Statistics for the Behavioral Sciences

Statistics for the Behavioral Sciences

Eighth Edition

Frederick J Gravetter
The College at Brockport, State University of New York

Larry B. Wallnau
The College at Brockport, State University of New York

WADSWORTH
CENGAGE Learning

Australia · Brazil · Japan · Korea · Mexico · Singapore · Spain · United Kingdom · United States

WADSWORTH
CENGAGE Learning

Statistics for the Behavioral Sciences
Frederick J Gravetter and Larry B. Wallnau

Sponsoring Editor: Jane Potter
Assistant Editor: Rebecca Rosenberg
Editorial Assistant: Ryan Patrick
Media Editor: Amy Cohen
Marketing Manager: Tierra Morgan
Marketing Coordinator: Molly Felz
Marketing Communications Manager:
    Talia Wise
Content Project Manager:
    Charlene Carpentier
Creative Director: Rob Hugel
Art Director: Vernon Boes
Print Buyer: Linda Hsu
Text Permissions Acquisitions Manager:
    Bob Kauser
Production Service: Carol O'Connell,
    Graphic World Inc.
Text Designer: Cheryl Carrington
Cover Designer: Cheryl Carrington
Cover Image: Rural Decay by Deborah Batt
Compositor: Graphic World Inc.

For product information and technology assistance, contact us at
Cengage Learning Customer & Sales Support, 1-800-354-9706.
For permission to use material from this text or product,
submit all requests online at **www.cengage.com/permissions.**
Further permissions questions can be e-mailed to
permissionrequest@cengage.com.

Library of Congress Control Number: 2008932407

ISBN-13: 9-780-495-60220-0

ISBN-10: 0-495-60220-5

Wadsworth
10 Davis Drive
Belmont, CA 94002-3098 USA

Cengage Learning is a leading provider of customized learning solutions with office locations around the globe, including Singapore, the United Kingdom, Australia, Mexico, Brazil, and Japan. Locate your local office at **www.cengage.com/international.**

Cengage Learning products are represented in Canada by Nelson Education, Ltd.

To learn more about Wadsworth, visit **www.cengage.com/wadsworth**

Purchase any of our products at your local college store or at our preferred online store **www.ichapters.com.**

Printed in the United States of America
1 2 3 4 5 6 7 12 11 10 09 08

Contents in Brief

Chapter 1   Introduction to Statistics   1

Chapter 2   Frequency Distributions   35

Chapter 3   Central Tendency   70

Chapter 4   Variability   104

Chapter 5 z-Scores: Location of Scores and Standardized Distributions   137

Chapter 6   Probability   163

Chapter 7   Probability and Samples: The Distribution of Sample Means   198

Chapter 8   Introduction to Hypothesis Testing   229

Chapter 9   Introduction to the t Statistic   280

Chapter 10   The t Test for Two Independent Samples   307

Chapter 11   The t Test for Two Related Samples   339

Chapter 12   Estimation   365

Chapter 13   Introduction to Analysis of Variance   392

Chapter 14   Repeated-Measures Analysis of Variance (ANOVA)   444

Chapter 15   Two-Factor Analysis of Variance (Independent Measures) 477

Chapter 16   Correlation   519

Chapter 17   Introduction to Regression   562

Chapter 18   The Chi-Square Statistic: Tests for Goodness of Fit
and Independence   604

Chapter 19   The Binomial Test   644

Chapter 20   Statistical Techniques for Ordinal Data: Mann-Whitney,
Wilcoxon, Kruskal-Wallis, and Friedman Tests   664

Contents

Chapter 1 Introduction to Statistics   1

Preview   2
1.1 Statistics, Science, and Observations   2
1.2 Populations and Samples   3
1.3 Data Structures, Research Methods, and Statistics   10
1.4 Variables and Measurement   18
1.5 Statistical Notation   24
Summary   28
Focus on Problem Solving   30
Demonstration 1.1   31
Problems   32

Chapter 2 Frequency Distributions   35

Preview   36
2.1 Introduction to Frequency Distributions   37
2.2 Frequency Distribution Tables   37
2.3 Frequency Distribution Graphs   44
2.4 The Shape of a Frequency Distribution   50
2.5 Percentiles, Percentile Ranks, and Interpolation   52
2.6 Stem and Leaf Displays   58
Summary   61
Focus on Problem Solving   63
Demonstrations 2.1 and 2.2   64
Problems   66

Chapter 3 Central Tendency   70

Preview   71
3.1 Overview   71
3.2 The Mean   73
3.3 The Median   82
3.4 The Mode   88
3.5 Selecting a Measure of Central Tendency   90
3.6 Central Tendency and the Shape of the Distribution   96
Summary   98
Focus on Problem Solving   100
Demonstration 3.1   100
Problems   101

Chapter 4   Variability   104

    Preview   105
4.1   Overview   105
4.2   The Range and Interquartile Range   107
4.3   Standard Deviation and Variance for a Population   109
4.4   Standard Deviation and Variance for Samples   117
4.5   More about Variance and Standard Deviation   122
4.6   Comparing Measures of Variability   129
    Summary   130
    Focus on Problem Solving   132
    Demonstration 4.1   133
    Problems   134

Chapter 5 z-Scores: Location of Scores and
Standardized Distributions   137

    Preview   138
5.1   Introduction to z-Scores   139
5.2 z-Scores and Location in a Distribution   141
5.3   Using z-Scores to Standardize a Distribution   146
5.4   Other Standardized Distributions Based on z-Scores   150
5.5   Computing z-Scores for Samples   153
5.6   Looking Ahead to Inferential Statistics   154
    Summary   157
    Focus on Problem Solving   159
    Demonstrations 5.1 and 5.2   159
    Problems   160

Chapter 6   Probability   163

    Preview   164
6.1   Introduction to Probability   164
6.2   Probability and the Normal Distribution   170
6.3   Probabilities and Proportions for Scores from
    a Normal Distribution   177
6.4   Probability and the Binomial Distribution   183
6.5   Looking Ahead to Inferential Statistics   189
    Summary   191
    Focus on Problem Solving   192
    Demonstrations 6.1 and 6.2   193
    Problems   195

Chapter 7   Probability and Samples: The Distribution
of Sample Means   198

    Preview   199
7.1   Samples and Populations   199
7.2   The Distribution of Sample Means   200
7.3   Probability and the Distribution of Sample Means   210

7.4   More about Standard Error   214
7.5   Looking Ahead to Inferential Statistics   219
       Summary   223
       Focus on Problem Solving   224
       Demonstration 7.1   225
       Problems   226

Chapter 8   Introduction to Hypothesis Testing   229

       Preview   230
8.1   The Logic of Hypothesis Testing   230
8.2   Uncertainty and Errors in Hypothesis Testing   241
8.3   An Example of a Hypothesis Test   246
8.4   Directional (One-Tailed) Hypothesis Tests   255
8.5   The General Elements of Hypothesis Testing: A Review   258
8.6   Concerns about Hypothesis Testing: Measuring Effect Size   260
8.7   Statistical Power   265
       Summary   271
       Focus on Problem Solving   273
       Demonstrations 8.1 and 8.2   274
       Problems   275

Chapter 9   Introduction to the t Statistic   280

       Preview   281
9.1   The t Statistic: An Alternative to z 281
9.2   Hypothesis Tests with the t Statistic   287
9.3   Measuring Effect Size for the t Statistic   291
9.4   Directional Hypotheses and One-Tailed Tests   297
       Summary   300
       Focus on Problem Solving   301
       Demonstrations 9.1 and 9.2   302
       Problems   303

Chapter 10   The t Test for Two Independent Samples   307

        Preview   308
10.1   Introduction to Independent-Measures Designs   308
10.2   The t Statistic for an Independent-Measures Research Design   310
10.3   Hypothesis Tests and Effect Size with the Independent-Measures t Statistic   318
10.4   Assumptions Underlying the Independent-Measures t Formula   327
        Summary   330
        Focus on Problem Solving   333
        Demonstrations 10.1 and 10.2   333
        Problems   335

Chapter 11   The *t* Test for Two Related Samples   339

Preview   340
11.1   Introduction to Repeated-Measures Designs   340
11.2   The *t* Statistic for a Repeated-Measures Research Design   342
11.3   Hypothesis Tests and Effect Size for the Repeated-Measures Design   346
11.4   Uses and Assumptions for Repeated-Measures *t* Tests   353
Summary   356
Focus on Problem Solving   358
Demonstrations 11.1 and 11.2   358
Problems   360

Chapter 12   Estimation   365

Preview   366
12.1   An Overview of Estimation   366
12.2   Estimation with the *t* Statistic   372
12.3   A Final Look at Estimation   382
Summary   383
Focus on Problem Solving   385
Demonstrations 12.1 and 12.2   385
Problems   388

Chapter 13   Introduction to Analysis of Variance   392

Preview   393
13.1   Introduction   394
13.2   The Logic of Analysis of Variance   398
13.3   ANOVA Notation and Formulas   403
13.4   The Distribution of *F*-Ratios   412
13.5   Examples of Hypothesis Testing and Effect Size with ANOVA   414
13.6   Post Hoc Tests   425
13.7   The Relationship Between ANOVA and *t* Tests   431
Summary   433
Focus on Problem Solving   435
Demonstrations 13.1 and 13.2   436
Problems   438

Chapter 14   Repeated-Measures Analysis of Variance (ANOVA)   444

Preview   445
14.1   Overview   445
14.2   Testing Hypotheses with the Repeated-Measures ANOVA   451
14.3   An Alternative View of the Repeated-Measures ANOVA   459
14.4   Advantages and Disadvantages of the Repeated-Measures Design   461
14.5   Individual Differences and the Consistency of the Treatment Effects   463

Summary   465
Focus on Problem Solving   468
Demonstrations 14.1 and 14.2   469
Problems   471

Chapter 15   Two-Factor Analysis of Variance (Independent Measures)   477

Preview   478
15.1   Overview   478
15.2   Main Effects and Interactions   480
15.3   Notation and Formulas   488
15.4   Interpreting the Results from a Two-Factor ANOVA   497
15.5   An Alternative View of the Two-Factor ANOVA   501
15.6   Assumptions for the Two-Factor ANOVA   505
Summary   506
Focus on Problem Solving   509
Demonstrations 15.1 and 15.2   510
Problems   513

Chapter 16   Correlation   519

Preview   520
16.1   Overview   520
16.2   The Pearson Correlation   525
16.3   Understanding and Interpreting the Pearson Correlation   530
16.4   Hypothesis Tests with the Pearson Correlation   537
16.5   The Spearman Correlation   540
16.6   Other Measures of Relationship   547
Summary   551
Focus on Problem Solving   554
Demonstrations 16.1 and 16.2   555
Problems   557

Chapter 17   Introduction to Regression   562

Preview   563
17.1   Introduction to Linear Equations and Regression   563
17.2   Testing the Significance of the Regression Equation:
        Analysis of Regression   577
17.3   Introduction to Multiple Regression with Two
        Predictor Variables   580
17.4   Evaluating the Contribution of Each Predictor Variable   586
Summary   593
Focus on Problem Solving   597
Demonstrations 17.1 and 17.2   597
Problems   600

Chapter 18   The Chi-Square Statistic: Tests for Goodness
of Fit and Independence   604

Preview   605
18.1   Parametric and Nonparametric Statistical Tests   606
18.2   The Chi-Square Test for Goodness of Fit   607
18.3   The Chi-Square Test for Independence   617
18.4   Measuring Effect Size for the Chi-Square
Test for Independence   626
18.5   Assumptions and Restrictions for Chi-Square Tests   628
18.6   Special Applications of the Chi-Square Tests   629
Summary   633
Focus on Problem Solving   636
Demonstration 18.1   637
Problems   639

Chapter 19   The Binomial Test   644

Preview   645
19.1   Overview   645
19.2   The Binomial Test   649
19.3   The Relationship Between Chi-Square and the Binomial Test   653
19.4   The Sign Test   654
Summary   657
Focus on Problem Solving   659
Demonstration 19.1   660
Problems   660

Chapter 20   Statistical Techniques for Ordinal Data: Mann-Whitney,
Wilcoxon, Kruskal-Wallis, and Friedman Tests   664

Preview   665
20.1   Data from an Ordinal Scale   665
20.2   The Mann-Whitney U-Test: An Alternative to the Independent-
Measures t Test   668
20.3   The Wilcoxon Signed-Ranks Test: An Alternative to the Repeated-
Measures t Test   675
20.4   The Kruskal-Wallis Test: An Alternative to the Independent-
Measures ANOVA   680
20.5   The Friedman Test: An Alternative to Repeated-Measures
ANOVA   684
Summary   688
Focus on Problem Solving   692
Demonstrations 20.1, 20.2, 20.3, and 20.4   692
Problems   697

Appendix A   Basic Mathematics Review   703

A.1   Symbols and Notation   705
A.2   Proportions: Fractions, Decimals, and Percentages   707
A.3   Negative Numbers   713
A.4   Basic Algebra: Solving Equations   715
A.5   Exponents and Square Roots   718

Appendix B   Statistical Tables   725

Appendix C   Solutions for Odd-Numbered Problems   741

Appendix D   General Instructions for Using SPSS   759

Statistics Organizer   762

References   774

Index   779

Preface

Many students in the behavioral sciences view the required statistics course as an intimidating obstacle that has been placed in the middle of an otherwise interesting curriculum. They want to learn about human behavior—not about math and science. As a result, the statistics course is seen as irrelevant to their education and career goals. However, as long as the behavioral sciences are founded in science, a knowledge of statistics will be necessary. Statistical procedures provide researchers with objective and systematic methods for describing and interpreting their research results. Scientific research is the system that we use to gather information, and statistics are the tools that we use to distill the information into sensible and justified conclusions. The goal of this book is not only to teach the methods of statistics but also to convey the basic principles of objectivity and logic that are essential for science and valuable in everyday life.

Those of you who are familiar with previous editions of *Statistics for the Behavioral Sciences* will notice that some changes have been made. These changes are summarized in the section entitled "To the Instructor." In revising this text, our students have been foremost in our minds. Over the years, they have provided honest and useful feedback. Their hard work and perseverance has made our writing and teaching most rewarding. We sincerely thank them. Students who are using this edition should please read the section of the preface entitled "To the Student."

ANCILLARIES

Ancillaries for this edition include the following:

- *Enhanced WebAssign:* New to this edition, Enhanced WebAssign allows instructors to assign additional homework and study problems to students so they can get even more practice on the most difficult concepts. Original guided tutorials and simulations lead students through step-by-step problem-solving. Algorithmic versions of the end-of-chapter problems offer opportunities for additional practice.

- *Study Guide:* Contains chapter overviews, learning objectives, new terms and concepts, new formulas, step-by-step procedures for problem solving, study hints and cautions, self-tests, and review. The Study Guide contains answers to the self-test questions.

- *Book Companion Website:* Additional, free study resources are available online at the book companion website. Practice and reinforce statistical concepts using the Statistics Workshops, chapter quizzes, chapter objectives, interactive flash cards, Web links, and more! Visit www.cengage.com/psychology/gravetter.

- *Instructor's Manual with Test Bank:* Contains chapter outlines, annotated learning objectives, lecture suggestions, test items, and solutions to all end-of-chapter problems in the text. Test items are also available as a Word download or for ExamView computerized test bank software.

- *PowerLecture with JoinIn and ExamView:* This CD includes the instructor's manual, test bank, lecture slides with book figures, and more. Featuring automatic grading, **ExamView**®, also available within PowerLecture, allows you to create, deliver, and customize tests and study guides (both print and online) in minutes. See assessments onscreen exactly as they will print or display online.

Build tests of up to 250 questions using up to 12 question types and enter an unlimited number of new questions or edit existing questions. The **JoinIn**™ content (for use with most "clicker" systems) available within PowerLecture delivers instant classroom assessment and active learning.

- *WebTutor on Blackboard and WebCT:* Load a WebTutor cartridge into your Course Management System to access text-specific content, media assets, quizzing, Web links, and discussion topics. Easily blend, add, edit, reorganize, or delete content.

ACKNOWLEDGMENTS

It takes a lot of good, hard-working people to produce a book. Our friends at Wadsworth have made enormous contributions to this textbook. We thank: Jane Potter, Psychology Editor; Amy Cohen, Media Editor; Rebecca Rosenberg, Assistant Editor; Ryan Patrick, Editorial Assistant; Talia Wise, Marketing Communications Manager; Tierra Morgan, Marketing Manager; Molly Felz, Marketing Coordinator; Charlene Carpentier, Content Project Manager; and Vernon Boes, Art Director. Reviewers play a very important role in the development of a manuscript. Accordingly, we offer our appreciation to the following colleagues for their assistance with the eighth edition: Sonia Ghumman, Michigan State University; Eric Olofson, University of Oregon; and Paul Voglewede, Syracuse University.

TO THE INSTRUCTOR

Those of you familiar with the previous edition of *Statistics for the Behavioral Sciences* will notice a number of changes in the eighth edition. Throughout the book, research examples have been updated and the end-of-chapter problems have been extensively revised. The Preview sections have been reformatted to emphasize the fact that each new statistical technique is intended to serve a specific purpose or solve a problem. Also, the Demonstration problems at the end of each chapter have been simplified to focus on the content of the chapter and minimize tedious computations.

Other examples of specific and noteworthy revisions include:

Chapter 1   The distinction between true experiments and other research strategies has been clarified, and there is expanded discussion of continuous and discrete variables.

Chapter 2   Minor editing simplifies examples and clarifies the idea that interpolation produces estimated values.

Chapter 3   The distinction between discrete and continuous variables has been expanded, especially in the context of computing the median.

Chapter 4   New discussion relates measures of variability to the scales of measurement introduced in Chapter 1. Specifically, most measures of variability are measures of distance and require an interval or ratio scale. Also, the sections discussing biased and unbiased statistics and the concept of degrees of freedom have been expanded.

Chapter 5   There is increased emphasis on the definition and concept of z-scores rather than formulas, encouraging students to estimate answers (based on the definition) before they begin calculations. A new section demonstrates that many z-score problems can be solved by sketching distributions rather than manipulating formulas and numbers. The section on z-scores for samples has been expanded.

Chapter 6   New text clarifies the distinction between the discrete binomial distribution and the normal approximation which is continuous.

Chapter 7   The discussion of the basic definition and purpose of the standard error has been expanded. A new section emphasizes that we are now dealing with three distributions: the original population, the sample, and the theoretical distribution of sample means.

Chapter 8   A new analogy compares hypothesis testing with the process of a jury trial. The section on statistical power has been expanded, including new discussion relating power to Type II errors, new examples demonstrating the calculation of power, and a new discussion concerning power calculations and determining an optimum sample size for a research study.

Chapter 9   New discussion emphasizes the fact that the shape and proportions of the t distribution depend on the degrees of freedom. Also, a new section simplifies locating the critical region for directional tests.

Chapter 10   A new section discusses the interpretation of the standard error for a sample mean difference.

Chapter 11   A new figure illustrates the null and alternative hypotheses for the repeated-measures t test. New text discusses the variability of the difference scores as a reflection of the consistency of the treatment effect. The section discussing the potential for order effects to be a disadvantage of the repeated-measures design has been revised.

Chapter 12   The comparison of hypothesis testing and estimation has been simplified, and the section on the interpretation of a confidence interval has been expanded.

Chapter 13   There is greater emphasis on concepts and less on computation for all three chapters covering analysis of variance. A new Box demonstrates how $SS_{between\ treatments}$ measures the variability of the treatment means. A new section expands the concept of effect size (percentage of variance accounted for) by demonstrating how the mean differences between treatments contribute to the overall variability of the scores.

Chapter 14   New text clarifies the distinction between systematic differences (caused by the treatments and by individual differences) and random, unsystematic differences that form the error term of the F-ratio. A new section demonstrates how $SS_{between\ treatments}$ and $SS_{between\ subjects}$ can be obtained by eliminating the mean differences between treatments and between subjects, respectively, and observing how the variability is reduced.

Chapter 15   A new section demonstrates how the SS values for both main effects and the interaction can be obtained by eliminating the corresponding mean differences in the data and noting how the overall variability is reduced for the resulting scores.

Chapter 16   Minor editing updates examples and clarifies the calculation of the Pearson correlation using z-scores from either samples or populations.

Chapter 17   A new section introduces the concept that a regression equation predicts deviations from the mean. The section on multiple regression has been greatly expanded, including a new section on partial correlations.

Chapter 18   An updated section discusses the relationship between chi-square tests and other statistical procedures such as correlations and the independent-measures *t* test and ANOVA.

Chapter 19   A new section discusses the distinction between the discrete binomial distribution and the continuous normal distribution and includes the concept of real limits that was introduced in Chapter 6 when the binomial distribution was first mentioned.

Chapter 20   The chapter has been simplified and shortened by editing and eliminating unnecessary examples.

TO THE STUDENT

A primary goal of this book is to make the task of learning statistics as easy and painless as possible. Among other things, you will notice that the book provides you with a number of opportunities to practice the techniques you will be learning in the form of learning checks, examples, demonstrations, and end-of-chapter problems. We encourage you to take advantage of these opportunities. Read the text rather than just memorize the formulas. We have taken care to present each statistical procedure in a conceptual context that explains why the procedure was developed and when it should be used. If you read this material and gain an understanding of the basic concepts underlying a statistical formula, you will find that learning the formula and how to use it will be much easier. In the following section, "Study Hints," we provide advice that we give our own students. Ask your instructor for advice as well; we are sure that other instructors will have ideas of their own.

Over the years, the students in our classes and other students using our book have given us valuable feedback. If you have any suggestions or comments about this book, you can write to either Professor Emeritus Frederick Gravetter or Professor Emeritus Larry Wallnau at the Department of Psychology, SUNY College at Brockport, 350 New Campus Drive, Brockport, New York 14420. You can also contact Professor Emeritus Gravetter directly at fgravett@brockport.edu.

Study Hints   You may find some of these tips helpful, as our own students have reported.

- The key to success in a statistics course is to keep up with the material. Each new topic builds on previous topics. If you have learned the previous material, then the new topic is just one small step forward. Without the proper background, however, the new topic can be a complete mystery. If you find that you are falling behind, get help immediately.

- You will learn (and remember) much more if you study for short periods several times per week rather than try to condense all of your studying into one long session. For example, it is far more effective to study half an hour every night than to have a single $3\frac{1}{2}$-hour study session once a week. We cannot even work on *writing* this book without frequent rest breaks.

- Do some work before class. Keep a little ahead of the instructor by reading the appropriate sections before they are presented in class. Although you may not fully understand what you read, you will have a general idea of the topic, which will make the lecture easier to follow. Also, you can identify material that is particularly confusing and then be sure the topic is clarified in class.

- Pay attention and think during class. Although this advice seems obvious, often it is not practiced. Many students spend so much time trying to write down every example presented or every word spoken by the instructor that they do not actually understand and process what is being said. Check with your instructor—there may not be a need to copy every example presented in class, especially if there are many examples like it in the text. Sometimes, we tell our students to put their pens and pencils down for a moment and just listen.

- Test yourself regularly. Do not wait until the end of the chapter or the end of the week to check your knowledge. After each lecture, work some of the end-of-chapter problems and do the Learning Checks. Review the Demonstration Problems, and be sure you can define the Key Terms. If you are having trouble, get your questions answered *immediately* (reread the section, go to your instructor, or ask questions in class). By doing so, you will be able to move ahead to new material.

- Do not kid yourself! Avoid denial. Many students observe their instructor solve problems in class and think to themselves, "This looks easy, I understand it." Do you really understand it? Can you really do the problem on your own without having to leaf through the pages of a chapter? Although there is nothing wrong with using examples in the text as models for solving problems, you should try working a problem with your book closed to test your level of mastery.

- We realize that many students are embarassed to ask for help. It is our biggest challenge as instructors. You must find a way to overcome this aversion. Perhaps contacting the instructor directly would be a good starting point, if asking questions in class is too anxiety-provoking. You could be pleasantly surprised to find that your instructor does not yell, scold, or bite! Also, your instructor might know of another student who can offer assistance. Peer tutoring can be very helpful.

Frederick J Gravetter
Larry B. Wallnau

About the Authors

Frederick J Gravetter is Professor Emeritus of Psychology at the State University of New York College at Brockport. While teaching at Brockport, Dr. Gravetter specialized in statistics, experimental design, and cognitive psychology. He received

his bachelor's degree in mathematics from M.I.T. and his Ph.D. in psychology from Duke University. In addition to publishing this textbook and several research articles, Dr. Gravetter co-authored *Research Methods for the Behavioral Sciences* and *Essentials of Statistics for the Behavioral Sciences*.

Nicki and Fred

Larry B. Wallnau is Professor Emeritus of Psychology at the State University of New York College at Brockport. While teaching at Brockport, he published numerous research articles, primarily on the behavioral effects of psychotropic drugs. With Dr. Gravetter, he co-authored *Essentials of Statistics for the Behavioral Sciences*. He also has provided editorial consulting for numerous publishers and journals. He routinely gives lectures and demonstrations on service dogs for those with disabilities.

Ed Berns

Ben and Larry

Statistics for the
Behavioral Sciences

CHAPTER

1

Introduction to Statistics

Preview

1.1 Statistics, Science, and Observations

1.2 Populations and Samples

1.3 Data Structures, Research Methods, and Statistics

1.4 Variables and Measurement

1.5 Statistical Notation

Summary

Focus on Problem Solving

Demonstration 1.1

Problems

Preview

Before we begin our discussion of statistics, we ask you to read the following paragraph taken from the philosophy of Wrong Shui (Candappa, 2000).

> The Journey to Enlightenment
> In Wrong Shui life is seen as a cosmic journey, a struggle to overcome unseen and unexpected obstacles at the end of which the traveler will find illumination and enlightenment. Replicate this quest in your home by moving light switches away from doors and over to the far side of each room.*

Why did we begin a statistics book with a bit of twisted philosophy? Actually, the paragraph is an excellent (and humorous) counterexample for the purpose of this book. Specifically, our goal is to help you avoid stumbling around in the dark by providing lots of easily available light switches and plenty of illumination as you journey through the world of statistics. To accomplish this, we try to present sufficient background and a clear statement of purpose before we introduce a new statistical procedure. If you have an appropriate background, it is much easier to fit new material into memory and to recall it later. In this book we begin each chapter with a preview. The purpose for the preview is to provide the background or context for the new material in the chapter. As you read each preview section, you should gain a general overview of the chapter content. Remember that all statistical procedures were developed to serve a purpose. If you understand why a new procedure is needed, you will find it much easier to learn the procedure.

The objectives for this first chapter are to provide an introduction to the topic of statistics and to give you some background for the rest of the book. We discuss the role of statistics within the general field of scientific inquiry, and we introduce some of the vocabulary and notation that are necessary for the statistical methods that follow. In some respects, this chapter serves as a preview section for the rest of the book.

As you read through the following chapters, keep in mind that the general topic of statistics follows a well-organized, logically developed progression that leads from basic concepts and definitions to increasingly sophisticated techniques. Thus, the material presented in the early chapters of this book will serve as a foundation for the material that follows. The content of the first nine chapters, for example, provides an essential background and context for the statistical methods presented in Chapter 10. If you turn directly to Chapter 10 without reading the first nine chapters, you will find the material confusing and incomprehensible. However, if you learn and use the background material, you will have a good frame of reference for understanding and incorporating new concepts as they are presented.

---

*Candappa, R. (2000) *The Little Book of Wrong Shui*. Kansas City: Andrews McMeel Publishing. Reprinted by permission.

1.1 STATISTICS, SCIENCE, AND OBSERVATIONS

DEFINITIONS OF STATISTICS By one definition, *statistics* consist of facts and figures such as average income, crime rate, birth rate, average snowfall, and so on. These statistics are usually informative and time-saving because they condense large quantities of information into a few simple figures. Later in this chapter we return to the notion of calculating statistics (facts and figures) but, for now, we concentrate on a much broader definition of statistics. Specifically, we use the term *statistics* to refer to a set of mathematical procedures. In this case, we are using the term *statistics* as a shortened version of *statistical procedures*. For example, you are probably using this book for a statistics course in which you will learn about the statistical techniques that are used for research in the behavioral sciences.

Research in psychology (and other fields) involves gathering information. To determine, for example, whether violence on TV has any effect on children's behavior, you would need to gather information about children's behaviors. When researchers finish the task of gathering information, they typically find themselves with pages and pages of measurements such as IQ scores, personality scores, reaction time scores, and so on. The role of statistics is to help researchers make sense of this information. Specifically, statistics serve two general purposes:

1. Statistics are used to organize and summarize the information so that the researcher can see what happened in the research study and can communicate the results to others.

2. Statistics help the researcher to answer the general questions that initiated the research by determining exactly what conclusions are justified based on the results that were obtained.

DEFINITION

The term **statistics** refers to a set of mathematical procedures for organizing, summarizing, and interpreting information.

Statistical procedures help ensure that the information or observations are presented and interpreted in an accurate and informative way. In somewhat grandiose terms, statistics help researchers bring order out of chaos. In addition, statistics provide researchers with a set of standardized techniques that are recognized and understood throughout the scientific community. Thus, the statistical methods used by one researcher will be familiar to other researchers, who can accurately interpret the statistical analyses with a full understanding of how the analysis was done and what the results signify.

1.2   POPULATIONS AND SAMPLES

WHAT ARE THEY?   Scientific research typically begins with a general question about a specific group (or groups) of individuals. For example, a researcher may be interested in the effect of divorce on the self-esteem of preteen children. Or a researcher may want to examine the political attitudes for men compared to women. In the first example, the researcher is interested in the group of *preteen children*. In the second example, the researcher wants to compare the group of *men* with the group of *women*. In statistical terminology, the entire group that a researcher wishes to study is called a *population*.

DEFINITION

A **population** is the set of all the individuals of interest in a particular study.

As you can well imagine, a population can be quite large—for example, the entire set of women on the planet Earth. A researcher might be more specific, limiting the population for study to women who are registered voters in the United States. Perhaps the investigator would like to study the population consisting of women who are heads of state. Populations can obviously vary in size from extremely large to very small, depending on how the investigator defines the population. The population being studied should always be identified by the researcher. In addition, the population need not consist of people—it could be a population of rats, corporations, parts produced in a factory, or anything else an investigator wants to study. In practice, populations are typically very large, such as the population of fourth-grade children in the United States or the population of small businesses.

Although research questions concern an entire population, it usually is impossible for a researcher to examine every individual in the population of interest. Therefore, researchers typically select a smaller, more manageable group from the population and limit their studies to the individuals in the selected group. In statistical terms, a set of individuals selected from a population is called a *sample*. A sample is intended to be representative of its population, and a sample should always be identified in terms of the population from which it was selected.

DEFINITION

A **sample** is a set of individuals selected from a population, usually intended to represent the population in a research study.

Just as we saw with populations, samples can vary in size. For example, one study might examine a sample of only 10 children in a preschool program, and another study might use a sample of over 1000 registered voters representing the population of a major city.

So far we have talked about a sample being selected from a population. However, this is actually only half of the full relationship between a sample and its population. Specifically, when a researcher finishes examining the sample, the goal is to generalize the results back to the entire population. Remember that the research started with a general question about the population. To answer the question, a researcher studies a sample and then generalizes the results from the sample to the population. The full relationship between a sample and a population is shown in Figure 1.1.

VARIABLES AND DATA

Typically, researchers are interested in specific characteristics of the individuals in the population (or in the sample), or they are interested in outside factors that may influence the individuals. For example, a researcher may be interested in the influence

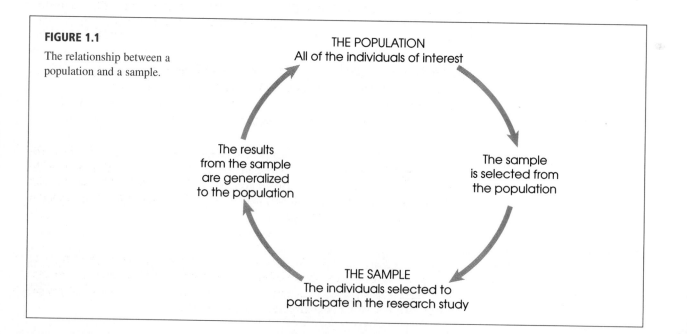

FIGURE 1.1

The relationship between a population and a sample.

THE POPULATION
All of the individuals of interest

The sample
is selected from
the population

THE SAMPLE
The individuals selected to
participate in the research study

The results
from the sample
are generalized
to the population

of the weather on people's moods. As the weather changes, do people's moods also change? Something that can change or have different values is called a *variable.*

DEFINITION

> A **variable** is a characteristic or condition that changes or has different values for different individuals.

Once again, variables can be characteristics that differ from one individual to another, such as height, weight, gender, or personality. Also, variables can be environmental conditions that change such as temperature, time of day, or the size of the room in which the research is being conducted.

In order to demonstrate changes in variables, it is necessary to make measurements of the variables being examined. The measurement obtained for each individual is called a *datum,* or more commonly, a *score* or *raw score.* The complete set of scores is called the *data set* or simply the *data.*

DEFINITIONS

> **Data** (plural) are measurements or observations. A **data set** is a collection of measurements or observations. A **datum** (singular) is a single measurement or observation and is commonly called a **score** or **raw score**.

Before we move on, we should make one more point about samples, populations, and data. Earlier, we defined populations and samples in terms of *individuals.* For example, we discussed a population of preteen children and a sample of preschool children. Be forewarned, however, that we will also refer to populations or samples of *scores.* Because research typically involves measuring each individual to obtain a score, every sample (or population) of individuals produces a corresponding sample (or population) of scores.

PARAMETERS AND STATISTICS

When describing data it is necessary to distinguish whether the data come from a population or a sample. A characteristic that describes a population—for example, the population average—is called a *parameter.* On the other hand, a characteristic that describes a sample is called a *statistic.* Thus, the average score for a sample is an example of a statistic. Typically, the research process begins with a question about a population parameter. However, the actual data come from a sample and are used to compute sample statistics.

DEFINITIONS

> A parameter is a value, usually a numerical value, that describes a population. A parameter is usually derived from measurements of the individuals in the population. A statistic is a value usually a numerical value, that describes a sample. A statistic is usually derived from measurements of the individuals in the sample.

Typically, every population parameter has a corresponding sample statistic, and much of this book is concerned with the relationship between sample statistics and the corresponding population parameters. In Chapter 7, for example, we examine the relationship between the mean obtained for a sample and the mean for the population from which the sample was obtained.

DESCRIPTIVE AND INFERENTIAL STATISTICAL METHODS

Although researchers have developed a variety of different statistical procedures to organize and interpret data, these different procedures can be classified into two general categories. The first category, *descriptive statistics,* consists of statistical procedures that are used to simplify and summarize data.

DEFINITION

Descriptive statistics are statistical procedures used to summarize, organize, and simplify data.

Descriptive statistics are techniques that take raw scores and organize or summarize them in a form that is more manageable. Often the scores are organized in a table or a graph so that it is possible to see the entire set of scores. Another common technique is to summarize a set of scores by computing an average. Note that even if the data set has hundreds of scores, the average provides a single descriptive value for the entire set.

The second general category of statistical techniques is called *inferential statistics*. Inferential statistics are methods that use sample data to make general statements about a population.

DEFINITION

Inferential statistics consist of techniques that allow us to study samples and then make generalizations about the populations from which they were selected.

It usually is not possible to measure everyone in the population. Because populations are typically very large, a sample is selected to represent the population. By analyzing the results from the sample, we hope to make general statements about the population. Typically, researchers use sample statistics as the basis for drawing conclusions about population parameters.

One problem with using samples, however, is that a sample provides only limited information about the population. Although samples are generally *representative* of their populations, a sample is not expected to give a perfectly accurate picture of the whole population. There usually is some discrepancy between a sample statistic and the corresponding population parameter. This discrepancy is called *sampling error*, and it creates another problem to be addressed by inferential statistics (see Box 1.1).

DEFINITION

Sampling error is the discrepancy, or amount of error, that exists between a sample statistic and the corresponding population parameter.

The concept of sampling error is illustrated in Figure 1.2. The figure shows a population of 1000 college students and two samples, each with 5 students, that have been

BOX 1.1 THE MARGIN OF ERROR BETWEEN STATISTICS AND PARAMETERS

One common example of sampling error is the error associated with a sample proportion. For example, in newspaper articles reporting results from political polls, you frequently find statements such as this:

Candidate Brown leads the poll with 51% of the vote. Candidate Jones has 42% approval, and the remaining 7% are undecided. This poll was taken from a sample of registered voters and has a margin of error of plus-or-minus 4 percentage points.

The "margin of error" is the sampling error. In this case, the percentages that are reported were obtained from a sample and are being generalized to the whole population. As always, you do not expect the statistics from a sample to be perfect. There always will be some "margin of error" when sample statistics are used to represent population parameters.

FIGURE 1.2

A demonstration of sampling error. Two samples are selected from the same population. Notice that the sample statistics are different from one sample to another, and all of the sample statistics are different from the corresponding population parameters. The natural differences that exist, by chance, between a sample statistic and a population parameter are called sampling error.

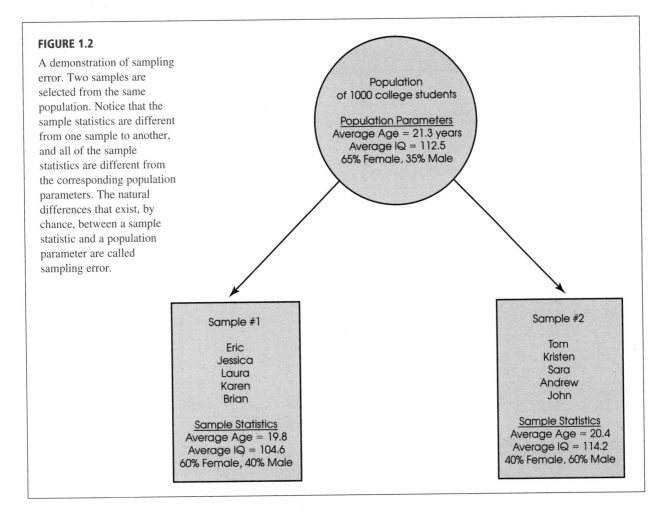

selected from the population. Notice that each sample contains different individuals who have different characteristics. Because the characteristics of each sample depend on the specific people in the sample, statistics will vary from one sample to another. In addition, neither sample has statistics that are exactly the same as the population parameters. You should also realize that Figure 1.2 shows only two of the hundreds of possible samples. Each sample would contain different individuals and would produce different statistics. This is the basic concept of sampling error: sample statistics vary from one sample to another and typically are different from the corresponding population parameters.

As a further demonstration of sampling error, imagine that your statistics class is separated into two groups by drawing a line from front to back through the middle of the room. Now imagine that you compute the average age (or height, or IQ) for each group. Will the two groups have exactly the same average? Almost certainly they will not. No matter what you chose to measure, you will probably find some difference between the two groups. Should you interpret this difference as meaning that there is a general tendency for older people to sit on the right-hand side of a room? Again, the answer is probably not. It is unlikely that the difference is "caused" by some systematic tendency that leads older people to gravitate to the right (or left) side of a room. Instead, the difference is probably the result of chance. The unpredictable, unsystematic differences that exist from one sample to another are an example of sampling error.

STATISTICS IN THE CONTEXT OF RESEARCH   The following example shows the general stages of a research study and demonstrates how descriptive statistics and inferential statistics are used to organize and interpret the data. At the end of the example, note how sampling error can affect the interpretation of experimental results, and consider why inferential statistical methods are needed to deal with this problem.

EXAMPLE 1.1   Figure 1.3 shows an overview of a general research situation and demonstrates the roles that descriptive and inferential statistics play. The purpose of the research study

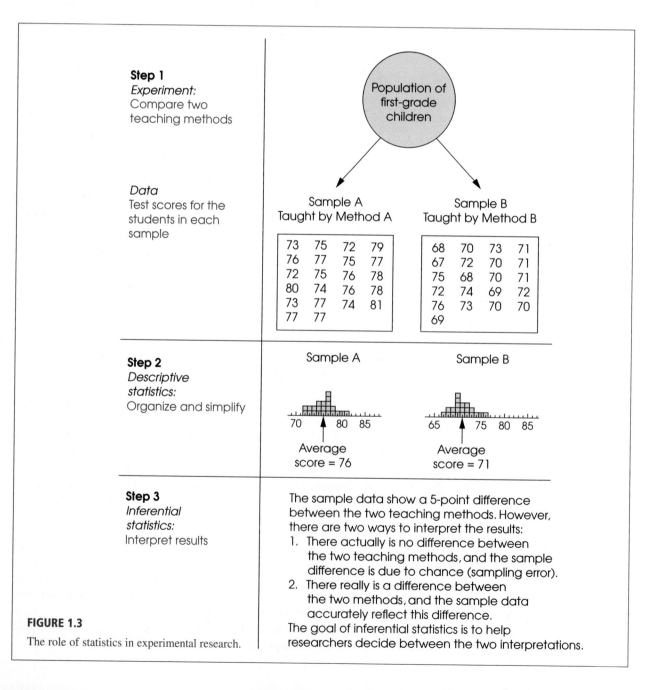

FIGURE 1.3

The role of statistics in experimental research.

is to evaluate the difference between two methods for teaching reading to first-grade children. Two samples are selected from the population of first-grade children. The children in sample A are assigned to teaching method A and the children in sample B are assigned to method B. After 6 months, all of the students are given a standardized reading test. At this point, the researcher has two sets of data: the scores for sample A and the scores for sample B (see Figure 1.3). Now is the time to begin using statistics.

First, descriptive statistics are used to simplify the pages of data. For example, the researcher could draw a graph showing the scores for each sample or compute the average score for each sample. Note that descriptive methods provide a simplified, organized description of the scores. In this example, the students taught by method A averaged 76 on the standardized test, and the students taught by method B averaged only 71.

Once the researcher has described the results, the next step is to interpret the outcome. This is the role of inferential statistics. In this example, the researcher has found a 5-point difference between the two samples (sample A averaged 76 and sample B averaged 71). The problem for inferential statistics is to differentiate between the following two interpretations:

1. There is no real difference between the two teaching methods, and the 5-point difference between the samples is just an example of sampling error (like the samples in Figure 1.2).

2. There really is a difference between the two teaching methods, and the 5-point difference between the samples was caused by the different methods of teaching.

In simple English, does the 5-point difference between samples provide convincing evidence of a difference between the two teaching methods, or is the 5-point difference just chance? The purpose of inferential statistics is to answer this question.

LEARNING CHECK

1. A researcher is interested in the reading skill of fourth-grade students in the state of Texas. The average reading score for the entire group of fourth-grade students would be an example of a _____.

2. A researcher is interested in the effect of humidity on the eating behavior of rats. A group of 30 rats is selected to participate in a research study. The group of 30 rats is an example of a _____.

3. A researcher is interested in the effect of humidity on the eating behavior of rats and selects a group of 30 rats to participate in a research study. The amount of food eaten in one day is measured for each rat and the researcher computes the average score for the 30 rats. The average score is an example of a _____.

4. Statistical techniques are classified into two general categories. What are the two categories called, and what is the general purpose for the techniques in each category?

5. Briefly define the concept of sampling error.

ANSWERS   1. parameter   2. sample   3. statistic

4. The two categories are descriptive statistics and inferential statistics. Descriptive techniques are intended to organize, simplify, and summarize data. Inferential techniques use sample data to reach general conclusions about populations.

5. Sampling error is the error or discrepancy between the value obtained for a sample statistic and the value for the corresponding population parameter.

1.3   DATA STRUCTURES, RESEARCH METHODS, AND STATISTICS

RELATIONSHIPS BETWEEN VARIABLES

Some research studies are conducted simply to describe individual variables as they exist naturally. For example, a college official may conduct a survey to describe the eating, sleeping, and study habits of a group of college students. Most research, however, is intended to examine the relationship between variables. For example, is there a relationship between the amount of violence that children see on television and the amount of aggressive behavior they display? Is there a relationship between the quality of breakfast and academic performance for elementary school children? Is there a relationship between the number of hours of sleep and grade point average for college students? To establish the existence of a relationship, researchers must make observations—that is, measurements of the two variables. The resulting measurements can be classified into two distinct data structures that also help to classify different research methods and different statistical techniques. In the following section we identify and discuss these two data structures.

I. Measuring two variables for each individual: the correlational method One method for examining the relationship between variables is to observe the two variables as they exist naturally for a set of individuals. That is, simply measure the two variables for each individual. For example, research has demonstrated a relationship between sleep habits, especially wake-up time, and academic performance for college students (Trockel, Barnes, and Egget, 2000). The researchers used a survey to measure wake-up time and school records to measure academic performance for each student. Figure 1.4 shows an example of the kind of data obtained in the study. The researchers then look for consistent patterns in the data to provide evidence for a relationship between variables. For example, as wake-up time changes from one student to another, is there also a tendency for academic performance to change?

Consistent patterns in the data are often easier to see if the scores are presented in a graph. Figure 1.4 also shows the scores for the eight students in a graph called a scatter

(a)

Child	Wake-up Time	Academic Performance
A	11	2.4
B	9	3.6
C	9	3.2
D	12	2.2
E	7	3.8
F	10	2.2
G	10	3.0
H	8	3.0

FIGURE 1.4

One of two data structures for studies evaluating the relationship between variables. Note that there are two separate measurements for each individual (wake-up time and academic performance). The same scores are shown in a table (a) and in a graph (b).

plot. In the scatter plot, each individual is represented by a point so that the horizontal position corresponds to the student's wake-up time and the vertical position corresponds to the student's academic performance score. The scatter plot shows a clear relationship between wake-up time and academic performance: as wake-up time increases, academic performance decreases.

A research study that simply measures two different variables for each individual and produces the kind of data shown in Figure 1.4 is an example of the *correlational method,* or the *correlational research strategy.* The relationship between variables is usually measured and described using a statistic called a correlation. Correlations and the correlational method are discussed in detail in Chapter 16.

DEFINITION

In the **correlational method,** two different variables are observed to determine whether there is a relationship between them.

Occasionally, the correlational method produces scores that are not numerical values. For example, a researcher could measure home location (city or suburb) and attitude toward a new budget proposal (for or against) for a group of registered voters. Note that the researcher has two measurements for each individual but neither of the measurements is a numerical score. This type of data is typically summarized in a table showing how many individuals are classified into each of the possible categories. Table 1.1 shows an example of this kind of summary table. The relationship between variables for non-numerical data, such as the data in Table 1.1, is evaluated using a statistical technique known as a *chi-square test.* Chi-square tests are presented in Chapter 18.

The results from a correlational study can demonstrate the existence of a relationship between two variables, but they do not provide an explanation for the relationship. In particular, a correlational study cannot demonstrate a cause-and-effect relationship. For example, the data in Figure 1.4 show a systematic relationship between wake-up time and academic performance for a group of college students; those who sleep late tend to have lower performance scores than those who wake early. However, there are many possible explanations for the relationship and we do not know exactly what factor (or factors) is responsible for late sleepers having lower grades. In particular, we cannot conclude that waking students up earlier would cause their academic performance to improve, or that studying more would cause students to wake up earlier. To demonstrate a cause-and-effect relationship between two variables, researchers must use the experimental method, which is discussed in the next section.

II. Comparing two (or more) groups of scores: experimental and nonexperimental methods The second method for examining the relationship beween two variables involves the comparison of two or more groups of scores. In this situation, the relationship between variables is examined by using one of the variables to define the groups, then measuring the second variable to obtain scores for each group. For example, one group of elementary school children is shown a 30-minute action/adventure television program involving numerous instances of violence, and a second group is shown a 30-minute comedy that includes no violence. Both groups are then observed on the playground and a researcher records the number of aggressive acts committed by each child. An example of the resulting data is shown in Figure 1.5. The researcher will compare the scores for the violence group with the scores for the no-violence group. A systematic difference between the two groups provides evidence for a relationship between viewing television violence and aggressive behavior for elementary school children.

TABLE 1.1

Correlational data consisting of nonnumerical scores. Note that there are two measurements for each individual: home location and attitude. The numbers indicate how many people are in each category. For example, out of the 50 people living in the city, 40 are for the proposal.

Attitude toward budget proposal

		For	Against	
Home location	City	40	10	50
	Suburb	20	30	50
		60	40	

FIGURE 1.5

The second data structure for studies evaluating the relationship between variables. Note that one variable is used to define the groups and the second variable is measured to obtain scores within each group.

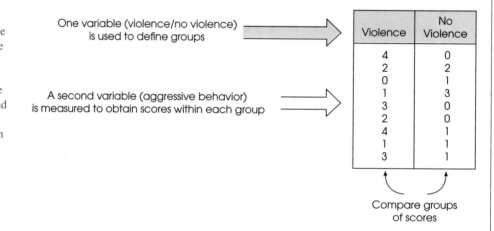

One variable (violence/no violence) is used to define groups

A second variable (aggressive behavior) is measured to obtain scores within each group

Violence	No Violence
4	0
2	2
0	1
1	3
3	0
2	0
4	1
1	1
3	1

Compare groups of scores

When the measurement procedure produces numerical scores, the statistical evaluation typically involves computing the average score for each group and then comparing the averages. The process of computing averages is presented in Chapter 3, and a variety of statistical techniques for comparing averages are presented in Chapters 8–15. If the measurement process simply classifies individuals into nonnumerical categories, the statistical evaluation usually consists of computing proportions for each group and then comparing proportions. Previously, in Table 1.1, we presented an example of nonnumerical data examining the relationship between attitude and home location. The same data can be used to compare the proportions for the two groups. For example, 80% of the people from the city are in favor of the new budget, whereas only 40% of people from the suburbs are in favor. As before, these data are evaluated using a chi-square test, which is presented in Chapter 18.

THE EXPERIMENTAL METHOD   One specific research method that involves comparing groups of scores is known as the *experimental method* or the *experimental research strategy*. The goal of an experimental study is to demonstrate a cause-and-effect relationship between two variables. Specifically, an experiment attempts to show that changing the value of one variable will cause changes to occur in the second variable. To accomplish this goal, the experimental method has two characteristics that differentiate experiments from other types of research studies:

1. **Manipulation**   The researcher manipulates one variable by changing its value from one level to another. A second variable is observed (measured) to determine whether the manipulation causes changes to occur.

2. **Control**   The researcher must exercise control over the research situation to ensure that other, extraneous variables do not influence the relationship being examined.

In more complex experiments, a researcher may systematically manipulate more than one variable and may observe more than one variable. Here we are considering the simplest case, in which only one variable is manipulated and only one variable is observed.

To demonstrate these two characteristics, consider an experiment in which a researcher is examining the effects of room temperature on memory performance. The purpose of the experiment is to determine whether changes in room temperature *cause* changes in memory performance.

The researcher manipulates temperature by creating two or more different treatment conditions. For example, our researcher could set the temperature at 70° for one condition and then change the temperature to 90° for a second condition. The researcher then measures memory performance for a group of individuals (called subjects, or participants) in the 70° room and compares their scores with another group that is tested in the 90° room. A systematic difference between the two groups of scores provides evidence that changing the temperature from 70° to 90° also caused a change in memory performance. The structure of this experiment is shown in Figure 1.6.

To be able to say that differences in memory performance are caused by temperature, the researcher must rule out any other possible explanations for the difference. That is, any other variables that might affect memory performance must be controlled. There are two general categories of variables that researchers must consider:

1. **Participant Variables**   These are characteristics such as age, gender, and intelligence that vary from one individual to another. Whenever an experiment compares different groups of participants (one group in treatment A and a different group in treatment B), researchers must ensure that participant variables do not differ from one group to another. For example, the experiment shown in Figure 1.6 compares memory scores for two different temperature

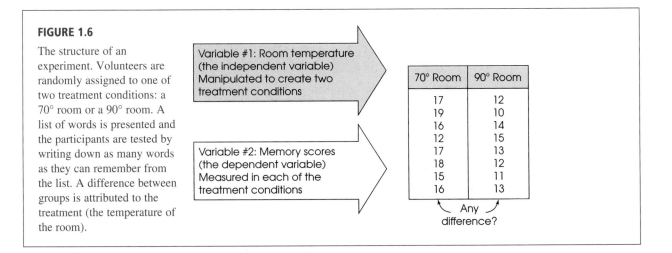

FIGURE 1.6

The structure of an experiment. Volunteers are randomly assigned to one of two treatment conditions: a 70° room or a 90° room. A list of words is presented and the participants are tested by writing down as many words as they can remember from the list. A difference between groups is attributed to the treatment (the temperature of the room).

Variable #1: Room temperature (the independent variable) Manipulated to create two treatment conditions

Variable #2: Memory scores (the dependent variable) Measured in each of the treatment conditions

70° Room	90° Room
17	12
19	10
16	14
12	15
17	13
18	12
15	11
16	13

Any difference?

conditions. Suppose, however, that the participants in the 70° room have higher IQs than those in the 90° room. In this case, there is an alternative explanation for any difference in memory scores that exists between the two groups. Specifically, it is possible that the difference in memory was caused by temperature, but it also is possible that the difference was caused by the participants' intelligence. Whenever a research study allows more than one explanation for the results, the study is said to be *confounded* because it is impossible to reach an unambiguous conclusion.

2. **Environmental Variables**   These are characteristics of the environment such as lighting, time of day, and weather conditions. A researcher must ensure that the individuals in treatment A are tested in the same environment as the individuals in treatment B. Using the temperature experiment (see Figure 1.6) as an example, suppose that the individuals in the 70° room were all tested in the morning and the individuals in the 90° room were all tested in the afternoon. Again, this would produce a confounded experiment because the researcher could not determine whether the differences in memory scores were caused by temperature or caused by the time of day.

Researchers typically use three basic techniques to control other variables. First, the researcher could use *random assignment,* which means that each participant has an equal chance of being assigned to each of the treatment conditions. The goal of random assignment is to distribute the participant characteristics evenly between the two groups so that neither group is noticeably smarter (or older, or faster) than the other. Random assignment can also be used to control environmental variables. For example, participants could be randomly assigned to be tested either in the morning or in the afternoon. Second, the researcher can use *matching* to ensure equivalent groups or equivalent environments. For example, the researcher could measure each particpant's IQ and then assign individuals to groups so that all of the groups have roughly the same average IQ. Finally, the researcher can control variables by *holding them constant.* For example, if an experiment uses only 10-year-old children as particpants (holding age constant), then the researcher can be certain that one group is not noticeably older than another.

DEFINITION

In the **experimental method**, one variable is manipulated while another variable is observed and measured. To establish a cause-and-effect relationship between the two variables, an experiment attempts to control all other variables to prevent them from influencing the results.

Terminology in the experimental method   Specific names are used for the two variables that are studied by the experimental method. The variable that is manipulated by the experimenter is called the *independent variable*. It can be identified as the treatment conditions to which participate are assigned. For the example in Figure 1.6, temperature is the independent variable. The variable that is observed and measured to obtain scores within each condition is the *dependent variable*. For the example in Figure 1.6, memory is the dependent variable.

DEFINITIONS

The **independent variable** is the variable that is manipulated by the researcher. In behavioral research, the independent variable usually consists of the two (or more) treatment conditions to which subjects are exposed. The independent variable consists of the *antecedent* conditions that were manipulated *prior* to observing the dependent variable.

The **dependent variable** is the one that is observed to assess the effect of the treatment.

Control conditions in an experiment   An experimental study evaluates the relationship between two variables by manipulating one variable (the independent variable) and measuring one variable (the dependent variable). Note that in an experiment only one variable is actually measured. You should realize that this is very different from a correlational study, in which both variables are measured and the data consist of two separate scores for each individual.

Often an experiment will include a condition in which the subjects do not receive any treatment. The scores from these subjects are then compared with scores from subjects who do receive the treatment. The goal of this type of study is to demonstrate that the treatment has an effect by showing that the scores in the treatment condition are substantially different from the scores in the no-treatment condition. In this kind of research, the no-treatment condition is called the *control condition*, and the treatment condition is called the *experimental condition*.

DEFINITIONS

Individuals in a **control condition** do not receive the experimental treatment. Instead, they either receive no treatment or they receive a neutral, placebo treatment. The purpose of a control condition is to provide a baseline for comparison with the experimental condition.

Individuals in the **experimental condition** do receive the experimental treatment.

Note that the independent variable always consists of at least two values. (Something must have at least two different values before you can say that it is "variable.") For the temperature experiment (see Figure 1.6), the independent variable is temperature (using values of 90° and 70°). For an experiment with an experimental group and a control group, the independent variable would be treatment versus no treatment.

NONEXPERIMENTAL METHODS: NONQUIVALENT GROUPS AND PRE–POST STUDIES

In informal conversation, there is a tendency for people to use the term *experiment* to refer to any kind of research study. You should realize, however, that the term only applies to studies that satisfy the specific requirements outlined earlier. In particular, a real experiment must include manipulation of an independent variable and rigorous control of other, extraneous variables. As a result, there are a number of other research designs that are not true experiments but still examine the relationship between variables by comparing groups of scores. Two examples are shown in Figure 1.7 and are discussed in the following paragraphs. This type of research study is classified as nonexperimental.

The top part of Figure 1.7 shows an example of a *nonequivalent groups* study comparing boys and girls. Notice that this study involves comparing two groups of scores (like an experiment). However, the researcher has no ability to control which participants go into which group—all the males must be in the boy group and all the females must be in the girl group. Because this type of research compares preexisting groups, the researcher cannot control the assignment of participants to groups and cannot ensure equivalent groups. Other examples of nonequivalent group studies include

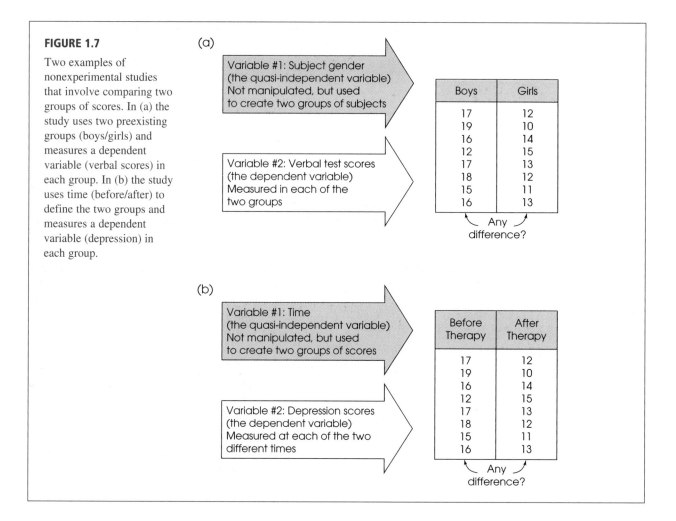

FIGURE 1.7

Two examples of nonexperimental studies that involve comparing two groups of scores. In (a) the study uses two preexisting groups (boys/girls) and measures a dependent variable (verbal scores) in each group. In (b) the study uses time (before/after) to define the two groups and measures a dependent variable (depression) in each group.

comparing 8-year-old children and 10-year-old children, people with an eating disorder and those with no disorder; and comparing children from a single-parent home and those from a two-parent home. Because it is impossible to use techniques like random assignment to control participant variables and ensure equivalent groups, this type of research is not a true experiment.

The bottom part of Figure 1.7 shows an example of a *pre–post* study comparing depression scores before therapy and after therapy. Again, the study is comparing two groups of scores (like an experiment). In this study, however, the researcher has no control over the passage of time. The "before" scores are always measured earlier than the "after" scores. Although a difference between the two groups of scores may be caused by the treatment, it is always possible that the scores simply change as time goes by. For example, the depression scores may decrease over time in the same way that the symptoms of a cold disappear over time. In a pre–post study the researcher also has no control over other variables that change with time. For example, the weather could change from dark and gloomy before therapy to bright and sunny after therapy. In this case, the depression scores could improve because of the weather and not because of the therapy. Because the researcher cannot control the passage of time or other variables related to time, this study is not a true experiment.

You may have noticed that a pre–post study is similar to a correlational study in that both designs measure two scores for each individual. In a correlational study, however, the two scores correspond to two different variables, such as height and weight. In a pre–post design the two scores are obtained by measuring the same variable twice under two different conditions at two different times. In the study shown in Figure 1.7(b) for example, depression is measured twice for each participant, once before treatment and again after treatment.

Terminology in nonexperimental research   Although the two research studies shown in Figure 1.7 are not true experiments, you should notice that they produce the same kind of data that are found in an experiment (see Figure 1.6). In each case, one variable is used to create groups, and a second variable is measured to obtain scores within each group. In an experiment, the groups are created by manipulating the independent variable and the scores are the dependent variable. The same terminology is often used to identify the two variables in nonexperimental studies. That is, the variable that is used to create groups is the independent variable and the scores are the dependent variable. For example, the top part of Figure 1.7, gender (boy/girl), is the independent variable and the verbal test scores are the dependent variable. However, you should realize that gender (boy/girl) is not a true independent variable because it is not manipulated. For this reason, the "independent variable" in a nonexperimental study is often called a *quasi-independent variable.*

DEFINITION
> In a nonexperimental study, the "independent variable" that is used to create the different groups of scores is often called the **quasi-independent variable.**

Most of the statistical procedures presented in this book are designed for research studies that compare sets of scores like the experimental study in Figure 1.6 and the nonexperimental studies in Figure 1.7. Specifically, we examine descriptive statistics that summarize and describe the scores in each group, and we examine inferential statistics that allow us to use the groups, or samples, to generalize to the entire population.

LEARNING CHECK

1. Researchers have observed that high school students who watched educational television programs as young children then to have higher grades than their peers who did not watch educational television. Is this study an example of an experiment? Explain why or why not.

2. What two elements are necessary for a research study to be an experiment?

3. The results from an experiment indicate that increasing the amount of indoor lighting during the winter months results in significantly lower levels of depression. Identify the independent variable and the dependent variable for this study.

ANSWERS

1. This study could be correlational or nonexperimental, but it is definitely not an example of a true experiment. The researcher is simply observing, not manipulating, the amount of educational television.

2. First, the researcher must manipulate one of the two variables being studied. Second, all other variables that might influence the results must be controlled.

3. The independent variable is the amount of indoor lighting, and the dependent variable is a measure of depression for each individual.

1.4   VARIABLES AND MEASUREMENT

The scores that make up the data from a research study are the result of observing and measuring variables. For example, a researcher may finish a study with a set of IQ scores, personality scores, or reaction time scores. In this section, we take a closer look at the variables that are being measured and the process of measurement.

CONSTRUCTS AND OPERATIONAL DEFINITIONS

Some variables, such as height, weight, and eye color are well-defined, concrete entities that can be observed and measured directly. On the other hand, many variables studied by behavioral scientists are internal characteristics that people use to help understand and explain behavior. For example, we say that a child does well in school because he or she is *intelligent*. Or we say that someone is anxious in social situations because he or she has low *self-esteem*. Variables like intelligence and self-esteem are called *constructs*, and because they are intangible and cannot be directly observed, they are often called *hypothetical constructs*.

Although constructs such as intelligence are internal characteristics that cannot be directly observed, it is possible to observe and measure behaviors that are representative of the construct. For example, we cannot "see" intelligence but we can see examples of intelligent behavior. The external behaviors can then be used to create an operational definition for the construct. An *operational definition* defines a construct in terms of external behaviors that can be observed and measured. For example, your intelligence is measured and defined by your performance on an IQ test.

DEFINITIONS

Constructs are internal attributes or characteristics that cannot be directly observed but are useful for describing and explaining behavior.

An **operational definition** identifies a measurement procedure (a set of operations) for measuring an external behavior and uses the resulting measurements as a definition and a measurement of a hypothetical construct. Note that an operational definition has two components: First, it describes a set of operations for measuring a construct. Second, it defines the construct in terms of the resulting measurements.

DISCRETE AND CONTINUOUS VARIABLES

The variables in a study can be characterized by the type of values that can be assigned to them. A *discrete variable* consists of separate, indivisible categories. For this type of variable, there are no intermediate values between two adjacent categories. Consider the values displayed when dice are rolled. Between neighboring values—for example, seven dots and eight dots—no other values can ever be observed.

DEFINITION

A **discrete variable** consists of separate, indivisible categories. No values can exist between two neighboring categories.

Discrete variables are commonly restricted to whole countable numbers—for example, the number of children in a family or the number of students attending class. If you observe class attendance from day to day, you may count 18 students one day and 19 students the next day. However, it is impossible ever to observe a value between 18 and 19. A discrete variable may also consist of observations that differ qualitatively. For example, a psychologist observing patients may classify some as having panic disorders, others as having dissociative disorders, and some as having psychotic disorders. The type of disorder is a discrete variable because there are distinct and finite categories that can be observed.

On the other hand, many variables are not discrete. Variables such as time, height, and weight are not limited to a fixed set of separate, indivisible categories. You can measure time, for example, in hours, minutes, seconds, or fractions of seconds. These variables are called *continuous* because they can be divided into an infinite number of fractional parts.

DEFINITION

For a **continuous variable**, there are an infinite number of possible values that fall between any two observed values. A continuous variable is divisible into an infinite number of fractional parts.

Suppose, for example, that a researcher is measuring weights for a group of individuals participating in a diet study. Because weight is a continuous variable, it can be pictured as a continuous line (Figure 1.8). Note that there are an infinite number of possible points on the line without any gaps or separations between neighboring points. For any two different points on the line, it is always possible to find a third value that is between the two points.

Two other factors apply to continuous variables:

1. When measuring a continuous variable, it should be very rare to obtain identical measurements for two different individuals. Because a continuous variable has an infinite number of possible values, it should be almost impossible for two people to have exactly the same score. If the data show a substantial number of tied scores, then you should suspect that the measurement procedure is very crude or that the variable is not really continuous.

2. When measuring a continuous variable, each measurement category is actually an *interval* that must be defined by boundaries. For example, two people who both claim to weigh 150 pounds are probably not *exactly* the same weight. However, they are both around 150 pounds. One person may actually weigh 149.6 and the other 150.3. Thus, a score of 150 is not a specific point on the scale but instead is an interval (see Figure 1.8). To differentiate a score of 150 from a

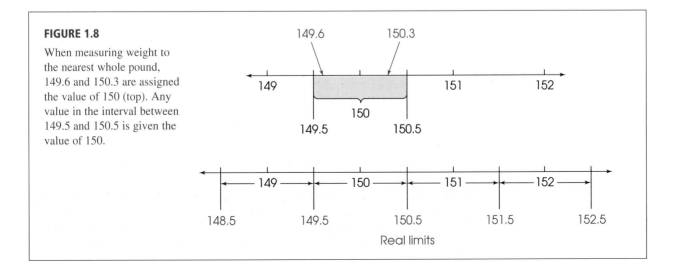

FIGURE 1.8

When measuring weight to the nearest whole pound, 149.6 and 150.3 are assigned the value of 150 (top). Any value in the interval between 149.5 and 150.5 is given the value of 150.

score of 149 or 151, we must set up boundaries on the scale of measurement. These boundaries are called *real limits* and are positioned exactly halfway between adjacent scores. Thus, a score of $X = 150$ pounds is actually an interval bounded by a *lower real limit* of 149.5 at the bottom and an *upper real limit* of 150.5 at the top. Any individual whose weight falls between these real limits will be assigned a score of $X = 150$.

DEFINITION

Real limits are the boundaries of intervals for scores that are represented on a continuous number line. The real limit separating two adjacent scores is located exactly halfway between the scores. Each score has two real limits. The **upper real limit** is at the top of the interval, and the **lower real limit** is at the bottom.

The concept of real limits applies to any measurement of a continuous variable, even when the score categories are not whole numbers. For example, if you were measuring time to the nearest tenth of a second, the measurement categories would be 31.0, 31.1, 31.2, and so on. Each of these categories represents an interval on the scale that is bounded by real limits. For example, a score of $X = 31.1$ seconds indicates that the actual measurement is in an interval bounded by a lower real limit of 31.05 and an upper real limit of 31.15. Remember that the real limits are always halfway between adjacent categories.

Later in this book, real limits are used for constructing graphs and for various calculations with continuous scales. For now, however, you should realize that real limits are a necessity whenever you make measurements of a continuous variable.

Finally, we should warn you that the term *continuous* applies to the variable that is being measured and not to the scores that are obtained from the measurement. For example, measuring people's heights to the nearest inch produces scores of 60, 61, 62, and so on. Although the scores may appear to be discrete numbers, the underlying variable is continuous. One key to determining whether a variable is continuous or discrete is that a continuous variable can be divided into any number of fractional parts. Height can be measured to the nearest inch, the nearest 0.5 inch, or the nearest 0.1 inch. Similarly, a professor evaluating students' knowledge could use a pass/fail system that classifies students into two broad categories. However, the professor could choose to use a 10-point quiz that divides student knowledge into 11 categories corresponding to quiz scores from 0 to 10. Or the professor could use a 100-point exam that potentially divides student knowledge into 101 categories from 0 to 100. Whenever you are free to choose the degree of precision or the number of categories for measuring a variable, the variable must be continuous.

SCALES OF MEASUREMENT

It should be obvious by now that data collection requires that we make measurements of our observations. Measurement involves assigning individuals or events to categories. The categories can simply be names such as male/female or employed/unemployed, or they can be numerical values such as 68 inches or 175 pounds. The complete set of categories makes up a *scale of measurement,* and the relationships between the categories determine different types of scales. The distinctions among the scales are important because they identify the limitations of certain types of measurements and because certain statistical procedures are appropriate for data collected on some scales but not on others. If you were interested in people's heights, for example, you could measure a group of individuals by simply classifying them into three categories: tall, medium, and short. However, this simple classification would not tell you much about the actual heights of the individuals, and these measurements would not give you enough information to calculate an average height for the group. Although the simple classification would be adequate for some purposes, you would need more sophisticated measurements before you could answer more detailed questions. In this section, we examine four different scales of measurement, beginning with the simplest and moving to the most sophisticated.

THE NOMINAL SCALE

The word *nominal* means "having to do with names." Measurement on a nominal scale involves classifying individuals into categories that have different names but are not related to each other in any systematic way. For example, if you were measuring the academic majors for a group of college students, the categories would be art, biology, business, chemistry, and so on. Each student would be classified in one category according to his or her major. The measurements from a nominal scale allow us to determine whether two individuals are different, but they do not identify either the direction or the size of the difference. If one student is an art major and another is a biology major we can say that they are different, but we cannot say that art is "more than" or "less than" biology and we cannot specify how much difference there is between art and biology. Other examples of nominal scales include classifying people by race, gender, or occupation.

DEFINITION

> A **nominal scale** consists of a set of categories that have different names. Measurements on a nominal scale label and categorize observations, but do not make any quantitative distinctions between observations.

Although the categories on a nominal scale are not quantitative values, they are occasionally represented by numbers. For example, the rooms or offices in a building may be identified by numbers. You should realize that the room numbers are simply names and do not reflect any quantitative information. Room 109 is not necessarily bigger than Room 100 and certainly not 9 points bigger. It also is fairly common to use numerical values as a code for nominal categories when data are entered into computer programs. For example, the data from a survey may code males with a 0 and females with a 1. Again, the numerical values are simply names and do not represent any quantitative difference. The scales that follow do reflect an attempt to make quantitative distinctions.

THE ORDINAL SCALE

The categories that make up an *ordinal scale* not only have different names (as in a nominal scale) but also are organized in a fixed order corresponding to differences of magnitude.

DEFINITION

> An **ordinal scale** consists of a set of categories that are organized in an ordered sequence. Measurements on an ordinal scale rank observations in terms of size or magnitude.

Often, an ordinal scale consists of a series of ranks (first, second, third, and so on) like the order of finish in a horse race. Occasionally, the categories are identified by verbal labels like small, medium, and large drink sizes at a fast-food restaurant. In either case, the fact that the categories form an ordered sequence means that there is a directional relationship between categories. With measurements from an ordinal scale, you can determine whether two individuals are different and you can determine the direction of difference. However, ordinal measurements do not allow you to determine the magnitude of the difference between two individuals. For example, if Billy is placed in the low-reading group and Tim is placed in the high-reading group, you know that Tim is a better reader, but you do not know how much better. Other examples of ordinal scales include socioeconomic class (upper, middle, lower) and T-shirt sizes (small, medium, large). In addition, ordinal scales are often used to measure variables for which it is difficult to assign numerical scores. For example, people can rank their food preferences but might have trouble explaining "how much" they prefer chocolate ice cream to steak.

THE INTERVAL AND RATIO SCALES

Both an *interval scale* and a *ratio scale* consist of series of ordered categories (like an ordinal scale) with the additional requirement that the categories form a series of intervals that are all exactly the same size. Thus, the scale of measurement consists of a series of equal intervals, such as inches on a ruler. Other examples of interval and ratio scales are the measurement of time in seconds, weight in pounds, and temperature in degrees Fahrenheit. Note that, in each case, one interval (1 inch, 1 second, 1 pound, 1 degree) is the same size, no matter where it is located on the scale. The fact that the intervals are all the same size makes it possible to determine both the size and the direction of the difference between two measurements. For example, you know that a measurement of 80° Fahrenheit is higher than a measure of 60°, and you know that it is exactly 20° higher.

The factor that differentiates an interval scale from a ratio scale is the nature of the zero point. An interval scale has an arbitrary zero point. That is, the value 0 is assigned to a particular location on the scale simply as a matter of convenience or reference. In particular, a value of zero does not indicate a total absence of the variable being measured. For example a temperature of 0 degrees Fahrenheit does not mean that there is no temperature, and it does not prohibit the temperature from going even lower. Interval scales with an arbitrary zero point are relatively rare. The two most common examples are the Fahrenheit and Celsius temperature scales. Other examples include golf scores (above and below par) and relative measures such as above and below average rainfall.

A ratio scale is anchored by a zero point that is not arbitrary but rather is a meaningful value representing none (a complete absence) of the variable being measured. The existence of an absolute, nonabitrary zero point means that we can measure the absolute amount of the variable; that is, we can measure the distance from 0. This makes it possible to compare measurements in terms of ratios. For example, an individual who requires 10 seconds to solve a problem (10 more than 0) has taken twice as much time as an individual who finishes in only 5 seconds (5 more than 0). With a ratio scale, we can measure the direction and the size of the difference between two measurements and we can describe the difference in terms of a ratio. Ratio scales are quite common and include physical measures such as height and weight, as well as variables such as reaction time or the number of errors on a test. The distinction between an interval scale and a ratio scale is demonstrated in Example 1.2.

DEFINITIONS

An **interval scale** consists of ordered categories that are all intervals of exactly the same size. Equal differences between numbers on scale reflect equal differences in magnitude. However, the zero point on an interval scale is arbitrary and does not indicate a zero amount of the variable being measured.

A **ratio scale** is an interval scale with the additional feature of an absolute zero point. With a ratio scale, ratios of numbers do reflect ratios of magnitude.

EXAMPLE 1.2

A researcher obtains measurements of height for a group of 8-year-old boys. Initially, the researcher simply records each child's height in inches, obtaining values such as 44, 51, 49, and so on. These initial measurements constitute a ratio scale. A value of zero represents no height (absolute zero). Also, it is possible to use these measurements to form ratios. For example, a child who is 60 inches tall is one and a half times taller than a child who is 40 inches tall.

Now suppose that the researcher converts the initial measurement into a new scale by calculating the difference between each child's actual height and the average height for this age group. A child who is 1 inch taller than average now gets a score of +1; a child 4 inches taller than average gets a score of +4. Similarly, a child who is 2 inches shorter than average gets a score of −2. The new scores constitute an interval scale of measurement. A score of zero no longer indicates an absence of height; now it simply means average height.

Notice that both sets of scores involve measurement in inches, and you can compute differences, or distances, on either scale. For example, there is a 6-inch difference in height between two boys who measure 57 and 51 inches tall on the first scale. Likewise, there is a 6-inch difference between two boys who measure +9 and +3 on the second scale. However, you should also notice that ratio comparisons are not possible on the second scale. For example, a boy who measures +9 is not three times as tall as a boy who measures +3.

For our purposes, scales of measurement are important because they influence the kind of statistics that can and cannot be used. For example, if you measure IQ scores for a group of students, it is possible to add the scores together and calculate a mean score for the group. On the other hand, if you measure the academic major for each student, you cannot compute the mean. (What is the mean of three psychology majors, an English major, and two chemistry majors?) The vast majority of the statistical techniques presented in this book are designed for numerical scores from an interval or a ratio scale. For most statistical applications, the distinction between an interval scale and a ratio scale is not important because both scales produce numerical values that permit us to compute differences between scores, to add scores, and to calculate mean scores. On the other hand, measurements from nominal or ordinal scales are typically not numerical values and are not compatible with many basic arithmetic operations. Therefore, alternative statistical techniques are necessary for data from nominal or ordinal scales of measurement (for example, the median and the mode in Chapter 3, the Spearman correlation in Chapter 16, and the chi-square tests in Chapter 18).

LEARNING CHECK

1. A survey asks people to identify their age, annual income, and marital status (single married, divorced, etc.). For each of these three variables, identify the scale of measurement that probably is used.

2. An educational psychologist classifies students as high, medium, and low intelligence. What kind of scale is being used?

3. In a study on perception of facial expressions, participants must classify the emotions displayed in photographs of people as anger, sadness, joy, disgust, fear, or surprise. Emotional expression is measured on a(n) _____ scale.

4. A researcher studies the factors that determine how many children couples decide to have. The variable, number of children, is a _____ (discrete/continuous) variable.

5. An investigator studies how concept-formation ability changes with age. Age is a _____ (discrete/continuous) variable.

6. **a.** When measuring height to the nearest inch, what are the real limits for a score of 68 inches?

   b. When measuring height to the nearest half inch, what are the real limits for a score of 68 inches?

ANSWERS

1. Age and annual income are measured on ratio scales. Marital status is measured on a nominal scale.

2. ordinal   **3.** nominal   **4.** discrete   **5.** continuous

6. **a.** 67.5 and 68.5   **b.** 67.75 and 68.25

1.5 STATISTICAL NOTATION

The measurements obtained in research studies provide the data for statistical analysis. Most of the statistical analyses will use the same general mathematical operations, notation, and basic arithmetic that you have learned during previous years of school. In case you are unsure of your mathematical skills, there is a mathematics review section in Appendix A at the back of this book. The appendix also includes a skills assessment exam (p. 704) to help you determine whether you need the basic mathematics review. In this section, we introduce some of the specialized notation that is used for statistical calculations. In later chapters, additional statistical notation is introduced as it is needed.

SCORES

Measuring a variable in research study typically yields a value or a score for each individual. Raw scores are the original, unchanged scores obtained in the study. Scores for a particular variable are represented by the letter X. For example, if performance in your statistics course is measured by tests and you obtain a 35 on the first test, then we could state that $X = 35$. A set of scores can be presented in a column that is headed by X. For example, a list of quiz scores from your class might be presented as shown in the margin (the single column on the left).

When observations are made for two variables, there will be two scores for each individual. The data can be presented as two lists labeled X and Y for the two variables.

X	X	Y
37	72	165
35	68	151
35	67	160
30	67	160
25	68	146
17	70	160
16	66	133

For example, observations for people's height in inches (variable X) and weight in pounds (variable Y) can be presented as shown in the margin (the double column on the right). Each pair X, Y represents the observations made of a single participant.

It is also useful to specify how many scores are in a set. We will use an uppercase letter N to represent the number of scores in a population and a lowercase letter n to represent the number of scores in a sample. Throughout the remainder of the book you will notice that we often use notational differences to distinguish between samples and populations. For the height and weight data in the preceding table, $n = 7$ for both variables. Note that by using a lowercase letter n, we are implying that these data are a sample.

SUMMATION NOTATION

Many of the computations required in statistics involve adding a set of scores. Because this procedure is used so frequently, a special notation is used to refer to the sum of a set of scores. The Greek letter *sigma*, or Σ, is used to stand for summation. The expression ΣX means to add all the scores for variable X. The summation sign Σ can be read as "the sum of." Thus, ΣX is read "the sum of the scores." For the following set of quiz scores,

$$10, \quad 6, \quad 7, \quad 4$$

$\Sigma X = 27$ and $N = 4$.

To use summation notation correctly, keep in mind the following two points:

1. The summation sign Σ is always followed by a symbol or mathematical expression. The symbol or expression identifies exactly which values are to be added. To compute ΣX, for example, the symbol following the summation sign is X, and the task is to find the sum of the X values. On the other hand, to compute $\Sigma(X - 1)^2$, the summation sign is followed by a relatively complex mathematical expression, so your first task is to calculate all of the $(X - 1)^2$ values and then add the results.

2. The summation process is often included with several other mathematical operations, such as multiplication or squaring. To obtain the correct answer, it is essential that the different operations be done in the correct sequence. Following is a list showing the correct *order of operations* for performing mathematical operations. Most of this list should be familiar, but you should note that we have inserted the summation process as the fourth operation in the list.

Order of Mathematical Operations

More information on the order of operations for mathematics is available in the Math Review appendix, page 705.

1. Any calculation contained within parentheses is done first.

2. Squaring (or raising to other exponents) is done second.

3. Multiplying and/or dividing is done third. A series of multiplication and/or division operations should be done in order from left to right.

4. Summation using the Σ notation is done next.

5. Finally, any other addition and/or subtraction is done.

The following examples demonstrate how summation notation is used in most of the calculations and formulas we will present in this book.

EXAMPLE 1.3

A set of four scores consists of values 3, 1, 7, and 4. We will compute ΣX, ΣX^2, and $(\Sigma X)^2$ for these scores. To help demonstrate the calculations, we will use a computational table showing the original scores (the X values) in the first column. Additional columns can then be added to show additional steps in the series of operations. You should notice that the first three operations in the list (parentheses, squaring, multiplying) all create a new column of values. The last two operations, however, produce a single value corresponding to the sum.

X	X^2
3	9
1	1
7	49
4	16

The table to the left shows the original scores (the X values) and the squared scores (the X^2 values) that are needed to compute ΣX^2.

The first calculation, ΣX, does not include any parentheses, squaring, or multiplication, so we go directly to the summation operation. The X values are listed in the first column of the table, and we simply add the values in this column:

$$\Sigma X = 3 + 1 + 7 + 4 = 15$$

To compute ΣX^2, the correct order of operations is to square each score and then find the sum of the squared values. The computational table shows the original scores and the results obtained from squaring (the first step in the calculation). The second step is to find the sum of the squared values, so we simply add the numbers in the X^2 column.

$$\Sigma X^2 = 9 + 1 + 49 + 16 = 75$$

The final calculation, $(\Sigma X)^2$, includes parentheses, so the first step is to perform the calculation inside the parentheses. Thus, we first find ΣX and then square this sum. Earlier, we computed $\Sigma X = 15$, so

$$(\Sigma X)^2 = (15)^2 = 225$$

EXAMPLE 1.4 Use the same set of four scores from Example 1.3 and compute $\Sigma(X - 1)$ and $\Sigma(X - 1)^2$. The following computational table helps demonstrate the calculations.

X	$(X - 1)$	$(X - 1)^2$	
3	2	4	The first column lists the
1	0	0	original scores. A second
7	6	36	column lists the $(X - 1)$
4	3	9	values, and a third column
			shows the $(X - 1)^2$ values.

To compute $\Sigma(X - 1)$, the first step is to perform the operation inside the parentheses. Thus, we begin by subtracting one point from each of the X values. The resulting values are listed in the middle column of the table. The next step is to add the $(X - 1)$ values.

$$\Sigma(X - 1) = 2 + 0 + 6 + 3 = 11$$

The calculation of $\Sigma(X - 1)^2$ requires three steps. The first step (inside parentheses) is to subtract 1 point from each X value. The results from this step are shown in the middle column of the computational table. The second step is to square each of the

$(X - 1)$ values. The results from this step are shown in the third column of the table. The final step is to add the $(X - 1)^2$ values to obtain

$$\Sigma(X - 1)^2 = 4 + 0 + 36 + 9 = 49$$

Notice that this calculation requires squaring before adding. A common mistake is to add the $(X - 1)$ values and then square the total. Be careful!

EXAMPLE 1.5   In both of the preceding examples, and in many other situations, the summation operation is the last step in the calculation. According to the order of operations, parentheses, exponents, and multiplication all come before summation. However, there are situations in which extra addition and subtraction are completed after the summation. For this example, use the same scores that appeared in the previous two examples, and compute $\Sigma X - 1$.

With no parentheses, exponents, or multiplication, the first step is the summation. Thus, we begin by computing ΣX. Earlier we found $\Sigma X = 15$. The next step is to subtract one point from the total. For these data,

$$\Sigma X - 1 = 15 - 1 = 14$$

EXAMPLE 1.6   For this example, each individual has two scores. The first score is identified as X, and the second score is Y. With the help of the following computational table, compute ΣX, ΣY, and ΣXY.

Person	X	Y	XY
A	3	5	15
B	1	3	3
C	7	4	28
D	4	2	8

To find ΣX, simply add the values in the X column.

$$\Sigma X = 3 + 1 + 7 + 4 = 15$$

Similarly, ΣY is the sum of the Y values.

$$\Sigma Y = 5 + 3 + 4 + 2 = 14$$

To compute ΣXY, the first step is to multiply X times Y for each individual. The resulting products (XY values) are listed in the third column of the table. Finally, we add the products to obtain

$$\Sigma XY = 15 + 3 + 28 + 8 = 54$$

1. Calculate each value requested for the following scores: 4, 2, 5, 1.
   a. ΣX
   b. $(\Sigma X)^2$
   c. ΣX^2
   d. $\Sigma X - 1$
   e. $\Sigma(X - 1)$
   f. $\Sigma(X - 1)^2$

2. Use summation notation to express each of the following calcualtions
   a. Add 2 points to each score, then add the resulting values.
   b. Square each score, then add the squared values.

ANSWERS

1. a. $\Sigma X = 12$
   b. $(\Sigma X)^2 = 144$
   c. $\Sigma X^2 = 46$
   d. $\Sigma X - 1 = 11$
   e. $\Sigma(X - 1) = 8$
   f. $\Sigma(X - 1)^2 = 26$

2. a. $\Sigma(X + 2)$
   b. ΣX^2

SUMMARY

1. The term *statistics* is used to refer to methods for organizing, summarizing, and interpreting data.

2. Scientific questions usually concern a population, which is the entire set of individuals one wishes to study. Usually, populations are so large that it is impossible to examine every individual, so most research is conducted with samples. A sample is a group selected from a population, usually for purposes of a research study.

3. A characteristic that describes a sample is called a statistic, and a characteristic that describes a population is called a parameter. Although sample statistics are usually representative of corresponding population parameters, there is typically some discrepancy between a statistic and a parameter. The naturally occurring difference between a statistic and a parameter is called sampling error.

4. Statistical methods can be classified into two broad categories: descriptive statistics, which organize and summarize data, and inferential statistics, which use sample data to draw inferences about populations.

5. The correlational method examines relationships between variables by measuring two different variables for each individual. This method allows researchers to measure and describe relationships, but cannot produce a cause-and-effect explanation for the relationship.

6. The experimental method examines relationships between variables by manipulating an independent

variable to create different treatment conditions and then measuring a dependent variable to obtain a group of scores in each condition. The groups of scores are then compared. A systematic difference between groups provides evidence that changing the independent variable from one condition to another also caused a change in the dependent variable. All other variables are controlled to prevent them from influencing the relationship. The intent of the experimental method is to demonstrate a cause-and-effect relationship between variables.

7. Nonexperimental studies also examine relationships between variables by comparing groups of scores, but they do not have the rigor of true experiments and cannot produce cause-and-effect explanations. Instead of manipulating a variable to create different groups, a nonexperimental study uses a preexisting participant characteristic (such as male/female) or the passage of time (before/after) to create the groups being compared.

8. A measurement scale consists of a set of categories that are used to classify individuals. A nominal scale consists of categories that differ only in name and are not differentiated in terms of magnitude or direction. In an ordinal scale, the categories are differentiated in terms of direction, forming an ordered series. An interval scale consists of an ordered series of categories that are all equal-sized intervals. With an interval scale, it is possible to differentiate direction and magnitude (or distance) between categories. Finally, a ratio scale

is an interval scale for which the zero point indicates none of the variable being measured. With a ratio scale, ratios of measurements reflect ratios of magnitude.

9. A discrete variable consists of indivisible categories, often whole numbers that vary in countable steps. A continuous variable consists of categories that are infinitely divisible. Each score corresponds to an interval on the scale. The boundaries that separate intervals are called real limits and are located exactly halfway between adjacent scores.

10. The letter X is used to represent scores for a variable. If a second variable is used, Y represents its scores.

The letter N is used as the symbol for the number of scores in a population; n is the symbol for a sample.

11. The Greek letter sigma (Σ) is used to stand for summation. Therefore, the expression ΣX is read "the sum of the scores." Summation is a mathematical operation (like addition or multiplication) and must be performed in its proper place in the order of operations; summation occurs after parentheses, exponents, and multiplying/dividing have been completed.

KEY TERMS

statistics	inferential statistics	operational definition
population	sampling error	discrete variable
sample	correlational method	continuous variable
variable	experimental method	real limits
data	independent variable	upper real limit
data set	dependent variable	lower real limit
datum	control condition	nominal scale
raw score	experimental condition	ordinal scale
parameter	quasi-independent variable	interval scale
statistic	construct	ratio scale
descriptive statistics		

RESOURCES

There is a book companion website containing practice quizzes and other learning aids for every chapter in this book as well as a series of workshops and other resources corresponding to the main topic areas. Go to www.cengage.com/psychology/gravetter Scroll down until you find the cover image for this book and click on *Student Companion Site*. In the left-hand column you will find a variety of learning exercises for Chapter 1, including a tutorial quiz. Also in the left-hand column, under Book Resources, you will find a link to the Workshops. For Chapter 1, there is a workshop that reviews the scales of measurement. To get there, click on the Workshop link, then click on *Scales of Measurement*. To find materials for other chapters, you begin by selecting the desired chapter at the top of the page. Note that the Workshops were not developed specifically for this book but are used by several different books written by different authors. As a result, you may find that some of the notation or terminology is different from that which you learned in this text.

At the end of each chapter we will remind you about the Web resources. Again, there is a tutorial quiz for every chapter, and we will notify you whenever there is a workshop that is related to the chapter content.

ENHANCED WebAssign

Guided interactive tutorials, end-of-chapter problems, and related testbank items may be assigned online at WebAssign.

WebTUTOR

For those using WebTutor along with this book, there is a WebTutor section corresponding to this chapter. The WebTutor contains a brief summary of Chapter 1, hints for learning the new material and for avoiding common errors, and sample exam items including solutions.

SPSS

The Statistical Package for the Social Sciences, known as SPSS, is a computer program that performs most of the statistical calculations that are presented in this book, and is commonly available on college and university computer systems. Appendix D contains a general introduction to SPSS. In the Resource section at the end of each chapter for which SPSS is applicable, there are step-by-step instructions for using SPSS to perform the statistical operations presented in the chapter.

FOCUS ON PROBLEM SOLVING

1. It may help to simplify summation notation if you observe that the summation sign is always followed by a symbol or symbolic expression—for example, ΣX or $\Sigma(X + 3)$. This symbol specifies which values you are to add. If you use the symbol as a column heading and list all the appropriate values in the column, your task is simply to add up the numbers in the column. To find $\Sigma(X + 3)$ for example, start a column headed with $(X + 3)$ next to the column of Xs. List all the $(X + 3)$ values; then find the total for the column.

2. Often, summation notation is part of a relatively complex mathematical expression that requires several steps of calculation. The series of steps must be performed according to the order of mathematical operations (see page 25). The best procedure is to use a computational table that begins with the original X values listed in the first column. Except for summation, each step in the calculation creates a new column of values. For example, computing $\Sigma(X - 1)^2$ involves three steps and produces a computional table with three columns. The final step is to add the values in the third column (see Example 1.4).

DEMONSTRATION 1.1

SUMMATION NOTATION

A set of data consists of the following scores:

$$7 \quad 3 \quad 9 \quad 5 \quad 4$$

For these data, find the following values:
   a. ΣX **b.** $(\Sigma X)^2$ **c.** ΣX^2 **d.** $\Sigma X + 5$ **e.** $\Sigma(X - 2)$

Compute ΣX   To compute ΣX, we simply add all of the scores in the group. For these data, we obtain

$$\Sigma X = 7 + 3 + 9 + 5 + 4 = 28$$

Compute $(\Sigma X)^2$ The key to determining the value of $(\Sigma X)^2$ is the presence of parentheses. The rule is to perform the operations that are inside the parentheses first.

STEP 1   Find the sum of the scores, ΣX.

STEP 2   Square the total.
   We have already determined that ΣX is 28. Squaring this total, we obtain

$$(\Sigma X)^2 = (28)^2 = 784$$

Compute ΣX^2   Calculating the sum of the squared scores, ΣX^2, involves two steps.

STEP 1   Square each score.

STEP 2   Add the squared values.
   These steps are most easily accomplished by constructing a computational table. The first column has the heading X and lists the scores. The second column is labeled X^2 and contains the squared value for each score. For this example, the table is presented in the margin. To find the value for ΣX^2, we add the values in the X^2 column.

X	X^2
7	49
3	9
9	81
5	25
4	16

$$\Sigma X^2 = 49 + 9 + 81 + 25 + 16 = 180$$

Compute $\Sigma X + 5$ In this expression, there are no parentheses. Thus, the summation sign is applied only to the X values.

STEP 1   Find the sum of X.

STEP 2   Add the constant 5 to the total from step 1.
   Earlier we found that the sum of the scores is 28. For $\Sigma X + 5$, we obtain the following:

$$\Sigma X + 5 = 28 + 5 = 33$$

Compute $\Sigma(X - 2)$ The summation sign is followed by an expression with parentheses. In this case, $X - 2$ is treated as a single expression, and the summation sign applies to the $(X - 2)$ values.

STEP 1   Subtract 2 from every score.

STEP 2   Add these new values.
This problem can be done by using a computational table with two columns, headed X and $X - 2$, respectively.

X	$X - 2$
7	5
3	1
9	7
5	3
4	2

To determine the value for $\Sigma(X - 2)$, we add the numbers in the $X - 2$ column.

$$\Sigma(X - 2) = 5 + 1 + 7 + 3 + 2 = 18$$

PROBLEMS

*1. A researcher is interested in the effect of an electrolytic sports drink on the endurance of adolescent boys. A group of 30 boys is selected and half are given a treadmill endurance test while consuming the sports drink and the other half take the test while drinking water. For this study,
a. Identify the population.
b. Identify the sample.

2. Define the terms parameter and statistic. Be sure that the concepts of population and sample are included in your definitions.

3. Statistical methods are classified into two major categories: descriptive and inferential. Describe the general purpose for the statistical methods in each category.

4. A researcher plans to compare two treatment conditions by measuring one sample in treatment 1 and a second sample in treatment 2. The researcher will then compare the scores for the two treatments. If there is a difference between the two groups of scores,
a. Briefly explain how the difference may have been caused by the treatments.
b. Briefly explain how the difference simply may be sampling error.

5. Describe the data for a correlational research study. Explain how correlational research is different from

other research evaluating the relationship between two variables.

6. Describe the goal of an experimental research study and identify the two elements that are necessary for an experiment to achieve its goal.

7. Strack, Martin, and Stepper (1988) found that people rated cartoons as funnier when holding a pen in their teeth (which forced them to smile) than when holding a pen in their lips (which forced them to frown). For this study, identify the independent variable and the dependent variable.

8. Downs and Abwender (2002) found neurological deficits in soccer players who are routinely hit on the head with soccer balls compared to swimmers, who are also athletes but who are not regularly hit in the head. Is this an example of an experimental or a nonexperimental study?

9. A researcher is interested in the relationship between caffeine consumption and activity level for elementary school children. A sample of 20 children is obtained and each child is interviewed to determine a level of caffeine consumption. The researcher then records the level of activity for each child during a 30-minute session on the playground. Is this an experimental or a correlational study? Explain your answer.

---

*Solutions for odd-numbered problems are provided in Appendix C.

10. A researcher is interested in the relationship between caffeine consumption and activity level for elementary school children. A sample of 20 children is obtained and the children are separated into two roughly equivalent groups. The children in one group are given a drink containing 300 mg of caffeine and the other group gets a group with no caffeine. The researcher then records the level of activity for each child during a 30-minute session on the playground. Is this an experimental or a correlational study? Explain your answer.

11. A research study comparing college alcohol use in the United States and Canada reports that more Canadian students drink but American students drink more (Kuo, Adlaf, Lee, Gliksman, Demers, and Wechsler, 2002). Is this study an example of an experiment? Explain why or why not.

12. Oxytocin is a naturally occurring brain chemical that is nicknamed the "love hormone" because it seems to play a role in the formation of social relationships such as mating pairs and parent-child bonding. A recent study demonstrated that oxytocin appears to increase people's tendency to trust others (Kosfeld, Heinrichs, Zak, Fischbacher, and Fehr, 2005). Using an investment game, the study demonstrated that people who inhaled oxytocin were more likely to give their money to a trustee compared to people who inhaled an inactive placebo. For this experimental study, identify the independent variable and the dependent variable.

13. Explain the difference between a discrete variable and a continuous variable. Give an example of each.

14. Four scales of measurement were introduced in this chapter: nominal, ordinal, interval, and ratio.
  a. What additional information is obtained from measurements on an ordinal scale compared to measurements on a nominal scale?
  b. What additional information is obtained from measurements on an interval scale compared to measurements on an ordinal scale?
  c. What additional information is obtained from measurements on a ratio scale compared to measurements on an interval scale?

15. In an experiment examining the effects of humor on memory, Schmidt (1994) showed participants a list of sentences, half of which were humorous and half were nonhumorous. The participants consistently recalled more of the humorous sentences than the nonhumorous sentences.
  a. Identify the independent variable for this study.
  b. What scale of measurement is used for the independent variable?.
  c. Identify the dependent variable for this study.
  d. What scale of measurement is used for the dependent variable?

16. Define or give an example of a hypothetical construct. Explain why operational definitions are needed to define and to measure constructs.

17. Research has shown that supplements high in antioxidants (strawberry, spinach, and especially blueberry) can actually reverse some effects of aging (Joseph et al., 1999). In a typical study, one group of elderly rats receives the supplement and a second group is given an inactive placebo. Both groups are then tested on a variety of motor-behavior tasks. One common task is the Morris water maze, which measures how much time it takes for a rat to find a submerged platform in a pool of opaque water. The results generally show that the older rats receiving the supplements are able to locate the platform much faster than their peers who are not receiving supplements.
  a. Identify the independent and dependent variables for this study.
  b. What scale of measurement is used for the dependent variable?

18. For the following scores, find the value of each expression:

	X
a. ΣX	
b. ΣX^2	
c. $(\Sigma X)^2$	6
d. $\Sigma(X - 1)$	1
	3
	4
	2

19. For the following set of scores, find the value of each expression:

	X
a. ΣX	
b. ΣX^2	2
c. $\Sigma(X + 1)$	5
d. $\Sigma(X + 1)^2$	0
	4

20. For the following set of scores, find the value of each expression:

	X
a. ΣX	
b. ΣX^2	
c. $\Sigma(X + 3)$	5
	-1
	0
	-3
	-1

21. Two scores, X and Y, are recorded for each of $n = 5$ subjects. For these scores, find the value of each expression.

    a. ΣX

    b. ΣY

    c. ΣXY

Subject	X	Y
A	3	7
B	5	4
C	0	3
D	8	1
E	6	5

22. Use summation notation to express each of the following calculations:

    a. Add 1 point to each score, then add the resulting values.

    b. Add 1 point to each score and square the result, then add the squared values.

    c. Add the scores and square the sum, then subtract 3 points from the squared value.

23. For the following set of scores, find the value of each expression:

    a. ΣX^2

    b. $(\Sigma X)^2$

    c. $\Sigma(X - 3)$

    d. $\Sigma(X - 3)^2$

X
4
0
0
8

CHAPTER

2

Frequency Distributions

Tools You Will Need

The following items are considered essential background material for this chapter. If you doubt your knowledge of any of these items, you should review the appropriate chapter or section before proceeding.

- Proportions (math review, Appendix A)
  - Fractions
  - Decimals
  - Percentages
- Scales of measurement (Chapter 1): Nominal, ordinal, interval, and ratio
- Continuous and discrete variables (Chapter 1)
- Real limits (Chapter 1)

Preview

2.1   Introduction to Frequency Distributions

2.2   Frequency Distribution Tables

2.3   Frequency Distribution Graphs

2.4   The Shape of a Frequency Distribution

2.5   Percentiles, Percentile Ranks, and Interpolation

2.6   Stem and Leaf Displays

Summary

Focus on Problem Solving

Demonstrations 2.1 and 2.2

Problems

Preview

If at first you don't succeed, you are probably not related to the boss.

Did we make you chuckle or, at least, smile a little? The use of humor is a common technique to capture attention and to communicate ideas. Advertisers, for example, will often try to make a commercial funny so that people will notice it and, perhaps, remember the product. After-dinner speakers will always put a few jokes into the speech in an effort to maintain the audience's interest. Although humor seems to capture our attention, does it actually affect our memory?

In an attempt to answer this question, Stephen Schmidt (1994) conducted a series of experiments examining the effects of humor on memory for sentences. Humorous sentences were collected from a variety of sources and then a nonhumorous version was constructed for each sentence. For example, the nonhumorous version of our opening sentence was:

> People who are related to the boss often succeed the very first time.

Participants were then presented with a list containing half humorous and half nonhumorous sentences. Later, each person was asked to recall as many sentences as possible. The researcher measured the number of humorous sentences and the number of nonhumorous sentences recalled by each participant. Data similar to the results obtained by Schmidt are shown in Table 2.1.

TABLE 2.1

Memory scores for a sample of 16 participants. The scores represent the number of sentences recalled from each category.

Humorous Sentences				Nonhumorous Sentences			
4	5	2	4	5	2	4	2
6	7	6	6	2	3	1	6
2	5	4	3	3	2	3	3
1	3	5	5	4	1	5	3

The Problem: It is difficult to see any clear pattern simply by looking at the list of numbers. Can you tell whether the memory scores for one type of sentence are generally higher than those for the other type?

The Solution: A *frequency distribution* provides an overview of the entire group of scores making it easy to see the general level of performance for each type of sentence. For example, the same memory scores that are shown in Table 2.1 have been organized in a frequency distribution graph in Figure 2.1. In the figure, each individual is represented by a block that is placed above that individual's score. The resulting pile of blocks shows a picture of how the individual scores are distributed. For this example, it is now easy to see that the scores for the humorous sentences are generally higher than the scores for the nonhumorous sentences; on average, participants recalled around 5 humorous sentences but only about 3 of the nonhumorous sentences.

In this chapter we present techniques for organizing data into tables and graphs so that an entire set of scores can be presented in a relatively simple display or illustration.

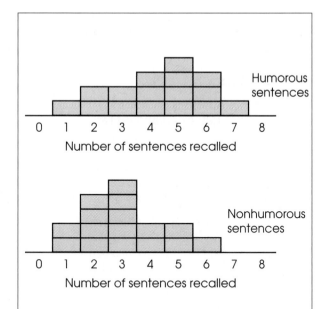

FIGURE 2.1

Hypothetical data showing the number of humorous sentences and the number of nonhumorous sentences recalled by participants in a memory experiment.

2.1   INTRODUCTION TO FREQUENCY DISTRIBUTIONS

When the data collection phase of a research study is completed, the results usually consist of pages of numbers. The immediate problem for the researcher is to organize the scores into some comprehensible form so that any patterns in the data can be seen easily and communicated to others. This is the job of descriptive statistics: to simplify the organization and presentation of data. One of the most common procedures for organizing a set of data is to place the scores in a *frequency distribution.*

DEFINITION   A **frequency distribution** is an organized tabulation of the number of individuals located in each category on the scale of measurement.

A frequency distribution takes a disorganized set of scores and places them in order from highest to lowest, grouping together all individuals who have the same score. If the highest score is $X = 10$, for example, the frequency distribution groups together all the 10s, then all the 9s, then the 8s, and so on. Thus, a frequency distribution allows the researcher to see "at a glance" the entire set of scores. It shows whether the scores are generally high or low, whether they are concentrated in one area or spread out across the entire scale, and generally provides an organized picture of the data. In addition to providing a picture of the entire set of scores, a frequency distribution allows you to see the location of any individual score relative to all of the other scores in the set.

A frequency distribution can be structured either as a table or as a graph, but in either case the distribution presents the same two elements:

1. The set of categories that make up the original measurement scale.

2. A record of the frequency, or number of individuals in each category.

Thus, a frequency distribution presents a picture of how the individual scores are distributed on the measurement scale—hence the name *frequency distribution.*

2.2   FREQUENCY DISTRIBUTION TABLES

It is customary to list categories from highest to lowest, but this is an arbitrary arrangement. Many computer programs list categories from lowest to highest.

The simplest frequency distribution table presents the measurement scale by listing the different measurement categories (X values) in a column from highest to lowest. Beside each X value, we indicate the frequency, or the number of times that particular measurement occurred in the data. It is customary to use an X as the column heading for the scores and an f as the column heading for the frequencies. An example of a frequency distribution table follows.

EXAMPLE 2.1   The following set of $N = 20$ scores was obtained from a 10-point statistics quiz. We will organize these scores by constructing a frequency distribution table. Scores:

8,　9,　8,　7,　10,　9,　6,　4,　9,　8,

7,　8,　10,　9,　8,　6,　9,　7,　8,　8

X	f
10	2
9	5
8	7
7	3
6	2
5	0
4	1

1. The highest score is $X = 10$, and the lowest score is $X = 4$. Therefore, the first column of the table lists the categories that make up the scale of measurement (X values) from 10 down to 4. Notice that all of the possible values are listed in the table. For example, no one had a score of $X = 5$, but this value is included. With an ordinal, interval, or ratio scale, the categories are listed in order (usually highest to lowest). For a nominal scale, the categories can be listed in any order.

2. The frequency associated with each score is recorded in the second column. For example, two people had scores of $X = 6$, so there is a 2 in the f column beside $X = 6$.

Because the table organizes the scores, it is possible to see very quickly the general quiz results. For example, there were only two perfect scores, but most of the class had high grades (8s and 9s). With one exception (the score of $X = 4$), it appears that the class has learned the material fairly well.

Notice that the X values in a frequency distribution table represent the scale of measurement, *not* the actual set of scores. For example, the X column lists the value 10 only one time, but the frequency column indicates that there are actually two values of $X = 10$. Also, the X column lists a value of $X = 5$, but the frequency column indicates that no one actually had a score of $X = 5$.

You also should notice that the frequencies can be used to find the total number of scores in the distribution. By adding up the frequencies, you obtain the total number of individuals:

$$\Sigma f = N$$

OBTAINING ΣX FROM A FREQUENCY DISTRIBUTION TABLE

There may be times when you need to compute the sum of the scores, ΣX, or perform other computations for a set of scores that has been organized into a frequency distribution table. Such calculations can present a problem because many students tend to disregard the frequencies and use only the values listed in the X column of the table. However, it is essential that you use the information in both the X column and the f column to obtain the full set of scores.

When it is necessary to perform calculations for scores that have been organized into a frequency distribution table, the safest procedure is to take the individual scores out of the table before you begin any computations. This process is demonstrated in the following example.

EXAMPLE 2.2

X	f
5	1
4	2
3	3
2	3
1	1

Consider the frequency distribution table shown in the margin. The table shows that the distribution has one 5, two 4s, three 3s, three 2s, and one 1. If you simply list all the individual scores, you can safely proceed with calculations such as finding ΣX or ΣX^2. Note that the complete set contains $N = \Sigma f = 10$ scores and your list should contain 10 values. For example, to compute ΣX you simply add all 10 scores:

$$\Sigma X = 5 + 4 + 4 + 3 + 3 + 3 + 2 + 2 + 2 + 1$$

For the distribution in this table, you should obtain $\Sigma X = 29$. Try it yourself. Similarly, to compute ΣX^2 you square each of the 10 scores and then add the squared values.

$$\Sigma X^2 = 5^2 + 4^2 + 4^2 + 3^2 + 3^2 + 3^2 + 2^2 + 2^2 + 2^2 + 1^2$$

This time you should obtain $\Sigma X^2 = 97$.

An alternative way to get ΣX from a frequency distribution table is to multiply each X value by its frequency and then add these products. This sum may be expressed in symbols as $\Sigma f X$. The computation is summarized as follows for the data in Example 2.2:

Caution: Doing calculations within the table works well for ΣX but can lead to errors for more complex formulas.

X	f	fX	
5	1	5	(the one 5 totals 5)
4	2	8	(the two 4s total 8)
3	3	9	(the three 3s total 9)
2	3	6	(the three 2s total 6)
1	1	1	(the one 1 totals 1)

$$\Sigma X = 29$$

No matter which method you use to find ΣX, the important point is that you must use the information given in the frequency column in addition to the information in the X column.

PROPORTIONS AND PERCENTAGES

In addition to the two basic columns of a frequency distribution, there are other measures that describe the distribution of scores and can be incorporated into the table. The two most common are proportion and percentage.

Proportion measures the fraction of the total group that is associated with each score. In Example 2.2, there were two individuals with $X = 4$. Thus, 2 out of 10 people had $X = 4$, so the proportion would be $\frac{2}{10} = 0.20$. In general, the proportion associated with each score is

$$\text{proportion} = p = \frac{f}{N}$$

Because proportions describe the frequency (f) in relation to the total number (N), they often are called *relative frequencies*. Although proportions can be expressed as fractions (for example, $\frac{2}{10}$), they more commonly appear as decimals. A column of proportions, headed with a p, can be added to the basic frequency distribution table (see Example 2.3).

In addition to using frequencies (f) and proportions (p), researchers often describe a distribution of scores with percentages. For example, an instructor might describe the results of an exam by saying that 15% of the class earned As, 23% Bs, and so on. To compute the percentage associated with each score, you first find the proportion (p) and then multiply by 100:

$$\text{percentage} = p(100) = \frac{f}{N}(100)$$

Percentages can be included in a frequency distribution table by adding a column headed with % (see Example 2.3).

EXAMPLE 2.3   The frequency distribution table from Example 2.2 is repeated here. This time we have added columns showing the proportion (p) and the percentage (%) associated with each score.

X	f	$p = f/N$	$\% = p(100)$
5	1	1/10 = 0.10	10%
4	2	2/10 = 0.20	20%
3	3	3/10 = 0.30	30%
2	3	3/10 = 0.30	30%
1	1	1/10 = 0.10	10%

LEARNING CHECK

1. Construct a frequency distribution table for the following set of scores.

Scores: 3,   4,   3,   2,   5,   3,   2,   1,   2,   5,   3,   4,   4,   2,   3

2. Find each of the following values for the sample in the following frequency distribution table.

   a. n
   b. ΣX
   c. ΣX^2

X	f
5	1
4	2
3	4
2	3
1	1

[handwritten] $5^2+4^2+4^2+3^2+3^2+3^2+3^2+2^2+2^2+2^2+1^2$ $25+$

ANSWERS   1.

X	f
5	2
4	3
3	5
2	4
1	1

2. **a.** $n = 11$ **b.** $\Sigma X = 32$ **c.** $\Sigma X^2 = 106$ (square then add all 11 scores)

GROUPED FREQUENCY DISTRIBUTION TABLES

When a set of data covers a wide range of values, it is unreasonable to list all the individual scores in a frequency distribution table. For example, a set of exam scores ranges from a low of $X = 41$ to a high of $X = 96$. These scores cover a *range* of more than 50 points.

If we were to list all the individual scores from $X = 96$ down to $X = 41$, it would take 56 rows to complete the frequency distribution table. Although this would organize and simplify the data, the table would be long and cumbersome. Remember:

When the scores are whole numbers, the total number of rows for a regular table can be obtained by finding the difference between the highest and the lowest scores and adding 1:

rows = highest − lowest + 1

The purpose for constructing a table is to obtain a relatively simple, organized picture of the data. This can be accomplished by grouping the scores into intervals and then listing the intervals in the table instead of listing each individual score. For example, we could construct a table showing the number of students who had scores in the 90s, the number with scores in the 80s, and so on. The result is called a *grouped frequency distribution table* because we are presenting groups of scores rather than individual values. The groups, or intervals, are called *class intervals*.

There are several rules that help guide you in the construction of a grouped frequency distribution table. These rules should be considered as guidelines rather than absolute requirements, but they do help produce a simple, well-organized, and easily understood table.

RULE 1   The grouped frequency distribution table should have about 10 class intervals. If a table has many more than 10 intervals, it becomes cumbersome and defeats the purpose of a frequency distribution table. On the other hand, if you have too few intervals, you begin to lose information about the distribution of the scores. At the extreme, with only one interval, the table would not tell you anything about how the scores are distributed. Remember that the purpose of a frequency distribution is to help a researcher see the data. With too few or too many intervals, the table will not provide a clear picture. You should note that 10 intervals is a general guide. If you are constructing a table on a blackboard, for example, you probably want only 5 or 6 intervals. If the table is to be printed in a scientific report, you may want 12 or 15 intervals. In each case, your goal is to present a table that is relatively easy to see and understand.

RULE 2   The width of each interval should be a relatively simple number. For example, 2, 5, 10, or 20 would be a good choice for the interval width. Notice that it is easy to count by 5s or 10s. These numbers are easy to understand and make it possible for someone to see quickly how you have divided the range.

RULE 3   The bottom score in each class interval should be a multiple of the width. If you are using a width of 10 points, for example, the intervals should start with 10, 20, 30, 40, and so on. Again, this makes it easier for someone to understand how the table has been constructed.

RULE 4   All intervals should be the same width. They should cover the range of scores completely with no gaps and no overlaps, so that any particular score belongs in exactly one interval.

The application of these rules is demonstrated in Example 2.4.

EXAMPLE 2.4

Remember, when the scores are whole numbers, the number of rows is determined by

highest − lowest + 1

An instructor has obtained the set of $N = 25$ exam scores shown here. To help organize these scores, we will place them in a frequency distribution table. The scores are:

82, 75, 88, 93, 53, 84, 87, 58, 72, 94, 69, 84, 61,
91, 64, 87, 84, 70, 76, 89, 75, 80, 73, 78, 60

The first step is to determine the range of scores. For these data, the smallest score is $X = 53$ and the largest score is $X = 94$, so a total of 42 rows would be needed for a table that lists each individual score. Because 42 rows would not provide a simple table, we have to group the scores into class intervals.

The best method for finding a good interval width is a systematic trial-and-error approach that uses rules 1 and 2 simultaneously. According to rule 1, we want about

10 intervals; according to rule 2, we want the interval width to be a simple number. For this example, the scores cover a range of 42 points, so we will try several different interval widths to see how many intervals are needed to cover this range. For example, if we try a width of 2, how many intervals would it take to cover the range of scores? With each interval only 2 points wide, we would need 21 intervals to cover the range. This is too many. What about an interval width of 5? What about a width of 10? The following table shows how many intervals would be needed for these possible widths:

Because the bottom interval usually extends below the lowest score and the top interval extends beyond the highest score, you often will need slightly more than the computed number of intervals.

Width	Number of Intervals Needed to Cover a Range of 42 Points	
2	21	(too many)
5	9	(OK)
10	5	(too few)

Notice that an interval width of 5 will result in about 10 intervals, which is exactly what we want.

The next step is to actually identify the intervals. The lowest score for these data is $X = 53$, so the lowest interval should contain this value. Because the interval should have a multiple of 5 as its bottom score, the interval should begin at 50. The interval has a width of 5, so it should contain 5 values: 50, 51, 52, 53, and 54. Thus, the bottom interval is 50–54. The next interval would start at 55 and go to 59. The complete frequency distribution table showing all of the class intervals is presented in Table 2.2.

Once the class intervals are listed, you complete the table by adding a column of frequencies, proportions, or percentages. The values in the frequency column indicate the number of individuals whose scores are located in that class interval. For this example, there were three students with scores in the 60–64 interval, so the frequency for this class interval is $f = 3$ (see Table 2.2). The basic table can be extended by adding columns showing the proportion and percentage associated with each class interval.

Finally, you should note that after the scores have been placed in a grouped table, you lose information about the specific value for any individual score. For example, Table 2.2 shows that one person had a score between 65 and 69, but the table does not identify the exact value for the score. In general, the wider the class intervals are, the more information is lost. In Table 2.2 the interval width is 5 points, and the table shows that there are three people with scores in the lower 60s and one person with a score in the upper 60s. This information would be lost if the interval width were

TABLE 2.2

A grouped frequency distribution table showing the data from Example 2.4. The original scores range from a high of $X = 94$ to a low of $X = 53$. This range has been divided into 9 intervals with each interval exactly 5 points wide. The frequency column (f) lists the number of individuals with scores in each of the class intervals.

X	f
90–94	3
85–89	4
80–84	5
75–79	4
70–74	3
65–69	1
60–64	3
55–59	1
50–54	1

increased to 10 points. With an interval width of 10, all of the 60s would be grouped together into one interval labeled 60–69. The table would show a frequency of four people in the 60–69 interval, but it would not tell whether the scores were in the upper 60s or the lower 60s.

REAL LIMITS AND FREQUENCY DISTRIBUTIONS

Recall from Chapter 1 that a continuous variable has an infinite number of possible values and can be represented by a number line that is continuous and contains an infinite number of points. However, when a continuous variable is measured, the resulting measurements correspond to *intervals* on the number line rather than single points. For example, a score of $X = 8$ for a continuous variable actually represents an interval bounded by the real limits 7.5 and 8.5. Thus, a frequency distribution table showing a frequency of $f = 3$ individuals all assigned a score of $X = 8$ does not mean that all three individuals had exactly the same measurement. Instead, you should realize that the three measurements are simply located in the same interval between 7.5 and 8.5.

The concept of real limits also applies to the class intervals of a grouped frequency distribution table. For example, a class interval of 40–49 contains scores from $X = 40$ to $X = 49$. These values are called the *apparent limits* of the interval because it appears that they form the upper and lower boundaries for the class interval. But $X = 40$ is actually an interval from 39.5 to 40.5. Similarly, $X = 49$ is an interval from 48.5 to 49.5. Therefore, the real limits of the interval are 39.5 (the lower real limit) and 49.5 (the upper real limit). Notice that the next higher class interval is 50–59, which has a lower real limit of 49.5. Thus, the two intervals meet at the real limit 49.5, so there are no gaps in the scale. You also should notice that the width of each class interval becomes easier to understand when you consider the real limits of an interval. For example, the interval 50–54 has real limits of 49.5 and 54.5. The distance between these two real limits (5 points) is the width of the interval.

LEARNING CHECK

1. A set of scores ranges from a high of $X = 142$ to a low of $X = 65$.
   a. Explain why it would not be reasonable to display these scores in a *regular* frequency distribution table.
   b. Determine what interval width is most appropriate for a grouped frequency distribution for this set of scores.
   c. What range of values would form the bottom interval for the grouped table?

2. Using only the frequency distribution table presented in Table 2.2, how many individuals had a score of $X = 73$?

ANSWERS

1. a. It would require a table with 78 rows to list all the individual score values. This is too many rows.

   b. An interval width of 10 points is probably best. With a width of 10, it would require 8 intervals to cover the range of scores. In some circumstances, an interval width of 5 points (requiring 16 intervals) might be appropriate.

   c. With a width of 10, the bottom interval would be 60–69.

2. After a set of scores has been summarized in a grouped table, you cannot determine the frequency for any specific score. There is no way to determine how many individuals had $X = 73$ from the table alone. (You can say that *at most* three people had $X = 73$.)

2.3   FREQUENCY DISTRIBUTION GRAPHS

A frequency distribution graph is basically a picture of the information available in a frequency distribution table. We will consider several different types of graphs, but all start with two perpendicular lines called *axes*. The horizontal line is called the *X*-axis, or the abscissa. The vertical line is called the *Y*-axis, or the ordinate. The measurement scale (set of *X* values) is listed along the *X*-axis with values increasing from left to right. The frequencies are listed on the *Y*-axis with values increasing from bottom to top. As a general rule, the point where the two axes intersect should have a value of zero for both the scores and the frequencies. A final general rule is that the graph should be constructed so that its height (*Y*-axis) is approximately two-thirds to three-quarters of its length (*X*-axis). Violating these guidelines can result in graphs that give a misleading picture of the data (see Box 2.1).

GRAPHS FOR INTERVAL OR RATIO DATA

When the data consist of numerical scores that have been measured on an interval or ratio scale, there are two options for constructing a frequency distribution graph. The two types of graph are called *histograms* and *polygons*.

Histograms   To construct a histogram, you first list the numerical scores (the categories of measurement) along the *X*-axis. Then you draw a bar above each *X* value so that

    a. The height of the bar corresponds to the frequency for that category.

    b. The width of the bar extends to the real limits of the category.

Because the bars extend to the real limits for each category, adjacent bars touch so that there are no spaces or gaps between bars. An example of a histogram is shown in Figure 2.2.

When data have been grouped into class intervals, you can construct a frequency distribution histogram by drawing a bar above each interval so that the width of the bar extends to the real limits of the interval (the lower real limit of the lowest score and the upper real limit of the highest score in the interval). This process is demonstrated in Figure 2.3.

For the two histograms shown in Figures 2.2 and 2.3, notice that the values on both the vertical and horizontal axes are clearly marked and that both axes are labeled. Also note that, whenever possible, the units of measurement are specified; for example, Figure 2.3 shows a distribution of heights measured in inches. Finally, notice that the horizontal axis in Figure 2.3 does not list all of the possible heights starting from zero and going up to 48 inches. Instead, the graph clearly shows a break between zero and 30, indicating that some scores have been omitted.

A modified histogram   A slight modification to the traditional histogram produces a very easy to draw and simple to understand sketch of a frequency distribution. Instead of drawing a bar above each score, the modification consists of drawing a stack of blocks. Each block represents one individual, so the number of blocks above each score corresponds to the frequency for that score. You may recall that we used this pile-of-blocks display in the Preview section of this chapter. Another example is shown in Figure 2.4.

Note that the number of blocks in each stack makes it very easy to see the absolute frequency for each category. In addition it is easy to see the exact difference in frequency from one category to another. In Figure 2.4, for example, it is easy to see that

FIGURE 2.2

An example of a frequency distribution histogram. The same set of quiz scores is presented in a frequency distribution table and in a histogram.

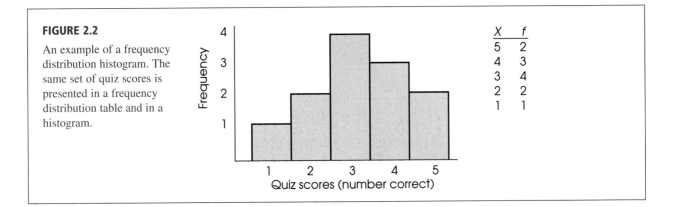

X	f
5	2
4	3
3	4
2	2
1	1

FIGURE 2.3

An example of a frequency distribution histogram for grouped data. The same set of children's heights is presented in a frequency distribution table and in a histogram.

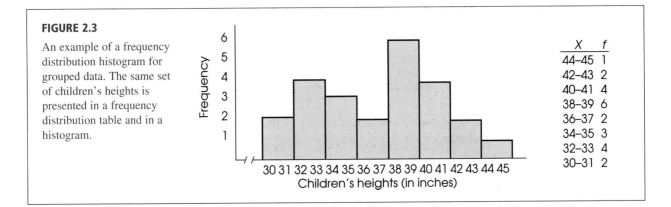

X	f
44–45	1
42–43	2
40–41	4
38–39	6
36–37	2
34–35	3
32–33	4
30–31	2

FIGURE 2.4

A frequency distribution in which each individual is represented by a block placed directly above the individual's score. For example, three people had scores of $X = 2$.

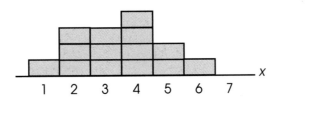

there are exactly two more people with scores of $X = 2$ than with scores of $X = 1$. Because the frequencies are clearly displayed by the number of blocks, this type of display eliminates the need for a vertical line (the Y-axis) showing frequencies. In general, this kind of graph provides a simple and concrete picture of the distribution for a sample of scores. Note that we will be using this kind of graph to show sample data throughout the rest of the book. You should also note, however, that this kind of display simply provides a sketch of the distribution and is not a substitute for an accurately drawn histogram with two labeled axes.

Polygons   The second option for graphing a distribution of numerical scores from an interval or ratio scale of measurement is called a polygon. To construct a polygon, you begin by listing the numerical scores (the categories of measurement) along the *X*-axis. Then,

a. A dot is centered above each score so that the vertical position of the dot corresponds to the frequency for the category.

b. A continuous line is drawn from dot to dot to connect the series of dots.

c. The graph is completed by drawing a line down to the *X*-axis (zero frequency) at each end of the range of scores. The final lines are usually drawn so that they reach the *X*-axis at a point that is one category below the lowest score on the left side and one category above the highest score on the right side. An example of a polygon is shown in Figure 2.5.

A polygon also can be used with data that have been grouped into class intervals. For a grouped distribution, you position each dot directly above the midpoint of the class interval. The midpoint can be found by averaging the highest and the lowest scores in the interval. For example, a class interval that is listed as 20–29 would have a midpoint of 24.5.

$$\text{midpoint} = \frac{20 + 29}{2} = \frac{49}{2} = 24.5$$

An example of a frequency distribution polygon with grouped data is shown in Figure 2.6.

GRAPHS FOR NOMINAL OR ORDINAL DATA

When the scores are measured on a nominal or ordinal scale (usually nonnumerical values), the frequency distribution can be displayed in a *bar graph.*

Bar graphs   A bar graph is essentially the same as a histogram, except that spaces are left between adjacent bars. For a nominal scale, the space between bars emphasizes that

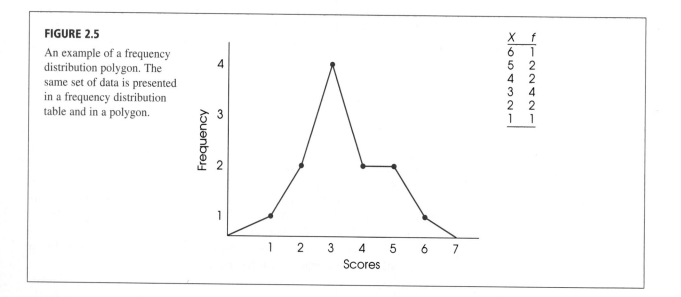

FIGURE 2.5

An example of a frequency distribution polygon. The same set of data is presented in a frequency distribution table and in a polygon.

X	f
6	1
5	2
4	2
3	4
2	2
1	1

FIGURE 2.6

An example of a frequency distribution polygon for grouped data. The same set of data is presented in a grouped frequency distribution table and in a polygon.

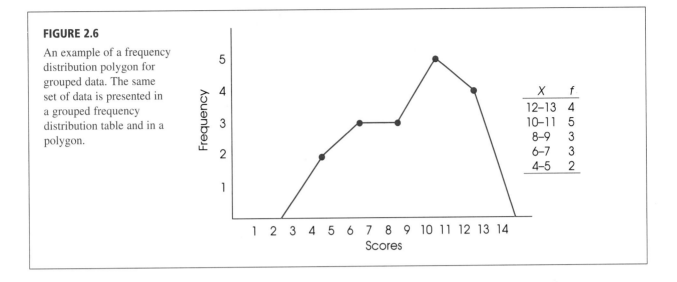

X	f
12–13	4
10–11	5
8–9	3
6–7	3
4–5	2

the scale consists of separate, distinct categories. For ordinal scales, separate bars are used because you cannot assume that the categories are all the same size.

To construct a bar graph, list the categories of measurement along the X-axis and then draw a bar above each category so that the height of the bar correponds to the frequency for the category. An example of a bar graph is shown in Figure 2.7.

GRAPHS FOR POPULATION DISTRIBUTIONS

When you can obtain an exact frequency for each score in a population, you can construct frequency distribution graphs that are exactly the same as the histograms, polygons, and bar graphs that are typically used for samples. For example, if a population is defined as a specific group of $N = 50$ people, we could easily determine how many have IQs of $X = 110$. However, if we are interested in the entire population of adults in the United States, it would be impossible to obtain an exact count of the number of people with an IQ of 110. Although it is still possible to construct graphs showing frequency distributions for extremely large populations, the graphs usually involve two special features: relative frequencies and smooth curves.

Relative frequencies   Although you usually cannot find the absolute frequency for each score in a population, you very often can obtain *relative frequencies*. For

FIGURE 2.7

A bar graph showing the distribution of personality types in a sample of college students. Because personality type is a discrete variable measured on a nominal scale, the graph is drawn with space between the bars.

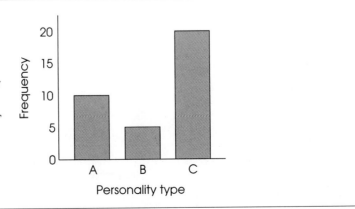

example, you may not know exactly how many fish are in the lake, but after years of fishing you do know that there are twice as many bluegill as there are bass. You can represent these relative frequencies in a bar graph by making the bar above bluegill two times taller than the bar above bass (Figure 2.8). Notice that the graph does not show the absolute number of fish. Instead, it shows the relative number of bluegill and bass.

Smooth curves   When a population consists of numerical scores from an interval or a ratio scale, it is customary to draw the distribution with a smooth curve instead of the jagged, step-wise shapes that occur with histograms and polygons. The smooth curve indicates that you are not connecting a series of dots (real frequencies) but instead are showing the relative changes that occur from one score to the next. One commonly occurring population distribution is the normal curve. The word *normal* refers to a specific shape that can be precisely defined by an equation. Less precisely, we can describe a normal distribution as being symmetrical, with the greatest frequency in the middle and relatively smaller frequencies as you move toward either extreme. A good example of a normal distribution is the population distribution for IQ scores shown in Figure 2.9. Because normal-shaped distributions occur commonly and because this

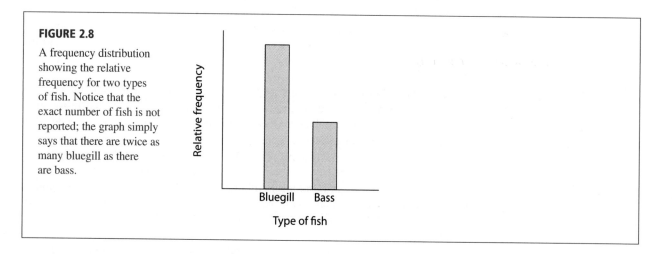

FIGURE 2.8

A frequency distribution showing the relative frequency for two types of fish. Notice that the exact number of fish is not reported; the graph simply says that there are twice as many bluegill as there are bass.

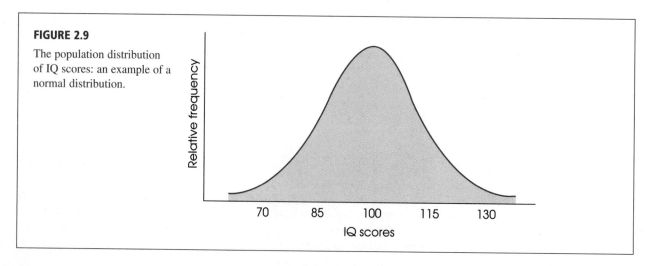

FIGURE 2.9

The population distribution of IQ scores: an example of a normal distribution.

BOX 2.1   THE USE AND MISUSE OF GRAPHS

Although graphs are intended to provide an accurate picture of a set of data, they can be used to exaggerate or misrepresent a set of scores. These misrepresentations generally result from failing to follow the basic rules for graph construction. The following example demonstrates how the same set of data can be presented in two entirely different ways by manipulating the structure of a graph.

For the past several years, the city has kept records of the number of homicides. The data are summarized as follows:

Year	Number of Homicides
2005	42
2006	44
2007	47
2008	49

These data are shown in two different graphs in Figure 2.10. In the first graph, we have exaggerated the height and started numbering the Y-axis at 40 rather than at zero. As a result, the graph seems to indicate a rapid rise in the number of homicides over the 4-year period. In the second graph, we have stretched out the X-axis and used zero as the starting point for the Y-axis. The result is a graph that shows little change in the homicide rate over the 4-year period.

Which graph is correct? The answer is that neither one is very good. Remember that the purpose of a graph is to provide an accurate display of the data. The first graph in Figure 2.10 exaggerates the differences between years, and the second graph conceals the differences. Some compromise is needed. Also note that in some cases a graph may not be the best way to display information. For these data, for example, showing the numbers in a table would be better than either graph.

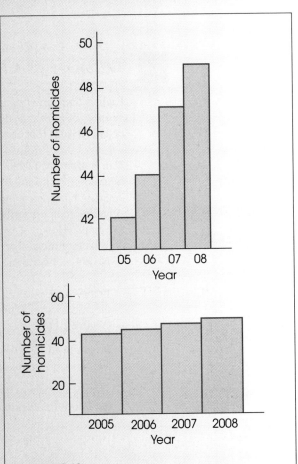

FIGURE 2.10

Two graphs showing the number of homicides in a city over a 4-year period. Both graphs show exactly the same data. However, the first graph gives the appearance that the homicide rate is high and rising rapidly. The second graph gives the impression that the homicides rate is low and has not changed over the 4-year period.

shape is mathematically guaranteed in certain situations, we give it extensive attention throughout this book.

In the future, we will be referring to *distributions of scores*. Whenever the term *distribution* appears, you should conjure up an image of a frequency distribution graph. The graph provides a picture showing exactly where the individual scores are located. To make this concept more concrete, you might find it useful to think of the graph as showing a pile of individuals just like we showed a pile of blocks in Figure 2.4. For the population of IQ scores shown in Figure 2.9, the pile is highest at an IQ score around 100 because most people have average IQs. There are only a few individuals piled up at an IQ of 130; it must be lonely at the top.

2.4   THE SHAPE OF A FREQUENCY DISTRIBUTION

Rather than drawing a complete frequency distribution graph, researchers often simply describe a distribution by listing its characteristics. There are three characteristics that completely describe any distribution: shape, central tendency, and variability. In simple terms, central tendency measures where the center of the distribution is located. Variability tells whether the scores are spread over a wide range or are clustered together. Central tendency and variability will be covered in detail in Chapters 3 and 4. Technically, the shape of a distribution is defined by an equation that prescribes the exact relationship between each X and Y value on the graph. However, we will rely on a few less-precise terms that serve to describe the shape of most distributions.

Nearly all distributions can be classified as being either symmetrical or skewed.

DEFINITIONS

In a **symmetrical distribution**, it is possible to draw a vertical line through the middle so that one side of the distribution is a mirror image of the other (Figure 2.11).

In a **skewed distribution**, the scores tend to pile up toward one end of the scale and taper off gradually at the other end (see Figure 2.11).

The section where the scores taper off toward one end of a distribution is called the **tail** of the distribution.

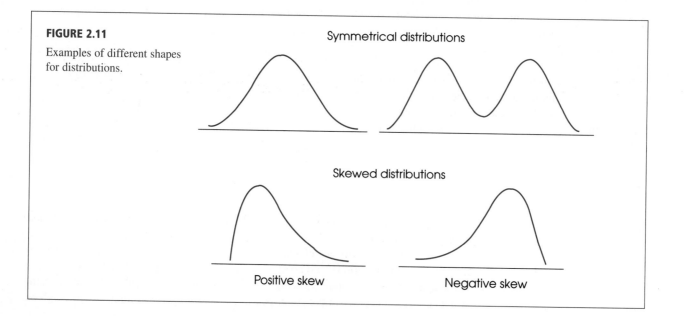

FIGURE 2.11

Examples of different shapes for distributions.

Symmetrical distributions

Skewed distributions

Positive skew   Negative skew

A skewed distribution with the tail on the right-hand side is said to be **positively skewed** because the tail points toward the positive (above-zero) end of the *X*-axis. If the tail points to the left, the distribution is said to be **negatively skewed** (see Figure 2.11).

For a very difficult exam, most scores tend to be low, with only a few individuals earning high scores. This produces a positively skewed distribution. Similarly, a very easy exam tends to produce a negatively skewed distribution, with most of the students earning high scores and only a few with low values.

LEARNING CHECK

1. Sketch a frequency distribution histogram and a frequency distribution polygon for the data in the following table:

X	f
5	4
4	6
3	3
2	1
1	1

2. Describe the shape of the distribution in Exercise 1.

3. A researcher surveys a group of 400 college students and asks each person to identify his or her favorite movie from the preceding year. What type of graph should be used to show the distribution of responses?

4. A college reports that 20% of the registered students are older than 25. What is the shape of the distribution of ages for these students?

ANSWERS

1. The graphs are shown in Figure 2.12.

2. The distribution is negatively skewed.

3. A bar graph is used for nominal data.

4. It would be positively skewed with most of the distribution around 18–21 and a few scores scattered at 25 and higher.

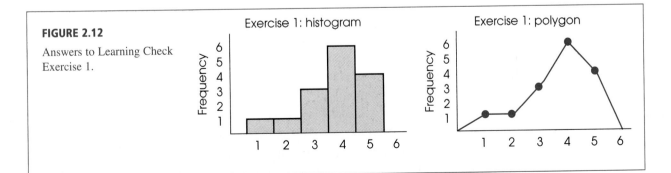

FIGURE 2.12

Answers to Learning Check Exercise 1.

2.5 PERCENTILES, PERCENTILE RANKS, AND INTERPOLATION

Although the primary purpose of a frequency distribution is to provide a description of an entire set of scores, it also can be used to describe the position of an individual within the set. Individual scores, or X values, are called raw scores. By themselves, raw scores do not provide much information. For example, if you are told that your score on an exam is $X = 43$, you cannot tell how well you did. To evaluate your score, you need more information, such as the average score or the number of people who had scores above and below you. With this additional information, you would be able to determine your relative position in the class. Because raw scores do not provide much information, it is desirable to transform them into a more meaningful form. One transformation that we will consider changes raw scores into percentiles.

DEFINITIONS

The **rank** or **percentile rank** of a particular score is defined as the percentage of individuals in the distribution with scores at or below the particular value.

When a score is identified by its percentile rank, the score is called a **percentile**.

Suppose, for example, that you have a score of $X = 43$ on an exam and that you know that exactly 60% of the class had scores of 43 or lower. Then your score $X = 43$ has a percentile rank of 60%, and your score would be called the 60th percentile. Notice that *percentile rank* refers to a percentage and that *percentile* refers to a score. Also notice that your rank or percentile describes your exact position within the distribution.

CUMULATIVE FREQUENCY AND CUMULATIVE PERCENTAGE

To determine percentiles or percentile ranks, the first step is to find the number of individuals who are located at or below each point in the distribution. This can be done most easily with a frequency distribution table by simply counting the number who are in or below each category on the scale. The resulting values are called *cumulative frequencies* because they represent the accumulation of individuals as you move up the scale.

EXAMPLE 2.5

In the following frequency distribution table, we have included a cumulative frequency column headed by *cf*. For each row, the cumulative frequency value is obtained by adding up the frequencies in and below that category. For example, the score $X = 3$ has a cumulative frequency of 14 because exactly 14 individuals had scores of $X = 3$ or less.

X	f	cf
5	1	20
4	5	19
3	8	14
2	4	6
1	2	2

The cumulative frequencies show the number of individuals located at or below each score. To find percentiles, we must convert these frequencies into percentages. The resulting values are called *cumulative percentages* because they show the percentage of individuals who are accumulated as you move up the scale.

EXAMPLE 2.6 This time we have added a cumulative percentage column ($c\%$) to the frequency distribution table from Example 2.5. The values in this column represent the percentage of individuals who are located in and below each category. For example, 70% of the individuals (14 out of 20) had scores of $X = 3$ or lower. Cumulative percentages can be computed by

$$c\% = \frac{cf}{N}(100\%)$$

X	f	cf	c%
5	1	20	100%
4	5	19	95%
3	8	14	70%
2	4	6	30%
1	2	2	10%

The cumulative percentages in a frequency distribution table give the percentage of individuals with scores at or below each X value. However, you must remember that the X values in the table are usually measurements of a continuous variable and, therefore, represent intervals on the scale of measurement (see page 19). A score of $X = 2$, for example, means that the measurement was somewhere between the real limits of 1.5 and 2.5. Thus, when a table shows that a score of $X = 2$ has a cumulative percentage of 30%, you should interpret this as meaning that 30% of the individuals have been accumulated by the time you reach the top of the interval for $X = 2$. Notice that each cumulative percentage value is associated with the upper real limit of its interval. This point is demonstrated in Figure 2.13, which shows the same data that were used in Example 2.6. Figure 2.13 shows that two people, or 10%, had scores of $X = 1$; that is,

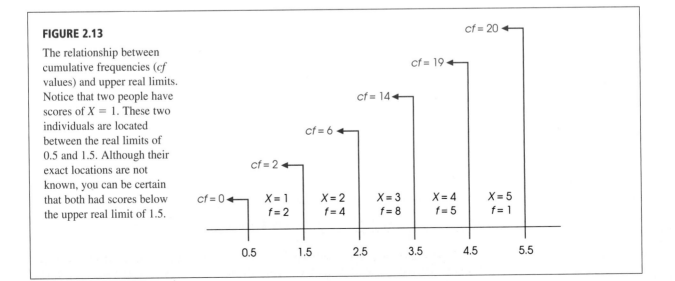

FIGURE 2.13

The relationship between cumulative frequencies (*cf* values) and upper real limits. Notice that two people have scores of $X = 1$. These two individuals are located between the real limits of 0.5 and 1.5. Although their exact locations are not known, you can be certain that both had scores below the upper real limit of 1.5.

two people had scores between 0.5 and 1.5. You cannot be sure that both individuals have been accumulated until you reach 1.5, the upper real limit of the interval. Similarly, a cumulative percentage of 30% is reached at 2.5 on the scale, a percentage of 70% is reached at 3.5, and so on.

INTERPOLATION   It is possible to determine some percentiles and percentile ranks directly from a frequency distribution table, provided the percentiles are upper real limits and the ranks are percentages that appear in the table. Using the table in Example 2.6, for example, you should be able to answer the following questions:

1. What is the 95th percentile? (Answer: $X = 4.5$.)
2. What is the percentile rank for $X = 3.5$? (Answer: 70%.)

However, there are many values that do not appear directly in the table, and it is impossible to determine these values precisely. Referring to the table in Example 2.6 again,

1. What is the 50th percentile?
2. What is the percentile rank for $X = 4$?

Because these values are not specifically reported in the table, you cannot answer the questions. However, it is possible to obtain estimates of these intermediate values by using a standard procedure known as *interpolation*.

Before we apply the process of interpolation to percentiles and percentile ranks, we will use a simple, commonsense example to introduce this method. Suppose that Bob walks to work each day. The total distance is 2 miles and the trip takes Bob 40 minutes. What is your estimate of how far Bob has walked after 20 minutes? To help, we have created a table showing the time and distance for the start and finish of Bob's trip.

Time	Distance
0	0
40	2

If you estimated that Bob walked 1 mile in 20 minutes, you have done interpolation. You probably went through the following logical steps:

1. The total time is 40 minutes.
2. The total distance is 2 miles.
3. 20 minutes represents half of the total time.
4. Assuming that Bob walks at a steady pace, he should walk half the distance in half the time, and 1 mile is half of the total distance.

The process of interpolation is pictured in Figure 2.14. Using the figure, try answering the following questions about time and distance.

1. How much time does it take for Bob to walk 1.5 miles?
2. How far has Bob walked after 10 minutes?

If you got answers of 30 minutes and $\frac{1}{2}$ mile, you have mastered the process of interpolation.

Notice that interpolation provides a method for finding intermediate values—that is, values that are located between two specified numbers. This is exactly the problem we faced with percentiles and percentile ranks. Some values are given in the table, but

FIGURE 2.14

A graphic representation of the process of interpolation. The same interval is shown on two separate scales, time and distance. Only the endpoints of the scales are known – Bob starts at 0 for both time and distance, and he ends at 40 minutes and 2 miles. Interpolation is used to estimate values within the interval by assuming that fractional portions of one scale correspond to the same fractional portions of the other. For example, it is assumed that halfway through the time scale corresponds to halfway through the distance scale.

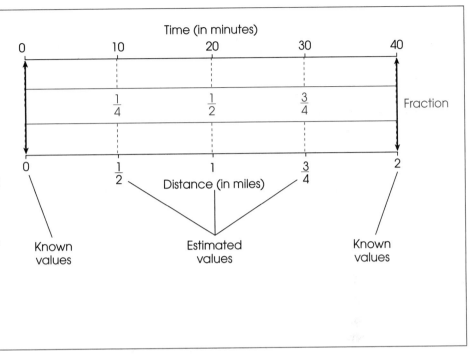

others are not. Also notice that interpolation only *estimates* the intermediate values. In the time and distance example, we do not know how far Bob has walked after 10 minutes. He may have been walking faster at the beginning of his trip and slowed down at the end. The basic assumption underlying interpolation is that there is a constant rate of change from one end of the interval to the other. In this example, we assume that Bob is walking at a constant rate of 1 mile per 20 minutes (3 miles an hour) for the entire trip. Because interpolation is based on this assumption, the values we calculate are only estimates. The general process of interpolation can be summarized as follows:

1. A single interval is measured on two separate scales (for example, time and distance). The endpoints of the interval are known for each scale.

2. You are given an intermediate value on one of the scales. The problem is to find the corresponding intermediate value on the other scale.

3. The interpolation process requires four steps:

   a. Find the width of the interval on both scales.

   b. Locate the position of the intermediate value in the interval. This position corresponds to a fraction of the whole interval:

   $$\text{fraction} = \frac{\text{distance from the top of the interval}}{\text{interval width}}$$

   c. Use this fraction to determine the distance from the top of the interval on the other scale:

   $$\text{distance} = (\text{fraction}) \times (\text{width})$$

   d. Use the distance from the top to determine the position on the other scale.

The following examples demonstrate the process of interpolation as it is applied to percentiles and percentile ranks. The key to successfully working these problems is that each cumulative percentage in the table is associated with the upper real limit of its score interval.

You may notice that in each of these problems we use interpolation working from the *top* of the interval. However, this choice is arbitrary, and you should realize that interpolation can be done just as easily working from the bottom of the interval.

EXAMPLE 2.7   Using the following distribution of scores, we will find the percentile rank corresponding to $X = 7.0$:

X	f	cf	c%
10	2	25	100%
9	8	23	92%
8	4	15	60%
7	6	11	44%
6	4	5	20%
5	1	1	4%

Notice that $X = 7.0$ is located in the interval bounded by the real limits of 6.5 and 7.5. The cumulative percentages corresponding to these real limits are 20% and 44%, respectively. These values are shown in the following table:

For interpolation problems, it is always helpful to create a table showing the range on both scales.

Scores (X)	Percentages
7.5	44%
7.0 ------------------- ?	
6.5	20%

STEP 1   For the scores, the width of the interval is 1 point. For the percentages, the width is 24 points.

STEP 2   Our particular score is located 0.5 point from the top of the interval. This is exactly halfway down in the interval.

STEP 3   Halfway down on the percentage scale would be

$$\frac{1}{2}(24 \text{ points}) = 12 \text{ points}$$

STEP 4   For the percentages, the top of the interval is 44%, so 12 points down would be

$$44\% - 12\% = 32\%$$

This is the answer. A score of $X = 7.0$ corresponds to a percentile rank of 32%.

This same interpolation procedure can be used with data that have been grouped into class intervals. Once again, you must remember that the cumulative percentage values are associated with the upper real limits of each interval. The following example demonstrates the calculation of percentiles and percentile ranks using data in a grouped frequency distribution.

EXAMPLE 2.8   Using the following distribution of scores, we will use interpolation to find the 50th percentile:

X	f	cf	c%
20–24	2	20	100%
15–19	3	18	90%
10–14	3	15	75%
5–9	10	12	60%
0–4	2	2	10%

A percentage value of 50% is not given in the table; however, it is located between 10% and 60%, which are given. These two percentage values are associated with the upper real limits of 4.5 and 9.5, respectively. These values are shown in the following table:

Scores (X)	Percentages
9.5	60%
? ------------------	50%
4.5	10%

STEP 1   For the scores, the width of the interval is 5 points. For the percentages, the width is 50 points.

STEP 2   The value of 50% is located 10 points from the top of the percentage interval. As a fraction of the whole interval, this is 10 out of 50, or $\frac{1}{5}$ of the total interval.

STEP 3   Using this same fraction for the scores, we obtain a distance of

$$\frac{1}{5}(5 \text{ points}) = 1 \text{ point}$$

The location we want is 1 point down from the top of the score interval.

STEP 4   Because the top of the interval is 9.5, the position we want is

$$9.5 - 1 = 8.5$$

This is the answer. The 50th percentile is $X = 8.5$.

1. On a statistics exam, would you rather score at the 80th percentile or at the 40th percentile?

2. For the distribution of scores presented in the following table,
   a. Find the 60th percentile.
   b. Find the percentile rank for $X = 39.5$.

X	f	cf	c%
40–49	4	25	100%
30–39	6	21	84%
20–29	10	15	60%
10–19	3	5	20%
0–9	2	2	8%

3. Using the distribution of scores from Exercise 2 and interpolation,
   a. Find the 40th percentile.
   b. Find the percentile rank for $X = 32$.

1. The 80th percentile is the higher score.

2. a. $X = 29.5$ is the 60th percentile.   b. $X = 39.5$ has a rank of 84%.

3. a. Because 40% is between the values of 20% and 60% in the table, you must use interpolation. The score corresponding to a rank of 40% is $X = 24.5$.

   b. Because $X = 32$ is between the real limits of 29.5 and 39.5, you must use interpolation. The percentile rank for $X = 32$ is 66%.

2.6   STEM AND LEAF DISPLAYS

The general term *display* is used because a stem and leaf display combines the elements of a table and a graph.

In 1977, J.W. Tukey presented a technique for organizing data that provides a simple alternative to a frequency distribution table or graph (Tukey, 1977). This technique, called a *stem and leaf display,* requires that each score be separated into two parts: The first digit (or digits) is called the *stem,* and the last digit (or digits) is called the *leaf.* For example, $X = 85$ would be separated into a stem of 8 and a leaf of 5. Similarly, $X = 42$ would have a stem of 4 and a leaf of 2. To construct a stem and leaf display for a set of data, the first step is to list all the stems in a column. For the data in Table 2.3, for example, the lowest scores are in the 30s and the highest scores are in the 90s, so the list of stems would be

Stems
3
4
5
6
7
8
9

TABLE 2.3

A set of $N = 24$ scores presented as raw data and organized in a stem and leaf display.

Data			Stem and Leaf Display	
83	82	63	3	23
62	93	78	4	26
71	68	33	5	6279
76	52	97	6	283
85	42	46	7	1643846
32	57	59	8	3521
56	73	74	9	37
74	81	76		

The next step is to go through the data, one score at a time, and write the leaf for each score beside its stem. For the data in Table 2.3, the first score is $X = 83$, so you would write 3 (the leaf) beside the 8 in the column of stems. This process is continued for the entire set of scores. The complete stem and leaf display is shown with the original data in Table 2.3.

COMPARING STEM AND LEAF DISPLAYS WITH FREQUENCY DISTRIBUTIONS

Notice that the stem and leaf display is very similar to a grouped frequency distribution. Each of the stem values corresponds to a class interval. For example, the stem 3 represents all scores in the 30s—that is, all scores in the interval 30–39. The number of leaves in the display shows the frequency associated with each stem. It also should be clear that the stem and leaf display has several advantages over a traditional frequency distribution:

1. The stem and leaf display is very easy to construct. By going through the data only one time, you can construct a complete display.

2. The stem and leaf display allows you to identify every individual score in the data. In the display shown in Table 2.3, for example, you know that there were three scores in the 60s and that the specific values were 62, 68, and 63. A frequency distribution would tell you only the frequency, not the specific values.

3. The stem and leaf display provides both a listing of the scores and a picture of the distribution. If a stem and leaf display is viewed from the side, it is essentially the same as a frequency distribution histogram (Figure 2.15).

FIGURE 2.15

A grouped frequency distribution histogram and a stem and leaf display showing the distribution of scores from Table 2.3. The stem and leaf display is placed on its side to demonstrate that the display gives the same information provided in the histogram.

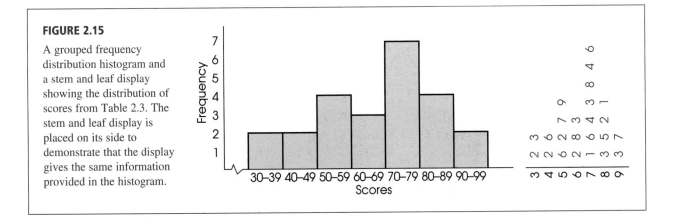

4. Because the stem and leaf display presents the actual value for each score, it is easy to modify a display if you want a more-detailed picture of a distribution. The modification simply requires that each stem be split into two (or more) parts. For example, Table 2.4 shows the same data that were presented in Table 2.3, but now we have split each stem in half. Notice that each stem value is now listed twice in the display. The first half of each stem is associated with the lower leaves (values 0–4), and the second half is associated with the upper leaves (values 5–9). In essence, we have regrouped the distribution using an interval width of 5 points instead of a width of 10 points in the original display.

Although stem and leaf displays are quite useful, they are considered to be only a preliminary means for organizing data. Typically, a researcher would use a stem and leaf display to get a first look at experimental data. The final, published report normally would present the distribution of scores in a traditional frequency distribution table or graph.

TABLE 2.4

A stem and leaf display with each stem split into two parts. Note that each stem value is listed twice: The first occurrence is associated with the lower leaf values (0–4), and the second occurrence is associated with the upper leaf values (5–9). The data shown in this display are taken from Table 2.3.

```
3 | 23
3 |
4 | 2
4 | 6
5 | 2
5 | 679
6 | 23
6 | 8
7 | 1434
7 | 686
8 | 321
8 | 5
9 | 3
9 | 7
```

LEARNING CHECK

1. Use a stem and leaf display to organize the following set of scores:

86, 114, 94, 107, 96, 100, 98, 118, 107,

132, 106, 127, 124, 108, 112, 119, 125, 115

2. Explain how a stem and leaf display contains more information than a grouped frequency distribution.

ANSWERS

1. The stem and leaf display for these data is

```
 8 | 6
 9 | 468
10 | 70768
11 | 48295
12 | 745
13 | 2
```

2. A grouped frequency distribution table tells only the number of scores in each interval; it does not identify the exact value for each score. The stem and leaf display gives the individual scores as well as the number in each interval.

SUMMARY

1. The goal of descriptive statistics is to simplify the organization and presentation of data. One descriptive technique is to place the data in a frequency distribution table or graph that shows exactly how many individuals (or scores) are located in each category on the scale of measurement.

2. A frequency distribution table lists the categories that make up the scale of measurement (the X values) in one column. Beside each X value, in a second column, is the frequency or number of individuals in that category. The table may include a proportion column showing the relative frequency for each category:

$$\text{proportion} = p = \frac{f}{n}$$

The table may include a percentage column showing the percentage associated with each X value:

$$\text{percentage} = p(100) = \frac{f}{n}(100)$$

3. It is recommended that a frequency distribution table have a maximum of 10 to 15 rows to keep it simple. If the scores cover a range that is wider than this suggested maximum, it is customary to divide the range into sections called class intervals. These intervals are then listed in the frequency distribution table along with the frequency or number of individuals with scores in each interval. The result is called a grouped frequency distribution. The guidelines for constructing a grouped frequency distribution table are as follows:
   a. There should be about 10 intervals.
   b. The width of each interval should be a simple number (e.g., 2, 5, or 10).
   c. The bottom score in each interval should be a multiple of the width.
   d. All intervals should be the same width, and they should cover the range of scores with no gaps.

4. A frequency distribution graph lists scores on the horizontal axis and frequencies on the vertical axis. The type of graph used to display a distribution depends on the scale of measurement used. For interval or ratio scales, you should use a histogram or a polygon. For a histogram, a bar is drawn above each score so that the height of the bar corresponds to the frequency. Each bar extends to the real limits of the score, so that adjacent bars touch. For a polygon, a dot is placed above the midpoint of each score or class interval so that the height of the dot corresponds to the frequency; then lines are drawn to connect the dots. Bar graphs are used with nominal or ordinal scales. Bar graphs are similar to histograms except that gaps are left between adjacent bars.

5. Shape is one of the basic characteristics used to describe a distribution of scores. Most distributions can be classified as either symmetrical or skewed. A skewed distribution that tails off to the right is said to be positively skewed. If it tails off to the left, it is negatively skewed.

6. The cumulative percentage is the percentage of individuals with scores at or below a particular point in the distribution. The cumulative percentage values are associated with the upper real limits of the corresponding scores or intervals.

7. Percentiles and percentile ranks are used to describe the position of individual scores within a distribution. Percentile rank gives the cumulative percentage associated with a particular score. A score that is identified by its rank is called a percentile.

8. When a desired percentile or percentile rank is located between two known values, it is possible to estimate the desired value using the process of interpolation. Interpolation assumes a regular linear change between the two known values.

9. A stem and leaf display is an alternative procedure for organizing data. Each score is separated into a stem (the first digit or digits) and a leaf (the last digit or digits). The display consists of the stems listed in a column with the leaf for each score written beside its stem. A stem and leaf display combines the characteristics of a table and a graph and produces a concise, well-organized picture of the data.

KEY TERMS

frequency distribution
range
grouped frequency distribution
class interval
apparent limits
histogram

polygon
bar graph
relative frequency
symmetrical distribution
tail(s) of a distribution
positively skewed distribution
negatively skewed distribution
percentile

percentile rank
cumulative frequency (cf)
cumulative percentage ($c\%$)
interpolation
stem and leaf display

RESOURCES

There are a tutorial quiz and other learning exercises for Chapter 2 at the Wadsworth website, www.cengage.com/psychology/gravetter. See page 29 for more information about accessing items on the website.

ENHANCED
WebAssign

Guided interactive tutorials, end-of-chapter problems, and related testbank items may be assigned online at WebAssign.

WebTUTOR

For those using WebTutor along with this book, there is a WebTutor section corresponding to this chapter. The WebTutor contains a brief summary of Chapter 2, hints for learning the new material and for avoiding common errors, and sample exam items including solutions.

SPSS

General instructions for using SPSS are presented in Appendix D. Following are detailed instructions for using SPSS to produce **Frequency Distribution Tables or Graphs.**

Frequency Distribution Tables

Data Entry

1. Enter all the scores in one column of the data editor, probably var00001.

Data Analysis

1. Click **Analyze** on the tool bar, select **Descriptive Statistics,** and click on **Frequencies.**
2. Highlight the column label for the set of scores (var00001) in the left box and click the arrow to move it into the **Variable** box.

3. Be sure that the option to **Display Frequency Table** is selected.
4. Click **OK.**

SPSS Output

The frequency distribution table will list the score values in a column from smallest to largest, with the percentage and cumulative percentage also listed for each score. Score values that do not occur (zero frequencies) are not included in the table, and the program does not group scores into class intervals (all values are listed).

Frequency Distribution Histograms or Bar Graphs

Data Entry

1. Enter all the scores in one column of the data editor, probably var00001.

Data Analysis

1. Click **Analyze** on the tool bar, select **Descriptive Statistics,** and click on **Frequencies.**
2. Highlight the column label for the set of scores (var00001) in the left box and click the arrow to move it into the **Variable** box.
3. Click **Charts.**
4. Select either **Bar Graphs** or **Histogram.**
5. Click **Continue.**
6. Click **OK.**

SPSS Output

After a brief delay, SPSS will display a frequency distribution table and a graph. Note that SPSS often produces a histogram that groups the scores in unpredictable intervals. A bar graph usually produces a clearer picture of the actual frequency associated with each score.

FOCUS ON PROBLEM SOLVING

1. The reason for constructing frequency distributions is to put a disorganized set of raw data into a comprehensible, organized format. Because several different types of frequency distribution tables and graphs are available, one problem is deciding which type should be used. Tables have the advantage of being easier to construct, but graphs generally give a better picture of the data and are easier to understand.
   To help you decide exactly which type of frequency distribution is best, consider the following points:
   a. What is the range of scores? With a wide range, you need to group the scores into class intervals.
   b. What is the scale of measurement? With an interval or a ratio scale, you can use a polygon or a histogram. With a nominal or an ordinal scale, you must use a bar graph.

2. When using a grouped frequency distribution table, a common mistake is to calculate the interval width by using the highest and lowest values that define each interval. For example, some students are tricked into thinking that an interval identified as 20–24 is only 4 points wide. To determine the correct interval width, you can
   a. Count the individual scores in the interval. For this example, the scores are 20, 21, 22, 23, and 24 for a total of 5 values. Thus, the interval width is 5 points.

b. Use the real limits to determine the real width of the interval. For example, an interval identified as 20–24 has a lower real limit of 19.5 and an upper real limit of 24.5 (halfway to the next score). Using the real limits, the interval width is

$$24.5 - 19.5 = 5 \text{ points}$$

3. Percentiles and percentile ranks are intended to identify specific locations within a distribution of scores. When solving percentile problems, especially with interpolation, it is helpful to sketch a frequency distribution graph. Use the graph to make a preliminary estimate of the answer before you begin any calculations. For example, to find the 60th percentile, you would want to draw a vertical line through the graph so that slightly more than half (60%) of the distribution is on the left-hand side of the line. Locating this position in your sketch will give you a rough estimate of what the final answer should be. When doing interpolation problems, you should keep several points in mind:

   a. Remember that the cumulative percentage values correspond to the upper real limits of each score or interval.

   b. You should always identify the interval with which you are working. The easiest way to do this is to create a table showing the endpoints on both scales (scores and cumulative percentages). This is illustrated in Example 2.7 on page 56.

   c. The word *interpolation* means *between two poles*. Remember: Your goal is to find an intermediate value between the two ends of the interval. Check your answer to be sure that it is located between the two endpoints. If it is not, then check your calculations.

DEMONSTRATION 2.1

A GROUPED FREQUENCY DISTRIBUTION TABLE

For the following set of $N = 20$ scores, construct a grouped frequency distribution table using an interval width of 5 points. The scores are:

14, 8, 27, 16, 10, 22, 9, 13, 16, 12,
10, 9, 15, 17, 6, 14, 11, 18, 14, 11

STEP 1 Set up the class intervals.

The largest score in this distribution is $X = 27$, and the lowest is $X = 6$. Therefore, a frequency distribution table for these data would have 22 rows and would be too large. A grouped frequency distribution table would be better. We have asked specifically for an interval width of 5 points, and the resulting table has five rows.

X
25–29
20–24
15–19
10–14
5–9

Remember that the interval width is determined by the real limits of the interval. For example, the class interval 25–29 has an upper real limit of 29.5 and a lower real limit of 24.5. The difference between these two values is the width of the interval—namely, 5.

STEP 2 Determine the frequencies for each interval.

Examine the scores, and count how many fall into the class interval of 25–29. Cross out each score that you have already counted. Record the frequency for this class interval. Now repeat this process for the remaining intervals. The result is the following table:

X	f	
25–29	1	(the score X = 27)
20–24	1	(X = 22)
15–19	5	(the scores X = 16, 16, 15, 17, and 18)
10–14	9	(X = 14, 10, 13, 12, 10, 14, 11, 14, and 11)
5–9	4	(X = 8, 9, 9, and 6)

DEMONSTRATION 2.2

USING INTERPOLATION TO FIND PERCENTILES AND PERCENTILE RANKS

Find the 50th percentile for the set of scores in the grouped frequency distribution table that was constructed in Demonstration 2.1.

STEP 1 Find the cumulative frequency (cf) and cumulative percentage values, and add these values to the basic frequency distribution table.

Cumulative frequencies indicate the number of individuals located in or below each category (class interval). To find these frequencies, begin with the bottom interval, and then accumulate the frequencies as you move up the scale. For this example, there are 4 individuals who are in or below the 5–9 interval ($cf = 4$). Moving up the scale, the 10–14 interval contains an additional 9 people, so the cumulative value for this interval is $9 + 4 = 13$ (simply add the 9 individuals in the interval to the 4 individuals below). Continue moving up the scale, cumulating frequencies for each interval.

Cumulative percentages are determined from the cumulative frequencies by the relationship

$$c\% = \left(\frac{cf}{N}\right)100\%$$

For example, the cf column shows that 4 individuals (out of the total set of $N = 20$) have scores in or below the 5–9 interval. The corresponding cumulative percentage is

$$c\% = \left(\frac{4}{20}\right)100\% = \left(\frac{1}{5}\right)100\% = 20\%$$

The complete set of cumulative frequencies and cumulative percentages is shown in the following table:

X	f	cf	c%
25–29	1	20	100%
20–24	1	19	95%
15–19	5	18	90%
10–14	9	13	65%
5–9	4	4	20%

STEP 2 Locate the interval that contains the value that you want to calculate.

We are looking for the 50th percentile, which is located between the values of 20% and 65% in the table. The scores (upper real limits) corresponding to these two percentages are 9.5 and 14.5, respectively. The interval, measured in terms of scores and percentages, is shown in the following table:

X	c%
14.5	65%
??--------50%	
9.5	20%

STEP 3 Locate the intermediate value as a fraction of the total interval.

Our intermediate value is 50%, which is located in the interval between 65% and 20%. The total width of the interval is 45 points ($65 - 20 = 45$), and the value of 50% is located 15 points down from the top of the interval. As a fraction, the 50th percentile is located $\frac{15}{45} = \frac{1}{3}$ down from the top of the interval.

STEP 4 Use the fraction to determine the corresponding location on the other scale.

Our intermediate value, 50%, is located $\frac{1}{3}$ of the way down from the top of the interval. Our goal is to find the score, the X value, that also is located $\frac{1}{3}$ of the way down from the top of the interval.

On the score (X) side of the interval, the top value is 14.5, and the bottom value is 9.5, so the total interval width is 5 points ($14.5 - 9.5 = 5$). The position we are seeking is $\frac{1}{3}$ of the way from the top of the interval. One-third of the total interval is

$$\left(\frac{1}{3}\right)5 = \frac{5}{3} = 1.67 \text{ points}$$

To find this location, begin at the top of the interval, and come down 1.67 points:

$$14.5 - 1.67 = 12.83$$

This is our answer. The 50th percentile is $X = 12.83$.

PROBLEMS

1. Construct a frequency distribution table for the following set of scores. Include columns for proportion and percentage in your table.

Scores: 5, 7, 8, 4, 7, 9, 6, 6, 5, 3
9, 6, 4, 7, 7, 8, 6, 7, 8, 5

2. Find each value requested for the distribution of scores in the following table.

a. n
b. ΣX
c. ΣX^2

X	f
5	2
4	2
3	4
2	0
1	1

3. Find each value requested for the distribution of scores in the following table.

a. n
b. ΣX
c. ΣX^2

X	f
5	1
4	2
3	4
2	3
1	1

4. For the following scores, the lowest value is $X = 13$. Place the scores in a frequency distribution table using
   a. An interval width of 5.
   b. An interval width of 10.

   21, 40, 18, 37, 32, 52, 33, 24, 13

   57, 41, 47, 32, 43, 58, 16, 38, 31

   47, 29, 49, 54, 22, 39, 34, 45, 20

5. Construct a grouped frequency distribution table to organize the following set of scores, which range from 88 to 470.

   319, 241, 104, 165,   97, 452, 280

    88, 225, 396, 271, 233, 470, 356

   159, 266, 112, 293, 174, 268, 153

6. Under what circumstances should you use a grouped frequency distribution instead of a regular frequency distribution?

7. What information is available about the data in a regular frequency distribution table that is not available in a grouped table?

8. Sketch a histogram and a polygon showing the distribution of scores presented in the following table:

X	f
7	1
6	1
5	3
4	6
3	4
2	1

9. Find each of the following values for the distribution of scores shown in the frequency distribution polygon.

   a. N
   b. ΣX

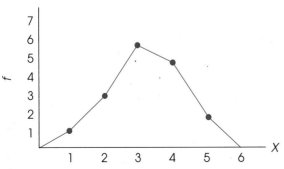

10. A survey given to a sample of 200 college students contained questions about the following variables. For each variable, identify the kind of graph that should be used to display the distribution of scores (histogram, polygon, or bar graph).
    a. number of pizzas consumed during the previous week
    b. size of T-shirt worn (S, M, L, XL)
    c. gender (male/female)
    d. grade point average for the previous semester
    e. college class (freshman, sophomore, junior, senior)

11. The following table shows the frequency distribution of birth-order position for a sample of $n = 24$ students.
    a. What kind of graph would be appropriate for showing this distribution?
    b. Sketch the frequency distribution graph.

Birth-Order Position	f
1st born	12
2nd born	5
3rd born	6
4th born	0
5th born	1

12. An instructor at the state college recorded the academic major for each student in a psychology class and obtained the following results:

Psych	Bio	Soc	Engl	Psych
Health	Phys Ed	Psych	Art	Psych
Soc	Health	Psych	Phys Ed	Soc
Hist	Psych	Art	Health	Psych
Nurs	Soc	Psych	Psych	Pol Sci

    a. Construct a frequency distribution table for these data.
    b. What type of graph would be appropriate to show the frequency distribution for these data?

13. For the following set of scores

    Scores: 7, 8, 5, 7, 6, 8, 9, 7, 4, 8
              6   9, 7, 7, 8, 6, 7, 8, 7

    a. Place the scores in a frequency distribution table.
    b. Identify the shape of the distribution.

14. For the following set of scores:

    Scores: 2, 3, 2, 4, 5, 2, 4, 2, 1, 7
              1, 3, 3, 2, 4, 3, 2, 1, 3, 2

    a. Construct a frequency distribution table.
    b. Sketch a polygon showing the distribution.

   c. Describe the distribution using the following
   characteristics:
   - **(1)** What is the shape of the distribution?
   - **(2)** What score best identifies the center (average)
     for the distribution?
   - **(3)** Are the scores clustered together, or are they
     spread out across the scale?

15. A psychologist would like to examine the effects of prior
experience on learning a new task. Two groups of rats
were compared. One group had participated in a variety
of learning experiments over the past 4 months, and the
other group had just arrived at the psychology department
with no prior research experience. All the rats were tested
on a discrimination learning problem and the psychologist
recorded the number of errors each animal made before it
solved the problem. The data are as follows:

Experienced rats:
9,   8,   8,   10,   7,   9,   11,   7,   9,   10
10,   8,   9,   9,   8,   10,   9,   8,   9,   8

Inexperienced rats:
13,   11,   12,   10,   14,   12,   13,   10,   9,   13
9,   12,   12,   10,   13,   12,   11,   12,   10,   14

16. Mental imagery has been shown to be very effective
for improving memory. In one demonstration study,
two groups of participants were presented with a list of
20 words. One group was given instructions to use
mental images to help them remember the words, and
the second group was given no special instructions.
After the list was presented, each person was asked to
recall as many words as possible. The scores for the
two groups are as follows:

Imagery group				
18	19	18	17	19
13	20	17	18	18
20	18	14	19	16

No-instruction group				
11	12	16	12	10
13	14	12	10	8
13	10	11	14	12

   a. Sketch a histogram showing the distribution of
   scores for each group (two separate graphs).
   b. Describe each graph by identifying the shape of the
   distribution and the approximate center of the
   distribution.

   c. Based on the appearance of the two distributions,
   what is the effect of using mental images on the
   level of performance for this memory task?

17. The following data are quiz scores from two different
sections of an introductory statistics course:

Section I			Section II		
9	6	8	4	7	8
10	8	3	6	3	7
7	8	8	4	6	5
7	5	10	10	3	6
9	6	7	7	4	6

   a. Organize the scores from each section in a
   frequency distribution histogram.
   b. Describe the general differences between the two
   distributions.

18. Complete the final two columns in the following
frequency distribution table and then find the
percentiles and percentile ranks requested:

X	f	cf	c%
7	2		
6	3		
5	4		
4	6		
3	4		
2	1		

   a. What is the percentile rank for $X = 2.5$?
   b. What is the percentile rank for $X = 5.5$?
   c. What is the 25th percentile?
   d. What is the 75th percentile?

19. Complete the final two columns in the following
frequency distribution table and then find the
percentiles and percentile ranks requested:

X	f	cf	c%
30–34	2		
25–29	4		
20–24	7		
15–19	5		
10–14	4		
5–9	3		

   a. What is the percentile rank for $X = 14.5$?
   b. What is the percentile rank for $X = 29.5$?

c. What is the 48th percentile?
d. What is the 92nd percentile?

20. Complete the final two columns in the following frequency distribution table and then use interpolation to find the percentiles and percentile ranks requested:

X	f	cf	c%
10	2		
9	5		
8	8		
7	15		
6	10		
5	6		
4	4		

a. What is the percentile rank for $X = 6$?
b. What is the percentile rank for $X = 9$?
c. What is the 25th percentile?
d. What is the 90th percentile?

21. Use interpolation to find the requested percentiles and percentile ranks requested for the following distribution of scores:

X	f	cf	c%
14–15	2	25	100
12–13	3	23	92
10–11	4	20	80
8–9	6	16	64
6–7	5	10	40
4–5	4	5	20
2–3	1	1	4

a. What is the percentile rank for $X = 5$?
b. What is the percentile rank for $X = 12$?
c. What is the 30th percentile?
d. What is the 72nd percentile?

22. The following frequency distribution presents a set of exam scores for a class of $N = 20$ students.

X	f	cf	c%
90–99	4	20	100
80–89	7	16	80
70–79	4	9	45
60–69	3	5	25
50–59	2	2	10

a. Find the 30th percentile.
b. Find the 88th percentile.
c. What is the percentile rank for $X = 77$?
d. What is the percentile rank for $X = 90$?

23. Using the following frequency distribution, find each of the percentiles and percentile ranks requested:

X	f	cf	c%
95–99	2	50	100
90–94	5	48	96
85–89	12	43	86
80–84	15	31	62
75–79	10	16	32
70–74	2	6	12
65–69	2	4	8
60–64	1	2	4
55–59	1	1	2

a. Find the 30th percentile.
b. Find the 80th percentile.
c. Find the percentile rank of $X = 60$
d. Find the percentile rank for $X = 78$.

24. Construct a stem and leaf display for the data in problem 4 using one stem for the scores in the 50s, one for scores in the 40s, and so on.

25. A set of scores has been organized into the following stem and leaf display. For this set of scores:
a. How many scores are in the 70s?
b. Identify the individual scores in the 70s.
c. How many scores are in the 40s?
d. Identify the individual scores in the 40s.

```
3 | 8
4 | 60
5 | 734
6 | 81469
7 | 2184
8 | 247
```

26. Use a stem and leaf display to organize the following distribution of scores. Use six stems with each stem corresponding to a 10-point interval.

Scores: 16, 42, 53, 41, 69
34, 33, 40, 61, 23
53, 49, 55, 29, 10
44, 64, 51, 21, 39
36, 58, 60, 27, 47
14, 44, 38, 31, 56

C H A P T E R

Central Tendency

Tools You Will Need

The following items are considered essential background material for this chapter. If you doubt your knowledge of any of these items, you should review the appropriate chapter or section before proceeding.

- Summation notation (Chapter 1)
- Frequency distributions (Chapter 2)

Preview

3.1   Overview

3.2   The Mean

3.3   The Median

3.4   The Mode

3.5   Selecting a Measure of Central Tendency

3.6   Central Tendency and the Shape of the Distribution

Summary

Focus on Problem Solving

Demonstration 3.1

Problems

Preview

In a classic study examining the relationship between heredity and intelligence, Tryon (1940) used a selective breeding program to develop separate strains of "smart" and "dumb" rats. Starting with a large sample of rats, Tryon tested each animal on a maze-learning problem. Based on their error scores for the maze, the brightest rats and the dullest rats were selected from this sample. The brightest males were mated with the brightest females. Similarly, the dullest rats were interbred. This process of testing and selectively breeding was continued for several generations until Tryon had established a line of maze-bright rats and a separate line of maze-dull rats. The results obtained by Tryon are shown in Figure 3.1.

Notice that after seven generations, there is an obvious difference between the two groups. As a rule, the maze-bright animals outperform the maze-dull animals. **The Problem:** Although it seems obvious that the maze-dull rats commit substantially more errors than the maze-bright rats, this conclusion is based on a general impression, or a subjective interpretation, of the figure. In fact, this conclusion is not always true. For example, there is overlap between the two groups so that some of the animals bred for brightness actually make more errors than some of the animals bred for dullness. What we need is a method to summarize each group as a whole so that we can objectively describe how much difference exists between the two groups.

The Solution: A measure of *central tendency* will identify the average or typical score to serve as a representative value for each group. Then we can use the two averages to describe the two groups and to measure the difference between them. The average maze-bright rat really is a faster learner than the average maze-dull rat.

As a footnote, you should know that later research with Tryon's rats revealed that the maze-bright rats were not really more intelligent than the maze-dull rats. Although the bright rats had developed specific abilities that are useful in mazes, the dull rats proved to be just as smart when tested on a variety of other tasks.

Tryon, R. C. (1940). "Genetic differences in maze-learning ability in rats." *The Thirty-ninth Yearbook of the National Society for the Study of Education*, 111–119. Adapted and reprinted with permission of the National Society for the Study of Education.

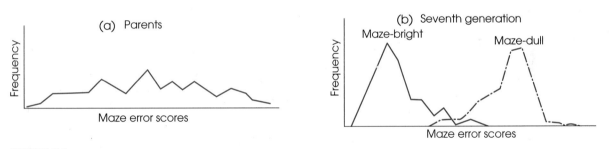

FIGURE 3.1

Distribution of error scores (a) for the original sample of rats (parents) and (b) for the two separate lines that were selectively bred for either good or poor maze performance (maze-bright and maze-dull).

3.1   OVERVIEW

The general purpose of descriptive statistical methods is to organize and summarize a set of scores. Perhaps the most common method for summarizing and describing a distribution is to find a single value that defines the average score and can serve as a representative for the entire distribution. In statistics, the concept of an average or representative score is called *central tendency*. The goal in measuring central tendency is to describe a distribution of scores by determining a single value that identifies the center of the distribution. Ideally, this central value will be the score that is the best representative value for all of the individuals in the distribution.

DEFINITION

Central tendency is a statistical measure to determine a single score that defines the center of a distribution. The goal of central tendency is to find the single score that is most typical or most representative of the entire group.

In everyday language, the goal of central tendency is to identify the "average" or "typical" individual. This average value can then be used to provide a simple description of an entire population or a sample. For example, the average price of a home in Greenwich, Connecticut, is more than $1,900,000. Obviously, not every home in Greenwich costs this much, but the average should give you a good idea of the neighborhood. Measures of central tendency are also useful for making comparisons between groups of individuals or between sets of figures. For example, weather data indicate that for Seattle, Washington, the average yearly temperature is 53° and the average annual precipitation is 34 inches. By comparison, the average temperature in Phoenix, Arizona, is 71° and the average precipitation is 7.4 inches. The point of these examples is to demonstrate the great advantage of being able to describe a large set of data with a single, representative number. Central tendency characterizes what is typical for a large population and in doing so makes large amounts of data more digestible. Statisticians sometimes use the expression "number crunching" to illustrate this aspect of data description. That is, we take a distribution consisting of many scores and "crunch" them down to a single value that describes them all.

Unfortunately, there is no single, standard procedure for determining central tendency. The problem is that no single measure will always produce a central, representative value in every situation. The three distributions shown in Figure 3.2 should help demonstrate this fact. Before we discuss the three distributions, take a moment to look at the figure and try to identify the "center" or the "most representative score" for each distribution.

1. The first distribution [Figure 3.2(a)] is symmetrical, with the scores forming a distinct pile centered around $X = 5$. For this type of distribution, it is easy to identify the "center," and most people would agree that the value $X = 5$ is an appropriate measure of central tendency.

2. In the second distribution [Figure 3.2(b)], however, problems begin to appear. Now the scores form a negatively skewed distribution, piling up at the high end of the scale around $X = 8$, but tapering off to the left all the way down to $X = 1$. Where is the "center" in this case? Some people might select $X = 8$ as the center because more individuals had this score than any other single value. However, $X = 8$ is clearly not in the middle of the distribution. In fact, the majority of the scores (10 out of 16) have values less than 8, so it seems reasonable that the "center" should be defined by a value that is less than 8.

3. Now consider the third distribution [Figure 3.2(c)]. Again, the distribution is symmetrical, but now there are two distinct piles of scores. Because the distribution is symmetrical with $X = 5$ as the midpoint, you may choose $X = 5$ as the "center." However, none of the scores is located at $X = 5$ (or even close), so this value is not particularly good as a representative score. On the other hand, because there are two separate piles of scores with one group

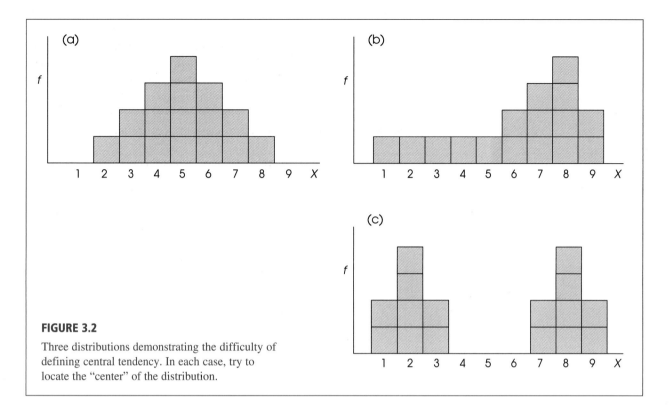

FIGURE 3.2

Three distributions demonstrating the difficulty of defining central tendency. In each case, try to locate the "center" of the distribution.

centered at $X = 2$ and the other centered at $X = 8$, it is tempting to say that this distribution has two centers. But can one distribution have two centers?

Clearly, there are problems defining the "center" of a distribution. Occasionally, you will find a nice, neat distribution like the one shown in Figure 3.2(a), for which everyone will agree on the center. But you should realize that other distributions are possible and that there may be different opinions concerning the definition of the center. To deal with these problems, statisticians have developed three different methods for measuring central tendency: the mean, the median, and the mode. They are computed differently and have different characteristics. To decide which of the three measures is best for any particular distribution, you should keep in mind that the general purpose of central tendency is to find the single most representative score. Each of the three measures we present has been developed to work best in a specific situation. We examine this issue in more detail after we introduce the three measures.

3.2 THE MEAN

The *mean,* commonly known as the arithmetic average, is computed by adding all the scores in the distribution and dividing by the number of scores. The mean for a population will be identified by the Greek letter mu, μ (pronounced "mew"), and the mean for a sample is identified by M or $\overline{X}$ (read "x-bar").

For years, the convention in statistics textbooks was to use $\overline{X}$ to represent the mean for a sample. However, in manuscripts and in published research reports the letter M is the standard notation for a sample mean. Because you will encounter the letter M when reading research reports and because you should use the letter M when writing research reports, we have decided to use the same notation in this text. Keep in mind that the $\overline{X}$ notation is still appropriate for identifying a sample mean, and you may find it used on occasion, especially in textbooks.

DEFINITION

The **mean** for a distribution is the sum of the scores divided by the number of scores.

The formula for the *population mean* is

$$\mu = \frac{\Sigma X}{N} \tag{3.1}$$

First, add all the scores in the population, and then divide by N. For a sample, the computation is done the same way, but the formula for the *sample mean* uses symbols that signify sample values:

$$\text{sample mean} = M = \frac{\Sigma X}{n} \tag{3.2}$$

In general, we will use Greek letters to identify characteristics of a population and letters of our own alphabet to stand for sample values. If a mean is identified with the symbol M, you should realize that we are dealing with a sample. Also note that n is used as the symbol for the number of scores in the sample.

EXAMPLE 3.1   For a population of $N = 4$ scores,

$$3, \quad 7, \quad 4, \quad 6$$

the mean is

$$\mu = \frac{\Sigma X}{N} = \frac{20}{4} = 5$$

ALTERNATIVE DEFINITIONS FOR THE MEAN

Although the procedure of adding the scores and dividing by the number provides a useful definition of the mean, there are two alternative definitions that may give you a better understanding of this important measure of central tendency.

The first alternative is to think of the mean as the amount each individual would get if the total (ΣX) were divided equally among all the individuals (N) in the distribution. This somewhat socialistic viewpoint is particularly useful in problems for which you know the mean and must find the total. Consider the following example.

EXAMPLE 3.2   A group of $n = 6$ boys buys a box of baseball cards at a garage sale and discovers that the box contains a total of 180 cards. If the boys divide the cards equally among themselves, how many cards will each boy get? You should recognize that this problem represents the standard procedure for computing the mean. Specifically, the total (ΣX) is divided by the number (n) to produce the mean, $\frac{180}{6} = 30$ cards for each boy.

You should also recognize that this example demonstrates that it is ⌐ define the mean as the amount that each individual gets when the total is ⌐ equally. This new definition can be useful for some problems involving the meᴀ. Consider the following example.

This time we have a group of $n = 4$ boys and we measure the amount of money that each boy has. The data produce a mean of $M = \$5$. Given this information, what is the total amount of money for the whole group? Although you do not know exactly how much money each boy has, the new definition of the mean tells you that if they pool their money together and then distribute the total equally, each boy will get $5. For each of $n = 4$ boys to get $5, the total must be $4(\$5) = \$20$. To check this answer, use the formula for the mean:

$$M = \frac{\Sigma X}{n} = \frac{\$20}{4} = \$5$$

The second alternative definition of the mean describes the mean as a balance point for a distribution. Consider the population consisting of $N = 4$ scores (2, 2, 6, 10). For this population, $\Sigma X = 20$ and $N = 4$, so $\mu = \frac{20}{4} = 5$. Figure 3.3 shows this population drawn as a histogram, with each score represented by a box that is sitting on a seesaw. If the seesaw is positioned so that it pivots at a point equal to the mean, then it will be balanced and will rest level.

The reason the seesaw is balanced over the mean becomes clear when we measure the distance of each box (score) from the mean:

Score	Distance from the Mean
$X = 2$	3 points below the mean
$X = 2$	3 points below the mean
$X = 6$	1 point above the mean
$X = 10$	5 points above the mean

Notice that the mean balances the distances. That is, the total distance below the mean is the same as the total distance above the mean:

3 points below + 3 points below = 6 points below

1 point above + 5 points above = 6 points above

FIGURE 3.3

The frequency distribution shown as a seesaw balanced at the mean.

Based on G. H. Weinberg, J. A. Schumaker, & D. Oltman (1981). *Statistics: An Intuitive Approach* (p. 14). Belmont, Calif.: Wadsworth.

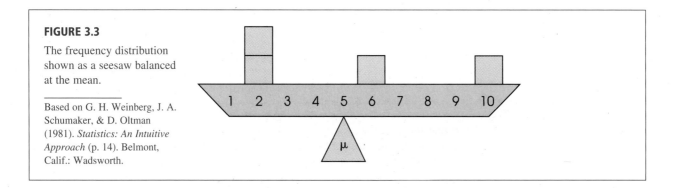

Because the mean serves as a balance point, the value of the mean will always be located somewhere between the highest score and the lowest score; that is, the mean can never be outside the range of scores. If the lowest score in a distribution is $X = 8$ and the highest is $X = 15$, then the mean *must* be between 8 and 15. If you calculate a value that is outside this range, then you have made an error.

The image of a seesaw with the mean at the balance point is also useful for determining how a distribution is affected if a new score is added or if an existing score is removed. For the distribution in Figure 3.3, for example, what would happen to the mean (balance point) if a new score were added at $X = 10$?

THE WEIGHTED MEAN

Often it is necessary to combine two sets of scores and then find the overall mean for the combined group. Suppose that we begin with two separate samples. The first sample has $n = 12$ scores and $M = 6$. The second sample has $n = 8$ and $M = 7$. If the two samples are combined, what is the mean for the total group?

To calculate the overall mean, we need two values:

1. the overall sum of the scores for the combined group (ΣX), and

2. the total number of scores in the combined group (n).

To determine the overall sum, we find the sum of the scores for the first sample (ΣX_1) and the sum for the second sample (ΣX_2) and then add the two sums together. Similarly, the total number of scores in the combined group can be found by adding the number of scores in the first sample (n_1) and the number in the second sample (n_2). With these two values, we can compute the mean using the basic equation

$$\text{overall mean} = M = \frac{\Sigma X \text{ (overall sum for the combined group)}}{n \text{ (total number in the combined group)}}$$

$$= \frac{\Sigma X_1 + \Sigma X_2}{n_1 + n_2}$$

The first sample has $n = 12$ and $M = 6$. Therefore, the total for this sample must be $\Sigma X = 72$. (Remember that when the total is distributed equally, each person gets the mean. For each of 12 people to get $M = 6$, the total must be 72.) In the same way, the second sample has $n = 8$ and $M = 7$, so the total must be $\Sigma X = 56$. Using these values, we obtain an overall mean of

$$\text{overall mean} = M = \frac{\Sigma X_1 + \Sigma X_2}{n_1 + n_2} = \frac{72 + 56}{12 + 8} = \frac{128}{20} = 6.4$$

The following table summarizes the calculations.

First Sample	Second Sample	Combined Sample
$n = 12$	$n = 8$	$n = 20$ $(12 + 8)$
$\Sigma X = 72$	$\Sigma X = 56$	$\Sigma X = 128$ $(72 + 56)$
$M = 6$	$M = 7$	$M = 6.4$

Note that the overall mean is not halfway between the original two sample means. Because the samples are not the same size, one makes a larger contribution to the total group and therefore carries more weight in determining the overall mean. For this reason, the overall mean we have calculated is called the *weighted mean*. In this example, the overall mean of $M = 6.4$ is closer to the value of $M = 6$ (the larger sample) than it is to $M = 7$ (the smaller sample).

In summary, when two samples are combined, the weighted mean is obtained as follows:

STEP 1   Determine the grand total of all the scores in both samples. This sum is obtained by adding the sum of the scores for the first sample (ΣX_1) and the sum of the scores for the second sample (ΣX_2).

STEP 2   Determine the total number of scores in both samples. This value is obtained by adding the number in the first sample (n_1) and the number in the second sample (n_2).

STEP 3   Divide the sum of all the scores (step 1) by the total number of scores (step 2). Expressed as an equation,

$$\text{weighted mean} = \frac{\text{combined sum}}{\text{combined } n} = \frac{\Sigma X_1 + \Sigma X_2}{n_1 + n_2}$$

BOX 3.1   **AN ALTERNATIVE PROCEDURE FOR FINDING THE WEIGHTED MEAN**

In the text, the weighted mean was obtained by first determining the total number of scores (n) for the two combined samples and then determining the overall sum (ΣX) for the two combined samples. The following example demonstrates how the same result can be obtained using a slightly different conceptual approach.

We begin with the same two samples that were used in the text: One sample has $M = 6$ for $n = 12$ students, and the second sample has $M = 7$ for $n = 8$ students. The goal is to determine the mean for the overall group when the two samples are combined.

Logically, when these two samples are combined, the larger sample (with $n = 12$ scores) will make a greater contribution to the combined group than the smaller sample (with $n = 8$ scores). Thus, the larger sample will carry more weight in determining the mean for the combined group. We will accommodate this fact by assigning a weight to each sample mean so that the weight is determined by the size of the sample. To determine how much weight should be assigned to each sample mean, you simply consider the sample's contribution to the combined group. When the two

samples are combined, the resulting group will have a total of 20 scores ($n = 12$ from the first sample and $n = 8$ from the second). The first sample contributes 12 out of 20 scores and, therefore, is assigned a weight of $\frac{12}{20}$. The second sample contributes 8 out of 20 scores, and its weight is $\frac{8}{20}$. Each sample mean is then multiplied by its weight, and the results are added to find the weighted mean for the combined sample. For this example,

$$\text{weighted mean} = \left(\frac{12}{20}\right)(6) + \left(\frac{8}{20}\right)(7)$$

$$= \frac{72}{20} + \frac{56}{20}$$

$$= 3.6 + 2.8$$

$$= 6.4$$

Note that this is the same result obtained using the method described in the text.

COMPUTING THE MEAN FROM A FREQUENCY DISTRIBUTION TABLE

When a set of scores has been organized in a frequency distribution table, the calculation of the mean is usually easier if you first remove the individual scores from the table. Table 3.1 shows a distribution of quiz scores organized in a frequency distribution table. To compute the mean for this distribution you must be careful to use both the X values in the first column and the frequencies in the second column. The values in the table show that the distribution consists of one 10, two 9s, four 8s, and one 6, for a total of $n = 8$ scores. Remember that you can determine the number of scores by adding the frequencies, $n = \Sigma f$. To find the sum of the scores, you must be careful to add all eight scores:

$$\Sigma X = 10 + 9 + 9 + 8 + 8 + 8 + 8 + 6 = 66$$

Note that you can also find the sum of the scores by computing ΣfX as we demonstrated in Chapter 2 (p. 39). Once you have found ΣX and n, you compute the mean as usual. For these data,

$$M = \frac{\Sigma X}{n} = \frac{66}{8} = 8.25$$

TABLE 3.1

Statistics quiz scores for a section of $n = 8$ students.

Quiz Score (X)	f	fX
10	1	10
9	2	18
8	4	32
7	0	0
6	1	6

LEARNING CHECK

1. Find the mean for the following sample of $n = 6$ scores: 4, 10, 7, 5, 9, 7

2. A sample of $n = 7$ scores has a mean of $M = 5$. What is the value of ΣX for this sample?

3. One sample has $n = 6$ scores with a mean of $M = 4$. A second sample has $n = 4$ scores with a mean of $M = 9$. If the two samples are combined, what is the mean for the combined sample?

4. A sample of $n = 6$ scores has a mean of $M = 40$. One new score is added to the sample and the new mean is found to be $M = 42$. What can you conclude about the value of the new score?
   a. It must be greater 40.
   b. It must be less than 40

5. Find the values for n, ΣX, and M for the sample that is summarized in the following frequency distribution table.

X	f
5	1
4	2
3	2
2	4
1	1

CHARACTERISTICS OF THE MEAN

The mean has many characteristics that will be important in future discussions. In general, these characteristics result from the fact that every score in the distribution contributes to the value of the mean. Specifically, every score adds to the total (ΣX) and every score contributes one point to the number of scores (n). These two values (ΣX and n) determine the value of the mean. We now discuss four of the more important characteristics of the mean.

Changing a score   Changing the value of any score will change the mean. For example, a sample of quiz scores for a psychology lab section consists of 9, 8, 7, 5, and 1. Note that the sample consists of $n = 5$ scores with $\Sigma X = 30$. The mean for this sample is

$$M = \frac{\Sigma X}{n} = \frac{30}{5} = 6.00$$

Now suppose that the score of $X = 1$ is changed to $X = 8$. Note that we have added 7 points to this individual's score, which will also add 7 points to the total (ΣX). After changing the score, the new distribution consists of

9, 8, 7, 5, 8

There are still $n = 5$ scores, but now the total is $\Sigma X = 37$. Thus, the new mean is

$$M = \frac{\Sigma X}{n} = \frac{37}{5} = 7.40$$

Notice that changing a single score in the sample has produced a new mean. You should recognize that changing any score also changes the value of ΣX (the sum of the scores), and thus always changes the value of the mean.

Introducing a new score or removing a score   In general, the mean is determined by two values: ΣX and N (or n). Whenever either of these values is changed, the mean also is changed. In the preceding example, the value of one score was changed. This produced a change in the total (ΣX) and therefore changed the mean. If you add a new score (or take away a score), you will change both ΣX and n, and you must compute the new mean using the changed values.

Usually, but not always, adding a new score or removing an existing score will change the mean. The exception is when the new score (or the removed score) is exactly equal to the mean. It is easy to visualize the effect of adding or removing a score if you remember that the mean is defined as the balance point for the distribution. Figure 3.4 shows a distribution of scores represented as boxes on a seesaw that is balanced at the mean, $\mu = 7$. Imagine what would happen if we added a new score (a new box) at $X = 10$. Clearly, the seesaw would tip to the right and we would need to change the pivot point (the mean) higher to restore balance.

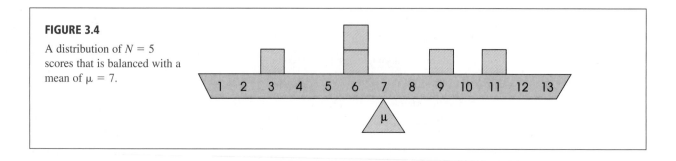

FIGURE 3.4

A distribution of $N = 5$ scores that is balanced with a mean of $\mu = 7$.

Now imagine what would happen if we removed the score (the box) at $X = 9$. This time the seesaw would tip to the left and, once again, we would need to change the mean to restore balance.

Finally, consider what would happen if we added a new score of $X = 7$, exactly equal to the mean. It should be clear that the seesaw would not tilt in either direction, so the mean would stay in exactly the same place. Also note that if we remove the new score at $X = 7$, the seesaw will remain balanced and the mean will not change.

In general, adding a new score or removing an existing score will cause the mean to change unless the new score (or existing score) is located exactly at the mean. Adding or removing a score that is equal to the mean will not cause a change in the mean.

The following example demonstrates exactly how the new mean is computed when a new score is added to an existing sample.

EXAMPLE 3.3   Adding a score (or removing a score) has the same effect on the mean whether the original set of scores is a sample or a population. To demonstrate the calculation of the new mean, we will use the set of scores that is shown in Figure 3.4. This time, however, we will treat the scores as a sample with $n = 5$ and $M = 7$. Note that this sample must have $\Sigma X = 35$. What will happen to the mean if a new score of $X = 13$ is added to the sample?

To find the new sample mean, we must determine how the values for n and ΣX will be changed by a new score. We begin with the original sample and then consider the effect of adding the new score. The original sample had $n = 5$ scores, so adding one new score will produce $n = 6$. Similarly, the original sample had $\Sigma X = 35$. Adding a score of $X = 13$ will produce a new sum of $\Sigma X = 35 + 13 = 48$. Finally, the new mean is computed using the new values for ΣX and n.

$$M = \frac{\Sigma X}{n} = \frac{48}{6} = 8$$

The entire process can be summarized as follows:

Original Sample	New Sample, Adding $X = 13$
$n = 5$	$n = 6$
$\Sigma X = 35$	$\Sigma X = 48$
$M = 35/5 = 7$	$M = 48/6 = 8$

Adding or subtracting a constant from each score   If a constant value is added to every score in a distribution, the same constant will be added to the mean. Similarly, if you subtract a constant from every score, the same constant will be subtracted from the mean.

In the Preview section for Chapter 2, we discussed a research study in which participants consistently had better memory for humorous sentences than for nonhumorous sentences (Schmidt, 1994). Table 3.2 shows the results for a sample of $n = 6$ participants. The first column shows their memory scores for nonhumorous sentences. Note that the total number of sentences recalled is $\Sigma X = 17$ for a sample of $n = 6$ participants, so the mean is $M = 17/6 = 2.83$. Now suppose that the effect of humor is to add a constant amount (2 points) to each individual's memory score. The resulting scores for humorous sentences are shown in the second column of the table. For these scores, the 6 participants recalled a total of $\Sigma X = 29$ sentences, so the mean is $M = 29/6 = 4.83$. Adding 2 points to each score has also added 2 points to the mean, from $M = 2.83$ to $M = 4.83$. (It is important to note that experimental effects are usually not as simple as adding or subtracting a constant amount. Nonetheless, the concept of adding a constant to every score is important and will be addressed in later chapters when we are using statistics to evaluate the effects of experimental manipulations.)

Multiplying or dividing each score by a constant   If every score in a distribution is multiplied by (or divided by) a constant value, the mean will change in the same way.

Multiplying (or dividing) each score by a constant value is a common method for changing the unit of measurement. To change a set of measurements from minutes to seconds, for example, you multiply by 60; to change from inches to feet, you divide by 12. One common task for researchers is converting measurements into metric units to conform to international standards. For example, publication guidelines of the American Psychological Association call for metric equivalents to be reported in parentheses when most nonmetric units are used. Table 3.3 shows how a sample of $n = 5$ scores measured in inches would be transformed to a set of scores measured in centimeters. (Note that one inch equals 2.54 centimeters.) The first column shows the original scores that total $\Sigma X = 50$ with $M = 10$ inches. In the second column, each of the original scores has been multiplied by 2.54 (to convert from inches to centimeters) and the resulting values total $\Sigma X = 127$, with $M = 25.4$. Multiplying each score by 2.54 has also caused the mean to be multiplied by 2.54. You should realize, however, that although the numerical values for the individual scores and the sample mean have changed, the actual measurements are not changed.

TABLE 3.2

Number of sentences recalled for humorous and nonhumorous sentences.

Participant	Nonhumorous Sentences	Humorous Sentences
A	4	6
B	2	4
C	3	5
D	3	5
E	2	4
F	3	5
	$\Sigma X = 17$	$\Sigma X = 29$
	$M = 2.83$	$M = 4.83$

TABLE 3.3

Measurements transformed from inches to centimeters.

Original Measurement in inches	Conversion to Centimeters (multiply by 2.54)
10	25.40
9	22.86
12	30.48
8	20.32
11	27.94
$\Sigma X = 50$	$\Sigma X = 127.00$
$M = 10$	$M = 25.40$

LEARNING CHECK

1. Adding a new score to a distribution will always change the mean. (True or false?)

2. Changing the value of a score in a distribution will always change the mean. (True or false?)

3. A population has a mean of $\mu = 80$.
   a. If 6 points were added to every score, what would be the value for the new mean?
   b. If every score were multiplied by 2, what would be the value for the new mean?

4. A sample of $n = 4$ scores has a mean of 9. If one new person with a score of $X = 14$ is added to the sample, what is the value for the new sample mean?

ANSWERS

1. False. If the score is equal to the mean, it will not change the mean.

2. True.

3. **a.** The new mean would be 86. **b.** The new mean would be 160.

4. The original sample has $n = 4$ and $\Sigma X = 36$. The new sample has $n = 5$ scores that total $\Sigma X = 50$. The new mean is $M = 10$.

3.3 THE MEDIAN

The second measure of central tendency we will consider is called the *median*. The median is the score that divides a distribution in half. If the scores are listed in order from smallest to largest, the median is the midpoint of the list.

DEFINITION

The **median** is the score that divides a distribution in half so that 50% of the individuals in a distribution have scores at or below the median.

By *midpoint* we mean that the scores are being divided into two equal-sized groups. We are not locating the midpoint between the highest and lowest X values.

Earlier, when we introduced the mean, specific symbols and notation were used to identify the mean and to differentiate a sample mean and a population mean. For the median, however, there are no symbols or notation. Instead, the median is simply identified by the word *median*. In addition, the definition and the computations for the median are identical for a sample and for a population.

The goal of the median is to determine the midpoint of the distribution. This commonsense goal is demonstrated in the following two examples which show how the median for most distributions can be found simply by counting scores.

EXAMPLE 3.4

This example demonstrates the calculation of the median when n is an odd number. With an odd number of scores, you list the scores in order (lowest to highest), and the median is the middle score in the list. Consider the following set of $N = 5$ scores, which have been listed in order:

3,   5,   8,   10,   11

The middle score is $X = 8$, so the median is equal to 8. If the scores are measurements of a continuous variable, the median corresponds to the precise point $X = 8.00$. In a frequency distribution graph, this point divides the area of the graph exactly in half (Figure 3.5). The amount of area above the median consists of $2\frac{1}{2}$ "boxes," the same as the area below the median (shaded portion).

EXAMPLE 3.5

This example demonstrates the calculation of the median when n is an even number. With an even number of scores in the distribution, you list the scores in order (lowest to highest) and then locate the median by finding the average of the middle two scores. Consider the following population:

3,   3,   4,   5,   7,   8

Now we select the middle pair of scores (4 and 5), add them together, and divide by 2:

$$\text{median} = \frac{4 + 5}{2} = \frac{9}{2} = 4.5$$

In terms of a graph, we see again that the median divides the area of the distribution exactly in half (Figure 3.6). There are three scores (or boxes) above the median and three below the median.

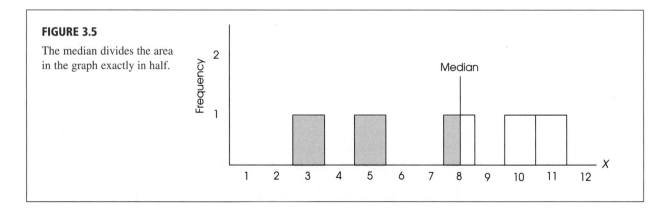

FIGURE 3.5

The median divides the area in the graph exactly in half.

FIGURE 3.6

The median divides the area in the graph exactly in half.

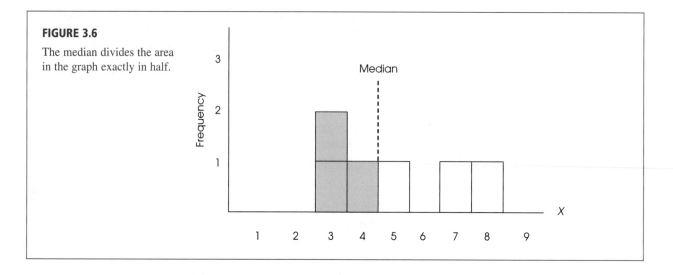

The technique of averaging the middle two scores to find the median will either produce a whole-number or a decimal value ending in .50 (as in this example). A whole-number answer is a sensible result for either a continuous variable, such as time, or a discrete variable, such as the number of children in a family. On the other hand, a decimal value is only sensible for a continuous variable. For example, it is reasonable to have a median time of $X = 4.5$ seconds but it is impossible to have a family with $X = 4.5$ children. Nonetheless, the decimal value provides an accurate description of the midpoint of the distribution. If the scores in this example correspond to the number of children in $n = 6$ families, it is accurate to say that 50% of the families have more than 4.5 children and 50% have fewer than 4.5 children. Thus, this method is used to find the median for either a continuous or a discrete variable.

Continuous and discrete variables were discussed in Chapter 1 on pages 18–19.

FINDING THE PRECISE MEDIAN FOR A CONTINUOUS VARIABLE

The simple technique of listing scores is sufficient to determine the median for most distributions. However, in some situations involving a continuous variable, it is possible to divide a distribution in half by splitting one of the measurement categories into fractional parts so that *exactly* 50% of a distribution is above (and below) a specific point. The following example demonstrates this process.

EXAMPLE 3.6

For this example, we will find the median for the following sample of $n = 10$

1,   2,   2,   3,   4,   4,   4,   4,   4,   5

The frequency distribution for this sample is also shown in Figure 3.7(a). With an even number of scores, you normally would average the middle pair to determine the median. Using this method produces a median of 4. However, if you look closely at the frequency distribution graph, you probably get the clear impression that $X = 4$ is not the exact midpoint. The problem comes from the tendency to interpret a score of 4 as meaning exactly 4.00 instead of meaning an interval from 3.5 to 4.5. The simple method of computing the median has determined that the value we want is located in this interval.

To find the median with greater precision, start at the left-hand side of the distribution, and move up the scale of measurement (the X-axis), counting boxes as

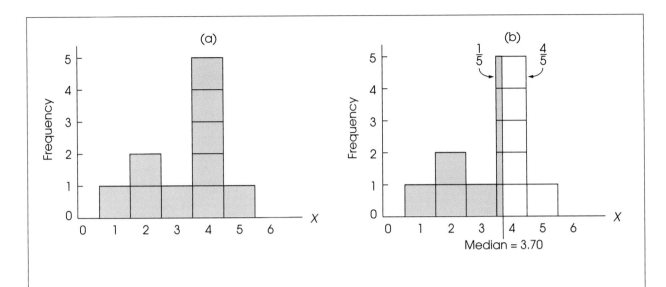

FIGURE 3.7

A distribution with several scores clustered at the median. The median for this distribution is positioned so that each of the five boxes above $X = 4$ is divided into two sections, with $\frac{1}{5}$ of each box below the median (to the left) and $\frac{4}{5}$ of each box above the median (to the right).

you go along. The vertical line corresponding to the median should be drawn at the point where you have counted exactly 5 boxes (50% of the total of 10 boxes). By the time you reach a value of 3.5 on the X-axis, you will have gathered a total of 4 boxes, so that only one more box is needed to give you exactly 50% of the distribution. The problem is that there are 5 boxes in the next interval. The solution is to take only a fraction of each of the 5 boxes so that the fractions combine to give you 1 more box. If you take $\frac{1}{5}$ of each block, the five fifths will combine to make one whole box. This solution is shown in Figure 3.7(b).

Notice that we have drawn a line separating each of the 5 boxes so that $\frac{1}{5}$ is on the left-hand side of the line and $\frac{4}{5}$ are on the right-hand side. Thus, the line should be drawn exactly $\frac{1}{5}$ of the way into the interval containing the 5 boxes. This interval extends from the lower real limit of 3.5 to the upper real limit of 4.5 on the X-axis and has a width of 1.00 point. One-fifth of the interval is 0.20 points ($\frac{1}{5}$ of 1.00). Therefore, the line should be drawn at the point where $X = 3.70$ (the lower real limit of 3.5 + $\frac{1}{5}$ of the interval, or 0.20). This value, $X = 3.70$, is the median and divides the distribution exactly in half.

The process demonstrated in Example 3.6 can be summarized in the following five steps, which can be generalized to any situation in which several scores are tied at the median.

STEP 1   Determine the real limits of the interval that contains the precise midpoint.

STEP 2   Count the number of scores (boxes) below the identified interval.

STEP 3   Find the number of additional scores (boxes) needed to reach exactly 50%.

STEP 4   Calculate a fraction $= \dfrac{\text{number of additional scores needed (step 2)}}{\text{total number of scores in the internal}}$

STEP 5   Add the fraction to the lower real limit of the interval.
When these steps are incorporated into a formula, we obtain

$$\text{median} = X_{\text{LRL}} + \left(\frac{0.5n - f_{\text{below LRL}}}{f_{\text{Interval}}} \right)$$

where X_{LRL} is the lower real limit of the interval, $f_{\text{below LRL}}$ is the frequency of number of scores with values less than X_{LRL}, and f_{Interval} is the number of scores in the interval. For the data in Example 3.6, we obtain,

$$\text{median} = 3.5 + \left(\frac{0.5(10) - 4}{5} \right)$$

$$= 3.5 + \left(\frac{5 - 4}{5} \right)$$

$$= 3.5 + \frac{1}{5}$$

$$= 3.5 + 0.2$$

$$= 3.7$$

Finding the precise midpoint by dividing scores into fractional parts is sensible for a continuous variable, however, it is not appropriate for a discrete variable. For example, finding a median time of 3.70 seconds is reasonable, but a median family size of 3.70 children is not. For a discrete variable, simply list the scores in order and find the middle score. If the distribution in Example 3.6 (see Figure 3.7) represented a discrete variable, the median would be reported as $X = 4$.

THE MEDIAN, THE MEAN, AND THE MIDDLE

Earlier, we defined the mean as the "balance point" for a distribution because the distances above the mean must have the same total as the distances below the mean. One consequence of this definition is that the mean is always located inside the group of scores; that is, there always will be at least one score above the mean and at least one score below the mean. You should notice, however, that the concept of a balance point focuses on distances rather than scores. In particular, it is possible to have a distribution in which the vast majority of the scores are located on one side of the mean. Figure 3.8 shows a distribution of $N = 6$ scores in which 5 out of 6 scores have values less than the mean. In this figure, the total of the distances above the mean is 8 points and the total of the distances below the mean is 8 points. Although it is reasonable to

FIGURE 3.8

A population of $N = 6$ scores with a mean of $\mu = 4$. Notice that the mean does not necessarily divide the scores into two equal groups. In this example, 5 out of the 6 scores have values less than the mean.

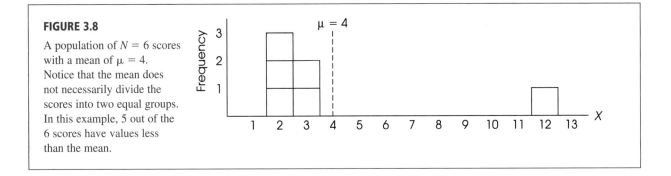

say that the mean is located in the middle of the distribution if you use the concept of distance to define the term "middle," you should realize that the mean is not necessarily located at the exact center of the group of scores.

The median, on the other hand, defines the middle of the distribution in terms of scores. In particular, the median is located so that half of the scores are on one side and half are on the other side. For the distribution in Figure 3.8, for example, the median is located at $X = 2.5$, with exactly 3 scores above this value and exactly 3 scores below. Thus, it is possible to claim that the median is located in the middle of the distribution, provided that the term "middle" is defined by the number of scores.

In summary, the mean and the median are both methods for defining and measuring central tendency. Although they both define the middle of the distribution, they use different definitions of the term "middle."

LEARNING CHECK

1. Find the median for each distribution of scores:
   a. 3, 10, 8, 4, 10, 7, 6
   b. 13, 8, 10, 11, 12, 10

2. If you have a score of 52 on an 80-point exam, then you definitely scored above the median. (True or false?)

3. It is possible for more than 50% of the scores in a distribution to have values greater than the mean. (True or false?)

4. It is possible for more than 50% of the scores in a distribution to have values greater than the median. (True or false?)

ANSWERS

1. **a.** The median is $X = 7$.   **b.** The median is $X = 10.5$.

2. False. The value of the median would depend on where the scores are located.

3. True

4. False

THE MODE

The final measure of central tendency that we will consider is called the *mode*. In its common usage, the word *mode* means "the customary fashion" or "a popular style." The statistical definition is similar in that the mode is the most common observation among a group of scores.

DEFINITION

In a frequency distribution, the **mode** is the score or category that has the greatest frequency.

As with the median, there are no symbols or special notation used to identify the mode or to differentiate between a sample mode and a population mode. In addition, the definition of the mode is the same for a population and for a sample distribution.

The mode is a useful measure of central tendency because it can be used to determine the typical or average value for any scale of measurement, including a nominal scale (see Chapter 1). Consider, for example, the data shown in Table 3.4. These data were obtained by asking a sample of 100 students to name their favorite restaurants in town. The result is a sample of $n = 100$ scores with the score for each student corresponding to the restaurant he or she named. For these data, the mode is Luigi's, the restaurant (score) that was named most frequently as a favorite place. Although we can identify a modal response for these data, you should notice that it would be impossible to compute a mean or a median. For example, you cannot add the scores to determine a mean (How much is 5 College Grills plus 42 Luigi's?). Also, it is impossible to list the scores in order because the restaurants do not form any natural order. For example, the College Grill is not "more than" or "less than" the Oasis Diner, they are simply two different restaurants. Thus, it is impossible to obtain the median by finding the midpoint of the list. In general, the mode is the only measure of central tendency that can be used with data from a nominal scale of measurement.

In a frequency distribution graph, the greatest frequency will appear as the tallest part of the figure. To find the mode, you simply identify the score located directly beneath the highest point in the distribution.

Although a distribution will have only one mean and only one median, it is possible to have more than one mode. Specifically, it is possible to have two or more scores that have the same highest frequency. In a frequency distribution graph, the different modes will correspond to distinct, equally high peaks. A distribution with two modes is said to be *bimodal,* and a distribution with more than two modes is called *multimodal.* Occasionally, a distribution with several equally high points is said to have no mode.

Incidentally, a bimodal distribution is often an indication that two separate and distinct groups of individuals exist within the same population (or sample). For example, if you measured height for each person in a set of 100 college students, the resulting

TABLE 3.4

Favorite restaurants named by a sample of $n = 100$ students. *Caution:* The mode is a score or category, not a frequency. For this example, the mode is Luigi's, not $f = 42$.

Restaurant	f
College Grill	5
George & Harry's	16
Luigi's	42
Oasis Diner	18
Roxbury Inn	7
Sutter's Mill	12

distribution would probably have two modes, one corresponding primarily to the males in the group and one corresponding primarily to the females.

Technically, the mode is the score with the absolute highest frequency. However, the term *mode* is often used more casually to refer to scores with relatively high frequencies— that is, scores that correspond to peaks in a distribution even though the peaks are not the absolute highest points. For example, Athos, Levinson, Kistler, Zemansky, Bostrom, Freimer, and Gitschier (2007) asked people to identify the pitch for both pure tones and piano tones. Participants were presented with a series of tones and had to name the note corresponding to each tone. Nearly half the participants (44%) had extraordinary pitch-naming ability (absolute pitch), and most of the others performed around chance level, apparently guessing the pitch names randomly. Figure 3.9 shows a distribution of scores that is consistent with the results of the study. There are two distinct peaks in the distribution, one located at $X = 2$ (chance performance) and the other located near $X = 10$ (perfect performance). Each of these values is a mode in the distribution. Note, however, that the two modes do not have identical frequencies. Eight people scored at $X = 2$ and only seven had scores of $X = 10$. Nonetheless, both of these points are called modes. When two modes have unequal frequencies, researchers occasionally differentiate the two values by calling the taller peak the *major mode*, and the shorter one the *minor mode*.

LEARNING CHECK

1. An instructor recorded the number of absences for each student in a class of 20 and obtained the frequency distribution shown in the following table.

Number of Absences (X)	f
7	1
6	0
5	3
4	3
3	3
2	5
1	4
0	1

a. Using the mean to define central tendency, what is the average number of absences for this class?

b. Using the median to define central tendency, what is the average number of absences for this class?

c. Using the mode to define central tendency, what is the average number of absences for this class?

2. In a recent survey comparing picture quality for three brands of color televisions, 63 people preferred brand A, 29 people preferred brand B, and 58 people preferred brand C. What is the mode for this distribution?

3. It is possible for a distribution to have more than one mode. (True or false?)

ANSWERS   **1. a.** The mean is 2.85.   **b.** The median is 2.50.   **c.** The mode is 2.

2. The mode is brand A.

3. True

FIGURE 3.9

A frequency distribution for tone identification scores. An example of a bimodal distribution.

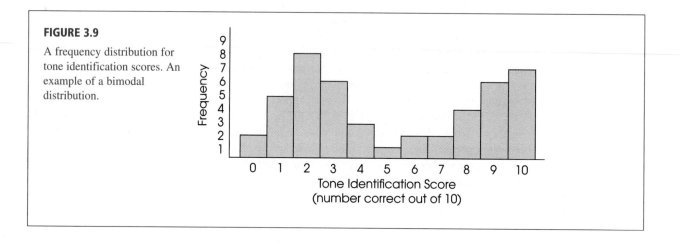

3.5   **SELECTING A MEASURE OF CENTRAL TENDENCY**

How do you decide which measure of central tendency to use? The answer to this question depends on several factors. Before we discuss these factors, however, note that you usually can compute two or even three measures of central tendency for the same set of data. Although the three measures often produce similar results, there are situations in which they are very different (see Section 3.6). Also note that the mean is usually the preferred measure of central tendency. Because the mean uses every score in the distribution, it typically produces a good representative value. Remember that the goal of central tendency is to find the single value that best represents the entire distribution. Besides being a good representative, the mean has the added advantage of being closely related to variance and standard deviation, the most common measures of variability (Chapter 4). This relationship makes the mean a valuable measure for purposes of inferential statistics. For these reasons, and others, the mean generally is considered to be the best of the three measures of central tendency. But there are specific situations in which it is impossible to compute a mean or in which the mean is not particularly representative. It is in these situations that the mode and the median are used.

WHEN TO USE THE MEDIAN   We will consider four situations in which the median serves as a valuable alternative to the mean. In the first three cases, the data consist of numerical values (interval or ratio scales) for which you would normally compute the mean. However, each case also involves a special problem so that it is either impossible to compute the mean, or the calculation of the mean produces a value that is not central or not representative of the distribution. The fourth situation involves measuring central tendency for ordinal data.

Extreme scores or skewed distributions   When a distribution has a few extreme scores, scores that are very different in value from most of the others, then the mean may not be a good representative of the majority of the distribution. The problem comes from the fact that one or two extreme values can have a large influence and cause the mean to be displaced. In this situation, the fact that the mean uses all of the scores equally can be a disadvantage. For example, suppose that a sample of $n = 10$ rats is tested in a T-maze for a food reward. The animals must choose the correct arm of the T (right or left) to find the food in the goal box. The experimenter records the number of errors each rat makes before it solves the maze. Hypothetical data are presented in Figure 3.10.

The mean for this sample is

$$M = \frac{\Sigma X}{n} = \frac{203}{10} = 20.3$$

Notice that the mean is not very representative of any score in this distribution. Most of the scores are clustered between 10 and 13. The extreme score of $X = 100$ (a slow learner) inflates the value of ΣX and distorts the mean.

The median, on the other hand, is not easily affected by extreme scores. For this sample, $n = 10$, so there should be five scores on either side of the median. The median is 11.50. Notice that this is a very representative value. Also note that the median would be unchanged even if the slow learner made 1000 errors instead of only 100. Because it is relatively unaffected by extreme scores, the median commonly is used when reporting the average value for a skewed distribution. For example, the distribution of personal incomes is very skewed, with a small segment of the population earning incomes that are astronomical. These extreme values distort the mean, so that it is not very representative of the salaries that most of us earn. As in the previous example, the median is the preferred measure of central tendency when extreme scores exist.

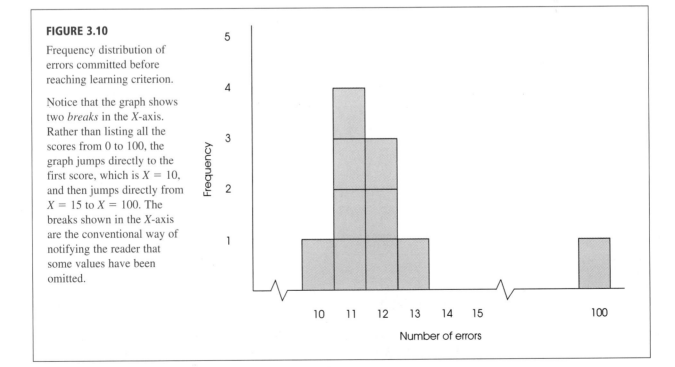

FIGURE 3.10

Frequency distribution of errors committed before reaching learning criterion.

Notice that the graph shows two *breaks* in the X-axis. Rather than listing all the scores from 0 to 100, the graph jumps directly to the first score, which is $X = 10$, and then jumps directly from $X = 15$ to $X = 100$. The breaks shown in the X-axis are the conventional way of notifying the reader that some values have been omitted.

Undetermined values   Occasionally, you will encounter a situation in which an individual has an unknown or undetermined score. In psychology, this often occurs in learning experiments in which you are measuring the number of errors (or amount of time) required for an individual to solve a particular problem. For example, suppose that participants are asked to assemble a wooden puzzle as quickly as possible. The experimenter records how long (in minutes) it takes each individual to arrange all the pieces to complete the puzzle. Table 3.5 presents results for a sample of $n = 6$ people.

Notice that person 6 never completed the puzzle. After an hour, this person still showed no sign of solving the puzzle, so the experimenter stopped him or her. This person has an undetermined score. (There are two important points to be noted. First, the experimenter should not throw out this individual's score. The whole purpose for using a sample is to gain a picture of the population, and this individual tells us that part of the population cannot solve the puzzle. Second, this person should not be given a score of $X = 60$ minutes. Even though the experimenter stopped the individual after 1 hour, the person did not finish the puzzle. The score that is recorded is the amount of time needed to finish. For this individual, we do not know how long this is.)

It is impossible to compute the mean for these data because of the undetermined value. We cannot calculate the ΣX part of the formula for the mean. However, it is possible to compute the median. For these data, the median is 12.5. Three scores are below the median, and three scores (including the undetermined value) are above the median.

Number of Pizzas (X)	f
5 or more	3
4	2
3	2
2	3
1	6
0	4

Open-ended distributions   A distribution is said to be *open-ended* when there is no upper limit (or lower limit) for one of the categories. The table at the left provides an example of an open-ended distribution, showing the number of pizzas eaten during a 1 month period for a sample of $n = 20$ high school students. The top category in this distribution shows that three of the students consumed "5 or more" pizzas. This is an open-ended category. Notice that it is impossible to compute a mean for these data because you cannot find ΣX (the total number of pizzas for all 20 students). However, you can find the median. Listing the 20 scores in order produces $X = 1$ and $X = 2$ as the middle two scores. For these data, the median is 1.5.

Ordinal scale   Many researchers believe that it is not appropriate to use the mean to describe central tendency for ordinal data. When scores are measured on an ordinal scale, the median is always appropriate and is usually the preferred measure of central tendency.

You should recall that ordinal measurements allow you to determine direction (greater than or less than) but do not allow you to determine distance. The median is compatible with this type of measurement because it is defined by direction: half of the stores are above the median and half are below the median. The mean, on the other hand, defines central tendency in terms of distance. Remember that the mean is the

TABLE 3.5

Amount of time to complete puzzle.

Person	Time (Min.)
1	8
2	11
3	12
4	13
5	17
6	Never finished

balance point for the distribution, so that the distances above the mean are exactly balanced by the distances below the mean. Because the mean is defined in terms of distances, and because ordinal scales do not measure distance, it is not appropriate to compute a mean for scores from an ordinal scale.

WHEN TO USE THE MODE

We will consider three situations in which the mode is commonly used as an alternative to the mean, or is used in conjunction with the mean to describe central tendency.

Nominal scales   The primary advantage of the mode is that it can be used to measure and describe central tendency for data that are measured on a nominal scale. Recall that the categories that make up a nominal scale are differentiated only by name. Because nominal scales do not measure quantity (distance or direction), it is impossible to compute a mean or a median for data from a nominal scale. Therefore, the mode is the only option for describing central tendency for nominal data.

Discrete variables   Recall that discrete variables are those that exist only in whole, indivisible categories. Often, discrete variables are numerical values, such as the number of children in a family or the number of rooms in a house. When these variables produce numerical scores, it is possible to calculate means. In this situation, the calculated means will usually be fractional values that cannot actually exist. For example, computing means will generate results such as "the average family has 2.4 children and a house with 5.33 rooms." On the other hand, the mode always identifies the most typical case and, therefore, it produces more sensible measures of central tendency. Using the mode, our conclusion would be "the typical, or modal, family has 2 children and a house with 5 rooms." In many situations, especially with discrete variables, people are more comfortable using the realistic, whole-number values produced by the mode.

Describing shape   Because the mode requires little or no calculation, it is often included as a supplementary measure along with the mean or median as a no-cost extra. The value of the mode (or modes) in this situation is that it gives an indication of the shape of the distribution as well as a measure of central tendency. Remember that the mode identifies the location of the peak (or peaks) in the frequency distribution graph. For example, if you are told that a set of exam scores has a mean of 72 and a mode of 80, you should have a better picture of the distribution than would be available from the mean alone (see Section 3.6).

IN THE LITERATURE
REPORTING MEASURES OF CENTRAL TENDENCY

Measures of central tendency are commonly used in the behavioral sciences to summarize and describe the results of a research study. For example, a researcher may report the sample means from two different treatments or the median score for a large sample. These values may be reported in verbal descriptions of the results, in tables, or in graphs.

In reporting results, many behavioral science journals use guidelines adopted by the American Psychological Association (APA), as outlined in the *Publication Manual of the American Psychological Association* (2001). We will refer to the APA manual from time to time in describing how data and research results are reported in the scientific literature. The APA style typically uses the letter M as the symbol for the sample mean. Thus, a study might state

The treatment group showed fewer errors ($M = 2.56$) on the task than the control group ($M = 11.76$).

When there are many means to report, tables with headings provide an organized and more easily understood presentation. Table 3.6 illustrates this point.

The median can be reported using the abbreviation *Mdn,* as in "Mdn = 8.5 errors," or it can simply be reported in narrative text, as follows:

The median number of errors for the treatment group was 8.5, compared to a median of 13 for the control group.

There is no special symbol or convention for reporting the mode. If mentioned at all, the mode is usually just reported in narrative text.

The modal hair color was blonde, with 23 out of 50 participants in this category.

PRESENTING MEANS AND MEDIANS IN GRAPHS

Graphs also can be used to report and compare measures of central tendency. Usually, graphs are used to display values obtained for sample means, but occasionally you will see sample medians reported in graphs (modes are rarely, if ever, shown in a graph). The value of a graph is that it allows several means (or medians) to be shown simultaneously so it is possible to make quick comparisons between groups or treatment conditions. When preparing a graph, it is customary to list the values for the different groups or treatment conditions on the horizontal axis. Typically, these are the different values that make up the independent variable or the quasi-independent variable. Values for the dependent variable (the scores) are listed on the vertical axis. The means (or medians) are then displayed using a *line graph,* a *histogram,* or a *bar graph,* depending on the scale of measurement used for the independent variable.

Figure 3.11 shows an example of a graph displaying the relationship between drug dose (the independent variable) and food consumption (the dependent variable). In this study, there were five different drug doses (treatment conditions) and they are listed along the horizontal axis. The five means appear as points in the graph. To construct this graph, a point was placed above each treatment condition so that the vertical position of the point corresponds to the mean score for the treatment condition. The points are then connected with straight lines and the resulting graph is called a *line graph.* A line graph is used when the values on the horizontal axis are measured on an interval or a ratio scale. An alternative to the line graph is a *histogram.* For this example, the histogram would show a bar above each drug dose so that the height of each bar corresponds to the mean food consumption for that group, with no space between adjacent bars.

Figure 3.12 shows a bar graph displaying the median selling price for single-family homes in different regions of the United States. Bar graphs are used to present means (or medians) when the groups or treatments shown on the horizontal

TABLE 3.6

The mean number of errors made on the task for treatment and control groups according to gender.

	Treatment	Control
Females	1.45	8.36
Males	3.83	14.77

FIGURE 3.11

The relationship between an independent variable (drug dose) and a dependent variable (food consumption). Because drug dose is a continuous variable, a continuous line is used to connect the different dose levels.

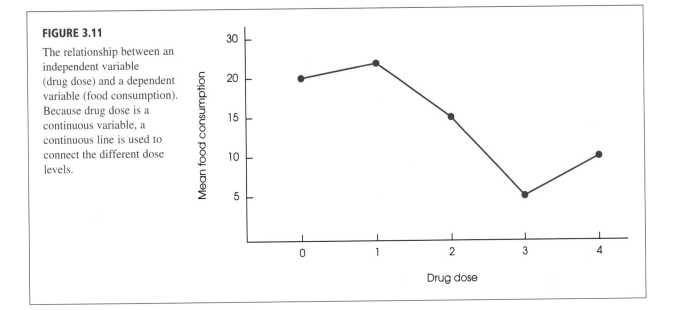

axis are measured on a nominal or an ordinal scale. To construct a bar graph, you simply draw a bar directly above each group or treatment so that the height of the bar corresponds to the mean (or median) for that group or treatment. For a bar graph, a space is left between adjacent bars to indicate that the scale of measurement is nominal or ordinal.

When constructing graphs of any type, you should recall the basic rules we introduced in Chapter 2:

1. The height of a graph should be approximately two-thirds to three-quarters of its length.

2. Normally, you start numbering both the X-axis and the Y-axis with zero at the point where the two axes intersect. However, when a value of zero is part of the

FIGURE 3.12

Median cost of a new, single-family home by region.

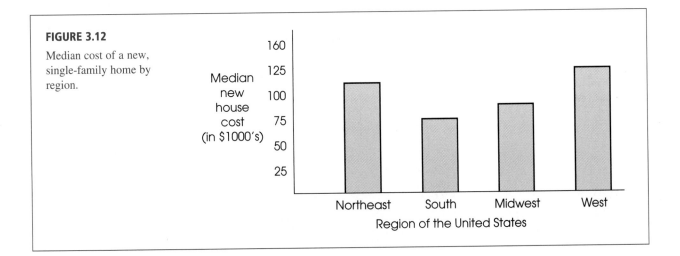

data, it is common to move the zero point away from the intersection so that the graph does not overlap the axes (see Figure 3.11).

Following these rules will help produce a graph that provides an accurate presentation of the information in a set of data. Although it is possible to construct graphs that distort the results of a study (see Box 2.1), researchers have an ethical responsibility to present an honest and accurate report of their research results. ❑

3.6   CENTRAL TENDENCY AND THE SHAPE OF THE DISTRIBUTION

We have identified three different measures of central tendency, and often a researcher calculates all three for a single set of data. Because the mean, the median, and the mode are all trying to measure the same thing (central tendency), it is reasonable to expect that these three values should be related. In fact, there are some consistent and predictable relationships among the three measures of central tendency. Specifically, there are situations in which all three measures will have exactly the same value. On the other hand, there are situations in which the three measures are guaranteed to be different. In part, the relationships among the mean, median, and mode are determined by the shape of the distribution. We will consider two general types of distributions.

SYMMETRICAL DISTRIBUTIONS

For a *symmetrical distribution,* the right-hand side of the graph is a mirror image of the left-hand side. If a distribution is perfectly symmetrical, the median is exactly at the center because exactly half of the area in the graph will be on either side of the center. The mean also is exactly at the center of a perfectly symmetrical distribution because each score on the left side of the distribution is balanced by a corresponding score (the mirror image) on the right side. As a result, the mean (the balance point) is located at the center of the distribution. Thus, for a perfectly symmetrical distribution, the mean and the median are the same (Figure 3.13). If a distribution is roughly symmetrical, but not perfect, the mean and median will be close together in the center of the distribution.

If a symmetrical distribution has only one mode, it will also be in the center of the distribution. Thus, for a perfectly symmetrical distribution with one mode, all three

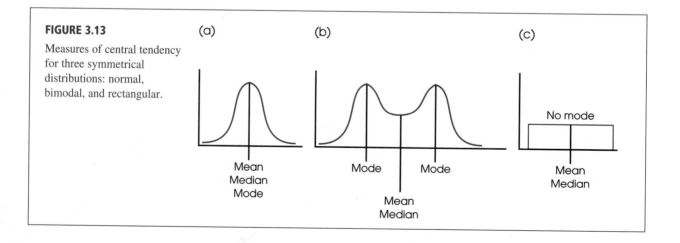

FIGURE 3.13

Measures of central tendency for three symmetrical distributions: normal, bimodal, and rectangular.

measures of central tendency, the mean, the median, and the mode, have the same value. For a roughly symmetrical distribution, the three measures are clustered together in the center of the distribution. On the other hand, a bimodal distribution that is symmetrical [see Figure 3.13(b)] will have the mean and median together in the center with the modes on each side. A rectangular distribution [see Figure 3.13(c)] has no mode because all *X* values occur with the same frequency. Still, the mean and the median are in the center of the distribution.

SKEWED DISTRIBUTIONS

The positions of the mean, median, and mode are not as consistently predictable in distributions of discrete variables (see Von Hippel, 2005).

Distributions are not always symmetrical; quite often, they are lopsided, or *skewed.* For example, Figure 3.14(a) shows a *positively skewed* distribution. In *skewed distributions,* especially distributions for continuous variables, there is a strong tendency for the mean, median, and mode to be located in predictable positions. In Figure 3.14(a), for example, the peak (highest frequency) is on the left-hand side. This is the position of the mode. However, it should be clear that the vertical line drawn at the mode does not divide the distribution into two equal parts. To have exactly 50% of the distribution on each side, the median must be located to the right of the mode. Finally, the mean is located to the right of the median because it is influenced most by extreme scores and is displaced farthest to the right by the scores in the tail. Therefore, in a positively skewed distribution, the order of the three measures of central tendency from smallest to largest (left to right) is the mode, the median, and the mean.

Negatively skewed distributions are lopsided in the opposite direction, with the scores piling up on the right-hand side and the tail tapering off to the left. The grades on an easy exam, for example, tend to form a negatively skewed distribution [see Figure 3.14(b)]. For a distribution with negative skew, the mode is on the right-hand side (with the peak), while the mean is displaced toward the left by the extreme scores in the tail. As before, the median is located between the mean and the mode. In order from smallest value to largest value (left to right), the three measures of central tendency are be the mean, the median, and the mode for a negatively skewed distribution.

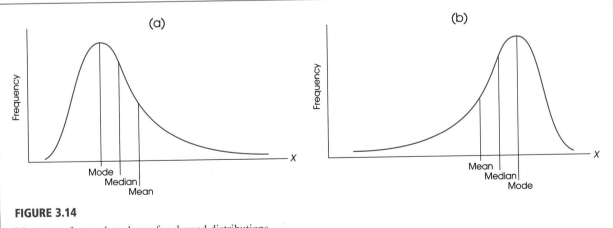

FIGURE 3.14

Measures of central tendency for skewed distributions.

1. Which measure of central tendency is most likely to be affected by one or two extreme scores in a distribution? (mean, median, or mode)

2. The mode is the correct way to measure central tendency for data from a nominal scale. (True or false?)

3. If a graph is used to show the means obtained from an experiment, the different treatment conditions (the independent variable) should be listed on the vertical axis. (True or false?)

4. A distribution has a mean of 75 and a median of 70. This distribution is probably positively skewed. (True or false?)

ANSWERS   1. mean   2. True   3 False   4. True

SUMMARY

1. The purpose of central tendency is to determine the single value that identifies the center of the distribution and best represents the entire set of scores. The three standard measures of central tendency are the mode, the median, and the mean.

2. The mean is the arithmetic average. It is computed by adding all the scores and then dividing by the number of scores. Conceptually, the mean is obtained by dividing the total (ΣX) equally among the number of individuals (N or n). The mean can also be defined as the balance point for the distribution. The distances above the mean are exactly balanced by the distances below the mean. Although the calculation is the same for a population or a sample mean, a population mean is identified by the symbol μ, and a sample mean is identified by M. In most situations with numerical scores from an interval or a ratio scale, the mean is the preferred measure of central tendency.

3. Changing any score in the distribution causes the mean to be changed. When a constant value is added to (or subtracted from) every score in a distribution, the same constant value is added to (or subtracted from) the mean. If every score is multiplied by a constant, the mean is multiplied by the same constant.

4. The median is the value that divides a distribution exactly in half. The median is the preferred measure of central tendency when a distribution has a few extreme scores that displace the value of the mean. The median also is used when there are undetermined (infinite) scores that make it impossible to compute a mean. Finally, the median is the preferred measure of central tendency for data from an ordinal scale.

5. The mode is the most frequently occurring score in a distribution. It is easily located by finding the peak in a frequency distribution graph. For data measured on a nominal scale, the mode is the appropriate measure of central tendency. It is possible for a distribution to have more than one mode.

6. For symmetrical distributions, the mean will equal the median. If there is only one mode, then it will have the same value, too.

7. For skewed distributions, the mode will be located toward the side where the scores pile up, and the mean will be pulled toward the extreme scores in the tail. The median is usually located between these two values.

KEY TERMS

central tendency	weighted mean	bimodal	minor mode	skewed distribution
population mean (μ)	median	multimodal	line graph	positive skew
sample mean (M)	mode	major mode	symmetrical distribution	negative skew

RESOURCES

There are a tutorial quiz and other learning exercises for Chapter 3 at the Wadsworth website, www.cengage.com/psychology/gravetter. The site also includes a workshop titled *Central Tendency and Variability* that reviews the basic concept of the mean and introduces the concept of variability that will be presented in Chapter 4. See page 29 for more information about accessing items on the website.

ENHANCED WebAssign

Guided interactive tutorials, end-of-chapter problems, and related testbank items may be assigned online at WebAssign.

WebTUTOR

For those using WebTutor along with this book, there is a WebTutor section corresponding to this chapter. The WebTutor contains a brief summary of Chapter 3, hints for learning the new material, cautions about common errors, and sample exam items including solutions.

SPSS

General instructions for using SPSS are presented in Appendix D. Following are detailed instructions for using SPSS to compute the **Mean and ΣX** for a set of scores.

Data Entry

1. Enter all of the scores in one column of the data editor, probably VAR00001.

Data Analysis

1. Click **Analyze** on the tool bar, select **Descriptive Statistics,** and click on **Descriptives.**
2. Highlight the column label for the set of scores (VAR00001) in the left box and click the arrow to move it into the **Variable** box.
3. If you want ΣX as well as the mean, click on the **Options** box, select **Sum,** then click **Continue.**
4. Click **OK.**

SPSS Output

SPSS will produce a summary table listing the number of scores (N), the maximum and minimum scores, the sum of the scores (if you selected this option), the mean, and the standard deviation. Note: The standard deviation is a measure of variability that is presented in Chapter 4.

FOCUS ON PROBLEM SOLVING

1. Because there are three different measures of central tendency, your first problem is to decide which one is best for your specific set of data. Usually the mean is the preferred measure, but the median may provide a more representative value if you are working with a skewed distribution. With data measured on a nominal scale, you must use the mode.

2. Although the three measures of central tendency appear to be very simple to calculate, there is always a chance for errors. The most common sources of error are listed next.

X	f
4	1
3	4
2	3
1	2

   a. Many students find it very difficult to compute the mean for data presented in a frequency distribution table. They tend to ignore the frequencies in the table and simply average the score values listed in the X column. You must use the frequencies *and* the scores! Remember that the number of scores is found by $N = \Sigma f$, and the sum of all N scores is found by ΣfX. For the distribution shown in the margin, the mean is $\frac{24}{10} = 2.40$.

   b. The median is the midpoint of the distribution of scores, not the midpoint of the scale of measurement. For a 100-point test, for example, many students incorrectly assume that the median must be $X = 50$. To find the median, you must have the *complete set* of individual scores. The median separates the individuals into two equal-sized groups.

   c. The most common error with the mode is for students to report the highest frequency in a distribution rather than the score with the highest frequency. Remember that the purpose of central tendency is to find the most representative score. For the distribution in the margin, the mode is $X = 3$, not $f = 4$.

DEMONSTRATION 3.1

COMPUTING MEASURES OF CENTRAL TENDENCY

For the following sample data, find the mean, median, and mode. The scores are:

   5,   6,   9,   11,   5,   11,   8,   14,   2,   11

Compute the mean   Calculating the mean involves two steps:

1. Obtain the sum of the scores, ΣX.
2. Divide the sum by the number of scores, n.

For these data, the sum of the scores is as follows:

   $\Sigma X = 5 + 6 + 9 + 11 + 5 + 11 + 8 + 14 + 2 + 11 = 82$

We can also observe that $n = 10$. Therefore, the mean of this sample is obtained by

   $$M = \frac{\Sigma X}{n} = \frac{82}{10} = 8.2$$

Find the median   The median divides the distribution in half, in that half of the scores are above or equal to the median and half are below or equal to it. In this demonstration, $n = 10$. Thus, the median should be a value that has 5 scores above it and 5 scores below it.

> **STEP 1**   Arrange the scores in order.
>
> $$2, \ 5, \ 5, \ 6, \ 8, \ 9, \ 11, \ 11, \ 11, \ 14$$
>
> **STEP 2**   With an even number of scores, compute the average of the middle two scores. The middle scores are $X = 8$ and $X = 9$.
>
> $$\text{median} = \frac{8 + 9}{2} = \frac{17}{2} = 8.5$$

Find the mode   The mode is the X value that has the highest frequency. Looking at the data, we can readily determine that $X = 11$ is the score that occurs most frequently.

PROBLEMS

1. What general purpose is served by a good measure of central tendency?

2. Why is it necessary to have more than one method for measuring central tendency?

3. Find the mean, median, and mode for the following sample of scores:

   $$2, \ 3, \ 3, \ 1, \ 2, \ 5, \ 2, \ 3, \ 4, \ 2$$

4. Find the mean, median, and mode for the following sample of scores:

   $$7, \ 5, \ 9, \ 7, \ 7, \ 8, \ 6, \ 8, \ 10, \ 7,$$
   $$4, \ 6$$

5. Find the mean, median, and mode for the set of scores in the following frequency distribution table:

X	f
5	1
4	5
3	4
2	1
1	1

6. Find the mean, median, and mode for the set of scores in the following frequency distribution table:

X	f
10	2
9	3
8	5
7	6
6	3
5	1

7. For the following sample
   a. Assume that the scores are measurements of a continuous variable and find the median by locating the precise midpoint of the distribution.
   b. Assume that the scores are measurements of a discrete variable and find the median.
   Scores: 1, 2, 2, 3, 4

8. In 2007, the American golfer Tiger Woods competed in 16 PGA golf tournaments. He finished 1st seven times, 2nd three times, and one time each in 6th, 9th, 12th, 15th, 22nd, and 37th. Find his median finishing position for the year. (Assume that finishing position is a discrete variable.)

9. A sample of $n = 8$ scores has a mean of $M = 6$. What is the value of ΣX for this sample?

10. A population of $N = 25$ scores has a mean of $\mu = 10$. What is the value of ΣX for this population?

11. A sample of $n = 8$ scores has a mean of $M = 10$. If one new person with a score of $X = 1$ is added to the sample, what will be the value for the new mean?

12. A sample of $n = 20$ scores has a mean of $M = 6$. If one new person with a score of $X = 27$ is added to the sample, what will be the value for the new mean?

13. A population of $N = 10$ scores has a mean of $\mu = 24$. If one person with a score of $X = 42$ is removed from the sample, what will be the value for the new mean?

14. A sample of $n = 5$ scores has a mean of $M = 12$. If one person with a score of $X = 8$ is removed from the sample, what will be the value for the new mean?

15. A sample of $n = 6$ scores has a mean of $M = 10$. If one score is changed from $X = 14$ to $X = 2$, what will be the value for the new sample mean?

16. A population of $N = 8$ scores has a mean of $\mu = 20$. If one score is changed from $X = 14$ to $X = 30$, what will be the value for the new mean?

17. A sample of $n = 7$ scores has a mean of $M = 5$. After one new score is added to the sample, the new mean is found to be $M = 6$. What is the value of the new score? (Hint: Compare the values for ΣX before and after the score was added.)

18. A population of $N = 10$ scores has a mean of $\mu = 9$. One score is removed from the sample and the new mean is found to be $\mu = 8$. What is the value of the score that was removed? (Hint: Compare the values for ΣX before and after the score was removed.)

19. One sample has a mean of $M = 4$ and a second sample has a mean of $M = 8$. The two samples are combined into a single set of scores.
 a. What is the mean for the combined set if both of the original samples have $n = 7$ scores?
 b. What is the mean for the combined set if the first sample has $n = 3$ and the second sample has $n = 7$?
 c. What is the mean for the combined set if the first sample has $n = 7$ and the second sample has $n = 3$?

20. One sample has a mean of $M = 6$ and a second sample has a mean of $M = 12$. The two samples are combined into a single set of scores.
 a. What is the mean for the combined set if both of the original samples have $n = 5$ scores?

 b. What is the mean for the combined set if the first sample has $n = 4$ scores and the second sample has $n = 8$?
 c. What is the mean for the combined set if the first sample has $n = 8$ scores and the second sample has $n = 4$?

21. A sample of $n = 7$ scores has a mean of $M = 8$. A second sample has $n = 3$ scores with a mean of $M = 12$. If the two samples are combined, what is the mean for for the combined sample?

22. One sample of $n = 20$ scores has a mean of $M = 50$. A second sample of $n = 5$ scores has a mean of $M = 10$. If the two samples are combined, what is the mean for the combined sample?

23. Explain why the mean is often not a good measure of central tendency for a skewed distribution.

24. Identify the circumstances in which the median rather than the mean is the preferred measure of central tendency.

25. For each of the following situations, identify the measure of central tendency (mean, median, or mode) that would provide the best description of the "average" score:
 a. The college recorded the age of each student who graduated last spring. Although the majority of the students were in their early 20s, a few mature students were in their 40s, 50s, and 60s.
 b. A researcher asked each participant to select his or her favorite color from a set of eight color patches.
 c. A professor uses a standardized test to measure verbal skill for a sample of new college freshmen.

26. One question on a student survey asks: In a typical week, how many times do you eat at a fast food restaurant? The following frequency distribution table summarizes the results for a sample of $n = 20$ students.

Number of times per week	f
5 or more	2
4	2
3	3
2	6
1	4
0	3

 a. Find the mode for this distribution.
 b. Find the median for the distribution.
 c. Explain why you cannot compute the mean number of times using the data in the table.

27. To evaluate the effectiveness of a treatment, researchers often measure individuals before and after treatment, and record the amount of difference between the two scores. If the differences average around zero, it is an indication that the treatment has no effect. However, if the differences are consistently positive (or consistently negative), it suggests that the treatment does have an effect. For example, Kerr, Walsh, and Baxter (2003) evaluated the effectiveness of acupuncture treatment by measuring pain for a group of participants before and after a 6-week treatment program. Hypothetical data, similar to the results of the study, are as follows. Note that each score measures the difference in pain level after treatment compared to the pain level before treatment.

Note: A positive score indicates more pain before treatment than after treatment.

$$+4, +5, +2, \quad 0, +7, +4, +3, +3, +2, -2$$
$$+1, +6, \quad 0, +4, +7, -1, +3, +5, +4, +2$$

a. Sketch a histogram showing the frequency distribution for the difference scores.
b. Compute the mean for the sample of $n = 20$ difference scores.
c. Does it appear that the acupuncture treatment had an effect on pain?

28. Does it ever seem to you that the weather is nice during the work week, but lousy on the weekend? Cerveny and Balling (1988) have confirmed that this is not your imagination—pollution accumulating during the work week most likely spoils the weekend weather for people on the Atlantic coast. Consider the following hypothetical data showing the daily amount of rainfall for 10 weeks during the summer.

Week	Average Daily Rainfall on Weekdays (Mon.–Fri.)	Average Daily Rainfall on Weekends (Sat.–Sun.)
1	1.2	1.5
2	0.6	2.0
3	0.0	1.8
4	1.6	1.5
5	0.8	2.2
6	2.1	2.4
7	0.2	0.8
8	0.9	1.6
9	1.1	1.2
10	1.4	1.7

a. Calculate the average daily rainfall (the mean) during the week, and the average daily rainfall for weekends.
b. Based on the two means, does there appear to be a pattern in the data?

CHAPTER

4

Variability

Tools You Will Need

The following items are considered essential background material for this chapter. If you doubt your knowledge of any of these items, you should review the appropriate chapter or section before proceeding.

- Summation notation (Chapter 1)

- Central tendency (Chapter 3)
  - Mean
  - Median

Preview

4.1 Overview

4.2 The Range and Interquartile Range

4.3 Standard Deviation and Variance for a Population

4.4 Standard Deviation and Variance for Samples

4.5 More about Variance and Standard Deviation

4.6 Comparing Measures of Variability

Summary

Focus on Problem Solving

Demonstration 4.1

Problems

Preview

Although measures of central tendency, such as the mean and median, are handy ways to summarize large sets of data, these measures do not tell the whole story. Specifically, not everyone is average. Many people may perform near average, but others demonstrate performance that is far above (or below) average. In simple terms, people are different.

The differences that exist from one person to another are often called *diversity*. Researchers comparing cognitive skills for younger adults and older adults typically find that differences between people tend to increase as people age. For example, Morse (1993) reviewed hundreds of research studies published in *Psychology and Aging* and in the *Journal of Gerontology* from 1986 to 1990, and found increased diversity in older adults on measures of reaction time, memory, and some measures of intelligence. One possible explanation for the increased diversity is that different people respond differently to the aging process; some are essentially unchanged and others show a rapid decline. As a rresult, the differences from one person to another are larger for older people than for those who are younger.

It also is possible to measure differences in performance for the same person. These differences provide a measure of *consistency*. Often, large differences from trial to trial for the same person are viewed as evidence of poor performance. For example, the ability to consistently hit a target is an indication of skilled performance in many sports,whereas inconsistent performance indicates a lack of skill. Researchers in the field of aging have also found that older participants tend to have larger differences from trial to trial than younger participants. That is, older people seem to lose the ability to perform consistently on many tasks. For example, in a study comparing older and younger women, Wegesin and Stern (2004) found lower consistency for older women on a recognition memory task.

The Problem: To study phenomena such as diversity and consistency, it is necessary to devise a method to measure and objectively describe the differences that exist within a set of scores.

The Solution: A measure of *variability* provides an objective description of the differences between the scores in a distribution by measuring the degree to which the scores are spread out or are clustered together.

4.1   OVERVIEW

The term *variability* has much the same meaning in statistics as it has in everyday language; to say that things are variable means that they are not all the same. In statistics, our goal is to measure the amount of variability for a particular set of scores, a distribution. In simple terms, if the scores in a distribution are all the same, then there is no variability. If there are small differences between scores, then the variability is small, and if there are large differences between scores, then the variability is large.

DEFINITION

> **Variability** provides a quantitative measure of the degree to which scores in a distribution are spread out or clustered together.

Figure 4.1 shows two distributions of familiar values: Part (a) shows the distribution of adult male heights (in inches), and part (b) shows the distribution of adult male weights (in pounds). Notice that the two distributions differ in terms of central tendency. The mean height is 70 inches (5 feet, 10 inches) and the mean weight is 170 pounds. In addition, notice that the distributions differ in terms of variability. For example, most heights are clustered close together, within 5 or 6 inches of the mean. On the other hand, weights are spread over a much wider range. In the weight distribution it is not unusual

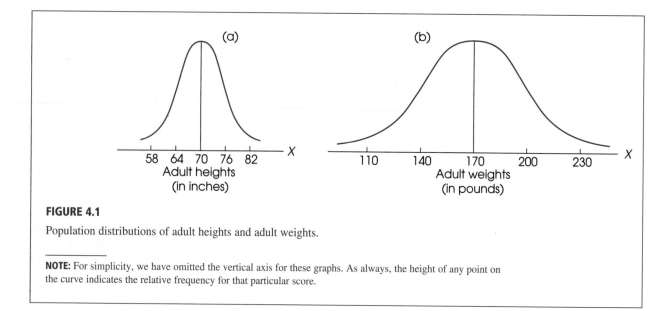

FIGURE 4.1

Population distributions of adult heights and adult weights.

NOTE: For simplicity, we have omitted the vertical axis for these graphs. As always, the height of any point on the curve indicates the relative frequency for that particular score.

to find individuals who are located more than 30 pounds away from the mean, and it would not be surprising to find two individuals whose weights differ by more than 30 or 40 pounds. The purpose for measuring variability is to obtain an objective measure of how the scores are spread out in a distribution. In general, a good measure of variability serves two purposes:

1. Variability describes the distribution. Specifically, it tells whether the scores are clustered close together or are spread out over a large distance. Usually, variability is defined in terms of *distance*. It tells how much distance to expect between one score and another, or how much distance to expect between an individual score and the mean. For example, we know that most adults' heights are clustered close together, within 5 or 6 inches of the average. Although more extreme heights exist, they are relatively rare.

2. Variability measures how well an individual score (or group of scores) represents the entire distribution. This aspect of variability is very important for inferential statistics where relatively small samples are used to answer questions about populations. For example, suppose that you selected a sample of one person to represent the entire population. Because most adult heights are within a few inches of the population average (the distances are small), there is a very good chance that you would select someone whose height is within 6 inches of the population mean. On the other hand, the scores are much more spread out (greater distances) in the distribution of adult weights. In this case, you probably would *not* obtain someone whose weight was within 6 pounds of the population mean. Thus, variability provides information about how much error to expect if you are using a sample to represent a population.

In this chapter, we consider three different measures of variability: the range, the interquartile range, and the standard deviation. Of these three, the standard deviation (and the related measure of variance) is by far the most important.

Finally, we have computed the mean and the variance for each sample. Note that we have computed the sample variance two different ways. First, we examine what would happen if there was no correction for bias and the sample variance was computed by simply dividing *SS* by *n*. Second, we examine the real sample variances where *SS* is divided by $n - 1$ to produce an unbiased measure of variance. You should verify our calculations by computing one or two of the values for yourself. The complete set of sample means and sample variances is presented in Table 4.1.

Now, direct your attention to the column of sample means. For this example, the original population has a mean of $\mu = 4$. Although none of the samples has a mean exactly equal to 4, if you consider the complete set of sample means, you will find that the 9 sample means add up to a total of 36, so the average of the sample means is $\frac{36}{9} = 4$. Note that the average of the sample means is exactly equal to the population mean. This is what is meant by the concept of an unbiased statistic. On average, the sample values provide an accurate representation of the population. In this example, the average of the 9 sample means is exactly equal to the population mean.

Next, consider the column of biased sample variances where we divided by *n*. For the original population the variance is $\sigma^2 = 14$. The 9 sample variances, however, add up to a total of 63, which produces an average variance of $\frac{63}{9} = 7$. Note that the average of these sample variances is *not* equal to the population variance. If the sample variance is computed by dividing by *n,* the resulting values will not produce an accurate estimate of the population variance. On average, these sample variances underestimate the population variance and, therefore, are biased statistics.

Finally, consider the column of sample variances that are computed using $n - 1$. Although the population has a variance of $\sigma^2 = 14$, you should notice that none of the samples has a variance exactly equal to 14. However, if you consider the complete set of sample variances, you will find that the 9 values add up to a total of 126, which produces an average variance of $\frac{126}{9} = 14.00$. Thus, the average of the sample variances is exactly equal to the population variance. On average, the sample variance (computed using $n - 1$) produces an accurate, unbiased estimate of the population variance.

In summary, both the sample mean and the sample variance (using $n - 1$) are examples of unbiased statistics. This fact makes the sample mean and sample variance extremely valuable for use as inferential statistics. Although no individual sample is likely to have a mean and variance exactly equal to the population values, both the sample mean and the sample variance, on average, do provide accurate estimates of the corresponding population values.

STANDARD DEVIATION AND DESCRIPTIVE STATISTICS

Because standard deviation requires extensive calculations, there is a tendency to get lost in the arithmetic and forget what standard deviation is and why it is important. Standard deviation is primarily a descriptive measure; it describes how variable, or how spread out, the scores are in a distribution. Behavioral scientists must deal with the variability that comes from studying people and animals. People are not all the same; they have different attitudes, opinions, talents, IQs, and personalities. Although we can calculate the average value for any of these variables, it is equally important to describe the variability. Standard deviation describes variability by measuring *distance from the mean.* In any distribution, some individuals will be close to the mean, and others will be relatively far from the mean. Standard deviation provides a measure of the typical, or standard, distance from the mean.

In addition to describing an entire distribution, standard deviation also allows us to interpret individual scores. At the beginning of this chapter we presented a figure showing the distribution of adult male heights and adult male weights. With a mean weight of $\sigma = 170$ pounds, a person who weights 180 is above average but not exceptionally heavy; with a standard deviation of $\sigma = 30$, a person who is 10 pounds above average is not extreme. If we are comparing heights, on the other hand, a person who is 10 inches taller than average is exceptional; with a small standard deviation of only 5 or 6 inches, a 10-point difference is extreme.

The mean and the standard deviation are the most common values used to describe a set of data. A research report, for example, typically will not list all of the individual scores but rather will summarize the data by reporting the mean and standard deviation. When you are given these two descriptive statistics, you should be able to visualize the entire set of data. For example, consider a sample with a mean of $M = 36$ and a standard deviation of $s = 4$. Although there are several different ways to picture the data, one simple technique is to imagine (or sketch) a histogram in which each score is represented by a box in the graph. For this sample, the data can be pictured as a pile of boxes (scores) with the center of the pile located at a value of $M = 36$. The individual scores or boxes are scattered on both sides of the mean with some of the boxes relatively close to the mean and some farther away. As a rule of thumb, roughly 70% of the scores in a distribution are located within a distance of one standard deviation from the mean, and almost all of the scores (roughly 95%) are within two standard deviations of the mean. In this example, the standard distance from the mean is $s = 4$ points so your image should have most of the boxes within 4 points of the mean, and nearly all of the boxes within 8 points. One possibility for the resulting image is shown in Figure 4.8.

Notice that Figure 4.8 not only shows the mean and the standard deviation, but also uses these two values to reconstruct the underlying scale of measurement (the X values along the horizontal line). The scale of measurement helps complete the picture of the entire distribution and helps to relate each individual score to the rest of the group. In this example, you should realize that a score of $X = 34$ is located near the center of the distribution, only slightly below the mean. On the other hand, a score of $X = 45$ is an extremely high score, located far out in the right-hand tail of the distribution.

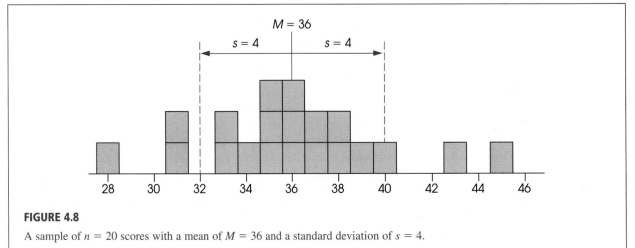

FIGURE 4.8

A sample of $n = 20$ scores with a mean of $M = 36$ and a standard deviation of $s = 4$.

The general point of this discussion is that the mean and standard deviation are not simply abstract concepts or mathematical equations. Instead, these two values should be concrete and meaningful, especially in the context of a set of scores. The mean and standard deviation are central concepts for most of the statistics that will be presented in the following chapters. A good understanding of these two statistics will help you with the more complex procedures that follow. (See Box 4.1.)

TRANSFORMATIONS OF SCALE

Occasionally a set of scores is transformed by adding a constant to each score or by multiplying each score by a constant value. This happens, for example, when exposure to a treatment adds a fixed amount to each participant's score or when you want to change the unit of measurement (to convert from minutes to seconds, multiply each X by 60). What happens to the standard deviation when the scores are transformed in this manner?

The easiest way to determine the effect of a transformation is to remember that the standard deviation is a measure of distance. If you select any two scores and see what happens to the distance between them, you also will find out what happens to the standard deviation.

1. **Adding a constant to each score does not change the standard deviation**   If you begin with a distribution that has $\mu = 40$ and $\sigma = 10$, what happens to σ if you add 5 points to every score? Consider any two scores in this distribution: Suppose, for example, that these are exam scores and that you had $X = 41$ and your friend had $X = 43$. The distance between these two scores is $43 - 41 = 2$ points. After adding the constant, 5 points, to each score, your score would be $X = 46$, and your friend would have $X = 48$. The distance between scores is still 2 points. Adding a constant to every score does not affect any of the distances and, therefore, does not change the standard deviation. This fact can be seen clearly if you imagine a frequency distribution graph. If, for example, you add 10 points to each score, then every score in the graph is moved 10 points to the right. The result is that the entire distribution is shifted to a new position 10 points up the scale. Note that the mean moves along with the scores and is

---

BOX 4.1

AN ANALOGY FOR THE MEAN AND THE STANDARD DEVIATION

Although the basic concepts of the mean and the standard deviation are not overly complex, the following analogy often helps students gain a more complete understanding of these two statistical measures.

In our local community, the site for a new high school was selected because it provides a central location. An alternative site on the western edge of the community was considered, but this site was rejected because it would require extensive busing for students living on the east side. In this example, the location of the high school is analogous to the concept of the mean; just as the high school is located in the center of the community, the mean is located in the center of the distribution of scores.

For each student in the community, it is possible to measure the distance between home and the new high school. Some students live only a few blocks from the new school and others live as much as 3 miles away. The average distance that a student must travel to school was calculated to be 0.80 miles. The average distance from the school is analogous to the concept of the standard deviation; that is, the standard deviation measures the standard distance from an individual score to the mean.

increased by 10 points. However, the variability does not change because each of the deviation scores $(X - \mu)$ does not change.

2. Multiplying each score by a constant causes the standard deviation to be multiplied by the same constant   Consider the same distribution of exam scores we looked at earlier. If $\mu = 40$ and $\sigma = 10$, what would happen to σ if each score were multiplied by 2? Again, we will look at two scores, $X = 41$ and $X = 43$, with a distance between them equal to 2 points. After all the scores have been multiplied by 2, these scores become $X = 82$ and $X = 86$. Now the distance between scores is 4 points, twice the original distance. Multiplying each score causes each distance to be multiplied, so the standard deviation also is multiplied by the same amount.

IN THE LITERATURE
REPORTING THE STANDARD DEVIATION

In reporting the results of a study, the researcher often provides descriptive information for both central tendency and variability. The dependent variables in psychology research frequently involve measures taken on interval or ratio scales. Thus, the mean (central tendency) and the standard deviation (variability) are commonly reported together. In many journals, especially those following APA style, the symbol SD is used for the sample standard deviation. For example, the results might state:

> Children who viewed the violent cartoon displayed more aggressive responses ($M = 12.45$, $SD = 3.7$) than those who viewed the control cartoon ($M = 4.22$, $SD = 1.04$).

When reporting the descriptive measures for several groups, the findings may be summarized in a table. Table 4.2 illustrates the results of hypothetical data.

TABLE 4.2

The number of aggressive responses in male and female children after viewing cartoons.

	Type of Cartoon	
	Violent	Control
Males	$M = 15.72$	$M = 6.94$
	$SD = 4.43$	$SD = 2.26$
Females	$M = 3.47$	$M = 2.61$
	$SD = 1.12$	$SD = 0.98$

Sometimes the table will also indicate the sample size, n, for each group. You should remember that the purpose of the table is to present the data in an organized, concise, and accurate manner.   ❏

VARIANCE AND INFERENTIAL STATISTICS

In very general terms, the goal of inferential statistics is to detect meaningful and significant patterns in research results. The basic question is whether the sample data reflect patterns that exist in the population, or the sample data simply show random fluctuations that occur by chance. Variability plays an important role in the inferential process because the variability in the data influences how easy it is to see patterns. In general, low variability means that existing patterns can be seen clearly, whereas high variability tends to obscure any patterns that might exist. The following two samples provide a simple demonstration of how variance can influence the perception of patterns. Your task is to examine each sample briefly and then estimate the sample mean.

Sample 1 X	Sample 2 X
34	26
35	10
36	64
35	40

After a few seconds of examining sample 1, you should realize that the sample mean is $M = 35$, and you should be very confident that this is an accurate estimate. With sample 2, on the other hand, the task is much more difficult. Although both samples have a mean of $M = 35$, it is much easier to see the mean in sample 1 than it is in sample 2.

The difference between the two samples is variability. In the first sample the scores are all clustered close together and variability is small. In this situation it is easy to see the sample mean. However, the scores in the second sample are spread out over a wide range and variability is large. With high variability, it is not easy to identify the location of the mean. In general, high variability makes it difficult to see any patterns in the data.

The preceding example demonstrates how variability can affect the ability to identify the mean for a single sample. In most research studies, however, the goal is to compare means for two (or more) sets of data. For example:

Is the mean level of depression lower after therapy than it was before therapy?

Is the mean attitude score for men different from the mean score for women?

Is the mean reading achievement score higher for students in a special program than for students in regular classrooms?

In each of these situations, the goal is to find a clear difference between two means that would demonstrate a significant, meaningful pattern in the results. Once again variability ability plays an important role in determining whether a clear pattern exists. Consider the following data representing hypothetical results from two experiments, each comparing two treatment conditions. For both experiments, your task is to determine whether there appears to be any consistent difference between the scores in treatment 1 and the scores in treatment 2.

Experiment A		Experiment B	
Treatment 1	Treatment 2	Treatment 1	Treatment 2
35	39	31	46
34	40	15	21
36	41	57	61
35	40	37	32

For each experiment, the data have been constructed so that there is a 5-point mean difference between the two treatments: On average, the scores in treatment 2 are 5 points higher than the scores in treatment 1. The 5-point difference is relatively easy to see in Experiment A, where the variability is low, but the same 5-point difference is difficult to see in Experiment B, where the variability is large. Again, high variability tends to obscure any patterns in the data. This general fact is perhaps even more convincing when the data are presented in a graph. Figure 4.9 shows the two sets of data from Experiments A and B. Notice that the results from Experiment A clearly show the

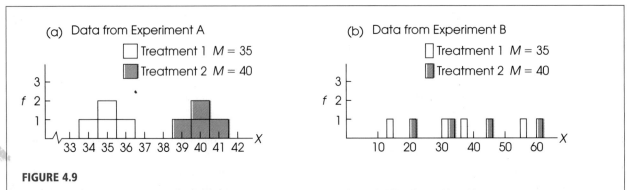

FIGURE 4.9

Graphs showing the results from two experiments. In Experiment A, the variability is small and it is easy to see the 5-point mean difference between the two treatments. In Experiment B, however, the 5-point mean difference between treatments is obscured by the large variability.

5-point difference between treatments. One group of scores piles up around 35 and the second group piles up around 40. On the other hand, the scores from Experiment B [Figure 4.9(b)] seem to be mixed together randomly with no clear difference between the two treatments.

In the context of inferential statistics, the variance that exists in a set of sample data is often classified as *error variance*. This term is used to indicate that the sample variance represents unexplained and uncontrolled differences between scores. As the error variance increases, it becomes more difficult to see any systematic differences or patterns that might exist in the data. An analogy is to think of variance as the static that appears on a radio station or a cell phone when you enter an area of poor reception. In general, variance makes it difficult to get a clear signal from the data. High variance can make it difficult or impossible to see a mean difference between two sets of scores, or to see any other meaningful patterns in the results from a research study.

LEARNING CHECK

1. What is the difference between a biased and an unbiased statistic?

2. In a population with a mean of $\mu = 50$ and a standard deviation of $\sigma = 10$, would a score of $X = 58$ be considered an extreme value (far out in the tail of the distribution)? What if the standard deviation were $\sigma = 3$?

3. A population has a mean of $\mu = 70$ and a standard deviation of $\sigma = 5$.
   a. If 10 points were added to every score in the population, what would be the new values for the population mean and standard deviation?
   b. If every score in the population were multiplied by 2, what would be the new values for the population mean and standard deviation?

ANSWERS

1. If a statistic is biased, it means that the average value of the statistic does not accurately represent the corresponding population parameter. Instead, the average value of the statistic

either overestimates or underestimates the parameter. If a statistic is unbiased, it means that the average value of the statistic is an accurate representation of the corresponding population parameter.

2. With $\sigma = 10$, a score of $X = 58$ would be located in the central section of the distribution (within one standard deviation). With $\sigma = 3$, a score of $X = 58$ would be an extreme value, located more than two standard deviations above the mean.

3. **a.** The new mean would be $\mu = 80$ but the standard deviation would still be $\sigma = 5$.

   b. The new mean would be $\mu = 140$ and the new standard deviation would be $\sigma = 10$.

4.6 COMPARING MEASURES OF VARIABILITY

By far the most commonly used measure of variability is standard deviation (together with the related measure of variance). Nonetheless, there are situations for which the range or the semi-interquartile range may be preferred. The advantages and disadvantages of each of these three measures will be discussed next.

In simple terms, two considerations determine the value of any statistical measurement:

1. The measure should provide a stable and reliable description of the scores. Specifically, it should not be greatly affected by minor details in the set of data.

2. The measure should have a consistent and predictable relationship with other statistical measurements.

We will examine each of these considerations separately.

FACTORS THAT AFFECT VARIABILITY

1. **Extreme scores**   Of the three measures of variability, the range is most affected by extreme scores. A single extreme value will have a large influence on the range. In fact, the range is determined exclusively by the two extremes of the distribution. Standard deviation and variance also are influenced by extreme scores. Because these measures are based on squared deviations, a single extreme value can have a disproportionate effect. For example, a score that is 10 points away from the mean will contribute $10^2 = 100$ points to the SS. For this reason, standard deviation and variance should be interpreted carefully in distributions with one or two extreme values. Because the semi-interquartile range focuses on the middle of the distribution, it is least affected by extreme values. For this reason, the semi-interquartile range often provides the best measure of variability for distributions that are very skewed or that have a few extreme scores.

2. **Sample size**   As you increase the number of scores in a sample, you also tend to increase the range because each additional score has the potential to replace the current highest or lowest value in the set. Thus, the range is directly related to sample size. This relationship between sample size and variability is unacceptable. A researcher should not be able to influence variability by manipulating sample size. Standard deviation, variance, and the semi-interquartile range are relatively unaffected by sample size and, therefore, provide better measures.

3. Stability under sampling   If you take several different samples from the same population, you should expect the samples to be similar. Specifically, if you compute variability for each of the separate samples, you should expect to obtain similar values. Because all of the samples come from the same source, it is reasonable that there should be some "family resemblance." When standard deviation and variance are used to measure variability, the samples tend to have similar variability. For this reason, standard deviation and variance are said to be stable under sampling. The semi-interquartile range also provides a reasonably stable measure of variability. The range, however, changes unpredictably from sample to sample and is said to be unstable under sampling.

4. Open-ended distributions   When a distribution does not have any specific boundary for the highest score or the lowest score, it is open-ended. This can occur when you have infinite or undetermined scores. For example, a subject who cannot solve a problem has taken an undetermined or infinite amount of time to reach the solution. In an open-ended distribution, you cannot compute the range, the standard deviation, or the variance. In this situation, the only available measure of variability is the semi-interquartile range.

RELATIONSHIP WITH OTHER STATISTICAL MEASURES

As noted earlier, variance and standard deviation are computed from squared deviation scores. Because they are based on squared distances, these measures fit into a coherent system of mathematical relationships that underlies many of the statistical techniques we will examine in this book. Although we generally will not present the underlying mathematics, you will notice that variance and standard deviation appear repeatedly. This is because they are valuable measures of variability. Also, you should notice that variance and standard deviation have a direct relationship to the mean (they are based on deviations from the mean). Therefore, the mean and standard deviation tend to be reported together. Because the mean is the most common measure of central tendency, the standard deviation is the most common measure of variability.

Because the median and the semi-interquartile range are both based on percentiles, they share a common foundation and tend to be associated. When the median is used to report central tendency, the semi-interquartile range is commonly used to report variability.

The range has no direct relationship to any other statistical measure. For this reason, it is rarely used in conjunction with other statistical techniques.

SUMMARY

1. The purpose of variability is to determine how spread out the scores are in a distribution. There are four basic measures of variability: the range, the interquartile range, the variance, and the standard deviation.

The range is the distance from the highest score to the lowest score in a distribution and is defined as the difference between the upper real limit of the largest X and the lower real limit of the smallest X. The interquartile range is the distance covered by the middle 50% of the distribution and is defined as the difference between the third quartile ($Q3$) and the first quartile ($Q1$). Standard deviation and variance are the most commonly used measures of variability. Both of these measures are based on the idea that each score can be described in terms of its deviation or distance from the mean. The variance is the mean of the squared deviations. The standard deviation is the square root of the variance and provides a measure of the standard distance from the mean.

2. To calculate variance or standard deviation, you first need to find the sum of the squared deviations, *SS*. Except for minor changes in notation, the calculation of *SS* is identical for samples and populations. There are two methods for calculating *SS:*

I. By definition, you can find *SS* using the following steps:

    a. Find the deviation $(X - \mu)$ for each score.

    b. Square each deviation.

    c. Add the squared deviations.

This process can be summarized in a formula as follows:

definitional formula: $SS = \Sigma(X - \mu)^2$

II. The sum of the squared deviations can also be found using a computational formula, which is especially useful when the mean is not a whole number:

computational formula: $SS = \Sigma X^2 - \dfrac{(\Sigma X)^2}{N}$

3. Variance is the mean squared deviation and is obtained by finding the sum of the squared deviations and then dividing by the number. For a population, variance is

$$\sigma^2 = \frac{SS}{N}$$

For a sample, only $n - 1$ of the scores are free to vary (degrees of freedom or $df = n - 1$), so sample variance is

$$s^2 = \frac{SS}{n - 1} = \frac{SS}{df}$$

Using $n - 1$ in the sample formula makes the sample variance an accurate and unbiased estimate of the population variance.

4. Standard deviation is the square root of the variance. For a population, this is

$$\sigma = \sqrt{\frac{SS}{N}}$$

Sample standard deviation is

$$s = \sqrt{\frac{SS}{n - 1}} = \sqrt{\frac{SS}{df}}$$

5. Adding a constant value to every score in a distribution will not change the standard deviation. Multiplying every score by a constant, however, will cause the standard deviation to be multiplied by the same constant.

KEY TERMS

variability	deviation score	sum of squares (*SS*)	degrees of freedom (df)
range	population variance (σ^2)	sample variance (s^2)	biased statistic
interquartile range	population standard	sample standard deviation (s)	unbiased statistic
semi-interquartile range	deviation (σ)		

RESOURCES

There are a tutorial quiz and other learning exercises for Chapter 4 at the book companion website, www.cengage.com/psychology/gravetter. The site also includes a workshop titled *Central Tendency and Variability* that examines the basic concepts of variability and the standard deviation, and reviews the concept of central tendency that was covered in Chapter 3. See page 29 for more information about accessing items on the website.

WebAssign

Guided interactive tutorials, end-of-chapter problems, and related testbank items may be assigned online at WebAssign.

WebTUTOR

For those using WebTutor along with this book, there is a WebTutor section corresponding to this chapter. The WebTutor contains a brief summary of Chapter 4, hints for learning the concepts and formulas for variability, cautions about common errors, and sample exam items including solutions.

SPSS

General instructions for using SPSS are presented in Appendix D. Following are detailed instructions for using SPSS to compute the **Standard Deviation and Variance** for a sample of scores**.**

Data Entry

  1. Enter all of the scores in one column of the data editor, probably VAR00001.

Data Analysis

  1. Click **Analyze** on the tool bar, select **Descriptive Statistics,** and click on **Descriptives.**
  2. Highlight the column label for the set of scores (VAR00001) in the left box and click the arrow to move it into the **Variable** box.
  3. If you want the variance reported along with the standard deviation, click on the **Options** box, select **Variance,** then click **Continue.**
  4. Click **OK.**

SPSS Output

SPSS will produce a summary table listing the number of scores the maximum and minimum scores, the mean, the standard deviation, and the variance. Caution: SPSS computes the *sample* standard deviation and *sample* variance using $n - 1$. If your scores are intended to be a population, you can multiply the sample standard deviation by the square root of $(n - 1)/n$ to obtain the population standard deviation.

  Note: You can also obtain the mean and standard deviation for a sample if you use SPSS to display the scores in a frequency distribution histogram (see the SPSS section at the end of Chapter 2). The mean and standard deviation are displayed beside the graph.

FOCUS ON PROBLEM SOLVING

  1. The purpose of variability is to provide a measure of how spread out the scores are in a distribution. Usually this is described by the standard deviation. Because the calculations are relatively complicated, it is wise to make a preliminary estimate of the standard deviation before you begin. Remember that standard deviation provides a measure of the typical, or standard, distance from the mean. Therefore, the standard deviation must have a value somewhere between the largest and the smallest deviation scores. As a rule of thumb, the standard deviation should be about one-fourth of the range.

2. Rather than trying to memorize all the formulas for SS, variance, and standard deviation, you should focus on the definitions of these values and the logic that relates them to each other:

SS is the sum of squared deviations.

Variance is the mean squared deviation.

Standard deviation is the square root of variance.

The only formula you should need to memorize is the computational formula for SS.

3. A common error is to use $n - 1$ in the computational formula for SS when you have scores from a sample. Remember that the SS formula always uses n (or N). After you compute SS for a sample, you must correct for the sample bias by using $n - 1$ in the formulas for variance and standard deviation.

DEMONSTRATION 4.1

COMPUTING MEASURES OF VARIABILITY

For the following sample data, compute the variance and standard deviation. The scores are:

$$10, \quad 7, \quad 6, \quad 10, \quad 6, \quad 15$$

STEP 1   **Compute SS, the sum of squared deviations**   We will use the definitional formula: $SS = \Sigma(X - M)^2$. For this sample, the mean is $M = \dfrac{\Sigma X}{n} = \dfrac{54}{6} = 9$. The following table shows the deviation and the squared deviation for each score.

X	$X - M$	$(X - M)^2$
10	+1	1
7	-2	4
6	-3	9
10	+1	1
6	-3	9
15	+6	36

SS is the sum of the squared deviations:

$$SS = \Sigma(X - M)^2 = 1 + 4 + 9 + 1 + 9 + 36 = 60$$

STEP 2   **Compute the sample variance**   For sample variance, SS is divided by the degrees of freedom, $df = n - 1$.

$$s^2 = \frac{SS}{n - 1} = \frac{60}{5} = 12$$

STEP 3   Compute the sample standard deviation   Standard deviation is simply the square root of the variance.

$$s = \sqrt{12} = 3.46$$

PROBLEMS

1. The following $n = 12$ scores are listed in order from smallest to largest. For this sample,

    1, 2, 2, 3, 3, 3, 3, 4, 4, 5, 6, 7,

    a. Find the range and the interquartile range.
    b. Now, change the score of $X = 7$ to $X = 27$ and find the new range and interquartile range.
    c. Describe how one extreme score influences the range and the interquartile range.

2. In words, explain what is measured by each of the following:
    a. SS
    b. Variance
    c. Standard deviation

3. A population has $\mu = 100$ and $\sigma = 20$. If you select a single score from this population, on the average, how close would it be to the population mean? Explain your answer.

4. Can SS ever have a value less than zero? Explain your answer.

5. What does it mean for a sample to have a standard deviation of zero? Describe the scores in such a sample.

6. What does it mean for a sample to have a standard deviation of $s = 5$? Describe the scores in such a sample. (Describe where the scores are located relative to the sample mean.)

7. For the following population of $N = 8$ scores:

    3, 3, 5, 1, 4, 3, 2, 3

    a. Calculate the range, the interquartile range, and the standard deviation.
    b. Add 2 points to every score, then compute the range, the interquartile range, and the standard deviation again. How is variability affected by adding a constant to every score?

8. Explain what it means to say that the sample variance provides an *unbiased* estimate of the population variance.

9. For the data in the following sample:

    1, 4, 3, 6, 2, 7, 8, 3, 7, 2, 4, 3

    a. Find the median and the semi-interquartile range. (It may help to sketch the frequency distribution histogram.)
    b. Now change the score of $X = 8$ to $X = 18$, and find the new median and semi-interquartile range.
    c. Describe how one extreme score influences the median and semi-interquartile range.

10. For the data in the following sample:

    8, 1, 5, 1, 5

    a. Find the mean and the standard deviation.
    b. Now change the score of $X = 8$ to $X = 18$, and find the new mean and standard deviation.
    c. Describe how one extreme score influences the mean and standard deviation.

11. A sample of $n = 20$ scores has a mean of $M = 30$.
    a. If the sample standard deviation is $s = 10$, would a score of $X = 38$ be considered an extreme value (out in the tail of the distribution)?
    b. If the sample standard deviation is $s = 2$, would a score of $X = 38$ be considered an extreme value (out in the tail of the distribution)?

12. On an exam with $\mu = 75$, you obtain a score of $X = 80$.
    a. Would you prefer that the exam distribution had $\sigma = 2$ or $\sigma = 10$? (Sketch each distribution, and locate the position of $X = 80$ in each one.)
    b. If your score is $X = 70$, would you prefer $\sigma = 2$ or $\sigma = 10$? (Again, sketch each distribution to determine how the value of σ affects your position relative to the rest of the class.)

13. A population has a mean of $\mu = 30$ and a standard deviation of $\sigma = 5$.
    a. If 5 points were added to every score in the population, what would be the new values for the mean and standard deviation?
    b. If every score in the population were multiplied by 3, what would be the new values for the mean and standard deviation?

14. A student was asked to compute the mean and standard deviation for the following sample of $n = 5$ scores: 81, 87, 89, 86, and 87. To simplify the arithmetic, the student first subtracted 80 points from each score to obtain a new sample consisting of 1, 7, 9, 6, and 7. The mean and standard deviation for the new sample were then calculated to be $M = 6$ and $s = 3$. What are the values of the mean and standard deviation for the original sample?

15. There are two different formulas or methods that can be used to calculate SS.
   a. Under what circumstances is the definitional formula easy to use?
   b. Under what circumstances is the computational formula preferred?

16. Calculate the mean for each of the following samples and then decide (yes or no) whether it would be easy to use the definitional formula to calculate the value for SS.

   Sample A: 1,   4,   7,   5
   Sample B: 3,   0,   9,   4

17. For the following scores:

   3, 2, 5, 0, 1, 2, 8

   a. Calculate the mean. (Note that the value of the mean does not depend on whether the set of scores is considered to be a sample or a population.)
   b. Find the deviation for each score, and check that the deviations add up to zero.
   c. Square each deviation, and compute SS. (Again, note that the value of SS is independent of whether the set of scores is a sample or a population.)

18. For the following set of scores:

   3,   1,   1,   5

   a. Calculate ΣX and ΣX^2.
   b. Use the two sums from part a and the computational formula to compute SS for the scores.

19. For the following population of $N = 6$ scores:

   8,   6,   2,   6,   3,   5

   a. Sketch a histogram showing the population distribution.
   b. Locate the value of the population mean in your sketch, and make an estimate of the standard deviation (as done in Example 4.2).
   c. Compute SS, variance, and standard deviation for the population. (How well does your estimate compare with the actual value of σ?)

20. For the following population of $N = 10$ scores:

   14, 6, 8, 3, 13, 8, 9, 12, 1, 6

   a. Sketch a histogram showing the population distribution.
   b. Locate the value of the population mean in your sketch, and make an estimate of the standard deviation (as done in Example 4.2).
   c. Compute SS, variance, and standard deviation for the sample. (How well does your estimate compare with the actual value of σ?)

21. For the following population of $N = 6$ scores:

   5,   0,   9,   3,   8,   5

   a. Sketch a histogram showing the population distribution.
   b. Locate the value of the population mean in your sketch, and make an estimate of the standard deviation (as done in Example 4.2).
   c. Compute SS, variance, and standard deviation for the population. (How well does your estimate compare with the actual value of σ?)

22. For the following sample of $n = 7$ scores:

   12,   1,   10,   6,   3,   3,   7

   a. Sketch a histogram showing the sample distribution.
   b. Locate the value of the sample mean in your sketch, and make an estimate of the standard deviation (as done in Example 4.5).
   c. Compute SS, variance, and standard deviation for the sample. (How well does your estimate compare with the actual value of s?)

23. Calculate SS, variance, and standard deviation for the following sample of $n = 6$ scores: 11, 0, 8, 2, 4, 5. (*Note:* The definitional formula for SS works well with these scores.)

24. Calculate SS, variance, and standard deviation for the following sample of $n = 5$ scores: 10, 3, 8, 0, 4. (*Note:* The definitional formula works well with these scores.)

25. Calculate SS, variance, and standard deviation for the following sample of $n = 4$ scores: 4, 0, 1, 1. (*Note:* The computational formula for SS works well with these scores.)

26. Calculate SS, variance, and standard deviation for the following sample of $n = 4$ scores: 3, 1, 1, 1. (*Note:* The computational formula works best with these scores.)

27. Calculate SS, variance, and standard deviation for the following sample of $N = 8$ scores: 0, 1, 0, 3, 6, 0, 2, 0. (*Note:* The computational formula for SS works best with these scores.)

28. In an extensive study involving thousands of British children, Arden and Plomin (2006) found significantly higher variance in the intelligence scores for males than for females. Following are hypothetical data, similar to the results obtained in the study. Note that the scores are not regular IQ scores but have been standardized so that the entire sample has a mean of $M = 10$ and a standard deviation of $s = 2$.

a. Calculate the mean and the standard deviation for the sample of $n = 8$ females and for the sample of $n = 8$ males.

b. Based on the means and the standard deviations, describe the differences in intelligence scores for males and females.

Female	Male
9	8
11	10
10	11
13	12
8	6
9	10
11	14
9	9

29. In the Preview section at the beginning of this chapter, we discussed a study by Wegesin and Stern (2004) that reported greater consistency (less variability) in memory scores for younger women than for older women. The following data represent scores for two women, one older and one younger, obtained over a series of memory trials.

a. Calculate the variance of the scores for each woman.

b. Are the scores for the younger woman more consistent (less variable)?

Younger	Older
8	7
6	5
6	8
7	5
8	7
7	6
8	8
8	5

C H A P T E R

z-Scores: Location of Scores and Standardized Distributions

Tools You Will Need

The following items are considered essential background material for this chapter. If you doubt your knowledge of any of these items, you should review the appropriate chapter and section before proceeding.

- The mean (Chapter 3)
- The standard deviation (Chapter 4)
- Basic algebra (math review, Appendix A)

Preview

5.1   Introduction to z-Scores

5.2   z-Scores and Location in a Distribution

5.3   Using z-Scores to Standardize a Distribution

5.4   Other Standardized Distributions Based on z-Scores

5.5   Computing z-Scores for Samples

5.6   Looking Ahead to Inferential Statistics

Summary

Focus on Problem Solving

Demonstrations 5.1 and 5.2

Problems

Preview

A common test of cognitive ability requires participants to search through a visual display and respond to specific targets as quickly as possible. This kind of test is called a perceptual-speed test. Measures of perceptual speed are commonly used for predicting performance on jobs that demand a high level of speed and accuracy. Although many different tests are used, a typical example is shown in Figure 5.1. This task requires the participant to search through the display of digit pairs as quickly as possible and circle each pair that adds up 10. Your score is determined by the amount of time required to complete the task with a correction for the number of errors that you make. One complaint about this kind of paper-and-pencil test is that it is tedious and time consuming to score because a researcher must also search through the entire display to identify errors to determine the participant's level of accuracy. An alternative, proposed by Ackerman and Beier (2007), is a computerized version of the task. The computer version presents a series of digit pairs and participants respond on a touch-sensitive monitor. The computerized test is very reliable and the scores are equivalent to the paper-and-pencil tests in terms of assessing cognitive skill. The advantage of the computerized test is that the computer produces a test score immediately when a participant finishes the test.

Suppose that you took Ackerman and Beier's test and your combined time and errors produced a score of 92. How did you do? Are you faster than average, fairly normal in perceptual speed, or does your score indicate a serious deficit in cognitive skill?

FIGURE 5.1

An example of a perceptual speed task. The participant is asked to search through the display as quickly as possible and circle each pair of digits that sum to 10.

Circle every pair of adjacent numbers that add up to 10.

64	23	19	31	19	46	31	91	83	82	82	46	19	87
11	42	94	87	64	44	19	55	82	46	57	98	39	46
78	73	72	66	63	71	67	42	62	73	45	22	62	99
73	91	52	37	55	97	91	51	44	23	46	64	97	62
97	31	21	49	93	91	89	46	73	82	55	98	12	56
73	82	37	55	89	83	73	27	83	82	73	46	97	62
57	96	46	55	46	19	13	67	73	26	58	64	32	73
23	94	66	55	91	73	67	73	82	55	64	62	46	39
87	11	99	73	56	73	63	73	91	82	63	33	16	88
19	42	62	91	12	82	32	92	73	46	68	19	11	64
93	91	32	82	63	91	46	46	36	55	19	92	62	71

The Problem: Without any frame of reference, a simple raw score provides relatively little information. Specifically, you have no idea how your test score of 92 compares with others who took the same test. **The Solution:** Transforming your test score into a z-*score* will identify exactly where you are located in the distribution of scores. In this case, the distribution of scores has a mean of 86.75 and a standard deviation of 10.50. With this additional information, you should realize that your score is somewhat higher than average but not extreme. The z-score combines all this information (your score, the mean, and the standard deviation) into a single number that precisely describes your location relative to the other scores in the distribution.

5.1   INTRODUCTION TO z-SCORES

In the previous two chapters, we introduced the concepts of the mean and the standard deviation as methods for describing an entire distribution of scores. Now we will shift attention to the individual scores within a distribution. In this chapter, we introduce a statistical technique that uses the mean and the standard deviation to transform each score (X value) into a *z-score* or a *standard score*. The purpose of z-scores, or standard scores, is to identify and describe the exact location of every score in a distribution.

The following example demonstrates why z-scores are useful and introduces the general concept of transforming X values into z-scores.

EXAMPLE 5.1   Suppose you received a score of $X = 76$ on a statistics exam. How did you do? It should be clear that you need more information to predict your grade. Your score of $X = 76$ could be one of the best scores in the class, or it might be the lowest score in the distribution. To find the location of your score, you must have information about the other scores in the distribution. It would be useful, for example, to know the mean for the class. If the mean were $\mu = 70$, you would be in a much better position than if the mean were $\mu = 85$. Obviously, your position relative to the rest of the class depends on the mean. However, the mean by itself is not sufficient to tell you the exact location of your score. Suppose you know that the mean for the statistics exam is $\mu = 70$ and your score is $X = 76$. At this point, you know that your score is 6 points above the mean, but you still do not know exactly where it is located. Six points may be a relatively big distance and you may have one of the highest scores in the class, or 6 points may be a relatively small distance and you are only slightly above the average. Figure 5.2 shows two possible distributions of exam scores. Both distributions have a mean of $\mu = 70$, but for one distribution, the standard deviation is $\sigma = 3$, and for the other, $\sigma = 12$. The location of $X = 76$ is highlighted in each of the two distributions. When the standard deviation is $\sigma = 3$, your score of $X = 76$ is in the extreme right-hand tail, one of the highest scores in the distribution. However, in the other distribution, where $\sigma = 12$, your score is only slightly above average. Thus, the relative location of your score within the distribution depends on the standard deviation as well as the mean.

The purpose of the preceding example is to demonstrate that a score *by itself* does not necessarily provide much information about its position within a distribution. These original, unchanged scores that are the direct result of measurement are often called *raw scores*. To make raw scores more meaningful, they are often transformed into new values that contain more information. This transformation is one purpose for z-scores. In particular, we transform X values into z-scores so that the resulting z-scores tell exactly where the original scores are located.

A second purpose for z-scores is to *standardize* an entire distribution. A common example of a standardized distribution is the distribution of IQ scores. Although there are several different tests for measuring IQ, the tests usually are standardized so that they have a mean of 100 and a standard deviation of 15. Because all the different tests are standardized, it is possible to understand and compare IQ scores even though they come from different tests. For example, we all understand that an IQ score of 95 is a little below average, *no matter which IQ test was used*. Similarly, an IQ of 145 is extremely high, *no matter which IQ test was used*. In general terms, the process of standardizing takes different distributions and makes them equivalent. The advantage

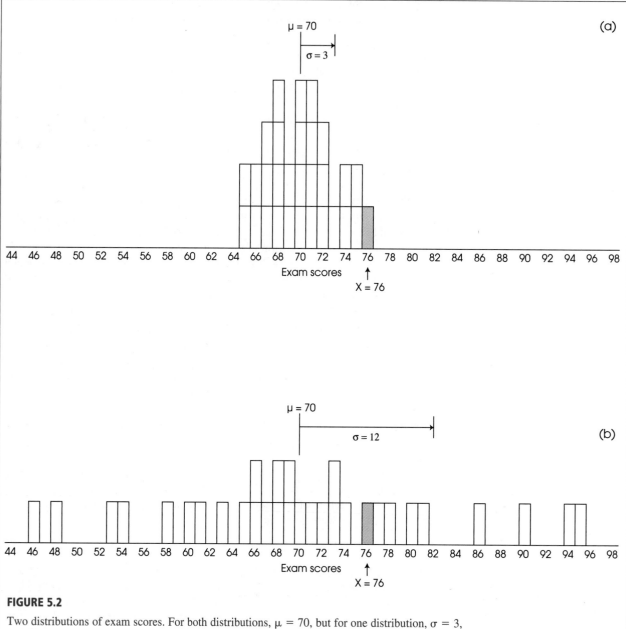

FIGURE 5.2

Two distributions of exam scores. For both distributions, $\mu = 70$, but for one distribution, $\sigma = 3$, and for the other, $\sigma = 12$. The relative position of $X = 76$ is very different for the two distributions.

of this process is that it is possible to compare distributions even though they may have been quite different before standardization.

In summary, the process of transforming X values into z-scores serves two useful purposes:

1. Each z-score tells the exact location of the original X value within the distribution.

2. The z-scores form a standardized distribution that can be directly compared to other distributions that also have been transformed into z-scores.

Each of these purposes will be discussed in the following sections.

5.2   z-SCORES AND LOCATION IN A DISTRIBUTION

One of the primary purposes of a z-score is to describe the exact location of a score within a distribution. The z-score accomplishes this goal by transforming each X value into a signed number (+ or −) so that

1. The *sign* tells whether the score is located above (+) or below (−) the mean, and
2. The *number* tells the distance between the score and the mean in terms of the number of standard deviations.

Thus, in a distribution of IQ scores with $\mu = 100$ and $\sigma = 15$, a score of $X = 130$ would be transformed into $z = +2.00$. The z value indicates that the score is located above the mean (+) by a distance of 2 standard deviations (30 points).

DEFINITION

A **z-score** specifies the precise location of each X value within a distribution. The sign of the z-score (+ or −) signifies whether the score is above the mean (positive) or below the mean (negative). The numerical value of the z-score specifies the distance from the mean by counting the number of standard deviations between X and μ.

Whenever you are working with z-scores, you should imagine or draw a picture similar to Figure 5.3. Although you should realize that not all distributions are normal, we will use the normal shape as an example when showing z-scores for populations.

Notice that a z-score always consists of two parts: a sign (+ or −) and a magnitude. Both parts are necessary to describe completely where a raw score is located within a distribution.

Figure 5.3 shows a population distribution with various positions identified by their z-score values. Notice that all z-scores above the mean are positive and all z-scores below the mean are negative. The sign of a z-score tells you immediately whether the score is located above or below the mean. Also, note that a z-score of $z = +1.00$

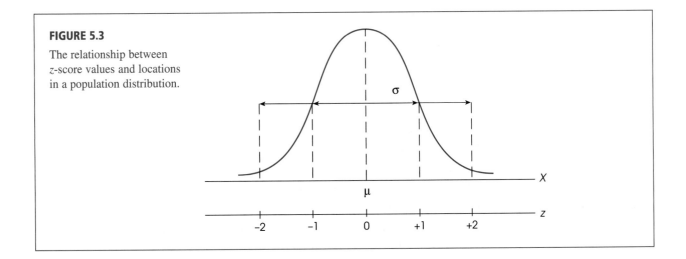

FIGURE 5.3

The relationship between z-score values and locations in a population distribution.

corresponds to a position exactly 1 standard deviation above the mean. A z-score of $z = +2.00$ is always located exactly 2 standard deviations above the mean. The numerical value of the z-score tells you the number of standard deviations from the mean. Finally, you should notice that Figure 5.3 does not give any specific values for the population mean or the standard deviation. The locations identified by z-scores are the same for *all distributions,* no matter what mean or standard deviation the distributions may have.

Now we can return to the two distributions shown in Figure 5.2 and use a z-score to describe the position of $X = 76$ within each distribution as follows:

In Figure 5.2(a), the score $X = 76$ corresponds to a z-score of $z = +2.00$. That is, the score is located above the mean by exactly 2 standard deviations.

In Figure 5.2(b), the score $X = 76$ corresponds to a z-score of $z = +0.50$. In this distribution, the score is located above the mean by exactly $\frac{1}{2}$ standard deviation.

LEARNING CHECK

1. Identify the z-score value corresponding to each of the following locations in a distribution.

   a. Below the mean by 1 standard deviation.

   b. Above the mean by $1\frac{1}{2}$ standard deviation

   c. Below the mean by $\frac{3}{4}$ standard deviation

2. For a population with $\mu = 20$ and $\sigma = 4$, find the z-score for each of the following scores:

   a. $X = 18$ **b.** $X = 28$ **c.** $X = 20$

3. For a population with $\mu = 60$ and $\sigma = 20$, find the X value corresponding to each of the following z-scores:

   a. $z = -0.25$ **b.** $z = 2.00$ **c.** $z = 0.50$

ANSWERS

1. **a.** $z = -1.00$ **b.** $z = 1.50$ **c.** $z = -0.75$

2. **a.** $z = -0.50$ **b.** $z = 2.00$ **c.** $z = 0$

3. **a.** $X = 55$ **b.** $X = 100$ **c.** $X = 70$

THE z-SCORE FORMULA

The z-score definition is adequate for transforming back and forth from X values to z-scores as long as the arithmetic is easy to do in your head. For more complicated values, it is best to have an equation to help structure the calculations. Fortunately, the relationship between X values and z-scores is easily expressed in a formula. The formula for transforming scores into z-scores is

$$z = \frac{X - \mu}{\sigma}$$

(5.1)

The numerator of the equation, $X - \mu$, is a *deviation score* (Chapter 4, page 110); it measures the distance in points between X and μ and indicates whether X is located above or below the mean. The deviation score is then divided by σ because we want the z-score to measure distance in terms of standard deviation units. The formula performs exactly the same arithmetic that is used with the z-score definition, and it provides a structured equation to organize the calculations when the numbers are more difficult. The following examples demonstrate the use of the z-score formula.

EXAMPLE 5.2   A distribution of scores has a mean of $\mu = 100$ and a standard deviation of $\sigma = 10$. What z-score corresponds to a score of $X = 120$ in this distribution?

According to the definition, the z-score will have a value of $+2$ because the score is located above the mean by exactly 2 standard deviations. Using the z-score formula, we obtain

$$z = \frac{X - \mu}{\sigma} = \frac{120 - 100}{10} = \frac{20}{10} = 2.00$$

The formula produces exactly the same result that is obtained using the z-score definition.

EXAMPLE 5.3   A distribution of scores has a mean of $\mu = 86$ and a standard deviation of $\sigma = 7$. What z-score corresponds to a score of $X = 95$ in this distribution?

Note that this problem is not particularly easy, especially if you try to use the z-score definition and perform the calculations in your head. However, the z-score formula organizes the numbers and allows you to finish the final arithmetic with your calculator. Using the formula, we obtain

$$z = \frac{X - \mu}{\sigma} = \frac{95 - 86}{7} = \frac{9}{7} = 1.29$$

According to the formula, a score of $X = 95$ corresponds to $z = 1.29$. The z-score indicates a location that is above the mean (positive) by slightly more than $1\frac{1}{4}$ standard deviations.

When you use the z-score formula, it can be useful to pay attention to the definition of a z-score as well. For example, we used the formula in Example 5.3 to calculate the z-score corresponding to $X = 95$, and obtained $z = 1.29$. Using the z-score definition, we note that $X = 95$ is located above the mean by 9 points, which is slightly more than one standard deviation ($\sigma = 7$). Therefore, the z-score should be positive and have a value slightly greater than 1.00. In this case, the answer predicted by the definition is in perfect agreement with the calculation. However, if the calculations produce a different value, for example $z = 0.78$, you should realize that this answer is not consistent with the definition of a z-score. In this case, an error has been made and you should double check the calculations.

DETERMINING A RAW SCORE (X) FROM A z-SCORE   Although the z-score equation (Formula 5.1) works well for transforming X values into z-scores, it can be awkward when you are trying to work in the opposite direction and change z-scores back into X values. In general it is easier to use the definition of a z-score, rather than a formula, when you are changing z-scores into X values. Remember, the z-score describes exactly where the score is located by identifying the direction and distance from the mean. It is possible, however, to express this definition as a formula, and we will use a sample problem to demonstrate how the formula can be created.

For a distribution with a mean of $\mu = 60$ and a standard deviation of $\sigma = 5$, what X value corresponds to a z-score of $z = -2.00$?

To solve this problem, we will use the z-score definition and carefully monitor the step-by-step process. The value of the z-score indicates that X is located 2 standard deviations below the mean. Thus, the first step in the calculation is to determine the distance corresponding to 2 standard deviations. For this problem, standard deviation is $\sigma = 5$ points, so 2 standard deviations is $2(5) = 10$ points. The next step is to start at the mean and go down by 10 points to find the value of X. In symbols,

$$X = \mu - 10 = 60 - 10 = 50$$

The two steps can be combined to form a single formula:

$$X = \mu + z\sigma \tag{5.2}$$

In the formula, the value of $z\sigma$ is the *deviation* of X and determines both the direction and the size of the distance from the mean. In this problem, $z\sigma = (-2)(5) = -10$, or 10 points below the mean. Formula 5.2 simply combines the mean and the deviation from the mean to determine the exact value of X.

Finally, you should realize that Formula 5.1 and Formula 5.2 are actually two different versions of the same equation. If you begin with either formula and use algebra to shuffle the terms around, you will soon end up with the other formula. We will leave this as an exercise for those who want to try it.

OTHER RELATIONSHIPS BETWEEN z, X, μ, AND σ

In most cases, we simply transform scores (X values) into z-scores, or change z-scores back into X values. However, you should realize that a z-score establishes a relationship between the score, the mean, and the standard deviation. This relationship can be used to answer a variety of different questions about scores and the distributions in which they are located. The following two examples demonstrate some possibilities.

EXAMPLE 5.4

In a population with a mean of $\mu = 65$, a score of $X = 59$ corresponds to $z = -2.00$. What is the standard deviation for the population?

To answer the question, we begin with the z-score value. A z-score of -2.00 indicates that the corresponding score is located below the mean by a distance of 2 standard deviations. By simple subtraction, you can also determine that the score ($X = 59$) is located below the mean ($\mu = 65$) by a distance of 6 points. Thus, 2 standard deviations correspond to a distance of 6 points, which means that 1 standard deviation must be $\sigma = 3$.

EXAMPLE 5.5

In a population with a standard deviation of $\sigma = 4$, a score of $X = 33$ corresponds to $z = +1.50$. What is the mean for the population?

Again, we begin with the z-score value. In this case, a z-score of $+1.50$ indicates that the score is located above the mean by a distance corresponding to 1.50 standard deviations. With a standard deviation of $\sigma = 4$, this distance is $(1.50)(4) = 6$ points. Thus, the score is located 6 points above the mean. The score is $X = 33$, so the mean must be $\mu = 27$.

identify locations. Recall from the definition of z-scores that all positive values are above the mean and all negative values are below the mean. A value of zero makes the mean a convenient reference point.

3. The Standard Deviation. The distribution of z-scores will *always* have a standard deviation of 1. In Figure 5.5, the original distribution of X values has $\mu = 100$ and $\sigma = 10$. In this distribution, a value of $X = 110$ is above the mean by exactly 10 points or 1 standard deviation. When $X = 110$ is transformed, it becomes $z = +1.00$, which is above the mean by exactly 1 point in the z-score distribution. Thus, the standard deviation corresponds to a 10-point distance in the X distribution and is transformed into a 1-point distance in the z-score distribution. The advantage of having a standard deviation of 1 is that the numerical value of a z-score is exactly the same as the number of standard deviations from the mean. For example, a z-score value of 2 is exactly 2 standard deviations from the mean.

In Figure 5.5, we showed the z-score transformation as a process that changed a distribution of X values into a new distribution of z-scores. In fact, there is no need to create a whole new distribution. Instead, you can think of the z-score transformation as simply *relabeling* the values along the X-axis. That is, after a z-score transformation, you still have the same distribution, but now each individual is labeled with a z-score instead of an X value. Figure 5.6 demonstrates this concept with a single distribution that has two sets of labels: the X values along one line and the corresponding z-scores along another line. Notice that the mean for the distribution of z-scores is zero and the standard deviation is 1.

When *any* distribution (with any mean or standard deviation) is transformed into z-scores, the resulting distribution will always have a mean of $\mu = 0$ and a standard deviation of $\sigma = 1$. Because all z-score distributions have the same mean and the same standard deviation, the z-score distribution is called a *standardized distribution*.

FIGURE 5.6

Following a z-score transformation, the X-axis is relabeled in z-score units. The distance that is equivalent to 1 standard deviation on the X-axis ($\sigma = 10$ points in this example) corresponds to 1 point on the z-score scale.

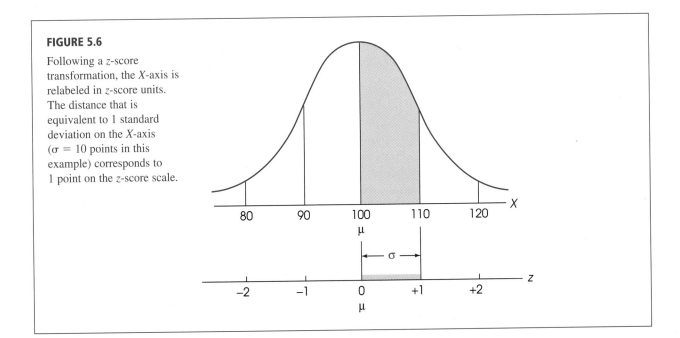

DEFINITION

A **standardized distribution** is composed of scores that have been transformed to create predetermined values for μ and σ. Standardized distributions are used to make dissimilar distributions comparable.

A z-score distribution is an example of a standardized distribution with $\mu = 0$ and $\sigma = 1$. That is, when any distribution (with any mean or standard deviation) is transformed into z-scores, the transformed distribution will always have $\mu = 0$ and $\sigma = 1$.

DEMONSTRATION OF A
z-SCORE TRANSFORMATION

Although the basic characteristics of a z-score distribution have been explained logically, the following example provides a concrete demonstration that a z-score transformation will create a new distribution with a mean of zero, a standard deviation of 1, and the same shape as the original population.

EXAMPLE 5.6

We begin with a population of $N = 6$ scores consisting of the following values: 0, 6, 5, 2, 3, 2. This population has a mean of $\mu = \frac{18}{6} = 3$ and a standard deviation of $\sigma = 2$ (check the calculations for yourself).

Each of the X values in the original population is then transformed into a z-score as summarized in the following table.

$X = 0$	Below the mean by $1\frac{1}{2}$ standard deviations	$z = -1.50$
$X = 6$	Above the mean by $1\frac{1}{2}$ standard deviations	$z = +1.50$
$X = 5$	Above the mean by 1 standard deviation	$z = +1.00$
$X = 2$	Below the mean by $\frac{1}{2}$ standard deviation	$z = -0.50$
$X = 3$	Exactly equal to the mean—no deviation	$z = 0$
$X = 2$	Below the mean by $\frac{1}{2}$ standard deviation	$z = -0.50$

The frequency distribution for the original population of X values is shown in Figure 5.7(a) and the corresponding distribution for the z-scores is shown in Figure 5.7(b). A simple comparison of the two distributions demonstrates the results of a z-score transformation.

1. The two distributions have exactly the same shape. Each individual has exactly the same relative position in the X distribution and in the z-score distribution.

2. After the transformation to z-scores, the mean of the distribution becomes $\mu = 0$. The individual with a score of $X = 3$ is located exactly at the mean in the X distribution and this individual is transformed into $z = 0$, exactly at the mean in the z-score distribution.

3. After the transformation, the standard deviation becomes $\sigma = 1$. For example, in the original distribution of scores, the individual with $X = 5$ is located above the mean by 2 points, which is a distance of exactly 1 standard deviation. After transformation, this same individual is located above the mean by 1 point, which is exactly 1 standard deviation in the z-score distribution.

USING z-SCORES FOR
MAKING COMPARISONS

One advantage of standardizing distributions is that it makes it possible to compare different scores or different individuals even though they come from completely different distributions. Normally, if two scores come from different distributions, it is

FIGURE 5.7

Transforming a distribution of raw scores (a) into *z*-scores (b) will not change the shape of the distribution.

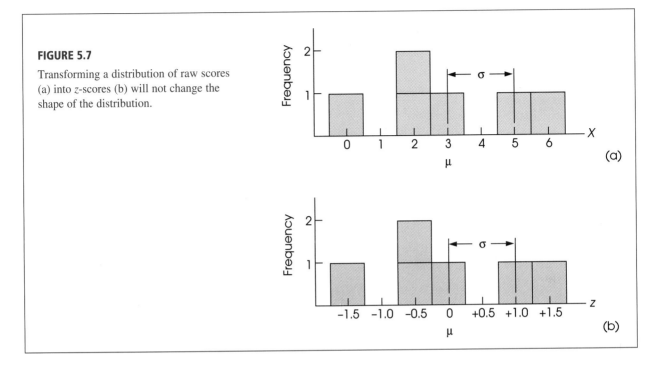

impossible to make any direct comparison between them. Suppose, for example, Bob received a score of X = 60 on a psychology exam and a score of X = 56 on a biology test. For which course should Bob expect the better grade?

Because the scores come from two different distributions, you cannot make any direct comparison. Without additional information, it is even impossible to determine whether Bob is above or below the mean in either distribution. Before you can begin to make comparisons, you must know the values for the mean and standard deviation for each distribution. Suppose the biology scores had $\mu = 48$ and $\sigma = 4$, and the psychology scores had $\mu = 50$ and $\sigma = 10$. With this new information, you could sketch the two distributions, locate Bob's score in each distribution, and compare the two locations.

Instead of drawing the two distributions to determine where Bob's two scores are located, we simply can compute the two *z*-scores to find the two locations. For psychology, Bob's *z*-score is

Be sure to use the μ and σ values for the distribution to which *X* belongs.

$$z = \frac{X - \mu}{\sigma} = \frac{60 - 50}{10} = \frac{10}{10} = +1.0$$

For biology, Bob's *z*-score is

$$z = \frac{56 - 48}{4} = \frac{8}{4} = +2.0$$

Note that Bob's *z*-score for biology is +2.0, which means that his test score is 2 standard deviations above the class mean. On the other hand, his *z*-score is +1.0 for psychology, or 1 standard deviation above the mean. In terms of relative class standing, Bob is doing much better in the biology class.

Notice that we cannot compare Bob's two exam scores ($X = 60$ and $X = 56$) because the scores come from different distributions with different means and standard deviations. However, we can compare the two z-scores because all distributions of z-scores have the same mean ($\mu = 0$) and the same standard deviation ($\sigma = 1$).

LEARNING CHECK

1. A normal-shaped distribution with $\mu = 40$ and $\sigma = 8$ is transformed into z-scores. The resulting distribution of z-scores will have a mean of _____ and a standard deviation of _____.

2. What is the advantage of having a mean of $\mu = 0$ for a distribution of z-scores?

3. A distribution of English exam scores has $\mu = 70$ and $\sigma = 4$. A distribution of history exams has $\mu = 65$ and $\sigma = 15$. For which exam would a score of $X = 78$ have a higher standing? Explain your answer.

ANSWERS

1. A z-score distribution always has a mean of 0 and a standard deviation of 1.

2. With a mean of zero, all positive scores are above the mean and all negative scores are below the mean.

3. For the English exam, $X = 78$ corresponds to $z = 2.00$, which is a higher standing than $z = \frac{13}{15} = 0.87$ for the history exam.

5.4   OTHER STANDARDIZED DISTRIBUTIONS BASED ON z-SCORES

TRANSFORMING z-SCORES TO A DISTRIBUTION WITH A PREDETERMINED μ AND σ

Although z-score distributions have distinct advantages, many people find them cumbersome because they contain negative values and decimals. For this reason, it is common to standardize a distribution by transforming the scores into a new distribution with a predetermined mean and standard deviation that are whole round numbers. The goal is to create a new (standardized) distribution that has "simple" values for the mean and standard deviation but does not change any individual's location within the distribution. Standardized scores of this type are frequently used in psychological or educational testing. For example, raw scores of the Scholastic Aptitude Test (SAT) are transformed to a standardized distribution that has $\mu = 500$ and $\sigma = 100$. For intelligence tests, raw scores are frequently converted to standard scores that have a mean of 100 and a standard deviation of 15. Because most IQ tests are standardized so that they have the same mean and standard deviation, it is possible to compare IQ scores even though they may come from different tests.

The procedure for standardizing a distribution to create new values for μ and σ involves two-steps:

1. The original raw scores are transformed into z-scores.

2. The z-scores are then transformed into new X values so that the specific μ and σ are attained.

This procedure ensures that each individual has exactly the same z-score location in the new distribution as in the original distribution. The following example demonstrates the standardization procedure.

EXAMPLE 5.7   An instructor gives an exam to a psychology class. For this exam, the distribution of raw scores has a mean of $\mu = 57$ with $\sigma = 14$. The instructor would like to simplify the distribution by transforming all scores into a new, standardized distribution with $\mu = 50$ and $\sigma = 10$. To demonstrate this process, we will consider what happens to two specific students: Maria, who has a raw score of $X = 64$ in the original distribution; and Joe, whose original raw score is $X = 43$.

STEP 1   Transform each of the original raw scores into z-scores. For Maria, $X = 64$, so her z-score is

$$z = \frac{X - \mu}{\sigma} = \frac{64 - 57}{14} = +0.5$$

Remember: The values of μ and σ are for the distribution from which X was taken.

For Joe, $X = 43$, and his z-score is

$$z = \frac{X - \mu}{\sigma} = \frac{43 - 57}{14} = -1.0$$

STEP 2   Change each z-score into an X value in the new standardized distribution that has a mean of $\mu = 50$ and a standard deviation of $\sigma = 10$.

Maria's z-score, $z = +0.50$, indicates that she is located above the mean by $\frac{1}{2}$ standard deviation. In the new, standardized distribution, this location corresponds to $X = 55$ (above the mean by 5 points).

Joe's z-score, $z = -1.00$, indicates that he is located below the mean by exactly 1 standard deviation. In the new distribution, this location corresponds to $X = 40$ (below the mean by 10 points).

The results of this two-step transformation process are summarized in Table 5.1. Note that Joe, for example, has exactly the same z-score ($z = -1.00$) in both the original distribution and the new standardized distribution. This means that Joe's position relative to the other students in the class has not changed.

Figure 5.8 provides another demonstration of the concept that standardizing a distribution does not change the individual positions within the distribution. The figure shows the original exam scores from Example 5.7, with a mean of $\mu = 57$ and a standard deviation of $\sigma = 14$. In the original distribution, Joe is located at a score of $X = 43$. In addition to the original scores, we have included a second scale showing the z-score value for each location in the distribution. In terms of z-scores, Joe is located at a value of $z = -1.00$. Finally, we have added a third scale showing the *standardized scores* where the mean is $\mu = 50$ and the standard deviation is $\sigma = 10$. For the standardized scores, Joe is located at $X = 40$. Note that Joe is always in the same place in the distribution. The only thing that changes is the number that is assigned to Joe: For the original scores, Joe is at 43; for the z-scores, Joe is at -1.00; and for the standardized scores, Joe is at 40.

TABLE 5.1

A demonstration of how two individual scores are changed when a distribution is standardized. See Example 5.7.

	Original Scores $\mu = 57$ and $\sigma = 14$		z-Score Location		Standardized Scores $\mu = 50$ and $\sigma = 10$
Maria	$X = 64$	$\longrightarrow$	$z = +0.50$	$\longrightarrow$	$X = 55$
Joe	$X = 43$	$\longrightarrow$	$z = -1.00$	$\longrightarrow$	$X = 40$

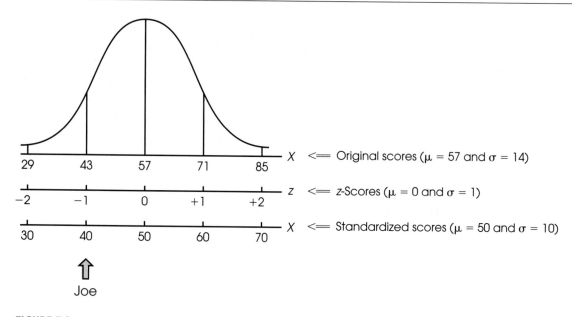

FIGURE 5.8

The distribution of exam scores from Example 5.7. The original distribution was standardized to produce a new distribution with $\mu = 50$ and $\sigma = 10$. Note that each individual is identified by an original score, a z-score, and a new, standardized score. For example, Joe has an original score of 43, a z-score of -1.00, and a standardized scorre of 40.

LEARNING CHECK

1. A population of scores has $\mu = 73$ and $\sigma = 8$. If the distribution is standardized to create a new distribution with $\mu = 100$ and $\sigma = 20$, what are the new values for each of the following scores from the original distribution?

   a. $X = 65$ **b.** $X = 71$ **c.** $X = 81$ **d.** $X = 83$

2. A population of $N = 6$ scores has $\mu = 8$ and $\sigma = 5$. The six scores are 14, 8, 0, 11, 3, and 12.

   a. Using z-scores, transform the population into a new distribution with $\mu = 100$ and $\sigma = 20$. (Find the new score corresponding to each of the original scores.)

   b. Compute the mean and standard deviation for the new scores. (You should obtain $\mu = 100$ and $\sigma = 20$.)

ANSWERS

1. **a.** $z = -1.00, X = 80$ **b.** $z = -0.25, X = 95$
   c. $z = 1.00, X = 120$ **d.** $z = 1.25, X = 125$

2. **a.** The original scores correspond to z-scores of 1.20, 0, -1.60, 0.60, -1.00, and 0.80. These values are transformed into new scores of 124, 100, 68, 112, 80, and 116.

   b. The new scores add to $\Sigma X = 600$ so the new mean is $\mu = 100$. The SS for the transformed scores is $SS = 2400$, the variance is 400, and the new standard deviation is $\sigma = 20$.

5.5 COMPUTING z-SCORES FOR SAMPLES

Although z-scores are most commonly used in the context of a population, the same principles can be used to identify individual locations within a sample. The definition of a z-score is the same for a sample as for a population, provided that you use the sample mean and the sample standard deviation to specify each z-score location. Thus, for a sample, each X value is transformed into a z-score so that

1. The sign of the z-score indicates whether the X value is above (+) or below (−) the sample mean, and

2. The numerical value of the z-score identifies the distance between the score and the sample mean in terms of the sample standard deviation.

Expressed as a formula, each X value in a sample can be transformed into a z-score as follows:

$$z = \frac{X - M}{s} \qquad (5.3)$$

Similarly, each z-score can be transformed back into an X value, as follows:

$$X = M + zs \qquad (5.4)$$

EXAMPLE 5.8   In a sample with a mean of $M = 40$ and a standard deviation of $s = 10$, what is the z-score corresponding to $X = 35$ and what is the X value corresponding to $z = +2.00$?

The score, $X = 35$, is located below the mean by 5 points, which is exactly half of a standard deviation. Therefore, the corresponding z-score is $z = -0.50$. The z-score, $z = +2.00$, corresponds to a location above the mean by 2 standard deviations. With a standard deviation of $s = 10$, this is distance of 20 points. The score that is located 20 points above the mean is $X = 60$. Note that it is possible to find these answers using either the z-score definition or one of the equations (5.3 or 5.4).

STANDARDIZING A SAMPLE DISTRIBUTION

If all the scores in a sample are transformed into z-scores, the result is a sample of z-scores. The transformation will have the same properties that exist when a population of X value is transformed into z-scores. Specifically,

1. The sample of z-scores will have the same shape as the original sample of scores.
2. The sample of z-scores will have a mean of $M = 0$.
3. The sample of z-scores will have a standard deviation of $s = 1$.

Note that the set of z-scores is still considered to be a sample (just like the set of X values) and the sample formulas must be used to compute variance and standard deviation. The following example demonstrate the process of transforming scores in a sample into z-scores.

EXAMPLE 5.9   We begin with a sample of $n = 5$ scores: 0, 2, 4, 4, 5. With a few simple calculations, you should be able to verify that the sample mean is $M = 3$, the sample variance is $s^2 = 4$, and the sample standard deviation is $s = 2$. Using the sample mean and sample standard deviation, we can convert each X value into a z-score. For example, $X = 5$ is located above the mean by 2 points. Thus, $X = 5$ is above the mean by exactly 1 standard deviation and has a z-score of $z = +1.00$. The z-scores for the entire sample are shown in the following table.

X	z
0	-1.50
2	-0.50
4	$+0.50$
4	$+0.50$
5	$+1.00$

Again, a few simple calculations demonstrate that the sum of the z-score values is $\Sigma z = 0$, so the mean is $M_z = 0$.

Because the mean is zero, each z-score value is its own deviation from the mean. Therefore, the sum of the squared deviations is simply the sum of the squared z-scores. For this sample of z-scores,

$$SS = \Sigma z^2 = (-1.50)^2 + (-0.50)^2 + (+0.50)^2 + (+0.50)^2 + (+1.00)^2$$
$$= 2.25 + 0.25 + 0.25 + 0.25 + 1.00$$
$$= 4.00$$

The variance for the sample of z-scores is

Notice that the set of z-scores is considered to be a sample and the variance is computed using the sample formula with $df = n - 1$.

$$s^2 = \frac{SS}{n-1} = \frac{4}{4} = 1.00$$

Finally, the sample standard deviation is $s = \sqrt{1.00} = 1.00$. As always, the distribution of z-scores has a mean of 0 and a standard deviation of 1.

5.6   LOOKING AHEAD TO INFERENTIAL STATISTICS

Recall that inferential statistics are techniques that use the information from samples to answer questions about populations. In later chapters, we will use inferential statistics to help interpret the results from a research study. A typical research study begins with a question about how a treatment will affect the individuals in a population. Because it is usually impossible to study an entire population, the researcher selects a sample and administers the treatment to the individuals in the sample. To evaluate the effect of the treatment, the researcher simply compares the treated sample with the original population (Figure 5.9). If the individuals in the sample are noticeably different from the individuals in the original population, the researcher has evidence that the treatment has had an effect. On the other hand, if the sample is not noticeably different from the original population, it would appear that the treatment has no effect.

Notice that the interpretation of the research results depends on whether the sample is *noticeably different* from the population. One technique for deciding whether a sample is noticeably different is to use z-scores. For example, an individual with a z-score near 0 is located in the center of the population and would be considered to be

FIGURE 5.9

A diagram of a research study. The goal of the study is to evaluate the effect of a treatment. A sample is selected from the population and the treatment is administered to the sample. If, after treatment, the individuals in the sample are noticeably different from the individuals in the original population, then we have evidence that the treatment does have an effect.

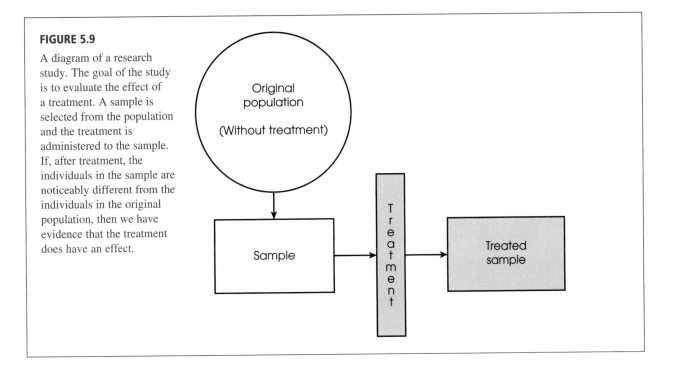

a fairly typical or representative individual. However, an individual with an extreme z-score, beyond +2.00 or −2.00 for example, would be considered "noticeably different" from most of the individuals in the population. Thus, we can use z-scores to help decide whether the treatment has caused a change. Specifically, if the individuals who receive the treatment in a research study tend to have extreme z-scores, we can conclude that the treatment does appear to have an effect. The following example demonstrates this process.

EXAMPLE 5.10   A researcher is evaluating the effect of a new growth hormone. It is known that regular adult rats weigh an average of $\mu = 400$ grams. The weights vary from rat to rat, and the distribution of weights is normal with a standard deviation of $\sigma = 20$ grams. The population distribution is shown in Figure 5.10. The researcher selects one newborn rat and injects the rat with the growth hormone. When the rat reaches maturity, it is weighed to determine whether there is any evidence that the hormone has an effect.

First, assume that the hormone-injected rat weighs $X = 418$ grams. Although this is more than the average nontreated rat ($\mu = 400$ grams), is it convincing evidence that the hormone has an effect? If you look at the distribution in Figure 5.10, you should realize that a rat weighing 418 grams is not noticeably different from the regular rats that did not receive any hormone injection. Specifically, our injected rat would be located near the center of the distribution for regular rats with a z-score of

$$z = \frac{X - \mu}{\sigma} = \frac{418 - 400}{20} = \frac{18}{20} = 0.90$$

FIGURE 5.10

The distribution of weights for the population of adult rats. Note that individuals with z-scores near 0 are typical or representative. However, individuals with z-scores beyond +2.00 or −2.00 are extreme and noticeably different from most of the others in the distribution.

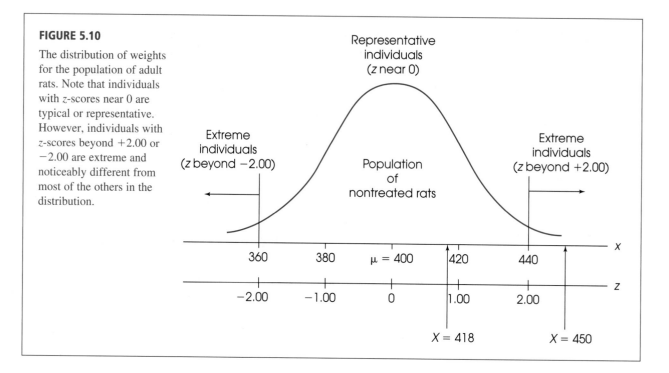

Because the injected rat still looks the same as a regular, nontreated rat, the conclusion is that the hormone does not appear to have an effect.

Now, assume that our injected rat weighs $X = 450$ grams as an adult. In the distribution of regular rats (see Figure 5.10), this animal would have a z-score of

$$z = \frac{X - \mu}{\sigma} = \frac{450 - 400}{20} = \frac{50}{20} = 2.50$$

In this case, the hormone-injected rat is substantially bigger than most ordinary rats, and it would be reasonable to conclude that the hormone does have an effect on weight.

In the preceding example, we used z-scores to help interpret the results obtained from a sample. Specifically, if the individuals who receive the treatment in a research study have extreme z-scores compared to those who do not receive the treatment, we can conclude that the treatment does appear to have an effect. The example, however, used rather vague definitions to determine which z-score values are noticeable different and which are not. Although it is reasonable to describe individuals with z-scores near 0 as "highly representative" of the population, and individuals with z-scores beyond ±2.00 as "extreme," you should realize that these z-score boundaries are arbitrary numbers. In the following chapter we will introduce *probability*, which will give us a rationale for deciding exactly where to set the boundaries.

LEARNING CHECK

1. For a sample with a mean of $M = 20$ and a standard deviation of $s = 4$, find the z-score corresponding to each of the following X values.

$$X = 12 \qquad X = 14 \qquad X = 19$$
$$X = 20 \qquad X = 22 \qquad X = 30$$

2. For a sample with a mean of $M = 80$ and a standard deviation of $s = 10$, find the X value corresponding to each of the following z-scores.

$$z = -1.00 \qquad z = -0.50 \qquad z = -0.20$$
$$z = 1.50 \qquad z = 0.80 \qquad z = 1.40$$

3. For a sample with a mean of $M = 36$, a score of $X = 40$ corresponds to $z = 0.50$. What is the standard deviation for the sample?

4. For a sample with a standard deviation of $s = 12$, a score of $X = 83$ corresponds to $z = -0.25$. What is the mean for the sample?

5. A sample has a mean of $M = 30$ and a standard deviation of $s = 8$.

   a. Would a score of $X = 36$ be considered a central score or an extreme score in the sample?

   b. If the standard deviation were $s = 2$, would $X = 36$ be central or extreme?

ANSWERS

1. $z = -2.00 \qquad z = -1.50 \qquad z = -0.25$
   $z = 0 \qquad z = 0.50 \qquad z = 2.50$

2. $X = 70 \qquad X = 75 \qquad X = 78$
   $X = 95 \qquad X = 88 \qquad X = 94$

3. $s = 8$

4. $M = 86$

5. a. $X = 36$ is a central score corresponding to $z = 0.75$.

   b. $X = 36$ is an extreme score corresponding to $z = 3.00$.

SUMMARY

1. Each X value can be transformed into a z-score that specifies the exact location of X within the distribution. The sign of the z-score indicates whether the location is above (positive) or below (negative) the mean. The numerical value of the z-score specifies the number of standard deviations between X and μ.

2. The z-score formula is used to transform X values into z-scores. For a population:

$$z = \frac{X - \mu}{\sigma}$$

For a sample:

$$z = \frac{X - M}{s}$$

3. To transform z-scores back into X values, it usually is easier to use the z-score definition rather than a formula. However, the z-score formula can be transformed into a new equation. For a population:

$$X = \mu + z\sigma$$

For a sample: $X = M + zs$

4. When an entire distribution of *X* values is transformed into *z*-scores, the result is a distribution of *z*-scores. The *z*-score distribution will have the same shape as the distribution of raw scores, and it always will have a mean of 0 and a standard deviation of 1.

5. When comparing raw scores from different distributions, it is necessary to standardize the distributions with a *z*-score transformation. The distributions will then be comparable because they will have the same parameters ($\mu = 0$, $\sigma = 1$). In practice, it is necessary to transform only those raw scores that are being compared.

6. In certain situations, such as psychological testing, a distribution may be standardized by converting the original *X* values into *z*-scores and then converting the *z*-scores into a new distribution of scores with predetermined values for the mean and the standard deviation.

7. In inferential statistics, *z*-scores provide an objective method for determining how well a specific score represents its population. A *z*-score near 0 indicates that the score is close to the population mean and therefore is representative. A *z*-score beyond $+2.00$ (or -2.00) indicates that the score is extreme and is noticeably different from the other scores in the distribution.

KEY TERMS

raw score

z-score

deviation score

z-score transformation

standardized distribution

standardized score

RESOURCES

There are a tutorial quiz and other learning exercises for Chapter 5 at the book companion website, www.cengage.com/psychology/gravetter. The site also includes a workshop titled *z*-scores that examines the basic concepts and calculations underlying *z*-scores. See page 29 for more information about accessing items on the website.

ENHANCED
WebAssign

Guided interactive tutorials, end-of-chapter problems, and related testbank items may be assigned online at WebAssign.

WebTUTOR

For those using WebTutor along with this book, there is a WebTutor section corresponding to this chapter. The WebTutor contains a brief summary of Chapter 5, hints for learning about *z*-scores, cautions about common errors, and sample exam items including solutions.

SPSS

General instructions for using SPSS are presented in Appendix D. Following are detailed instructions for using SPSS to **Transform *X* Values into *z*-Scores for a Sample.**

Data Entry

1. Enter all of the scores in one column of the data editor, probably VAR00001.

Data Analysis

1. Click **Analyze** on the tool bar, select **Descriptive Statistics,** and click on **Descriptives.**
2. Highlight the column label for the set of scores (VAR0001) in the left box and click the arrow to move it into the **Variable** box.
3. Click the box to **Save standardized values as variables** at the bottom of the **Descriptives** screen.
4. Click **OK.**

SPSS Output

The program will produce the usual output display listing the number of scores (*N*), the maximum and minimum scores, the mean, and the standard deviation. However, if you go back to the Data Editor (use the tool bar at the bottom of the screen), SPSS will have produced a new column showing the *z*-score corresponding to each of the original *X* values.

Caution: The SPSS program computes the *z*-scores using the sample standard deviation instead of the population standard deviation. If your set of scores is intended to be a population, SPSS will not produce the correct *z*-score values. You can convert the SPSS values into population *z*-scores by multiplying each *z*-score value by the square root of $n/(n-1)$.

FOCUS ON PROBLEM SOLVING

1. When you are converting an *X* value to a *z*-score (or vice versa), do not rely entirely on the formula. You can avoid careless mistakes if you use the definition of a *z*-score (sign and numerical value) to make a preliminary estimate of the answer before you begin computations. For example, a *z*-score of $z = -0.85$ identifies a score located *below* the mean by almost 1 standard deviation. When computing the *X* value for this *z*-score, be sure that your answer is smaller than the mean, and check that the distance between *X* and μ is slightly less than the standard deviation.

2. When comparing scores from distributions that have different standard deviations, it is important to be sure that you use the correct value for σ in the *z*-score formula. Use the σ value for the distribution from which the raw score in question was taken.

3. Remember that a *z*-score specifies a relative position within the context of a specific distribution. A *z*-score is a relative value, not an absolute value. For example, a *z*-score of $z = -2.0$ does not necessarily suggest a very low raw score—it simply means that the raw score is among the lowest within that specific group.

DEMONSTRATION 5.1

TRANSFORMING *X* VALUES INTO *z*-SCORES

A distribution of scores has a mean of $\mu = 60$ with $\sigma = 12$. Find the *z*-score for $X = 75$.

STEP 1   Determine the sign of the *z*-score.

First, determine whether *X* is above or below the mean. This will determine the sign of the *z*-score. For this demonstration, *X* is larger than (above) μ, so the *z*-score will be positive.

STEP 2   Convert the distance between X and μ into standard deviation units.

For $X = 75$ and $\mu = 60$, the distance between X and μ is 15 points. With $\sigma = 12$ points, this distance corresponds to $\frac{15}{12} = 1.25$ standard deviations.

STEP 3   Combine the sign from step 1 with the numerical value from step 2.

The score is above the mean (+) by a distance of 1.25 standard deviations. Thus,

$$z = +1.25$$

STEP 4   Confirm the answer using the z-score formula.

For this example, $X = 75$, $\mu = 60$, and $\sigma = 12$.

$$z = \frac{X - \mu}{\sigma} = \frac{75 - 60}{12} = \frac{+15}{12} = +1.25$$

DEMONSTRATION 5.2

CONVERTING z-SCORES TO X VALUES

For a population with $\mu = 60$ and $\sigma = 12$, what is the X value corresponding to $z = -0.50$?

Notice that in this situation we know the z-score and must find X.

STEP 1   Locate X in relation to the mean.

A z-score of -0.50 indicates a location below the mean by half of a standard deviation.

STEP 2   Convert the distance from standard deviation units to points.

With $\sigma = 12$, half of a standard deviation is 6 points.

STEP 3   Identify the X value.

The value we want is located below the mean by 6 points. The mean is $\mu = 60$, so the score must be $X = 54$

PROBLEMS

1. What information is provided by the sign (+/−) of a z-score? What information is provided by the numerical value of the z-score?

2. A distribution has a standard deviation of $\sigma = 8$. Find the z-score for each of the following locations in the distribution.
   a. Above the mean by 4 points.
   b. Above the mean by 16 points.
   c. Below the mean by 8 points.
   d. Below the mean by 12 points.

3. A distribution has a standard deviation of $\sigma = 4$. Describe the location of each of the following z-scores in terms of its position relative to the mean. For example, $z = +1.00$ is a location that is 4 points above the mean.
   a. $z = +2.00$
   b. $z = +0.50$
   c. $z = -2.00$
   d. $z = -0.50$

4. For a population with $\mu = 50$ and $\sigma = 10$,
   a. Find the z-score for each of the following X values. (*Note:* You should be able to find these values using the definition of a z-score. You should not need to use a formula or do any serious calculations.)

$$X = 55 \qquad X = 60 \qquad X = 75$$
$$X = 45 \qquad X = 30 \qquad X = 35$$

   b. Find the score (X value) that corresponds to each of the following z-scores. (Again, you should be able to find these values without any formula or serious calculations.)

$$z = 1.00 \qquad z = 0.80 \qquad z = 1.50$$
$$z = -0.50 \qquad z = -0.30 \qquad z = -1.50$$

5. For a population with $\mu = 60$ and $\sigma = 9$, find the z-score for each of the following X values. (*Note:* You probably will need to use a formula and a calculator to find these values.)

$$X = 65 \qquad X = 70 \qquad X = 53$$
$$X = 48 \qquad X = 75 \qquad X = 58$$

6. For a population with a mean of $\mu = 100$ and a standard deviation of $\sigma = 10$,
   a. Find the z-score for each of the following X values.

$$X = 105 \qquad X = 120 \qquad X = 130$$
$$X = 90 \qquad X = 85 \qquad X = 60$$

   b. Find the score (X value) that corresponds to each of the following z-scores.

$$z = -1.00 \qquad z = -0.50 \qquad z = 2.00$$
$$z = 0.70 \qquad z = 1.50 \qquad z = -1.50$$

7. A population has a mean of $\mu = 40$ and a standard deviation of $\sigma = 12$.
   a. For this population, find the z-score for each of the following X values.

$$X = 43 \qquad X = 49 \qquad X = 52$$
$$X = 34 \qquad X = 28 \qquad X = 64$$

   b. For the same population, find the score (X value) that corresponds to each of the following z-scores.

$$z = 0.75 \qquad z = 1.50 \qquad z = -2.00$$
$$z = -0.25 \qquad z = -0.50 \qquad z = 1.25$$

8. A sample has a mean of $M = 40$ and a standard deviation of $s = 4$. Find the z-score for each of the following X values from this sample.

$$X = 44 \qquad X = 42 \qquad X = 48$$
$$X = 28 \qquad X = 50 \qquad X = 46$$

9. A sample has a mean of $M = 80$ and a standard deviation of $s = 5$. For this sample, find the X value corresponding to each of the following z-scores.

$$z = 0.80 \qquad z = 1.20 \qquad z = 2.00$$
$$z = -0.40 \qquad z = -0.60 \qquad z = -1.80$$

10. Find the z-score corresponding to a score of $X = 60$ for each of the following distributions.
   a. $\mu = 50$ and $\sigma = 10$
   b. $\mu = 50$ and $\sigma = 5$
   c. $\mu = 70$ and $\sigma = 20$
   d. $\mu = 70$ and $\sigma = 5$

11. Find the X value corresponding to $z = 1.50$ for each of the following distributions.
   a. $\mu = 40$ and $\sigma = 6$
   b. $\mu = 40$ and $\sigma = 20$
   c. $\mu = 80$ and $\sigma = 4$
   d. $\mu = 80$ and $\sigma = 8$

12. A score that is 6 points below the mean corresponds to a z-score of $z = -2.00$. What is the population standard deviation?

13. A score that is 4 points above the mean corresponds to a z-score of $z = 0.50$. What is the population standard deviation?

14. For a population with a standard deviation of $\sigma = 10$, a score of $X = 44$ corresponds to $z = -0.50$. What is the population mean?

15. For a sample with a standard deviation of $s = 6$, a score of $X = 65$ corresponds to $z = 1.50$. What is the sample mean?

16. For a sample with a mean of $M = 85$, a score of $X = 90$ corresponds to $z = 0.50$. What is the sample standard deviation?

17. For a population with a mean of $\mu = 70$, a score of $X = 64$ corresponds to $z = -2.00$. What is the population standard deviation?

18. In a population of exam scores, a score of $X = 88$ corresponds to $z = +2.00$ and a score of $X = 79$ corresponds to $z = -1.00$. Find the mean and standard deviation for the population. (*Hint:* Sketch the distribution and locate the two scores on your sketch.)

19. In a distribution of scores, $X = 62$ corresponds to $z = +0.50$, and $X = 52$ corresponds to $z = -2.00$. Find the mean and standard deviation for the distribution.

20. For each of the following populations, would a score of $X = 50$ be considered a central score (near the middle of the distribution) or an extreme score (far out in the tail of the distribution)?
  a. $\mu = 45$ and $\sigma = 10$
  b. $\mu = 40$ and $\sigma = 2$
  c. $\mu = 55$ and $\sigma = 2$
  d. $\mu = 60$ and $\sigma = 20$

21. Suppose that you have a score of $X = 75$ on an exam with $\mu = 70$. Which standard deviation would give you a better grade: $\sigma = 2$ or $\sigma = 10$?

22. Suppose that you have a score of $X = 75$ on an exam with $\mu = 80$. Which standard deviation would give you a better grade: $\sigma = 2$ or $\sigma = 10$?

23. Which of the following exam scores should lead to the better grade?
  a. A score of $X = 43$ on an exam with $\mu = 40$ and $\sigma = 2$.
  b. A score of $X = 60$ on an exam with $\mu = 50$ and $\sigma = 20$.
  Explain your answer.

24. Which of the following exam scores should lead to the better grade?
  a. A score of $X = 55$ on an exam with $\mu = 60$ and $\sigma = 5$.
  b. A score of $X = 40$ on an exam with $\mu = 50$ and $\sigma = 20$.
  Explain your answer.

25. A distribution with a mean of $\mu = 64$ and a standard deviation of $\sigma = 4$ is being transformed into a standardized distribution with $\mu = 100$ and $\sigma = 20$. Find the new, standardized score for each of the following values from the original population.
  a. $X = 60$ **b.** $X = 52$
  c. $X = 72$ **d.** $X = 66$

26. A distribution with a mean of $\mu = 38$ and a standard deviation of $\sigma = 20$ is being transformed into a standardized distribution with $\mu = 50$ and $\sigma = 10$. Find the new, standardized score for each of the following values from the original population.
  a. $X = 48$ **b.** $X = 40$
  c. $X = 30$ **d.** $X = 18$

27. A population consists of the following $N = 6$ scores: 6, 1, 3, 4, 7, and 3.
  a. Compute μ and σ for the population.
  b. Find the z-score for each score in the population.
  c. Transform the original population into a new population of $N = 6$ scores with a mean of $\mu = 50$ and a standard deviation of $\sigma = 10$.

28. A population consists of the following $N = 5$ scores: 0, 6, 4, 3, and 12.
  a. Compute μ and σ for the population.
  b. Find the z-score for each score in the population.
  c. Transform the original population into a new population of $N = 5$ scores with a mean of $\mu = 60$ and a standard deviation of $\sigma = 8$.

29. The following sample of $n = 6$ scores has a mean of $M = 6$ and a standard deviation of $s = 4$. Scores: 10, 0, 4, 8, 4, and 10.
  a. Find the z-score for each X value.
  b. For the sample of $n = 6$ z-scores, verify that the mean is zero and the standard deviation is 1.

C H A P T E R

6

Probability

Tools You Will Need

The following items are considered essential background material for this chapter. If you doubt your knowledge of any of these items, you should review the appropriate chapter or section before proceeding.

- Proportions (math review, Appendix A)
  - Fractions
  - Decimals
  - Percentages
- Basic algebra (math review, Appendix A)
- Upper and lower real limits (Chapters 1 and 2)
- Percentiles and percentile ranks (Chapter 2)
- z-Scores (Chapter 5)

Preview

6.1   Introduction to Probability

6.2   Probability and the Normal Distribution

6.3   Probabilities and Proportions for Scores from a Normal Distribution

6.4   Probability and the Binomial Distribution

6.5   Looking Ahead to Inferential Statistics

Summary

Focus on Problem Solving

Demonstrations 6.1 and 6.2

Problems

Preview

Background: If you open a dictionary and randomly pick one word, which are you more likely to select:

1. A word beginning with the letter *K*?
2. A word with a *K* as its third letter?

If you think about this question and answer honestly, you probably will decide that words beginning with a *K* are more probable.

A similar question was asked a group of participants in an experiment reported by Tversky and Kahneman (1973). Their participants estimated that words beginning with *K* are twice as likely as words with a *K* as the third letter. In truth, the relationship is just the opposite. There are more than twice as many words with a *K* in the third position as there are words beginning with a *K*. How can people be so wrong? Do they completely misunderstand probability?

When you were deciding which type of *K* words are more likely, you probably searched your memory and tried to estimate which words are more common. How many words can you think of that start with the letter *K*? How many words can you think of that have a *K* as the third letter? Because you have had years of practice alphabetizing

words according to their first letter, you should find it much easier to search your memory for words beginning with a *K* than to search for words with a *K* in the third position. Consequently, you would conclude that first-letter *K* words are more common.

If you had searched for words in a dictionary (instead of those in your memory), you would have found more third-letter *K* words, and you would have concluded (correctly) that these words are more common.

The Problem: If you open a dictionary and randomly pick one word, it is impossible to predict exactly which word you will get. In the same way, when researchers recruit people to participate in research studies, it is impossible to predict exactly which individuals will be obtained.

The Solution: Although it is impossible to predict exactly which word will be picked from a dictionary, or which person will participate in a research study, you can use *probability* to demonstrate that some outcomes are more likely than others. For example, it is more likely that you will pick a third-letter *K* word than a first-letter *K* word. Similarly, it is more likely that you will obtain a person with an IQ around 100 than a person with an IQ around 150.

INTRODUCTION TO PROBABILITY

In Chapter 1, we introduced the idea that research studies begin with a general question about an entire population, but the actual research is conducted using a sample. In this situation, the role of inferential statistics is to use the sample data as the basis for answering questions about the population. To accomplish this goal, inferential procedures are typically built around the concept of probability. Specifically, the relationships between samples and populations are usually defined in terms of probability. Suppose, for example, you are selecting a single marble from a jar that contains 50 black and 50 white marbles. (In this example, the jar of marbles is the *population* and the single marble to be selected is the *sample*.) Although you cannot guarantee the exact outcome of your sample, it is possible to talk about the potential outcomes in terms of probabilities. In this case, you have a 50-50 chance of getting either color. Now consider another jar (population) that has 90 black and only 10 white marbles. Again, you cannot specify the exact outcome of a sample, but now you know that the sample probably will be a black marble. By knowing the makeup of a population, we can determine the probability of obtaining specific samples. In this way, probability gives us a connection between populations and samples, and this connection will be the foundation for the inferential statistics to be presented in the chapters that follow.

You may have noticed that the preceding examples begin with a population and then use probability to describe the samples that could be obtained. This is exactly backward from what we want to do with inferential statistics. Remember that the goal of inferential statistics is to begin with a sample and then answer general questions about the population. We will reach this goal in a two-stage process. In the first stage, we develop probability as a bridge from populations to samples. This stage involves identifying the types of samples that probably would be obtained from a specific population. Once this bridge is established, we simply reverse the probability rules to allow us to move from samples to populations (Figure 6.1). The process of reversing the probability relationship can be demonstrated by considering again the two jars of marbles we looked at earlier. (Jar 1 has 50 black and 50 white marbles; jar 2 has 90 black and only 10 white marbles.) This time, suppose you are blindfolded when the sample is selected, so you do not know which jar is being used. Your task is to look at the sample and then choose which jar is most likely. If you select a sample of $n = 4$ marbles and all are black, which jar would you choose? It should be clear that it would be relatively unlikely (low probability) to obtain this sample from jar 1; in four draws, you almost certainly would get at least 1 white marble. On the other hand, this sample would have a high probability of coming from jar 2, where nearly all the marbles are black. Your decision therefore is that the sample probably came from jar 2. Note that you now are using the sample to make an inference about the population.

PROBABILITY DEFINITION   Probability is a huge topic that extends far beyond the limits of introductory statistics, and we will not attempt to examine it all here. Instead, we will concentrate on the few concepts and definitions that are needed for an introduction to inferential statistics. We begin with a relatively simple definition of probability.

D E F I N I T I O N   For a situation in which several different outcomes are possible, the **probability** for any specific outcome is defined as a fraction or a proportion of all the possible outcomes. If the possible outcomes are identified as A, B, C, D, and so on, then

$$\text{probability of } A = \frac{\text{number of outcomes classified as } A}{\text{total number of possible outcomes}}$$

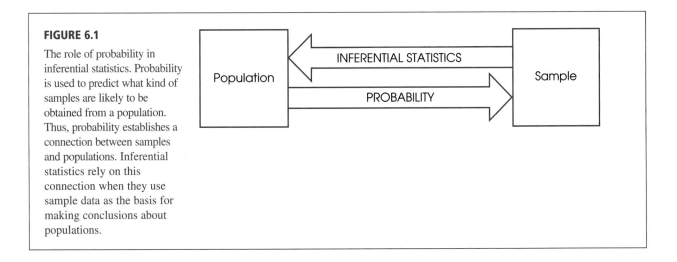

FIGURE 6.1

The role of probability in inferential statistics. Probability is used to predict what kind of samples are likely to be obtained from a population. Thus, probability establishes a connection between samples and populations. Inferential statistics rely on this connection when they use sample data as the basis for making conclusions about populations.

For example, if you are selecting a card from a complete deck, there are 52 possible outcomes. The probability of selecting the king of hearts is $p = \frac{1}{52}$. The probability of selecting an ace is $p = \frac{4}{52}$ because there are 4 aces in the deck.

To simplify the discussion of probability, we use a notation system that eliminates a lot of the words. The probability of a specific outcome is expressed with a p (for probability) followed by the specific outcome in parentheses. For example, the probability of selecting a king from a deck of cards is written as $p(\text{king})$. The probability of obtaining heads for a coin toss is written as $p(\text{heads})$.

Note that probability is defined as a proportion, or a part of the whole. This definition makes it possible to restate any probability problem as a proportion problem. For example, the probability problem "What is the probability of selecting a king from a deck of cards?" can be restated as "What proportion of the whole deck consists of kings?" In each case, the answer is $\frac{4}{52}$, or "4 out of 52." This translation from probability to proportion may seem trivial now, but it will be a great aid when the probability problems become more complex. In most situations, we are concerned with the probability of obtaining a particular sample from a population. The terminology of *sample* and *population* will not change the basic definition of probability. For example, the whole deck of cards can be considered as a population, and the single card we select is the sample.

The definition we are using identifies probability as a fraction or a proportion. If you work directly from this definition, the probability values you obtain will be expressed as fractions. For example, if you are selecting a card at random,

$$p(\text{spade}) = \frac{13}{52} = \frac{1}{4}$$

Of if you are tossing a coin,

$$p(\text{heads}) = \frac{1}{2}$$

You should be aware that these fractions can be expressed equally well as either decimals or percentages:

If you are unsure how to convert from fractions to decimals or percentages, you should review the section on proportions in the math review, Appendix A.

$$p = \frac{1}{4} = 0.25 = 25\%$$
$$p = \frac{1}{2} = 0.50 = 50\%$$

By convention, probability values most often are expressed as decimal values. But you should realize that any of these three forms is acceptable.

You also should note that all the possible probability values are contained in a limited range. At one extreme, when an event never occurs, the probability is zero, or 0%. At the other extreme, when an event always occurs, the probability is 1, or 100%. Thus, all probability values are contained in a range from 0 to 1 (Box 6.1). For example, suppose that you have a jar containing 10 white marbles. The probability of randomly selecting a black marble is

$$p(\text{black}) = \frac{0}{10} = 0$$

The probability of selecting a white marble is

$$p(\text{white}) = \frac{10}{10} = 1$$

BOX 6.1	ZERO PROBABILITY

An event that never occurs has a probability of zero. However, the opposite of this statement is not always true: A probability of zero does not mean that the event is guaranteed never to occur. Whenever there are an extremely large number of possible events, the probability of any specific event is assigned the value zero. This is done because the probability value tends toward zero as the number of possible events get large. Consider, for example, the series

$$\frac{1}{10} \quad \frac{1}{100} \quad \frac{1}{1000} \quad \frac{1}{10,000} \quad \frac{1}{100,000}$$

Note that the value of the fraction is getting smaller and smaller, headed toward zero. At the far extreme, when the number of possible events is so large that it cannot be specified, the probability of a single, specific event is said to be zero.

$$\frac{1}{\text{infinite number}} = 0$$

Consider, for example, the fish in the ocean. If there were only 10 fish, then the probability of selecting any particular one would be $p = \frac{1}{10}$. Note that if you add up the probabilities for all 10 fish, you get a total of 1.00. In reality, the number of fish in the ocean is essentially infinite, and, for all practical purposes, the probability of catching any specific fish is $p = 0$. However, this does not mean that you are doomed to fail whenever you go fishing. The zero probability simply means that you cannot predict in advance which fish you will catch. Note that each individual fish has a probability of being caught that is near or, practically speaking, equal to zero. However, there are so many fish that when you add up all the "zeros," you still get a total of 1.00. For large populations, a probability value of zero does not mean never. But, practically speaking, it does mean very, very close to never.

RANDOM SAMPLING   For the preceding definition of probability to be accurate, it is necessary that the outcomes be obtained by a process called *random sampling*.

DEFINITION

A **random sample** requires that each individual in the population has an *equal chance* of being selected. A second requirement, necessary for many statistical formulas, states that the probabilities must *stay constant* from one selection to the next if more than one individual is selected.

Each of the two requirements for random sampling has some interesting consequences. The first assures that there is no bias in the selection process. For a population with N individuals, each individual must have the same probability, $p = \frac{1}{N}$, of being selected. This means, for example, that you would not get a random sample of people in your city by selecting names from a yacht club membership list. Similarly, you would not get a random sample of college students by selecting individuals from your psychology classes. You also should note that the first requirement of random sampling prohibits you from applying the definition of probability to situations where the possible outcomes are not equally likely. Consider, for example, the question of whether you will win a million dollars in the lottery tomorrow. There are only two possible alternatives.

1. You will win.

2. You will not win.

According to our simple definition, the probability of winning would be one out of two, or $p = \frac{1}{2}$. However, the two alternatives are not equally likely, so the simple definition of probability does not apply.

The second requirement also is more interesting than may be apparent at first glance. Consider, for example, the selection of $n = 2$ cards from a complete deck. For the first draw, the probability of obtaining the jack of diamonds is

$$p(\text{jack of diamonds}) = \frac{1}{52}$$

After selecting one card for the sample, you are ready to draw the second card. What is the probability of obtaining the jack of diamonds this time? Assuming that you still are holding the first card, there are two possibilities:

$$p(\text{jack of diamonds}) = \frac{1}{51} \text{ if the first card was not the jack of diamonds}$$

or

$$p(\text{jack of diamonds}) = 0 \text{ if the first card was the jack of diamonds}$$

In either case, the probability is different from its value for the first draw. This contradicts the requirement for random sampling, which says that the probability must stay constant. To keep the probabilities from changing from one selection to the next, it is necessary to return each individual to the population before you make the next selection. This process is called *sampling with replacement*. The second requirement for random samples (constant probability) demands that you sample with replacement.

(*Note:* The definition that we are using identifies one type of random sampling, often called a *simple random sample* or an *independent random sample*. This kind of sampling is important for the mathematical foundation of many of the statistics we will encounter later. However, you should realize that other definitions exist for the concept of random sampling. In particular, it is very common to define random sampling without the requirement of constant probabilities—that is, without replacement. In addition, there are many different sampling techniques that are used when researchers are selecting individuals to participate in research studies.)

PROBABILITY AND FREQUENCY DISTRIBUTIONS

The situations in which we are concerned with probability usually will involve a population of scores that can be displayed in a frequency distribution graph. If you think of the graph as representing the entire population, then different portions of the graph represent different portions of the population. Because probabilities and proportions are equivalent, a particular portion of the graph corresponds to a particular probability in the population. Thus, whenever a population is presented in a frequency distribution graph, it will be possible to represent probabilities as proportions of the graph. The relationship between graphs and probabilities is demonstrated in the following example.

EXAMPLE 6.1

We will use a very simple population that contains only $N = 10$ scores with values 1, 1, 2, 3, 3, 4, 4, 4, 5, 6. This population is shown in the frequency distribution graph in Figure 6.2. If you are taking a random sample of $n = 1$ score from this population,

FIGURE 6.2

A frequency distribution histogram for a population that consists of $N = 10$ scores. The shaded part of the figure indicates the portion of the whole population that corresponds to scores greater than $X = 4$. The shaded portion is two-tenths $\left(p = \frac{2}{10}\right)$ of the whole distribution.

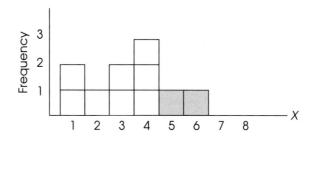

what is the probability of obtaining an individual with a score greater than 4? In probability notation,

$$p(X > 4) = ?$$

Using the definition of probability, there are 2 scores that meet this criterion out of the total group of $N = 10$ scores, so the answer would be $p = \frac{2}{10}$. This answer can be obtained directly from the frequency distribution graph if you recall that probability and proportion measure the same thing. Looking at the graph (see Figure 6.2), what proportion of the population consists of scores greater than 4? The answer is the shaded part of the distribution—that is, 2 squares out of the total of 10 squares in the distribution. Notice that we now are defining probability as a proportion of *area* in the frequency distribution graph. This provides a very concrete and graphic way of representing probability.

Using the same population once again, what is the probability of selecting an individual with a score less than 5? In symbols,

$$p(X < 5) = ?$$

Going directly to the distribution in Figure 6.2, we now want to know what part of the graph is not shaded. The unshaded portion consists of 8 out of the 10 blocks $\left(\frac{8}{10}\right.$ of the area of the graph$\left.\right)$, so the answer is $p = \frac{8}{10}$.

LEARNING CHECK

1. The animal colony in the psychology department contains 20 male rats and 30 female rats. Of the 20 males, 15 are white and 5 are spotted. Of the 30 females, 15 are white, and 15 are spotted. Suppose that you randomly select 1 rat from this colony.

   a. What is the probability of obtaining a female?

   b. What is the probability of obtaining a white male?

   c. Which selection is more likely, a spotted male or a spotted female?

2. A jar contains 10 red marbles and 30 blue marbles.

   a. If you randomly select 1 marble from the jar, what is the probability of obtaining a red marble?

    b. If you take a *random sample* of $n = 3$ marbles from the jar and the first two marbles are both blue, what is the probability that the third marble will be red?

3. Suppose that you are going to select a random sample of $n = 1$ score from the distribution in Figure 6.2. Find the following probabilities:

    a. $p(X > 2)$ **b.** $p(X > 5)$ **c.** $p(X < 3)$

ANSWERS   **1. a.** $p = \frac{30}{50} = 0.60$ **b.** $p = \frac{15}{50} = 0.30$

        c. A spotted female ($p = 0.30$) is more likely than a spotted male ($p = 0.10$).

  2. a. $p = \frac{10}{40} = 0.25$

       b. $p = \frac{10}{40} = 0.25$ Remember that random sampling requires sampling with replacement.

  3. a. $p = \frac{7}{10} = 0.70$ **b.** $p = \frac{1}{10} = 0.10$ **c.** $p = \frac{3}{10} = 0.30$

6.2   PROBABILITY AND THE NORMAL DISTRIBUTION

The normal distribution was first introduced in Chapter 2 as an example of a commonly occurring shape for population distributions. An example of a normal distribution is shown in Figure 6.3.

    Note that the normal distribution is symmetrical, with the highest frequency in the middle and frequencies tapering off as you move toward either extreme. Although the exact shape for the normal distribution is defined by an equation (see Figure 6.3),

FIGURE 6.3

The normal distribution. The exact shape of the normal distribution is specified by an equation relating each X value (score) with each Y value (frequency). The equation is

$$Y = \frac{1}{\sqrt{2\pi\sigma^2}}\, e^{-(X-\mu)^2/2\sigma^2}$$

(π and e are mathematical constants.) In simpler terms, the normal distribution is symmetrical with a single mode in the middle. The frequency tapers off as you move farther from the middle in either direction.

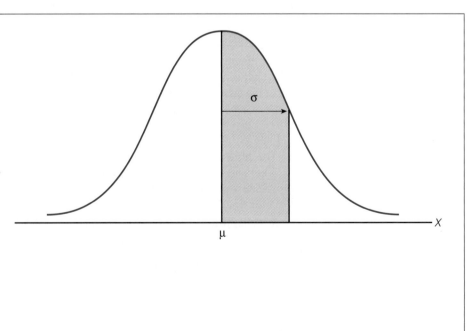

the normal shape can also be described by the proportions of area contained in each section of the distribution. Statisticians often identify sections of a normal distribution by using *z*-scores. Figure 6.4 shows a normal distribution with several sections marked in *z*-score units. You should recall that *z*-scores measure positions in a distribution in terms of standard deviations from the mean. (Thus, $z = +1$ is 1 standard deviation above the mean, $z = +2$ is 2 standard deviations above the mean, and so on.) The graph shows the percentage of scores that fall in each of these sections. For example, the section between the mean ($z = 0$) and the point that is 1 standard deviation above the mean ($z = 1$) contains 34.13% of the scores. Similarly, 13.59% of the scores are located in the section between 1 and 2 standard deviations above the mean. In this way it is possible to define a normal distribution in terms of its proportions; that is, a distribution is normal if and only if it has all the right proportions.

There are two additional points to be made about the distribution shown in Figure 6.4. First, you should realize that the sections on the left side of the distribution have exactly the same areas as the corresponding sections on the right side because the normal distribution is symmetrical. Second, because the locations in the distribution are identified by *z*-scores, the percentages shown in the figure apply to *any normal distribution* regardless of the values for the mean and the standard deviation. Remember: When any distribution is transformed into *z*-scores, the mean becomes zero and the standard deviation becomes one.

Because the normal distribution is a good model for many naturally occurring distributions and because this shape is guaranteed in some circumstances (as you will see in Chapter 7), we will devote considerable attention to this particular distribution.

The process of answering probability questions about a normal distribution is introduced in the following example.

EXAMPLE 6.2   Assume that the population of adult heights forms a normal-shaped distribution with a mean of $\mu = 68$ inches and a standard deviation of $\sigma = 6$ inches. Given this information about the population and the known proportions for a normal distribution (see Figure 6.4), we can determine the probabilities associated with specific samples.

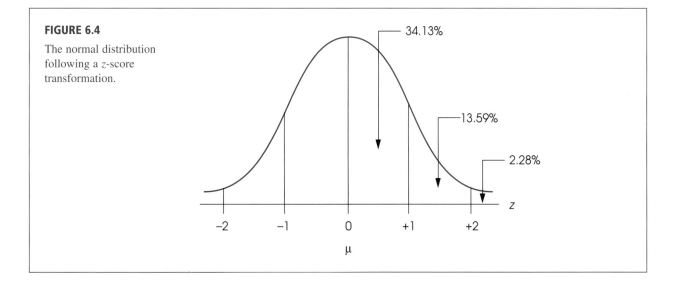

FIGURE 6.4

The normal distribution following a *z*-score transformation.

34.13%

13.59%

2.28%

z

−2   −1   0   +1   +2

μ

For example, what is the probability of randomly selecting an individual from this population who is taller than 6 feet 8 inches ($X = 80$ inches)?

Restating this question in probability notation, we get

$$p(X > 80) = ?$$

We will follow a step-by-step process to find the answer to this question.

1. First, the probability question is translated into a proportion question: Out of all possible adult heights, what proportion is greater than 80 inches?

2. The set of "all possible adult heights" is simply the population distribution. This population is shown in Figure 6.5(a). The mean is $\mu = 68$, so the score $X = 80$ is to the right of the mean. Because we are interested in all heights greater than 80, we shade in the area to the right of 80. This area represents the proportion we are trying to determine.

3. Identify the exact position of $X = 80$ by computing a z-score. For this example,

$$z = \frac{X - \mu}{\sigma} = \frac{80 - 68}{6} = \frac{12}{6} = 2.00$$

That is, a height of $X = 80$ inches is exactly 2 standard deviations above the mean and corresponds to a z-score of $z = +2.00$ [see Figure 6.5(b)].

4. The proportion we are trying to determine may now be expressed in terms of its z-score:

$$p(z > 2.00) = ?$$

According to the proportions shown in Figure 6.4, all normal distributions, regardless of the values for μ and σ, will have 2.28% of the scores in the tail beyond $z = +2.00$. Thus, for the population of adult heights,

$$p(X > 80) = p(z > +2.00) = 2.28\%$$

THE UNIT NORMAL TABLE   Before we attempt any more probability questions, we must introduce a more useful tool than the graph of the normal distribution shown in Figure 6.4. The graph shows proportions for only a few selected z-score values. A more complete listing of z-scores

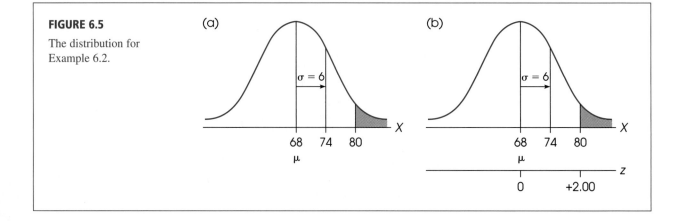

FIGURE 6.5

The distribution for Example 6.2.

and proportions is provided in the *unit normal table*. This table lists proportions of the normal distribution for a full range of possible *z*-score values.

The complete unit normal table is provided in Appendix B Table B.1, and part of the table is reproduced in Figure 6.6. Notice that the table is structured in a four-column format. The first column (A) lists *z*-score values corresponding to different positions in a normal distribution. If you imagine a vertical line drawn through a normal distribution, then the exact location of the line can be described by one of the *z*-score values listed in column A. You should also realize that a vertical line will separate the distribution into two sections: a larger section called the *body* and a smaller section called the *tail*. Columns B and C in the table identify the proportion of the distribution in each of the two sections. Column B presents the proportion in the body (the larger portion), and column C presents the proportion in the tail. Finally, we have added a fourth column, column D, that identifies the proportion of the distribution that is located *between* the mean and the *z*-score. Using the portion of the table shown in Figure 6.6, you should verify that a vertical line drawn through a normal distribution at $z = +0.25$ separates the distribution into two sections with the larger section containing 0.5987 (59.87%) of the distribution and the smaller section containing 0.4013 (40.13%) of the distribution. Also, there is exactly 0.0987 (9.87%) of the distribution between the mean and $z = +0.25$.

To make full use of the unit normal table, there are a few facts to keep in mind:

1. The *body* always corresponds to the larger part of the distribution whether it is on the right-hand side or the left-hand side. Similarly, the *tail* is always the smaller section whether it is on the right or the left.

FIGURE 6.6

A portion of the unit normal table. This table lists proportions of the normal distribution corresponding to each *z*-score value. Column A of the table lists *z*-scores. Column B lists the proportion in the body of the normal distribution up to the *z*-score value. Column C lists the proportion of the normal distribution that is located in the tail of the distribution beyond the *z*-score value. Column D lists the proportion between the mean and the *z*-score value.

(A) z	(B) Proportion in body	(C) Proportion in tail	(D) Proportion between mean and z
0.00	.5000	.5000	.0000
0.01	.5040	.4960	.0040
0.02	.5080	.4920	.0080
0.03	.5120	.4880	.0120
0.21	.5832	.4168	.0832
0.22	.5871	.4129	.0871
0.23	.5910	.4090	.0910
0.24	.5948	.4052	.0948
0.25	.5987	.4013	.0987
0.26	.6026	.3974	.1026
0.27	.6064	.3936	.1064
0.28	.6103	.3897	.1103
0.29	.6141	.3859	.1141
0.30	.6179	.3821	.1179
0.31	.6217	.3783	.1217
0.32	.6255	.3745	.1255
0.33	.6293	.3707	.1293
0.34	.6331	.3669	.1331

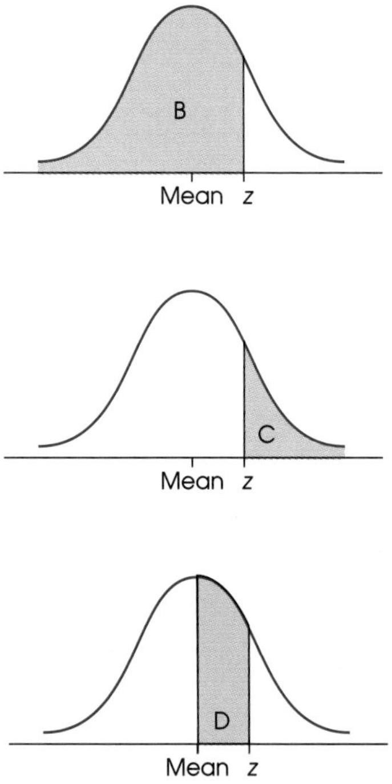

2. Because the normal distribution is symmetrical, the proportions on the right-hand side are exactly the same as the corresponding proportions on the left-hand side. For example, the proportion in the right-hand tail beyond $z = +1.00$ is exactly the same as the proportion in the left-hand tail beyond $z = -1.00$. Note that the table does not list negative z-score values. To find proportions for negative z-scores, you must look up the corresponding proportions for the positive value of z.

3. Although the z-score values change signs ($+$ and $-$) from one side to the other, the proportions are always positive. Thus, column C in the table always lists the proportion in the tail whether it is the right-hand tail or the left-hand tail.

PROBABILITIES, PROPORTIONS, AND z-SCORES

The unit normal table lists relationships between z-score locations and proportions in a normal distribution. For any z-score location, you can use the table to look up the corresponding proportions. Similarly, if you know the proportions, you can use the table to look up the specific z-score location. Because we have defined probability as equivalent to proportion, you can also use the unit normal table to look up probabilities for normal distributions. The following examples demonstrate a variety of different ways that the unit normal table can be used.

Finding proportions/probabilities for specific z-score values For each of the following examples, we begin with a specific z-score value and then use the unit normal table to find probabilities or proportions associated with the z-score.

EXAMPLE 6.3A

What proportion of the normal distribution corresponds to z-score values greater than $z = 1.00$? First, you should sketch the distribution and shade in the area you are trying to determine. This is shown in Figure 6.7(a). In this case, the shaded portion is the tail of the distribution beyond $z = 1.00$. To find this shaded area, you simply look up $z = 1.00$ in column A of the unit normal table. Then you read column C (tail) for the proportion. Using the table in Appendix B, you should find that the answer is 0.1587.

You also should notice that this same problem could have been phrased as a probability question. Specifically, we could have asked, "For a normal distribution, what is the probability of selecting a z-score value greater than $z = +1.00$?" Again, the answer is $p(z > 1.00) = 0.1587$ (or 15.87%).

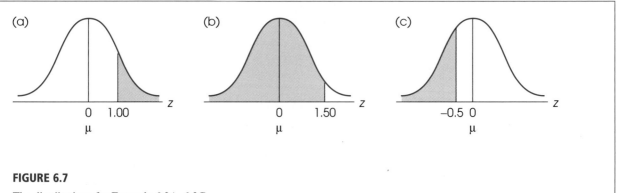

FIGURE 6.7

The distributions for Example 6.3A–6.3C.

EXAMPLE 6.3B

For a normal distribution, what is the probability of selecting a z-score less than $z = 1.50$? In symbols, $p(z < 1.50) = ?$ Our goal is to determine what proportion of the normal distribution corresponds to z-scores less than 1.50. A normal distribution is shown in Figure 6.7(b) and $z = 1.50$ is located in the distribution. Note that we have shaded all the values to the left of (less than) $z = 1.50$. This is the portion we are trying to find. Clearly the shaded portion is more than 50% so it corresponds to the body of the distribution. Therefore, we find $z = 1.50$ in the unit normal table and read the proportion from column B. The answer is $p(z < 1.50) = 0.9332$ (or 93.32%).

EXAMPLE 6.3C

Moving to the left on the X-axis results in smaller X values and smaller z-scores. Thus, a z-score of -3.00 reflects a smaller value than a z-score of -1.

Many problems will require that you find proportions for negative z-scores. For example, what proportion of the normal distribution is contained in the tail beyond $z = -0.50$? That is, $p(z < -0.50)$. This portion has been shaded in Figure 6.7(c). To answer questions with negative z-scores, simply remember that the normal distribution is symmetrical with a z-score of zero at the mean, positive values to the right, and negative values to the left. The proportion in the left tail beyond $z = -0.50$ is identical to the proportion in the right tail beyond $z = +0.50$. To find this proportion, look up $z = 0.50$ in column A, and find the proportion in column C (tail). You should get an answer of 0.3085 (30.85%).

Finding the z-score location that corresponds to specific proportions   The preceding examples all involved using a z-score value in column A to look up a proportion in column B or C. You should realize, however, that the table also allows you to begin with a known proportion and then look up the corresponding z-score. In general, the unit normal table can be used for two purposes:

1. If you know a specific location (z-score) in a normal distribution, you can use the table to look up the corresponding proportions.

2. If you know a specific proportion (or proportions), you can use the table to look up the exact z-score location in the distribution.

The following examples demonstrate how the table can be used to find specific z-scores if you begin with known proportions.

EXAMPLE 6.4A

For a normal distribution, what z-score separates the top 10% from the remainder of the distribution? To answer this question, we have sketched a normal distribution [Figure 6.8(a)] and drawn a vertical line that separates the highest 10% (approximately)

FIGURE 6.8

The distributions for Examples 6.4A and 6.4B.

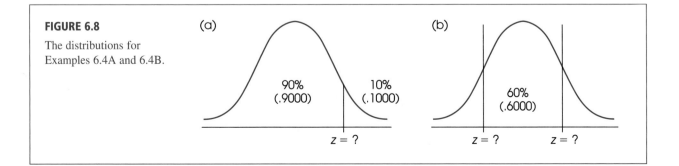

from the rest. The problem is to locate the exact position of this line. For this distribution, we know that the tail contains 0.1000 (10%) and the body contains 0.9000 (90%). To find the z-score value, you simply locate 0.1000 in column C or 0.9000 in column B of the unit normal table. Note that you probably will not find the exact proportion, but you can use the closest value listed in the table. For example, you will not find 0.1000 listed in column C but you can use 0.1003, which is listed. Once you have found the correct proportion in the table, simply read across the row to find the corresponding z-score value in column A.

For this example, the z-score that separates the extreme 10% in the tail is $z = 1.28$. At this point you must be careful because the table does not differentiate between the right-hand tail and the left-hand tail of the distribution. Specifically, the final answer could be either $z = +1.28$, which separates 10% in the right-hand tail, or $z = -1.28$, which separates 10% in the left-hand tail. For this problem we want the right-hand tail (the highest 10%), so the z-score value is $z = +1.28$.

EXAMPLE 6.4B   For a normal distribution, what z-score values form the boundaries that separate the middle 60% of the distribution from the rest of the scores?

Again, we have sketched a normal distribution [Figure 6.8(b)] and drawn vertical lines in the approximate locations. For this example, we want slightly more than half of the distribution in the central section, with the remainder split equally between the two tails. The problem is to find the z-score values that define the exact locations for the lines. To find the z-score values, we begin with the known proportions: 0.6000 in the center and 0.4000 divided equally between the two tails. Although these proportions can be used in several different ways, this example provides an opportunity to demonstrate how column D in the table can be used to solve problems. For this problem, the 0.6000 in the center can be divided in half with exactly 0.3000 to the right of the mean and exactly 0.3000 to the left. Each of these sections corresponds to the proportion listed in column D. Looking in column D for a value of 0.3000, you will discover that this exact proportion is not in the table, but the closest value is 0.2995. Reading across the row to column A, you should find a z-score value of $z = 0.84$. Looking again at the sketch [Figure 6.8(b)], you can see that the right-hand line is located at $z = +0.84$ and the left-hand line is located at $z = -0.84$.

You may have noticed that we have sketched distributions for each of the preceding problems. As a general rule, you should always sketch a distribution, locate the mean with a vertical line, and shade in the portion you are trying to determine. Look at your sketch. It will indicate which columns to use in the unit normal table. If you make a habit of drawing sketches, you will avoid careless errors when using the table.

LEARNING CHECK   1. Find the proportion of the normal distribution that is associated with the following sections of a graph:

   a. $z < +1.00$ **b.** $z > +0.80$ **c.** $z < -2.00$ **d.** $z > -0.33$

   e. $z > -0.50$

2. For a normal distribution, find the *z*-score location that separates the distribution as follows:

   a. Separate the highest 30% from the rest of the distribution.

   b. Separate the lowest 40% from the rest of the distribution.

   c. Separate the highest 75% from the rest of the distribution.

ANSWERS   **1. a.** 0.8413   **b.** 0.2119   **c.** 0.0228   **d.** 0.6293   **e.** 0.6915

   2. a. $z = 0.52$ **b.** $z = -0.25$ **c.** $z = -0.67$

6.3

PROBABILITIES AND PROPORTIONS FOR SCORES FROM A NORMAL DISTRIBUTION

In the preceding section, we used the unit normal table to find probabilities and proportions corresponding to specific *z*-score values. In most situations, however, it will be necessary to find probabilities for specific *X* values. Consider the following example:

It is known that IQ scores form a normal distribution with $\mu = 100$ and $\sigma = 15$. Given this information, what is the probability of randomly selecting an individual with an IQ score less than 130?

This problem is asking for a specific probability or proportion of a normal distribution. However, before we can look up the answer in the unit normal table, we must first transform the IQ scores (*X* values) into *z*-scores. Thus, to solve this new kind of probability problem, we must add one new step to the process. Specifically, to answer probability questions about scores (*X* values) from a normal distribution, you must use the following two-step procedure:

Caution: The unit normal table can be used only with normal-shaped distributions. If a distribution is not normal, transforming to *z*-scores will not make it normal.

1. Transform the *X* values into *z*-scores.

2. Use the unit normal table to look up the proportions corresponding to the *z*-score values.

This process is demonstrated in the following examples. Once again, we suggest that you sketch the distribution and shade the portion you are trying to find in order to avoid careless mistakes.

EXAMPLE 6.5   We will now answer the probability question about IQ scores that we presented earlier. Specifically, what is the probability of randomly selecting an individual with an IQ score less than 130?

$$p(X < 130) = ?$$

Restated in terms of proportions, we want to find the proportion of the IQ distribution that corresponds to scores less than 130. The distribution is drawn in Figure 6.9, and the portion we want has been shaded.

The first step is to change the *X* values into *z*-scores. In particular, the score of $X = 130$ is changed to

$$z = \frac{X - \mu}{\sigma} = \frac{130 - 100}{15} = \frac{30}{15} = 2.00$$

FIGURE 6.9

The distribution of IQ scores. The problem is to find the probability or proportion of the distribution corresponding to scores less than 130.

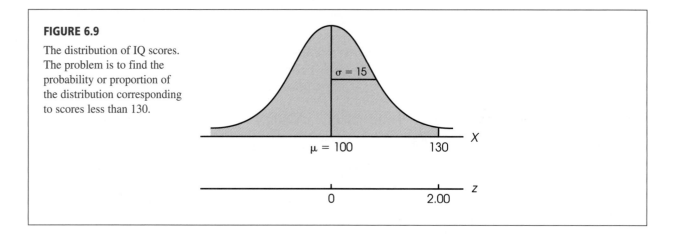

Thus, an IQ score of $X = 130$ corresponds to a z-score of $z = +2.00$, and IQ scores less than 130 correspond to z-scores less than 2.00.

Next, look up the z-score value in the unit normal table. Because we want the proportion of the distribution in the body to the left of $X = 130$ (see Figure 6.9), the answer will be found in column B. Consulting the table, we see that a z-score of 2.00 corresponds to a proportion of 0.9772. The probability of randomly selecting an individual with an IQ less than 130 is 0.9772:

$$p(X < 130) = p(z < 2.00) = 0.9772 \text{ (or 97.72\%)}$$

Finally, notice that we phrased this question in terms of a *probability*. Specifically, we asked, "What is the probability of selecting an individual with an IQ less than 130?" However, the same question can be phrased in terms of a *proportion:* "What proportion of the individuals in the population have IQ scores less than 130?" Both versions ask exactly the same question and produce exactly the same answer. A third alternative for presenting the same question is introduced in Box 6.2.

Finding proportions/probabilities located between two scores   The next example demonstrates the process of finding the probability of selecting a score that is located *between* two specific values. Although these problems can be solved using the proportions of columns B and C (body and tail), they are often easier to solve with the proportions listed in column D.

EXAMPLE 6.6   The highway department conducted a study measuring driving speeds on a local section of interstate highway. They found an average speed of $\mu = 58$ miles per hour with a standard deviation of $\sigma = 10$. The distribution was approximately normal. Given this information, what proportion of the cars are traveling between 55 and 65 miles per hour? Using probability notation, we can express the problem as

$$p(55 < X < 65) = \text{?}$$

The distribution of driving speeds is shown in Figure 6.10 with the appropriate area shaded. The first step is to determine the z-score corresponding to the X value at each end of the interval.

BOX 6.2

PROBABILITIES, PROPORTIONS, AND PERCENTILE RANKS

Thus far we have discussed parts of distributions in terms of proportions and probabilities. However, there is another set of terminology that deals with many of the same concepts. Specifically, in Chapter 2 we defined the *percentile rank* for a specific score as the percentage of the individuals in the distribution who have scores that are less than or equal to the specific score. For example, if 70% of the individuals have scores of $X = 45$ or lower, then $X = 45$ has a percentile rank of 70%. Also, when a score is referred to by its percentile rank, the score is called a *percentile*. For example, a score with a percentile rank of 70% is called the 70th percentile.

Using this terminology, it is possible to rephrase some of the probability problems that we have been working. In Example 6.5, the problem is presented as "What is the probability of randomly selecting an individual with an IQ of less than 130?" Exactly the same question could be phrased as "What is the percentile rank for an IQ score of 130?" In each case, we are looking for the proportion of the distribution corresponding to scores equal to or less than 130. Similarly, Example 6.8 asks "What is the minimum score necessary to be in the top 15% of the SAT distribution?" Because this score separates the top 15% from the bottom 85%, the same question could be rephrased as "What is the 85th percentile for the distribution of SAT scores?"

FIGURE 6.10

The distribution for Example 6.6.

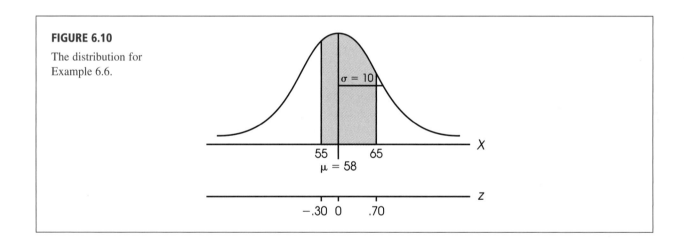

$$\text{For } X = 55: \quad z = \frac{X - \mu}{\sigma} = \frac{55 - 58}{10} = \frac{-3}{10} = -0.30$$

$$\text{For } X = 65: \quad z = \frac{X - \mu}{\sigma} = \frac{65 - 58}{10} = \frac{7}{10} = 0.70$$

Looking again at Figure 6.10, we see that the proportion we are seeking can be divided into two sections: (1) the area left of the mean, and (2) the area right of the mean. The first area is the proportion between the mean and $z = -0.30$ and the second is the proportion between the mean and $z = +0.70$. Using column D of the unit normal table, these two proportions are 0.1179 and 0.2580. The total proportion is obtained by adding these two sections:

$$p(55 < X < 65) = p(-0.30 < z < +0.70) = 0.1179 + 0.2580 = 0.3759$$

EXAMPLE 6.7   Using the same distribution of driving speeds from the previous example, what proportion of cars are traveling between 65 and 75 miles per hour?

$$p(65 < X < 75) = ?$$

The distribution is shown in Figure 6.11 with the appropriate area shaded. Again, we start by determining the z-score corresponding to each end of the interval.

For $X = 65$: $z = \dfrac{X - \mu}{\sigma} = \dfrac{65 - 58}{10} = \dfrac{7}{10} = 0.70$

For $X = 75$: $z = \dfrac{X - \mu}{\sigma} = \dfrac{75 - 58}{10} = \dfrac{17}{10} = 1.70$

According to column D in the unit normal table, the proportion between the mean and $z = 1.70$ is $p = 0.4554$. Note that this proportion includes the section that we want, but it also includes an extra, unwanted section between the mean and $z = 0.70$. Again, using column D, we see that the unwanted section is $p = 0.2580$. To obtain the correct answer, we subtract the unwanted portion from the total proportion between the mean and $z = 1.70$.

$$p(65 < X < 75) = p(0.70 < z < 1.70) = 0.4554 - 0.2580 = 0.1974$$

Finding scores corresponding to specific proportions or probabilities   In the previous two examples, the problem was to find the proportion or probability corresponding to specific X values. The two-step process for finding these proportions is shown in Figure 6.12. Thus far, we have only considered examples that move in a clockwise direction around the triangle shown in the figure; that is, we start with an X value that is transformed into a z-score, and then we use the unit normal table to look up the appropriate proportion. You should realize, however, that it is possible to reverse this two-step process so that we move backward, or counterclockwise, around the triangle. This reverse process will allow us to find the score (X value) corresponding to a specific proportion in the distribution. Following the lines in Figure 6.12, we will begin with a specific proportion, use the unit normal table to look up the corresponding z-score, and then transform the z-score into an X value. The following example demonstrates this process.

FIGURE 6.11

The distribution for Example 6.7.

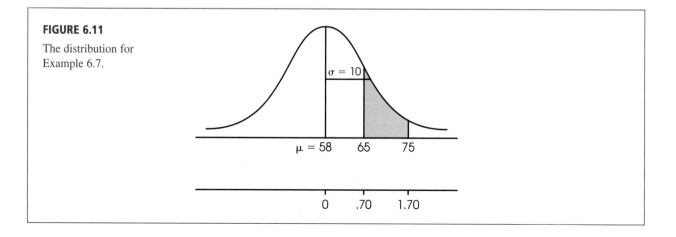

FIGURE 6.12

Determining probabilities or proportions for a normal distribution is shown as a two-step process with z-scores as an intermediate stop along the way. Note that you cannot move directly along the dashed line between X values and probabilities or proportions. Instead, you must follow the solid lines around the corner.

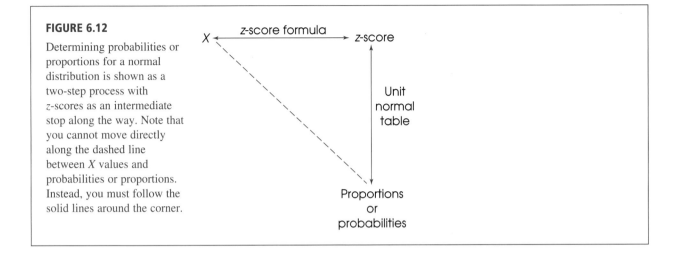

EXAMPLE 6.8   Scores on the SAT form a normal distribution with $\mu = 500$ and $\sigma = 100$. What is the minimum score necessary to be in the top 15% of the SAT distribution? (An alternative form of the same question is presented in Box 6.2.) This problem is shown graphically in Figure 6.13.

In this problem, we begin with a proportion (15% = 0.15), and we are looking for a score. According to the map in Figure 6.12, we can move from p (proportion) to X (score) via z-scores. The first step is to use the unit normal table to find the z-score that corresponds to a proportion of 0.15. Because the proportion is located beyond z in the tail of the distribution, we will look in column C for a proportion of 0.1500. Note that you may not find 0.1500 exactly, but locate the closest value possible. In this case, the closest value in the table is 0.1492, and the z-score that corresponds to this proportion is $z = 1.04$.

The next step is to determine whether the z-score is positive or negative. Remember that the table does not specify the sign of the z-score. Looking at the graph in

FIGURE 6.13

The distribution of SAT scores. The problem is to locate the score that separates the top 15% from the rest of the distribution. A line is drawn to divide the distribution roughly into 15% and 85% sections.

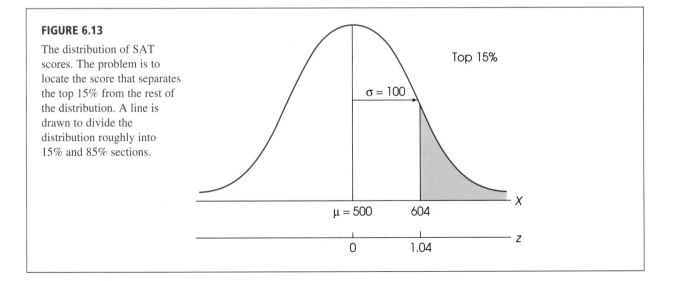

Figure 6.13, you should realize that the score we want is above the mean, so the z-score is positive, $z = +1.04$.

Now you are ready for the last stage of the solution—that is, changing the z-score into an X value. By definition, a z-score of $z = +1.04$ corresponds to an X value that is located above the mean $(+)$ by 1.04 standard deviations. One standard deviation is $\sigma = 100$ points, so 1.04 standard deviations is

$$1.04\sigma = 1.04(100) = 104 \text{ points}$$

Thus, our score is located 104 points above the mean. With a mean of $\mu = 500$, the score is $X = 500 + 104 = 604$. The conclusion for this example is that you must have an SAT score of at least 604 to be in the top 15% of the distribution.

EXAMPLE 6.9

An alternative for solving this problem is to note that 80% in the middle leaves 10% or 0.1000 in each tail. Looking up 0.1000 in column C also produces a z-score of $z = 1.28$.

This time the problem is to find the range of values that defines the middle 80% of the distribution of SAT scores. The distribution is shown in Figure 6.14 with the appropriate section shaded.

The 80% (0.8000) in the middle of the distribution can be split in half with 0.4000 to the right of the mean and 0.4000 to the left. Looking up 0.4000 in column D of the unit normal table, you will discover that the exact proportion is not listed but the closest value is 0.3997. Reading across to column A, the corresponding z-score is $z = 1.28$. Thus, the z-score at the right-hand boundary is $z = +1.28$ and the z-score at the left boundary is $z = -1.28$. In either case, a z-score of 1.28 indicates a location that is 1.28 standard deviations away from the mean. For the distribution of SAT scores, 1.28 standard deviations corresponds to

$$1.28\sigma = 1.28(100) = 128 \text{ points}$$

Therefore, the right-hand boundary ($z = +1.28$) is $X = 500 + 128 = 628$. The left-hand boundary ($z = -1.28$) is $X = 500 - 128 = 372$. The middle 80% of the distribution corresponds to X values between 372 and 628.

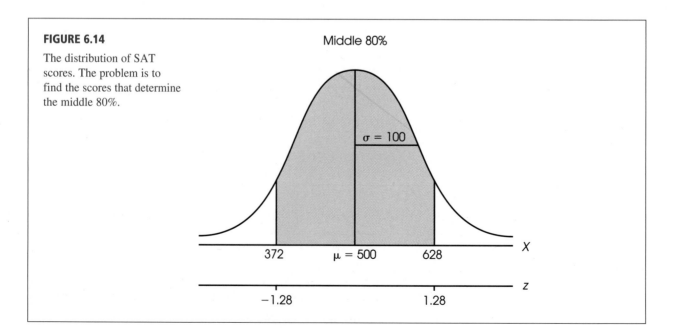

FIGURE 6.14

The distribution of SAT scores. The problem is to find the scores that determine the middle 80%.

Middle 80%

$\sigma = 100$

372    $\mu = 500$ 628   X

-1.28 1.28   z

LEARNING CHECK

1. For a normal distribution with a mean of 80 and a standard deviation of 10, find each probability value requested.

   a. $p(X > 85) = ?$ **b.** $p(X < 95) = ?$

   c. $p(X > 70) = ?$ **d.** $p(75 < X < 100) = ?$

2. For a normal distribution with a mean of 100 and a standard deviation of 20, find each value requested.

   a. What score separates the top 40% from the bottom 60% of the distribution?

   b. What is the minimum score needed to be in the top 5% of the distribution?

   c. What scores form the boundaries for the middle 60% of the distribution?

3. What is the probability of selecting a score greater than 45 from a positively skewed distribution with $\mu = 40$ and $\sigma = 10$? (Be careful.)

ANSWERS

1. **a.** $p = 0.3085$ (30.85%)   **b.** $p = 0.9332$ (93.32%)   **c.** $p = 0.8413$ (84.13%)

   d. $p = 0.6687$ (66.87%)

2. **a.** $z = +0.25; X = 105$ **b.** $z = +1.64; X = 132.8$

   c. $z = \pm 0.84$; boundaries are 83.2 and 116.8

3. You cannot obtain the answer. The unit normal table cannot be used to answer this question because the distribution is not normal.

6.4 PROBABILITY AND THE BINOMIAL DISTRIBUTION

A variable that is classified in exactly two categories is also called a *dichotomous* variable.

When a variable is measured on a scale consisting of exactly two categories, the resulting data are called binomial. The term *binomial* can be loosely translated as "two names," referring to the two categories on the measurement scale.

Binomial data can occur when a variable naturally exists with only two categories. For example, people can be classified as male or female, and a coin toss results in either heads or tails. It also is common for a researcher to simplify data by collapsing the scores into two categories. For example, a psychologist may use personality scores to classify people as either high or low in aggression.

In binomial situations, the researcher often knows the probabilities associated with each of the two categories. With a balanced coin, for example, $p(\text{heads}) = p(\text{tails}) = \frac{1}{2}$. The question of interest is the number of times each category occurs in a series of trials or in a sample of individuals. For example:

What is the probability of obtaining 15 heads in 20 tosses of a balanced coin?

What is the probability of obtaining more than 40 introverts in a sampling of 50 college freshmen?

As we shall see, the normal distribution serves as an excellent model for computing probabilities with binomial data.

THE BINOMIAL DISTRIBUTION

To answer probability questions about binomial data, we need to examine the binomial distribution. To define and describe this distribution, we first introduce some notation.

1. The two categories are identified as *A* and *B*.

2. The probabilities (or proportions) associated with each category are identified as

$$p = p(A) = \text{the probability of } A$$

$$q = p(B) = \text{the probability of } B$$

Notice that $p + q = 1.00$ because A and B are the only two possible outcomes.

3. The number of individuals or observations in the sample is identified by n.

4. The variable X refers to the number of times category A occurs in the sample.

Notice that X can have any value from 0 (none of the sample is in category A) to n (all the sample is in category A).

DEFINITION

Using the notation presented here, the **binomial distribution** shows the probability associated with each value of X from $X = 0$ to $X = n$.

A simple example of a binomial distribution is presented next.

EXAMPLE 6.10

Figure 6.15 shows the binomial distribution for the number of heads obtained in 2 tosses of a balanced coin. This distribution shows that it is possible to obtain as many as 2 heads or as few as 0 heads in 2 tosses. The most likely outcome (highest probability) is to obtain exactly 1 head in 2 tosses. The construction of this binomial distribution is discussed in detail next.

For this example, the event we are considering is a coin toss. There are two possible outcomes, heads and tails. We assume the coin is balanced, so

$$p = p(\text{heads}) = \frac{1}{2}$$

$$q = p(\text{tails}) = \frac{1}{2}$$

We are looking at a sample of $n = 2$ tosses, and the variable of interest is

$$X = \text{the number of heads}$$

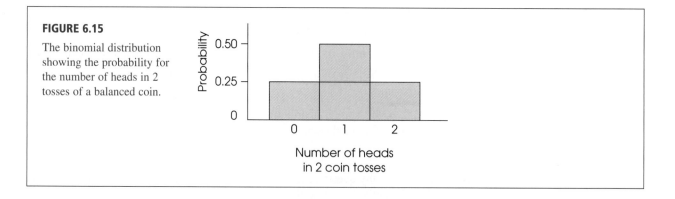

FIGURE 6.15

The binomial distribution showing the probability for the number of heads in 2 tosses of a balanced coin.

To construct the binomial distribution, we will look at all the possible outcomes from tossing a coin 2 times. The complete set of 4 outcomes is listed below.

1st Toss	2nd Toss	
Heads	Heads	(both heads)
Heads	Tails	(each sequence has
Tails	Heads	exactly 1 head)
Tails	Tails	(no heads)

Notice that there are 4 possible outcomes when you toss a coin 2 times. Only 1 of the 4 outcomes has 2 heads, so the probability of obtaining 2 heads is $p = \frac{1}{4}$. Similarly, 2 of the 4 outcomes have exactly 1 head, so the probability of 1 head is $p = \frac{2}{4} = \frac{1}{2}$. Finally, the probability of no heads ($X = 0$) is $p = \frac{1}{4}$. These are the probabilities shown in Figure 6.15.

Note that this binomial distribution can be used to answer probability questions. For example, what is the probability of obtaining at least 1 head in 2 tosses? According to the distribution shown in Figure 6.15, the answer is $\frac{3}{4}$.

Similar binomial distributions have been constructed for the number of heads in 4 tosses of a balanced coin and in 6 tosses of a coin (Figure 6.16). It should be obvious from the binomial distributions shown in Figures 6.15 and 6.16 that the binomial distribution tends toward a normal shape, especially when the sample size (n) is relatively large.

It should not be surprising that the binomial distribution tends to be normal. With $n = 10$ coin tosses, for example, the most likely outcome would be to obtain around $X = 5$ heads. On the other hand, values far from 5 would be very unlikely—you would

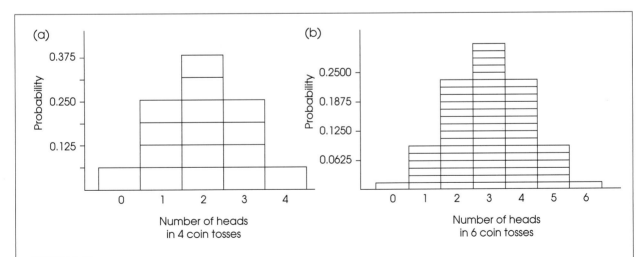

FIGURE 6.16

Binomial distributions showing probabilities for the number of heads (a) in 4 tosses of a balanced coin and (b) in 6 tosses of a balanced coin.

not expect to get all 10 heads or all 10 tails (0 heads) in 10 tosses. Notice that we have described a normal-shaped distribution: The probabilities are highest in the middle (around $X = 5$), and they taper off as you move toward either extreme.

THE NORMAL APPROXIMATION TO THE BINOMIAL DISTRIBUTION

We have stated that the binomial distribution tends to approximate a normal distribution, particularly when n is large. To be more specific, the binomial distribution will be a nearly perfect normal distribution when pn and qn are both equal to or greater than 10. Under these circumstances, the binomial distribution will approximate a normal distribution with the following parameters:

Mean: $\mu = pn$ (6.1)

standard deviation: $\sigma = \sqrt{npq}$ (6.2)

Within this normal distribution, each value of X has a corresponding z-score,

$$z = \frac{X - \mu}{\sigma} = \frac{X - pn}{\sqrt{npq}}$$ (6.3)

The value of 10 for pn or qn is a general guide, not an absolute cutoff. Values slightly less than 10 still provide a good approximation. However, with smaller values the normal approximation becomes less accurate as a substitute for the binomial distribution.

The fact that the binomial distribution tends to be normal in shape means that we can compute probability values directly from z-scores and the unit normal table.

It is important to remember that the normal distribution is only an approximation of a true binomial distribution. Binomial values, such as the number of heads in a series of coin tosses, are *discrete*. The normal distribution is *continuous*. However, the *normal approximation* provides an extremely accurate model for computing binomial probabilities in many situations. Figure 6.17 shows the difference between the true binomial distribution, the discrete histogram, and the normal curve that approximates the binomial distribution. Although the two distributions are slightly different, the area under the distributions is nearly equivalent. *Remember, it is the area under the distribution that is used to find probabilities.*

Coin tosses produce discrete events. In a series of coin tosses, you may observe 1 head, 2 heads, 3 heads, and so on, but no values between them are possible (p. 18).

To gain maximum accuracy when using the normal approximation, you must remember that each X value in the binomial distribution actually corresponds to a bar in the histogram. In the histogram in Figure 6.17, for example, the score $X = 6$ is represented by a bar that is bounded by real limits of 5.5 and 6.5. The actual probability of $X = 6$ is determined by the area contained in this bar. To approximate this probability using the normal distribution, you should find the area that is contained between the two real limits. Similarly, if you are using the normal approximation to find the probability of obtaining

FIGURE 6.17

The relationship between the binomial distribution and the normal distribution. The binomial distribution is always a discrete histogram, and the normal distribution is a continuous, smooth curve. Each X value is represented by a bar in the histogram or a section of the normal distribution.

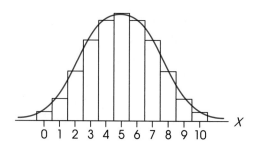

a score greater than $X = 6$, you should use the area beyond the real limit boundary of 6.5. The following two examples demonstrate how the normal approximation to the binomial distribution is used to compute probability values.

EXAMPLE 6.11A   Suppose that you are testing for ESP (extra-sensory perception) by asking people to predict the suit of a card that is randomly selected from a complete deck. What is the probability that an individual would predict the suit correctly for exactly 14 out of 48 trials?

On each trial, there are two possible outcomes: correct or incorrect. Because there are four different suits, the probability of a correct prediction (assuming that there is no ESP) is $p = \frac{1}{4}$ and the probability of an incorrect prediction is $q = \frac{3}{4}$. With a series of $n = 48$ trials, this situation meets the criteria for the normal approximation to the binomial distribution:

$$pn = \frac{1}{4}(48) = 12 \qquad qn = \frac{3}{4}(48) = 36 \qquad \text{Both are greater than 10.}$$

Thus, the distribution of correct predictions will form a normal shaped distribution with a mean of $\mu = pn = 12$ and a standard deviation of $\sigma = \sqrt{npq} = \sqrt{9} = 3$.

This distribution is shown in Figure 6.18. We want the proportion of this distribution that corresponds to $X = 14$, that is, the portion bounded by $X = 13.5$ and $X = 14.5$. This portion is shaded in Figure 6.18. To find the shaded portion, we first convert each of the boundaries into a z-score. For $X = 13.5$,

Remember to use the real limits when determining probabilities from the normal approximation to the binomial distribution.

$$z = \frac{X - \mu}{\sigma} = \frac{13.5 - 12}{3} = 0.50$$

For $X = 14.5$,

$$z = \frac{X - \mu}{\sigma} = \frac{14.5 - 12}{3} = 0.83$$

The area we want is contained between these two z-scores. Looking up the z-score values in the unit normal table, we find

a. The area in the tail beyond $z = 0.50$ is 0.3085.

b. The area in the tail beyond $z = 0.83$ is 0.2033.

FIGURE 6.18

The binomial distribution (normal approximation) for the number of correct predictions of the suit of a card in a series of $n = 48$ trials. The shaded area corresponds to the probability of obtaining exactly 14 correct predictions. Note that the score $X = 14$ is defined by its real limits.

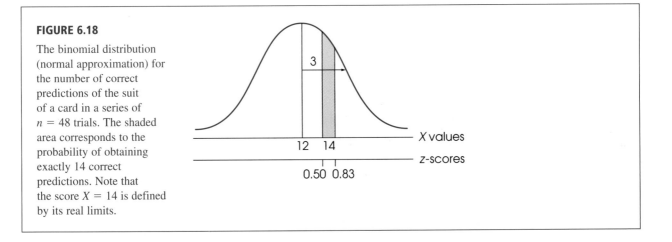

Therefore, the area between the two z-scores is

$$0.3085 - 0.2033 = 0.1052$$

You should remember that this value is an approximation to the exact probability value. However, it is a reasonably accurate approximation. To illustrate the accuracy of the normal approximation, the exact probability of getting $X = 14$ correct from the binomial distribution is 0.1015. Notice that the normal approximation value of 0.1052 is very close to the exact probability of 0.1015. As we mentioned earlier, the normal distribution provides an excellent approximation for finding binomial probabilities.

EXAMPLE 6.11B

Caution: If the question had asked for the probability of *20 or more* correct predictions, we would find the area beyond $X = 19.5$. Read the question carefully.

Again, suppose that you are testing for ESP by having people predict the suit of a card that is randomly selected from a deck. This time, what is the probability that an individual would predict correctly more than 20 times in a series of 48 trials?

As before, the distribution of correct predictions forms a normal shaped distribuion with a mean of $\mu = pn = 12$ and a standard deviation of $\sigma = \sqrt{npq} = \sqrt{9} = 3$. Because we want the probability of obtaining *more than* 20 correct predictions, we must find the area in the tail of the distribution beyond $X = 20.5$. (Remember that a score of 20 corresponds to an interval from 19.5 to 20.5. We want scores beyond this interval.) The first step is to find the z-score corresponding to $X = 20.5$.

$$z = \frac{X - \mu}{\sigma} = \frac{20.5 - 12}{3} = \frac{8.5}{3} = 2.83$$

Next, look up the probability in the unit normal table. In this case, we want the proportion in the tail beyond $z = 2.83$. The value from the table is $p = 0.0023$. This is the answer we want. The probability of correctly predicting the suit of a card more than 20 times in a series of 48 trials is only $p = 0.0023$ or 0.23%.

LEARNING CHECK

1. Under what circumstances is the normal distribution an accurate approximation to the binomial distribution?

2. A multiple-choice test consists of 48 questions with 4 possible answers for each question. What is the probability that you would get more than 15 questions correct just by guessing?

3. If you toss a balanced coin 36 times, you would expect, on the average, to get 18 heads and 18 tails. What is the probability of obtaining exactly 18 heads in 36 tosses?

ANSWERS

1. When pn and qn are both greater than 10

2. $p = \frac{1}{4}$, $q = \frac{3}{4}$; $p(X > 15.5) = p(z > 1.17) = 0.1210$

3. $z = \pm 0.17$, $p = 0.1350$

6.5   LOOKING AHEAD TO INFERENTIAL STATISTICS

Probability forms a direct link between samples and the populations from which they come. As we noted at the beginning of this chapter, this link will be the foundation for the inferential statistics in future chapters. The following example provides a brief preview of how probability will be used in the context of inferential statistics.

We ended Chapter 5 with a demonstration of how inferential statistics are used to help interpret the results of a research study. A general research situation was shown in Figure 5.9 and is repeated here in Figure 6.19. The research begins with a population that forms a normal distribution with a mean of $\mu = 400$ and a standard deviation of $\sigma = 20$. A sample is selected from the population and a treatment is administered to the sample. The goal for the study is to evaluate the effect of the treatment.

To determine whether the treatment has an effect, the researcher simply compares the treated sample with the original population. If the individuals in the sample have scores around 400 (the original population mean), then we must conclude that the treatment appears to have no effect. On the other hand, if the treated individuals have scores that are noticeably different from 400, then the researcher has evidence that the treatment does have an effect. Notice that the study is using a sample to help answer a question about a population; this is the essence of inferential statistics.

The problem for the researcher is determining exactly what is meant by "noticeably different" from 400. If a treated individual has a score of $X = 415$, is that enough to say that the treatment has an effect? What about $X = 420$ or $X = 450$? In Chapter 5, we suggested that z-scores provide one method for solving this problem. Specifically, we suggested that a z-score value beyond $z = 2.00$ (or -2.00) was an extreme value and therefore noticeably different. However, the choice of $z = \pm 2.00$ was purely arbitrary. Now we have another tool, *probability*, to help us decide exactly where to set the boundaries.

FIGURE 6.19

A diagram of a research study. A sample is selected from the population and receives a treatment. The goal is to determine whether the treatment has an effect.

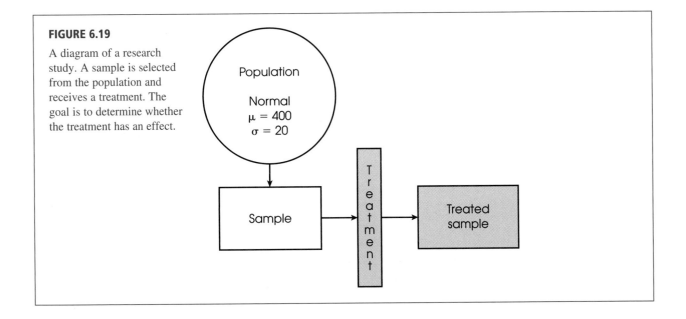

Figure 6.20 shows the original population from our hypothetical research study. Note that most of the scores are located close to $\mu = 400$. Also note that we have added boundaries separating the middle 95% of the distribution from the extreme 5% or 0.0500 in the two tails. Dividing the 0.0500 in half produces a proportion of 0.0250 in the right-hand tail and 0.0250 in the left-hand tail. Using column C of the unit normal table, the z-score boundaries for the right and left tails are $z = +1.96$ and $z = -1.96$, respectively.

The boundaries set at $z = \pm 1.96$ provide objective criteria for deciding whether our sample provides evidence that the treatment has an effect. Specifically, we will use the sample data to help decide between the following two alternatives:

1. The treatment has no effect and the scores still average $\mu = 400$.

2. The treatment does have an effect that changes the scores so they no longer average $\mu = 400$.

If the first alternative is correct, then the sample should be close to 400. According to Figure 6.20, any sample between the ± 1.96 boundaries is considered close to 400 and will support this alternative. On the other hand, a sample beyond the ± 1.96 boundaries is significantly different from $\mu = 400$. Specifically, this kind of sample is very unlikely to occur (probability less than 5%) if the population mean is $\mu = 400$. Therefore, if we obtain a sample that is beyond the ± 1.96 boundaries, we reject the first alternative and conclude that the population mean is no longer equal to 400. In terms of the research study, we conclude that the treatment does have a significant effect.

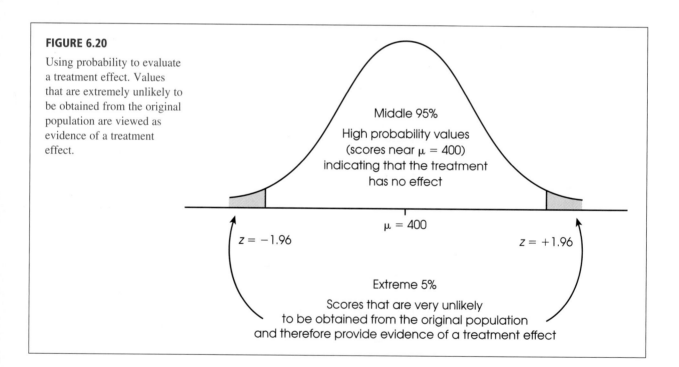

FIGURE 6.20

Using probability to evaluate a treatment effect. Values that are extremely unlikely to be obtained from the original population are viewed as evidence of a treatment effect.

Middle 95%
High probability values
(scores near $\mu = 400$)
indicating that the treatment
has no effect

$\mu = 400$

$z = -1.96$

$z = +1.96$

Extreme 5%
Scores that are very unlikely
to be obtained from the original population
and therefore provide evidence of a treatment effect

SUMMARY

1. The probability of a particular event A is defined as a fraction or proportion:

$$p(A) = \frac{\text{number of outcomes classified as } A}{\text{total number of possible outcomes}}$$

2. Our definition of probability is accurate only for random samples. There are two requirements that must be satisfied for a random sample:
   a. Every individual in the population has an equal chance of being selected.
   b. When more than one individual is being selected, the probabilities must stay constant. This means there must be sampling with replacement.

3. All probability problems can be restated as proportion problems. The "probability of selecting a king from a deck of cards" is equivalent to the "proportion of the deck that consists of kings." For frequency distributions, probability questions can be answered by determining proportions of area. The "probability of selecting an individual with an IQ greater than 108" is equivalent to the "proportion of the whole population that consists of IQs above 108."

4. For normal distributions, probabilities (proportions) can be found in the unit normal table. The table provides a listing of the proportions of a normal distribution that correspond to each z-score value. With the table, it is possible to move between X values and probabilities using a two-step procedure:
   a. The z-score formula (Chapter 5) allows you to transform X to z or to change z back to X.
   b. The unit normal table allows you to look up the probability (proportion) corresponding to each z-score or the z-score corresponding to each probability.

5. Percentiles and percentile ranks measure the relative standing of a score within a distribution (see Box 6.2). Percentile rank is the percentage of individuals with scores at or below a particular X value. A percentile is an X value that is identified by its rank. The percentile rank always corresponds to the proportion to the left of the score in question.

6. The binomial distribution is used whenever the measurement procedure simply classifies individuals into exactly two categories. The two categories are identified as A and B, with probabilities of

$$p(A) = p \quad \text{and} \quad p(B) = q$$

7. The binomial distribution gives the probability for each value of X, where X equals the number of occurrences of category A in a series of n events. For example, X equals the number of heads in $n = 10$ tosses of a coin.

8. When pn and qn are both at least 10, the binomial distribution is closely approximated by a normal distribution with

$$\mu = pn$$
$$\sigma = \sqrt{npq}$$

9. In the normal approximation to the binomial distribution, each value of X has a corresponding z-score:

$$z = \frac{X - \mu}{\sigma} = \frac{X - pn}{\sqrt{npq}}$$

With the z-score and the unit normal table, you can find probability values associated with any value of X. For maximum accuracy, you should use the appropriate real limits for X when computing z-scores and probabilities.

KEY TERMS

probability	unit normal table	binomial distribution
random sample	percentile rank	normal approximation (binomial)
sampling with replacement	percentile	

RESOURCES

There are a tutorial quiz and other learning exercises for Chapter 6 at the Wadsworth website, www.cengage.com/psychology/gravetter. See page 29 for more information about accessing items on the website.

ENHANCED
WebAssign

Guided interactive tutorials, end-of-chapter problems, and related testbank items may be assigned online at WebAssign.

WebTUTOR

For those using WebTutor along with this book, there is a WebTutor section corresponding to this chapter. The WebTutor contains a brief summary of Chapter 6, hints for learning about probability, cautions about common errors, and sample exam items including solutions.

SPSS

The statistics computer package SPSS is not structured to compute probabilities. However, the program does report probability values as part of the inferential statistics that we will examine later in this book. In the context of inferential statistics, the probabilities are called *significance levels,* and they warn researchers about the probability of misinterpreting their research results.

FOCUS ON PROBLEM SOLVING

1. We have defined probability as being equivalent to a proportion, which means that you can restate every probability problem as a proportion problem. This definition is particularly useful when you are working with frequency distribution graphs in which the population is represented by the whole graph and probabilities (proportions) are represented by portions of the graph. When working problems with the normal distribution, you always should start with a sketch of the distribution. You should shade the portion of the graph that reflects the proportion you are looking for.

2. Remember that the unit normal table shows only positive z-scores in column A. However, since the normal distribution is symmetrical, the proportions in the table apply to both positive and negative z-score values.

3. A common error for students is to use negative values for proportions on the left-hand side of the normal distribution. Proportions (or probabilities) are always positive: 10% is 10% whether it is in the left or right tail of the distribution.

4. The proportions in the unit normal table are accurate only for normal distributions. If a distribution is not normal, you cannot use the table.

5. For maximum accuracy when using the normal approximation to the binomial distribution, you must remember that each X value is an interval bounded by real limits. For example, to find the probability of obtaining an X value greater than 10, you should use the real limit 10.5 in the z-score formula. Similarly, to find the probability of obtaining an X value less than 10, you should use the real limit 9.5.

DEMONSTRATION 6.1

FINDING PROBABILITY FROM THE UNIT NORMAL TABLE

A population is normally distributed with a mean of $\mu = 45$ and a standard deviation of $\sigma = 4$. What is the probability of randomly selecting a score that is greater than 43? In other words, what proportion of the distribution consists of scores greater than 43?

STEP 1   Sketch the distribution.
    You should always start by sketching the distribution, identifying the mean and standard deviation ($\mu = 45$ and $\sigma = 4$ in this example). You should also find the approximate location of the specified score and draw a vertical line through the distribution at that score. The score of $X = 43$ is lower than the mean, and, therefore, it should be placed somewhere to the left of the mean. Figure 6.21(a) shows the preliminary sketch.

STEP 2   Shade in the appropriate portion of the distribution.
    For this problem, we want the proportion corresponding to scores greater than $X = 43$. Thus, we shade the area to the right of this score [Figure 6.21(b)]. Note that the shaded area covers more than half of the distribution and, therefore, corresponds to the body of the distribution.

STEP 3   Transform the X value to a z-score.

$$z = \frac{X - \mu}{\sigma} = \frac{43 - 45}{4} = \frac{-2}{4} = -0.5$$

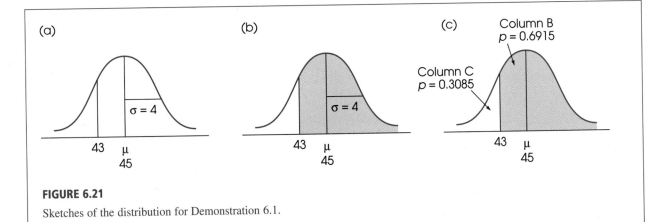

FIGURE 6.21

Sketches of the distribution for Demonstration 6.1.

STEP 4   Consult the unit normal table.

Look up the z-score (ignoring the sign) in the unit normal table. Find the proportions in the table that are associated with the z-score, and write the proportions in the appropriate regions on the figure. Remember that column B gives the proportion of area in the body of the distribution (more than 50%), and column C gives the area in the tail beyond z. Figure 6.21(c) shows the proportions for different regions of the normal distribution.

STEP 5   Determine the probability.

If one of the proportions in step 4 corresponds exactly to the entire shaded area, then that proportion is your answer. Otherwise, you will need to do some additional arithmetic. In this example, the proportion in the shaded area is given in column B:

$$p(X > 43) = 0.6915$$

DEMONSTRATION 6.2

PROBABILITY AND THE BINOMIAL DISTRIBUTION

Suppose that you forgot to study for a quiz and now must guess on every question. It is a true-false quiz with $n = 40$ questions. What is the probability that you will get at least 26 questions correct just by chance? Stated in symbols,

$$p(X \geq 26) = ?$$

STEP 1   Identify p and q.

This problem is a binomial situation, in which

$$p = \text{probability of guessing correctly} = 0.50$$

$$q = \text{probability of guessing incorrectly} = 0.50$$

With $n = 40$ quiz items, both pn and qn are greater than 10. Thus, the criteria for the normal approximation to the binomial distribution are satisfied:

$$pn = 0.50(40) = 20$$

$$qn = 0.50(40) = 20$$

STEP 2   Identify the parameters, and sketch the distribution.

The normal approximation will have a mean and a standard deviation as follows:

$$\mu = pn = 0.5(40) = 20$$

$$\sigma = \sqrt{npq} = \sqrt{10} = 3.16$$

Figure 6.22 shows the distribution. We are looking for the probability of getting $X = 26$ or more questions correct. Remember that when determining probabilities from the binomial distribution, we must use real limits. Because we are interested in scores equal to or greater than 26, we will use the lower real limit for 26 (25.5). By using the lower real limit, we include the entire interval (25.5 to 26.5) that corresponds to $X = 26$. In Figure 6.22, everything to the right of $X = 25.5$ is shaded.

STEP 3   Compute the z-score, and find the probability.

The z-score for $X = 25.5$ is calculated as follows:

$$z = \frac{X - pn}{\sqrt{npq}} = \frac{25.5 - 20}{3.16} = +1.74$$

The shaded area in Figure 6.22 corresponds to column C of the unit normal table. For the z-score of 1.74, the proportion of the shaded area is $P = 0.0409$. Thus, the probability of getting at least 26 questions right just by guessing is

$$p(X \geq 26) = 0.0409 \text{ (or 4.09\%)}$$

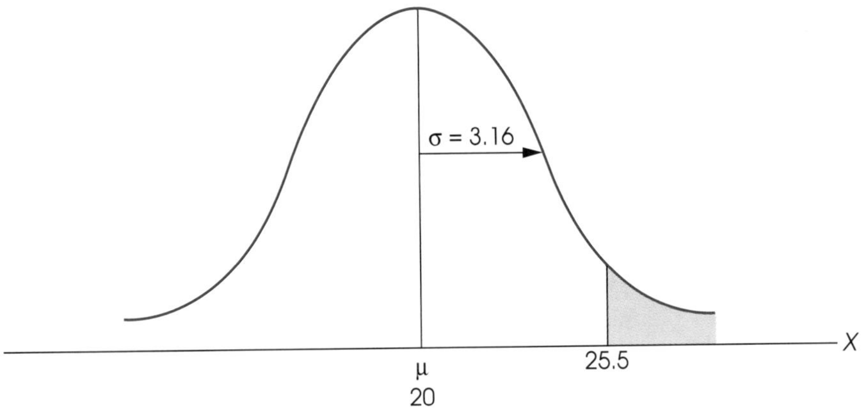

FIGURE 6.22

The normal approximation to a binomial distribution with $\mu = 20$ and $\sigma = 3.16$. The proportion of all scores equal to or greater than 26 is shaded. Notice that the real lower limit (25.5) for $X = 26$ is used.

PROBLEMS

1. A jar contains 20 red marbles, of which 10 are large and 10 are small, and 30 blue marbles, of which 10 are large and 20 are small. If 1 marble is randomly selected from the jar,
   a. What is the probability of obtaining a blue marble?
   b. What is the probability of obtaining a large marble?
   c. What is the probability of obtaining a large blue marble?

2. A kindergarten class consists of 14 boys and 11 girls. If the teacher selects children from the class using *random sampling,*
   a. What is the probability that the first child selected will be a girl?
   b. If the teacher selects a random sample of $n = 3$ children and the first two children are both boys, what is the probability that the third child selected will be a girl?

3. What requirements must be satisfied to have a *random sample?*

4. What is sampling with replacement, and why is it used?

5. For each of the following z-scores, sketch a normal distribution and draw a vertical line at the location of the z-score. Then, determine whether the tail is to the right or the left of the line and find the proportion in the tail.
   a. $z = 1.50$
   b. $z = -1.75$
   c. $z = 0.20$
   d. $z = -0.50$

6. For each of the following z-scores, sketch a normal distribution and draw a vertical line at the location of the z-score. Then, determine whether the body is to the right or the left of the line and find the proportion in the body.
   a. $z = 0.75$
   b. $z = -1.40$
   c. $z = 0.85$
   d. $z = -2.10$

7. Find each of the following probabilities for a normal distribution.
   a. $p(z > 1.00)$
   b. $p(z > 0.80)$
   c. $p(z < -0.25)$
   d. $p(z < -1.25)$

8. Find each of the following probabilities for a normal distribution.

  a. $p(z > -1.00)$ **b.** $p(z > -0.80)$
  c. $p(z < 0.25)$ **d.** $p(z < 1.25)$

9. For a normal distribution, identify the z-score location that would separate the distribution into two sections so that there is
  a. 10% in the tail on the right-hand side.
  b. 20% in the tail on the right-hand side.
  c. 20% in the tail on the left-hand side.
  d. 30% in the tail on the left-hand side.

10. For a normal distribution, identify the z-score location that would separate the distribution into two sections so that there is
  a. 70% in the body on the right-hand side.
  b. 80% in the body on the right-hand side.
  c. 75% in the body on the left-hand side.
  d. 90% in the body on the left-hand side.

11. What proportion of a normal distribution is located
  a. Between $z = 0.50$ and $z = -0.50$.
  b. Between $z = 1.00$ and $z = -1.00$.
  c. Between $z = 0$ and $z = -1.50$.
  d. Between $z = 1.75$ and $z = -0.25$.

12. For a normal distribution, find the z-score values that separate
  a. The middle 60% of the distribution from the 40% in the tails.
  b. The middle 70% of the distribution from the 30% in the tails.
  c. The middle 80% of the distribution from the 20% in the tails.
  d. The middle 90% of the distribution from the 10% in the tails.

13. A normal distribution has a mean of $\mu = 50$ and a standard deviation of $\sigma = 12$. For each of the following scores, indicate whether the tail is to the right or left of the score and find the proportion of the distribution located in the tail.
  a. $X = 53$ **b.** $X = 44$
  c. $X = 68$ **d.** $X = 38$

14. The distribution of IQ scores is normal with $\mu = 100$ and $\sigma = 15$. What proportion of the population has IQ scores
  a. Greater than 115?
  b. Greater than 130?
  c. Greater than 145?

15. The distribution of scores on the SAT is approximately normal with a mean of $\mu = 500$ and a standard deviation of $\sigma = 100$. For the population of students who have taken the SAT,
  a. What proportion have SAT scores greater than 700?
  b. What proportion have SAT scores greater than 550?

  c. What is the minimum SAT score needed to be in the highest 10% of the population?
  d. If the state college only accepts students from the top 60% of the SAT distribution, what is the minimum SAT score needed to be accepted?

16. The distribution of SAT scores is normal with $\mu = 500$ and $\sigma = 100$.
  a. What SAT score, X value, separates the top 15% of the distribution from the rest?
  b. What SAT score, X value, separates the top 20% of the distribution from the rest?
  c. What SAT score, X value, separates the top 25% of the distribution from the rest?

17. A normal distribution has a mean of $\mu = 60$ and a standard deviation of $\sigma = 20$. For each of the following scores, indicate whether the tail of the distribution is to the left or right of the score, and find the proportion of the distribution in the tail.
  a. $X = 75$ **b.** $X = 65$
  c. $X = 50$ **d.** $X = 30$

18. A normal distribution has a mean of $\mu = 80$ and a standard deviation of $\sigma = 10$. For each of the following scores, indicate whether the body of the distribution is to the left or right of the score, and find the proportion of the distribution in the body.
  a. $X = 85$ **b.** $X = 92$
  c. $X = 78$ **d.** $X = 72$

19. Information from the Department of Motor Vehicles indicates that the average age of licensed drivers is $\mu = 39.7$ years with a standard deviation of $\sigma = 12.5$ years. Assuming that the distribution of drivers' ages is approximately normal,
  a. What proportion of licensed drivers are more than 50 years old?
  b. What proportion of licensed drivers are less than 30 years old?

20. A consumer survey indicates that the average household spends $\mu = \$155$ on groceries each week. The distribution of spending amounts is approximately normal with a standard deviation of $\sigma = \$25$. Based on this distribution,
  a. What proportion of the population spends more than $175 per week on groceries?
  b. What is the probability of randomly selecting a family that spends less than $100 per week on groceries?
  c. How much money do you need to spend on groceries each week to be in the top 20% of the distribution?

21. Laboratory rats commit an average of $\mu = 40$ errors before they solve a standardized maze problem. The distribution of error scores is approximately normal with a standard deviation of $\sigma = 8$. A researcher is testing the effect of a new dietary supplement on intelligence. A newborn rat is selected and is given the supplement daily until it reaches maturity. The rat is then tested on the maze and finishes with a total of $X = 24$ errors.

a. What is the probability that a regular rat (without the supplement) would solve the maze with a score less than or equal to $X = 24$ errors?

b. Is it reasonable to conclude that the rat with the supplement is smarter than the vast majority of regular rats?

c. Does it appear that the supplement has an effect on intelligence? Explain your answer.

22. A multiple-choice test has 48 questions, each with four response choices. If a student is simply guessing at the answers,

a. What is the probability of guessing correctly for any question?

b. On average, how many questions would a student get correct for the entire test?

c. What is the probability that a student would get more than 18 answers correct simply by guessing?

d. What is the probability that a student would get 18 or more answers correct simply by guessing?

23. A true/false test has 50 questions. If a students is simply guessing at the answers,

a. What is the probability of guessing correctly for any question?

b. On average, how many questions would a student get correct for the entire test?

c. What is the probability that a student would get more than 30 answers correct simply by guessing?

d. What is the probability that a student would get 30 or more answers correct simply by guessing?

24. A true-false test has 36 questions. If a passing grade on this test is $X = 24$ (or more) correct, what is the probability of obtaining a passing grade just by guessing?

25. One test for ESP involves using Zener cards. Each card shows one of five different symbols (square, circle, star, cross, wavy lines), and the person being tested has to predict the shape on each card before it is selected. Find each of the probabilities requested for a person who has no ESP and is just guessing.

a. What is the probability of correctly predicting 20 cards in a series of 100 trials?

b. What is the probability of correctly predicting more than 30 cards in a series of 100 trials?

c. What is the probability of correctly predicting 50 or more cards in a series of 200 trials?

26. A trick coin has been weighted so that heads occurs with a probability of $p = \frac{2}{3}$, and $p(\text{tails}) = \frac{1}{3}$. If you toss this coin 72 times:

a. How many heads would you expect to get on average?

b. What is the probability of getting more than 50 heads?

c. What is the probability of getting exactly 50 heads?

27. For a balanced coin:

a. What is the probability of getting more than 30 heads in 50 tosses?

b. What is the probability of getting more than 60 heads in 100 tosses?

c. Parts a and b both asked for the probability of getting more than 60% heads in a series of coin tosses $\left(\frac{30}{50} = \frac{60}{100} = 60\%\right)$. Why do you think the two probabilities are different.

28. A disease affects 10% of the individuals in a population and a sample of 100 people is selected from the population.

a. What is the probability of finding the disease in at least 15 people? (Be careful. "At least 15" means "15 or more.")

b. What is the probability of finding the disease in fewer than 5 people. (Be careful. "Fewer than 5" means "4 or less.")

CHAPTER

7

Probability and Samples: The Distribution of Sample Means

Tools You Will Need

The following items are considered essential background material for this chapter. If you doubt your knowledge of any of these items, you should review the appropriate chapter and section before proceeding.

- Random sampling (Chapter 6)
- Probability and the normal distribution (Chapter 6)
- *z*-Scores (Chapter 5)

Preview

7.1   Samples and Populations
7.2   The Distribution of Sample Means
7.3   Probability and the Distribution of Sample Means
7.4   More about Standard Error
7.5   Looking Ahead to Inferential Statistics

Summary

Focus on Problem Solving

Demonstration 7.1

Problems

Preview

Now that you have some understanding of probability, consider the following problem.

Imagine a jar filled with marbles. Two-thirds of the marbles are one color and the remaining one-third are a second color. One individual selects 5 marbles from the jar and finds that 4 are red and 1 is white. Another individual selects 20 marbles and finds that 12 are red and 8 are white. Which of these two individuals should feel more confident that the jar contains two-thirds red marbles and one-third white marbles, rather than the opposite?*

When Tversky and Kahneman (1974) presented this problem to a group of experimental participants, they found that most people felt that the first sample (4 out of 5) provided much stronger evidence and, therefore, should give more confidence. At first glance, it may appear that this is the correct decision. After all, the first sample contained $\frac{4}{5} = 80\%$ red marbles, and the second sample contained only $\frac{12}{20} = 60\%$ red marbles. However, you should also notice that the two samples differ in another important respect: the sample size. One sample contains only $n = 5$, and the other sample contains $n = 20$. The correct answer to the problem is that the larger sample (12 out of 20) gives a much stronger justification for concluding that the marbles in the jar are predominantly red. It appears that most people tend to focus on the sample proportion and pay very little attention to the sample size.

The importance of sample size may be easier to appreciate if you approach the jar problem from a different perspective. Suppose that you are the individual assigned responsibility for selecting a sample and then deciding which color is in the majority. Before you select your sample, you are offered a choice of selecting either a sample of 5 marbles or a sample of 20 marbles. Which would you prefer? It should be clear that the larger sample would be better. With a small number, you risk obtaining an unrepresentative sample. By chance, you could end up with 3 white marbles and 2 red marbles even though the reds outnumber the whites by 2 to 1. The larger sample is much more likely to provide an accurate representation of the population. This is an example of the *law of large numbers,* which states that large samples will be representative of the population from which they are selected. One final example should help demonstrate this law. If you were tossing a coin, you probably would not be greatly surprised to obtain 3 heads in a row. However, if you obtained a series of 20 heads in a row, you almost certainly would suspect a trick coin. The large sample has more authority.

The Problem: Whether you are picking marbles from a jar or recruiting people to participate in a research study, it usually is possible to obtain thousands or even millions of different samples from the same population. Under these circumstances, how can we determine the probability for obtaining any specific sample? Additionally, although it is intuitively reasonable that sample size is an important factor in determining how accurately a sample represents its population, we need to quantify exactly how sample size influences the probabilities.

The Solution: *The distribution of sample means* describes the entire set of all the possible sample means for any sized sample. Because we can describe the entire set, we can find probabilities associated with specific sample means. (Recall from Chapter 6 that probabilities are equivalent to proportions of the entire distribution.) Also, because the distribution of sample means tends to be normal, it is possible to find probabilities using z-scores and the unit normal table. Although it is impossible to predict exactly which sample will be obtained, the probabilities allow researchers to determine which samples are likely (and which are very unlikely).

---

*Adapted from Tversky, A., and Kahneman, D. (1974). Judgments under uncertainty: Heuristics and biases. *Science, 185,* 1124–1131. Copyright 1974 by the AAAS.

7.1 SAMPLES AND POPULATIONS

The preceding two chapters presented the topics of z-scores and probability. Whenever a score is selected from a population, you should be able to compute a z-score that describes exactly where the score is located in the distribution. And, if the population is normal, you should be able to determine the probability value for obtaining any individual score. In a normal distribution, for example, any score located in the tail of the distribution beyond $z = +2.00$ is an extreme value, and a score this large has a probability of only $p = 0.0228$.

However, the z-scores and probabilities that we have considered so far are limited to situations in which the sample consists of a single score. Most research studies involve much larger samples such as $n = 25$ preschool children or $n = 100$ laboratory rats. In these situations, the sample mean, rather than a single score, is used to answer questions about the population. In this chapter we extend the concepts of z-scores and probability to cover situations with larger samples. In particular, we introduce a procedure for transforming a sample mean into a z-score. Thus, a researcher is able to compute a z-score that describes an entire sample. As always, a z-score value near zero indicates a central, representative sample; a z-value beyond $+2.00$ or -2.00 indicates an extreme sample. Thus, it is possible to describe how any specific sample is related to all the other possible samples. In addition, we can use the z-score values to look up probabilities for obtaining certain samples, no matter how many scores the sample contains.

In general, the difficulty of working with samples is that a sample provides an incomplete picture of the population. Suppose, for example, a researcher selects a sample of $n = 25$ students from the state college. Although the sample should be representative of the entire student population, there will almost certainly be some segments of the population that are not included in the sample. In addition, any statistics that are computed for the sample will not be identical to the corresponding parameters for the entire population. For example, the average IQ for the sample of 25 students will not be the same as the overall mean IQ for the entire population. This difference, or *error* between sample statistics and the corresponding population parameters, is called *sampling error* and was illustrated in Figure 1.2 (page 7).

DEFINITION

Sampling error is the natural discrepancy, or amount of error, between a sample statistic and its corresponding population parameter.

Furthermore, samples are variable; they are not all the same. If you take two separate samples from the same population, the samples will be different. They will contain different individuals, they will have different scores, and they will have different sample means. How can you tell which sample gives the best description of the population? Can you even predict how well a sample will describe its population? What is the probability of selecting a sample with specific characteristics? These questions can be answered once we establish the rules that relate samples and populations.

7.2   THE DISTRIBUTION OF SAMPLE MEANS

As noted, two separate samples probably will be different even though they are taken from the same population. The samples will have different individuals, different scores, different means, and so on. In most cases, it is possible to obtain thousands of different samples from one population. With all these different samples coming from the same population, it may seem hopeless to try to establish some simple rules for the relationships between samples and populations. Fortunately, however, the huge set of possible samples forms a relatively simple and orderly pattern that makes it possible to predict the characteristics of a sample with some accuracy. The ability to predict sample characteristics is based on the *distribution of sample means*.

DEFINITION

The **distribution of sample means** is the collection of sample means for all the possible random samples of a particular size (n) that can be obtained from a population.

Notice that the distribution of sample means contains *all the possible samples*. It is necessary to have all the possible values in order to compute probabilities. For example, if the entire set contains exactly 100 samples, then the probability of obtaining any specific sample is 1 out of 100: $p = \frac{1}{100}$ (Box 7.1).

Also, you should notice that the distribution of sample means is different from distributions we have considered before. Until now we always have discussed distributions of scores; now the values in the distribution are not scores, but statistics (sample means). Because statistics are obtained from samples, a distribution of statistics is referred to as a *sampling distribution.*

DEFINITION

A **sampling distribution** is a distribution of statistics obtained by selecting all the possible samples of a specific size from a population.

Thus, the distribution of sample means is an example of a sampling distribution. In fact, it often is called the sampling distribution of *M*.

If you actually wanted to construct the distribution of sample means, you would first select a random sample of a specific size (*n*) from a population, calculate the sample mean, and place the sample mean in a frequency distribution. Then you select another random sample with the same number of scores. Again, you calculate the sample mean and add it to your distribution. You continue selecting samples and calculating means, over and over, until you have the complete set of all the possible random samples. At this point, your frequency distribution will show the distribution of sample means.

We demonstrate the process of constructing a distribution of sample means in Example 7.1, but first we use common sense and a little logic to predict the general characteristics of the distribution.

1. The sample means should pile up around the population mean. Samples are not expected to be perfect but they are representative of the population. As a result, most of the sample means should be relatively close to the population mean.

BOX 7.1	**PROBABILITY AND THE DISTRIBUTION OF SAMPLE MEANS**

I have a bad habit of losing playing cards. This habit is compounded by the fact that I always save the old deck in the hope that someday I will find the missing cards. As a result, I have a drawer filled with partial decks of playing cards. Suppose that I take one of these almost-complete decks, shuffle the cards carefully, and then randomly select one card. What is the probability that I will draw a king?

You should realize that it is impossible to answer this probability question. To find the probability of selecting a king, you must know how many cards are in the deck and exactly which cards are missing. (It is crucial that you know whether or not any kings are missing.) The point of this simple example is that any probability question requires that you have complete information about the population from which the sample is being selected. In this case, you must know all the possible cards in the deck before you can find the probability for selecting any specific card.

In this chapter, we are examining probability and sample means. To find the probability for any specific sample mean, you first must know *all the possible sample means.* Therefore, we begin by defining and describing the set of all possible sample means that can be obtained from a particular population. Once we have specified the complete set of all possible sample means (i.e., the distribution of sample means), we will be able to find the probability of selecting any specific sample mean.

2. The pile of sample means should tend to form a normal-shaped distribution. Logically, most of the samples should have means close to μ, and it should be relatively rare to find sample means that are substantially different from μ. As a result, the sample means should pile up in the center of the distribution (around μ) and the frequencies should taper off as the distance between M and μ increases. This describes a normal-shaped distribution.

3. In general, the larger the sample size, the closer the sample means should be to the population mean, μ. Logically, a large sample should be a better representative than a small sample. Thus, the sample means obtained with a large sample size should cluster relatively close to the population mean; the means obtained from small samples should be more widely scattered.

As you will see, each of these three commonsense characteristics is an accurate description of the distribution of sample means. The following example demonstrates the process of constructing the distribution of sample means by repeatedly selecting samples from a population.

EXAMPLE 7.1

Consider a population that consists of only 4 scores: 2, 4, 6, 8. This population is pictured in the frequency distribution histogram in Figure 7.1.

We are going to use this population as the basis for constructing the distribution of sample means for $n = 2$. Remember: This distribution is the collection of sample means from all the possible random samples of $n = 2$ from this population. We begin by looking at all the possible samples. For this example, there are 16 different samples, and they are all listed in Table 7.1. Notice that the samples are listed systematically. First, we list all the possible samples with $X = 2$ as the first score, then all the possible samples with $X = 4$ as the first score, and so on. In this way, we are sure that we have all of the possible random samples.

Next, we compute the mean, M, for each of the 16 samples (see the last column of Table 7.1). The 16 means are then placed in a frequency distribution histogram in Figure 7.2. This is the distribution of sample means. Note that the distribution in Figure 7.2 demonstrates two of the characteristics that we predicted for the distribution of sample means.

Remember that random sampling requires sampling with replacement.

1. The sample means pile up around the population mean. For this example, the population mean is $\mu = 5$, and the sample means are clustered around a value of 5. It should not surprise you that the sample means tend to approximate the population mean. After all, samples are supposed to be representative of the population.

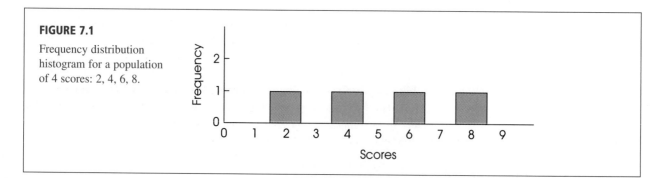

FIGURE 7.1

Frequency distribution histogram for a population of 4 scores: 2, 4, 6, 8.

TABLE 7.1

All the possible samples of $n = 2$ scores that can be obtained from the population presented in Figure 7.1. Notice that the table lists *random samples*. This requires sampling with replacement, so it is possible to select the same score twice.

| Sample | Scores | | Sample Mean |
	First	Second	(M)
1	2	2	2
2	2	4	3
3	2	6	4
4	2	8	5
5	4	2	3
6	4	4	4
7	4	6	5
8	4	8	6
9	6	2	4
10	6	4	5
11	6	6	6
12	6	8	7
13	8	2	5
14	8	4	6
15	8	6	7
16	8	8	8

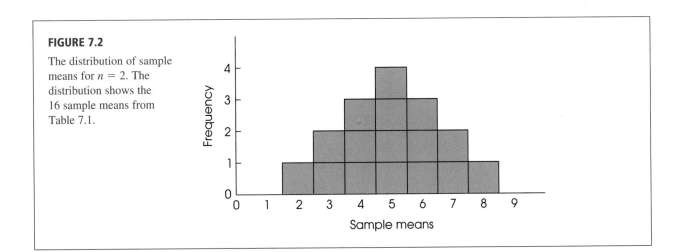

FIGURE 7.2

The distribution of sample means for $n = 2$. The distribution shows the 16 sample means from Table 7.1.

2. The distribution of sample means is approximately normal in shape. This is a characteristic that will be discussed in detail later and will be extremely useful because we already know a great deal about probabilities and the normal distribution (Chapter 6).

Finally, you should notice that we can use the distribution of sample means to answer probability questions about sample means (Box 7.1). For example, if you take a sample of $n = 2$ scores from the original population, what is the probability of obtaining a sample mean greater than 7? In symbols,

$$p(M > 7) = ?$$

Remember that our goal in this chapter is to answer probability questions about samples with $n > 1$.

Because probability is equivalent to proportion, the probability question can be restated as follows: What proportion of all the possible sample means have values greater than 7? In this form, the question is easily answered by looking at the distribution of sample means. All the possible sample means are pictured (see Figure 7.2), and only 1 out of the 16 means has a value greater than 7. The answer, therefore, is 1 out of 16, or $p = \frac{1}{16}$.

THE CENTRAL LIMIT THEOREM

Example 7.1 demonstrates the construction of the distribution of sample means for an overly simplified situation with a very small population and samples that each contain only $n = 2$ scores. In more realistic circumstances, with larger populations and larger samples, the number of possible samples will increase dramatically and it will be virtually impossible to actually obtain every possible random sample. Fortunately, it is possible to determine exactly what the distribution of sample means will look like without taking hundreds or thousands of samples. Specifically, a mathematical proposition known as the *central limit theorem* provides a precise description of the distribution that would be obtained if you selected every possible sample, calculated every sample mean, and constructed the distribution of the sample mean. This important and useful theorem serves as a cornerstone for much of inferential statistics. Following is the essence of the theorem.

> **Central limit theorem:** For any population with mean μ and standard deviation σ, the distribution of sample means for sample size n will have a mean of μ and a standard deviation of $\sigma/\sqrt{n}$ and will approach a normal distribution as n approaches infinity.

The value of this theorem comes from two simple facts. First, it describes the distribution of sample means for *any population,* no matter what shape, mean, or standard deviation. Second, the distribution of sample means "approaches" a normal distribution very rapidly. By the time the sample size reaches $n = 30$, the distribution is almost perfectly normal.

Note that the central limit theorem describes the distribution of sample means by identifying the three basic characteristics that describe any distribution: shape, central tendency, and variability. We will examine each of these.

THE SHAPE OF THE DISTRIBUTION OF SAMPLE MEANS

It has been observed that the distribution of sample means tends to be a normal distribution. In fact, this distribution will be almost perfectly normal if either of the following two conditions is satisfied:

1. The population from which the samples are selected is a normal distribution.
2. The number of scores (n) in each sample is relatively large, around 30 or more.

(As n gets larger, the distribution of sample means will closely approximate a normal distribution. When $n > 30$, the distribution is almost normal regardless of the shape of the original population.)

As we noted earlier, the fact that the distribution of sample means tends to be normal is not surprising. Whenever you take a sample from a population, you expect the sample mean to be near to the population mean. When you take lots of different samples, you expect the sample means to "pile up" around μ, resulting in a normal-shaped distribution. You can see this tendency emerging (although it is not yet normal) in Figure 7.2.

THE MEAN OF THE DISTRIBUTION OF SAMPLE MEANS: THE EXPECTED VALUE OF *M*

In Example 7.1, the distribution of sample means is centered around the mean of the population from which the samples were obtained. In fact, the average value of all the sample means is exactly equal to the value of the population mean. This fact should be intuitively reasonable; the sample means are expected to be close to the population mean, and they do tend to pile up around μ. The formal statement of this phenomenon is that the mean of the distribution of sample means always will be identical to the population mean. This mean value is called the *expected value of* M.

In commonsense terms, a sample mean is "expected" to be near its population mean. When all of the possible sample means are obtained, the average value will be identical to μ.

The fact that the average value of *M* is equal to μ was first introduced in Chapter 4 (page 123) in the context of *biased* versus *unbiased* statistics. The sample mean is an example of an unbiased statistic, which means that on average the sample statistic produces a value that is exactly equal to the corresponding population parameter. In this case, the average value of all the sample means is exactly equal to μ.

DEFINITION

> The mean of the distribution of sample means is equal to the mean of the population of scores, μ, and is called the **expected value of *M***.

THE STANDARD ERROR OF *M*

So far, we have considered the shape and the central tendency of the distribution of sample means. To completely describe this distribution, we need one more characteristic, variability. The value we will be working with is the standard deviation for the distribution of sample means, This standard deviation is identified by the symbol σ_M and is called the *standard error of* M.

When the standard deviation was first introduced in Chapter 4, we noted that this measure of variability serves two general purposes. First, the standard deviation describes the distribution by telling whether the individual scores are clustered close together or scattered over a wide range. Second, the standard deviation measures how well any individual score represents the population by providing a measure of how much distance is reasonable to expect between a score and the population mean. The standard error serves the same two purposes for the distribution of sample means.

1. The standard error describes the distribution of sample means. It provides a measure of how much difference is expected from one sample to another. When the standard error is small, all the sample means are close together and have similar values. If the standard error is large, the sample means are scattered over a wide range and there are big differences from one sample to another.

2. Standard error measures how well an individual sample mean represents the entire distribution. Specifically, it provides a measure of how much distance is reasonable to expect between a sample mean and the overall mean for the distribution of sample means. However, because the overall mean is equal to μ, the standard error also provides a measure of how much distance to expect between a sample mean (*M*) and the mean of the population of scores, μ.

Remember that a sample is not expected to provide a perfectly accurate reflection of its population. Although a sample mean should be representative of the population mean, there typically is some error between the sample and the population. The standard error measures exactly how much difference should be expected on average between a sample mean, M and the population mean, μ.

DEFINITION

The standard deviation of the distribution of sample means, σ_M, is called the **standard error of M**. The standard error provides a measure of how much distance is expected on average between a sample mean (M) and the population mean (μ).

Once again, the symbol for the standard error is σ_M. The σ indicates that this value is a standard deviation, and the subscript M indicates that it is the standard deviation for the distribution of sample means. Similarly, it is common to use the symbol μ_M to represent the mean of the distribution of sample means. However, μ_M is always equal to μ and our primary interest in inferential statistics is to compare sample means (M) with their population means (μ). Therefore, we simply use the symbol μ to refer to the mean of the distribution of sample means.

The standard error is an extremely valuable measure because it specifies precisely how well a sample mean estimates its population mean—that is, how much error you should expect, on the average, between M and μ. Remember that one basic reason for taking samples is to use the sample data to answer questions about the population. However, you do not expect a sample to provide a perfectly accurate picture of the population. There always will be some discrepancy or error between a sample statistic and the corresponding population parameter. Now we are able to calculate exactly how much error to expect. For any sample size (n), we can compute the standard error, which measures the average distance between a sample mean and the population mean.

The magnitude of the standard error is determined by two factors: (1) the size of the sample and (2) the standard deviation of the population from which the sample is selected. We will examine each of these factors.

The sample size   Earlier we predicted, based on commonsense, that the size of a sample should influence how accurately the sample represents its population. Specifically, a large sample should be more accurate than a small sample. In general, as the sample size increases, the error between the sample mean and the population mean should decrease. This rule is also known as the *law of large numbers*.

DEFINITION

The **law of large numbers** states that the larger the sample size (n), the more probable it is that the sample mean will be close to the population mean.

The population standard deviation   As we noted earlier, there is an inverse relationship between the sample size and the standard error: bigger samples have smaller error, and smaller samples have bigger error. At the extreme, the smallest possible sample (and the largest standard error) occurs when the sample consists of $n = 1$ score. At this extreme, the sample is a single score, X, and the sample mean is simply the value of X. In this case, the standard error is measuring the standard distance between a score X and the population mean μ. However, the standard distance between X and μ is the standard deviation, σ. Thus, when $n = 1$, the standard error and the standard deviation are identical.

When $n = 1$, standard error $= \sigma_M = \sigma =$ standard deviation

You can think of the standard deviation as the "starting point" for standard error. When $n = 1$, the standard error and the standard deviation are the same: $\sigma_M = \sigma$. As sample size increases beyond $n = 1$, the sample becomes a more accurate representative of the population, and the standard error decreases. The formula for standard error expresses this relationship between standard deviation and sample size (n).

$$\text{standard error} = \sigma_M = \frac{\sigma}{\sqrt{n}} \qquad (7.1)$$

Note that the formula satisfies all of the requirements for the standard error. Specifically,

   a. As sample size (n) increases, the size of the standard error decreases. (Larger samples are more accurate.)
   b. When the sample consists of a single score ($n = 1$), the standard error is the same as the standard deviation ($\sigma_M = \sigma$).

In Equation 7.1 and in most of the preceding discussion, we have defined standard error in terms of the population standard deviation. However, the population standard deviation (σ) and the population variance (σ^2) are directly related, and it is easy to substitute variance into the equation for standard error. Using the simple equality $\sigma = \sqrt{\sigma^2}$, the equation for standard error can be rewritten as follows:

$$\text{standard error} = \sigma_M = \frac{\sigma}{\sqrt{n}} = \frac{\sqrt{\sigma^2}}{\sqrt{n}} = \sqrt{\frac{\sigma^2}{n}} \qquad (7.2)$$

Throughout the rest of this chapter (and in Chapter 8), we will continue to define standard error in terms of the standard deviation (Equation 7.1). However, in later chapters (starting in Chapter 9) the formula based on variance (Equation 7.2) will become more useful.

Figure 7.3 illustrates the general relationship between standard error and sample size. (The calculations for the data points in Figure 7.3 are presented in Table 7.2.)

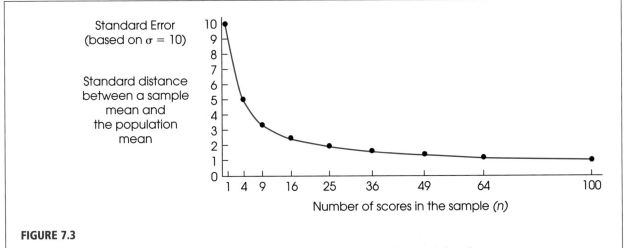

FIGURE 7.3

The relationship between standard error and sample size. As the sample size is increased, there is less error between the sample mean and the population mean.

TABLE 7.2

Calculations for the points shown in Figure 7.3. The size of the standard error decreases as the size of the sample increases.

Sample size (n)	Standard error	
1	$\sigma_M = 10 \sqrt{1}$	$= 10.00$
4	$\sigma_M = 10 \sqrt{4}$	$= 5.00$
9	$\sigma_M = 10 \sqrt{9}$	$= 3.33$
16	$\sigma_M = 10 \sqrt{16}$	$= 2.50$
25	$\sigma_M = 10 \sqrt{25}$	$= 2.00$
49	$\sigma_M = 10 \sqrt{49}$	$= 1.43$
64	$\sigma_M = 10 \sqrt{64}$	$= 1.25$

Again, the basic concept is that the larger a sample is, the more accurately it represents its population. Also note that the standard error decreases in relation to the *square root* of the sample size. As a result, researchers can substantially reduce error by increasing sample size up to around $n = 30$. However, increasing sample size beyond $n = 30$ does not produce much additional improvement in how well the sample represents the population.

THREE DIFFERENT DISTRIBUTIONS

Before we move forward with our discussion of the distribution of sample means, we will pause for a moment to emphasize the idea that we are now dealing with three different but interrelated distributions.

1. First, we have the original population of scores. This population contains the scores for thousands or millions of individual people, and it has its own shape, mean, and standard deviation. For example, the population of IQ scores consists of millions of individual IQ scores that form a normal distribution with a mean of $\mu = 100$ and a standard deviation of $\sigma = 15$. An example of a population is shown in Figure 7.4(a).

2. Next, we have a sample that is selected from the population. The sample consists of a small set of scores for a few people who have been selected to represent the entire population. For example, we could select a sample of $n = 25$ people, measure each individual's IQ score. The 25 scores could be organized in a frequency distribution and we could calculate the sample mean and the sample standard deviation. Note that the sample also has its own shape, mean, and standard deviation. An example of a sample is shown in Figure 7.4(b).

3. The third distribution is the distribution of sample means. This is a theoretical distribution consisting of the sample means obtained from all the possible random samples of a specific size. For example, the distribution of sample means for samples of $n = 25$ IQ scores would be normal with a mean (expected value) of $\mu = 100$ and a standard deviation (standard error) of $\sigma_M = 15\sqrt{25} = 3$. This distribution, shown in Figure 7.4(c), also has its own shape, mean, and standard deviation.

Note that the scores for the sample were taken from the original population and that the mean for the sample is one of the values contained in the distribution of sample means. Thus, the three distribution are all connected, but they are all distinct.

FIGURE 7.4

The distribution. Part (a) shows the population of IQ scores. Part (b) shows a sample of $n = 25$ IQ scores. Part (c) shows the distribution of sample means for samples of $n = 25$ scores. Note that the sample mean from part (b) is one of the thousands of sample means in the part (c) distribution.

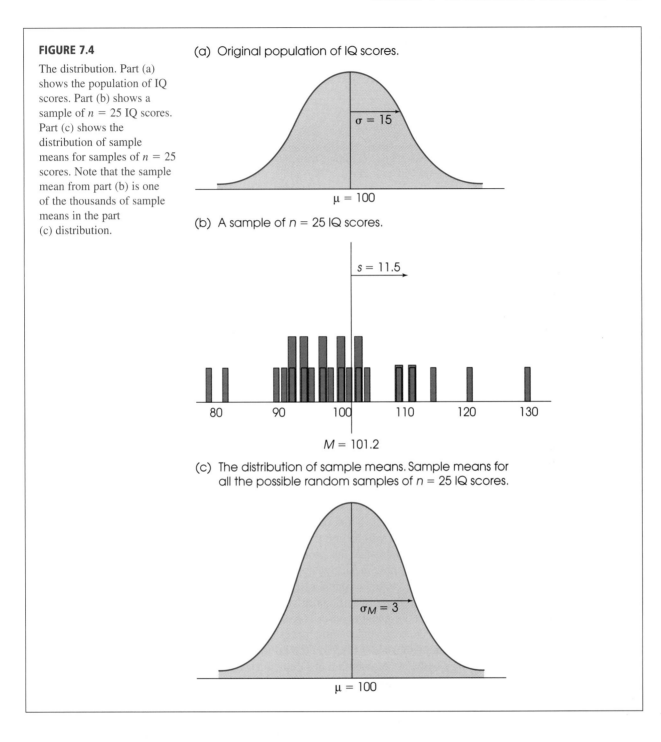

(a) Original population of IQ scores.

(b) A sample of $n = 25$ IQ scores.

(c) The distribution of sample means. Sample means for all the possible random samples of $n = 25$ IQ scores.

LEARNING CHECK

1. A population of scores is normal with $\mu = 80$ and $\sigma = 20$.

   a. Describe the distribution of sample means for samples of size $n = 16$ selected from this population. (Describe shape, central tendency, and variability for the distribution.)

b. How would the distribution of sample means be changed if the sample size were $n = 100$ instead of $n = 16$?

2. As sample size increases, the value of the standard error also increases. (True or false?)

3. Under what circumstances will the distribution of sample means be a normal-shaped distribution?

ANSWERS
1. **a.** The distribution of sample means is normal with a mean (expected value) of $\mu = 80$ and a standard error of $\sigma_M = 20/\sqrt{16} = 5$.

   b. The distribution would still be normal with a mean of 80 but the standard error would be reduced to $\sigma_M = 20/\sqrt{100} = 2$.

2. False. Standard error decreases as n increases.

3. The distribution of sample means will be normal if the sample size is relatively large (around $n = 30$ or more) or if the population of scores is normal.

7.3	PROBABILITY AND THE DISTRIBUTION OF SAMPLE MEANS

The primary use of the distribution of sample means is to find the probability associated with any specific sample. Recall that probability is equivalent to proportion. Because the distribution of sample means presents the entire set of all possible Ms, we can use proportions of this distribution to determine probabilities. The following example demonstrates this process.

EXAMPLE 7.2

The population of scores on the SAT forms a normal distribution with $\mu = 500$ and $\sigma = 100$. If you take a random sample of $n = 25$ students, what is the probability that the sample mean will be greater than $M = 540$?

Caution: Whenever you have a probability question about a sample mean, you must use the distribution of sample means.

First, you can restate this probability question as a proportion question: Out of all the possible sample means, what proportion have values greater than 540? You know about "all the possible sample means"; this is the distribution of sample means. The problem is to find a specific portion of this distribution.

Although we cannot construct the distribution of sample means by repeatedly taking samples and calculating means (as in Example 7.1), we know exactly what the distribution looks like based on the information from the central limit theorem. Specifically, the distribution of sample means has the following characteristics:

   a. The distribution is normal because the population of SAT scores is normal.

   b. The distribution has a mean of 500 because the population mean is $\mu = 500$.

   c. For $n = 25$, the distribution has a standard error of $\sigma_M = 20$:

$$\sigma_M = \frac{\sigma}{\sqrt{n}} = \frac{100}{\sqrt{25}} = \frac{100}{5} = 20$$

This distribution of sample means is shown in Figure 7.5.

We are interested in sample means greater than 540 (the shaded area in Figure 7.5), so the next step is to use a z-score to locate the exact position of

FIGURE 7.5

The distribution of sample means for $n = 25$. Samples were selected from a normal population with $\mu = 500$ and $\sigma = 100$.

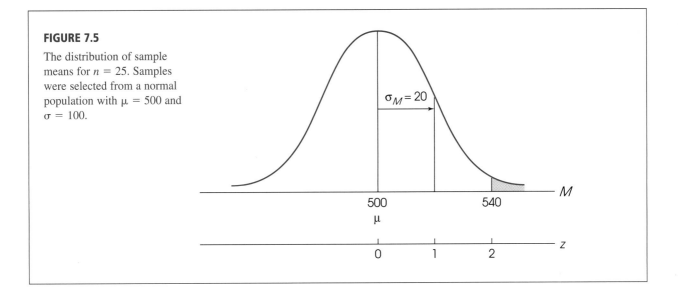

$M = 540$ in the distribution. The value 540 is located above the mean by 40 points, which is exactly 2 standard deviations (in this case, exactly 2 standard errors). Thus, the z-score for $M = 540$ is $z = +2.00$.

Because this distribution of sample means is normal, you can use the unit normal table to find the probability associated with $z = +2.00$. The table indicates that 0.0228 of the distribution is located in the tail of the distribution beyond $z = +2.00$. Our conclusion is that it is very unlikely, $p = 0.0228$ (2.28%), to obtain a random sample of $n = 25$ students with an average SAT score greater than 540.

A z-SCORE FOR SAMPLE MEANS

As demonstrated in Example 7.2, it is possible to use a z-score to describe the exact location of any specific sample mean within the distribution of sample means. The z-score tells exactly where the sample mean is located in relation to all the other possible sample means that could have been obtained. As defined in Chapter 5, a z-score identifies the location with a signed number so that

1. The sign tells whether the location is above ($+$) or below ($-$) the mean.

2. The number tells the distance between the location and the mean in terms of the number of standard deviations.

However, we are now finding a location within the distribution of sample means. Therefore, we must use the notation and terminology appropriate for this distribution. The result will produce some changes in the appearance of the z-score formula. First, we are finding the location for a sample mean (M) rather than a score (X). Second, the standard deviation for the distribution of sample means is the standard error, σ_M. With these changes, the z-score formula for locating a sample mean is

$$z = \frac{M - \mu}{\sigma_M} \tag{7.3}$$

Caution: When computing z for a single score, use the standard deviation, σ. When computing z for a sample mean, you must use the standard error, σ_M. (see Box 7.2).

Every sample mean has a z-score that describes its position in the distribution of sample means. Using z-scores and the unit normal table, it is possible to find the probability associated with any specific sample mean (as in Example 7.2). Example 7.3 demonstrates that it also is possible to make quantitative predictions about the kinds of samples that should be obtained from any population.

EXAMPLE 7.3

Once again, the distribution of SAT scores forms a normal distribution with a mean of $\mu = 500$ and a standard deviation of $\sigma = 100$. For this example, we are going to determine what kind of sample mean is likely to be obtained as the average SAT score for a random sample of $n = 25$ students. Specifically, we will determine the exact range of values that is expected for the sample mean 80% of the time.

We begin with the distribution of sample means for $n = 25$. As demonstrated in Example 7.2, this distribution is normal with an expected value of $\mu = 500$ and a standard error of $\sigma_M = 20$ (Figure 7.6). Our goal is to find the range of values that make up the middle 80% of the distribution. Because the distribution is normal we can use the unit normal table. First, the 80% in the middle is split in half, with 40% (0.4000) on each side of the mean. Looking up 0.4000 in column D (the proportion between the mean and z), we find a corresponding z-score of $z = 1.28$. Thus, the z-score boundaries for the middle 80% are $z = +1.28$ and $z = -1.28$. By definition, a z-score of 1.28 represents a location that is 1.28 standard deviations (or standard errors) from the mean. With a standard error of 20 points, the distance from the mean is $1.28(20) = 25.6$ points. The mean is $\mu = 500$, so a distance of 25.6 in both directions produces a range of values from 474.4 to 525.6.

BOX 7.2   **THE DIFFERENCE BETWEEN STANDARD DEVIATION AND STANDARD ERROR**

A constant source of confusion for many students is the difference between standard deviation and standard error. Remember that standard deviation measures the standard distance between a *score* and the population mean, $X - \mu$. Whenever you are working with a distribution of scores, the standard deviation is the appropriate measure of variability. Standard error, on the other hand, measures the standard distance between a *sample mean* and the population mean, $M - \mu$. Whenever you have a question concerning a sample, the standard error is the appropriate measure of variability.

If you still find the distinction confusing, there is a simple solution. Namely, if you always use standard error, you always will be right. Consider the formula for standard error:

$$\text{standard error} = \sigma_M = \frac{\sigma}{\sqrt{n}}$$

If you are working with a single score, then $n = 1$, and the standard error becomes

$$\text{standard error} = \sigma_M = \frac{\sigma}{\sqrt{n}} = \frac{\sigma}{\sqrt{1}}$$

$$= \sigma = \text{standard deviation}$$

Thus, standard error always measures the standard distance from the population mean, whether you have a sample of $n = 1$ or $n = 100$.

FIGURE 7.6

The middle 80% of the distribution of sample means for $n = 25$. Sample were selected from a normal population with $\mu = 500$ and $\sigma = 100$.

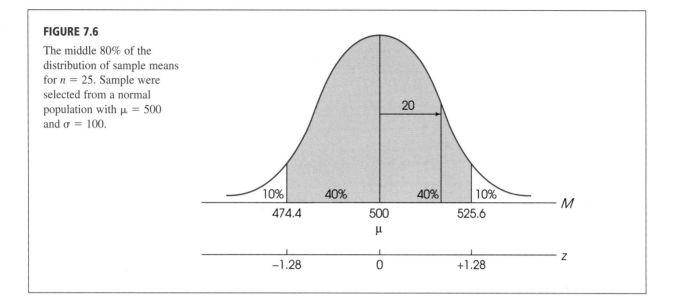

Thus, 80% of all the possible sample means are contained in a range between 474.4 and 525.6. If we select a sample of $n = 25$ students, we can be 80% confident that the mean SAT score for the sample will be in this range.

The point of Example 7.3 is that the distribution of sample means makes it possible to predict the value that ought to be obtained for a sample mean. We know, for example, that a sample of $n = 25$ students ought to have a mean SAT score around 500. More specifically, we are 80% confident that the value of the sample mean will be between 474.4 and 525.6. The ability to predict sample means in this way will be a valuable tool for the inferential statistics that follow.

LEARNING CHECK

1. A population has $\mu = 80$ and $\sigma = 12$. Find the z-score corresponding to each of the following sample means:
   a. $M = 84$ for a sample of $n = 9$ scores
   b. $M = 74$ for a sample of $n = 16$ scores
   c. $M = 81$ for a sample of $n = 36$ scores

2. A random sample of $n = 25$ scores is selected from a normal population with $\mu = 90$ and $\sigma = 10$.
   a. What is the probability that the sample mean will have a value greater than 94?
   b. What is the probability that the sample mean will have a value less than 91?

3. A positively skewed distribution has $\mu = 60$ and $\sigma = 8$.
   a. What is the probability of obtaining a sample mean greater than $M = 62$ for a sample of $n = 4$ scores? (Be careful. This is a trick question.)

b. What is the probability of obtaining a sample mean greater than $M = 62$ for a sample of $n = 64$ scores?

ANSWERS **1. a.** The standard error is $12/\sqrt{9} = 4$. The z-score is $z = +1.00$.

b. The standard error is $12/\sqrt{16} = 3$. The z-score is $z = -2.00$.

c. The standard error is $12/\sqrt{36} = 2$. The z-score is $z = +0.50$.

2. a. The standard error is $10/\sqrt{25} = 2$. A mean of $M = 94$ corresponds to $z = +2.00$, and $p = 0.0228$.

b. A mean of $M = 91$ corresponds to $z = +0.50$, and $p = 0.6915$.

3. a. The distribution of sample means does not satisfy either of the criteria for being normal. Therefore, you cannot use the unit normal table, and it is impossible to find the probability.

b. With $n = 64$, the distribution of sample means is nearly normal. The standard error is $8/\sqrt{64} = 1$, the z-score is $+2.00$, and the probability is 0.0228.

7.4   MORE ABOUT STANDARD ERROR

At the beginning of this chapter, we introduced the idea that it is possible to obtain thousands of different samples from a single population. Each sample will have its own individuals, its own scores, and its own sample mean. The distribution of sample means provides a method for organizing all of the different sample means into a single picture that shows how the sample means are related to each other and how they are related to the overall population mean. Figure 7.7 shows a prototypical distribution of sample means. To emphasize the fact that the distribution contains many different samples, we have constructed this figure so that the distribution is made up of hundreds of small

FIGURE 7.7

An example of a typical distribution of sample means. Each of the small boxes represents the mean obtained for one sample.

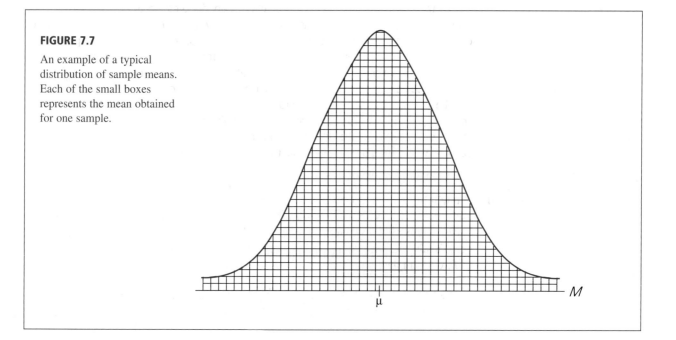

boxes, each box representing a single sample mean. Also notice that the sample means tend to pile up around the population mean (μ), forming a normal-shaped distribution as predicted by the central limit theorem.

The distribution shown in Figure 7.7 provides a concrete example for reviewing the general concepts of sampling error and standard error. Although the following points may seem obvious, they are intended to provide you with a better understanding of these two statistical concepts.

1. Sampling Error.   The general concept of sampling error is that a sample typically will not provide a perfectly accurate representation of its population. More specifically, there typically will be some discrepancy (or error) between a statistic computed for a sample and the corresponding parameter for the population. As you look at Figure 7.7, notice that the individual sample means are not exactly equal to the population mean. In fact, 50% of the samples have means that are smaller than μ (the entire left-hand side of the distribution). Similarly, 50% of the samples produce means that overestimate the true population mean. In general, there will be some discrepancy, or *sampling error*, between the mean for a sample and the mean for the population from which the sample was obtained.

2. Standard Error.   Again, looking at Figure 7.7, notice that most of the sample means are relatively close to the population mean (those in the center of the distribution). These samples provide a fairly accurate representation of the population. On the other hand, some samples will produce means that are out in the tails of the distribution, relatively far way from the population mean. These extreme sample means do not accurately represent the population. For each individual sample, you can measure the error (or distance) between the sample mean and the population mean. For some samples, the error will be relatively small, but for other samples, the error will be relatively large. The *standard error* provides a way to measure the "average", or standard, distance between a sample mean and the population mean.

Thus, the standard error provides a method for defining and measuring sampling error. Knowing the standard error gives researchers a good indication of how accurately their sample data represent the populations they are studying. In most research situations, for example, the population mean is unknown, and the researcher selects a sample to help obtain information about the unknown population. Specifically, the sample mean provides information about the value of the unknown population mean. The sample mean is not expected to give a perfectly accurate representation of the population mean; there will be some error, and the standard error tells *exactly how much error*, on average, should exist between the sample mean and the unknown population mean. The following example demonstrates the use of the standard error and provides additional information about the relationship between standard error and standard deviation.

EXAMPLE 7.4   We will begin with a population that is normally distributed with a mean of $\mu = 80$ and a standard deviation of $\sigma = 20$. Next, we will take a sample from this population and examine how accurately the sample mean represents the population mean. More specifically, we will examine how sample size affects accuracy by considering three different samples: one with $n = 1$ score, one with $n = 4$ scores, and one with $n = 100$ scores.

Figure 7.8 shows the distributions of sample means based on samples of $n = 1$, $n = 4$, and $n = 100$. Each distribution shows the collection of all possible sample means that could be obtained for that particular sample size. Notice that all three

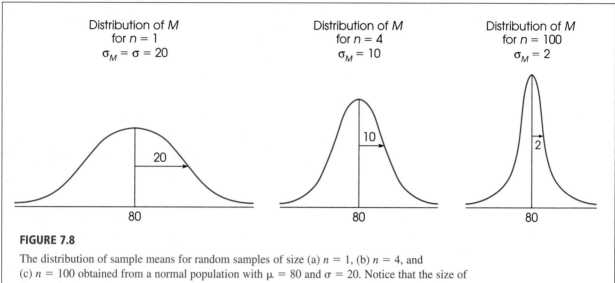

FIGURE 7.8

The distribution of sample means for random samples of size (a) $n = 1$, (b) $n = 4$, and (c) $n = 100$ obtained from a normal population with $\mu = 80$ and $\sigma = 20$. Notice that the size of the standard error decreases as the sample size increases.

sampling distributions are normal (because the original population is normal), and all three have the same mean, $\mu = 80$, which is the expected value of M. However, the three distributions differ greatly with respect to variability. We will consider each one separately.

The smallest sample size is $n = 1$. When a sample consists of a single score, the mean for the sample will equal the value of that score, $M = X$. Thus, when $n = 1$, the distribution of sample means is identical to the original population of scores. In this case, the standard error for the distribution of sample means is equal to the standard deviation for the original population. Equation 7.1 confirms this observation.

$$\sigma_M = \frac{\sigma}{\sqrt{n}} = \frac{20}{\sqrt{1}} = 20$$

As we noted earlier, the population standard deviation is the "starting point" for the standard error. With the smallest possible sample, $n = 1$, the standard error is equal to the standard deviation [see Figure 7.8(a)].

As the sample size increases, however, the standard error gets smaller. For $n = 4$, the standard error is

$$\sigma_M = \frac{\sigma}{\sqrt{n}} = \frac{20}{\sqrt{4}} = \frac{20}{2} = 10$$

That is, the typical (or standard) distance between M and μ is 10 points. Figure 7.8(b) illustrates this distribution. Notice that the sample means in this distribution approximate the population mean more closely than in the previous distribution where $n = 1$.

With a sample of $n = 100$, the standard error is still smaller.

$$\sigma_M = \frac{\sigma}{\sqrt{n}} = \frac{20}{\sqrt{100}} = \frac{20}{10} = 2$$

A sample of $n = 100$ scores should provide a sample mean that is a much better estimate of μ than you would obtain with a sample of $n = 4$ or $n = 1$. As shown in Figure 7.8(c), there is very little error between M and μ (you would expect only a 2-point error, on average). The sample means pile up very close to μ.

In summary, this example illustrates that with the smallest possible sample ($n = 1$), the standard error and the population standard deviation are the same. When sample size is increased, the standard error gets smaller, and the sample means tend to approximate μ more closely. Thus, standard error defines the relationship between sample size and the accuracy with which M represents μ.

IN THE LITERATURE
REPORTING STANDARD ERROR

As we will see later, standard error plays a very important role in inferential statistics. Because of its crucial role, the standard error for a sample mean, rather than the sample standard deviation, is often reported in scientific papers. Scientific journals vary in how they refer to the standard error, but frequently the symbols *SE* and *SEM* (for standard error of the mean) are used. The standard error is reported in two ways. Much like the standard deviation, it may be reported in a table along with the sample means (Table 7.3). Alternatively, the standard error may be reported in graphs.

Figure 7.9 illustrates the use of a bar graph to display information about the sample mean and the standard error. In this experiment, two samples (groups A and B) are given different treatments, and then the subjects' scores on a dependent variable are recorded. The mean for group A is $M = 15$, and for group B, it is $M = 30$. For both samples, the standard error of M is $\sigma_M = 4$. Note that the mean is represented by the height of the bar, and the standard error is depicted on the graph by brackets at the top of each bar. Each bracket extends 1 standard error above and 1 standard error below the sample mean. Thus, the graph illustrates the mean for each group plus or minus 1 standard error ($M \pm SE$). When you glance at Figure 7.9, not only do you get a "picture" of the sample means, but also you get an idea of how much error you should expect for those means.

Figure 7.10 shows how sample means and standard error are displayed in a line graph. In this study, two samples representing different age groups are tested on a task for four trials. The number of errors committed on each trial is recorded for all participants. The graph shows the mean (M) number of errors committed for each group on each trial. The brackets show the size of the standard error for each sample mean. Again, the brackets extend 1 standard error above and below the value of the mean. ❏

TABLE 7.3

The mean self-consciousness scores for participants who were working in front of a video camera and those who were not (controls)

	n	Mean	*SE*
Control	17	32.23	2.31
Camera	15	45.17	2.78

FIGURE 7.9

The mean ($\pm SE$) score for treatment groups A and B.

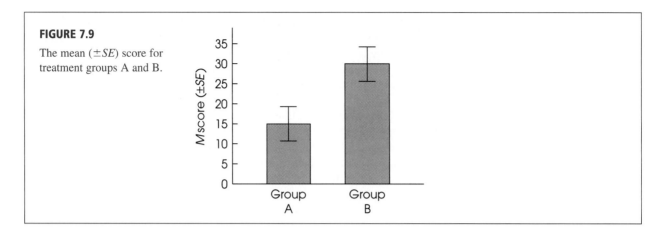

FIGURE 7.10

The mean ($\pm SE$) number of mistakes made for groups A and B on each trial.

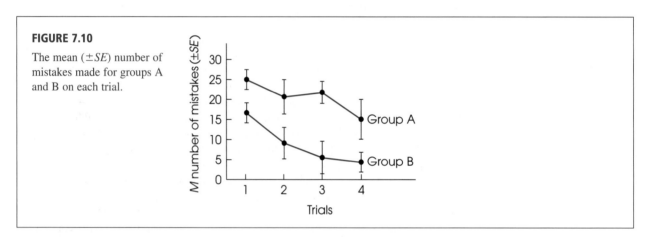

LEARNING CHECK

1. A population has a standard deviation of $\sigma = 20$.

   a. If a single score is selected from this population, how close, on average, would you expect the score to be to the population mean?

   b. If a sample of $n = 4$ scores is selected from the population, how close, on average, would you expect the sample mean to be to the population mean?

   c. If a sample of $n = 25$ scores is selected from the population, how close, on average, would you expect the sample mean to be to the population mean?

2. Can the magnitude of the standard error ever be larger than the population standard deviation? Explain your answer.

3. A researcher plans to select a random sample of $n = 36$ individuals from a population with a standard deviation of $\sigma = 18$. On average, how much error should the researcher expect between the sample mean and the population mean?

4. A researcher plans to select a random sample from a population with a standard deviation of $\sigma = 20$.

   a. How large a sample is needed to have a standard error of 10 points or less?

   b. How large a sample is needed if the researcher wants the standard error to be 2 points or less?

ANSWERS   **1. a.** The standard deviation, $\sigma = 20$, measures the standard distance between a score and the mean.

      b. The standard error is $20/\sqrt{4} = 10$ points.

      c. The standard error is $20/\sqrt{25} = 4$ points.

2. No. The standard error is always less than or equal to the standard deviation. The two values are equal only when $n = 1$.

3. The standard error is $18/\sqrt{36} = 3$ points.

4. a. A sample of $n = 4$ or larger.

      b. A sample of $n = 100$ or larger.

7.5   LOOKING AHEAD TO INFERENTIAL STATISTICS

Inferential statistics are methods that use sample data as the basis for drawing general conclusions about populations. However, we have noted that a sample is not expected to give a perfectly accurate reflection of its population. In particular, there will be some error or discrepancy between a sample statistic and the corresponding population parameter. In this chapter, we have observed that a sample mean will not be exactly equal to the population mean. The standard error of M specifies how much difference is expected on average between the mean for a sample and the mean for the population.

The natural differences that exist between samples and populations introduce a degree of uncertainty and error into all inferential processes. For example, there is always a margin of error that must be considered whenever a researcher uses a sample mean as the basis for drawing a conclusion about a population mean. Remember that the sample mean is not perfect. In the next seven chapters we will introduce a variety of statistical methods that all use sample means to draw inferences about population means.

In each case, the distribution of sample means and the standard error will be critical elements in the inferential process. Before we begin this series of chapters, we pause briefly to demonstrate how the distribution of sample means, along with z-scores and probability, can help us use sample means to draw inferences about population means.

EXAMPLE 7.5   Suppose that a psychologist is planning a research study to evaluate the effect of a new growth hormone. It is known that regular, adult rats (with no hormone) weigh an average of $\mu = 400$ grams. Of course, not all rats are the same size, and the distribution of their weights is normal with $\sigma = 20$. The psychologist plans to select a sample of $n = 25$ newborn rats, inject them with the hormone, and then measure their weights when they become adults. The structure of this research study is shown in Figure 7.11.

The psychologist will make a decision about the effect of the hormone by comparing the sample of treated rats with the regular untreated rats in the original population. If the treated rats in the sample are noticeably different from untreated rats, then the researcher has evidence that the hormone has an effect. The problem is to determine exactly how much difference is necessary before we can say that the sample is *noticeably different*.

FIGURE 7.11

The structure of the research study described in Example 7.5. The purpose of the study is to determine whether the treatment (a growth hormone) has an effect on weight for rats.

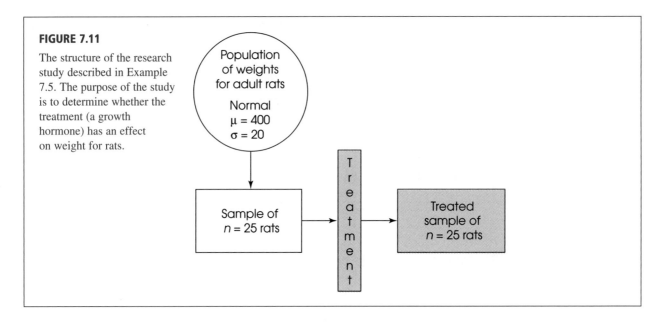

The distribution of sample means and the standard error can help researchers make this decision. In particular, the distribution of sample means can be used to show exactly what would be expected for a sample of rats who do not receive any hormone injections. This allows researchers to make a simple comparison between

a. The sample of treated rats (from the research study)

b. Samples of untreated rats (from the distribution of sample means)

If our treated sample is noticeably different from the untreated samples, then we will have evidence that the treatment has an effect. On the other hand, if our treated sample still looks like one of the untreated samples, then we must conclude that the treatment does not appear to have any effect.

We begin with the original population of untreated rats and consider the distribution of sample means for all the possible samples of $n = 25$ rats. The distribution of sample means has the following characteristics:

1. It will be a normal distribution, because the population of rat weights is normal.

2. It will have an expected value of 400, because the population mean for untreated rat is $\mu = 400$.

3. It will have a standard error of $\sigma_M = 20/\sqrt{25} = 20/5 = 4$, because the population standard deviation is $\sigma = 20$ and the sample size is $n = 25$.

The distribution of sample means is shown in Figure 7.12. Notice that a sample of $n = 25$ untreated rats (without the hormone) should have a mean weight around 400 grams. To be more precise, we can use z-scores to determine the middle 95% of all the possible sample means. As demonstrated in Chapter 6 (page 190), the middle 95% of a normal distribution is located between z-score boundaries of $z = +1.96$ and $z = -1.96$ (check the unit normal table). These z-score boundaries are shown in Figure 7.12. With a standard error of $\sigma = 4$ points, a z-score of $z = 1.96$ corresponds to a distance of $1.96(4) = 7.84$ points from the mean. Thus, the z-score boundaries of ± 1.96 correspond to sample means of 392.16 and 407.84.

Thus, a sample of untreated rats is almost guaranteed (95% probability) to have a sample mean between 392.16 and 407.84. If our sample has a mean within this range,

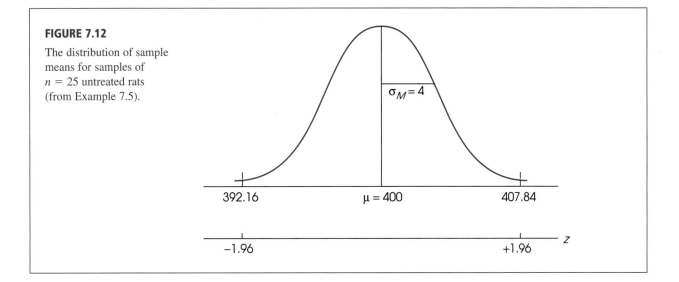

FIGURE 7.12

The distribution of sample means for samples of $n = 25$ untreated rats (from Example 7.5).

$\sigma_M = 4$

392.16 $\mu = 400$ 407.84

-1.96 $+1.96$ z

then we must conclude that our sample of treated rats does not look any different from a sample of untreated rats. In this case, we conclude that the hormone does not appear to have any effect.

On the other hand, if the mean for the treated sample is outside the 95% range, then we can conclude that our sample of treated rats is noticeably different from the samples that would be obtained for $n = 25$ rats without any treatment. In this case, the research results provide evidence that the treatment has an effect.

In Example 7.5 we used the distribution of sample means, together with z-scores and probability, to provide a description of what is reasonable to expect for an untreated sample. Then, we evaluated the effect of a treatment by determining whether the treated sample was noticeably different from an untreated sample. This procedure forms the foundation for the inferential technique known as *hypothesis testing* that is introduced in Chapter 8.

STANDARD ERROR AS A MEASURE OF RELIABILITY

The research situation shown in Figure 7.11 introduces one final issue concerning sample means and standard error. In Figure 7.11, as in most research studies, the researcher must rely on a *single* sample to provide an accurate representation of the population being investigated. As we have noted, however, if you take two different samples from the same population, you will get different individuals with different scores and different sample means. Thus, every researcher must face the nagging question, "If I had taken a different sample, would I have obtained different results?"

The importance of this question is directly related to the degree of similarity among all the different samples. For example, if there is a high level of consistency from one sample to another, then a researcher can be reasonably confident that the specific sample being studied will provide a good measurement of the population. That is, when all the samples are similar, then it does not matter which one you have selected. On the other hand, if there are big differences from one sample to another, then the researcher is left with some doubts about the accuracy of his/her specific sample. In this case, a different sample could have produced vastly different results.

In this context, the standard error can be viewed as a measure of the *reliability* of a sample mean. The term *reliability* refers to the consistency of different measurements of the same thing. More specifically, a measurement procedure is said to be reliable if you make two different measurements of the same thing and obtain identical (or nearly identical) values. If you view a sample as a "measurement" of a population, then a sample mean is a "measurement" of a population mean.

If the standard error is small, all the possible sample means are clustered close together and a researcher can be confident than any individual sample mean will provide a reliable measure of the population. On the other hand, a large standard error indicates that there are relatively large differences from one sample mean to another, and a researcher must be concerned that a different sample could produce a different conclusion. Fortunately, the size of the standard error can be controlled. In particular, if a researcher is concerned about a large standard error and the potential for big differences from one sample to another, the researcher has the option of reducing the standard error by selecting a larger sample. Thus, the ability to compute the value of the standard error provides researchers the ability to control the reliability of their samples.

Earlier, in Figure 7.8, we showed how the size of the standard error is related to sample size. The same figure can now be used to demonstrate how standard error provides a measure of reliability for a sample mean.

Look back at Figure 7.8(b), and see that it shows a distribution of sample means based on $n = 4$ scores. In this distribution, the standard error is $\sigma_M = 10$, and it is relatively easy to find two samples with means that differ by 10 or 20 points. In this situation, a researcher cannot be confident that the mean for any individual sample is reliable; that is, a different sample could provide a very different mean.

Now consider what happens when the sample size is increased. Figure 7.8(c) shows the distribution of sample means based on $n = 100$. For this distribution, the standard error is $\sigma_M = 2$. With a larger sample and a smaller standard error, all of the sample means are clustered close together. With $n = 100$, a researcher can be much more confident that the mean for any individual sample is reliable; that is, if a second sample were taken, it would probably produce a sample mean that is essentially equivalent to the mean for the first sample.

LEARNING CHECK

1. A population forms a normal distribution with a mean of $\mu = 50$ and a standard deviation of $\sigma = 10$.

   a. A sample of $n = 4$ scores from this population has a mean of $M = 55$. Would you describe this as a relatively typical sample or is the mean an extreme value? Explain your answer.

   b. If the sample from part a had $n = 25$ scores, would it be considered typical or extreme?

2. The scores from a depression questionnaire form a normal distribution with a mean of $\mu = 80$ and a standard deviation of $\sigma = 10$.

   a. If the questionnaire is given to a sample of $n = 25$ people, what range of values for the sample mean would be expected 95% of the time?

   b. What range of values would have a 95% probability if the sample size were $n = 100$?

3. An automobile manufacturer claims that a newly introduced model will average $\mu = 45$ miles per gallon with $\sigma = 2$. A sample of $n = 4$ cars is tested and averages only $M = 42$ miles per gallon. Is this sample mean likely to occur if the manufacturer's claim is true? Specifically, is the sample mean within the range of values that would be expected 95% of the time? (Assume that the distribution of mileage scores is normal.)

ANSWERS   **1. a.** With $n = 4$ the standard error is 5, and the sample mean corresponds to $z = 1.00$. This is a relatively typical value.

   b. With $n = 25$ the standard error is 2, and the sample mean corresponds to $z = 2.50$. This is an extreme value.

2. a. With $n = 25$ the standard error is $\sigma_M = 2$ points. Using $z = 1.96$, the 95% range extends from 76.08 to 83.92.

   b. With $n = 100$ the standard error is only 1 point and the range extends from 78.04 to 81.96.

3. With $n = 4$, the standard error is $\sigma_M = 1$. If the real mean is $\mu = 45$, then 95% of all sample means should be within $1.96(1) = 1.96$ points of $\mu = 45$. This is a range of values from 43.04 to 46.96. Our sample mean is outside this range, so it is not the kind of sample that ought to be obtained if the manufacturer's claim is true.

SUMMARY

1. The distribution of sample means is defined as the set of Ms for all the possible random samples for a specific sample size (n) that can be obtained from a given population. According to the *central limit theorem,* the parameters of the distribution of sample means are as follows:

   a. *Shape.* The distribution of sample means will be normal if either one of the following two conditions is satisfied:

   (1) The population from which the samples are selected is normal.

   (2) The size of the samples is relatively large (around $n = 30$ or more).

   b. *Central Tendency.* The mean of the distribution of sample means will be identical to the mean of the population from which the samples are selected. The mean of the distribution of sample means is called the expected value of M.

   c. *Variability.* The standard deviation of the distribution of sample means is called the standard error of M and is defined by the formula

$$\sigma_M = \frac{\sigma}{\sqrt{n}} \quad \text{or} \quad \sigma_M = \sqrt{\frac{\sigma^2}{n}}$$

Standard error measures the standard distance between a sample mean (M) and the population mean (μ).

2. One of the most important concepts in this chapter is standard error. The standard error is the standard deviation of the distribution of sample means. It measures the standard distance between a sample mean (M) and the population mean (μ). The standard error tells how much error to expect if you are using a sample mean to represent a population mean.

3. The location of each M in the distribution of sample means can be specified by a z-score:

$$z = \frac{M - \mu}{\sigma_M}$$

Because the distribution of sample means tends to be normal, we can use these z-scores and the unit normal table to find probabilities for specific sample means. In particular, we can identify which sample means are likely and which are very unlikely to be obtained from any given population. This ability to find probabilities for samples is the basis for the inferential statistics in the chapters ahead.

4. In general terms, the standard error measures how much discrepancy you should expect, between a sample statistic and a population parameter. Statistical inference involves using sample statistics to make a general conclusion about a population parameter. Thus, standard error plays a crucial role in inferential statistics.

KEY TERMS

sampling error

distribution of sample means

sampling distribution

central limit theorem

expected value of M

standard error of M

law of large numbers

RESOURCES

There are a tutorial quiz and other learning exercises for Chapter 7 at the Wadsworth website, www.cengage.com/psychology/gravetter. The site also provides access to two workshops entitled *Standard Error* and *Central Limit Theorem* that review the material covered in Chapter 7. See page 29 for more information about accessing items on the website.

ENHANCED WebAssign

Guided interactive tutorials, end-of-chapter problems, and related testbank items may be assigned online at WebAssign.

WebTUTOR

For those using WebTutor along with this book, there is a WebTutor section corresponding to this chapter. The WebTutor contains a brief summary of Chapter 7, hints for learning about the distribution of sample means, cautions about common errors, and sample exam items including solutions.

SPSS

The statistical computer package SPSS is not structured to compute the standard error or a z-score for a sample mean. In later chapters, however, we introduce new inferential statistics that are included in SPSS. When these new statistics are computed, SPSS typically includes a report of standard error that describes how accurately, on average, the sample represents its population.

FOCUS ON PROBLEM SOLVING

1. Whenever you are working probability questions about sample means, you must use the distribution of sample means. Remember that every probability question can be restated as a proportion question. Probabilities for sample means are equivalent to proportions of the distribution of sample means.

2. When computing probabilities for sample means, the most common error is to use standard deviation (σ) instead of standard error (σ_M) in the z-score formula. Standard deviation measures the typical deviation (or "error") for a single score. Standard error measures the typical deviation (or error) for a sample. Remember: The larger the sample is, the more accurately the sample represents the population. Thus, sample size (n) is a critical part of the standard error.

$$\text{Standard error} = \sigma_M = \frac{\sigma}{\sqrt{n}}$$

3. Although the distribution of sample means is often normal, it is not always a normal distribution. Check the criteria to be certain the distribution is normal before you use the unit normal table to find probabilities (see item 1a of the Summary). Remember that all probability problems with the normal distribution are easier if you sketch the distribution and shade in the area of interest.

DEMONSTRATION 7.1

PROBABILITY AND THE DISTRIBUTION OF SAMPLE MEANS

For a normally distributed population with $\mu = 60$ and $\sigma = 12$, what is the probability of selecting a random sample of $n = 36$ scores with a sample mean greater than 64?
In symbols, for $n = 36$

$$p(M > 64) = ?$$

Notice that we may rephrase the probability question as a proportion question. Out of all the possible sample means for $n = 36$, what proportion has values greater than 64?

STEP 1   Sketch the distribution.
We are looking for a specific proportion of *all possible sample means*. Therefore, we will have to sketch the distribution of sample means, including the expected value, $\mu = 60$, and the standard error

$$\sigma_M = \frac{\sigma}{\sqrt{n}} = \frac{12}{\sqrt{36}} = \frac{12}{6} = 2$$

The sample mean, $M = 64$, is located in the tail on the right-hand side of the distribution. Figure 7.13(a) shows the preliminary sketch.

STEP 2   Shade the appropriate area of the distribution.
In this example we are looking for the proportion greater than $M = 64$, so we shade the area on the right-hand side of $M = 64$. See Figure 7.13(b).

STEP 3   Compute the z-score for the sample mean.
Using the z-score formula for sample means, including the standard error in the denominator, we obtain

$$z = \frac{M - \mu}{\sigma_M} = \frac{64 - 60}{2} = \frac{4}{2} = 2.00$$

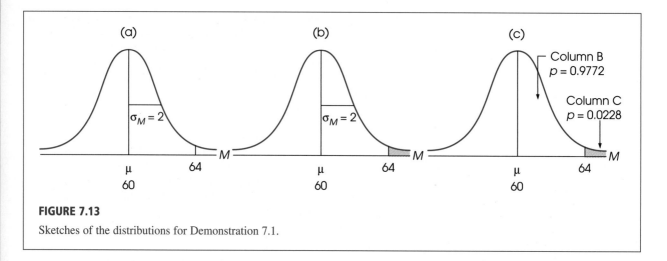

FIGURE 7.13

Sketches of the distributions for Demonstration 7.1.

STEP 4   Consult the unit normal table.

Look up the z-score in the unit normal table, and note the two proportions in columns B and C. Jot them down in the appropriate areas of the distribution [Figure 7.13(c)]. For this demonstration, the value in column C (the tail beyond z) corresponds exactly to the proportion we want (the shaded area). Thus, for $n = 36$,

$$p(M > 64) = p(z > +2.00) = 0.0228$$

PROBLEMS

1. Briefly define each of the following:
   a. Distribution of sample means
   b. Expected value of M
   c. Standard error of M

2. Describe the distribution of sample means (shape, expected value, and standard error) for samples of $n = 36$ selected from a population with a mean of $\mu = 100$ and a standard deviation of $\sigma = 12$.

3. A sample is selected from a population with a mean of $\mu = 80$ and a standard deviation of $\sigma = 20$.
   a. What is the expected value of M and the standard error of M for a sample of $n = 4$ scores?
   b. What is the expected value of M and the standard error of M for a sample of $n = 16$ scores?

4. The distribution of sample means is not always a normal distribution. Under what circumstances will the distribution of sample means *not* be normal?

5. A random sample is obtained from a population with a mean of $\mu = 100$ and a standard deviation of $\sigma = 20$.
   a. How much error would you expect between the sample mean and the population mean for a sample of $n = 4$ scores?

   b. How much error would you expect for a sample of $n = 16$ scores?
   c. How much error would be expected for a sample of $n = 100$ scores?

6. For a population with a mean of $\mu = 50$ and a standard deviation of $\sigma = 10$, how much error, on average, would you expect between the sample mean (M) and the population mean for:
   a. a sample of $n = 4$ scores
   b. a sample of $n = 16$ scores
   c. a sample of $n = 25$ scores

7. For a population with a standard deviation of $\sigma = 10$, how large a sample is necessary to have a standard error that is:
   a. less than or equal to 5 points?
   b. less than or equal to 2 points?
   c. less than or equal to 1 point?

8. If the population standard deviation is $\sigma = 8$, how large a sample is necessary to have a standard error that is
   a. less than 4 points?
   b. less than 2 points?
   c. less than 1 point?

9. For a sample of $n = 16$ scores, what is the value of the population standard deviation (σ) in order to have a standard error of
   a. $\sigma_M = 10$ points?
   b. $\sigma_M = 5$ points?
   c. $\sigma_M = 2$ points?

10. For a population with a mean of $\mu = 60$ and a standard deviation of $\sigma = 24$, find the z-score corresponding to each of the following samples.
   a. $M = 63$ for a sample of $n = 16$ scores
   b. $M = 63$ for a sample of $n = 36$ scores
   c. $M = 63$ for a sample of $n = 64$ scores

11. A sample of $n = 25$ scores has a mean of $M = 84$. Find the z-score for this sample:
   a. If it was obtained from a population with $\mu = 80$ and $\sigma = 10$.
   b. If it was obtained from a population with $\mu = 80$ and $\sigma = 20$.
   c. If it was obtained from a population with $\mu = 80$ and $\sigma = 40$.

12. A random sample is obtained from a population with a mean of $\mu = 50$ and a standard deviation of $\sigma = 12$.
   a. For a sample of $n = 4$ scores, would a sample mean of $M = 55$ be considered an extreme value or is it a fairly typical sample mean?
   b. For a sample of $n = 36$ scores, would a sample mean of $M = 55$ be considered an extreme value or is it a fairly typical sample mean?

13. A normally distributed population has a mean of $\mu = 80$ and a standard deviation of $\sigma = 20$. Calculate the z-score for each of the following samples. Based on the z-score value, determine whether the sample mean is a typical, representative value or an extreme value for samples of this size.
   a. A sample of $n = 4$ scores with a mean of $M = 90$
   b. A sample of $n = 25$ scores with a mean of $M = 84$
   c. A sample of $n = 100$ scores with a mean of $M = 84$

14. The population of IQ scores forms a normal distribution with a mean of $\mu = 100$ and a standard deviation of $\sigma = 15$. What is the probability of obtaining a sample mean greater than $M = 105$,
   a. for a random sample of $n = 9$ people?
   b. for a random sample of $n = 36$ people?

15. A population of scores forms a normal distribution with a mean of $\mu = 75$ and a standard deviation of $\sigma = 20$.
   a. What proportion of the scores in the population have values less than $X = 80$?
   b. If samples of $n = 4$ are selected from the population, what proportion of the samples will have means less than $M = 80$?

   c. If samples of $n = 16$ are selected from the population, what proportion of the samples will have means less than $M = 80$?

16. A population of scores forms a normal distribution with a mean of $\mu = 40$ and a standard deviation of $\sigma = 12$.
   a. What is the probability of randomly selecting a score less than $X = 34$?
   b. What is the probability of selecting a sample of $n = 9$ scores with a mean less than $M = 34$?
   c. What is the probability of selecting a sample of $n = 36$ scores with a mean less than $M = 34$?

17. A population of scores forms a normal distribution with a mean of $\mu = 80$ and a standard deviation of $\sigma = 10$.
   a. What proportion of the scores have values between 75 and 85?
   b. For samples of $n = 4$, what proportion of the samples will have means between 75 and 85?
   c. For samples of $n = 16$, what proportion of the samples will have means between 75 and 85?

18. The population of SAT scores forms a normal distribution with a mean of $\mu = 500$ and a standard deviation of $\sigma = 100$. If the average SAT score is calculated for a sample of $n = 25$ students,
   a. What is the probability that the sample mean will be greater than $M = 510$. In symbols, what is $p(M > 510)$?
   b. What is the probability that the sample mean will be greater than $M = 520$. In symbols, what is $p(M > 520)$?
   c. What is the probability that the sample mean will be between $M = 510$ and $M = 520$? In symbols, what is $p(510 < M < 520)$?

19. A random sample of $n = 4$ scores is selected from a normal population with a mean of $\mu = 90$ and a standard deviation of $\sigma = 12..$
   a. What is the probability that the sample mean is between $M = 87$ and $M = 93$?
   b. What is the probability that the sample mean is between $M = 84$ and $M = 96$?
   c. What is the probability that the sample mean is between $M = 81$ and $M = 99$?

20. A normal-shaped distribution has $\mu = 80$ and $\sigma = 15$.
   a. Sketch the distribution of sample means for samples of $n = 25$ scores from this population.
   b. What are the z-score values that form the boundaries for the middle 95% of the distribution of sample means?
   c. Compute the z-score for $M = 89$ for a sample of $n = 25$ scores. Is this sample mean in the middle 95% of the distribution?

d. Compute the z-score for $M = 84$ for a sample of $n = 25$ scores. Is this sample mean in the middle 95% of the distribution?

21. The average age for licensed drivers in the county is $\mu = 42.6$ years with a standard deviation of $\sigma = 12$. The distribution is approximately normal.
   a. A researcher obtained a sample of $n = 36$ drivers who received parking tickets. The average age for these drivers was $M = 40.7$. Is this a reasonable outcome for a sample of $n = 36$ or is this sample mean an extreme value?
   b. The same researcher also obtained a sample of $n = 16$ drivers who received speeding tickets. The average age for this sample was $M = 34.4$. Is this a reasonable outcome for a sample of $n = 16$, or is the sample mean very different from what would usually be expected? (Again, compute the z-score for the sample mean.)

22. People are selected to serve on juries by randomly picking names from the list of registered voters. The average age for registered voters in the county is $\mu = 39.7$ years with a standard deviation of $\sigma = 12.4$. A statistician randomly selects a sample of $n = 16$ people who are currently serving on juries. The average age for the individuals in the sample is $M = 48.9$ years.
   a. How likely is it to obtain a random sample of $n = 16$ jurors with an average age greater than or equal to 48.9?

b. Is it reasonable to conclude that this set of $n = 16$ people is not a representative random sample of registered voters?

23. Welsh, Davis, Burke, and Williams (2002) conducted a study to evaluate the effectiveness of a carbohydrate-electrolyte drink on sports performance and endurance. Experienced athletes were given either a carbohydrate-electrolyte drink or a placebo while they were tested on a series of high-intensity exercises. One measure was how much time it took for the athletes to run to fatigue. Data similar to the results obtained in the study are shown in the following table.

Time to Run to Fatigue (in minutes)		
	Mean	SE
Placebo	21.7	2.2
Carbohydrate-electrolyte	28.6	2.7

   a. Construct a bar graph that incorporates all of the information in the table.
   b. Looking at your graph, do you think that the carbohydrate-electrolyte drink helps performance?

C H A P T E R

8

Introduction to Hypothesis Testing

Tools You Will Need

The following items are considered essential background material for this chapter. If you doubt your knowledge of any of these items, you should review the appropriate chapter or section before proceeding.

- *z*-Scores (Chapter 5)

- Distribution of sample means (Chapter 7)
  - Expected value
  - Standard error
  - Probability and sample means

Preview

8.1   The Logic of Hypothesis Testing

8.2   Uncertainty and Errors in Hypothesis Testing

8.3   An Example of a Hypothesis Test

8.4   Directional (One-Tailed) Hypothesis Tests

8.5   The General Elements of Hypothesis Testing: A Review

8.6   Concerns about Hypothesis Testing: Measuring Effect Size

8.7   Statistical Power

Summary

Focus on Problem Solving

Demonstrations 8.1 and 8.2

Problems

Preview

In the spring of 1998, a group of parents in Rochester, New York, filed a $185-million lawsuit against a major corporation claiming that environmental pollutants from the company had caused a cancer cluster. A "cluster" is defined as a disproportionate number of similar illnesses that could have a common cause, occurring in a defined geographic area. The parents had gathered data showing an unusually large number of childhood cancers concentrated in a relatively small area of the city. The cancers, which are fairly rare, affect the brain and nervous system of children and are typically fatal. The fact that a large number of cancers were concentrated in a single neighborhood seemed to indicate that there must be some factor that was responsible for causing the cluster.

One part of the company's defense involved challenging the existence of a real cluster. They argued that the number of cases and the concentration of cases were not unusual, but were well within the boundaries of what would be normal to expect by chance. For an analogy, imagine that a huge map of the city is pasted on one wall and that you are standing across the room, blindfolded, with a handful of darts. After throwing 20 or 30 darts blindly at the map, you remove your blindfold and observe the pattern that you have created. What is almost guaranteed to happen is that the darts will *not* be spaced evenly across the map. Instead, there will be large sections with no darts at all, and other sections where darts are concentrated in clusters. In this example, the clusters are not caused by any external factor; they are simply the result of chance.

The Problem: There are always two possible explanations for patterns that appear in the sample data obtained from a research study:

1. The pattern was caused by some systematic factor and therefore represents a real, meaningful event.
2. The pattern is simply the result of random influences or chance occurrences and is not meaningful.

The dilemma for researchers is trying to distinguish between these two alternatives.

The Solution: A *hypothesis test* helps researchers differentiate between real and random patterns by first determining what kind of patterns might reasonably be produced by random factors. If the sample pattern falls into this category, the researchers must conclude that it has no significance. On the other hand, if the pattern does not fall into the randomly produced category, the researchers can conclude that the pattern is not random but rather represents a significant and meaningful pattern in the population.

8.1 THE LOGIC OF HYPOTHESIS TESTING

It usually is impossible or impractical for a researcher to observe every individual in a population. Therefore, researchers usually collect data from a sample and then use the sample data to help answer questions about the population. Hypothesis testing is a statistical procedure that allows researchers to use sample data to draw inferences about the population of interest.

Hypothesis testing is one of the most commonly used inferential procedures. In fact, most of the remainder of this book examines hypothesis testing in a variety of different situations and applications. Although the details of a hypothesis test change from one situation to another, the general process remains constant. In this chapter, we introduce the general procedure for a hypothesis test. You should notice that we use the statistical techniques that have been developed in the preceding three chapters—that is, we will combine the concepts of *z*-scores, probability, and the distribution of sample means to create a new statistical procedure known as a *hypothesis test*.

DEFINITION | A **hypothesis test** is a statistical method that uses sample data to evaluate a hypothesis about a population.

In very simple terms, the logic underlying the hypothesis-testing procedure is as follows:

1. First, we state a hypothesis about a population. Usually the hypothesis concerns the value of a population parameter. For example, we might hypothesize that the mean IQ for registered voters in the United States is $\mu = 110$.

2. Before we actually select a sample, we use the hypothesis to predict the characteristics that the sample should have. For example, if we hypothesize that the population mean IQ is $\mu = 110$, then we would predict that our sample should have a mean *around* 110. Remember: The sample should be similar to the population, but you should always expect a certain amount of error.

3. Next, we obtain a random sample from the population. For example, we might select a random sample of $n = 200$ registered voters and compute the mean IQ for the sample.

4. Finally, we compare the obtained sample data with the prediction that was made from the hypothesis. If the sample mean is consistent with the prediction, we conclude that the hypothesis is reasonable. But if there is a big discrepancy between the data and the prediction, we decide that the hypothesis is wrong.

A hypothesis test is typically used in the context of a research study. That is, a researcher completes a research study and then uses a hypothesis test to evaluate the results. Depending on the type of research and the type of data, the details of the hypothesis test change from one research situation to another. In later chapters, we will examine different versions of hypothesis testing that are used for different kinds of research. For now, however, we will focus on the basic elements that are common to all hypothesis tests. To accomplish this general goal, we will examine a hypothesis test as it applies to the simplest possible research study. In particular, we will look at the situation in which a researcher is using one sample to examine one unknown population.

The unknown population   Figure 8.1 shows the general research situation that we will use to introduce the process of hypothesis testing. Notice that the researcher begins with a known population. This is the set of individuals as they exist *before treatment*. For this example, we are assuming that the original set of scores forms a normal distribution with $\mu = 80$ and $\sigma = 20$. The purpose of the research is to determine the effect of the treatment on the individuals in the population. That is, the goal is to determine what happens to the population *after the treatment is administered*.

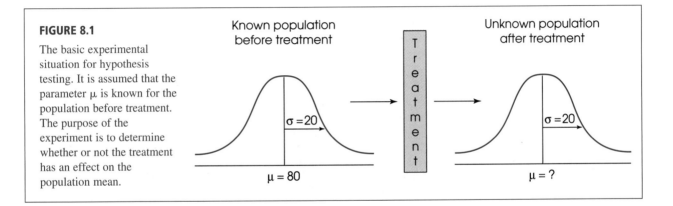

FIGURE 8.1

The basic experimental situation for hypothesis testing. It is assumed that the parameter μ is known for the population before treatment. The purpose of the experiment is to determine whether or not the treatment has an effect on the population mean.

To simplify the hypothesis-testing situation, one basic assumption is made about the effect of the treatment: If the treatment has any effect, it is simply to add a constant amount to (or subtract a constant amount from) each individual's score. You should recall from Chapters 3 and 4 that adding (or subtracting) a constant will change the mean but will not change the shape of the population, nor will it change the standard deviation. Thus, we will assume that the population after treatment has the same shape as the original population and the same standard deviation as the original population. This assumption is incorporated into the situation shown in Figure 8.1.

Note that the unknown population, after treatment, is the focus of the research question. Specifically, the purpose of the research is to determine what would happen if the treatment were administered to every individual in the population.

The sample in the research study   The goal of the hypothesis test is to determine whether the treatment has any effect on the individuals in the population (see Figure 8.1). Usually, however, we cannot administer the treatment to the entire population so the actual research study is conducted using a sample. Figure 8.2 shows the structure of the research study from the point of view of the hypothesis test. The original population, before treatment, is shown on the left-hand side. The unknown population, after treatment, is shown on the right-hand side. Note that the unknown population is actually *hypothetical* (the treatment is never administered to the entire population). Instead, we are asking *what if* the treatment were administered to the entire population. However, we do have a *real* sample that has received the treatment, and we will use this sample to evaluate a hypothesis about the unknown population mean. (Note that the sample is actually selected from the original population and then is given the treatment. Nonetheless, we end up with a treated sample that should be representative of the treated population.)

A hypothesis test is a formalized procedure that follows a standard series of operations. In this way, researchers have a standardized method for evaluating the results of their research studies. Other researchers will recognize and understand exactly how the data were evaluated and how conclusions were reached. To emphasize the formal structure of a hypothesis test, we will present hypothesis testing as a

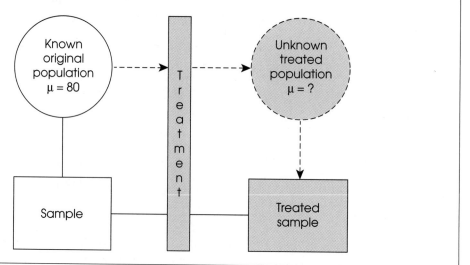

FIGURE 8.2

From the point of view of the hypothesis test, the entire population receives the treatment and then a sample is selected from the treated population. In the actual research study, a sample is selected from the original population and the treatment is administered to the sample. From either perspective, the result is a treated sample that represents the treated population.

four-step process that is used throughout the rest of the book. The following example provides a concrete foundation for introducing the hypothesis-testing procedure.

EXAMPLE 8.1   Researchers have noted a decline in cognitive functioning as people age (Bartus, 1990). However, the results from other research suggest that the antioxidants in foods such as blueberries can reduce and even reverse these age-related declines, at least in laboratory rats (Joseph, Shukitt-Hale, Denisova, et al., (1999). Based on these results, one might theorize that the same antioxidants might also benefit elderly humans. Suppose a researcher is interested in testing this theory.

The researcher plans to use a neuropsychological test (such as the perceptual speed task that was discussed in the Preview for Chapter 5) to measure cognitive function. It is known that the distribution of scores on this test is approximately normal and, for adults older than 65, the average score is $\mu = 80$ with a standard deviation of $\sigma = 20$. The researcher intends to obtain a sample of $n = 25$ adults who are older than 65, and give each participant a daily dose of a blueberry supplement that is very high in antioxidants. After taking the supplement for 6 months, the participants will be given the neuropsychological test to measure their level of cognitive function. If the mean score for the sample is noticeably different from the mean for the general population of elderly adults, the researcher can conclude that the supplement does appear to have an effect on cognitive function. On the other hand, if the sample mean is around 80 (the same as the general population mean), the researcher must conclude that the supplement does not appear to have any effect.

Figure 8.2 depicts the research situation that was described in the preceding example. Notice that the population after treatment is unknown. Specifically, we do not know what will happen to the mean score if the entire population of elderly adults is given the blueberry supplement. However, we do have a sample of $n = 25$ participants who have received the supplement and we can use this sample to help draw inferences about the unknown population. The following four steps outline the hypothesis-testing procedure that allows us to use sample data to answer questions about an unknown population.

STEP 1: STATE THE HYPOTHESES

As the name implies, the process of hypothesis testing begins by stating a hypothesis about the unknown population. Actually, we state two opposing hypotheses. Notice that both hypotheses are stated in terms of population parameters.

The first and most important of the two hypotheses is called the *null hypothesis*. The null hypothesis states that the treatment has no effect. In general, the null hypothesis states that there is no change, no effect, no difference—nothing happened, hence the name *null*. The null hypothesis is identified by the symbol H_0. (The *H* stands for *hypothesis*, and the zero subscript indicates that this is the *zero-effect*, null hypothesis.) For the study in Example 8.1, the null hypothesis states that the blueberry supplement has no effect on cognitive functioning for the population of adults who are more than 65 years old. In symbols, this hypothesis is

The goal of inferential statistics is to make general statements about the population by using sample data. Therefore, when testing hypotheses, we make our predictions about the population parameters.

$$H_0: \quad \mu_{\text{with supplement}} = 80 \qquad \text{(Even with the supplement, the mean test score still will be 80.)}$$

DEFINITION   The **null hypothesis** (H_0) states that in the general population there is no change, no difference, or no relationship. In the context of an experiment, H_0

predicts that the independent variable (treatment) *has no effect* on the dependent variable for the population.

The second hypothesis is simply the opposite of the null hypothesis, and it is called the *scientific,* or *alternative, hypothesis* (H_1). This hypothesis states that the treatment has an effect on the dependent variable.

DEFINITION

The **alternative hypothesis** (H_1) states that there is a change, a difference, or a relationship for the general population. In the context of an experiment, H_1 predicts that the independent variable (treatment) *does have an effect* on the dependent variable.

The null hypothesis and the alternative hypothesis are mutually exclusive and exhaustive. They cannot both be true. The data will determine which one should be rejected.

For this example, the alternative hypothesis states that the supplement does have an effect on cognitive functioning for the population and will cause a change in the mean score. In symbols, the alternative hypothesis is represented as

$$H_1: \quad \mu_{\text{with supplement}} \neq 80 \qquad \text{(With the supplement, the mean test score will be different from 80.)}$$

Notice that the alternative hypothesis simply states that there will be some type of change. It does not specify whether the effect will be increased or decreased test scores. In some circumstances, it is appropriate for the alternative hypothesis to specify the direction of the effect. For example, the researcher might hypothesize that the supplement will increase neuropsychological test scores ($\mu > 80$). This type of hypothesis results in a directional hypothesis test, which will be examined in detail later in this chapter. For now we concentrate on nondirectional tests, for which the hypotheses simply state that the treatment has some effect (H_1) or has no effect (H_0).

STEP 2: SET THE CRITERIA FOR A DECISION

The researcher will eventually use the data from the sample to evaluate the credibility of the null hypothesis. The data will either provide support for the null hypothesis or tend to refute the null hypothesis. In particular, if there is a big discrepancy between the data and the hypothesis, we will conclude that the hypothesis is wrong.

To formalize the decision process, we use the null hypothesis to predict the kind of sample mean that ought to be obtained. Specifically, we determine exactly what sample means are consistent with the null hypothesis and what sample means are at odds with the null hypothesis.

For our example, the null hypothesis states that the supplement has no effect and the population mean is still $\mu = 80$. If this is true, then the sample mean should have a value around 80. Therefore, a sample mean near 80 is consistent with the null hypothesis. On the other hand, a sample mean that is very different from 80 is not consistent with the null hypothesis. To determine exactly what values are "near" 80 and what values are "very different from" 80, we will examine all of the possible sample means that could be obtained if the null hypothesis is true. For our example, this is the distribution of sample means for $n = 25$. According to the null hypothesis, this distribution is centered at $\mu = 80$. The distribution of sample means is then divided into two sections:

1. Sample means that are likely to be obtained if H_0 is true; that is, sample means that are close to the null hypothesis

2. Sample means that are very unlikely to be obtained if H_0 is true; that is, sample means that are very different from the null hypothesis

The distribution of sample means divided into these two sections is shown in Figure 8.3. Notice that the high-probability samples are located in the center of the distribution and have sample means close to the value specified in the null hypothesis. On the other hand, the low-probability samples are located in the extreme tails of the distribution. After the distribution has been divided in this way, we can compare our sample data with the values in the distribution. Specifically, we can determine whether our sample mean is consistent with the null hypothesis (like the values in the center of the distribution) or whether our sample mean is very different from the null hypothesis (like the values in the extreme tails).

The alpha level   To find the boundaries that separate the high-probability samples from the low-probability samples, we must define exactly what is meant by "low" probability and "high" probability. This is accomplished by selecting a specific probability value, which is known as the *level of significance,* or the *alpha level,* for the hypothesis test. The alpha (α) value is a small probability that is used to identify the low-probability samples. By convention, commonly used alpha levels are $\alpha = .05$ (5%), $\alpha = .01$ (1%), and $\alpha = .001$ (0.1%). For example, with $\alpha = .05$, we separate the most unlikely 5% of the sample means (the extreme values) from the most likely 95% of the sample means (the central values).

With rare exceptions, an alpha level is never larger than .05.

The extremely unlikely values, as defined by the alpha level, make up what is called the *critical region*. These extreme values in the tails of the distribution define outcomes that are not consistent with the null hypothesis; that is, they are very unlikely to occur if the null hypothesis is true. Whenever the data from a research study produce a sample mean that is located in the critical region, we conclude that the data are not consistent with the null hypothesis, and we reject the null hypothesis.

DEFINITIONS

The **alpha level,** or the **level of significance,** is a probability value that is used to define the very unlikely sample outcomes if the null hypothesis is true.

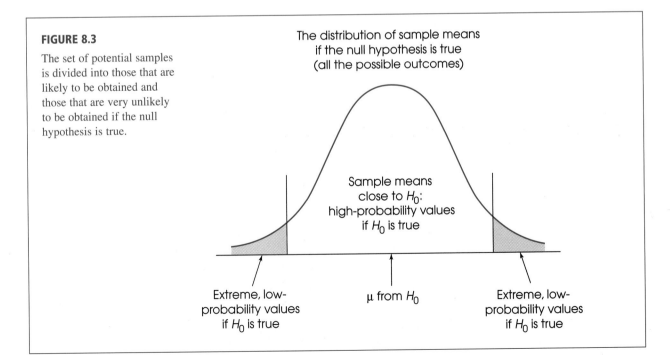

FIGURE 8.3

The set of potential samples is divided into those that are likely to be obtained and those that are very unlikely to be obtained if the null hypothesis is true.

The distribution of sample means
if the null hypothesis is true
(all the possible outcomes)

Sample means
close to H_0:
high-probability values
if H_0 is true

Extreme, low-
probability values
if H_0 is true

μ from H_0

Extreme, low-
probability values
if H_0 is true

The **critical region** is composed of extreme sample values that are very unlikely to be obtained if the null hypothesis is true. The boundaries for the critical region are determined by the alpha level. If sample data fall in the critical region, the null hypothesis is rejected.

Technically, the critical region is defined by sample outcomes that are *very unlikely* to occur if the treatment has no effect (that is, if the null hypothesis is true). Reversing the point of view, we can also define the critical region as sample values that provide *convincing evidence* that the treatment really does have an effect. For our example, the regular population of elderly adults has a mean test score of $\mu = 80$. We selected a sample from this population and administered a treatment (the blueberry supplement) to the individuals in the sample. What kind of sample mean would convince you that the treatment has an effect? It should be obvious that the most convincing evidence would be a sample mean that is really different from $\mu = 80$. In a hypothesis test, the critical region is determined by sample values that are "really different" from the original population.

The boundaries for the critical region   To determine the exact location for the boundaries that define the critical region, we will use the alpha-level probability and the unit normal table. In most cases, the distribution of sample means is normal, and the unit normal table will provide the precise z-score location for the critical region boundaries. With $\alpha = .05$, for example, the boundaries separate the extreme 5% from the middle 95%. Because the extreme 5% is split between two tails of the distribution, there is exactly 2.5% (or 0.0250) in each tail. In the unit normal table, you can look up a proportion of 0.0250 in column C (the tail) and find that the z-score boundary is $z = 1.96$. Thus, for any normal distribution, the extreme 5% is in the tails of the distribution beyond $z = 1.96$ and $z = -1.96$. These values define the boundaries of the critical region for a hypothesis test using $\alpha = .05$ (Figure 8.4).

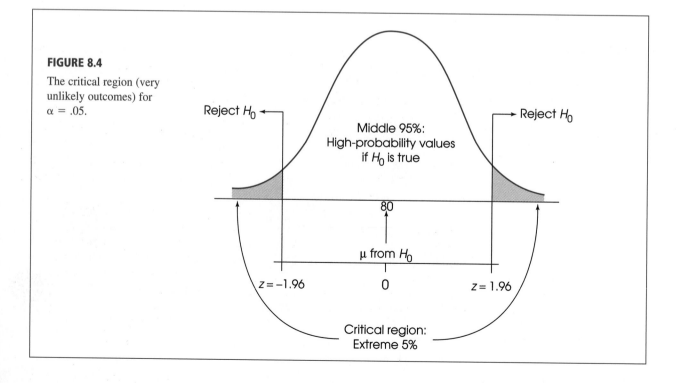

FIGURE 8.4

The critical region (very unlikely outcomes) for $\alpha = .05$.

Reject H_0

Middle 95%:
High-probability values
if H_0 is true

Reject H_0

80

μ from H_0

$z = -1.96$

0

$z = 1.96$

Critical region:
Extreme 5%

Similarly, an alpha level of $\alpha = .01$ means that 1% or .0100 is split between the two tails. In this case, the proportion in each tail is .0050, and the corresponding z-score boundaries are $z = \pm 2.58$ (± 2.57 is equally good). For $\alpha = .001$, the boundaries are located at $z = \pm 3.30$. You should verify these values in the unit normal table and be sure that you understand exactly how they are obtained.

LEARNING CHECK

1. A researcher is investigating the effectiveness of a new medication that is intended to lower cholesterol. In words, what would the null hypothesis say about the effectiveness of the medication?

2. If the alpha level is increased from $\alpha = .01$ to $\alpha = .05$, the size of the critical region also increases. (True or false?)

3. If a researcher conducted a hypothesis test with an alpha level of $\alpha = .04$, what z-score values would form the boundaries for the critical region?

ANSWERS

1. The null hypothesis would say that the medication has no effect.

2. True. A larger alpha means that the boundaries for the critical region move closer to the center of the distribution.

3. The .04 would be split between the two tails, with .02 in each tail. The z-score boundaries would be $z = +2.05$ and $z = -2.05$.

STEP 3: COLLECT DATA AND COMPUTE SAMPLE STATISTICS

At this time, we select a sample of adults who are more than 65 years old, and give each one a daily dose of the blueberry supplement. After 6 months, the neuropsychological test is used to measure cognitive function for the sample of participants. Notice that the data are collected *after* the researcher has stated the hypotheses and established the criteria for a decision. This sequence of events helps ensure that a researcher makes an honest, objective evaluation of the data and does not tamper with the decision criteria after the experimental outcome is known.

Next, the raw data from the sample are summarized with the appropriate statistics: For this example, the researcher would compute the sample mean. Now it is possible for the researcher to compare the sample mean (the data) with the null hypothesis. This is the heart of the hypothesis test: comparing the data with the hypothesis.

The comparison is accomplished by computing a z-score that describes exactly where the sample mean is located relative to the hypothesized population mean from H_0. In step 2, we constructed the distribution of sample means that would be expected if the null hypothesis were true—that is, the entire set of sample means that could be obtained if the treatment has no effect (see Figure 8.4). Now we will calculate a z-score that identifies where our sample mean is located in this hypothesized distribution. The z-score formula for a sample mean is

$$z = \frac{M - \mu}{\sigma_M}$$

In the formula, the value of the sample mean (M) is obtained from the sample data, and the value of μ is obtained from the null hypothesis. Thus, the z-score formula can be expressed in words as follows:

$$z = \frac{\text{sample mean} - \text{hypothesized population mean}}{\text{standard error between } M \text{ and } \mu}$$

Notice that the top of the z-score formula measures how much difference there is between the data and the hypothesis. The bottom of the formula measures the standard distance that ought to exist between a sample mean and the population mean.

STEP 4: MAKE A DECISION   In the final step, the researcher uses the z-score value obtained in step 3 to make a decision about the null hypothesis according to the criteria established in step 2. There are two possible decisions, and both are stated in terms of the null hypothesis (Box 8.1).

One possible decision is to *reject the null hypothesis*. This decision is made whenever the sample data fall in the critical region. By definition, a sample value in the critical region indicates that the sample is not consistent with the population defined by the null hypothesis. Specifically, this sample is very unlikely to occur if the null hypothesis is true, so our decision is to reject H_0. Remember, the null hypothesis states that there is no treatment effect, so rejecting H_0 means we are concluding that the treatment did have an effect. For the example we have been considering, suppose the sample produced a mean of $M = 92$ after taking the supplement for 6 months. With $n = 25$ and $\sigma = 20$, the standard error for the sample mean is

$$\sigma_M = \frac{\sigma}{\sqrt{n}} = \frac{20}{\sqrt{25}} = \frac{20}{5} = 4$$

B O X 8.1   REJECTING THE NULL HYPOTHESIS VERSUS PROVING THE ALTERNATIVE HYPOTHESIS

It may seem awkward to pay so much attention to the null hypothesis. After all, the purpose of most experiments is to show that a treatment does have an effect, and the null hypothesis states that there is no effect. The reason for focusing on the null hypothesis, rather than the alternative hypothesis, comes from the limitations of inferential logic. Remember that we want to use the sample data to draw conclusions, or inferences, about a population. Logically, it is much easier to demonstrate that a universal (population) hypothesis is false than to demonstrate that it is true. This principle is shown more clearly in a simple example. Suppose you make the universal statement "all dogs have four legs" and you intend to test this hypothesis by using a sample of ten dogs. If all the dogs in your sample have four legs, have you proved

the statement? It should be clear that ten four-legged dogs do not prove the general statement to be true. On the other hand, suppose one dog in your sample has only three legs. In this case, you have proved the statement to be false. Again, it is much easier to show that something is false than to prove that it is true.

Hypothesis testing uses this logical principle to achieve its goals. We state the null hypothesis, "the treatment has no effect," and try to show that it is false. The end result still is to demonstrate that the treatment does have an effect. That is, we find support for the alternative hypothesis by disproving (rejecting) the null hypothesis.

The z-score for the sample is

$$z = \frac{M - \mu}{\sigma_M} = \frac{92 - 80}{4} = \frac{12}{4} = 3.00$$

With an alpha level of $\alpha = .05$, this z-score is far beyond the boundary of 1.96. Because the sample z-score is in the critical region, we reject the null hypothesis and conclude that the blueberry supplement did have an effect on cognitive functioning.

The second possibility occurs when the sample data are not in the critical region. In this case, the data are reasonably close to the null hypothesis (in the center of the distribution). Because the data do not provide strong evidence that the null hypothesis is wrong, our conclusion is to *fail to reject the null hypothesis*. This conclusion means that the treatment does not appear to have an effect. For the research study examining the blueberry supplement, suppose our sample produced a mean test score of $M = 84$. As before, the standard error for a sample of $n = 25$ is $\sigma_M = 4$, and the null hypothesis states that $\mu = 80$. These values produce a z-score of

$$z = \frac{M - \mu}{\sigma_M} = \frac{84 - 80}{4} = \frac{4}{4} = 1.00$$

The z-score of 1.00 is not in the critical region. Therefore, we fail to reject the null hypothesis and conclude that the blueberry supplement does not appear to have an effect on cognitive functioning.

In general, the final decision is made by comparing our treated sample with the distribution of sample means for untreated samples. If the treated sample still looks much the same as other samples that did not receive the blueberry treatment, we conclude that the treatment does not appear to have any effect. On the other hand, if the treated sample is noticeably different from the majority of untreated samples, we conclude that the treatment did have an effect.

The two possible decisions may be easier to understand if you think of a research study as an attempt to gather evidence to prove that a treatment works. From this perspective, the research study has two possible outcomes:

1. You gather enough evidence to demonstrate convincingly that the treatment really works. That is, you reject the null hypothesis and conclude that the treatment does have an effect.

2. The evidence you gather from the research study is not convincing. In this case, all you can do is conclude that there is not enough evidence. The research study has failed to demonstrate that the treatment has an effect, and the statistical decision is to fail to reject the null hypothesis.

An Analogy for Hypothesis Testing   In many respects, the process of conducting a hypothesis test is similar to the process that takes place during a jury trial. For example,

1. The test begins with a null hypothesis stating that there is no treatment effect. The trial begins with a null hypothesis that there is no crime (innocent until proven guilty).

2. The research study gathers evidence to show that the treatment actually does have an effect, and the police gather evidence to show that there really is a crime. Note that both are trying to refute the null hypothesis.

3. If there is enough evidence, the researcher rejects the null hypothesis and concludes that there really is a treatment effect. If there is enough evidence, the jury rejects the hypothesis and concludes that the defendant is guilty of a crime.

4. If there is not enough evidence, the researcher fails to reject the null hypothesis. Note that the researcher does not conclude that there is no treatment effect, simply that there is not enough evidence to conclude that there is an effect. Similarly, if there is not enough evidence, the jury fails to find the defendant guilty. Note that the jury does not conclude that the defendant is innocent, simply that there is not enough evidence for a guilty verdict.

A CLOSER LOOK AT THE z-SCORE STATISTIC

The z-score statistic that is used in the hypothesis test is the first specific example of what is called a *test statistic*. The term *test statistic* simply indicates that the sample data are converted into a single, specific statistic that is used to test the hypotheses. In the chapters that follow, we will introduce several other test statistics that are used in a variety of different research situations. However, each of the new test statistics have the same basic structure and serve the same purpose as the z-score. We have already described the z-score equation as a formal method for comparing the sample data and the population hypothesis. In this section, we discuss the z-score from two other perspectives that may give you a better understanding of hypothesis testing and the role that z-scores play in this inferential technique. In each case, keep in mind that the z-score will serve as a general model for other test statistics that will come in future chapters.

The z-score formula as a recipe   The z-score formula, like any formula, can be viewed as a recipe. If you follow instructions and use all the right ingredients, the formula will produce a z-score. In the hypothesis-testing situation, however, you do not have all the necessary ingredients. Specifically, you do not know the value for the population mean (μ), which is one component or ingredient in the formula.

This situation is similar to trying to follow a cake recipe where one of the ingredients is not clearly listed. For example, the recipe may call for flour but there is a grease stain that makes it impossible to read how much flour. Faced with this situation, you might try the following steps:

1. Make a hypothesis about the amount of flour. For example, hypothesize that the correct amount is 2 cups.

2. To test your hypothesis, add the rest of the ingredients along with the hypothesized flour and bake the cake.

3. If the cake turns out to be good, you can reasonably conclude that your hypothesis was correct. But if the cake is terrible, you conclude that your hypothesis was wrong.

In a hypothesis test with z-scores, we do essentially the same thing. We have a formula (recipe) for z-scores but one ingredient is missing. Specifically, we do not know the value for the population mean, μ. Therefore, we try the following steps:

1. Make a hypothesis about the value of μ. This is the null hypothesis.

2. Plug the hypothesized value in the formula along with the other values (ingredients).

3. If the formula produces a z-score near zero (which is where z-scores are supposed to be), we conclude that the hypothesis was correct. On the other hand, if the formula produces an extreme value (a very unlikely result), we conclude that the hypothesis was wrong.

The z-score formula as a ratio In the context of a hypothesis test, the z-score formula has the following structure:

$$z = \frac{M - \mu}{\sigma_M} = \frac{\text{sample mean} - \text{hypothesized population mean}}{\text{standard error between } M \text{ and } \mu}$$

Notice that the numerator of the formula involves a direct comparison between the sample data and the null hypothesis. In particular, the numerator measures the obtained difference between the sample mean and the hypothesized population mean. The standard error in the denominator of the formula measures the standard amount of distance that exists naturally between a sample mean and the population mean without any treatment effect that causes the sample to be different. Thus, the z-score formula (and most other test statistics) forms a ratio

$$z = \frac{\text{actual difference between the sample } (M) \text{ and the hypothesis } (\mu)}{\text{standard difference between } M \text{ and } \mu \text{ with no treatment effect}}$$

Thus, for example, a z-score of $z = 3.00$ means that the obtained difference between the sample and the hypothesis is 3 times bigger than would be expected if the treatment had no effect.

In general, a large value for a test statistic like the z-score indicates a large discrepancy between the sample data and the null hypothesis. Specifically, a large value indicates that the sample data are very unlikely to have occurred by chance alone. Therefore, when we obtain a large value (in the critical region), we conclude that it must have been caused by a treatment effect.

LEARNING CHECK

1. If the null hypothesis is true, the mean for the population after treatment is identical to the mean for the population before treatment. (True or false?)

2. A small value (near zero) for the z-score statistic is evidence that the null hypothesis should be rejected. (True or false?)

3. A decision to reject the null hypothesis means you have demonstrated that the treatment has no effect. (True or false?)

ANSWERS

1. True. The null hypothesis says that the treatment has no effect.

2. False. A z-score near zero indicates that the data support the null hypothesis.

3. False. Rejecting the null hypothesis means you have concluded that the treatment does have an effect.

8.2 **UNCERTAINTY AND ERRORS IN HYPOTHESIS TESTING**

Hypothesis testing is an *inferential process*, which means that it uses limited information as the basis for reaching a general conclusion. Specifically, a sample provides only limited or incomplete information about the whole population, and yet a hypothesis test uses a sample to draw a conclusion about the population. In this situation, there is always the possibility that an incorrect conclusion will be made. Although sample data are usually representative of the population, there is always a chance that the sample is misleading and will cause a researcher to make the wrong decision about the research results. In a hypothesis test, there are two different kinds of errors that can be made.

TYPE I ERRORS

It is possible that the data will lead you to reject the null hypothesis when in fact the treatment has no effect. Remember: Samples are not expected to be identical to their populations, and some extreme samples can be very different from the populations they are supposed to represent. If a researcher selects one of these extreme samples by chance, then the data from the sample may give the appearance of a strong treatment effect, even though there is no real effect. In the previous section, for example, we discussed a research study examining how a food supplement that is high in antioxidants affects the cognitive functioning of elderly adults. Suppose the researcher selected a sample of $n = 25$ people who already had cognitive functioning that was well above average. When these people are tested after 6 months of taking the supplement, they will score higher than average on the neuropsychological test even though the blueberry supplement (the treatment) may have had no effect. In this case, the researcher is likely to conclude that the treatment has had an effect, when in fact it really did not. This is an example of what is called a *Type I error.*

DEFINITION

A **Type I error** occurs when a researcher rejects a null hypothesis that is actually true. In a typical research situation, a Type I error means the researcher concludes that a treatment does have an effect when in fact it has no effect.

You should realize that a Type I error is not a stupid mistake in the sense that a researcher is overlooking something that should be perfectly obvious. On the contrary, the researcher is looking at sample data that appear to show a clear treatment effect. The researcher then makes a careful decision based on the available information. The problem is that the information from the sample is misleading.

In most research situations, the consequences of a Type I error can be very serious. Because the researcher has rejected the null hypothesis and believes that the treatment has a real effect, it is likely that the researcher will report or even publish the research results. A Type I error, however, means that this is a false report. Thus, Type I errors lead to false reports in the scientific literature. Other researchers may try to build theories or develop other experiments based on the false results. A lot of precious time and resources may be wasted.

A Type I error occurs when a researcher unknowingly obtains an extreme, nonrepresentative sample. Fortunately, the hypothesis test is structured to minimize the risk that this will occur. Figure 8.4 shows the distribution of sample means and the critical region for the research study we have been discussing. This distribution contains all of the possible sample means for samples of $n = 25$ if the null hypothesis is true. Notice that most of the sample means are near the hypothesized population mean, $\mu = 80$.

For this test, we used an alpha level of $\alpha = .05$ and the critical region consists of the extreme sample means located in the tails of the distribution beyond $z = \pm 1.96$. Although these extreme sample means are very unlikely to occur if H_0 is true, they are not impossible. In fact, a sample mean in the extreme 5% of the distribution will occur 5% of the time. Thus, the alpha level determines the probability of obtaining a sample mean in the critical region when the null hypothesis is true. In other words, the alpha level determines the probability of a Type I error.

DEFINITION

The **alpha level** for a hypothesis test is the probability that the test will lead to a Type I error. That is, the alpha level determines the probability of obtaining sample data in the critical region even though the null hypothesis is true.

In summary, whenever the sample data are in the critical region, the appropriate decision for a hypothesis test is to reject the null hypothesis. Normally this is the correct decision because the treatment has caused the sample to be different from the original population; that is, the treatment effect has pushed the sample mean into the critical region. In this case, the hypothesis test has correctly identified a real treatment effect. Occasionally, however, sample data will be in the critical region just by chance, without any treatment effect. When this occurs, the researcher will make a Type I error; that is, the researcher will conclude that a treatment effect exists, when in fact it does not. Fortunately, the risk of a Type I error is small and is under the control of the researcher. Specifically, the probability of a Type I error is equal to the alpha level.

TYPE II ERRORS

Whenever a researcher rejects the null hypothesis, there is a risk of a Type I error. Similarly, whenever a researcher fails to reject the null hypothesis, there is a risk of a *Type II error*. By definition, a Type II error is the failure to reject a false null hypothesis. In more straightforward English, a Type II error means that a treatment effect really exists, but the hypothesis test fails to detect it.

DEFINITION

A **Type II error** occurs when a researcher fails to reject a null hypothesis that is really false. In a typical research situation, a Type II error means that the hypothesis test has failed to detect a real treatment effect.

A Type II error occurs when the sample mean is not in the critical region even though the treatment has had an effect on the sample. Often this happens when the effect of the treatment is relatively small. In this case, the treatment does influence the sample, but the magnitude of the effect is not big enough to move the sample mean into the critical region. Because the sample is not substantially different from the original population (it is not in the critical region), the statistical decision is to fail to reject the null hypothesis and to conclude that there is not enough evidence to say there is a treatment effect.

The consequences of a Type II error are usually not as serious as those of a Type I error. In general terms, a Type II error means that the research data do not show the results that the researcher had hoped to obtain. The researcher can accept this outcome and conclude that the treatment either has no effect or has only a small effect that is not worth pursuing, or the researcher can repeat the experiment (usually with some improvements) and try to demonstrate that the treatment really does work.

Unlike a Type I error, it is impossible to determine a single, exact probability value for a Type II error. Instead, the probability of a Type II error depends on a variety of factors and therefore is a function, rather than a specific number. Nonetheless, the probability of a Type II error is represented by the symbol β, the Greek letter *beta*.

In summary, a hypothesis test always leads to one of two decisions:

1. The sample data provide sufficient evidence to reject the null hypothesis and conclude that the treatment has an effect.

2. The sample data do not provide enough evidence to reject the null hypothesis. In this case, you fail to reject H_0 and conclude that the treatment does not appear to have an effect.

In either case, there is a chance that the data are misleading and the decision is wrong. The complete set of decisions and outcomes is shown in Table 8.1. The risk of an error is especially important in the case of a Type I error, which can lead to a false report. Fortunately, the probability of a Type I error is determined by the alpha level, which is completely under the control of the researcher. At the beginning of a hypothesis test, the researcher states the hypotheses and selects the alpha level, which immediately determines the risk that a Type I error will be made.

SELECTING AN ALPHA LEVEL

As you have seen, the alpha level for a hypothesis test serves two very important functions. First, alpha helps determine the boundaries for the critical region by defining the concept of "very unlikely" outcomes. More importantly, alpha determines the probability of a Type I error. When you select a value for alpha at the beginning of a hypothesis test, your decision influences both of these functions.

The primary concern when selecting an alpha level is to minimize the risk of a Type I error. Thus, alpha levels tend to be very small probability values. By convention, the largest permissible value is $\alpha = .05$. When there is no treatment effect, an alpha level of .05 means that there is still a 5% risk, or a 1-in-20 probability, of rejecting the null hypothesis and committing a Type I error. Because the consequences of a Type I error can be relatively serious, many individual researchers and many scientific publications prefer to use a more conservative alpha level such as .01 or .001 to reduce the risk that a false report is published and becomes part of the scientific literature. (For more information on the origins of the .05 level of significance, see the excellent short article by Cowles and Davis, 1982.)

At this point, it may appear that the best strategy for selecting an alpha level is to choose the smallest possible value to minimize the risk of a Type I error. However, there is a different kind of risk that develops as the alpha level is lowered. Specifically, a lower alpha level means less risk of a Type I error, but it also means that the hypothesis test demands more evidence from the research results.

The trade-off between the risk of a Type I error and the demands of the test is controlled by the boundaries of the critical region. For the hypothesis test to conclude that the treatment does have an effect, the sample data must be in the critical region. If the treatment really has an effect, it should cause the sample to be different from the original population; essentially, the treatment should push the sample into the critical region. However, as the alpha level is lowered, the boundaries for the critical region

TABLE 8.1

Possible outcomes of a statistical decision

		Actual Situation	
		No Effect, H_0 True	Effect Exists, H_0 False
EXPERIMENTER'S DECISION	Reject H_0	Type I error	Decision correct
	Retain H_0	Decision correct	Type II error

move farther out and become more difficult to reach. Figure 8.5 shows how the boundaries for the critical region move farther into the tails as the alpha level decreases. Notice that $z = 0$, in the center of the distribution, corresponds to the value of μ specified in the null hypothesis. The boundaries for the critical region determine how much distance between the sample mean and μ is needed to reject the null hypothesis. As the alpha level gets smaller, this distance gets larger.

Thus, an extremely small alpha level, such as .000001 (one in a million), would mean almost no risk of a Type I error but would push the critical region so far out that it would become essentially impossible to ever reject the null hypothesis; that is, it would require an enormous treatment effect before the sample data would reach the critical boundaries.

In general, researchers try to maintain a balance between the risk of a Type I error and the demands of the hypothesis test. Alpha levels of .05, .01, and .001 are considered reasonably good values because they provide a low risk of error without placing excessive demands on the research results.

FIGURE 8.5

The locations of the critical region boundaries for three different levels of significance: $\alpha = .05$, $\alpha = .01$, and $\alpha = .001$.

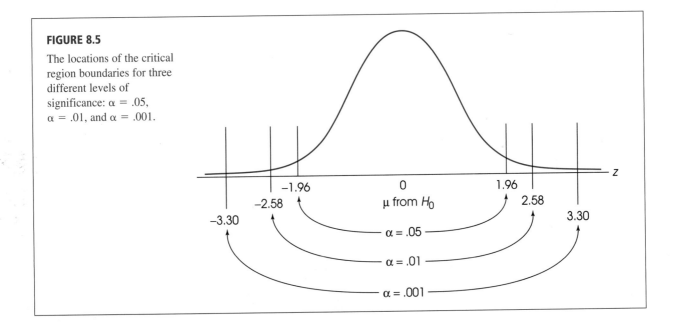

1. What is a Type I error?

2. Why is the consequence of a Type I error considered serious?

3. If a sample mean is in the critical region with $\alpha = .05$, it would still (always) be in the critical region if alpha were changed to $\alpha = .01$. (True or false?)

4. Define a Type II error.

5. What research situation is likely to lead to a Type II error?

ANSWERS   1. A Type I error is rejecting a true null hypothesis—that is, saying that the treatment has an effect when in fact it does not.

2. A Type I error often results in a false report. A researcher reports or publishes a treatment effect that does not exist.

3. False. With $\alpha = .01$, the boundaries for the critical region move farther out into the tails of the distribution. It is possible that a sample mean could be beyond the .05 boundary but not beyond the .01 boundary.

4. A Type II error is the failure to reject a false null hypothesis. In terms of a research study, a Type II error occurs when a study fails to detect a treatment effect that really exists.

5. A Type II error is likely to occur when the treatment effect is very small. In this case, a research study is more likely to fail to detect the effect.

8.3   AN EXAMPLE OF A HYPOTHESIS TEST

At this time, we have introduced all the elements of a hypothesis test. In this section, we present a complete example of the hypothesis-testing process and discuss how the results from a hypothesis test are presented in a research report. For purposes of demonstration, the following scenario will be used to provide a concrete background for the hypothesis-testing process.

EXAMPLE 8.2   Alcohol appears to be involved in a variety of birth defects, including low birth weight and retarded growth. A researcher would like to investigate the effect of prenatal alcohol on birth weight. A random sample of $n = 16$ pregnant rats is obtained. The mother rats are given daily doses of alcohol. At birth, one pup is selected from each litter to produce a sample of $n = 16$ newborn rats. The average weight for the sample is $M = 15$ grams. The researcher would like to compare the sample with the general population of rats. It is known that regular newborn rats (not exposed to alcohol) have an average weight of $\mu = 18$ grams. The distribution of weights is normal with $\sigma = 4$. Figure 8.6 shows the overall research situation. Notice that the researcher's question concerns the unknown population that is exposed to alcohol. Also notice that we have a sample representing the unknown population, and we have a hypothesis about the unknown population mean. Specifically, the null hypothesis says that the alcohol has no effect and the unknown mean is still $\mu = 18$. The goal of the hypothesis test is to determine whether the sample data are compatible with the hypothesis.

The following steps outline the hypothesis test that evaluates the effect of alcohol exposure on birth weight.

STEP 1   *State the hypotheses*, and select the alpha level. Both hypotheses concern the population that is exposed to alcohol, for which the population mean is unknown (the population on the right-hand side of Figure 8.6). The null hypothesis states that exposure to alcohol has no effect on birth weight. Thus, the population of rats with alcohol exposure should have the same mean birth weight as the regular, unexposed rats. In symbols,

$$H_0: \quad \mu_{\text{alcohol exposure}} = 18 \qquad \text{(Even with alcohol exposure, the rats still average 18 grams at birth.)}$$

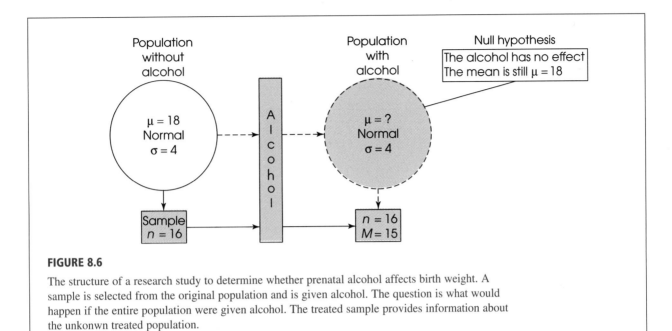

FIGURE 8.6

The structure of a research study to determine whether prenatal alcohol affects birth weight. A sample is selected from the original population and is given alcohol. The question is what would happen if the entire population were given alcohol. The treated sample provides information about the unkonwn treated population.

The alternative hypothesis states that alcohol exposure does affect birth weight, so the exposed population should be different from the regular rats. In symbols,

$$H_1: \quad \mu_{\text{alcohol exposure}} \neq 18 \qquad \text{(Alcohol exposure will change birth weight.)}$$

Notice that both hypotheses concern the unknown population. For this test, we will use an alpha level of $\alpha = .05$. That is, we are taking a 5% risk of committing a Type I error.

STEP 2   *Set the decision criteria by locating the critical region.* By definition, the critical region consists of outcomes that are very unlikely if the null hypothesis is true. To locate the critical region we will go through a three-stage process that is portrayed in Figure 8.7. We begin with the null hypothesis, which states that the alcohol has no effect on newborn rats. If H_0 is true, the population treated with alcohol is the same as the original population: that is, a normal distribution with $\mu = 18$ and $\sigma = 4$. Next, we consider all the possible outcomes for a sample of $n = 16$ newborn rats. This is the distribution of sample means for $n = 16$. For this example, the distribution of sample mean is normal, is centered at $\mu = 18$ (according to H_0), and has a standard error of $\sigma_M = 4/\sqrt{16} = 1$. Finally, we use the distribution of sample means to identify the critical region, which consists of those outcomes that are very unlikely if the null hypothesis is true. With $\alpha = .05$, the critical region consists of the extreme 5% of the distribution. As we saw earlier, for any normal distribution, z-scores of $z = \pm 1.96$ separate the middle 95% from the extreme 5% (a proportion of 0.0250 in each tail). Thus, we have identified the sample means that, according to the null hypothesis, are very unlikely to occur. It is the unlikely sample means, those with z-score values beyond ± 1.96, that form the critical region for the test.

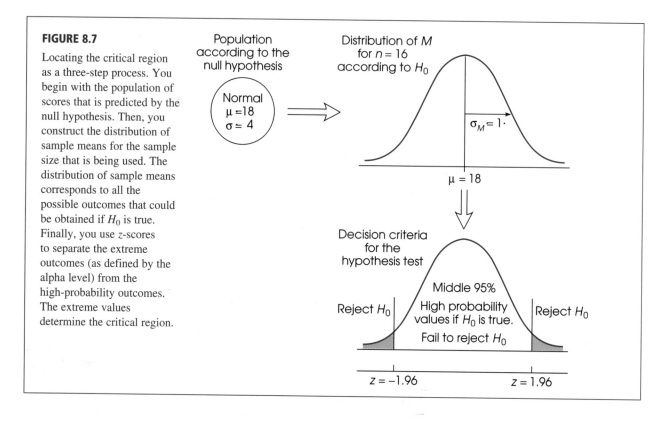

FIGURE 8.7

Locating the critical region as a three-step process. You begin with the population of scores that is predicted by the null hypothesis. Then, you construct the distribution of sample means for the sample size that is being used. The distribution of sample means corresponds to all the possible outcomes that could be obtained if H_0 is true. Finally, you use z-scores to separate the extreme outcomes (as defined by the alpha level) from the high-probability outcomes. The extreme values determine the critical region.

STEP 3 *Collect the data, and compute the test statistic.* At this point, we would select our sample of $n = 16$ pups whose mothers had received alcohol during pregnancy. The birth weight would be recorded for each pup and the sample mean computed. As noted, we obtained a sample mean of $M = 15$ grams. The sample mean is then converted to a z-score, which is our test statistic.

$$z = \frac{M - \mu}{\sigma_M} = \frac{15 - 18}{1} = \frac{-3}{1} = -3.00$$

STEP 4 *Make a decision.* The z-score computed in step 3 has a value of -3.00, which is beyond the boundary of -1.96. Therefore, the sample mean is located in the critical region. This is a very unlikely outcome if the null hypothesis is true, so our decision is to reject the null hypothesis. In addition to this statistical decision concerning the null hypothesis, it is customary to state a conclusion about the results of the research study. For this example, we conclude that prenatal exposure to alcohol does have a significant effect on birth weight.

IN THE LITERATURE
REPORTING THE RESULTS OF THE STATISTICAL TEST

A special jargon and notational system are used in published reports of hypothesis tests. When you are reading a scientific journal, for example, you typically will not be told explicitly that the researcher evaluated the data using a z-score as a test statistic with an alpha level of .05. Nor will you be told that "the null hypothesis is rejected." Instead, you will see a statement such as

> The treatment with alcohol had a significant effect on the birth weight of newborn rats, $z = 3.00$, $p < .05$.

Let us examine this statement piece by piece. First, what is meant by the word *significant*? In statistical tests, a *significant* result means that the null hypothesis has been rejected, which means that the result is very unlikely to have occurred merely by chance. For this example, the null hypothesis stated that the alcohol has no effect, however the data clearly indicated that the alcohol did have an effect. Specifically, it is very unlikely that the data would have been obtained if the alcohol did not have an effect.

DEFINITION

A result is said to be **significant** or **statistically significant** if it is very unlikely to occur when the null hypothesis is true. That is, the result is sufficient to reject the null hypothesis. Thus, a treatment has a significant effect if the decision from the hypothesis test is to reject H_0.

Next, what is the meaning of $z = 3.00$? The z indicates that a z-score was used as the test statistic to evaluate the sample data and that its value is 3.00. Finally, what is meant by $p < .05$? This part of the statement is a conventional way of specifying the alpha level that was used for the hypothesis test. It also acknowledges the possibility (and the probability) of a Type I error. Specifically, the researcher is reporting that the treatment had an effect but admits that this could be a false report. That is, it is possible that the sample mean was in the critical region even though the alcohol had no effect. However, the probability (p) of obtaining a sample mean in the critical region is extremely small (less than .05) if there is no treatment effect.

In circumstances where the statistical decision is to *fail to reject H_0*, the report might state that

The APA style does not use a leading zero in a probability value that refers to a level of significance.

> There was no evidence that the alcohol had an effect on birth weight, $z = 1.30$, $p > .05$.

In this case, we are saying that the obtained result, $z = 1.30$, is not unusual (not in the critical region) and has a relatively high probability of occurring (greater than .05) even if the null hypothesis is true and there is no treatment effect.

Sometimes students become confused trying to differentiate between $p < .05$ and $p > .05$. Remember that you reject the null hypothesis with extreme, low-probability values, located in the critical region in the tails of the distribution. Thus, a significant result that rejects the null hypothesis corresponds to $p < .05$ (Figure 8.8).

When a hypothesis test is conducted using a computer program, the printout often includes not only a z-score value but also an exact value for p, the probability that the result occurred without any treatment effect. In this case, researchers are encouraged to report the exact p value instead of using the less-than or greater-than notation. For example, a research report might state that the treatment effect was significant, with $z = 2.45$, $p = .0142$. When using exact values for p, however, you must still satisfy the traditional criterion for significance; specifically, the p value must be smaller than .05 to be considered statistically significant. Remember: The p value is the probability that the result would occur if H_0 were true (without any treatment effect), which is also the probability of a Type I error. It is essential that this probability be very small.

Finally, you should notice that in scientific reports, the researcher does not actually state that "the null hypothesis was rejected." Instead, the researcher reports that the

FIGURE 8.8

Sample means that fall in the critical region (shaded areas) have a probability *less than* alpha ($p < \alpha$). In this case, H_0 should be rejected. Sample means that do not fall in the critical region have a probability *greater than* alpha ($p > \alpha$).

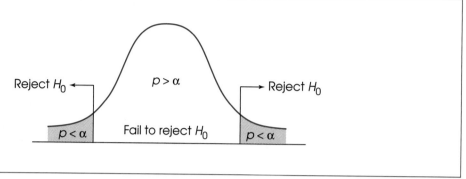

effect of the treatment was statistically significant. Likewise, when H_0 is not rejected, one simply states that the treatment effect was not statistically significant or that there was no evidence for a treatment effect. In fact, when you read scientific reports, you will note that the terms *null hypothesis* and *alternative hypothesis* are rarely mentioned. Nevertheless, H_0 and H_1 are part of the logic of hypothesis testing even if they are not formally stated in a scientific report. Because of their central role in the process of hypothesis testing, you should be able to identify and state these hypotheses. ❏

FACTORS THAT INFLUENCE A HYPOTHESIS TEST

In a hypothesis test, a large value for the z-score statistic is an indication that the sample mean, M, is very unlikely to have occurred if there is no treatment effect. Thus, a large value for z is grounds for concluding that the treatment has a significant effect. However, there are several factors that help determine whether the z-score will be large enough to reject H_0. In this section we examine three factors that can influence the outcome of a hypothesis test.

1. The size of the difference between the sample mean and the original population mean. This is the value that appears in the numerator of the z-score.

2. The variability of the scores, which is measured by either the standard deviation or the variance. The variability influences the size of the standard error in the denominator of the z-score.

3. The number of scores in the sample. This value also influences the size of the standard error in the denominator.

We will use the research study shown in Figure 8.9 to examine each of the three factors. The figure shows a population that is normally distributed with a mean of $\mu = 80$ and a standard deviation of $\sigma = 10$. The goal is to determine what would happen if a treatment was administered to the entire population. To help answer the question, a sample of $n = 25$ individuals is selected and the treatment is administered to the sample. After treatment, the sample mean is $M = 84$. As usual, the null hypothesis says that the treatment has no effect and that the population mean after treatment is still $\mu = 80$. The hypothesis test for this study produces a standard error of $\sigma_M = 10/\sqrt{25} = 2$ and a z-score of

$$z = \frac{M - \mu}{\sigma_M} = \frac{84 - 80}{2} = 2.00$$

The Research Study   The Hypothesis Test

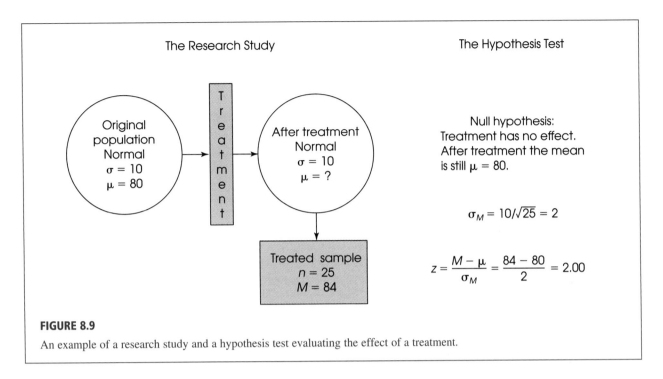

FIGURE 8.9

An example of a research study and a hypothesis test evaluating the effect of a treatment.

With $\alpha = .05$, this is sufficient to reject the null hypothesis and conclude that the treatment has a significant effect. Now consider what happens when we modify the size of the mean difference, the variability of the scores, and the number of scores in the sample.

The size of the mean difference   The most obvious factor influencing the outcome of a hypothesis test is the size of the mean difference. The amount of difference between the treated sample mean and the original population mean is the clearest indicator of a significant treatment effect. In Figure 8.9, there is a 4-point mean difference between $M = 84$ and $\mu = 80$, producing $z = 2.00$, which is just big enough to be significant. If the sample mean is changed to $M = 86$, increasing the mean difference to 6 points, the z-score becomes

$$z = \frac{M - \mu}{\sigma_M} = \frac{86 - 80}{2} = 3.00$$

Now the z-score is well into the critical region (even if α is reduced to .01) and we confidently reject the null hypothesis. In general, the larger the difference between the sample mean and the population mean is, the larger the z-score will be, and the greater the likelihood that we will find a significant treatment effect.

The variability of the scores   In Chapter 4 (page 128) we noted that high variability can make it very difficult to see any clear patterns in the results from a research study. In a hypothesis test, higher variability can reduce the chances of finding a significant treatment effect. For the study in Figure 8.9, the population standard deviation was $\sigma = 10$. With a sample of $n = 25$ individuals, this produced a standard error of $\sigma_M = 2$, and a significant z-score of $z = 200$. Now consider what happens if we increase the variability of the scores by increasing the standard deviation to $\sigma = 20$. With the increased variability, the standard error becomes $\sigma_M = 20/\sqrt{25} = 4$, and the z-score becomes

$$z = \frac{M - \mu}{\sigma_M} = \frac{84 - 80}{4} = 1.00$$

The z-score is no longer greater than the critical boundary of 1.96, so the statistical decision is to fail to reject the null hypothesis. The increased variability means that the sample data are no longer sufficient to conclude that the treatment has a significant effect. In general, increasing the variability of the scores will produce a larger standard error and a smaller value (closer to zero) for the z-score. If other factors are held constant, the larger the variability, the lower the likelihood of finding a significant treatment effect.

The number of scores in the sample   The final factor that influences the outcome of a hypothesis test is the number of scores in the sample. In simple terms, if you find a 4-point treatment effect with a sample of $n = 100$ people, it is more convincing evidence than if you find a 4-point effect with a sample of only $n = 4$ people. This relationship is demonstrated by once again examining the study shown in Figure 8.9. The study used a sample of $n = 25$ individuals and produced a significant z-score of $z = 2.00$. Now consider what happens if the sample size is increased to $n = 100$. With $n = 100$, the standard error becomes $\sigma_M = 10/\sqrt{100} = 1$, and the z-score for the hypothesis test becomes

$$z = \frac{M - \mu}{\sigma_M} = \frac{84 - 80}{1} = 4.00$$

This z-score is far into the critical region for $\alpha = .05$ or $\alpha = .01$ and we can confidently reject the null hypothesis and conclude that there is a significant treatment effect. In general, increasing the number of scores in the sample will produce a smaller standard error and a larger value for the z-score. If all other factors are held constant, the larger the sample size, the greater the likelihood of finding a significant treatment effect.

ASSUMPTIONS FOR HYPOTHESIS TESTS WITH z-SCORES   The mathematics that are used for a hypothesis test are based on a set of assumptions. When these assumptions are satisfied, you can be confident that the test produces a justified conclusion. However, if the assumptions are not satisfied, the hypothesis test may be compromised. In practice, researchers are not overly concerned with the assumptions underlying a hypothesis test because the tests usually work well even when the assumptions are violated. However, you should be aware of the fundamental conditions that are associated with each type of statistical test to ensure that the test is being used appropriately. The assumptions for hypothesis tests with z-scores are summarized as follows.

Random sampling   It is assumed that the subjects used to obtain the sample data were selected randomly. Remember, we wish to generalize our findings from the sample to the population. We accomplish this task when we use sample data to test a hypothesis about the population. Therefore, the sample must be representative of the population from which it has been drawn. Random sampling helps to ensure that it is representative.

Independent observations   The values in the sample must consist of *independent* observations. In everyday terms, two observations are independent if there is no consistent, predictable relationship between the first observation and the second. More precisely, two events (or observations) are independent if the occurrence of the first event has no effect on the probability of the second event. Specific examples of independence and nonindependence are examined in Box 8.2. Usually, this assumption is satisfied by using a *random* sample, which also helps ensure that the sample is representative of the population and that the results can be generalized to the population.

The value of σ is unchanged by the treatment   A critical part of the *z*-score formula in a hypothesis is the standard error, σ_M. To compute the value for the standard error, we must know the sample size (*n*) and the population standard deviation (σ). In a hypothesis test, however, the sample comes from an *unknown* population (see Figures 8.2

BOX 8.2	INDEPENDENT OBSERVATIONS

Independent observations are a basic requirement for nearly all hypothesis tests. The critical concern is that each observation or measurement is not influenced by any other observation or measurement. An example of independent observations is the set of outcomes obtained in a series of coin tosses. Assuming that the coin is balanced, each toss has a 50–50 chance of coming up either heads or tails. More important, each toss is *independent* of the tosses that came before. On the fifth toss, for example, there is a 50% chance of heads no matter what happened on the previous four tosses; the coin does not remember what happened earlier and is not influenced by the past. (*Note:* Many people fail to believe in the independence of events. For example, after a series of four tails in a row, it is tempting to think that the probability of heads must increase because the coin is overdue to come up heads. This is a mistake, called the "gambler's fallacy." Remember that the coin does not know what happened on the preceding tosses and cannot be influenced by previous outcomes.)

In most research situations, the requirement for independent observations is typically satisfied by using a random sample of separate, unrelated individuals. Thus, the measurement obtained for each individual is not influenced by other subjects in the sample. The following two situations demonstrate circumstances in which the observations are *not* independent.

1. A researcher is interested in examining television preferences for children. To obtain a sample of *n* = 20 children, the researcher selects 4 children from family A, 3 children from family B, 5 children from family C, 2 children from family D, and 6 children from family E.

   It should be obvious that the researcher does *not* have 20 independent observations. Within each family, the children probably share television preference (at least they watch the same shows). Thus, the response for each child is likely to be related to the responses of his or her siblings.

2. A researcher is interested in people's ability to judge distances. A sample of 20 people is obtained. All 20 participants are gathered together in the same room. The researcher asks the first person to estimate the distance between New York and Miami. The second person is asked the same question, then the third person, the fourth person, and so on.

   Again, the observations that the researcher obtains are not independent: The response of each participant is probably influenced by the responses of the previous participants. For example, consistently low estimates by the first few participants could create pressure to conform and thereby increase the probability that the following participants would also produce low estimates.

and 8.6). If the population is really unknown, it would suggest that we do not know the standard deviation and, therefore, we cannot calculate the standard error. To solve this dilemma, we have made an assumption. Specifically, we assume that the standard deviation for the unknown population (after treatment) is the same as it was for the population before treatment.

Actually, this assumption is the consequence of a more general assumption that is part of many statistical procedures. This general assumption states that the effect of the treatment is to add a constant amount to (or subtract a constant amount from) every score in the population. You should recall that adding (or subtracting) a constant changes the mean but has no effect on the standard deviation. You also should note that this assumption is a theoretical ideal. In actual experiments, a treatment generally will not show a perfect and consistent additive effect.

Normal sampling distribution To evaluate hypotheses with z-scores, we have used the unit normal table to identify the critical region. This table can be used only if the distribution of sample means is normal.

LEARNING CHECK

1. A researcher would like to know whether room temperature will affect eating behavior in rats. The lab temperature is usually kept at 72°, and under these conditions, the rats eat an average of $\mu = 10$ grams of food each day. The amount of food varies from one rat to another, forming a normal distribution with $\sigma = 2$. The researcher selects a sample of $n = 4$ rats and places them in a room where the temperature is kept at 65°. The daily food consumption for these rats averaged $M = 13$ grams. Do these data indicate that temperature has a significant effect on eating? Test at the .05 level of significance.

2. In a research report, the term *significant* is used when the null hypothesis is rejected. (True or false?)

3. In a research report, the results of a hypothesis test include the phrase "$p < .01$." This means that the test failed to reject the null hypothesis. (True or false?)

4. If other factors are held constant, increasing the size of the sample increases the likelihood of rejecting the null hypothesis. (True or false?)

5. If other factors are held constant, increasing the value of the standard deviation increases the likelihood of rejecting the null hypothesis. (True or false?)

ANSWERS

1. The null hypothesis states that temperature has no effect, or $\mu = 10$ grams even at the lower temperature. With $\alpha = .05$, the critical region consists of z-scores beyond $z = \pm 1.96$. The sample data produce $z = 3.00$. The decision is to reject H_0 and conclude that temperature has a significant effect on eating behavior.

2. True.

3. False. The probability is *less than* .01, which means it is very unlikely that the result occurred without any treatment effect. In this case, the data are in the critical region, and H_0 is rejected.

4. True. A larger sample produces a smaller standard error, which leads to a larger z-score.

5. False. A larger standard deviation produces a larger standard error, which leads to a smaller z-score.

8.4 DIRECTIONAL (ONE-TAILED) HYPOTHESIS TESTS

The hypothesis-testing procedure presented in Section 8.3 was the standard, or *two-tailed*, test format. The term *two-tailed* comes from the fact that the critical region is located in both tails of the distribution. This format is by far the most widely accepted procedure for hypothesis testing. Nonetheless, there is an alternative that is discussed in this section.

Usually a researcher begins an experiment with a specific prediction about the direction of the treatment effect. For example, a special training program is expected to *increase* student performance, or alcohol consumption is expected to *slow* reaction times. In these situations, it is possible to state the statistical hypotheses in a manner that incorporates the directional prediction into the statement of H_0 and H_1. The result is a directional test, or what commonly is called a *one-tailed test*.

DEFINITION

In a **directional hypothesis test**, or a **one-tailed test**, the statistical hypotheses (H_0 and H_1) specify either an increase or a decrease in the population mean score. That is, they make a statement about the direction of the effect.

The following example will be used to demonstrate the elements of a one-tailed hypothesis test.

EXAMPLE 8.3

Earlier, in Example 8.1, we discussed a research study that examined the effect of antioxidants (such as those found in blueberries) on the cognitive skills of elderly adults. In the study, each participant in a sample of $n = 25$ received a blueberry supplement every day for 6 months and then was given a standardized test to measure cognitive skill. For the general population of elderly adults (without any supplement), the test scores form a normal distribution with a mean of $\mu = 80$ and a standard deviation of $\sigma = 20$. For this example, the expected effect is that the blueberry supplement will improve cognitive performance. If the researcher obtains a sample mean of $M = 87$ for the $n = 25$ participants, is the result sufficient to conclude that the supplement really works?

THE HYPOTHESES FOR A DIRECTIONAL TEST

Because a specific direction is expected for the treatment effect, it is possible for the researcher to perform a directional test. The first step (and the most critical step) is to state the statistical hypotheses. Remember that the null hypothesis states that there is no treatment effect and that the alternative hypothesis says that there is an effect. For this example, the predicted effect is that the blueberry supplement will increase test scores. Thus, the two hypotheses would state:

H_0: Test scores are not increased.

H_1: Test scores are increased.

To express directional hypotheses in symbols, it usually is easier to begin with the alternative hypothesis (H_1). Again, we know that the general population has an average test score of $\mu = 80$, and H_1 states that test scores will be increased by the blueberry supplement. Therefore, expressed in symbols, H_1 states,

H_1: $\mu > 80$ (With the supplement, the average score is greater than 80.)

The null hypothesis states hat the supplement does not increase scores. In symbols,

H_0: $\mu \leq 80$ (With the supplement, the average score is not greater than 80.)

Note again that the two hypotheses are mutually exclusive and cover all of the possibilities.

THE CRITICAL REGION FOR DIRECTIONAL TESTS

The critical region is defined by sample outcomes that are very unlikely to occur if the null hypothesis is true (that is, if the treatment has no effect). Earlier (page 236), we noted that the critical region can also be defined in terms of sample values that provide *convincing evidence* that the treatment really does have an effect. For a directional test, the concept of "convincing evidence" is the simplest way to determine the location of the critical region. We begin with all the possible sample means that could be obtained if the null hypothesis is true. This is the distribution of sample means and it will be normal (because the population of test scores is normal), have an expected value of $\mu = 80$ (from H_0), and, for a sample of $n = 25$, will have a standard error of $\sigma_M = 20/\sqrt{25} = 4$. The distribution is shown in Figure 8.10.

For this example, the treatment is expected to increase test scores. If untreated adults average $\mu = 80$ on the test, then a sample mean that is substantially more than 80 would provide convincing evidence that the treatment worked. Thus, the critical region is located entirely in the right-hand tail of the distribution corresponding to sample means much greater than $\mu = 80$ (Figure 8.10). Because the critical region is contained in one tail of the distribution, a directional test is commonly called a *one-tailed* test. Also note that the proportion specified by the alpha level is not divided between two tails, but rather is contained entirely in one tail. Using $\alpha = .05$ for example, the whole 5% is located in one tail. In this case, the z-score boundary for the critical region is $z = 1.65$, which is obtained by looking up a proportion of .05 in column C (the tail) of the unit normal table.

If the prediction is that the treatment will produce a decrease *in scores, the critical region will be located entirely in the left-hand tail of the distribution.*

Notice that a directional (one-tailed) test requires two changes in the step-by-step hypothesis-testing procedure.

1. In the first step of the hypothesis test, the directional prediction is incorporated into the statement of the hypotheses.

2. In the second step of the process, the critical region is located entirely in one tail of the distribution.

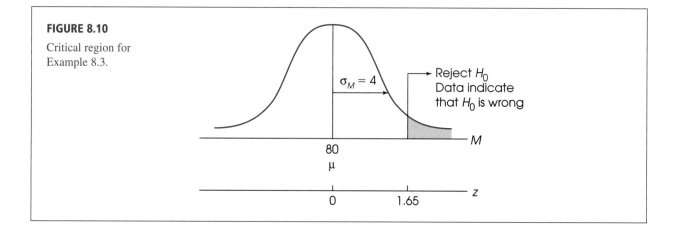

FIGURE 8.10

Critical region for Example 8.3.

After these two changes, the remainder of a one-tailed test proceeds exactly the same as a regular two-tailed test. Specifically, you calculate the z-score statistic and then make a decision about H_0 depending on whether the z-score is in the critical region.

For this example, the researcher obtained a mean of $M = 87$ for the 25 participants who received the blueberry supplement. This sample mean corresponds to a z-score of

$$z = \frac{M - \mu}{\sigma_M} = \frac{87 - 80}{4} = \frac{7}{4} = 1.75$$

A z-score of $z = 1.75$ is in the critical region for a one-tailed test (see Figure 8.10). This is a very unlikely outcome if H_0 is true. Therefore, we reject the null hypothesis and conclude that the blueberry supplement does produce a significant increase in cognitive performance scores.

COMPARISON OF ONE-TAILED VERSUS TWO-TAILED TESTS

The general goal of hypothesis testing is to determine whether a particular treatment has any effect on a population. The test is performed by selecting a sample, administering the treatment to the sample, and then comparing the result with the original population. If the treated sample is noticeably different from the original population, then we conclude that the treatment has an effect, and we reject H_0. On the other hand, if the treated sample is still similar to the original population, then we conclude that there is no convincing evidence for a treatment effect, and we fail to reject H_0. The critical factor in this decision is the *size of the difference* between the treated sample and the original population. A large difference is evidence that the treatment worked; a small difference is not sufficient to say that the treatment has any effect.

The major distinction between one-tailed and two-tailed tests is in the criteria they use for rejecting H_0. A one-tailed test allows you to reject the null hypothesis when the difference between the sample and the population is relatively small, provided the difference is in the specified direction. A two-tailed test, on the other hand, requires a relatively large difference independent of direction. This point is illustrated in the following example.

EXAMPLE 8.4

Consider again the study in Example 8.3 evaluating the effect of an antioxidant supplement. If we had used a standard two-tailed test, the hypotheses would be

H_0: $\mu = 80$ (The supplement has no effect on test scores.)

H_1: $\mu \neq 80$ (The supplement does have an effect on test scores.)

For a two-tailed test with $\alpha = .05$, the critical region consists of z-scores beyond ± 1.96. The data from Example 8.3 produced a sample mean of $M = 87$ and $z = 1.75$. For the two-tailed test, this z-score is not in the critical region, and we conclude that the supplement does not have a significant effect.

With the two-tailed test in Example 8.4, the 7-point difference between the sample mean and the hypothesized population mean ($M = 87$ and $\mu = 80$) was not big enough to reject the null hypothesis. However, with the one-tailed test in Example 8.3, the same 7-point difference was large enough to reject H_0 and conclude that the treatment had a significant effect.

All researchers agree that one-tailed tests are different from two-tailed tests. However, there are several ways to interpret the difference. One group of researchers contends that a two-tailed test is more rigorous and, therefore, more convincing than a one-tailed test. Remember that the two-tailed test demands more evidence to reject H_0 and thus provides a stronger demonstration that a treatment effect has occurred.

Other researchers feel that one-tailed tests are preferable because they are more sensitive. That is, a relatively small treatment effect may be significant with a one-tailed test but fail to reach significance with a two-tailed test. Also, there is the argument that one-tailed tests are more precise because they test hypotheses about a specific directional effect instead of an indefinite hypothesis about a general effect.

In general, two-tailed tests should be used in research situations when there is no strong directional expectation or when there are two competing predictions. For example, a two-tailed test would be appropriate for a study in which one theory predicts an increase in scores but another theory predicts a decrease. One-tailed tests should be used only in situations when the directional prediction is made before the research is conducted and there is a strong justification for making the directional prediction. In particular, you should never use a one-tailed test as a second attempt to salvage a significant result from a research study for which the first analysis with a two-tailed test failed to reach significance.

LEARNING CHECK

1. A researcher predicts that a treatment will increase scores. If this researcher uses a one-tailed hypothesis test with $\alpha = .01$, what z-score values will define the critical region?

2. A psychologist is examining the effects of early sensory deprivation on the development of perceptual discrimination. A sample of $n = 9$ newborn kittens is obtained. These kittens are raised in a completely dark environment for 4 weeks, after which they receive normal visual stimulation. At age 6 months, the kittens are tested on a visual discrimination task. The average score for this sample is $M = 32$. It is known that under normal circumstances kittens score an average of $\mu = 40$ on this task. The distribution of scores is normal with $\sigma = 12$. The researcher is predicting that the early sensory deprivation will reduce the kittens' performance on the discrimination task. Use a one-tailed test with $\alpha = .05$ to test this hypothesis.

ANSWERS

1. The critical region will be in the right-hand tail and consist of z-score values greater than $z = +2.33$.

2. The hypotheses are H_0: $\mu \geq 40$ and H_1: $\mu < 40$. The critical region consists of z-scores less than -1.65. For this sample, $z = -2.00$. Reject the null hypothesis and conclude that the deprived kittens scored significantly lower than regular kittens.

8.5   THE GENERAL ELEMENTS OF HYPOTHESIS TESTING: A REVIEW

The z-score hypothesis test presented in this chapter is one specific example of the general process of hypothesis testing. In later chapters, we examine many other hypothesis tests. Although the details of the hypothesis test will vary from one situation to another, all of the different tests use the same basic logic and consist of the same elements. In this section, we present a generic hypothesis test that outlines the basic elements and logic common to all tests. The better you understand the general process of a hypothesis test, the better you will understand inferential statistics.

Hypothesis tests consist of five basic elements organized in a logical pattern. The elements and their relationships to each other are as follows:

1. Hypothesized Population Parameter. The first step in a hypothesis test is to state a null hypothesis. You will notice that the null hypothesis typically provides a specific value for an unknown population parameter. The specific value predicted by the null hypothesis is the first element of hypothesis testing.

2. Sample Statistic. The sample data are used to calculate a sample statistic that corresponds to the hypothesized population parameter. Thus far we have considered only situations in which the null hypothesis specifies a value for the population mean, and the corresponding sample statistic is the sample mean.

3. Estimate of Error. To evaluate our findings, we must know what types of sample data are likely to occur simply due to chance. Typically, a measure of standard error is used to provide this information. The general purpose of standard error is to provide a measure of how much difference it is reasonable to expect between a sample statistic and the corresponding population parameter. Remember that samples (statistics) are not expected to provide a perfectly accurate picture of populations (parameters). There always will be some error or discrepancy between a statistic and a parameter; this is the concept of sampling error. The standard error tells how much error is expected.

In Chapter 7, we introduced the standard error of M. This is the first specific example of the general concept of standard error, and it provides a measure of how much error is expected between a sample mean (statistic) and the corresponding population mean (parameter). In later chapters, you will encounter other examples of standard error, each of which measures the standard distance (expected by chance) between a specific statistic and its corresponding parameter. Although the exact formula for standard error will change from one situation to another, you should recall that standard error is basically determined by two factors:

a. The variability of the scores. Standard error is directly related to the variability of the scores: The larger the variability, the larger the standard error.

b. The size of the sample. Standard error is inversely related to sample size: The larger the sample, the smaller the standard error.

4. The Test Statistic. In this chapter, the test statistic is a z-score:

$$z = \frac{M - \mu}{\sigma_M}$$

This z-score statistic provides a model for other test statistics that will follow. In particular, a test statistic typically forms a *ratio*. The numerator of the ratio is simply the obtained difference between the sample statistic and the hypothesized parameter. The denominator of the ratio is the standard error measuring how much difference is expected by chance. Thus, the generic form of a test statistic is

$$\text{test statistic} = \frac{\text{obtained difference}}{\text{difference expected by chance}}$$

In general, the purpose of a test statistic is to determine whether the result of an experiment (the obtained difference) is more than expected by chance alone. Thus, a test statistic that has a value greater than 1.00 indicates that the obtained result (numerator) is greater than chance (denominator). In addition, the test statistic provides a measure of *how much* greater than chance. A value of 3.00, for example, indicates that the obtained difference is three times greater than would be expected by chance.

5. The Alpha Level. The final element in a hypothesis test is the alpha level, or level of significance. The alpha level provides a criterion for interpreting the test statistic. As we noted, a test statistic greater than 1.00 indicates that the obtained result is greater than would be expected by chance. However, researchers are not satisfied with results that are simply "more than chance," instead they demand that results be *significantly more than chance*. The alpha level provides a criterion for "significance."

Earlier in this chapter (page 249), we noted that when a significant result is reported in the literature, it is always accompanied by a p value; for example, $p < .05$ means that it is *very unlikely* (a probability less than 5%) that the result would have been obtained without any treatment effect. In general, a significant result means that the researcher is very confident that the treatment effect is real and that the research results are not due to chance.

8.6   CONCERNS ABOUT HYPOTHESIS TESTING: MEASURING EFFECT SIZE

Although hypothesis testing is the most commonly used technique for evaluating and interpreting research data, a number of scientists have expressed a variety of concerns about the hypothesis testing procedure (for example see Loftus, 1996; Hunter, 1997; and Killeen, 2005).

Perhaps the most serious criticism of a hypothesis test concerns the interpretation of a significant result. Actually, there are two serious limitations with using a hypothesis test to establish the significance of a treatment effect. The first concern is that the focus of a hypothesis test is on the data rather than the hypothesis. Specifically, when the null hypothesis is rejected, we are actually making a strong probability statement about the sample data, not about the null hypothesis. A significant result permits the following conclusion: "This specific sample mean is very unlikely ($p < .05$) if the null hypothesis is true." Note that the conclusion does not make any definite statement about the probability of the null hypothesis being true or false. The fact that the data are very unlikely *suggests* that the null hypothesis is also very unlikely, but we do not have any solid grounds for making a probability statement about the null hypothesis. Rejecting the null hypothesis with $\alpha = .05$ does not justify a conclusion that the probability of the null hypothesis being true is less than 5%.

A second concern is that demonstrating a *significant* treatment effect does not necessarily indicate a *substantial* treatment effect. In particular, statistical significance does not provide any real information about the absolute size of a treatment effect. Instead, the hypothesis test has simply established that the results obtained in the research study are very unlikely to have occurred if there is no treatment effect. The hypothesis test reaches this conclusion by (1) calculating the standard error, which measures how much difference is reasonable to expect between M and μ, and (2) demonstrating that the obtained mean difference is substantially bigger than the standard error.

Notice that the test is making a *relative* comparison: the size of the treatment effect is being evaluated relative to the standard error. If the standard error is very small, then the treatment effect can also be very small and still be large enough to be significant. Thus, a significant effect does not necessarily mean a big effect.

The idea that a hypothesis test evaluates the relative size of a treatment effect, rather than the absolute size, is illustrated in the following example.

EXAMPLE 8.5   We begin with a population of scores that forms a normal distribution with $\mu = 50$ and $\sigma = 10$. A sample is selected from the population and a treatment is administered to the sample. After treatment, the sample mean is found to be $M = 51$. Does this sample provide evidence of a statistically significant treatment effect?

Although there is only a 1-point difference between the sample mean and the original population mean, the difference may be enough to be significant. In particular, the outcome of the hypothesis test depends on the sample size.

For example, with a sample of $n = 25$ the standard error is

$$\sigma_M = \frac{\sigma}{\sqrt{n}} = \frac{10}{\sqrt{25}} = \frac{10}{5} = 2.00$$

and the z-score for $M = 51$ is

$$z = \frac{M - \mu}{\sigma_M} = \frac{51 - 50}{2} = \frac{1}{2} = 0.50$$

For an alpha level of .05, the critical region would begin at $z = 1.96$. Our z-score fails to reach the critical region, so we fail to reject the null hypothesis. In this case, the 1-point difference between M and μ is not significant because it is being evaluated relative to a standard error of 2 points.

Now consider the outcome with a sample of $n = 400$. The standard error is

$$\sigma_M = \frac{\sigma}{\sqrt{n}} = \frac{10}{\sqrt{400}} = \frac{10}{20} = 0.50$$

and the z-score for $M = 51$ is

$$z = \frac{M - \mu}{\sigma_M} = \frac{51 - 50}{0.5} = \frac{1}{0.5} = 2.00$$

Now the z-score is beyond the 1.96 boundary, so we reject the null hypothesis and conclude that there is a signficant effect. In this case, the 1-point difference between M and μ is considered statistically significant because it is being evaluated relative to a standard error of only 0.5 points.

The point of Example 8.5 is that a small treatment effect can still be statistically significant. If the sample size is large enough, any treatment effect, no matter how small, can be enough for us to reject the null hypothesis.

MEASURING EFFECT SIZE   As noted in the previous section (point 2), one concern with hypothesis testing is that a hypothesis test does not really evaluate the absolute size of a treatment effect. To correct this problem, it is recommended that whenever researchers report a statistically significant effect, they also provide a report of the effect size (see the guidelines presented by L. Wilkinson and the APA Task Force on Statistical Inference, 1999). Therefore, as we present different hypothesis tests we will also present different options for measuring and reporting *effect size.*

D E F I N I T I O N

A measure of **effect size** is intended to provide an measurement of the absolute magnitude of a treatment effect, independent of the size of the sample(s) being used.

One of the simplest and most direct methods for measuring effect size is *Cohen's* d. Cohen (1988) recommended that effect size can be standardized by measuring the mean difference in terms of the standard deviation. The resulting measure of effect size is computed as

$$\text{Cohen's } d = \frac{\text{mean difference}}{\text{standard deviation}} = \frac{\mu_{\text{treatment}} - \mu_{\text{no treatment}}}{\sigma} \tag{8.1}$$

For the z-score hypothesis test, the mean difference is determined by the difference between the population mean before treatment and the population mean after treatment. However, the population mean after treatment is unknown. Therefore, we must use the mean for the treated sample in its place. Remember, the sample mean is expected to be representative of the population mean and provides the best measure of the treatment effect. Thus, the actual calculations are really estimating the value of Cohen's d as follows:

Cohen's d measures the distance between two means and is typically reported as a positive number even when the formula produces a negative value.

$$\text{estimated Cohen's } d = \frac{\text{mean difference}}{\text{standard deviation}} = \frac{M_{\text{treatment}} - \mu_{\text{no treatment}}}{\sigma} \tag{8.2}$$

The standard deviation is included in the calculation to standardize the size of the mean difference in much the same way that z-scores standardize locations in a distribution. For example, a 15-point mean difference can be a relatively large treatment effect or a relatively small effect depending on the size of the standard deviation. This phenomenon is demonstrated in Figure 8.11. The top portion of the figure (part a) shows the results of a treatment that produces a 15-point mean difference in SAT scores; before treatment, the average SAT score is 500, and after treatment the average is 515. Notice that the standard deviation for SAT scores is $\sigma = 100$, so the 15-point difference appears to be small. For this example, Cohen's d is

$$\text{Cohen's } d = \frac{\text{mean difference}}{\text{standard deviation}} = \frac{15}{100} = 0.15$$

Now consider the treatment effect shown in Figure 8.11(b). This time, the treatment produces a 15-point mean difference in IQ scores; before treatment the average IQ is 100, and after treatment the average is 115. Because IQ scores have a standard deviation of $\sigma = 15$, the 15-point mean difference now appears to be large. For this example, Cohen's d is

$$\text{Cohen's } d = \frac{\text{mean difference}}{\text{standard deviation}} = \frac{15}{15} = 1.00$$

Notice that Cohen's d measures the size of the treatment effect in terms of the standard deviation. For example, a value of $d = 0.50$ indicates that the treatment changed the

Sample size   One factor that has a huge influence on power is the size of the sample. In Example 8.7 we demonstrated power for an 4-point treatment effect using a sample of $n = 25$. If the researcher decided to conduct the study using a sample of $n = 50$, then the power would be dramatically different. With $n = 50$, the standard error for the sample means would be

$$\sigma_M = \frac{\sigma}{\sqrt{n}} = \frac{10}{\sqrt{50}} = \frac{10}{7.07} = 1.41$$

Figure 8.14 shows the two distributions of sample means with $n = 50$ and a standard error of $\sigma_M = 1.41$ points. The distribution on the left is centered at $\mu = 80$ and shows all the possible sample means if H_0 is true. As always, this distribution is used to

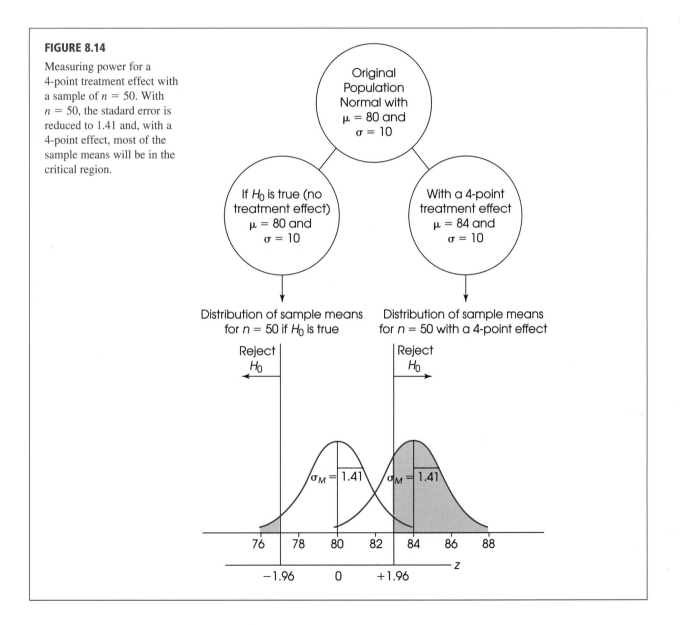

FIGURE 8.14

Measuring power for a 4-point treatment effect with a sample of $n = 50$. With $n = 50$, the stadard error is reduced to 1.41 and, with a 4-point effect, most of the sample means will be in the critical region.

locate the critical boundaries for the hypothesis test. The distribution on the right is centered at $\mu = 84$ and shows all the possible sample means if there is a 4-point treatment effect. Note that the majority of the treated sample means are now located beyond the $z = 1.96$ boundary. To find the exact proportion, we first identify the sample mean corresponding to $z = 1.96$ in the left-hand distribution. In this case, $M = 82.76$ [$80 + 1.96(1.41)$]. Next, find the z-score location of this sample mean in the right-hand (treated) distribution. In this case

$$z = \frac{M - \mu}{\sigma_M} = \frac{82.76 - 84}{1.41} = -0.88$$

Finally, use the unit normal table to find the proportion of the right-hand distribution corresponding to values greater than $z = -0.88$. Using column B (body) in the table we find $p = 0.8106$ or 81.06%. Thus, with $n = 50$, over 81% of all the possible samples will lead to rejecting H_0. By increasing the sample size from 25 to 50, the power has increased from 51.06% to 81.06%.

Because power is directly related to sample size, one of the primary reasons for computing power is to determine what sample size is necessary to achieve a reasonable probability for a successful research study. In Example, 8.7 we found that a sample of $n = 25$ gives the study only a 51% chance of detecting a 4-point treatment effect. However, if the sample size is increased to $n = 50$, the likelihood that the 4-point effect will be significant increases to 81%. In this way, researchers can determine the probability that their research will successfully reject the null hypothesis before a study is conducted. If the probability (power) is too small, they always have the option of increasing sample size to increase power.

Alpha level   Reducing the alpha level for a hypothesis test will also reduce the power of the test. For example, lowering α from .05 to .01 will lower the power of the hypothesis test. The effect of reducing the alpha level can be seen by referring again to Figure 8.14. In this figure, the boundaries for the critical region are drawn using $\alpha = .05$. Specifically, the critical region on the right-hand side begins at $z = 1.96$. If α were changed to .01, the boundary would be moved farther to the right, out to $z = 2.58$. It should be clear that moving the critical boundary to the right means that a smaller portion of the treatment distribution (the distribution on the right-hand side) will be in the critical region. Thus, there would be a lower probability of rejecting the null hypothesis and a lower value for the power of the test.

One-tailed versus two-tailed tests   Changing from a regular two-tailed test to a one-tailed test will increase the power of the hypothesis test. Again, this effect can be seen by referring to Figure 8.14. The figure shows the boundaries for the critical region using a two-tailed test with $\alpha = .05$ so that the critical region on the right-hand side begins at $z = 1.96$. Changing to a one-tailed test would move the critical boundary to the left to a value of $z = 1.65$. Moving the boundary to the left would cause a larger proportion of the treatment distribution to be in the critical region and, therefore, would increase the power of the test.

1. For a particular hypothesis test, the power is .50 for a 5-point treatment effect. Will the power be greater or less for a 10-point treatment effect?

2. As the power of a test increases, what happens to the probability of a Type II error?

3. If a researcher uses $\alpha = .01$ rather than $\alpha = .05$, then power will increase. (True or false?)

4. What is the effect of increasing sample size on power?

ANSWERS

1. The hypothesis test is more likely to detect a 10-point effect, so power will be greater.

2. As power increases, the probability of a Type II error decreases.

3. False.

4. Power increases as sample size gets larger.

SUMMARY

1. Hypothesis testing is an inferential procedure that uses the data from a sample to draw a general conclusion about a population. The procedure begins with a hypothesis about an unknown population. Then a sample is selected, and the sample data provide evidence that either supports or refutes the hypothesis.

2. In this chapter, we introduced hypothesis testing using the simple situation in which a sample mean is used to test a hypothesis about an unknown population mean. We begin with an unknown population, generally a population that has received a treatment. The question is to determine whether the treatment has an effect on the population mean (see Figure 8.1).

3. Hypothesis testing is structured as a four-step process that is used throughout the remainder of the book.
   a. State the null hypothesis (H_0), and select an alpha level. The null hypothesis states that there is no effect or no change. In this case, H_0 states that the mean for the treated population is the same as the mean before treatment. The alpha level, usually $\alpha = .05$ or $\alpha = .01$, provides a definition of the term *very unlikely* and determines the risk of a Type I error. Also state an alternative hypothesis (H_1), which is the exact opposite of the null hypothesis.
   b. Locate the critical region. The critical region is defined as sample outcomes that would be very unlikely to occur if the null hypothesis is true. The alpha level defines "very unlikely." For example,

with $\alpha = .05$, the critical region is defined as sample means in the extreme 5% of the distribution of sample means. When the distribution is normal, the extreme 5% corresponds to z-scores beyond $z = \pm 1.96$.
   c. Collect the data, and compute the test statistic. The sample mean is transformed into a z-score by the formula

$$z = \frac{M - \mu}{\sigma_M}$$

The value of μ is obtained from the null hypothesis. The z-score test statistic identifies the location of the sample mean in the distribution of sample means. Expressed in words, the z-score formula is

$$z = \frac{\text{sample mean} - \begin{array}{c}\text{hypothesized}\\\text{population mean}\end{array}}{\text{standard error}}$$

   d. Make a decision. If the obtained z-score is in the critical region, reject H_0 because it is very unlikely that these data would be obtained if H_0 were true. In this case, conclude that the treatment has changed the population mean. If the z-score is not in the critical region, fail to reject H_0 because the data are not significantly different from the null hypothesis. In this case, the data do not provide sufficient evidence to indicate that the treatment has had an effect.

4. Whatever decision is reached in a hypothesis test, there is always a risk of making the incorrect decision. There are two types of errors that can be committed.

   A Type I error is defined as rejecting a true H_0. This is a serious error because it results in falsely reporting a treatment effect. The risk of a Type I error is determined by the alpha level and therefore is under the experimenter's control.

   A Type II error is defined as the failure to reject a false H_0. In this case, the experiment fails to detect an effect that actually occurred. The probability of a Type II error cannot be specified as a single value and depends in part on the size of the treatment effect. It is identified by the symbol β (beta).

5. When a researcher expects that a treatment will change scores in a particular direction (increase or decrease), it is possible to do a directional or one-tailed test. The first step in this procedure is to incorporate the directional prediction into the hypotheses. For example, if the prediction is that a treatment will increase scores, the null hypothesis says that there is no increase and the alternative hypothesis states that there is an increase. To locate the critical region, you must determine what kind of data would refute the null hypothesis by demonstrating that the treatment worked as predicted. These outcomes will be located entirely in one tail of the distribution, so the entire critical region (5% or 1% depending on α) will be in one tail.

6. A one-tailed test is used when there is prior justification for making a directional prediction. These *a priori* reasons may be previous reports and findings

or theoretical considerations. In the absence of the *a priori* basis, a two-tailed test is appropriate. In this situation, you might be unsure of what to expect in the study, or you might be testing competing theories.

7. In addition to using a hypothesis test to evaluate the *significance* of a treatment effect, it is recommended that you also measure and report the *effect size*. One measure of effect size is Cohen's *d*, which is a standardized measure of the mean difference. Cohen's *d* is computed as

$$\text{Cohen's } d = \frac{\text{mean difference}}{\text{standard deviation}}$$

8. The power of a hypothesis test is defined as the probability that the test will correctly reject the null hypothesis.

9. To determine the power for a hypothesis test, you must first identify the treatment and null distributions. Also, you must specify the magnitude of the treatment effect. Next, you locate the critical region in the null distribution. The power of the hypothesis test is the portion of the treatment distribution that is located beyond the boundary (critical value) of the critical region.

10. As the size of the treatment effect increases, statistical power increases. Also, power is influenced by several factors that can be controlled by the experimenter:
    a. Increasing the alpha level will increase power.
    b. A one-tailed test will have greater power than a two-tailed test.
    c. A large sample will result in more power than a small sample.

KEY TERMS

hypothesis testing	alpha level	Type II error	one-tailed test
null hypothesis	critical region	beta	effect size
alternative hypothesis	test statistic	significant	Cohen's *d*
level of significance	Type I error	directional test	power

RESOURCES

There are a tutorial quiz and other learning exercises for Chapter 8 at the Wadsworth website, www.cengage.com/psychology/gravetter. The site also provides access to a workshop entitled *Hypothesis Testing* that reviews the concept and logic of hypothesis testing. See page 29 for more information about accessing items on the website.

Guided interactive tutorials, end-of-chapter problems, and related testbank items may be assigned online at WebAssign.

WebTUTOR™

For those using WebTutor along with this book, there is a WebTutor section corresponding to this chapter. The WebTutor contains a brief summary of Chapter 8, some hints for learning the new material and for avoiding common errors, and a review of the new vocabulary that accompanies hypothesis testing. As usual, WebTutor also presents some sample exam items including solutions.

SPSS

The statistical computer package SPSS is not structured to conduct hypothesis tests using z-scores. In truth, the z-score test presented in this chapter is rarely used in actual research situations. The problem with the z-score test is that it requires that you know the value of the population standard deviation, and this information is usually not available. Researchers rarely have detailed information about the populations that they wish to study. Instead, they must obtain information entirely from samples. In the following chapters we introduce new hypothesis-testing techniques that are based entirely on sample data. These new techniques are included in SPSS.

FOCUS ON PROBLEM SOLVING

1. Hypothesis testing involves a set of logical procedures and rules that enable us to make general statements about a population when all we have are sample data. This logic is reflected in the four steps that have been used throughout this chapter. Hypothesis-testing problems will become easier to tackle when you learn to follow the steps.

STEP 1 State the hypotheses and set the alpha level.

STEP 2 Locate the critical region.

STEP 3 Compute the test statistic (in this case, the z-score) for the sample.

STEP 4 Make a decision about H_0 based on the result of step 3.

2. Students often ask, "What alpha level should I use?" Or a student may ask, "Why is an alpha of .05 used?" as opposed to something else. There is no single correct answer to either of these questions. Keep in mind the aim of setting an alpha level in the first

place: *to reduce the risk of committing a Type I error.* Therefore, the maximum acceptable value is $\alpha = .05$. However, some researchers prefer to take even less risk and use alpha levels of .01 or smaller.

Most statistical tests are now done with computer programs that provide an exact probability (*p* value) for a Type I error. Because an exact value is available, most researchers simply report the *p* value from the computer printout rather than setting an alpha level at the beginning of the test. However, the same criterion still applies: A result is not significant unless the *p* value is less than .05.

3. Take time to consider the implications of your decision about the null hypothesis. The null hypothesis states that there is no effect. Therefore, if your decision is to reject H_0, you should conclude that the sample data provide evidence for a treatment effect. However, it is an entirely different matter if your decision is to fail to reject H_0. Remember that when you fail to reject the null hypothesis, the results are inconclusive. It is impossible to *prove* that H_0 is correct; therefore, you cannot state with certainty that "there is no effect" when H_0 is not rejected. At best, all you can state is that "there is insufficient evidence for an effect" (see Box 8.1).

4. It is very important that you understand the structure of the *z*-score formula (page 241). It will help you understand many of the other hypothesis tests that will be covered later.

5. When you are doing a directional hypothesis test, read the problem carefully, and watch for key words (such as increase or decrease, raise or lower, and more or less) that tell you which direction the researcher is predicting. The predicted direction will determine the alternative hypothesis (H_1) and the critical region. For example, if a treatment is expected to *increase* scores, H_1 would contain a *greater than* symbol, and the critical region would be in the tail associated with high scores.

DEMONSTRATION 8.1

HYPOTHESIS TEST WITH *z*

A researcher begins with a known population—in this case, scores on a standardized test that are normally distributed with $\mu = 65$ and $\sigma = 15$. The researcher suspects that special training in reading skills will produce a change in the scores for the individuals in the population. Because it is not feasible to administer the treatment (the special training) to everyone in the population, a sample of $n = 25$ individuals is selected, and the treatment is given to this sample. Following treatment, the average score for this sample is $M = 70$. Is there evidence that the training has an effect on test scores?

STEP 1   State the hypothesis and select an alpha level.
The null hypothesis states that the special training has no effect. In symbols,

$H_0: \mu = 65$ (After special training, the mean is still 65.)

The alternative hypothesis states that the treatment does have an effect.

$H_1: \mu \neq 65$ (After training, the mean is different from 65.)

At this time you also select the alpha level. For this demonstration, we will use $\alpha = .05$. Thus, there is a 5% risk of committing a Type I error if we reject H_0.

STEP 2   Locate the critical region.

With $\alpha = .05$, the critical region consists of sample means that correspond to z-scores beyond the critical boundaries of $z = \pm 1.96$.

STEP 3   Obtain the sample data, and compute the test statistic.

For this example, the distribution of sample means, according to the null hypothesis, will be normal with an expected value of $\mu = 65$ and a standard error of

$$\sigma_M = \frac{\sigma}{\sqrt{n}} = \frac{15}{\sqrt{25}} = \frac{15}{5} = 3$$

In this distribution, our sample mean of $M = 70$ corresponds to a z-score of

$$z = \frac{M - \mu}{\sigma_M} = \frac{70 - 65}{3} = \frac{5}{3} = +1.67$$

STEP 4   Make a decision about H_0, and state the conclusion.

The z-score we obtained is not in the critical region. This indicates that our sample mean of $M = 70$ is not an extreme or unusual value to be obtained from a population with $\mu = 65$. Therefore, our statistical decision is to *fail to reject H_0*. Our conclusion for the study is that the data do not provide sufficient evidence that the special training changes test scores.

DEMONSTRATION 8.2

EFFECT SIZE USING COHEN'S *d*

We will compute Cohen's *d* using the research situation and the data from Demonstration 8.1. Again, the original population mean was $\mu = 65$ and, after treatment (special training), the sample mean was $M = 70$. Thus, there is a 5-point mean difference. Using the population standard deviation, $\sigma = 15$, we obtain an effect size of

$$\text{Cohen's } d = \frac{\text{mean difference}}{\text{standard deviation}} = \frac{5}{15} = 0.33$$

According to Cohen's evaluation standards (Table 8.2), this is a medium treatment effect.

PROBLEMS

1. In the z-score formula as it is used in a hypothesis test,
   a. Explain what is measured by $M - \mu$ in the numerator.
   b. Explain what is measured by the standard error in the denominator.

2. The value of the z-score in a hypothesis test is influenced by a variety of factors. Assuming that all other variables are held constant, explain how the value of z is influenced by each of the following:
   a. Increasing the difference between the sample mean and the original population mean.
   b. Increasing the population standard deviation.
   c. Increasing the number of scores in the sample.

3. In words, define the alpha level and the critical region for a hypothesis test.

4. If the alpha level is changed from $\alpha = .05$ to $\alpha = .01$,

   a. What happens to the boundaries for the critical region?
   b. What happens to the probability of a Type I error?

5. Although there is a popular belief that herbal remedies such as Ginkgo biloba and Ginseng may improve learning and memory in healthy adults, these effects are usually not supported by well-controlled research (Peresson, Bringlov, Nilsson, & Nyberg, 2004). In a typical study, a researcher obtains a sample of $n = 36$ participants and has each person take the herbal supplements every day for 90 days. At the end of the 90 days, each person takes a standardized memory test. For the general population, scores from the test are normally distributed with a mean of $\mu = 80$ and a standard deviation of $\sigma = 18$. The sample of research participants had an average of $M = 84$.
   a. In a sentence, state the null hypothesis being tested.
   b. Using symbols, state the null hypothesis and the alternative hypothesis.
   c. Conduct the hypothesis test using a two-tailed test with $\alpha = .05$.

6. A researcher is investigating the effectiveness of a new study-skills training program for elementary school children. A sample of $n = 25$ third-grade children is selected to participate in the program and each child is given a standardized achievement test at the end of the year. For the regular population of third-grade children, scores on the test form a normal distribution with a mean of $\mu = 150$ and a standard deviation of $\sigma = 25$. The mean for the sample is $M = 158$.
   a. Identify the independent and the dependent variables for this study.
   b. Assuming a two-tailed test, state the null hypothesis in a sentence that includes the independent variable and the dependent variable.
   c. Using symbols, state the hypotheses (H_0 and H_1) for the two-tailed test.
   d. Sketch the appropriate distribution, and locate the critical region for $\alpha = .05$.
   e. Calculate the test statistic (z-score) for the sample.
   f. What decision should be made about the null hypothesis, and what decision should be made about the effect of the program?

7. A researcher is testing the effectiveness of a new drug that is intended to improve learning and memory performance. A sample of $n = 16$ rats is obtained, and the rats are administered the drug and then are tested on a standardized learning task. For the general population of rats (with no drug) the average score on the standardized task is $\mu = 50$ with a standard deviation of $\sigma = 10$. The distribution of test scores is normal.
   a. If the drug has a 4-point effect and produces a mean score of $M = 54$ for the sample, is this sufficient to conclude that there is a significant effect using a two-tailed test with $\alpha = .05$?
   b. If there is a 6-point effect with a sample mean of $M = 56$, is this sufficient to conclude that there is a significant effect using a two-tailed test with $\alpha = .05$?

8. Childhood participation in sports, cultural groups, and youth groups appears to be related to improved self-esteem for adolescents (McGee, Williams, Howden-Chapman, Martin, & Kawachi, 2006). In a representative study, a sample of $n = 100$ adolescents with a history of group participation is given a standardized self-esteem questionnaire. For the general population of adolescents, scores on this questionnaire form a normal distribution with a mean of $\mu = 40$ and a standard deviation of $\sigma = 12$. The sample of group-participation adolescents had an average of $M = 43.84$. Does this sample provide enough evidence to conclude that self-esteem scores for group-participation adolescents are significantly different from those of the general population? Use a two-tailed test with $\alpha = .01$.

9. A researcher is evaluating the effectiveness of a new physical education program for elementary school children. The program is designed to reduce competition and increase individual self-esteem. A sample of $n = 16$ children is selected and the children are placed in the new program. After 3 months, each child is given a standardized self-esteem test. For the general population of elementary school children, the scores on the self-esteem test form a normal distribution with $\mu = 40$ and $\sigma = 8$.
   a. If the researcher obtains a sample mean of $M = 42$, is this enough evidence to conclude that the program has a significant effect? Assume a two-tailed test with $\alpha = .05$.
   b. If the sample mean is $M = 44$, is this enough to demonstrate a significant effect? Again, assume a two-tailed test with $\alpha = .05$.
   c. Briefly explain why you reached different conclusions for part a and part b.

10. State College is evaluating a new English composition course for freshmen. A random sample of $n = 25$ freshmen is obtained and the students are placed in the course during their first semester. One year later, a writing sample is obtained for each student and the writing samples are graded using a standardized evaluation technique. The average score for the sample

is $M = 76$. For the general population of college students, writing scores form a normal distribution with a mean of $\mu = 70$.

a. If the writing scores for the population have a standard deviation of $\sigma = 20$, does the sample provide enough evidence to conclude that the new composition course has a significant effect? Assume a two-tailed test with $\alpha = .05$.

b. If the population standard deviation is $\sigma = 10$, is the sample sufficient to demonstrate a significant effect? Again, assume a two-tailed test with $\alpha = .05$.

c. Briefly explain why you reached different conclusions for part (a) and part (b).

11. To test the effectiveness of a treatment, a sample is selected from a normal population with a mean of $\mu = 40$ and a standard deviation of $\sigma = 12$. After the treatment is administered to the individuals in the sample, the sample mean is found to be $M = 46$.

a. If the sample consists of $n = 4$ individuals, is this result sufficient to conclude that there is a significant treatment effect? Use a two-tailed test with $\alpha = .05$.

b. If the sample consists of $n = 36$ individuals, is this result sufficient to conclude that there is a significant treatment effect? Use a two-tailed test with $\alpha = .05$.

c. Compute Cohen's d to measure effect size for both tests ($n = 4$ and $n = 36$).

d. Briefly describe how sample size influences the outcome of the hypothesis test. How does sample size influence measures of effect size?

12. To test the effectiveness of a treatment, a sample of $n = 25$ people is selected from a normal population with a mean of $\mu = 60$. After the treatment is administered to the individuals in the sample, the sample mean is found to be $M = 55$.

a. If the population standard deviation is $\sigma = 10$, can you conclude that the treatment has a significant effect? Use a two-tailed test with $\alpha = .05$.

b. If the population standard deviation is $\sigma = 20$, can you conclude that the treatment has a significant effect? Use a two-tailed test with $\alpha = .05$.

c. Compute Cohen's d to measure effect size for both tests ($\sigma = 20$ and $\sigma = 10$).

d. Briefly describe how the standard deviation influences the outcome of the hypothesis test. How does σ influence measures of effect size?

13. There is some evidence that REM sleep, associated with dreaming, may also play a role in learning and memory processing. For example, Smith and Lapp (1991) found increased REM activity for college students during exam periods. Suppose that REM activity for a sample of $n = 16$ students during the final exam period produced an average score of $M = 143$. Regular REM activity for the college population averages $\mu = 110$ with a standard deviation of $\sigma = 50$. The population distribution is approximately normal.

a. Do the data from this sample provide evidence for a significant increase in REM activity during exams? Use a one-tailed test with $\alpha = .01$.

b. Compute Cohen's d to estimate the size of the effect.

14. Under some circumstances a 6-point treatment effect can be very large, and in some circumstances it can be very small. Assume that a sample of $n = 16$ individuals is selected from a population with a mean of $\mu = 70$. A treatment is administered to the sample and, after treatment, the sample mean is found to be $M = 76$. Notice that the treatment appears to have increased scores by an average of 6 points.

a. If the population standard deviation is $\sigma = 20$, is the 6-point effect large enough to be statistically significant? Use a two-tailed test with $\alpha = .05$.

b. If the population standard deviation is $\sigma = 8$, is the 6-point effect large enough to be statistically significant? Use a two-tailed test with $\alpha = .05$.

15. A psychologist is investigating the hypothesis that children who grow up as the only child in the household develop different personality characteristics than those who grow up in larger families. A sample of $n = 30$ only children is obtained and each child is given a standardized personality test. For the general population, scores on the test from a normal distribution with a mean of $\mu = 50$ and a standard deviation of $\sigma = 15$. If the mean for the sample is $M = 58$, can the researcher conclude that there is a significant difference in personality between only children and the rest of the population? Use a two-tailed test with $\alpha = .05$.

16. A researcher is testing the hypothesis that consuming a sports drink during exercise will improve endurance. A sample of $n = 50$ male college students is obtained and each student is given a series of three endurance tasks and asked to consume 4 ounces of the drink during each break between tasks. The overall endurance score for this sample is $M = 53$. For the general population, without any sports drink, the scores for this task average $\mu = 50$ with a standard deviation of $\sigma = 12$.

a. Can the researcher conclude that endurance scores with the sports drink are significantly higher than scores without the drink? Use a one-tailed test with $\alpha = .05$.

b. Can the researcher conclude that endurance scores with the sports drink are significantly different than scores without the drink? Use a two-tailed test with $\alpha = .05$.

c. You should find that the two tests lead to different conclusions. Explain why.

17. Montarello and Martins (2005) found that fifth-grade students completed more mathematics problems correctly when simple problems were mixed in with their regular math assignments. To further explore this phenomenon, suppose that a researcher selects a standardized mathematics achievement test that produces a normal distribution of scores with a mean of $\mu = 100$ and a standard deviation of $\sigma = 18$. The researcher modifies the test by inserting a set of very easy problems among the standardized questions, and gives the modified test to a sample of $n = 36$ students. If the average test score for the sample is $M = 104$, is this result sufficient to conclude that inserting the easy questions improves student performance? Use a one-tailed test with $\alpha = .01$.

18. A sample of $n = 16$ individuals is selected from a normal population with a mean of $\mu = 48$ and a standard deviation of $\sigma = 12$. After receiving a treatment, the sample mean is found to be $M = 52$.

a. Compute Cohen's d to evaluate the size of the treatment effect.

b. If the sample size were $n = 36$, what value would be obtained for Cohen's d? How does sample size influence the measure of effect size?

c. If the population standard deviation were $\sigma = 24$, what value would be obtained for Cohen's d? How does standard deviation influence the measure of effect size?

d. If the sample mean were $M = 56$, what value would be obtained for Cohen's d? How does the size of the mean difference influence the measure of effect size?

19. Researchers have often noted increases in violent crimes when it is very hot. In fact, Reifman, Larrick, and Fein (1991) noted that this relationship even extends to baseball. That is, there is a much greater chance of a batter being hit by a pitch when the temperature increases. Consider the following hypothetical data. Suppose that over the past 30 years, during any given week of the major-league season, an average of $\mu = 12$ players are hit by wild pitches. Assume that the distribution is nearly normal with $\sigma = 3$. For a sample of $n = 4$ weeks in which the daily temperature was extremely hot, the weekly average of hit-by-pitch players was $M = 15.5$. Are players more likely to get hit by pitches during hot weeks? Set alpha to .05 for a one-tailed test.

20. A sample of $n = 9$ individuals was selected from a normally distributed population with a mean of $\mu = 25$ and a standard deviation of $\sigma = 6$. A treatment was administered to the sample and, after treatment, the nine individuals produced the following scores: 36, 27, 30, 33, 22, 30, 30, 24, 38.

a. Calculate the sample mean.

b. Sketch a histogram showing the sample distribution.

c. Does the sample appear to be centered around $\mu = 25$ or does the treatment appear to have moved the scores away from $\mu = 25$?

d. Use a hypothesis test with $\alpha = .05$ to determine whether the treatment had a significant effect.

e. Looking at the sample distribution from part b, determine how big the treatment effect is. Specifically, what is the distance between the sample mean and the original population mean?

f. Compute Cohen's d to measure the size of the treatment effect.

21. Briefly explain how increasing sample size influences each of the following. Assume that all other factors are held constant.

a. The size of the z-score in a hypothesis test.

b. The size of Cohen's d.

c. The power of a hypothesis test.

22. Explain how the power of a hypothesis test is influenced by each of the following. Assume that all other factors are held constant.

a. Increasing the alpha level from .01 to .05.

b. Changing from a one-tailed test to a two-tailed test.

23. A researcher is investigating the effectiveness of a new medication for lowering blood pressure for individuals with systolic pressure greater than 140. For this population, systolic scores average $\mu = 160$ with a standard deviation of $\sigma = 20$, and the scores form a normal shaped distribution. The researcher plans to select a sample of $n = 25$ individuals, and measure their systolic blood pressure after they take the medication for 60 days. If the researcher uses a two-tailed test with $\alpha = .05$,

a. What is the power of the test if the medication has a 5-point effect?

b. What is the power of the test if the medication has a 10-point effect?

24. A researcher is evaluating the influence of a treatment using a sample selected from a normally distributed population with a mean of $\mu = 80$ and a standard deviation of $\sigma = 20$. The researcher expects a 12-point treatment effect and plans to use a two-tailed hypothesis test with $\alpha = .05$.

a. Compute the power of the test if the researcher uses a sample of $n = 16$ individuals. (See Example 8.6.)

   b. Compute the power of the test if the researcher uses a sample of $n = 25$ individuals.

25. A researcher expects a treatment to increase scores by 5 points. The regular population, without treatment, averages $\mu = 40$ with a standard deviation of $\sigma = 8$, and the scores form a normal distribution. If the researcher uses a one-tailed test with $\alpha = .01$,

   a. What is the power of the test for a sample of $n = 16$?

   b. What is the power of the test for a sample of $n = 64$?

CHAPTER

9

Introduction to the *t* Statistic

Tools You Will Need

The following items are considered essential background material for this chapter. If you doubt your knowledge of any of these items, you should review the appropriate chapter or section before proceeding.

- Sample standard deviation (Chapter 4)
- Degrees of freedom (Chapter 4)
- Standard error (Chapter 7)
- Hypothesis testing (Chapter 8)

Preview

9.1 The *t* Statistic: An Alternative to *z*

9.2 Hypothesis Tests with the *t* Statistic

9.3 Measuring Effect Size for the *t* Statistic

9.4 Directional Hypotheses and One-Tailed Tests

Summary

Focus on Problem Solving

Demonstrations 9.1 and 9.2

Problems

Preview

Numerous accounts suggest that for many animals, including humans, direct stare from another animal is aversive (e.g., Cook, 1977). Try it out for yourself. Make direct eye contact with a stranger in a cafeteria. Chances are the person will display avoidance by averting his or her gaze or turning away from you. Some insects, such as moths, have even developed eye-spot patterns on the wings or body to ward off predators (mostly birds) who may have a natural fear of eyes (Blest, 1957). Suppose that a comparative psychologist is interested in determining whether or not the birds that feed on these insects show an avoidance of eye-spot patterns.

Using methods similar to those of Scaife (1976), the researcher performed the following experiment. A sample of $n = 16$ moth-eating birds is selected. The animals are tested in an apparatus that consists of a two-chambered box. The birds are free to roam from one side of the box to the other through a doorway in the partition that separates the two chambers. In one chamber, there are two eye-spot patterns painted on the wall. The other side of the box has plain walls. One at a time, the researcher tests each bird by placing it in the doorway between the chambers. Each subject is left in the apparatus for 60 minutes, and the amount of time spent in the plain chamber is recorded.

The null hypothesis for this study would state that eye-spot patterns have no effect on the behavior of moth-eating birds. If this is tue, then birds placed in the apparatus should wander randomly from side to side during the 60-minute period, averaging half of the time on each side. Therefore, for the general population of moth-eating birds, the null hypothesis states

$$H_0: \mu_{plain\ side} = 30\ minutes$$

The Problem: The researcher has most of the information needed to conduct a hypothesis test. In particular, the researchers has a hypothesis about the population ($\mu = 30$ minutes) and a sample of $n = 16$ scores that will produce a sample mean (M). However, the researcher does not know the population standard deviation (σ). This value is needed to compute the standard error for the sample mean (σ_M) that appears in the denominator of the z-score equation. Recall that the standard error measures how much difference is reasonable to expect between a sample mean (M) and the population mean (μ). The value of the standard error is critical to deciding whether the sample data are consistent with the null hypothesis or refute the null hypothesis. Without the standard error it is impossible to conduct a z-score hypothesis test.

The Solution: Because it is impossible to compute the standard error, a z-score cannot be used for the hypothesis test. However, it is possible to estimate the standard error using the sample data. The estimated standard error can then be used to compute a new statistic that is similar in structure to the z-score. The new statistic is called a *t statistic* and it can be used to conduct a new kind of hypothesis test.

9.1   THE *t* STATISTIC: AN ALTERNATIVE TO *z*

In the previous chapter, we presented the statistical procedures that permit researchers to use a sample mean to test hypotheses about an unknown population mean. These statistical procedures were based on a few basic concepts, which we summarize as follows:

Remember that the expected value of the distribution of sample means is μ, the population mean.

1. A sample mean (M) is expected more or less to approximate its population mean (μ). This permits us to use the sample mean to test a hypothesis about the population mean.

2. The standard error provides a measure of how well a sample mean approximates the population mean. Specifically, the standard error determines how much difference is reasonable to expect between a sample mean (M) and the population mean (μ).

$$\sigma_M = \frac{\sigma}{\sqrt{n}} \qquad \text{or} \qquad \sigma_M = \sqrt{\frac{\sigma^2}{n}}$$

3. To quantify our inferences about the population, we compare the obtained sample mean (M) with the hypothesized population mean (μ) by computing a z-score test statistic.

$$z = \frac{M - \mu}{\sigma_M} = \frac{\text{obtained difference between data and hypothesis}}{\text{standard distance between } M \text{ and } \mu}$$

The goal of the hypothesis test is to determine whether the obtained difference between the data and the hypothesis is significantly greater than would be expected by chance. When the z-scores form a normal distribution, we are able to use the unit normal table (Appendix B) to find the critical region for the hypothesis test.

THE PROBLEM WITH *z*-SCORES

The shortcoming of using a z-score as an inferential statistic is that the z-score formula requires more information than is usually available. Specifically, a z-score requires that we know the value of the population standard deviation (or variance), which is needed to compute the standard error. In most situations, however, the standard deviation for the population is not known. In fact, the whole reason for conducting a hypothesis test is to gain knowledge about an *unknown* population. This situation appears to create a paradox: You want to use a z-score to find out about an unknown population, but you must know about the population before you can compute a z-score. Fortunately, there is a relatively simple solution to this problem. When the variability for the population is not known, we use the sample variability in its place.

INTRODUCING THE *t* STATISTIC

In Chapter 4, the sample variance was developed specifically to provide an unbiased estimate of the corresponding population variance. Recall the formulas for sample variance and sample standard deviation as follows:

The concept of degrees of freedom, $df = n - 1$, was introduced in Chapter 4 (page 121) and is discussed later in this chapter (page 283).

$$\text{sample variance} = s^2 = \frac{SS}{n - 1} = \frac{SS}{df}$$

$$\text{sample standard deviation} = s = \sqrt{\frac{SS}{n - 1}} = \sqrt{\frac{SS}{df}}$$

Using the sample values, we can now *estimate* the standard error. Recall from Chapters 7 and 8 that the value of the standard error can be computed using either standard deviation or variance:

$$\text{standard error} = \sigma_M = \frac{\sigma}{\sqrt{n}} \quad \text{or} \quad \sigma_M = \sqrt{\frac{\sigma^2}{n}}$$

Now we will estimate the standard error by simply substituting the sample variance or standard deviation in place of the unknown population value:

$$\text{estimated standard error} = s_M = \frac{s}{\sqrt{n}} \quad \text{or} \quad s_M = \sqrt{\frac{s^2}{n}} \tag{9.1}$$

Notice that the symbol for the *estimated standard error of* M is s_M instead of σ_M, indicating that the estimated value is computed from sample data rather than from the actual population parameter.

D E F I N I T I O N

The **estimated standard error (s_M)** is used as an estimate of the real standard error σ_M when the value of σ is unknown. It is computed from the sample variance or sample standard deviation and provides an estimate of the standard distance between a sample mean M and the population mean μ.

Finally, you should recognize that we have shown formulas for standard error (actual or estimated) using both the standard deviation and the variance. In the past (Chapters 7 and 8), we concentrated on the formula using the standard deviation. At this point, however, we will shift our focus to the formula based on variance. Thus, throughout the remainder of this chapter, and in following chapters, the estimated standard error of *M* typically is presented and computed using

$$s_M = \sqrt{\frac{s^2}{n}}$$

There are two reasons for making this shift from standard deviation to variance:

1. In Chapter 4 (page 122) we saw that the sample variance is an *unbiased* statistic; on average, the sample variance (s^2) will provide an accurate and unbiased estimate of the population variance (σ^2). Therefore, the most accurate way to estimate the standard error is to use the sample variance to estimate the population variance.

2. In future chapters we will encounter other versions of the *t* statistic that require variance (instead of standard deviation) in the formulas for estimated standard error. To maximize the similarity from one version to another, we will use variance in the formula for *all* of the different *t* statistics. Thus, whenever we present a *t* statistic, the estimated standard error will be computed as

$$\text{estimated standard error} = \sqrt{\frac{\text{sample variance}}{\text{sample size}}}$$

Now we can substitute the estimated standard error in the denominator of the *z*-score formula. The result is a new test statistic called a *t statistic*:

$$t = \frac{M - \mu}{s_M} \tag{9.2}$$

<table>
<tr><td>D E F I N I T I O N</td><td>The ***t* statistic** is used to test hypotheses about an unknown population mean μ when the value of σ is unknown. The formula for the *t* statistic has the same structure as the *z*-score formula, except that the *t* statistic uses the estimated standard error in the denominator.</td></tr>
</table>

The only difference between the *t* formula and the *z*-score formula is that the *z*-score uses the actual population variance, σ^2 (or the standard deviation), and the *t* formula uses the corresponding sample variance (or standard deviation) when the population value is not known.

$$z = \frac{M - \mu}{\sigma_M} = \frac{M - \mu}{\sqrt{\sigma^2/n}} \qquad t = \frac{M - \mu}{s_M} = \frac{M - \mu}{\sqrt{s^2/n}}$$

DEGREES OF FREEDOM AND THE *t* STATISTIC

In this chapter, we have introduced the *t* statistic as a substitute for a *z*-score. The basic difference between these two is that the *t* statistic uses sample variance (s^2) and the *z*-score uses the population variance (σ^2). To determine how well a *t* statistic approximates a *z*-score, we must determine how well the sample variance approximates the population variance.

In Chapter 4, we introduced the concept of degrees of freedom (page 121). Reviewing briefly, you must know the sample mean before you can compute sample variance. This places a restriction on sample variability such that only $n - 1$ scores in a sample are independent and free to vary. The value $n - 1$ is called the *degrees of freedom* (or *df*) for the sample variance.

$$\text{degrees of freedom} = df = n - 1 \tag{9.3}$$

DEFINITION

Degrees of freedom describe the number of scores in a sample that are independent and free to vary. Because the sample mean places a restriction on the value of one score in the sample, there are $n - 1$ degrees of freedom for the sample (see Chapter 4).

The greater the value of *df* for a sample, the better s^2 represents σ^2, and the better the *t* statistic approximates the *z*-score. This should make sense because the larger the sample (*n*) is, the better the sample represents its population. Thus, the degrees of freedom associated with s^2 also describe how well *t* represents *z*.

THE *t* DISTRIBUTION

Every sample from a population can be used to compute a *z*-score or a *t* statistic. If you select all the possible samples of a particular size (*n*), and compute the *z*-score for each sample mean, then the entire set of *z*-scores will form a *z*-score distribution. As we saw in Chapter 7, the distribution of *z*-scores for sample means tends to be a normal distribution. In the same way, you can compute the *t* statistic for every sample and the entire set of *t* values will form a t *distribution*. The *t* distribution will approximate a normal distribution in the same way that a *t* statistic approximates a *z*-score. How well a *t* distribution approximates a normal distributor is determined by degrees of freedom. In general, the greater the sample size (*n*) is, the larger the degrees of freedom ($n - 1$) are, and the better the t distribution approximates the normal distribution (Figure 9.1).

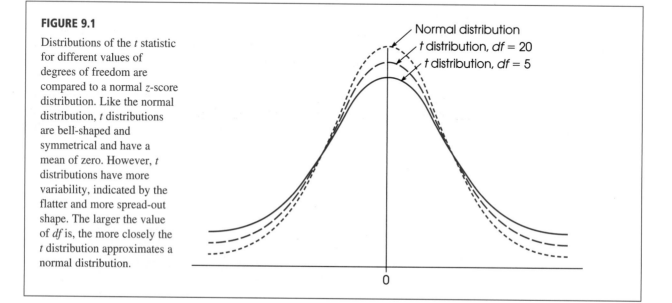

FIGURE 9.1

Distributions of the *t* statistic for different values of degrees of freedom are compared to a normal *z*-score distribution. Like the normal distribution, *t* distributions are bell-shaped and symmetrical and have a mean of zero. However, *t* distributions have more variability, indicated by the flatter and more spread-out shape. The larger the value of *df* is, the more closely the *t* distribution approximates a normal distribution.

Normal distribution
t distribution, *df* = 20
t distribution, *df* = 5

0

DEFINITION	A *t* **distribution** is the complete set of *t* values computed for every possible random sample for a specific sample size (*n*) or a specific degrees of freedom (*df*). The *t* distribution approximates the shape of a normal distribution.

THE SHAPE OF THE *t* DISTRIBUTION

The exact shape of a *t* distribution changes with degrees of freedom. In fact, statisticians speak of a "family" of *t* distributions. That is, there is a different sampling distribution of *t* (a distribution of all possible sample *t* values) for each possible number of degrees of freedom. As *df* gets very large, the *t* distribution gets closer in shape to a normal *z*-score distribution. A quick glance at Figure 9.1 reveals that distributions of *t* are bell-shaped and symmetrical and have a mean of zero. However, the *t* distribution has more variability than a normal *z* distribution, especially when *df* values are small (see Figure 9.1). The *t* distribution tends to be flatter and more spread out, whereas the normal *z* distribution has more of a central peak.

The reason that the *t* distribution is flatter and more variable than the normal *z*-score distribution becomes clear if you look at the structure of the formulas for *z* and *t*. For both formulas, *z* and *t*, the top of the formula, $M - \mu$, can take on different values because the sample mean (*M*) will vary from one sample to another. For *z*-scores, however, the bottom of the formula will not vary, provided that all of the samples are the same size and are selected from the same population. Specifically, all the *z*-scores will have the same standard error in the denominator, $\sigma_M = \sqrt{\sigma^2/n}$, because the population variance and the sample size are the same for every sample. For *t* statistics, on the other hand, the bottom of the formula will vary from one sample to another. Specifically, the sample variance (s^2) will change from one sample to the next, so the estimated standard error will also vary, $s_M = \sqrt{s^2/n}$. Thus, only the numerator of the *z*-score formula varies, but both the numerator and the denominator of the *t* statistic vary. As a result, *t* statistics are more variable than are *z*-scores, and the *t* distribution will be flatter and more spread out. As sample size and *df* increase however, the variability in the *t* distribution decreases, and it more closely resembles a normal distribution.

DETERMINING PROPORTIONS AND PROBABILITIES FOR *t* DISTRIBUTIONS

Just as we used the unit normal table to locate proportions associated with *z*-scores, we will use a *t* distribution table to find proportions for *t* statistics. The complete *t* distribution table is presented in Appendix B, page 729, and a portion of this table is reproduced in Table 9.1. The two rows at the top of the table show proportions of the *t* distribution contained in either one or two tails, depending on which row is used. The first column of

TABLE 9.1

A portion of the *t*-distribution table. The numbers in the table are the values of *t* that separate the tail from the main body of the distribution. Proportions for one or two tails are listed at the top of the table, and *df* values for *t* are listed in the first column.

		Proportion in One Tail				
	0.25	0.10	0.05	0.025	0.01	0.005
		Proportion in Two Tails Combined				
df	0.50	0.20	0.10	0.05	0.02	0.01
1	1.000	3.078	6.314	12.706	31.821	63.657
2	0.816	1.886	2.920	4.303	6.965	9.925
3	0.765	1.638	2.353	3.182	4.541	5.841
4	0.741	1.533	2.132	2.776	3.747	4.604
5	0.727	1.476	2.015	2.571	3.365	4.032
6	0.718	1.440	1.943	2.447	3.143	3.707

the table lists degrees of freedom for the *t* statistic. Finally, the numbers in the body of the table are the *t* values that mark the boundary between the tails and the rest of the *t* distribution.

For example, with *df* = 3, exactly 5% of the *t* distribution is located in the tail beyond *t* = 2.353 (Figure 9.2). The process of finding this value is highlighted in Table 9.1. Begin by finding *df* = 3 in the first column of the table. Then locate a proportion of 0.05 (5%) in the one-tail proportion row. When you line up these two values in the table, you should find *t* = 2.353. Similarly, 5% of the *t* distribution is located in the tail beyond *t* = −2.353 (see Figure 9.2). Finally, notice that a total of 10% is contained in the two tails beyond *t* = ±2.353 (check the proportion value in the "two-tails combined" row at the top of the table).

A close inspection of the *t* distribution table in Appendix B will demonstrate a point we made earlier: As the value for *df* increases, the *t* distribution becomes more similar to a normal distribution. For example, examine the column containing *t* values for a 0.05 proportion in two tails. You will find that when *df* = 1, the *t* values that separate the extreme 5% (0.05) from the rest of the distribution are *t* = ±12.706. As you read down the column, however, you should find that the critical *t* values become smaller and smaller, ultimately reaching ±1.96. You should recognize ±1.96 as the *z*-score values that separate the extreme 5% in a normal distribution. Thus, as *df* increases, the proportions in a *t* distribution become more like the proportions in a normal distribution. When the sample size (and degrees of freedom) is sufficiently large, the difference between a *t* distribution and the normal distribution becomes negligible.

Caution: The *t* distribution table printed in this book has been abridged and does not include entries for every possible *df* value. For example, the table lists *t* values for *df* = 40 and for *df* = 60, but does not list any entries for *df* values between 40 and 60. Occasionally, you will encounter a situation in which your *t* statistic has a *df* value that is not listed in the table. In these situations, you should look up the critical *t* for both of the surrounding *df* values listed and then use the *larger* value for *t*. If, for example, you have *df* = 53 (not listed), look up the critical *t* value for both *df* = 40 and *df* = 60 and *then use the larger t value.* If your sample *t* statistic is greater than the larger value listed, you can be certain that the data are in the critical region, and you can confidently reject the null hypothesis.

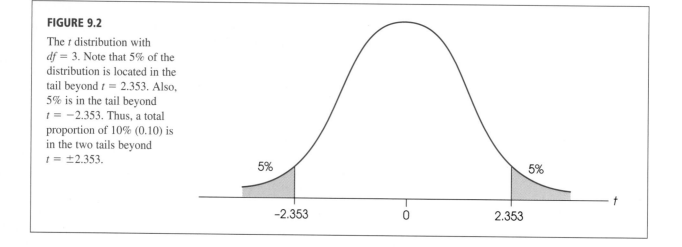

FIGURE 9.2

The *t* distribution with *df* = 3. Note that 5% of the distribution is located in the tail beyond *t* = 2.353. Also, 5% is in the tail beyond *t* = −2.353. Thus, a total proportion of 10% (0.10) is in the two tails beyond *t* = ±2.353.

LEARNING CHECK

1. Under what circumstances is a *t* statistic used instead of a *z*-score for a hypothesis test?

2. A sample of $n = 4$ scores has $SS = 108$.
   a. Compute the variance for the sample.
   b. Compute the estimated standard error for the sample mean.

3. In general, a distribution of *t* statistics is flatter and more spread out than the standard normal distribution. (True or false?)

4. A sample of $n = 15$ scores would produce a *t* statistic with $df = 16$. (True or false?)

5. For $df = 15$, find the value(s) of *t* associated with each of the following:
   a. The top 5% of the distribution.
   b. The middle 95% versus the extreme 5% of the distribution.
   c. The middle 99% versus the extreme 1% of the distribution.

ANSWERS

1. A *t* statistic is used instead of a *z*-score when the population standard deviation and variance are not known.

2. a. $s^2 = 36$
   b. $s_M = 3$

3. True.

4. False. With $n = 15$, the *df* value would be 14.

5. a. $t = +1.753$ b. $t = \pm 2.131$ c. $t = \pm 2.947$

9.2   HYPOTHESIS TESTS WITH THE *t* STATISTIC

In the hypothesis-testing situation, we begin with a population with an unknown mean and an unknown variance, often a population that has received some treatment (Figure 9.3). The goal is to use a sample from the treated population (a treated sample) as the basis for determining whether the treatment has any effect.

As always, the null hypothesis states that the treatment has no effect; specifically, H_0 states that the population mean is unchanged. Thus, the null hypothesis provides a specific value for the unknown population mean. The sample data provide a value for the sample mean. Finally, the variance and estimated standard error are computed from the sample data. When these values are used in the *t* formula, the result becomes

$$t = \frac{\underset{\text{(from the data)}}{\text{sample mean}} - \underset{\text{(hypothesized from } H_0\text{)}}{\text{population mean}}}{\underset{\text{(computed from the sample data)}}{\text{estimated standard error}}}$$

As with the *z*-score formula, the *t* statistic forms a ratio. The numerator measures the actual difference between the sample data (M) and the population hypothesis (μ). The

FIGURE 9.3

The basic experimental situation for using the *t* statistic or the *z*-score is presented. It is assumed that the parameter μ is known for the population before treatment. The purpose of the experiment is to determine whether the treatment has an effect. Note that the population after treatment has unknown values for the mean and the variance. We will use a sample to test a hypothesis about the population mean.

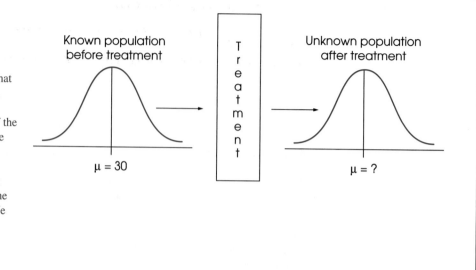

estimated standard error in the denominator measures how much difference is reasonable to expect between a sample mean and the population mean. When the obtained difference between the data and the hypothesis (numerator) is much greater than expected (denominator), we will obtain a large value for *t* (either large positive or large negative). In this case, we conclude that the data are not consistent with the hypothesis, and our decision is to "reject H_0." On the other hand, when the difference between the data and the hypothesis is small relative to the standard error, we will obtain a *t* statistic near zero, and our decision will be "fail to reject H_0."

HYPOTHESIS-TESTING EXAMPLE

The following research situation will be used to demonstrate the procedures of hypothesis testing with the *t* statistic.

EXAMPLE 9.1

Infants, even newborns, prefer to look at attractive faces compared to less attractive faces (Slater, et al., 1998). In the study, infants from 1 to 6 days old were shown two photographs of women's faces. Previously, a group of adults had rated one of the faces as significantly more attractive than the other. The babies were positioned in front of a screen on which the photographs were presented. The pair of faces remained on the screen until the baby accumulated a total of 20 seconds of looking at one or the other. The number of seconds looking at the attractive face was recorded for each infant. Suppose that the study used a sample of $n = 9$ infants and the data produced an average of $M = 13$ seconds for the attractive face with $SS = 72$. Note that all the available information comes from the sample. Specifically, we do not know the population mean or the population standard deviation.

STEP 1

State the hypotheses and select an alpha level. Although we have no information about the population of scores, it is possible to form a hypothesis about the value of μ. In this case, the null hypothesis states that the infants have no preference for either face. That is, they should average half of the 20 seconds looking at each of the two faces. In symbols, the null hypothesis states

$$H_0: \quad \mu_{attractive} = 10 \text{ seconds}$$

The alternative hypothesis states that there is a preference and one of the faces is preferred over the other. A directional, one-tailed test would specify which of the two faces is preferred, but the nondirectional alternative hypothesis is expressed as follows:

$$H_1: \quad \mu_{attractive} \neq 10 \text{ seconds}$$

We will set the level of significance at $\alpha = .05$ for two tails.

STEP 2 Locate the critical region. The test statistic is a *t* statistic because the population variance is not known. The exact shape of the *t* distribution and, therefore, the proportions under the *t* distribution depend on the number of degrees of freedom associated with the sample. To find the critical region, *df* must be computed:

$$df = n - 1 = 9 - 1 = 8$$

For a two-tailed test at the .05 level of significance and with 8 degrees of freedom, the critical region consists of *t* values greater than $+2.306$ or less than -2.306. Figure 9.4 depicts the critical region in this *t* distribution.

STEP 3 Calculate the test statistic. The *t* statistic typically requires much more computation than is necessary for a *z*-score. Therefore, we recommend that you divide the calculations into a three-stage process as follows:

  a. First, calculate the sample variance. Remember that the population variance is unknown, and you must use the sample value in its place. (This is why we are using a *t* statistic instead of a *z*-score.)

$$s^2 = \frac{SS}{n - 1} = \frac{SS}{df}$$

$$= \frac{72}{8} = 9$$

  b. Next, use the sample variance (s^2) to compute the estimated standard error. This value will be the denominator of the *t* statistic and measures how much

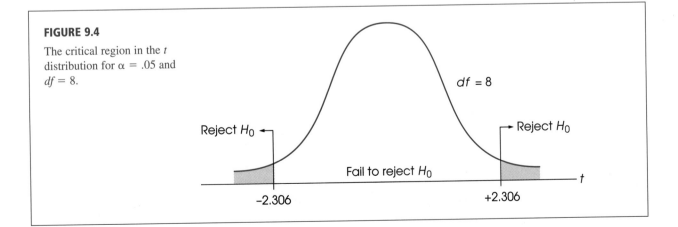

FIGURE 9.4

The critical region in the *t* distribution for $\alpha = .05$ and $df = 8$.

error is reasonable to expect by chance between a sample mean and the corresponding population mean.

$$s_M = \sqrt{\frac{s^2}{n}}$$

$$= \sqrt{\frac{9}{9}} = \sqrt{1} = 1$$

c. Finally, compute the *t* statistic for the sample data.

$$t = \frac{M - \mu}{s_M}$$

$$= \frac{13 - 10}{1} = 3.00$$

STEP 4   Make a decision regarding H_0. The obtained *t* statistic of 3.00 falls into the critical region on the right-hand side of the *t* distribution (see Figure 9.4). Our statistical decision is to reject H_0 and conclude that babies do show a preference when given a choice between an attractive and an unattractive face. Specifically, the average amount of time that the babies spent looking at the attractive face was significantly different from the 10 seconds that would be expected if there were no preference. As indicated by the sample mean, there is a tendency for the babies to spend more time looking at the attractive face.

ASSUMPTIONS OF THE *t* TEST   Two basic assumptions are necessary for hypothesis tests with the *t* statistic.

1. The values in the sample must consist of *independent* observations.
   In everyday terms, two observations are independent if there is no consistent, predictable relationship between the first observation and the second. More precisely, two events (or observations) are independent if the occurrence of the first event has no effect on the probability of the second event. We examined specific examples of independence and nonindependence in Box 8.2 (page 253).

2. The population sampled must be normal.
   This assumption is a necessary part of the mathematics underlying the development of the *t* statistic and the *t* distribution table. However, violating

LEARNING CHECK

1. A sample of $n = 16$ individuals is selected from a population with a mean of $\mu = 40$. A treatment is administered to the individuals in the sample and, after treatment, the sample mean is $M = 44$ with $SS = 540$. Is this sample sufficient to conclude that the treatment has a significant effect? Use a two-tailed test with $\alpha = .05$.

ANSWER

1. H_0: $\mu = 40$ even after the treatment. The sample variance is 36, the estimated standard error is 1.50, and $t = \frac{4}{1.50} = 2.67$. With $df = 15$, the critical boundaries are set at $t = \pm 2.131$. Reject H_0 and conclude that the treatment has a significant effect.

this assumption has little practical effect on the results obtained for a *t* statistic, especially when the sample size is relatively large. With very small samples, a normal population distribution is important. With larger samples, this assumption can be violated without affecting the validity of the hypothesis test. If you have reason to suspect that the population distribution is not normal, use a large sample to be safe.

9.3 MEASURING EFFECT SIZE FOR THE *t* STATISTIC

In Chapter 8 we noted that one criticism of a hypothesis test is that it does not really evaluate the size of the treatment effect. Instead, a hypothesis test simply determines whether the treatment effect is greater than chance, where "chance" is measured by the standard error. In particular, it is possible for a very small treatment effect to be "statistically significant," especially when the sample size is very large. To correct for this problem, it is recommended that the results from a hypothesis test be accompanied by a report of effect size such as Cohen's *d*.

ESTIMATING COHEN'S *d*   When Cohen's *d* was originally introduced (page 262), the formula was presented as

$$\text{Cohen's } d = \frac{\text{mean difference}}{\text{standard deviation}} = \frac{\mu_{treatment} - \mu_{no\ treatment}}{\sigma}$$

Cohen defined this measure of effect size in terms of the population mean difference and the population standard deviation. However, in most situations the population values are not known and you must substitute the corresponding sample values, in their place. When this is done, many researchers prefer to identify the calculated value as an "*estimated* d" or name the value after one of the statisticians who first substituted sample statistics into Cohen's formula (e.g., Glass's *g* or Hedges's *g*). For hypothesis tests using the *t* statistic, the population mean with no treatment is the value specified by the null hypothesis. However, the population mean with treatment and the standard deviation are both unknown. Therefore, we use the mean for the treated sample and the standard deviation for the sample after treatment as estimates of the unknown parameters. With these substitutions, the formula for estimating Cohen's *d* becomes

$$\text{estimated } d = \frac{\text{mean difference}}{\text{sample standard deviation}} = \frac{M - \mu}{s} \tag{9.4}$$

The numerator measures that magnitude of the treatment effect by finding the difference between the mean for the treated sample and the mean for the untreated population (μ from H_0). The sample standard deviation in the denominator standardizes the mean difference into standard deviation units. Thus, an estimated *d* of 1.00 indicates that the size of the treatment effect is equivalent to one standard deviation. The following example demonstrates how the estimated *d* is used to measure effect size for a hypothesis test using a *t* statistic.

EXAMPLE 9.2   For the infant face preference study in Example 9.1, the babies averaged $M = 13$ out of 20 seconds looking at the attractive face. If there were no preference (as stated by the null hypothesis), the population mean would be $\mu = 10$ seconds. Thus, the results show a 3-second difference between the mean with a preference ($M = 13$)

and the mean with no preference ($\mu = 10$). Also, for this study the sample standard deviation is

$$s = \sqrt{\frac{SS}{df}} = \sqrt{\frac{72}{8}} = \sqrt{9} = 3$$

Thus, Cohen's *d* for this example is estimated to be

$$\text{Cohen's } d = = \frac{M - \mu}{s} = \frac{13 - 10}{3} = 1.00$$

According to the standards suggested by Cohen (Table 8.2, page 264), this is a large treatment effect.

To help you visualize what is measured by Cohen's *d*, we have constructed a set of $n = 9$ scores with a mean of $M = 13$ and a standard deviation of $s = 3$ (the same values as in Examples 9.1 and 9.2). The set of scores is shown in Figure 9.5. Notice that the figure also includes an arrow that locates $\mu = 10$. Recall that $\mu = 10$ is the value specified by the null hypothesis and identifies what the mean ought to be if the treatment has no effect. Clearly, our sample is not centered around $\mu = 10$. Instead, the scores have been shifted to the right so that the sample mean is $M = 13$. This shift, from 10 to 13, is the 3-point mean difference that was caused by the treatment effect. Also notice that the 3-point mean difference is exactly equal to the standard deviation. Thus, the size of the treatment effect is equal to 1 standard deviation. In other words, Cohen's $d = 1.00$.

MEASURING THE PERCENTAGE OF VARIANCE EXPLAINED, r^2

An alternative method for measuring effect size is to determine how much of the variability in the scores is explained by the treatment effect. The concept behind this measure is that the treatment causes the scores to increase (or decrease), which means that the treatment is causing the scores to vary. If we can measure how much of the

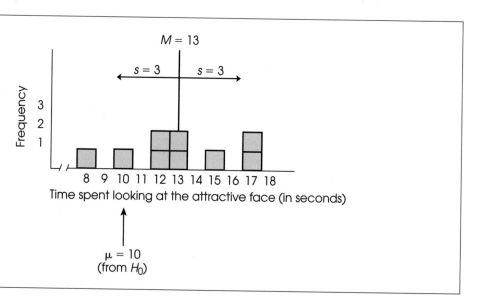

FIGURE 9.5

The sample distribution for the scores that were used in Examples 9.1 and 9.2. The population mean, $\mu = 10$ seconds, is the value that would be expected if attractiveness has no effect on thet infants' behavior. Note that the sample mean is displaced away from $\mu = 10$ by a distance equal to one standard deviation.

variability is explained by the treatment, we will obtain a measure of the size of the treatment effect.

To demonstrate this concept we will use the data from the hypothesis test in Example 9.1. Recall that the null hypothesis stated that the treatment (the attractiveness of the faces) has no effect on the infants' behavior. According to the null hypothesis, the infants should show no preference between the two photographs, and therefore should spend an average of $\mu = 10$ out of 20 seconds looking at the attractive face. However, if you look at the data in Figure 9.5, the scores are not centered around $\mu = 10$. Instead, the scores are shifted to the right so that they are centered around the sample mean, $M = 13$. This shift is the treatment effect. To measure the size of the treatment effect we calculate deviations from the mean and the sum of squared deviations, SS, two different ways.

Figure 9.6(a) shows the original set of scores. For each score, the deviation from $\mu = 10$ is shown as a colored line. Recall that $\mu = 10$ comes from the null hypothesis and represents the mean that should be obtained if the treatment has no effect. Note that almost all of the scores are located on the right-hand side of $\mu = 10$. This shift to the right is the treatment effect. Specifically, the preference for the attractive face has caused the infants to spend more time looking at the attractive photograph, which means that their scores are generally greater than 10. Thus, the treatment has pushed the scores away from $\mu = 10$ and has increased the size of the deviations.

Next, we will see what happens if the treatment effect is removed. In this example, the treatment has a 3-point effect (the average increases from $\mu = 10$ to $M = 13$). To remove the treatment effect, we will simply subtract 3 points from each score. The

FIGURE 9.6

Deviations from $\mu = 10$ (no treatment effect) for the scores in Example 9.1. The colored lines in part (a) show the deviations for the original scores, including the treatment effect. In part (b) the colored lines show the deviations for the adjusted scores after the treatment effect has been removed.

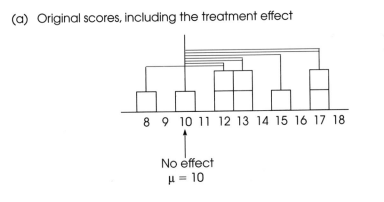

(a) Original scores, including the treatment effect

8 9 10 11 12 13 14 15 16 17 18

No effect
$\mu = 10$

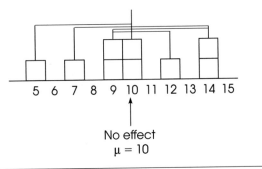

(b) Adjusted scores with the treatment effect removed

5 6 7 8 9 10 11 12 13 14 15

No effect
$\mu = 10$

adjusted scores are shown in Figure 9.6(b) and, once again, the deviations from $\mu = 10$ are shown as colored lines. First, notice that the adjusted scores are centered at $\mu = 10$, indicating that there is no treatment effect. Also notice that the deviations, the colored lines, are noticeably smaller when the treatment effect is removed.

To measure how much the variability is reduced when the treatment effect is removed, we will compute the sum of squared deviations, SS, for each set of scores. The left-hand columns of Table 9.2 show the calculations for the original scores [Figure 9.6(a)], and the right-hand columns show the calculations for the adjusted scores [Figure 9.6(b)]. Note that the total variability, including the treatment effect, is $SS = 153$. However, when the treatment effect is removed, the variability is reduced to $SS = 72$. The difference between these two values, $153 - 72 = 81$ points, is the amount of variability that is accounted for by the treatment effect. This value is usually reported as a proportion or percentage of the total variability:

$$\frac{\text{variability accounted for}}{\text{total variability}} = \frac{81}{153} = 0.5294 \quad (52.94\%)$$

Thus, removing the treatment effect reduces the variability by 52.94%. This value is called the *percentage of variance accounted for by the treatment* and is identified as r^2.

Rather than computing r^2 directly by comparing two different calculations for SS, the value can be found from a single equation based on the outcome of the t test.

The letter r is the traditional symbol used for a correlation, and the concept of r^2 will be discussed again when we consider correlations in Chapter 16. Also, in the context of t statistics, the percentage of variance that we are calling r^2 is often identified by the Greek letter omega squared (ω^2).

$$r^2 = \frac{t^2}{t^2 + df} \tag{9.5}$$

For the hypothesis test in Example 9.1, we obtained $t = 3.00$ with $df = 8$. These values produce

$$r^2 = \frac{3^2}{3^2 + 8} = \frac{9}{17} = 0.5294 \quad (52.94\%)$$

TABLE 9.2

Calculation of SS, the sum of squared deviations, for the data in Figure 9.7. The first three columns show the calculations for the original scores, including the treatment effect. The last three columns show the calculations for the adjusted scores after the treatment effect has been removed.

	Calculation of SS including the treatment effect			Calculation of SS after the treatment effect is removed		
Score	Deviation from $\mu = 10$	Squared Deviation		Adjusted Score	Deviation from $\mu = 10$	Squared Deviation
8	−2	4		8 − 3 = 5	−5	25
10	0	0		10 − 3 = 7	−3	9
12	2	4		12 − 3 = 9	−1	1
12	2	4		12 − 3 = 9	−1	1
13	3	9		13 − 3 = 10	0	0
13	4	16		13 − 3 = 10	0	0
15	5	25		15 − 3 = 12	2	4
17	7	49		17 − 3 = 14	4	16
17	7	49		17 − 3 = 14	4	16
		$SS = 153$				$SS = 72$

Note that this is exactly the same value we obtained with the direct calculation of the percentage of variability accounted for by the treatment.

Interpreting r^2   In addition to developing the Cohen's *d* measure of effect size, Cohen (1988) also proposed criteria for evaluating the size of a treatment effect that is measured by r^2. The criteria were actually suggested for evaluating the size of a correlation, *r*, but are easily extended to apply to r^2. Cohen's standards for interpreting r^2 are shown in Table 9.3.

According to these standards, the data we constructed for Examples 9.1 and 9.2 show a very large effect size with $r^2 = .5294$.

TABLE 9.3

Criteria for interpreting the value of r^2 as proposed by Cohen (1988).

Percentage of Variance Explained, r^2	
$r^2 = 0.01$	Small effect
$r^2 = 0.09$	Medium effect
$r^2 = 0.25$	Large effect

THE INFLUENCE OF SAMPLE SIZE AND SAMPLE VARIANCE

As we noted in Chapter 8 (page 250), a variety of factors can influence the outcome of a hypothesis test. In particular, the number of scores in the sample and the magnitude of the sample variance both have a large effect on the *t* statistic and thereby influence the statistical decision. The structure of the *t* formula makes these factors easier to understand.

$$t = \frac{M - \mu}{s_M} \qquad \text{where } s_M = \sqrt{s^2/n}$$

Because the estimated standard error, s_M, appears in the denominator of the formula, a larger value for s_M will produce a smaller value (closer to zero) for *t*. Thus, any factor that increases the standard error will also reduce the likelihood of rejecting the null hypothesis and finding a significant treatment effect. The two factors that determine the size of the standard error are the sample variance, s^2, and the sample size, *n*.

The estimated standard error is directly related to the sample variance so that the larger the variance, the larger the error. Thus, large variance means that you are less likely to obtain a significant treatment effect. In general, large variance is bad for inferential statistics. Large variance means that the scores are widely scattered, which makes it difficult to see any consistent patterns or trends in the data. In general, high variance reduces the likelihood of rejecting the null hypothesis and it reduces measures of effect size.

On the other hand, the estimated standard error is inversely related to the number of scores in the sample. The larger the sample is, the smaller the error is. If all other factors are held constant, large samples tend to produce bigger *t* statistics and therefore are more likely to produce significant results. For example, a 2-point mean difference with a sample of $n = 4$ may not be convincing evidence of a treatment effect. However, the same 2-point difference with a sample of $n = 100$ is much more compelling.

As a final note, we should remind you that although sample size affects the hypothesis test, this factor has little or no impact on measures of effect size. In particular, estimates of Cohen's *d* are not influenced at all by sample size, and measures of r^2 are only slightly affected by changes in the size of the sample.

IN THE LITERATURE
REPORTING THE RESULTS OF A *t* TEST

In Chapter 8, we noted the conventional style for reporting the results of a hypothesis test, according to APA format. First, recall that a scientific report typically uses the term *significant* to indicate that the null hypothesis has been rejected and the term *not significant* to indicate failure to reject H_0. Additionally, there is a prescribed format for reporting the calculated value of the test statistic, degrees of freedom, and alpha level for a *t* test. This format parallels the style introduced in Chapter 8 (page 248).

In Example 9.1 we calculated a *t* statistic of 3.00 with $df = 8$, and we decided to reject H_0 with alpha set at .05. Using the same data in Example 9.2, we obtained $r^2 = 52.94\%$ for the percentage of variance explained by the treatment effect. In a scientific report, this information is conveyed in a concise statement, as follows:

> The infants spent an average of $M = 13$ out of 20 seconds looking at the attractive face, with $SD = 3.00$. Statistical analysis indicates that the time spent looking at the attractive face was significantly greater than would be expected if there were no preference, $t(8) = 3.00, p < .05$, $r^2 = 0.5294$.

In the first statement, the mean ($M = 13$) and the standard deviation ($SD = 3$) are reported as previously noted (Chapter 4, page 126). The next statement provides the results of the statistical analysis. Note that the degrees of freedom are reported in parentheses immediately after the symbol *t*. The value for the obtained *t* statistic follows (3.00), and next is the probability of committing a Type I error (less than 5%). Finally, the effect size is reported, $r^2 = 52.94\%$.

Often, researchers will use a computer to perform a hypothesis test like the one in Example 9.1. In addition to calculating the mean, standard deviation, and the *t* statistic for the data, the computer usually calculates and reports the *exact probability* associated with the *t* value. In Example 9.1 we determined that any *t* value beyond ±2.306 has a probability of less than .05 (see Figure 9.4). Thus, the obtained *t* value, $t = 3.00$, is reported as being very unlikely, $p < .05$. A computer printout, however, would have included an exact probability for our specific *t* value.

Whenever a specific probability value is available, you are encouraged to use it in a research report. For example, a computer analysis of these data reports an exact *p* value of $p = .017$, and the research report would state "$t(8) = 3.00, p = .017$" instead of using the less specific "$p < .05$." As one final caution, we note that occasionally a *t* value is so extreme that the computer reports $p = 0.000$. The zero value does not mean that the probability is literally zero; instead, it means that the computer has rounded off the probability value to three decimal places and obtained a result of 0.000. In this situation, you do not know the exact probability value, but you can report $p < .001$. ❏

The statement $p < .05$ was explained in Chapter 8, page 249.

LEARNING CHECK

1. A sample of $n = 16$ individuals is selected from a population with a mean of $\mu = 80$. A treatment is administered to the sample and, after treatment, the sample mean is found to be $M = 86$ with a standard deviation of $s = 8$.

   a. Does the sample provide sufficient evidence to conclude that the treatment has a significant effect? Test with $\alpha = .05$. Also, compute Cohen's d and r^2 to measure the effect size.

   b. Now assume the sample has only $n = 4$ individuals, but the mean is still $M = 86$ and the standard deviation is still $s = 8$. Repeat the hypothesis test with $\alpha = .05$ and compute the new measures of effect size (d and r^2) for the sample of $n = 4$.

   c. Finally, assume that the standard deviation is increased to $s = 12$ for the original sample ($n = 16$ and $M = 86$). Once again, repeat the hypothesis test and compute the new measures of effect size.

   d. How does sample size influence the hypothesis test and the measures of effect size? How does the standard deviation influence the hypothesis test and the measures of effect size?

ANSWERS

1. a. The estimated standard error is 2 points and the data produce $t = 6/2 = 3.00$. With $df = 15$, the critical values are $t = \pm2.131$, so the decision is to reject H_0 and conclude that there is a significant treatment effect. For these data, $d = 6/8 = 0.75$ and $r^2 = 9/24 = 0.375$ or 37.5%.

   b. With $n = 4$, the estimated standard error is 4 points and $t = 6/4 = 1.50$. With $df = 3$, the critical boundaries are ±3.182, so the decision is to fail to reject H_0 and conclude that the treatment does not have a significant effect. The data produce $d = 6/8 = 0.75$ and $r^2 = 2.25/5.25 = 0.429$ (42.9%).

   c. With $s = 12$, the standard error is 3 and $t = 2.00$. Fail to reject the null hypothesis and conclude there is no significant effect. Cohen's $d = 0.50$ and $r^2 = 0.21$ or 21%.

   d. Increasing sample size increases the likelihood of rejecting the null hypothesis but has little or no effect on measures of effect size. Increasing the sample variance reduces the likelihood of rejecting the null hypothesis and reduces measures of effect size.

9.4 DIRECTIONAL HYPOTHESES AND ONE-TAILED TESTS

As noted in Chapter 8, the nondirectional (two-tailed) test is more commonly used than the directional (one-tailed) alternative. On the other hand, a directional test may be used in some research situations, such as exploratory investigations or pilot studies or when there is *a priori* justification (for example, a theory or previous findings). The following example demonstrates a directional hypothesis test with a t statistic, using the same experimental situation presented in Example 9.1.

EXAMPLE 9.3

The research question is whether attractiveness will affect the behavior of infants looking at photographs of women's faces. The researcher is expecting the infants to prefer the more attractive face. Therefore, the researcher predicts that the infants will spend most of the 20-second period looking at the attractive face. For this example we will use the same sample data that were used in the original hypothesis test in

Example 9.1. Specifically, the researcher tested a sample of $n = 9$ infants and obtained a mean of $M = 13$ seconds looking at the attractive face with $SS = 72$.

STEP 1   State the hypotheses, and select an alpha level. With most directional tests, it is usually easier to state the hypothesis in words, including the directional prediction, and then convert the words into symbols. For this example, the researcher is predicting that attractiveness will cause the infants to increase the amount of time they spend looking at the attractive face; that is, more than half of the 20 seconds should be spent looking at the attractive face. In general, the null hypothesis states that the predicted effect will not happen. For this study, the null hypothesis states that the infants will not spend more than half of the 20 seconds looking at the attractive face. In symbols,

H_0: $\mu_{attractive} \leq 10$ seconds   (Not more than half of the 20 seconds looking at the attractive face)

Similarly, the alternative states that the treatment will work. In this case, H_1 states that the infants will spend more than half of the time looking at the attractive face. In symbols,

H_1: $\mu_{attractive} > 10$ seconds   (More than half of the 20 seconds looking at the attractive face)

We will set the level of significance at $\alpha = .05$.

STEP 2   Locate the critical region. In this example, the researcher is predicting that the sample mean (M) will be greater than 10 seconds. Thus, if the infants average more than 10 seconds looking at the attractive face, the data will provide support for the researcher's prediction and will tend to refute the null hypothesis. However, we must still determine exactly how large a value is necessary to reject the null hypothesis. To find the critical value, you must look in the *t* distribution table using the one-tail proportions. With a sample of $n = 9$, the *t* statistic will have $df = 8$; using $\alpha = .05$, you should find a critical value of $t = 1.860$. Therefore, if we obtain a sample mean greater than 10 seconds and the sample mean produces a *t* statistic greater than 1.860, we will reject the null hypothesis and conclude that the infants show a significant preference for the attractive face. Figure 9.7 shows the one-tailed critical region for this test.

STEP 3   Calculate the test statistic. The computation of the *t* statistic is the same for either a one-tailed or a two-tailed test. Earlier (in Example 9.1), we found that the data for this experiment produce a test statistic of $t = 3.00$.

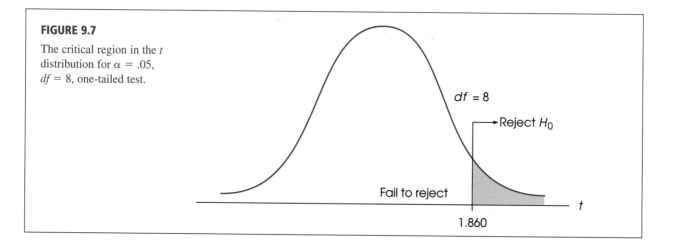

FIGURE 9.7

The critical region in the *t* distribution for $\alpha = .05$, $df = 8$, one-tailed test.

$df = 8$

Reject H_0

Fail to reject

t

1.860

STEP 4   Make a decision. The test statistic is in the critical region, so we reject H_0. In terms of the experimental variables, we have decided that the infants show a preference and spend significantly more time looking at the attractive face than they do looking at the unattractive face.

THE CRITICAL REGION FOR A ONE-TAILED TEST

In step 2 of Example 9.3, we determined that the critical region is in the right-hand tail of the distribution. However, it is possible to divide this step into two stages that eliminate the need to determine which tail (right or left) should contain the critical region. The first stage in this process is simply to determine whether the sample mean is in the direction predicted by the original research question. For this example, the researcher predicted that the infants would prefer the attractive face and spend more time looking at it. Specifically, the researcher expects the infants to spend more than 10 out of 20 seconds focused on the attractive face. The obtained sample mean, $M = 13$ seconds, is in the correct direction. This first stage eliminates the need to determine whether the critical region is in the left- or right-hand tail. Because we already have determined that the effect is in the correct direction, the sign of the t statistic ($+$ or $-$) no longer matters. The second stage of the process is to determine whether the effect is large enough to be significant. For this example, the requirement is that the sample produce a t statistic greater than 1.860. If the magnitude of the t statistic, independent of its sign, is greater than 1.860, the result is significant and H_0 is rejected.

LEARNING CHECK

1. A researcher would like to evaluate the effect of a new cold medication on reaction time. It is known that under regular circumstances the distribution of reaction times is normal with $\mu = 200$. A sample of $n = 9$ participants is obtained. Each person is given the new cold medication, and 1 hour later reaction time is measured for each individual. The average reaction time for this sample is $M = 206$ with $SS = 648$. The researcher would like to use a hypothesis test with $\alpha = .05$ to evaluate the effect of the medication. For the one-tailed test, assume that the medication is expected to increase reaction time.

    a. State the hypotheses for a one-tailed and a two-tailed test.

    b. Locate the critical region for a one-tailed and a two-tailed test.

    c. Calculate the test statistic (the t statistic).

    d. What is the decision for a one-tailed test? What is the decision for the two-tailed test?

ANSWERS

1. **a.** For the one-tailed test, H_0: $\mu \leq 200$ and H_1: $\mu > 200$. For the two-tailed test, H_0: $\mu = 200$ and H_1: $\mu \neq 200$.

    b. For the one-tailed test, the critical region consists of t values in the right-hand tail beyond $t = 1.860$. For the two-tailed test, the critical region consists of t values in the tails beyond $t = \pm 2.306$.

    c. The sample variance is 81 and the estimated standard error is 3. For this sample, $t = \frac{6}{3} = 2.00$.

    d. For the one-tailed test, t is in the critical region so we reject H_0 and conclude that the medication produced a significant increase in reaction time. For the two-tailed test, t is not in the critical region and we conclude that there is no significant treatment effect.

SUMMARY

1. The *t* statistic is used instead of a *z*-score for hypothesis testing when the population standard deviation (or variance) is unknown.

2. To compute the *t* statistic, you must first calculate the sample variance (or standard deviation) as a substitute for the unknown population value.

$$\text{sample variance} = s^2 = \frac{SS}{df}$$

Next, the standard error is *estimated* by substituting s^2 in the formula for standard error. The estimated standard error is calculated in the following manner:

$$\text{estimated standard error} = s_M = \sqrt{\frac{s^2}{n}}$$

Finally, a *t* statistic is computed using the estimated standard error. The *t* statistic is used as a substitute for a *z*-score that cannot be computed when the population variance or standard deviation is unknown.

$$t = \frac{M - \mu}{s_M}$$

3. The structure of the *t* formula is similar to that of the *z*-score in that

$$z \text{ or } t = \frac{\text{sample mean} - \text{population mean}}{\text{(estimated) standard error}}$$

For a hypothesis test, you hypothesize a value for the unknown population mean and plug the hypothesized value into the equation along with the sample mean and the estimated standard error, which are computed from the sample data. If the hypothesized mean produces an extreme value for *t*, you conclude that the hypothesis was wrong.

4. The *t* distribution is an approximation of the normal *z* distribution. To evaluate a *t* statistic that is obtained for a sample mean, the critical region must be located in a *t* distribution. There is a family of *t* distributions, with the exact shape of a particular distribution of *t* values depending on degrees of freedom $(n - 1)$. Therefore, the critical *t* values will depend on the value for *df* associated with the *t* test. As *df* increases, the shape of the *t* distribution approaches a normal distribution.

5. When a *t* statistic is used for a hypothesis test, Cohen's *d* can be computed to measure effect size. In this situation, the sample standard deviation is used in the formula to obtain an estimated value for *d*:

$$\text{estimated } d = \frac{\text{mean difference}}{\text{standard deviation}} = \frac{M - \mu}{s}$$

6. A second measure of effect size is r^2, which measures the percentage of the variability that is accounted for by the treatment effect. This value is computed as follows:

$$r^2 = \frac{t^2}{t^2 + df}$$

KEY TERMS

estimated standard error   degrees of freedom   estimated *d*

t statistic   *t* distribution   percentage of variance accounted for by the treatment (r^2)

RESOURCES

There are a tutorial quiz and other learning exercises for Chapter 9 at the Wadsworth website, www.cengage.com/psychology/gravetter. The site also provides access to a workshop entitled *Single-Sample t Test* that reviews the concepts and logic of hypothesis testing with the *t* statistic. See page 29 for more information about accessing items on the website.

ENHANCED
WebAssign

Guided interactive tutorials, end-of-chapter problems, and related testbank items may be assigned online at WebAssign.

WebTUTOR

For those using WebTutor along with this book, there is a WebTutor section corresponding to this chapter. The WebTutor contains a brief summary of Chapter 9, hints for learning the concepts and the formulas for the *t* score hypothesis test, cautions about common errors, and sample exam items including solutions.

SPSS

General instructions for using SPSS are presented in Appendix D. Following are detailed instructions for using SPSS to perform **the *t* Test** presented in this chapter.

Data Entry

1. Enter all of the scores from the sample in one column of the data editor, probably VAR00001.

Data Analysis

1. Click **Analyze** on the tool bar, select **Compare Means,** and click on **One-Sample T Test.**
2. Highlight the column label for the set of scores (VAR0001) in the left box and click the arrow to move it into the **Test Variable(s)** box.
3. In the **Test Value** box at the bottom of the One-Sample *t* Test window, enter the hypothesized value for the population mean from the null hypothesis. *Note:* The value is automatically set at zero until you type in a new value.
4. Click **OK.**

SPSS Output

The program will produce a table of One-Sample Statistics showing the number of scores, the sample mean and standard deviation, and the estimated standard error of the mean. A second table presents the results of the One-Sample Test, including the value of *t,* the degrees of freedom, the level of significance (the *p* value or alpha level for the test), and the size of the mean difference between the sample mean and the hypothesized population mean.

   Note: The output also includes the upper and lower boundaries for a 95% confidence interval for the mean difference. Confidence intervals are discussed in Chapter 12 and generally estimate the size of the mean difference.

FOCUS ON PROBLEM SOLVING

1. The first problem we confront in analyzing data is determining the appropriate statistical test. Remember that you can use a *z*-score for the test statistic only when the value for σ is known. If the value for σ is not provided, then you must use the *t* statistic.

2. For the *t* test, the sample variance is used to find the value for estimated standard error. Remember that when computing the sample variance, use $n - 1$ in the denominator (see Chapter 4). When computing estimated standard error, use *n* in the denominator.

DEMONSTRATION 9.1

A HYPOTHESIS TEST WITH THE *t* STATISTIC

A psychologist has prepared an "Optimism Test" that is administered yearly to graduating college seniors. The test measures how each graduating class feels about its future—the higher the score, the more optimistic the class. Last year's class had a mean score of $\mu = 15$. A sample of $n = 9$ seniors from this year's class was selected and tested. The scores for these seniors are 7, 12, 11, 15, 7, 8, 15, 9, and 6, which produce a sample mean of $M = 10$ with $SS = 94$.

On the basis of this sample, can the psychologist conclude that this year's class has a different level of optimism than last year's class?

Note that this hypothesis test will use a *t* statistic because the population variance (σ^2) is not known.

STEP 1   State the hypotheses, and select an alpha level.

The statements for the null hypothesis and the alternative hypothesis follow the same form for the *t* statistic and the *z*-score test.

$$H_0: \quad \mu = 15 \qquad \text{(There is no change.)}$$

$$H_1: \quad \mu \neq 15 \qquad \text{(This year's mean is different.)}$$

For this demonstration, we will use $\alpha = .05$, two tails.

STEP 2   Locate the critical region.

With a sample of $n = 9$ students, the *t* statistic will have $df = n - 1 = 8$. For a two-tailed test with $\alpha = .05$ and $df = 8$, the critical *t* values are $t = \pm 2.306$. These critical *t* values define the boundaries of the critical region. The obtained *t* value must be more extreme than either of these critical values to reject H_0.

STEP 3   Compute the test statistic. As we have noted, it is easier to separate the calculation of the *t* statistic into three stages.
Sample variance.

$$s^2 = \frac{SS}{n-1} = \frac{94}{8} = 11.75$$

Estimated standard error. The estimated standard error for these data is

$$s_M = \sqrt{\frac{s^2}{n}} = \sqrt{\frac{11.75}{9}} = 1.14$$

The t statistic. Now that we have the estimated standard error and the sample mean, we can compute the *t* statistic. For this demonstration,

$$t = \frac{M - \mu}{s_M} = \frac{10 - 15}{1.14} = \frac{-5}{1.14} = -4.39$$

STEP 4   Make a decision about H_0, and state a conclusion.

The *t* statistic we obtained ($t = -4.39$) is in the critical region. Thus, our sample data are unusual enough to reject the null hypothesis at the .05 level of significance. We can conclude that there is a significant difference in level of optimism between this year's and last year's graduating classes, $t(8) = -4.39$, $p < .05$, two-tailed.

DEMONSTRATION 9.2

EFFECT SIZE: ESTIMATING COHEN'S d AND COMPUTING r^2

We will estimate Cohen's d for the same data used for the hypothesis test in Demonstration 9.1. The mean optimism score for the sample from this year's class was 5 points lower than the mean from last year ($M = 10$ versus $\mu = 15$). In Demonstration 9.1 we computed a sample variance of $s^2 = 11.75$, so the standard deviation is $\sqrt{11.75} = 3.43$. With these values,

$$\text{estimated } d = \frac{\text{mean difference}}{\text{standard deviation}} = \frac{5}{3.43} = 1.46$$

To calculate the percentage of variance explained by the treatment effect, r^2, we need the value of t and the df value from the hypothesis test. In Demonstration 9.1 we obtained $t = -4.39$ with $df = 8$. Using these values in Equation 9.5, we obtain

$$r^2 = \frac{t^2}{t^2 + df} = \frac{(-4.39)^2}{(-4.39)^2 + 8} = \frac{19.27}{27.27} = 0.71$$

PROBLEMS

1. Under what circumstances is a t statistic used instead of a z-score for a hypothesis test?

2. A sample of $n = 16$ scores has a mean of $M = 83$ and a standard deviation of $s = 12$.
   a. Explain what is measured by the sample standard deviation.
   b. Compute the estimated standard error for the sample mean and explain what is measured by the standard error.

3. Find the estimated standard error for the sample mean for each of the following samples.
   a. $n = 9$ with $SS = 1152$
   b. $n = 25$ with $SS = 2400$
   c. $n = 36$ with $SS = 1260$

4. Explain why t distributions tend to be flatter and more spread out than the normal distribution.

5. What is the relationship between the value for degrees of freedom and the shape of the t distribution? What happens to the critical value of t for a particular alpha level when df increases in value?

6. Find the t values that form the boundaries of the critical region for a two-tailed test with $\alpha = .05$ for each of the following df values.
   a. $df = 8$
   b. $df = 15$
   c. $df = 24$

7. The following sample was obtained from a population with unknown parameters.

   Scores: 1, 5, 1, 1

   a. Compute the sample mean and standard deviation. (Note that these are descriptive values that summarize the sample data.)
   b. Compute the estimated standard error for M. (Note that this is an inferential value that describes how accurately the sample mean represents the unknown population mean.)

8. The following sample was obtained from a population with unknown parameters.

   Scores: 6, 12, 0, 3, 4

   a. Compute the sample mean and standard deviation. (Note that these are descriptive values that summarize the sample data.)
   b. Compute the estimated standard error for M. (Note that this is an inferential value that describes how accurately the sample mean represents the unknown population mean.)

9. A sample is randomly selected from a population with a mean of $\mu = 50$, and a treatment is administered to the individuals in the sample. After treatment, the sample is found to have a mean of $M = 56$ with a standard deviation of $s = 8$.
   a. If there are $n = 4$ individuals in the sample, are the data sufficient to reject H_0 and conclude that the treatment has a significant effect using a two-tailed test with $\alpha = .05$?
   b. If there are $n = 16$ individuals in the sample, are the data sufficient to conclude that the treatment has a significant effect? Again, assume a two-tailed test with $\alpha = .05$.

10. A sample of $n = 25$ individuals is randomly selected from a population with a mean of $\mu = 65$, and a treatment is administered to the individuals in the sample. After treatment, the sample mean is found to be $M = 70$.

    a. If the sample standard deviation is $s = 10$, are the data sufficient to conclude that the treatment has a significant effect using a two-tailed test with $\alpha = .05$?

    b. If the sample standard deviation is $s = 20$, are the data sufficient to conclude that the treatment has a significant effect using a two-tailed test with $\alpha = .05$?

11. A random sample of $n = 16$ individuals is selected from a population with $\mu = 70$, and a treatment is administered to each individual in the sample. After treatment, the sample mean is found to be $M = 76$ with $SS = 960$.

    a. How much difference is there between the mean for the treated sample and the mean for the original population? (*Note:* In a hypothesis test, this value forms the numerator of the t statistic.)

    b. How much difference is expected just by chance between the sample mean and its population mean? That is, find the standard error for M. (*Note:* In a hypothesis test, this value is the denominator of the t statistic.)

    c. Based on the sample data, does the treatment have a significant effect? Use a two-tailed test with $\alpha = .05$.

12. Last fall, a sample of $n = 36$ freshmen was selected to participate in a new 4-hour training program designed to improve study skills. To evaluate the effectiveness of the new program, the sample was compared with the rest of the freshman class. All freshmen must take the same English Language Skills course, and the mean score on the final exam for the entire freshman class was $\mu = 74$. The students in the new program had a mean score of $M = 79.4$ with a standard deviation of $s = 18$.

    a. On the basis of these data, can the college conclude that the students in the new program performed significantly better than the rest of the freshman class? Use a one-tailed test with $\alpha = .05$.

    b. Can the college conclude that the students in the new program are significantly different from the rest of the freshman class? Use a two-tailed test with $\alpha = .05$.

13. Numerous studies have shown that IQ scores have been increasing, generation by generation, for years (Flynn, 1984, 1999). The increase is called the Flynn Effect, and the data indicate that the increase appears to be about 7 points per decade. To demonstrate this phenomenon, a researcher obtains an IQ test that was written in 1980. At the time the test was prepared, it was standardized to produce a population mean of $\mu = 100$. The researcher administers the test to a random sample of $n = 16$ of today's high school students and obtains a sample mean IQ of $M = 121$ with $SS = 6000$. Is this result sufficient to conclude that today's sample scored significantly higher than would be expected from a population with $\mu = 100$? Use a one-tailed test with $\alpha = .01$.

14. In the Preview for this chapter, we discussed a study that examined the effect of eye-spot patterns on the behavior of moth-eating birds (Scaife, 1976). The birds were tested in a box with two chambers and were free to move from one chamber to another. In one chamber, two large eye-spots were painted on one wall. The other chamber had plain wall. The researcher recorded the amount of time each bird spent in the plain chamber during a 60-minute session. Suppose the study produced a mean of $M = 37$ minutes on the plain chamber with $SS = 288$ for a sample of $n = 9$ birds. (*Note:* If the eye spots have no effect, then the birds should spend an average of $\mu = 30$ minutes in each chamber.)

    a. Is this sample sufficient to conclude that the eye-spots have a significant influence on the birds' behavior? Use a two-tailed test with $\alpha = .05$.

    b. Compute the estimated Cohen's d to measure the size of the treatment effect.

15. The librarian at the local elementary school claims that, on average, the books in the library are more than 20 years old. To test this claim, a student takes a sample of $n = 30$ books and records the publication date for each. The sample produces an average age of $M = 23.8$ years with a variance of $s^2 = 67.5$. Use this sample to conduct a one-tailed test with $\alpha = .01$ to determine whether the average age of the library books is significantly greater than 20 years ($\mu > 20$).

16. In a classic study of infant attachment, Harlow (1959) placed infant monkeys in cages with two artificial surrogate mothers. One "mother" was made from bare wire mesh and contained a baby bottle from which the infants could feed. The other mother was made from soft terry cloth and did not provide any access to food. Harlow observed the infant monkeys and recorded how much time per day was spent with each mother. In a typical day, the infants spent a total of 18 hours clinging to one of the two mothers. If there were no preference between the two, you would expect the time to be divided evenly, with an average of $\mu = 9$ hours for each of the mothers. However, the typical monkey spent around 15 hours per day with the terry cloth mother, indicating a strong preference for the soft, cuddly mother. Suppose a sample of $n = 9$ infant monkeys averaged $M = 15.3$ hours per day with

SS = 216 with the terry cloth mother. Is this result sufficient to conclude that the monkeys spent significantly more time with the softer mother than would be expected if there were no preference? Use a two-tailed test with $\alpha = .05$.

17. Belsky, Weinraub, Owen, and Kelly (2001) reported on the effects of preschool childcare on the development of young children. One result suggests that children who spend more time away from their mothers are more likely to show behavioral problems in kindergarten. Using a standardized scale, the average rating of behavioral problems for kindergarten children is $\mu = 35$. A sample of $n = 16$ kindergarten children who had spent at least 20 hours per week in child care during the previous year produced a mean score of $M = 42.7$ with a standard deviation of $s = 6$. Are the data sufficient to conclude that children with a history of child care show significantly more behavioral problems than the average kindergarten child? Use a one-tailed test with $\alpha = .01$.

18. Other research examining the effects of preschool childcare has found that children who spent time in day care, especially high-quality day care, perform better on math and language tests than children who stay home with their mothers (Broberg, Wessels, Lamb, & Hwang, 1997). Typical results, for example, show that a sample of $n = 25$ children who attended day care before starting school had an average score of $M = 87$ with $SS = 1536$ on a standardized math test for which the population mean is $\mu = 81$. Is this sample sufficient to conclude that the children with a history of preschool day care are significantly different from the general population? Use a two-tailed test with $\alpha = .01$.

19. A random sample is obtained from a population with a mean of $\mu = 100$, and a treatment is administered to the sample. After treatment, the sample mean is found to be $M = 104$ and the sample variance is $s^2 = 400$.
   a. Assuming the sample contained $n = 16$ individuals, measure the size of the treatment effect by computing the estimated d and r^2.
   b. Assuming the sample contained $n = 25$ individuals, measure the size of the treatment effect by computing the estimated d and r^2.
   c. Comparing your answers from parts a and b, how does sample size influence measures of effect size?

20. A random sample of $n = 16$ scores is obtained from a population with a mean of $\mu = 80$, and a treatment is administered to the sample. After treatment, the sample mean is found to be $M = 82$.
   a. Assuming the sample variance is $s^2 = 64$, compute the estimated d and r^2 to measure the size of the treatment effect.

   b. Assuming the sample variance is $s^2 = 1600$, compute the estimated d and r^2 to measure the size of the treatment effect.
   c. Comparing your answers from parts a and b, how does sample variance influence measures of effect size?

21. An example of the vertical-horizontal illusion is shown in the figure. Although the two lines are exactly the same length, the vertical line appears to be much longer. To examine the strength of this illusion, a researcher prepared an example in which both lines were exactly 10 inches long. The example was shown to individual participants who were told that the horizontal line was 10 inches long and then were asked to estimate the length of the vertical line. For a sample of $n = 25$ participants, the average estimate was $M = 12.2$ inches with a standard deviation of $s = 1.00$.
   a. Use a one-tailed hypothesis test with $\alpha = .01$ to demonstrate that the individuals in the sample significantly overestimate the true length of the line. (*Note:* Accurate estimation would produce a mean of $\mu = 10$ inches.)
   b. Calculate the estimated d and r^2, the percentage of variance accounted for, to measure the size of this effect.

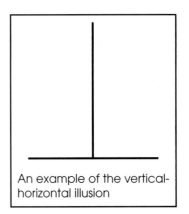

An example of the vertical-horizontal illusion

22. A researcher would like to examine the effects of humidity on eating behavior. It is known that laboratory rats normally eat an average of $\mu = 21$ grams of food each day. The researcher selects a random sample of $n = 16$ rats and places them in a controlled atmosphere room in which the relative humidity is maintained at 90%. The daily food consumption scores for the rats are as follows:

   14, 18, 21, 15, 18, 18, 21, 18
   16, 20, 17, 19, 20, 17, 17, 19

   a. Can the researcher conclude that humidity has a significant effect on eating behavior? Use a two-tailed test with $\alpha = .05$.

   b. Compute the estimated d and r^2 to measure the size of the treatment effect.

23. A psychologist would like to determine whether there is a relation between depression and aging. It is known that the general population averages $\mu = 40$ on a standardized depression test. The psychologist obtains a sample of $n = 9$ individuals who are all more than 70 years old. The depression scores for this sample are as follows: 37, 50, 43, 41, 39, 45, 49, 44, 48. On the basis of this sample, can the psychologist conclude that depression for elderly people is significantly different from depression in the general population? Use a two-tailed test at the .05 level of significance.

24. On a standardized spatial skills task, normative data reveal that people typically get $\mu = 15$ correct solutions. A psychologist tests $n = 7$ individuals who have brain injuries in the right cerebral hemisphere. For the following data, determine whether or not right-hemisphere damage results in significantly reduced performance on the spatial skills task. Test with alpha set at .05 with one tail. The data are as follows: 12, 16, 9, 8, 10, 17, 10.

25. People tend to evaluate the quality of their lives relative to others around them. In a demonstration of this phenomenon, Frieswijk, Buunk, Steverink, and Slaets (2004) conducted interviews with frail elderly people. In the interview, each person was compared with fictitious others who were worse off. After the interviews, the elderly people reported more satisfaction with their own lives. Following are hypothetical data similar to those obtained in the research study. The scores are measures on a life-satisfaction scale for a sample of $n = 9$ elderly people who completed the interview. Assume that the average score on this scale is $\mu = 20$. Are the data sufficient to conclude that the people in this sample are significantly more satisfied than others in the general population? Use a one-tailed test with $\alpha = .05$. The life-satisfaction scores for the sample are 18, 23, 24, 22, 19, 27, 23, 26, 25.

C H A P T E R

10

The *t* Test for Two Independent Samples

Tools You Will Need

The following items are considered essential background material for this chapter. If you doubt your knowledge of any of these items, you should review the appropriate chapter or section before proceeding.

- Sample variance (Chapter 4)

- Standard error formulas (Chapter 7)

- The *t* statistic (Chapter 9)
  - Distribution of *t* values
  - *df* for the *t* statistic
  - Estimated standard error

Preview

10.1   Introduction to Independent-Measures Designs

10.2   The *t* Statistic for an Independent-Measures Research Design

10.3   Hypothesis Tests and Effect Size with the Independent-Measures *t* Statistic

10.4   Assumptions Underlying the Independent-Measures *t* Formula

Summary

Focus on Problem Solving

Demonstrations 10.1 and 10.2

Problems

Preview

In a classic study in the area of problem solving, Katona (1940) compared the effectiveness of two methods of instruction. One group of participants was shown the exact, step-by-step procedure for solving a problem, and then these participants were required to memorize the solution. This method was called *learning by memorization* (later called the *expository* method). Participants in a second group were encouraged to study the problem and find the solution on their own. Although these participants were given helpful hints and clues, the exact solution was never explained. This method was called *learning by understanding* (later called the *discovery* method).

Katona's experiment included the problem shown in Figure 10.1. This figure shows a pattern of five squares made of matchsticks. The problem is to change the pattern into exactly four squares by moving only three matches. (All matches must be used, none can be removed, and all the squares must be the same size.) Two groups of participants learned the solution to this problem. One group learned by understanding, and the other group learned by memorization. After 3 weeks, both groups returned to be tested again. The two groups did equally well on the matchstick problem they had learned earlier. But when they were given two new problems (similar to the matchstick problem), the *understanding* group performed much better than the *memorization* group.

The Problem: Although the data a show a mean difference between the two groups in Katona's study, you cannot automatically conclude that the difference was caused by the method they used to solve the first problem. Specifically, the two groups consist of different people with different backgrounds, different skills, different IQs, and so on. Because the two different groups consist of different individuals, you should expect them to have different scores and different means. This issue was first presented in Chapter 1 when we introduced the concept of sampling error (see Figure 1.2 on page 7). Thus, there are two possible explanations for the difference between the two groups.

1. It is possible that there really is a difference between the two treatment conditions so that the method of understanding produces better learning than the method of memorization.

2. It is possible that there is no difference between the two treatment conditions and the mean difference obtained in the experiment is simply the result of sampling error.

A hypothesis test is necessary to determine which of the two explanations is most plausible. However, the hypothesis tests we have examined thus far are intended to evaluate the data from only one sample. In this study there are two separate samples.

The Solution: In this chapter we introduce the *independent-measures* t *test* which is a hypothesis test that uses two separate samples to evaluate the mean difference between two treatment conditions or between two different populations. Like the *t* test introduced in Chapter 9, the independent-measures test uses the sample variance to compute an estimated standard error. This test, however, combines the variance from the two separate samples to evaluate the difference between two separate sample means.

Incidentally, if you still have not discovered the solution to the matchstick problem, keep trying. According to Katona's results, it would be very poor teaching strategy for us to give you the answer.

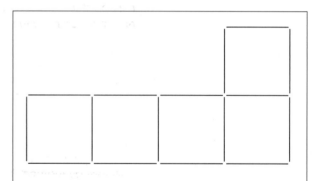

FIGURE 10.1

A pattern of five squares made of matchsticks. The problem is to change the pattern into exactly four squares by moving only three matchsticks.

Katona, G. (1940). *Organizing and Memorizing*. New York: Colombia University Press. Reprinted by permission.

10.1 INTRODUCTION TO INDEPENDENT-MEASURES DESIGNS

Until this point, all the inferential statistics we have considered involve using one sample as the basis for drawing conclusions about one population. Although these *single-sample* techniques are used occasionally in real research, most research studies require the comparison of two (or more) sets of data. For example, a social psychologist may want to compare men and women in terms of their political attitudes, an educational psychologist may want to compare two methods for teaching mathematics, or a clinical psychologist may want to evaluate a therapy technique by comparing depression scores for patients before therapy with their scores after therapy. In each case, the research question concerns a mean difference between two sets of data.

There are two general research strategies that can be used to obtain the two sets of data to be compared:

1. The two sets of data could come from two completely separate samples. For example, the study could involve a sample of men compared with a sample of women. Or the study could compare one sample of students taught by method A and a second sample taught by method B.

2. The two sets of data could both come from the same sample. For example, the researcher could obtain one set of scores by measuring depression for a sample of patients before they begin therapy and then obtain a second set of data by measuring the same individuals after 6 weeks of therapy.

The first research strategy, using completely separate samples, is called an *independent-measures* research design or a *between-subjects* design. These terms emphasize the fact that the design involves separate and independent samples and makes a comparison between two groups of individuals. The structure of an independent-measures research design is shown in Figure 10.2. Notice that the research study uses two separate samples to represent the two different populations (or two different treatments) being compared.

DEFINITION

> A research design that uses a separate sample for each treatment condition (or for each population) is called an **independent-measures** research design or a **between-subjects** design.

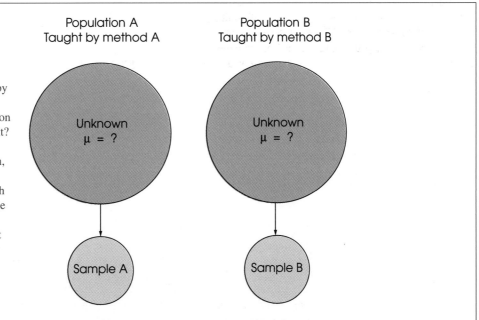

FIGURE 10.2

Do the achievement scores for children taught by method A differ from the scores for children taught by method B? In statistical terms, are the two population means the same or different? Because neither of the two population means is known, it will be necessary to take two samples, one from each population. The first sample will provide information about the mean for the first population, and the second sample will provide information about the second population.

Population A
Taught by method A

Unknown
$\mu = ?$

Sample A

Population B
Taught by method B

Unknown
$\mu = ?$

Sample B

In this chapter, we examine the statistical techniques used to evaluate the data from an independent-measures design. More precisely, we introduce the hypothesis test that allows researchers to use the data from two separate samples to evaluate the mean difference between two populations or between two treatment conditions.

The second research strategy, in which the two sets of data are obtained from the same sample, is called a *repeated-measures* research design or a *within-subjects* design. The statistics for evaluating the results from a repeated-measures design are introduced in Chapter 11. Also, at the end of Chapter 11, we discuss some of the advantages and disadvantages of independent-measures and repeated-measures designs.

10.2	THE *t* STATISTIC FOR AN INDEPENDENT-MEASURES RESEARCH DESIGN

Because an independent-measures study involves two separate samples, we will need some special notation to help specify which data go with which sample. This notation involves the use of subscripts, which are small numbers written beside a sample statistic. For example, the number of scores in the first sample would be identified by n_1; for the second sample, the number of scores is n_2. The sample means would be identified by M_1 and M_2. The sums of squares would be SS_1 and SS_2.

THE HYPOTHESES FOR AN INDEPENDENT-MEASURES TEST

The goal of an independent-measures research study is to evaluate the mean difference between two populations (or between two treatment conditions). Using subscripts to differentiate the two populations, the mean for the first population is μ_1, and the second population mean is μ_2. The difference between means is simply $\mu_1 - \mu_2$. As always, the null hypothesis states that there is no change, no effect, or, in this case, no difference. Thus, in symbols, the null hypothesis for the independent-measures test is

$$H_0: \quad \mu_1 - \mu_2 = 0 \qquad \text{(No difference between the population means)}$$

You should notice that the null hypothesis could also be stated as $\mu_1 = \mu_2$. However, the first version of H_0 produces a specific numerical value (zero) that will be used in the calculation of the *t* statistic. Therefore, we prefer to phrase the null hypothesis in terms of the difference between the two population means.

The alternative hypothesis states that there is a mean difference between the two populations,

$$H_1: \quad \mu_1 - \mu_2 \neq 0 \qquad \text{(There is a mean difference.)}$$

Equivalently, the alternative hypothesis can simply state that the two population means are not equal: $\mu_1 \neq \mu_2$.

THE FORMULAS FOR AN INDEPENDENT-MEASURES HYPOTHESIS TEST

The independent-measures hypothesis test will be based on another *t* statistic. The formula for this new *t* statistic will have the same general structure as the *t* statistic formula that was introduced in Chapter 9. To help distinguish between the two *t* formulas, we will refer to the original formula (Chapter 9) as the *single-sample* t *statistic* and we will refer to the new formula as the *independent-measures* t *statistic*. Because the new independent-measures *t* will include data from two separate samples and hypotheses about two populations, the formulas may appear to be a bit overpowering.

However, the new formulas will be easier to understand if you view them in relation to the single-sample *t* formulas from Chapter 9. In particular, there are two points to remember:

1. The basic structure of the *t* statistic is the same for both the independent-measures and the single-sample hypothesis tests. In both cases,

$$t = \frac{\text{sample statistic} - \text{hypothesized population parameter}}{\text{estimated standard error}}$$

2. The independent-measures *t* is basically a *two-sample* t *that doubles all the elements of the single-sample* t *formulas.*

To demonstrate the second point, we examine the two *t* formulas piece by piece.

The overall *t* formula   The single-sample *t* uses one sample mean to test a hypothesis about one population mean. The sample mean and the population mean appear in the numerator of the *t* formula, which measures how much difference there is between the sample data and the population hypothesis.

$$t = \frac{\substack{\text{sample} \\ \text{mean}} - \substack{\text{population} \\ \text{mean}}}{\text{estimated standard error}} = \frac{M - \mu}{s_M}$$

The independent-measures *t* uses the difference between *two* sample means to evaluate the difference between *two* population means. Thus, the independent-measures *t* formula is

$$t = \frac{\substack{\text{sample} \\ \text{mean difference}} - \substack{\text{population} \\ \text{mean difference}}}{\text{estimated standard error}} = \frac{(M_1 - M_2) - (\mu_1 - \mu_2)}{s_{(M_1 - M_2)}}$$

In this formula, the value of $M_1 - M_2$ is obtained from the sample data and the value for $\mu_1 - \mu_2$ comes from the null hypothesis. Finally, note that the null hypothesis sets the population mean difference equal to zero, so the independent measures *t* formula can be simplified further,

$$t = \frac{\text{sample mean difference}}{\text{estimated standard error}}$$

In this form, the *t* statistic is a simple ratio comparing the actual mean difference (numerator) with the difference that is expected by chance (denominator).

The estimated standard error   In each of the *t*-score formulas, the standard error in the denominator measures how accurately the sample statistic represents the population parameter. In the single-sample *t* formula, the standard error measures the amount of error expected for a sample mean and is represented by the symbol s_M. For the independent-measures *t* formula, the standard error measures the amount of error that is expected when you use a sample mean difference ($M_1 - M_2$) to represent a population mean difference ($\mu_1 - \mu_2$). The standard error for the sample mean difference is represented by the symbol $s_{(M_1 - M_2)}$.

Caution: Do not let the notation for standard error confuse you. In general, standard error measures how accurately a statistic represents a parameter. The symbol for standard error takes the form $s_{\text{statistic}}$. When the statistic is a sample mean, *M*, the symbol for standard error is s_M. For the independent-measures test, the statistic is a sample mean difference $(M_1 - M_2)$, and the symbol for standard error is $s_{(M_1-M_2)}$. In each case, the standard error tells how much discrepancy is reasonable to expect between the sample statistic and the corresponding population parameter.

Interpreting the estimated standard error   The *estimated standard error of* $M_1 - M_2$ that appears in the bottom of the independent-measures *t* statistic can be interpreted in two ways. First, the standard error is defined as a measure of the standard or average distance between a sample statistic $(M_1 - M_2)$ and the corresponding population parameter $(\mu_1 - \mu_2)$. As always, samples are not expected to be perfectly accurate and the standard error measures how much difference is reasonable to expect between a sample statistic and the population parameter.

Sample mean
difference
$(M_1 - M_2)$
   ⟵ estimated standard error (average distance) ⟶
Population mean
difference
$(\mu_1 - \mu_2)$

When the null hypothesis is true, however, the population mean difference is zero.

Sample mean
difference
$(M_1 - M_2)$
   ⟵ estimated standard error (average distance) ⟶
0 (If H_0 is true)

Now, the standard error is measuring how close the sample mean difference is to zero, which is equivalent to measuring how much difference there is between the two sample means.

M_1
   ⟵ estimated standard error (average distance) ⟶
M_2

This produces a second interpretation for the estimated standard error. Specifically, the standard error can be viewed as a measure of how much difference is reasonable to expect between two sample means if the null hypothesis is true.

The second interpretation of the estimated standard error produces a simplified version of the independent-measures *t* statistic.

$$t = \frac{\text{sample mean difference}}{\text{estimated standard error}}$$

$$= \frac{\text{actual difference between } M_1 \text{ and } M_2}{\text{standard difference (If } H_0 \text{ is true) between } M_1 \text{ and } M_2}$$

In this version, the numerator of the *t* statistic measures how much difference *actually exists* between the two sample means, including any difference that is caused by the different treatments. The denominator measures how much difference *should exist* between the two sample means if there is no treatment effect that causes them to be different. A large value for the *t* statistic is evidence for the existence of a treatment effect.

CALCULATING THE ESTIMATED STANDARD ERROR

To develop the formula for $s_{(M_1-M_2)}$ we will consider the following three points:

1. Each of the two sample means represents it own population mean, but in each case there is some error.

   M_1 approximates μ_1 with some error

   M_2 approximates μ_2 with some error

   Thus, there are two sources of error.

2. The amount of error associated with each sample mean is measured by the estimated standard error of *M*. Using Equation 9.1 (page 282), the estimated standard error for each sample mean is computed as follows:

   For M_1 $s_M = \sqrt{\dfrac{s_1^2}{n_1}}$ For M_2 $s_M = \sqrt{\dfrac{s_2^2}{n_2}}$

3. For the independent-measures *t* statistic, we want to know the total amount of error involved in using *two* sample means to approximate *two* population means. To do this, we will find the error from each sample separately and then add the two errors together. The resulting formula for standard error is

$$s_{(M_1-M_2)} = \sqrt{\frac{s_1^2}{n_1} + \frac{s_2^2}{n_2}} \tag{10.1}$$

Because the independent-measures *t* statistic uses two sample means, the formula for the estimated standard error simply combines the error for the first sample mean and the error for the second sample mean (Box 10.1).

POOLED VARIANCE

Although Equation 10.1 accurately presents the concept of standard error for the independent-measures *t* statistic, this formula is limited to situations in which the two samples are exactly the same size (that is, $n_1 = n_2$). In situations in which the two sample sizes are different, the formula is *biased* and, therefore, inappropriate. The bias comes from the fact that Equation 10.1 treats the two sample variances equally. However, when the sample sizes are different, the two sample variances are not equally good and should not be treated equally. In Chapter 7, we introduced the law of large numbers, which states that statistics obtained from large samples tend to be better (more accurate) estimates of population parameters than statistics obtained from small samples. This same fact holds for sample variances: The variance obtained from a large sample will be a more accurate estimate of σ^2 than the variance obtained from a small sample.

An alternative to computing pooled variance is presented in Box 10.2, page 329.

One method for correcting the bias in the sample variances is to combine the two sample variances into a single value called the *pooled variance*. The pooled variance is obtained by averaging or "pooling" the two sample variances using a procedure that allows the bigger sample to carry more weight in determining the final value.

You should recall that when there is only one sample, the sample variance is computed as

$$s^2 = \frac{SS}{df}$$

BOX 10.1   THE VARIABILITY OF DIFFERENCE SCORES

It may seem odd that the independent-measures *t* statistic *adds* together the two sample errors when it *subtracts* to find the difference between the two sample means. The logic behind this apparently unusual procedure is demonstrated here.

We begin with two populations, I and II (Figure 10.3). The scores in population I range from a high of 70 to a low of 50. The scores in population II range from 30 to 20. We will use the range as a measure of how spread out (variable) each population is:

For population I, the scores cover a range of 20 points.

For population II, the scores cover a range of 10 points.

If we randomly select a score from population I and a score from population II and compute the difference between these two scores ($X_1 - X_2$), what range of values is possible for these differences? To answer this

question, we need to find the biggest possible difference and the smallest possible difference. Look at Figure 10.2; the biggest difference occurs when $X_1 = 70$ and $X_2 = 20$. This is a difference of $X_1 - X_2 = 50$ points. The smallest difference occurs when $X_1 = 50$ and $X_2 = 30$. This is a difference of $X_1 - X_2 = 20$ points. Notice that the differences go from a high of 50 to a low of 20. This is a range of 30 points:

range for population I (X_1 scores) = 20 points

range for population II (X_2 scores) = 10 points

range for the differences ($X_1 - X_2$) = 30 points

The variability for the difference scores is found by *adding* together the variabilities for the two populations.

In the independent-measures *t* statistics, we are computing the variability (standard error) for a sample mean difference. To compute this value, we add together the variability for each of the two sample means.

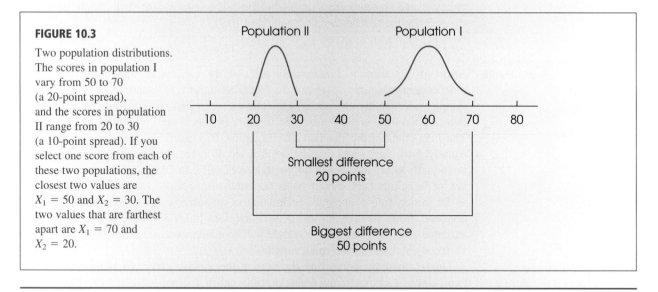

FIGURE 10.3

Two population distributions. The scores in population I vary from 50 to 70 (a 20-point spread), and the scores in population II range from 20 to 30 (a 10-point spread). If you select one score from each of these two populations, the closest two values are $X_1 = 50$ and $X_2 = 30$. The two values that are farthest apart are $X_1 = 70$ and $X_2 = 20$.

For the independent-measures *t* statistic, there are two *SS* values and two *df* values (one from each sample). The values from the two samples are combined to compute what is called the *pooled variance*. The pooled variance is identified by the symbol s_p^2 and is computed as

$$\text{pooled variance} = s_p^2 = \frac{SS_1 + SS_2}{df_1 + df_2} \tag{10.2}$$

With one sample, variance is computed as *SS* divided by *df*. With two samples, the two *SS*s are divided by the two *df*s to compute pooled variance.

As we mentioned earlier, the pooled variance is actually an average of the two sample variances, but the average is computed so that the larger sample carries more weight in determining the final value. The following examples demonstrate this point.

Equal samples sizes   We begin with two samples that are exactly the same size. The first sample has $n = 6$ scores with $SS = 50$, and the second sample has $n = 6$ scores with $SS = 30$. Individually, the two sample variances are

$$\text{Variance for sample 1:} \quad s^2 = \frac{SS}{df} = \frac{50}{5} = 10$$

$$\text{Variance for sample 2:} \quad s^2 = \frac{SS}{df} = \frac{30}{5} = 6$$

The pooled variance for these two samples is

$$s_p^2 = \frac{SS_1 + SS_2}{df_1 + df_2} = \frac{50 + 30}{5 + 5} = \frac{80}{10} = 8.00$$

Note that the pooled variance is exactly halfway between the two sample variances. Because the two samples are exactly the same size, the pooled variance is simply the average of the two individual sample variances.

Unequal samples sizes   Now consider what happens when the samples are not the same size. This time the first sample has $n = 3$ scores with $SS = 20$, and the second sample has $n = 9$ scores with $SS = 48$. Individually, the two sample variances are

$$\text{Variance for sample 1:} \quad s^2 = \frac{SS}{df} = \frac{20}{2} = 10$$

$$\text{Variance for sample 2:} \quad s^2 = \frac{SS}{df} = \frac{48}{8} = 6$$

The pooled variance for these two samples is

$$s_p^2 = \frac{SS_1 + SS_2}{df_1 + df_2} = \frac{20 + 48}{2 + 8} = \frac{68}{10} = 6.80$$

This time the pooled variance is not located halfway between the two sample variances. Instead, the pooled value is closer to the variance for the larger sample ($n = 9$ and $s^2 = 6$) than to the variance for the smaller sample ($n = 3$ and $s^2 = 10$). The larger sample carries more weight when the pooled variance is computed.

When computing the pooled variance, the weight for each of the individual sample variances is determined by its degrees of freedom. Because the larger sample has a larger *df* value, it will carry more weight when averaging the two variances. This produces an alternative formula for computing pooled variance.

$$\text{pooled variance} = s_p^2 = \frac{df_1 s_1^2 + df_2 s_2^2}{df_1 + df_2} \tag{10.3}$$

For example, if the first sample has $df_1 = 3$ and the second sample has $df_2 = 7$, then the formula instructs you to take 3 of the first sample variance and 7 of the second sample variance for a total of 10 variances. You then divide by 10 to obtain the average. The alternative formula is especially useful if the sample data are summarized as means and variances. Finally, you should note that because the pooled variance is an average of the two sample variances, the value obtained for the pooled variance will always be located between the two sample variances.

ESTIMATED STANDARD ERROR

Using the pooled variance in place of the individual sample variances, we can now obtain an unbiased measure of the standard error for a sample mean difference. The resulting formula for the independent-measures estimated standard error is

$$\text{estimated standard error of } M_1 - M_2 = s_{(M_1-M_2)} = \sqrt{\frac{s_p^2}{n_1} + \frac{s_p^2}{n_2}} \qquad (10.4)$$

Conceptually, this standard error measures how accurately the difference between two sample means represents the difference between the two population means. The formula combines the error for the first sample mean with the error for the second sample mean. Notice that the pooled variance from the two samples is used to compute the standard error for the two samples.

THE FINAL FORMULA AND DEGREES OF FREEDOM

The complete formula for the independent-measures *t* statistic is as follows:

$$t = \frac{(M_1 - M_2) - (\mu_1 - \mu_2)}{s_{(M_1-M_2)}}$$

$$= \frac{\text{sample mean difference} - \text{population mean difference}}{\text{estimated standard error}} \qquad (10.5)$$

In the formula, the estimated standard error in the denominator is calculated using Equation 10.4, and requires calculation of the pooled variance using either Equation 10.2. or 10.3.

The degrees of freedom for the independent-measures *t* statistic are determined by the *df* values for the two separate samples:

$$df \text{ for the } t \text{ statistic} = df \text{ for the first sample} + df \text{ for the second sample}$$

$$= df_1 + df_2$$

$$= (n_1 - 1) + (n_2 - 1) \qquad (10.6)$$

The independent-measures *t* statistic will be used for hypothesis testing. Specifically, we will use the difference between two sample means $(M_1 - M_2)$ as the basis for testing hypotheses about the difference between two population means $(\mu_1 - \mu_2)$. In this context, the overall structure of the *t* statistic can be reduced to the following:

$$t = \frac{\text{data} - \text{hypothesis}}{\text{error}}$$

This same structure is used for both the single-sample *t* from Chapter 9 and the new independent-measures *t* that was introduced in the preceding pages. Table 10.1 identifies each component of these two *t* statistics and should help reinforce the point that we made earlier in the chapter; that is, the independent-measures *t* statistic simply doubles each aspect of the single-sample *t* statistic.

TABLE 10.1

The basic elements of a *t* statistic for the single-sample *t* and the independent-measures *t*

	Sample Data	Hypothesized Population Parameter	Estimated Standard Error	Sample Variance
Single-sample *t* statistic	M	μ	$\sqrt{\dfrac{s^2}{n}}$	$s^2 = \dfrac{SS}{df}$
Independent-measures *t* statistic	$(M_1 - M_2)$	$(\mu_1 - \mu_2)$	$\sqrt{\dfrac{s_p^2}{n_1} + \dfrac{s_p^2}{n_2}}$	$s_p^2 = \dfrac{SS_1 + SS_2}{df_1 + df_2}$

LEARNING CHECK

1. Describe the general characteristics of an independent-measures research study.

2. An independent-measures *t* statistic is used to evaluate the mean difference between two treatments with $n = 8$ in one treatment and $n = 12$ in the other. What is the *df* value for the *t* statistic?

3. One sample from an independent-measures study has $n = 6$ with $SS = 400$. The other sample has $n = 9$ and $SS = 770$.
   a. Compute the pooled variance. (*Note:* Equation 10.2 works well with these data.)
   b. Compute the estimated standard error for the mean difference.

4. One sample from an independent-measures study has $n = 8$ with a variance of $s^2 = 21$. The other sample has $n = 4$ and $s^2 = 31$.
   a. Compute the pooled variance. (*Note:* Equation 10.3 works well with these data.)
   b. Compute the estimated standard error for the mean difference.

ANSWERS

1. An independent-measures study uses a separate sample to represent each of the populations or treatment conditions being compared.

2. $df = df_1 + df_2 = 7 + 11 = 18$.

3. a. The pooled variance is $\frac{1170}{13} = 90$.
   b. The estimated standard error is 5.

4. a. The pooled variance is $\frac{240}{10} = 24$.
   b. The estimated standard error is 3.

10.3   HYPOTHESIS TESTS AND EFFECT SIZE WITH THE INDEPENDENT-MEASURES *t* STATISTIC

The independent-measures *t* statistic uses the data from two separate samples to help decide whether there is a significant mean difference between two populations or between two treatment conditions. A complete example of a hypothesis test with two independent samples follows.

EXAMPLE 10.1   Research results suggest a relationship between the TV viewing habits of 5-year-old children and their future performance in high school. For example, Anderson, Huston, Wright, & Collins (1998) report that high school students who regularly watched Sesame Street as children had better grades in high school than their peers who did not watch Sesame Street. Suppose that a researcher intends to examine this phenomenon using a sample of 20 high school students.

The researcher first surveys the students' parents to obtain information on the family's TV viewing habits during the time that the students were 5 years old. Based on the survey results, the researcher selects a sample of $n = 10$ students with a history of watching Sesame Street and a sample of $n = 10$ students who did not watch the program. The average high school grade is recorded for each student and the data are as follows:

Average High School Grade			
Watched Sesame Street		Did Not Watch Sesame Street	
86	99	90	79
87	97	89	83
91	94	82	86
97	89	83	81
98	92	85	92
$n = 10$		$n = 10$	
$M = 93$		$M = 85$	
$SS = 200$		$SS = 160$	

Note that this is an independent-measures study using two separate samples representing two distinct populations of high school students. The researcher would like to know whether there is a significant difference between the two types of high school student.

STEP 1   State the hypotheses and select the alpha level.

Directional hypotheses could be used and would specify whether the students who watched Sesame Street should have higher or lower grades.

$$H_0: \quad \mu_1 - \mu_2 = 0 \quad \text{(No difference.)}$$

$$H_1: \quad \mu_1 - \mu_2 \neq 0 \quad \text{(There is a difference.)}$$

We will set $\alpha = .05$.

STEP 2   This is an independent-measures design. The *t* statistic for these data will have degrees of freedom determined by

$$df = df_1 + df_2$$

$$= (n_1 - 1) + (n_2 - 1)$$

$$= 9 + 9$$

$$= 18$$

The t distribution for $df = 18$ is presented in Figure 10.4. For $\alpha = .05$, the critical region consists of the extreme 5% of the distribution and has boundaries of $t = +2.101$ and $t = -2.101$.

STEP 3   Obtain the data and compute the test statistic. The data are as given, so all that remains is to compute the t statistic.

As with the single-sample t test in Chapter 9, we recommend that the calcuations be divided into three parts.

First, find the pooled variance for the two samples:

Caution: The pooled variance combines the two samples to obtain a single estimate of variance. In the formula, the two samples are combined in a single fraction.

$$s_p^2 = \frac{SS_1 + SS_2}{df_1 + df_2}$$

$$= \frac{200 + 160}{9 + 9}$$

$$= \frac{360}{18}$$

$$= 20$$

Second, use the pooled variance to compute the estimated standard error:

Caution: The standard error adds the errors from two separate samples. In the formula, these two errors are added as two separate fractions. In this case, the two errors are equal because the sample sizes are the same.

$$s_{(M_1 - M_2)} = \sqrt{\frac{s_p^2}{n_1} + \frac{s_p^2}{n_2}} = \sqrt{\frac{20}{10} + \frac{20}{10}}$$

$$= \sqrt{2 + 2}$$

$$= \sqrt{4}$$

$$= 2$$

FIGURE 10.4

The t distribution with $df = 18$. The critical region for $\alpha = .05$ is shown.

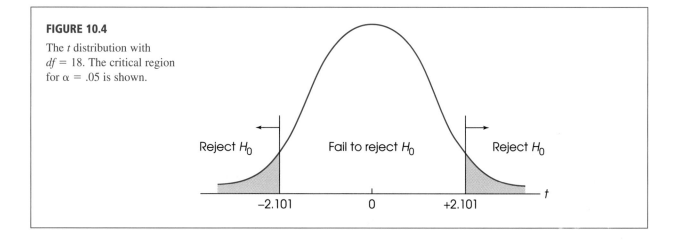

Reject H_0 Fail to reject H_0 Reject H_0

-2.101   0   +2.101 t

Third, compute the t statistic:

$$t = \frac{(M_1 - M_2) - (\mu_1 - \mu_2)}{s_{(M_1 - M_2)}} = \frac{(93 - 85) - 0}{2}$$

$$= \frac{8}{2}$$

$$= 4.00$$

S T E P   4   Make a decision. The obtained value ($t = 4.00$) is in the critical region. In this example, the obtained sample mean difference is four times greater than would be expected if there were no difference between the two populations. In other words, this result is very unlikely if H_0 is true. Therefore, we reject H_0 and conclude that there is a significant difference between the high school grades for students who watched Sesame Street and those who did not. Specifically, the students who watched Sesame Street had significantly higher grades than those who did not watch the program.

Note that the Sesame Street study in Example 10.1 is an example of nonexperimental research (see Chapter 1, pages 15–17). Specifically, the researcher did not manipulate the TV programs watched by the children, and did not control a variety of variables that could influence high school grades. As a result, we cannot conclude that watching Sesame Street *causes* higher high school grades. In particular, many other, uncontrolled factors, such as the parents' level of education or family economic status, might explain the difference between the two groups. Thus, we do not know exactly why there is a relationship between watching educational TV and high school grades, but we do know that a relationship exists.

MEASURING EFFECT SIZE FOR THE INDEPENDENT-MEASURES

As noted in Chapters 8 and 9, a hypothesis test is usually accompanied by a report of effect size to provide an indication of the absolute magnitude of the treatment effect. One technique for measuring effect size is Cohen's d, which produces a standardized measure of mean difference. In its general form, Cohen's d is defined as

$$d = \frac{\text{mean difference}}{\text{standard deviation}} = \frac{\mu_1 - \mu_2}{\sigma}$$

In the context of an independent-measures research study, the difference between the two sample means ($M_1 - M_2$) is used as the best estimate of the mean difference, and the pooled standard deviation (the square root of the pooled variance) is used to estimate the population standard deviation. Thus, the formula for estimating Cohen's d becomes

$$\text{estimated } d = \frac{\text{estimated mean difference}}{\text{estimated standard deviation}} = \frac{M_1 - M_2}{\sqrt{s_p^2}} \tag{10.7}$$

For the data from Example 10.1, the two sample means are 93 and 85, and the pooled variance is 20. The estimated d for these data is

$$d = \frac{M_1 - M_2}{\sqrt{s_p^2}} = \frac{93 - 85}{\sqrt{20}} = \frac{8}{4.47} = 1.79$$

Using the criteria established to evaluate Cohen's *d* (see Table 8.2 on page 264), this value indicates a very large treatment effect.

The independent-measures *t* hypothesis test also allows for measuring effect size by computing the percentage of variance accounted for, r^2. As we saw in Chapter 9, r^2 measures how much of the variability in the scores can be explained by the treatment effects. For example the participants in the Sesame Street study (Example 10.1) have different scores. However, some of the variability in the set of high scholol grades can be explained by knowing which group the student is in; students who watched Sesame Street tend to have higher grades and students who did not watch tend to have lower grades. By measuring exacly how much of the variability can be explained, we can obtain a measure of how big the treatment effect actually is. The calculation of r^2 for the independent-measures *t* is exactly the same as it was for the single-sample *t* in Chapter 9:

$$r^2 = \frac{t^2}{t^2 + df}$$

(10.8)

For the data in Example 10.1, we obtained $t = 4.00$ with $df = 18$. These values produce an r^2 of

$$r^2 = \frac{4^2}{4^2 + 18} = \frac{16}{16 + 18} = \frac{16}{34} = 0.47$$

According to the standards used to evaluate r^2 (see Table 9.3 on page 295), this value also indicates a very large treatment effect.

Although the value of r^2 is usually obtained by using Equation 10.8, it is possible to determine the percentage of variability directly by computing *SS* values for the set of scores. The following example demonstrates this process using the data from the Sesame Street study in Example 10.1.

EXAMPLE 10.2   The Sesame Street study described in Example 10.1 compared high school grades for two groups of students; one group who watched Sesame Street when they were children and one group who did not watch the program. If we assume that the null hypothesis is true and there is no difference between the two populations of students, there should be no systematic difference between the two samples. In this case, the two samples can be combined to form a single set of $n = 20$ scores with an overall mean of $M = 89$. The two samples are shown as a single distribution in Figure 10.5(a).

For this example, however, the result of the hypothesis test is that there is a significant difference between the two populations of students. This population mean difference should produce a corresponding mean difference between the two samples. Specifically, the Sesame Street sample had a mean score of $M = 93$ which is 4 points higher than the overall mean of $M = 89$. Similarly, the sample that did not watch Sesame Street had a mean score of $M = 85$ which is four points lower than the overall mean. Thus, the Sesame Street effect that differentiates the two populations causes one group of scores to move away from $M = 89$ toward the right, and causes the second group of scores to move away from $M = 89$ toward the left. Notice that the students who watched Sesame Street are clustered on the right side of the distribution in Figure 10.5(a), and the students who did not watch are clustered on the left. Thus, the effect of watching or not watching Sesame Street has shifted the scores away from the mean and has increased the variability.

To determine how much the treatment effect has increased the variability, we will remove the treatment effect and examine the resulting scores. To remove the effect,

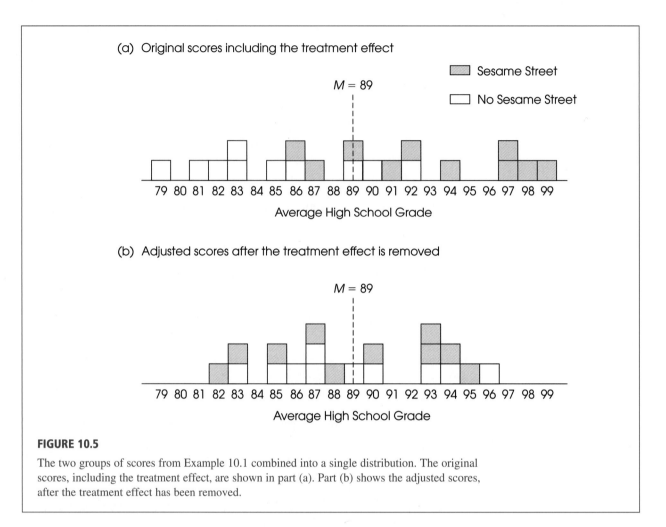

FIGURE 10.5

The two groups of scores from Example 10.1 combined into a single distribution. The original scores, including the treatment effect, are shown in part (a). Part (b) shows the adjusted scores, after the treatment effect has been removed.

we add 4 points to the score for each student who did not watch Sesame Street and we subtract 4 points for each student who did watch. This adjustment causes both groups to have a mean of $M = 89$ so there is no longer any mean difference between the two groups. The adjusted scores are shown in Figure 10.5(b).

It should be clear that the adjusted scores in Figure 10.5(b) are less variable (more closely clustered) than the original scores in Figure 10.5(a). That is, removing the treatment effect has reduced the variability. To determine exactly how much the treatment influences variability, we will compute *SS*, the sum of squared deviations, for each set of scores. The calculations are shown in Table 10.2. For the original set of scores, including the treatment effect, the total variability is $SS = 680$. When the treatment effect is removed, the variability is reduced to $SS = 360$. The difference between these two values, 320 points, is the amount of variability that is explained by the treatment effect. When expressed as a proportion of the total variability, we obtain

$$\frac{\text{variability explained by the treatment}}{\text{total variability}} = \frac{320}{680} = 0.47 = 47\%$$

TABLE 10.2

The calculation of *SS* for the two sets of scores shown in Figure 10.5. The calculations in the left-hand columns are for the original scores, including the treatment effect. The values in the right-hand columns are for the adjusted scores, after the treatment effect is removed.

		Calculation of *SS* Including the Treatment Effect			Calculation of *SS* After the Treatment Effect Is Removed		
	Score	Deviation from M = 89	Squared Deviation		Adjusted Score	Deviation from M = 89	Squared Deviation
	79	−10	100		79 + 4 = 83	−6	36
	81	−8	64		81 + 4 = 85	−4	16
	82	−7	49		82 + 4 = 86	−3	9
Students	83	−6	36		83 + 4 = 87	−2	4
Not	83	−6	36		83 + 4 = 87	−2	4
Watching	85	−4	16		85 + 4 = 89	0	0
Sesame	86	−3	9		86 + 4 = 90	1	1
Street	89	0	0		89 + 4 = 93	4	16
	90	1	1		90 + 4 = 94	5	25
	92	3	9		92 + 4 = 96	7	49
	86	−3	9		86 − 4 = 82	−7	49
	87	−2	4		87 − 4 = 83	−6	36
	89	0	0		89 − 4 = 85	−4	16
	91	2	4		91 − 4 = 87	−2	4
Students	92	3	9		92 − 4 = 88	−1	1
Watching	94	5	25		94 − 4 = 90	1	1
Sesame	97	8	64		97 − 4 = 93	4	16
Street	97	8	64		97 − 4 = 93	4	16
	98	9	81		98 − 4 = 94	5	25
	99	10	100		99 − 4 = 95	6	36
			SS = 680				*SS* = 360

You should recognize that this is exactly the same value we obtained for r^2 using Equation 10.8.

IN THE LITERATURE:
REPORTING THE RESULTS OF AN INDEPENDENT-MEASURES *t* TEST

In Chapter 4 (page 126), we demonstrated how the mean and the standard deviation are reported in APA format. In Chapter 9 (page 296), we illustrated the APA style for reporting the results of a *t* test. Now we will use the APA format to report the results of Example 10.1, an independent-measures *t* test. A concise statement might read as follows:

> The students who watched Sesame Street as children had higher high school grades (*M* = 93, *SD* = 4.71) than the students who did not watch the program (*M* = 85, *SD* = 4.22). The mean difference was significant, *t*(18) = 4.00, *p* < .05, *d* = 1.79.

You should note that standard deviation is not a step in the computations for the independent-measures *t* test, yet it is useful when providing descriptive statistics for each treatment group. It is easily computed when doing the *t* test because you need *SS* and *df* for both groups to determine the pooled variance. Note that the format for reporting *t* is exactly the same as that described in Chapter 9 (page 296) and that the measure of effect size is reported immediately after the results of the hypothesis test.

Also, as we noted in Chapter 9, if an exact probability is available from a computer analysis, it should be reported. For the data in Example 10.1, the computer analysis reports a probability value of $p = .001$ for $t = 4.00$ with $df = 18$. In the research report, this value would be included as follows:

The difference was significant, $t(18) = 4.00$, $p = .001$, $d = 1.79$. ❏

DIRECTIONAL HYPOTHESES AND ONE-TAILED TESTS

When planning an independent-measures study, a researcher usually has some expectation or specific prediction for the outcome. For the Sesame Street study in Example 10.1, the researcher clearly expects the students who watched Sesame Street to have higher grades than the students who did not watch. This kind of directional prediction can be incorporated into the statement of the hypotheses, resulting in a directional, or one-tailed, test. Recall from Chapter 8 that one-tailed tests can be less conservative than two-tailed tests, and should be used when clearly justified by theory or previous findings. The following example demonstrates the procedure for stating hypotheses and locating the critical region for a one-tailed test using the independent-measures *t* statistic.

EXAMPLE 10.3

We will use the same research situation that was described in Example 10.1. The researcher is using an independent-measures design to examine the relationship between watching educational TV as a child and academic performance as a high school student. The prediction is that high school students who watched Sesame Street regularly as 5-year-old children will have higher grades.

STEP 1

State the hypotheses and select the alpha level. As always, the null hypothesis says that there is no effect, and the alternative hypothesis says that there is an effect. For this example, the predicted effect is that the students who watched Sesame Street will have higher grades. Thus, the two hypotheses are as follows.

H_0: $\mu_{\text{Sesame Street}} \leq \mu_{\text{No Sesame Street}}$ (Grades are not higher with Sesame Street)

H_1: $\mu_{\text{Sesame Street}} > \mu_{\text{No Sesame Street}}$ (Grades are higher with Sesame Street)

Note that it is usually easier to state the hypotheses in words before you try to write them in symbols. Also, it usually is easier to begin with the alternative hypothesis (H_1), which states that the treatment works as predicted. Then the null hypothesis is just the opposite of H_1. Also note that the equal sign goes in the null hypothesis, indicating *no difference* between the two treatment conditions. The idea of zero difference is the essence of the null hypothesis, and the numerical value of zero will be used for $(\mu_1 - \mu_2)$ during the calculation of the *t* statistic.

STEP 2

Locate the critical region. For a directional test, the critical region will be located entirely in one tail of the distribution. Rather than trying to determine which tail, positive or negative, is the correct location, we suggest you identify the criteria for the critical region in a two-step process as follows. First, look at the data and determine whether the sample mean difference is in the direction that was predicted. If the answer is no, then the data obviously do not support the predicted treatment effect, and you can stop the analysis. On the other hand, if the difference is in the predicted direction, then the second step is to determine whether the difference is large enough to be significant. To test for significance, simply find the one-tailed critical value in the *t* distribution table. If the calculated *t* statistic is more extreme (either positive or negative) than the critical value, then the difference is significant.

For this example, the students who watched Sesame Street had higher grades, as predicted. With $df = 18$, the one-tailed critical value for $\alpha = .05$ is $t = 1.734$.

STEP 3   Collect the data and calculate the test statistic. The details of the calculations were shown in Example 10.1. The data produce a *t* statistic of $t = 4.00$.

STEP 4   Make a decision. The *t* statistic of $t = 4.00$ is well beyond the critical boundary of $t = 1.734$. Therefore, we reject the null hypothesis and conclude that grades for students who watched Sesame Street are significantly higher than grades for students who did not watch the program.

LEARNING CHECK

1. A researcher report states that there is a significant difference between treatments for an independent-measures design with $t(28) = 2.27$.

   a. How many individuals participated in the research study? (*Hint:* Start with the *df* value.)

   b. Should the report state that $p > .05$ or $p < .05$?

2. A developmental psychologist would like to examine the difference in verbal skills for 6-year-old girls compared with 6-year-old boys. A sample of $n = 5$ girls and $n = 5$ boys is obtained, and each child is given a standardized verbal abilities test.

   The data are as follows:

GIRLS	BOYS
$M = 83$	$M = 71$
$SS = 200$	$SS = 120$

   Are these data sufficient to indicate a significant difference in verbal skills for 6-year-old girls and 6-year-old boys? Test at the .05 level of significance.

3. Suppose that the psychologist in the previous question predicted that girls have higher verbal scores at age 5. What value of *t* would be necessary to conclude that girls score significantly higher using a one-tailed test with $\alpha = .01$?

4. Calculate the effect size for the data in Question 2. Compute Cohen's *d* and r^2.

ANSWERS

1. a. The $df = 28$, so the total number of participants is 30.

   b. A significant result is indicated by $p < .05$.

2. Pooled variance = 40, the standard error = 4, and $t = 3.00$. With $df = 8$, the decision is to reject the null hypothesis.

3. With $df = 8$ and $\alpha = .01$ the critical value is $t = 2.896$.

4. For these data, $d = 1.90$ (a very large effect size), and $r^2 = 0.53$ (or 53%).

THE ROLE OF SAMPLE VARIANCE IN THE INDEPENDENT-MEASURES *t* TEST

In Chapter 8 and in Chapter 9 (pages 250 and 295) we identified several factors that can influence the outcome of a hypothesis test. One factor that plays an important role is the variability of the scores. In general, high variability (measured by either sample variance or standard deviation) reduces the likelihood of obtaining a significant result. The independent-measures *t* test offers an opportunity to present a visual demonstration of this phenomenon.

EXAMPLE 10.4

We will use the data in Figure 10.6 to demonstrate the influence of sample variance. The figure shows the results from a research study comparing two treatments. Notice that the study uses the two separate samples, each with $n = 9$, and there is a 5-point mean difference between the two samples: $M = 8$ for treatment 1 and $M = 13$ for treatment 2. Also notice that there is no overlap between the two distributions; all of the scores for treatment 2 are higher than all of the scores for treatment 1. In this situation, there is a very clear and obvious difference between the two treatments.

For the hypothesis test, the data produce a pooled variance of 1.50 and an estimated standard error of 0.58. The *t* statistic is

$$t = \frac{\text{mean difference}}{\text{estimated standard error}} = \frac{5}{0.58} = 8.62$$

With $df = 16$, this value is far into the critical region (for $\alpha = .05$ or $\alpha = .01$), so we reject the null hypothesis and conclude that there is a significant difference between the two treatments.

Now consider the effect of increasing sample variance. Figure 10.7 shows the results from a second research study comparing two treatments. To create these scores we simply increased the variance for the two samples shown in Figure 10.6. Notice that there are still $n = 9$ scores in each sample, and the two sample means are still $M = 8$ and $M = 13$. However, the sample variances have been greatly increased: Each sample now has $s^2 = 44.25$ as compared with $s^2 = 1.5$ for the data in Figure 10.6. Notice that the increased variance means that the scores are now spread out over a wider range, with the result that the two samples are mixed together without any clear distinction between them.

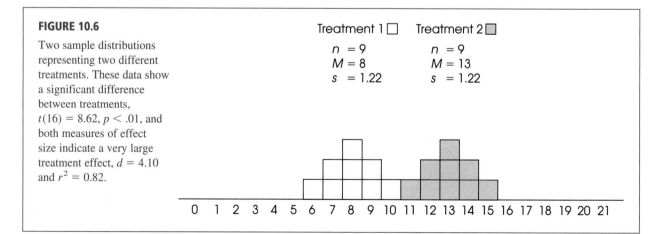

FIGURE 10.6

Two sample distributions representing two different treatments. These data show a significant difference between treatments, $t(16) = 8.62$, $p < .01$, and both measures of effect size indicate a very large treatment effect, $d = 4.10$ and $r^2 = 0.82$.

Treatment 1 □   Treatment 2 ▨

$n = 9$ $n = 9$
$M = 8$ $M = 13$
$s = 1.22$ $s = 1.22$

0 1 2 3 4 5 6 7 8 9 10 11 12 13 14 15 16 17 18 19 20 21

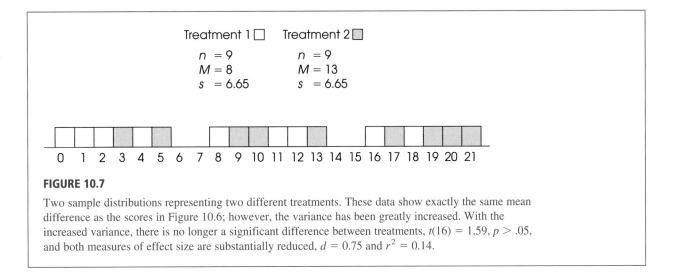

FIGURE 10.7

Two sample distributions representing two different treatments. These data show exactly the same mean difference as the scores in Figure 10.6; however, the variance has been greatly increased. With the increased variance, there is no longer a significant difference between treatments, $t(16) = 1.59$, $p > .05$, and both measures of effect size are substantially reduced, $d = 0.75$ and $r^2 = 0.14$.

The absence of a clear difference between the two samples is supported by the hypothesis test. The pooled variance is 44.25, the estimated standard error is 3.14, and the independent-measures *t* statistic is

$$t = \frac{\text{mean difference}}{\text{estimated standard error}} = \frac{5}{3.14} = 1.59$$

With $df = 16$ and $\alpha = .05$, this value is not in the critical region. Therefore, we fail to reject the null hypothesis and conclude that there is no significant difference between the two treatments. Although there is still a 5-point difference between sample means (as in Figure 10.6), the 5-point difference is not significant with the increased variance. In general, large sample variance can obscure any mean difference that exists in the data and reduces the likelihood of obtaining a significant difference in a hypothesis test.

10.4

ASSUMPTIONS UNDERLYING THE INDEPENDENT-MEASURES *t* FORMULA

There are three assumptions that should be satisfied before you use the independent-measures *t* formula for hypothesis testing:

1. The observations within each sample must be independent (see page 253).
2. The two populations from which the samples are selected must be normal.
3. The two populations from which the samples are selected must have equal variances.

The first two assumptions should be familiar from the single-sample *t* hypothesis test presented in Chapter 9. As before, the normality assumption is the less important of the two, especially with large samples. When there is reason to suspect that the populations are far from normal, you should compensate by ensuring that the samples are relatively large.

Remember: Adding a constant to (or subtracting a constant from) each score does not change the standard deviation.

The third assumption is referred to as *homogeneity of variance* and states that the two populations being compared must have the same variance. You may recall a similar assumption for the *z*-score hypothesis test in Chapter 8. For that test, we assumed that the effect of the treatment was to add a constant amount to (or subtract a constant amount from) each individual score. As a result, the population standard deviation after treatment was the same as it had been before treatment. We now are making essentially the same assumption but phrasing it in terms of variances.

Recall that the pooled variance in the *t*-statistic formula is obtained by averaging together the two sample variances. It makes sense to average these two values only if they both are estimating the same population variance—that is, if the homogeneity of variance assumption is satisfied. If the two sample variances are estimating different population variances, then the average is meaningless. (*Note:* If two people are asked to estimate the distance between Los Angeles and San Francisco, it would be sensible to average the two estimates. However, if one person estimates the distance from New York to Miami and another estimates the distance from Los Angeles to San Francisco, it would not be meaningful to average the two estimates.)

The importance of the homogeneity assumption increases when there is a large discrepancy between the sample sizes. With equal (or nearly equal) sample sizes, this assumption is less critical, but still important.

The homogeneity of variance assumption is quite important because violating this assumption can negate any meaningful interpretation of the data from an independent-measures experiment. Specifically, when you compute the *t* statistic in a hypothesis test, all the numbers in the formula come from the data except for the population mean difference, which you get from H_0. Thus, you are sure of all the numbers in the formula except one. If you obtain an extreme result for the *t* statistic (a value in the critical region), you conclude that the hypothesized value was wrong. But consider what happens when you violate the homogeneity of variance assumption. In this case, you have two questionable values in the formula (the hypothesized population value and the meaningless average of the two variances). Now if you obtain an extreme *t* statistic, you do not know which of these two values is responsible. Specifically, you cannot reject the hypothesis because it may have been the pooled variance that produced the extreme *t* statistic. Without satisfying the homogeneity of variance requirement, you cannot accurately interpret a *t* statistic, and the hypothesis test becomes meaningless.

If you suspect that the homogenity of variance assumption is not justified, you should not compute an independent-measure *t* statistic using pooled variance. However, there is an alternative formula for the *t* statistic that does not pool the two sample variances and does not require the homogeneity assumption. The alternative formula is presented in Box 10.2.

HARTLEY'S *F*-MAX TEST

How do you know whether the homogeneity of variance assumption is satisfied? One simple test involves just looking at the two sample variances. Logically, if the two population variances are equal, then the two sample variances should be very similar. When the two sample variances are reasonably close, you can be reasonably confident that the homogeneity assumption has been satisfied and proceed with the test. However, if one sample variance is more than three or four times larger than the other, then there is reason for concern. A more objective procedure involves a statistical test to evaluate the homogeneity assumption. Although there are many different statistical methods for determining whether the homogeneity of variance assumption has been satisfied, Hartley's *F*-max test is one of the simplest to compute and to understand. An additional advantage is that this test can also be used to check homogeneity of variance with more than two independent samples. Later, in Chapter 13, we will examine statistical methods for comparing several different samples, and Hartley's test will be useful again.

The following example demonstrates the *F*-max test for two independent samples.

BOX 10.2	AN ALTERNATIVE TO POOLED VARIANCE

Computing the independent-measures *t* statistic using pooled variance requires that the data satisfy the homogeneity of variance assumption. Specifically, the two distributions from which the samples are obtained must have equal variances. To avoid this assumption, many statisticians recommend an alternative formula for computing the independent-measures *t* statistic that does not require pooled variance or the homogeneity assumption. The alternative procedure consists of two steps:

1. The standard error is computed using the two separate sample variances as in Equation 10.1.

2. The value of degrees of freedom for the *t* statistic is adjusted using the following equation:

Decimal values for *df* should be rounded down to the next integer.

The adjustment to degrees of freedom will lower the value of *df*, which pushes the boundaries for the critical region farther out. Thus, the adjustment makes the test more demanding and therefore corrects for the same bias problem that the pooled variance attempts to avoid.

Note: Many computer programs that perform statistical analysis (such as SPSS) report two versions of the independent-measures *t* statistic; one using pooled variance (with equal variances assumed) and one using the adjustment shown here (with equal variances not assumed).

$$df = \frac{(V_1 + V_2)^2}{\dfrac{V_1^2}{n_1 - 1} + \dfrac{V_2^2}{n_2 - 1}} \quad \text{where } V_1 = \frac{s_1^2}{n_1} \quad \text{and} \quad V_2 = \frac{s_2^2}{n_2}$$

EXAMPLE 10.5

The *F*-max test is based on the principle that a sample variance provides an unbiased estimate of the population variance. Therefore, if the population variances are the same, the sample variances should be very similar. The procedure for using the *F*-max test is as follows:

1. Compute the sample variance, $s^2 = SS/df$, for each of the separate samples.

2. Select the largest and the smallest of these sample variances and compute

$$F\text{-max} = \frac{s^2(\text{largest})}{s^2(\text{smallest})}$$

A relatively large value for *F*-max indicates a large difference between the sample variances. In this case, the data suggest that the population variances are different and that the homogeneity assumption has been violated. On the other hand, a small value of *F*-max (near 1.00) indicates that the sample variances are similar and that the homogeneity assumption is reasonable.

3. The *F*-max value computed for the sample data is compared with the critical value found in Table B.3 (Appendix B). If the sample value is larger than the table value, you conclude that the variances are different and that the homogeneity assumption is not valid.

To locate the critical value in the table, you need to know

a. k = number of separate samples. (For the independent-measures *t* test, $k = 2$.)

b. $df = n - 1$ for each sample variance. The Hartley test assumes that all samples are the same size.

c. The alpha level. The table provides critical values for $\alpha = .05$ and $\alpha = .01$. Generally a test for homogeneity would use the larger alpha level.

Example: Two independent samples each have $n = 10$. The sample variances are 12.34 and 9.15. For these data,

$$F\text{-max} = \frac{s^2(\text{largest})}{s^2(\text{smallest})} = \frac{12.34}{9.15} = 1.35$$

With $\alpha = .05$, $k = 2$, and $df = n - 1 = 9$, the critical value from the table is 4.03. Because the obtained F-max is smaller than this critical value, you conclude that the data do not provide evidence that the homogeneity of variance assumption has been violated.

Note: The goal for most hypothesis tests is to reject the null hypothesis to demonstrate a significant difference or a significant treatment effect. However, when testing for homogeneity of variance, the preferred outcome is to fail to reject H_0. Failing to reject H_0 with the F-max test means that there is no significant difference between the two population variances and the homogeneity assumption is satisfied. In this case, you may proceed with the independent-measures t test using pooled variance.

LEARNING CHECK

1. A researcher is using an independent-measures design to compare two treatment conditions with $n = 8$ participants in each treatment. If the results are evaluated using a one-tailed test with $\alpha = .05$, what is the critical value for t?

2. The homogeneity of variance assumption requires that the two sample variances be equal. (True or false?)

3. When you are using an F-max test to evaluate the homogeneity of variance assumption, you usually do not want to find a significant difference between the variances. (True or false?)

ANSWERS

1. With $df = 14$, the critical $t = 1.761$.

2. False. The assumption is that the two population variances are equal.

3. True. If there is a significant difference between the two variances, you cannot do the t test.

SUMMARY

1. The independent-measures t statistic uses the data from two separate samples to draw inferences about the mean difference between two populations or between two different treatment conditions.

2. The formula for the independent-measures t statistic has the same structure as the original z-score or the single-sample t:

$$t = \frac{\text{sample statistic} - \text{population parameter}}{\text{estimated standard error}}$$

For the independent-measures t, the sample statistic is the sample mean difference $(M_1 - M_2)$. The population parameter is the population mean difference, $(\mu_1 - \mu_2)$. The estimated standard error for the sample mean difference is computed by combining the errors for the two sample means. The resulting formula is

$$t = \frac{(M_1 - M_2) - (\mu_1 - \mu_2)}{s_{(M_1-M_2)}}$$

where the estimated standard error is

$$s_{(M_1-M_2)} = \sqrt{\frac{s_p^2}{n_1} + \frac{s_p^2}{n_2}}$$

The pooled variance in the formula, s_p^2, is the weighted mean of the two sample variances:

$$s_p^2 = \frac{SS_1 + SS_2}{df_1 + df_2}$$

This t statistic has degrees of freedom determined by the sum of the df values for the two samples:

$$df = df_1 + df_2$$
$$= (n_1 - 1) + (n_2 - 1)$$

3. For hypothesis testing, the null hypothesis states that there is no difference between the two population means:

$$H_0: \mu_1 = \mu_2 \quad \text{or} \quad \mu_1 - \mu_2 = 0$$

4. When a hypothesis test with an independent-measures t statistic indicates a significant difference, it is recommended that you also compute a measure of the effect size. One measure of effect size is Cohen's d, which is a standardized measured of the mean

difference. For the independent-measures t statistic, Cohen's d is estimated as follows:

$$\text{estimated } d = \frac{M_1 - M_2}{\sqrt{s_p^2}}$$

A second common measure of effect size is the percentage of variance accounted for by the treatment effect. This measure is identified by r^2 and is computed as

$$r^2 = \frac{t^2}{t^2 + df}$$

5. Appropriate use and interpretation of the t statistic using pooled variance require that the data satisfy the homogeneity of variance assumption. This assumption stipulates that the two populations have equal variances. An informal test of the assumption can be made by verifying that the two sample variances are approximately equal. Hartley's F-max test provides a statistical technique for determining whether the data satisfy the homogeneity assumption. An alternative technique that avoids pooling variances and eliminates the need for the homogeneity assumption is presented in Box 10.2.

KEY TERMS

independent-measures research design

between-subjects research design

repeated-measures research design

within-subjects research design

independent-measures t statistic

estimated standard error of $M_1 - M_2$

pooled variance

homogeneity of variance

RESOURCES

There are a tutorial quiz and other learning exercises for Chapter 10 at the Wadsworth website, www.cengage.com/psychology/gravetter. The site also provides access to a workshop entitled *Independent vs. Repeated t-tests* that compares the t test presented in this chapter with the repeated-measures test presented in Chapter 11. See page 29 for more information about accessing items on the website.

Guided interactive tutorials, end-of-chapter problems, and related testbank items may be assigned online at WebAssign.

WebTUTOR

For those using WebTutor along with this book, there is a WebTutor section corresponding to this chapter. The WebTutor contains a brief summary of Chapter 10, hints for learning the concepts and the formulas for the independent-measures *t* test, cautions about common errors, and sample exam items including solutions.

SPSS

General instructions for using SPSS are presented in Appendix D. Following are detailed instructions for using SPSS to perform **The Independent-Meausres *t* Test** presented in this chapter.

Data Entry

1. The scores are entered in what is called *stacked format,* which means that all the scores from *both samples* are entered in one column of the data editor (probably VAR00001). Enter the scores for sample #2 directly beneath the scores from sample #1 with no gaps or extra spaces.
2. Values are then entered into a second column (VAR00002) to identify the sample or treatment condition corresponding to each of the scores. For example, enter a 1 beside each score from sample #1 and enter a 2 beside each score from sample #2.

Data Analysis

1. Click **Analyze** on the tool bar, select **Compare Means,** and click on **Independent-Samples T Test.**
2. Highlight the column label for the set of scores (VAR0001) in the left box and click the arrow to move it into the **Test Variable(s)** box.
3. Highlight the label from the column containing the sample numbers (VAR0002) in the left box and click the arrow to move it into the **Group Variable** box.
4. Click on **Define Groups.**
5. Assuming that you used the numbers 1 and 2 to identify the two sets of scores, enter the values 1 and 2 into the appropriate group boxes.
6. Click **Continue.**
7. Click **OK.**

SPSS Output

SPSS produces a summary table showing the number of scores, the mean, the standard deviation, and the standard error for each of the two samples. A separate table presents the results of the hypothesis test. SPSS first conducts a test for homogeneity of variance, using Levene's test. This test should *not* be significant (you do not want the two variances to be different), so you want the reported Sig. value to be greater than .05. Next, the results of the independent-measures *t* test are presented using two different assumptions. The top row shows the outcome assuming equal variances, using the pooled variance to compute *t*. The second row does not assume equal variances and computes the *t* statistic using the alternative method presented in Box 10.2. Each row reports the calculated *t* value, the degrees of freedom, the level of significance (the *p* value or alpha level for the test), and the size of the mean difference. Finally, the output includes a report of the standard error for the mean difference (the denominator of the *t* statistic) and a 95% confidence interval for the mean difference. Confidence intervals are presented in Chapter 12 and provide an estimate of the size of the mean difference.

FOCUS ON PROBLEM SOLVING

1. As you learn more about different statistical methods, one basic problem will be deciding which method is appropriate for a particular set of data. Fortunately, it is easy to identify situations in which the independent-measures t statistic is used. First, the data will always consist of two separate samples (two ns, two Ms, two SSs, and so on). Second, this t statistic is always used to answer questions about a mean difference: On the average, is one group different (better, faster, smarter) than the other group? If you examine the data and identify the type of question that a researcher is asking, you should be able to decide whether an independent-measures t is appropriate.

2. When computing an independent-measures t statistic from sample data, we suggest that you routinely divide the formula into separate stages rather than trying to do all the calculations at once. First, find the pooled variance. Second, compute the standard error. Third, compute the t statistic.

3. One of the most common errors for students involves confusing the formulas for pooled variance and standard error. When computing pooled variance, you are "pooling" the two samples together into a single variance. This variance is computed as a *single fraction*, with two SS values in the numerator and two df values in the denominator. When computing the standard error, you are adding the error from the first sample and the error from the second sample. These two separate errors add as *two separate fractions* under the square root symbol.

DEMONSTRATION 10.1

THE INDEPENDENT-MEASURES t TEST

In a study of jury behavior, two samples of participants were provided details about a trial in which the defendant was obviously guilty. Although group 2 received the same details as group 1, the second group was also told that some evidence had been withheld from the jury by the judge. Later the participants were asked to recommend a jail sentence. The length of term suggested by each participant is presented here. Is there a significant difference between the two groups in their responses?

Group 1	Group 2	
4	3	
4	7	
3	8	for Group 1: $M = 3$ and $SS = 16$
2	5	
5	4	for Group 2: $M = 6$ and $SS = 24$
1	7	
1	6	
4	8	

There are two separate samples in this study. Therefore, the analysis will use the independent-measures t test.

STEP 1   State the hypothesis, and select an alpha level.

H_0: $\mu_1 - \mu_2 = 0$ (For the population, knowing evidence has been
withheld has no effect on the suggested sentence.)

H_1: $\mu_1 - \mu_2 \neq 0$ (For the population, knowledge of withheld evidence has
an effect on the jury's response.)

We will set the level of significance to $\alpha = .05$, two tails.

STEP 2   Identify the critical region.
For the independent-measures *t* statistic, degrees of freedom are determined by

$$df = n_1 + n_2 - 2$$
$$= 8 + 8 - 2$$
$$= 14$$

The *t* distribution table is consulted, for a two-tailed test with $\alpha = .05$ and $df = 14$. The
critical *t* values are $+2.145$ and -2.145.

STEP 3   Compute the test statistic. As usual, we recommended that the calculation of the *t* statistic
be separated into three stages.

Pooled variance   For these data, the pooled variance equals

$$s_p^2 = \frac{SS_1 + SS_2}{df_1 + df_2} = \frac{16 + 24}{7 + 7} = \frac{40}{14} = 2.86$$

Estimated standard error   Now we can calculate the estimated standard error for mean
differences.

$$s_{(M_1 - M_2)} = \sqrt{\frac{s_p^2}{n_1} + \frac{s_p^2}{n_2}} = \sqrt{\frac{2.86}{8} + \frac{2.86}{8}} = \sqrt{0.358 + 0.358}$$
$$= \sqrt{0.716} = 0.85$$

The t *statistic*   Finally, the *t* statistic can be computed.

$$t = \frac{(M_1 - M_2) - (\mu_1 - \mu_2)}{s_{(M_1 - M_2)}} = \frac{(3 - 6) - 0}{0.85} = \frac{-3}{0.85}$$
$$= -3.53$$

STEP 4   Make a decision about H_0, and state a conclusion.
The obtained *t* value of -3.53 falls in the critical region of the left tail (critical
$t = \pm2.145$). Therefore, the null hypothesis is rejected. The participants that were
informed about the withheld evidence gave significantly longer sentences, $t(14) = -3.53$,
$p < .05$, two tails.

DEMONSTRATION 10.2

EFFECT SIZE FOR THE INDEPENDENT-MEASURES t

We will estimate Cohen's d and compute r^2 for the jury decision data in Demonstration 10.1. For these data, the two sample means are $M_1 = 3$ and $M_2 = 6$, and the pooled variance is 2.86. Therefore, our estimate of Cohen's d is

$$\text{estimated } d = \frac{M_1 - M_2}{\sqrt{s_p^2}} = \frac{3 - 6}{\sqrt{2.86}} = \frac{3}{1.69} = 1.78$$

With a t value of $t = 3.53$ and $df = 14$, the percentage of variance accounted for is

$$r^2 = \frac{t^2}{t^2 + df} = \frac{(3.53)^2}{(3.53)^2 + 14} = \frac{12.46}{26.46} = 0.47 \quad \text{(or 47\%)}$$

PROBLEMS

1. Describe the basic characteristics of an independent-measures, or a between-subjects, research study.

2. Describe what is measured by the estimated standard error in the bottom of the independent-measures t statistic.

3. If other factors are held constant, how does increasing the number of scores in each sample affect the value of the independent-measures t statistic and the likelihood of rejecting the null hypothesis?

4. If other factors are held constant, how does increasing the sample variance affect the value of the independent-measures t statistic and the likelihood of rejecting the null hypothesis?

5. Describe the homogeneity of variance assumption and explain why it is important for the independent-measures t test.

6. One sample has $SS = 48$ and a second sample has $SS = 42$.
   a. Assuming that $n = 4$ for both samples, find each of the sample variances, and calculate the pooled variance. You should find that the pooled variance is exactly halfway between the two sample variances.
   b. Now assume that $n = 4$ for the first sample and $n = 8$ for the second. Find each of the sample variances, and calculate the pooled variance. You should find that the pooled variance is closer to the variance for the larger sample ($n = 8$).

7. One sample has $SS = 35$ and a second sample has $SS = 45$.
   a. Assuming that $n = 6$ for both samples, calculate each of the sample variances, then calculate the pooled variance. Because the samples are the same size, you should find that the pooled variance is exactly halfway between the two sample variances.
   b. Now assume that $n = 6$ for the first sample and $n = 16$ for the second. Again, calculate the two sample variances and the pooled variance. You should find that the pooled variance is closer to the variance for the larger sample.

8. As noted on page 312, when the two population means are equal, the estimated standard error for the independent-measures t test provides a measure of how much difference to expect between two sample means. For each of the following situations, assume that $\mu_1 = \mu_2$ and calculate how much difference should be expected between the two sample means.
   a. One sample has $n = 8$ scores with $SS = 45$ and the second sample has $n = 4$ scores with $SS = 15$.
   b. One sample has $n = 8$ scores with $SS = 150$ and the second sample has $n = 4$ scores with $SS = 90$.
   c. In part b, the samples have larger variability (bigger SS values) than in part a, but the sample sizes are unchanged. How does larger variability affect the size of the standard error for the sample mean difference?

9. One sample has $n = 15$ with $SS = 1660$, and a second sample has $n = 15$ with $SS = 1700$.
   a. Find the pooled variance for the two samples.
   b. Compute the estimated standard error for the sample mean difference.
   c. If the sample mean difference is 8 points, is this enough to reject the null hypothesis and conclude that there is a significant difference for a two-tailed test at the .05 level?
   d. If the sample mean difference is 12 points, is this enough to indicate a significant difference for a two-tailed test at the .05 level?
   e. Calculate the percentage of variance accounted for (r^2) to measure the effect size for an 8-point mean difference and for a 12-point mean difference.

10. One sample has $n = 4$ scores with $SS = 100$ and a second sample has $n = 8$ with $SS = 140$.
   a. Calculate the pooled variance for the two samples.
   b. Calculate the estimated standard error for the sample mean difference.
   c. If the sample mean difference is 6 points, is this enough to reject the null hypothesis for a two-tailed test with $\alpha = .05$?
   d. If the sample mean difference is 9 points, is this enough to reject the null hypothesis for a two-tailed test with $\alpha = .05$?

11. For each of the following, assume that the two samples are selected from populations with equal means and calculate how much difference should be expected, on average, between the two sample means.
   a. Each sample has $n = 5$ scores with $s^2 = 38$ for the first sample and $s^2 = 42$ for the second. (*Note:* Because the two samples are the same size, the pooled variance is equal to the average of the two sample variances.)
   b. Each sample has $n = 20$ scores with $s^2 = 38$ for the first sample and $s^2 = 42$ for the second.
   c. In part b, the two samples are bigger than in part a, but the variances are unchanged. How does sample size affect the size of the standard error for the sample mean difference?

12. An independent-measures research study was used to compare two treatment conditions with $n = 12$ participants in each treatment. The first treatment had a mean of $M = 55$ with a variance of $s^2 = 8$, and the second treatment had $M = 52$ and $s^2 = 4$. Do these data indicate a significant difference between the two treatments? Use a two-tailed test with $\alpha = .05$. (*Note:* Because the two samples are the same size, the pooled variance is simply the average of the two sample variances.)

13. Suppose the research study in the previous problem produced larger variances of $s^2 = 27$ and $s^2 = 21$ for

the two samples. Assuming that the means and sample sizes are the same as in problem 12, are the data sufficient to conclude that there is a significant difference between the two treatments? Use a two-tailed test with $\alpha = .05$. Note how larger variances affect the outcome of the hypothesis test. (Again, the two samples are the same size, so the pooled variance is simply the average of the two sample variances.)

14. Do you view a chocolate bar as delicious or as fattening? Your attitude may depend on your gender. In a study of American college students, Rozin, Bauer, and Catanese (2003) examined the importance of food as a source of pleasure versus concerns about food associated with weight gain and health. The following results are similar to those obtained in the study. The scores are a measure of concern about the negative aspects of eating.

Males	Females
$n = 6$	$n = 9$
$M = 31$	$M = 45$
$SS = 490$	$SS = 680$

   a. Based on these results, is there a significant difference between the attitudes for males and for females? Use a two-tailed test with $\alpha = .05$.
   b. Compute r^2, the percentage of variance accounted for by the gender difference, to measure effect size for this study.

15. In a study comparing individual performance with group performance, Laughlin, Zander, Knievel, and Tan (2003) found that groups consistently outperformed the best of the individuals. The task was a letters-to-numbers problem in which an arithmetic problem was presented substituting a letter in place of each digit. Participants had to determine which letters corresponded to each number. Three-person groups competed with individuals and scores were compared for the groups and the best of the individuals. The dependent variable was the number of trials needed to solve each problem. The following data are similar to those obtained in the study.

Groups	Individuals
$n = 15$	$n = 15$
$M = 3.8$	$M = 4.7$
$SS = 15.4$	$SS = 18.2$

   a. Based on these results, is there a significant difference between the performance for individuals versus the performance for groups? Use a two-tailed test with $\alpha = .05$.

b. Compute the estimated value of Cohen's d to measure effect size for this study.

16. An independent-measures research study was used to compare two treatment conditions with $n = 4$ participants in each treatment. The first treatment had a mean of $M = 58$ with a variance of $s^2 = 28$, and the second treatment had $M = 52$ and $s^2 = 36$. Do these data indicate a significant difference between the two treatments? Use a two-tailed test with $\alpha = .05$. (*Note:* Because the two samples are the same size, the pooled variance is simply the average of the two sample variances.)

17. The independent-measures study from the previous problem was repeated using larger samples with $n = 16$ participants in each treatment. Assuming that the larger samples produced exactly the same sample means and variances as were obtained in problem 16, are the data sufficient to conclude that there is a significant difference between treatments? Use a two-tailed test with $\alpha = .05$. Note how larger sample sizes affect the outcome of the hypothesis test. (Again, the two samples are the same size so, the pooled variance is simply the average of the two sample variances.)

18. In 1974, Loftus and Palmer conducted a classic study demonstrating how the language used to ask a question can influence eyewitness memory. In the study, college students watched a film of an automobile accident and then were asked questions about what they saw. One group was asked, "About how fast were the cars going when they smashed into each other?" Another group was asked the same question except the verb was changed to "hit" instead of "smashed into." The "smashed into" group reported significantly higher estimates of speed than the "hit" group. Suppose a researcher repeats this study with a sample of today's college students and obtains the following results.

Estimated Speed	
Smashed into	Hit
$n = 15$	$n = 15$
$M = 40.8$	$M = 34.0$
$SS = 510$	$SS = 414$

Do the results indicate a significantly higher mean for the "smashed into" group? Use a one-tailed test with $\alpha = .01$.

19. Research indicates that adolescent boys report higher levels of self-esteem than is reported by adolescent girls (Kling, Hyde, & Buswell, 1999). To examine this phenomenon a researcher obtains a sample of 40 adolescents, 20 boys and 20 girls, and administers a standardized measure of self-esteem to each participant. The boys produce a mean score of $M = 84$ with $SS = 1940$, and the girls produce $M = 73$ with $SS = 1480$.

a. Do these results indicate a significant difference in the level of self-esteem reported by adolescent girls and boys? Use a two-tailed test with $\alpha = .01$.

a. Compute Cohen's d and r^2 to measure the size of the effect.

20. Harris, Schoen, and Hensley (1992) conducted a research study showing how cultural experience can influence memory. They presented participants with two different versions of stories. One version contained facts or elements that were consistent with a U.S. culture and the second version contained material consistent with a Mexican culture. The results showed that the participants tended to make memory errors for the information that was not consistent with their own culture. Specifically, participants from Mexico either forgot or distorted information that was unique to U.S. culture, and participants from the United States forgot or distorted information that was unique to Mexican culture. The following data represent results similar to those obtained by Harris, Schoen, and Hensley. Is there a significant difference between the two groups? Use a two-tailed test with $\alpha = .05$.

Number of Errors Recalling the Mexican Story	
Participants from Mexico	Participants from the United States
$n = 20$	$n = 20$
$M = 4.1$	$M = 6.9$
$SS = 180$	$SS = 200$

21. When people learn a new task, their performance usually improves when they are tested the next day, but only if they get at least 6 hours of sleep (Stickgold, Whidbee, Schirmer, Patel, & Hobson, 2000). The following data demonstrate this phenomenon. The participants learned a visual discrimination task on one day, and then were tested on the task the following day. Half of the participants were allowed to have at least 6 hours of sleep and the other half were kept awake all night. Is there a significant difference between the two conditions? Use a two-tailed test with $\alpha = .05$.

Performance Scores	
6-hours Sleep	No Sleep
$n = 8$	$n = 8$
$M = 72$	$M = 61$
$SS = 440$	$SS = 456$

22. Steven Schmidt (1994) conducted a series of experiments examining the effects of humor on memory. In one study, participants were given a mix of humorous and nonhumorous sentences and significantly more humorous sentences were recalled. However, Schmidt argued that the humorous sentences were not necessarily easier to remember, they were simply preferred when participants had a choice between the two types of sentence. To test this argument, he switched to an independent-measures design in which one group got a set of exclusively humorous sentences and another group got a set of exclusively nonhumorous sentences. The following data are similar to the results from the independent-measures study.

Humorous Sentences				Nonhumorous Sentences			
4	5	2	4	6	3	5	3
6	7	6	6	3	4	2	6
2	5	4	3	4	3	4	4
3	3	5	3	5	2	6	4

Do the results indicate a significant difference in the recall of humorous versus nonhumorous sentences? Use a two-tailed test with $\alpha = .05$.

23. Siegel (1990) found that elderly people who owned dogs were less likely to pay visits to their doctors after upsetting events than were those who did not own pets. Similarly, consider the following hypothetical data. A sample of elderly dog owners is compared to a similar group (in terms of age and health) who do not own dogs. The researcher records the number of visits to the doctor during the past year for each person. The data are as follows:

Control Group	Dog Owners
10	7
8	4
7	9
9	3
13	7
7	
6	
12	

a. Is the number of doctor visits significantly lower for the dog owners than for participants in the control group. Use a one-tailed test with $\alpha = .05$.
b. Calculate the value of r^2 (percentage of variance accounted for) for these data.

24. A researcher conducts an independent-measures research study and obtains $t = 2.070$ with $df = 28$.
a. How many individuals participated in the entire research study?
b. Using a two-tailed test with $\alpha = .05$, is there a significant difference between the two treatment conditions?
c. Compute r^2 to measure the percentage of variance accounted for by the treatment effect.

25. A researcher is comparing the effectiveness of two sets of instructions for assembling a child's bike. A sample of eight fathers is obtained. Half of the fathers are given one set of instructions and the other half receives the second set. The researcher measures how much time is needed for each father to assemble the bike. The scores are the number of minutes needed by each participant.

Instruction Set I	Instruction Set II
8	14
4	10
8	6
4	10

a. Is there a significant difference in time for the two sets of instructions? Use a two-tailed test at the .05 level of significance.
b. Calculate the estimated Cohen's d and r^2 to measure effect size for this study.

C H A P T E R

11

The *t* Test for Two Related Samples

Tools You Will Need

The following items are considered essential background material for this chapter. If you doubt your knowledge of any of these items, you should review the appropriate chapter or section before proceeding.

- Introduction to the *t* statistic (Chapter 9)
  - Estimated standard error
  - Degrees of freedom
  - *t* Distribution
  - Hypothesis tests with the *t* statistic
- Independent-measures design (Chapter 10)

Preview

11.1 Introduction to Repeated-Measures Designs

11.2 The *t* Statistic for a Repeated-Measures Research Design

11.3 Hypothesis Tests and Effect Size for the Repeated-Measures Design

11.4 Uses and Assumptions for Repeated-Measures *t* Tests

Summary

Focus on Problem Solving

Demonstrations 11.1 and 11.2

Problems

Preview

At the Olympic level of competition, even the smallest factors can make the difference between winning and losing. Pelton (1983) reports on one study demonstrating this point. The study recorded the scores obtained by Olympic marksmen when firing *during* heartbeats compared with the scores for shots fired *between* heartbeats. The results show the marksmen had significantly higher scores when they shot between heartbeats. Apparently the small vibrations caused by a heartbeat were enough to disturb the marksmen's aim.

To conduct this study, the researchers monitored each individual's heart during a shooting session (Figure 11.1). The shots fired during heartbeats were then separated from the shots fired between heartbeats, and two separate scores were computed for each marksman. Thus, the study involved only one sample of participants, but produced two sets of scores. The goal of the study is to look for a mean difference between the two sets of scores.

In the previous chapter, we introduced a statistical procedure for evaluating the mean difference between two sets of data (the independent-measures *t* statistic). However, the independent-measures *t* statistic is intended for research situations involving two separate and independent samples. You should realize that the two sets of shooting scores are not independent samples. In fact, for each participant, the score for shots fired during a heartbeat is *directly related*, one to one, to the score for shots fired between heartbeats, because both scores come from the same individual. This kind of study is called a repeated-measures (or related-samples) design because the data are obtained by literally repeating measurements, under different conditions, for the same sample. Because the two sets of measurements are related, rather than independent, a different statistical analysis is required.

The Problem: One concern for an independent-measures design (two separate samples) is that the individuals in one sample may have characteristics that are different from individuals in the other sample. For example, one group may be smarter or older than the other. If this occurs, the mean difference between the groups may be explained by the individual characteristics rather than a treatment effect.

As a result, researchers take steps to ensure that the two groups are as similar as possible. In a repeated-measures design, however, this is not a problem because the two sets of scores come from the same group of individuals. Thus, the individuals in one treatment are perfectly matched with the individuals in the other treatment. A new hypothesis testing procedure is needed to make full advantage of this perfect match.

The Solution: In this chapter, we introduce the *repeated-measures* t *statistic* which is used for hypothesis tests evaluating the mean difference between two sets of scores obtained from the same group of individuals. As you will see, however, this new *t* statistic is very similar to the original *t* statistic that was introduced in Chapter 9.

Finally, we should note that independent measures and repeated measures are two basic research strategies that are available whenever a researcher is planning a study that will compare treatment conditions. That is, a researcher may choose to use two separate samples, one for each of the treatments, or a researcher may choose to use one sample and measure each individual in both of the treatment conditions. Later in this chapter, we take a closer look at the differences between independent-measures and repeated-measures designs and discuss the advantages and disadvantages of each.

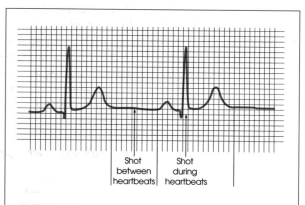

FIGURE 11.1

Illustration of the timing of shots taken between heartbeats or during a heartbeat.

11.1 INTRODUCTION TO REPEATED-MEASURES DESIGNS

In the previous chapter, we introduced the independent-measures research design as one strategy for comparing two treatment conditions or two populations. The independent-measures design is characterized by the fact that two separate samples are used to obtain the two sets of data that are to be compared. In this chapter, we examine an alternative strategy known as a *repeated-measures design,* or a *within-subjects design.* With a repeated-measures design, two sets of data are obtained from the same sample of individuals. For example, a group of patients could be measured before therapy and then measured again after therapy. Or a group of individuals could be tested in a quiet, comfortable environment and then tested again in a noisy, stressful environment to examine the effects of stress on performance. In each case, notice that the same variable is being measured twice for the same set of individuals; that is, we are literally repeating measurements on the same sample.

DEFINITION

> A **repeated-measures design,** or a **within-subject design,** is one in which a single sample of individuals is measured more than once on the same dependent variable. The same subjects are used in all of the treatment conditions.

The main advantage of a repeated-measures study is that it uses exactly the same individuals in all treatment conditions. Thus, there is no risk that the participants in one treatment are substantially different from the participants in another. With an independent-measures design, on the other hand, there is always a risk that the results are biased because the individuals in one sample are systematically different (smarter, faster, more extroverted, and so on) than the individuals in the other sample. At the end of this chapter, we present a more detailed comparison of repeated-measures studies and independent-measures studies, considering the advantages and disadvantages of both types of research.

Occasionally, researchers try to approximate the advantages of a repeated-measures design by using a technique known as *matched subjects.* A matched-subjects design involves two separate samples, but each individual in one sample is matched one-to-one with an individual in the other sample. Typically, the individuals are matched on one or more variables that are considered to be especially important for the study. For example, a researcher studying verbal learning might want to be certain that the two samples are matched in terms of IQ and gender. In this case, a male participant with an IQ of 120 in one sample would be matched with another male with an IQ of 120 in the other sample. Although the participants in one sample are not *identical* to the participants in the other sample, the matched-subjects design at least ensures that the two samples are equivalent (or matched) with respect to some specific variables.

DEFINITION

> In a **matched-subjects** study, each individual in one sample is matched with an individual in the other sample. The matching is done so that the two individuals are equivalent (or nearly equivalent) with respect to a specific variable that the researcher would like to control.

A matched-subjects study occasionally is called a *matched-samples design.* But the subjects in the samples must be matched one-to-one before you can use the statistical techniques in this chapter.

Of course, it is possible to match participants on more than one variable. For example, a researcher could match pairs of subjects on age, gender, race, and IQ. In this case, for example, a 22-year-old white female with an IQ of 115 who was in one sample would be matched with another 22-year-old white female with an IQ of 115 in the second sample. The more variables that are used, however, the more difficult it is to find matching pairs. The goal of the matching process is to simulate a repeated-measures design as closely as possible. In a repeated-measures design, the matching is perfect because the same individual is used in both conditions. In a matched-subjects design, however, the best you can get is a degree of match that is limited to the variable(s) that are used for the matching process.

In a repeated-measures design, or a matched-subjects design, the data consist of two sets of scores with the scores in one set directly related, one-to-one, with the scores in the second set. For this reason, the two research designs are statistically equivalent and are grouped together under the common name *related-samples* designs (or correlated-samples designs). In this chapter, we focus our discussion on repeated-measures designs because they are overwhelmingly the more common example of related-groups designs. However, you should realize that the statistical techniques used for repeated-measures studies can be applied directly to data from a matched-subjects study.

Now we will examine the statistical techniques that allow a researcher to use the sample data from a repeated-measures study to draw inferences about the general population.

11.2   THE *t* STATISTIC FOR A REPEATED-MEASURES RESEARCH DESIGN

The *t* statistic for a repeated-measures design is structurally similar to the other *t* statistics we have examined. As we shall see, it is essentially the same as the single-sample *t* statistic covered in Chapter 9. One major distinction of the related-samples *t* is that it is based on difference scores rather than raw scores (X values). In this section, we examine difference scores and develop the *t* statistic for related samples.

DIFFERENCE SCORES: THE DATA FOR A REPEATED-MEASURES STUDY

Table 11.1 presents hypothetical data from a repeated-measures study examining how reaction time is affected by a common, over-the-counter cold medication. Note that there is one sample of $n = 4$ participants, and that each indiviual is measured twice. The first score for each person (X_1) is a measurement of reaction time before the medication was administered. The second score (X_2) is a measure of reaction time 1 hour after taking the medication. Because we are interested in how the medication affects reaction time, we have computed the difference between the first score and the second score for each individual. The *difference scores*, or D values, are shown in the last column of the table. Notice that the difference scores measure the amount of change in reaction time for each person. Typically, the difference scores are obtained by subtracting the first score (before treatment) from the second score (after treatment) for each person:

$$\text{difference score} = D = X_2 - X_1 \tag{11.1}$$

TABLE 11.1

Reaction time measurements taken before and after taking an over-the-counter cold medication.

Person	Before Medication (X_1)	After Medication (X_2)	Difference D
A	215	210	−5
B	221	242	21
C	196	219	23
D	203	228	25

Note that M_D is the mean for the sample of D scores.

$$\Sigma D = 64$$

$$M_D = \frac{\Sigma D}{n} = \frac{64}{4} = 16$$

Note that the sign of each D score tells you the direction of the change. Person A, for example, shows a decrease in reaction time after taking the medication (a negative change), but person B shows an increase (a positive change).

The sample of difference scores (D values) will serve as the sample data for the hypothesis test. To compute the t statistic, we will use the number of D scores (n) as well as the sample mean (M_D) and the value of SS for the sample of D scores.

THE HYPOTHESES FOR A RELATED-SAMPLES TEST

The researcher's goal is to use the sample of difference scores to answer questions about the general population. In particular, the researcher would like to know whether there is any difference between the two treatment conditions for the general population. Note that we are interested in a population of *difference scores*. That is, we would like to know what would happen if every individual in the population were measured in two treatment conditions (X_1 and X_2) and a difference score (D) were computed for everyone. Specifically, we are interested in the mean for the population of difference scores. We identify this population mean difference with the symbol μ_D (using the subscript letter D to indicate that we are dealing with D values rather than X scores).

As always, the null hypothesis states that for the general population there is no effect, no change, or no difference. For a repeated-measures study, the null hypothesis states that the mean difference for the general population is zero. In symbols,

$$H_0: \quad \mu_D = 0$$

Again, this hypothesis refers to the mean for the entire population of difference scores. Figure 11.2(a) shows an example of a population of difference scores with a mean of $\mu_D = 0$. Although the population mean is zero, the individual scores in the population are not all equal to zero. Thus, even when the null hypothesis is true, we still expect some individuals to have positive difference scores and some to have negative difference scores. However, the positives and negatives are unsystematic and in the long run will balance out to $\mu_D = 0$. Also note that a sample selected from this population will probably not have a mean exactly equal to zero. As always, there will

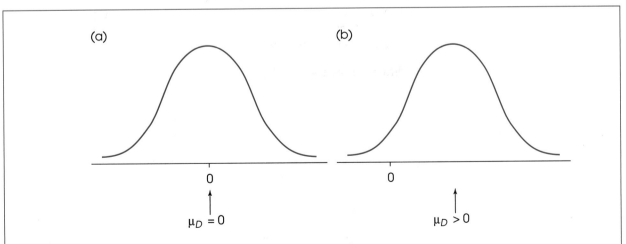

FIGURE 11.2

(a) A population of difference scores for which the mean is $\mu_D = 0$. Note that the typical difference score (D value) is not equal to zero. (b) A population of difference scores for which the mean is greater than zero. Note that most of the difference scores are also greater than zero.

be some error between a sample mean and the population mean, so even if $\mu_D = 0$ (H_0 is true), we do not expect M_D to be exactly equal to zero.

The alternative hypothesis states that there is a treatment effect that causes the scores in one treatment condition to be systematically higher (or lower) than the scores in the other condition. In symbols,

$$H_1: \quad \mu_D \neq 0$$

According to H_1, the difference scores for the individuals in the population tend to be systematically positive (or negative), indicating a consistent, predictable difference between the two treatments.

Figure 11.2(b) shows an example of a population of difference scores with a positive mean difference, $\mu_D > 0$. This time, most of the individuals in the population have difference scores that are greater than zero. Also, a sample selected from this population will contain primarily positive difference scores and will probably have a mean difference that is greater than zero, $M_D > 0$. See Box 11.1 for further discussion of H_0 and H_1.

THE *t* STATISTIC FOR RELATED SAMPLES

Figure 11.3 shows the general situation that exists for a repeated-measures hypothesis test. You may recognize that we are facing essentially the same situation that we encountered in Chapter 9. In particular, we have a population for which the mean and the standard deviation are unknown, and we have a sample that will be used to test a hypothesis about the unknown population. In Chapter 9, we introduced a *t* statistic that allowed us to use the sample mean as a basis for testing hypotheses about the population mean. This *t*-statistic formula will be used again here to develop the repeated-measures *t* test. To refresh your memory, the single-sample *t* statistic (Chapter 9) is defined by the formula

$$t = \frac{M - \mu}{s_M}$$

BOX 11.1	**ANALOGIES FOR H_0 AND H_1 IN THE REPEATED-MEASURES TEST**

An Analogy for H_0: Intelligence is a fairly stable characteristic; that is, you do not get noticeably smarter or dumber from one day to the next. However, if we gave you an IQ test every day for a week, we probably would get seven different numbers. The day-to-day changes in your IQ score are caused by random factors such as your health, your mood, and your luck at guessing answers you do not know. Some days your IQ score is slightly higher, and some days it is slightly lower. On average, the day-to-day changes in IQ should balance out to zero. This is the situation that is predicted by the null hypothesis for a repeated-measures test. According to H_0, any changes that occur either for an individual or for a sample are just due to chance, and in the long run, they will average out to zero.

An Analogy for H_1: On the other hand, suppose we evaluate the effects of a fitness training program by measuring the strength of the muscles in your right arm. We will measure your grip strength every day over a 4-week period. We probably will find small differences in your scores from one day to the next, just as we did with the IQ scores. However, the day-to-day changes in grip strength will not be random. Although your grip strength may decrease occasionally, there should be a general trend toward increased strength as you go through the training program. Thus, most of the day-to-day changes should show an increase. This is the situation that is predicted by the alternative hypothesis for the repeated-measures test. According to H_1, the changes that occur are systematic and predictable and will not average out to zero.

FIGURE 11.3

A sample of $n = 4$ people is selected from the population. Each individual is measured twice, once in treatment I and once in treatment II, and a difference score, *D*, is computed for each individual. This sample of difference scores is intended to represent the population. Note that we are using a sample of difference scores to represent a population of difference scores. Note that the mean for the population of difference scores is unknown. The null hypothesis states that for the general population there is no consistent or systematic difference between the two treatments, so the population mean difference is $\mu_D = 0$.

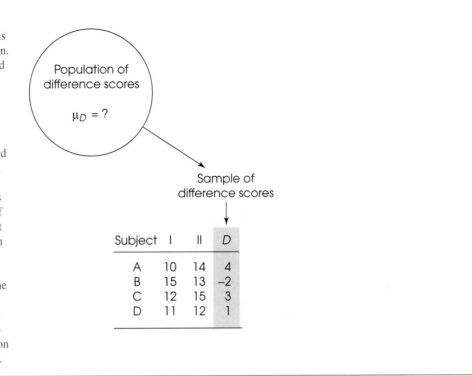

In this formula, the sample mean, *M*, is calculated from the data, and the value for the population mean, μ, is obtained from the null hypothesis. The estimated standard error, s_M, is also calculated from the data and provides a measure of how much difference it is reasonable to expect between a sample mean and the population mean.

For the repeated-measures design, the sample data are difference scores and are identified by the letter *D*, rather than *X*. Therefore, we will use *D*s in the formula to emphasize that we are dealing with difference scores instead of *X* values. Also, the population mean that is of interest to us is the population mean difference (the mean amount of change for the entire population), and we identify this parameter with the symbol μ_D. With these simple changes, the *t* formula for the repeated-measures design becomes

As noted, the repeated-measures *t* *formula is also used for matched-subjects designs.*

$$t = \frac{M_D - \mu_D}{s_{M_D}} \tag{11.2}$$

In this formula, the *estimated standard error*, s_{M_D}, is computed in exactly the same way as it is computed for the single-sample *t* statistic. To calculate the estimated standard error, the first step is to compute the variance (or the standard deviation) for the sample of *D* scores.

$$s^2 = \frac{SS}{n-1} = \frac{SS}{df} \qquad \text{or} \qquad s = \sqrt{\frac{SS}{df}}$$

The estimated standard error is then computed using the sample variance (or sample standard deviation) and the sample size, *n*.

$$s_{M_D} = \sqrt{\frac{s^2}{n}} \quad \text{or} \quad s_{M_D} = \frac{s}{\sqrt{n}}$$

Notice that all of the calculations are done using the difference scores (the *D* scores) and that there is only one *D* score for each subject. With a sample of *n* subjects, there will be exactly *n* *D* scores, and the *t* statistic will have $df = n - 1$. Remember that *n* refers to the number of *D* scores, not the number of *X* scores in the original data.

You should also note that the *repeated-measures t statistic* is conceptually similar to the *t* statistics we have previously examined:

$$t = \frac{\text{sample statistic} - \text{population parameter}}{\text{estimated standard error}}$$

In this case, the sample data are represented by the sample mean of the difference scores (M_D), the population parameter is the value predicted by H_0 ($\mu_D = 0$), and the amount of sampling error is measured by the standard error of the sample mean (s_{M_D}).

LEARNING CHECK

1. For a research study comparing two treatment conditions, what characteristic differentiates a repeated-measures design from an independent-measures design?

2. Describe the data used to compute the sample mean and the sample variance for the repeated-measures *t* statistic.

3. In words and in symbols, what is the null hypothesis for a repeated-measures *t* test?

ANSWERS

1. For a repeated-measures design, the same group of individuals is tested in both of the treatments. An independent-measures design uses a separate group for each treatment.

2. The two scores obtained for each individual are used to compute a difference score. The sample of difference scores is used to compute the mean and variance.

3. The null hypothesis states that for the general population, the average difference between the two conditions is zero. In symbols, $\mu_D = 0$.

11.3 HYPOTHESIS TESTS AND EFFECT SIZE FOR THE REPEATED-MEASURES DESIGN

In a repeated-measures study, we are interested in whether there is a systematic difference between the scores in the first treatment condition and the scores in the second treatment condition. The hypothesis test will use the difference scores obtained from a sample to evaluate the overall mean difference, μ_D, for the entire population. According to the null hypothesis, there is no consistent or systematic difference between treatments and $\mu_D = 0$. The alternative hypothesis, on the other hand, says that there is a systematic difference and $\mu_D \neq 0$. The purpose of the hypothesis test is to decide between these two options.

The hypothesis test is based on the repeated-measures *t* formula,

$$t = \frac{M_D - \mu_D}{s_{M_D}}$$

The numerator measures the actual difference between the data (M_D) and the hypothesis (μ_D), and the denominator measures the standard difference that is expected if H_0 is true. A sample mean that is close to the hypothesized population mean will produce a t statistic near zero, and we will conclude that the data support the null hypothesis. On the other hand, a large value for the t statistic (either positive or negative) indicates that the data do not fit the hypothesis and will result in rejecting H_0.

The hypothesis test with the repeated-measures t statistic follows the same four-step process that we have used for other tests. The complete hypothesis-testing procedure is demonstrated in Example 11.1.

EXAMPLE 11.1

One technique to help people deal with phobias is to have them counteract the feared objects by using imagination to move themselves to a place of safety. In an experimental test of this technique, patients sit in front of a screen and are instructed to relax. Then, they are shown an image of the feared object (for example, a picture of a spider). The patient signals the researcher as soon as feelings of anxiety reach a point at which viewing the image can no longer be tolerated. The researcher records the amount of time that the patient was able to endure looking at the image. The patient then spends 2 minutes imagining a "safe scene" such as a tropical beach or a familiar room before the image is presented again. As before, the patient signals when the level of anxiety becomes intolerable. If patients can tolerate the feared object longer after the imagination exercise, it is viewed as a reduction in the phobia. The data in Table 11.2 summarize the times recorded for a sample of $n = 7$ patients. Do the data indicate that the imagination technique effectively alters phobias?

STEP 1   State the hypotheses, and select the alpha level.

$$H_0: \quad \mu_D = 0 \qquad \text{(There is no change in the phobia.)}$$

$$H_1: \quad \mu_D \neq 0 \qquad \text{(There is a change.)}$$

For this test, we will use $\alpha = .01$ for the level of significance.

TABLE 11.2

The amount of time (in seconds) that patients are able to view a feared object before experiencing feelings of anxiety. Each individual is measured before and after imagination therapy.

Patient	Time Before Imagination	Time After Imagination	D	D^2
A	15	24	+9	81
B	10	23	+13	169
C	7	11	+4	16
D	18	25	+7	49
E	5	14	+9	81
F	9	14	+5	25
G	12	21	+9	81
			$\Sigma D = 56$	$\Sigma D^2 = 502$

Because the data consist of difference scores (D values) instead of X scores, the formulas for the mean and SS also use D in place of X. However, you should recognize the computational formula for SS that was introduced in Chapter 4.

$$M_D = \frac{\Sigma D}{n} = \frac{56}{7} = 8.00$$

$$SS = \Sigma D^2 - \frac{(\Sigma D)^2}{n} = 502 - \frac{(56)^2}{7}$$

$$= 502 - 448 = 54$$

STEP 2   Locate the critical region. For this example, $n = 7$, so the t statistic has $df = n - 1 = 6$. From the t distribution table, you should find that the critical value for $\alpha = .01$ is ± 3.707. The critical region is shown in Figure 11.4.

STEP 3   Calculate the t statistic. Table 11.2 shows the sample data and the calculations for $M_D = 8.00$ and $SS = 54$. As we have done with the other t statistics, we will present the calculation of the t statistic as a three-step process.
First, compute the sample variance.

$$s^2 = \frac{SS}{n-1} = \frac{54}{6} = 9.00$$

Next, use the sample variance to compute the estimated standard error.

$$s_{M_D} = \sqrt{\frac{s^2}{n}} = \sqrt{\frac{9.00}{7}} = 1.134$$

Finally, use the sample mean (M_D) and the hypothesized population mean (μ_D) along with the estimated standard error to compute the value for the t statistic.

$$t = \frac{M_D - \mu_D}{s_{M_D}} = \frac{8.0 - 0}{1.134} = 7.05$$

STEP 4   Make a decision. The t value we obtained falls in the critical region (see Figure 11.4). The researcher rejects the null hypothesis and concludes that the imagination technique does affect the onset of anxiety when patients are exposed to feared objects.

MEASURING EFFECT SIZE FOR THE REPEATED-MEASURES *t*

As we noted with other hypothesis tests, whenever a treatment effect is found to be statistically significant, it is recommended that you also report a measure of the absolute magnitude of the effect. The two commonly used measures of effect size are Cohen's d and r^2, the percentage of variance accounted for. Using the data from Example 11.1, we will demonstrate how these two measures are used to calculate effect size.

FIGURE 11.4

The critical region for the t distsribution with $df = 6$ and $\alpha = .01$.

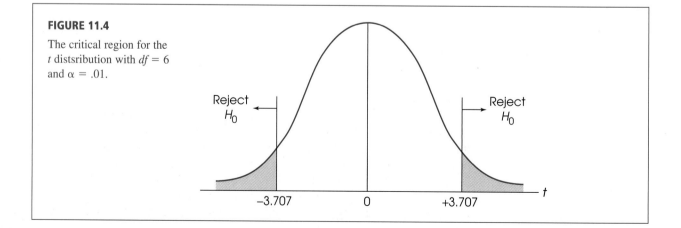

In Chapters 8 and 9 we introduced Cohen's d as a standardized measure of the mean difference between treatments. The standardization simply divides the population mean difference by the standard deviation. For a repeated-measures study, Cohen's d is defined as

$$d = \frac{\text{population mean difference}}{\text{standard deviation}} = \frac{\mu_D}{\sigma_D}$$

Because the population mean and standard deviation are unknown, we will use the sample values instead. The sample mean, M_D, is the best estimate of the actual mean difference, and the sample standard deviation (square root of sample variance) provides the best estimate of the actual standard deviation. Thus, we are able to estimate the value of d as follows:

Because we are measuring the size of the effect and not the direction, we have ignored the minus sign for the sample mean difference.

$$\text{estimated } d = \frac{\text{sample mean difference}}{\text{sample standard deviation}} = \frac{M_D}{s} \qquad (11.3)$$

For the repeated-measures study in Example 11.1, $M_D = 8$ and the sample variance was $s^2 = 9$, so the data produce

$$\text{estimated } d = \frac{M_D}{s} = \frac{8.00}{\sqrt{9}} = \frac{8.00}{3} = 2.67$$

Any value greater than 0.80 is considered to be a large effect, and these data are clearly in that category (see Table 8.2 on page 264).

Percentage of variance is computed using the obtained t value and the df value from the hypothesis test, exactly as was done for the independent-measures t (see page 321). For the data in Example 11.1, we obtain

$$r^2 = \frac{t^2}{t^2 + df} = \frac{(7.05)^2}{(7.05)^2 + 6} = \frac{49.70}{55.70} = 0.892 \quad \text{or} \quad 89.2\%$$

For these data, 89.2% of the variance in the scores is explained by the effect of the imagination therapy. More specifically, increased tolerance after the therapy caused the difference scores to be consistently positive. Thus, the deviations from zero are largely explained by the treatment.

IN THE LITERATURE:
REPORTING THE RESULTS OF A REPEATED-MEASURES t TEST

As we have seen in Chapters 9 and 10, the APA format for reporting the results of t tests consists of a concise statement that incorporates the t value, degrees of freedom, and alpha level. One typically includes values for means and standard deviations, either in a statement or a table (Chapter 4). For Example 11.1, we observed a mean difference of $M = 8.00$ with $s = 3$. Also, we obtained a t statistic of $t = 7.05$ with $df = 6$, and our decision was to reject the null hypothesis at the .01 level of significance. Finally, we measured effect size by computing the percentage of variance explained and obtained $r^2 = 0.892$. A published report of this study might summarize the results as follows:

Imagining a safe scene increased the amount of time that patients could tolerate a feared stimulus before feeling anxiety by an average of $M = 8.00$ with $SD = 3$. The treatment effect was statistically significant, $t(6) = 7.05, p < .01, r^2 = 0.892$.

When the hypothesis test is conducted with a computer program, the printout typically includes an exact probability for the level of significance. The *p*-value from the printout is then stated as the level of significance in the research report. However, the data from Example 11.1 produced a significance level of $p = .000$ in the computer printout. In this case, the probability was so small that the computer rounded it off to 3 decimal points and obtained a value of zero. In this situation you do not know the exact probability value and should report $p < .001$.   ❑

DESCRIPTIVE STATISTICS AND THE HYPOTHESIS TEST

Often, a close look at the sample data from a research study makes it easier to see the size of the treatment effect and to understand the outcome of the hypothesis test. In Example 11.1, we obtained a sample of $n = 7$ individuals who produced a mean difference of $M_D = 8.00$ with a standard deviation of $s = 3$ points. The sample mean and standard deviation describe a set of scores centered at $M_D = 8.00$ with most of the scores located within 3 points of the mean. Figure 11.5 shows the actual set of difference scores that were obtained in Example 11.1. In addition to showing the scores in the sample, we have highlighted the position of $\mu_D = 0$; that is, the value specified in the null hypothesis. Notice that the scores in the sample are displaced away from zero. Specifically, the data are not consistent with a population mean of $\mu_D = 0$, which is why we rejected the null hypothesis. In addition, note that the sample mean is located more than 2 standard deviations above zero. This distance corresponds to the effect size measured by Cohen's $d = 2.67$. For these data, the picture of the sample distribution (see Figure 11.5) should help you to understand the measure of effect size and the outcome of the hypothesis test.

VARIABILITY AS A MEASURE OF CONSISTENCY FOR THE TREATMENT EFFECT

In a repeated-measures study, the variability of the difference scores becomes a relatively concrete and easy-to-understand concept. In particular, the sample variability describes the *consistency* of the treatment effect. For example, if a treatment consistently adds 7 or 8 points to each individual's score, then the set of difference scores will be clustered

---

FIGURE 11.5

A sample of difference scores from Example 11.1. The mean is $M_D = 8$ and the standard deviation is $s = 3$. The data show a consistent increase in scores (positive change) and suggest that $\mu_D = 0$ (no effect) is not a reasonable hypothesis.

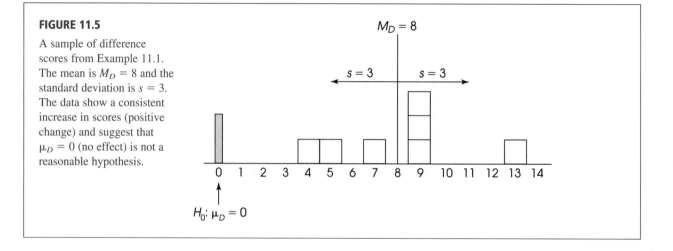

around a mean of $M = 8$ with relatively small variability. This is the situation that we observed in Example 11.1 (see Figure 11.5), in which all the patients increased their tolerance of a feared object after the imagination therapy. In this situation, with small variability, it is easy to see the treatment effect and the effect is likely to be significant.

Now consider what happens when the variability is large. Suppose that the imagination study in Example 11.1 produced a sample of $n = 7$ difference scores consisting of -10, -4, $+5$, $+8$, $+17$, $+17$, and $+23$. These difference scores also have a mean of $M_D = 8$, but now the variability is substantially increased so that $SS = 684$ and the standard deviation is now $s = 12$. Figure 11.6 shows the new set of difference scores. Again, we have highlighted the position of $\mu_D = 0$, which is the value specified in the null hypothesis. Notice that the high variability means that there is no consistent treatment effect. For some participants the treatment produces an increase in scores and for some it produces a decrease. Although these data have exactly the same mean difference ($M_D = 8$) that is shown in Figure 11.5, the mean difference is no longer significant. In the hypothesis test, the high variability increases the size of the estimated standard error and produces $t = 1.76$, which is not in the critical region. With these data, we fail to reject the null hypothesis and must conclude that there is no significant treatment effect. As we have noted several times, high variability can obscure patterns in the data. In general, the more variability there is in the sample data, the less likely you are to obtain a significant treatment effect.

DIRECTIONAL HYPOTHESES AND ONE-TAILED TESTS

In many repeated-measures and matched-subjects studies, the researcher has a specific prediction concerning the direction of the treatment effect. For example, in the study described in Example 11.1, the researcher expects the imagination exercise to increase the amount of time that phobia patients can tolerate viewing the feared object. This kind of directional prediction can be incorporated into the statement of the hypotheses, resulting in a directional, or one-tailed, hypothesis test. The following example demonstrates how the hypotheses and critical region are determined for a directional test.

EXAMPLE 11.2   We will reexamine the experiment presented in Example 11.1. The researcher is using a repeated-measures design to investigate the effect of imagining a safe scene on the anxiety felt by patients suffering from phobias. The researcher predicts that the

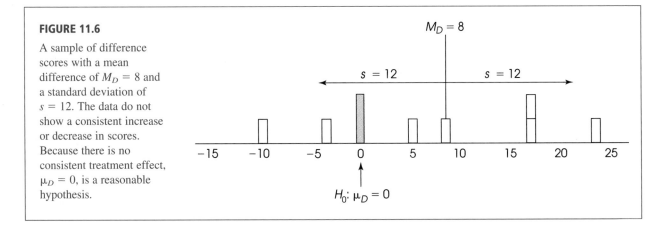

FIGURE 11.6

A sample of difference scores with a mean difference of $M_D = 8$ and a standard deviation of $s = 12$. The data do not show a consistent increase or decrease in scores. Because there is no consistent treatment effect, $\mu_D = 0$, is a reasonable hypothesis.

imagination technique will increase the amount of time that patients can tolerate viewing the feared stimulus.

STEP 1 State the hypothesis and select the alpha level. For this example, the researcher predicts that imagining a safe scene will increase the amount of time that patients can tolerate the stimulus. The null hypothesis, on the other hand, says that the amount of time will not be increased but rather will be unchanged or even reduced after imagining a safe scene. In symbols,

$$H_0: \quad \mu \leq 0 \quad \text{(The amount of time is not increased.)}$$

The alternative hypothesis says that the treatment does work. For this example, H_1 states that imagining a safe scene will increase the amount of time that patients can tolerate the feared stimulus.

$$H_1: \quad \mu > 0 \quad \text{(The amount of time is increased.)}$$

STEP 2 Locate the critical region. As we demonstrated with the independent-measures *t* statistic (page 324), the critical region for a one-tailed test can be located using a two-stage process. Rather than trying to determine which tail of the distribution contains the critical region, you first look at the sample mean difference to verify that it is in the predicted direction. If not, then the treatment clearly did not work as expected and you can stop the test. If the change is in the correct direction, then the question is whether it is large enough to be significant. For this example, change is in the predicted direction (the researcher predicted an increase in time and the sample mean shows an increase). With $n = 7$, we obtain $df = 6$ and a critical value of $t = 3.143$ for a one-tailed test with $\alpha = .01$. Thus, any *t* statistic beyond 3.143 (positive or negative) will be sufficient to reject H_0.

STEP 3 Compute the *t* statistic. We calculated the *t* statistic in Example 11.1, and obtained $t = 7.05$.

STEP 4 Make a decision. The obtained *t* statistic is well beyond the critical boundary. Therefore, we reject H_0 and conclude that the imagination technique did significantly increase the amount of time that patients could tolerate viewing the feared stimulus.

LEARNING CHECK

1. A researcher is investigating the effectiveness of acupuncture treatment for chronic back pain. A sample of $n = 4$ participants is obtained from a pain clinic. Each individual ranks the current level of pain and then begins a 6-week program of acupuncture treatment. At the end of the program, the pain level is rated again and the researcher records the amount of difference between the two ratings. For this sample, pain level decreased by an average of $M = 4.5$ points with $SS = 27$.

   a. Are the data sufficient to conclude that acupuncture has a significant effect on back pain? Use a two-tailed test with $\alpha = .05$.

   b. Can you conclude that acupuncture significantly reduces back pain? Use a one-tailed test with $\alpha = .05$.

2. Compute the effect size using both Cohen's *d* and r^2 acupuncture study in the previous question.

3. A computer printout for a repeated-measures *t* test reports a *p* value of $p = .021$.

   a. Can the researcher claim a significant effect with $\alpha = .01$?

   b. Is the effect significant with $\alpha = .05$?

ANSWERS   1. a. For these data, the sample variance is 9, the standard error is 1.50, and $t = 3.00$. With $df = 3$, the critical values are $t = \pm 3.182$. Fail to reject the null hypothesis.

   b. For a one-tailed test, the critical value is $t = 2.353$. Reject the null hypothesis and conclude that acupuncture treatment significantly reduces pain.

2. $d = 4.5/3 = 1.50$ and $r^2 = 9/12 = 0.75$.

3. a. The exact *p* value, $p = .021$, is not less than $\alpha = .01$. Therefore, the effect is not significant for $\alpha = .01$ ($p > .01$).

   b. The *p* value is less than .05, so the effect is significant with $\alpha = .05$.

11.4   USES AND ASSUMPTIONS FOR REPEATED-MEASURES *t* TESTS

REPEATED-MEASURES VERSUS INDEPENDENT-MEASURES DESIGNS

In many research situations, it is possible to use either a repeated-measures study or an independent-measures study to compare two treatment conditions. The independent-measures design would use two separate samples (one in each treatment condition) and the repeated-measures design would use only one sample with the same individuals in both treatments. The decision about which design to use is often made by considering the advantages and disadvantages of the two designs. In general, the repeated-measures design has most of the advantages.

Number of subjects   A repeated-measures design typically requires fewer subjects than an independent-measures design. The repeated-measures design uses the subjects more efficiently because each individual is measured in both of the treatment conditions. This can be especially important when there are relatively few subjects available (for example, when you are studying a rare species or individuals in a rare profession).

Study changes over time   The repeated-measures design is especially well suited for studying learning, development, or other changes that take place over time. Remember that this design involves measuring individuals at one time and then returning to measure the same individuals at a later time. In this way, a researcher can observe behaviors that change or develop over time.

Individual differences   The primary advantage of a repeated-measures design is that it reduces or eliminates problems caused by individual differences. *Individual differences* are characteristics such as age, IQ, gender, and personality that vary from one individual to another. These individual differences can influence the scores obtained in a research study, and they can affect the outcome of a hypothesis test. Consider the data in Table 11.3. The first set of data represents the results from a typical independent-measures study and the second set represents a repeated-measures study. Note that we have identified each participant by name to help demonstrate the effects of individual differences.

For the independent-measures data, note that every score represents a different person. For the repeated-measures study, on the other hand, the same participants are

TABLE 11.3

Hypothetical data showing the results from an independent-measures study and a repeated-measures study. The two sets of data use exactly the same numerical scores and they both show the same 5-point mean difference between treatments.

Independent-Measures Study (2 separate samples)		Repeated-Measures Study (same sample in both treatments)		
Treatment 1	Treatment 2	Treatment 1	Treatment 2	D
(John) $X = 18$	(Sue) $X = 15$	(John) $X = 18$	(John) $X = 15$	-3
(Mary) $X = 27$	(Tom) $X = 20$	(Mary) $X = 27$	(Mary) $X = 20$	-7
(Bill) $X = 33$	(Dave) $X = 28$	(Bill) $X = 33$	(Bill) $X = 28$	-5
$M = 26$	$M = 21$			$M_D = -5$
$SS = 114$	$SS = 86$			$SS = 8$

measured in both of the treatment conditions. This difference between the two designs has some important consequences.

1. We have constructed the data so that both research studies have exactly the same scores and they both show the same 5-point mean difference between treatments. In each case, the researcher would like to conclude that the 5-point difference was caused by the treatments. However, with the independent-measures design, there is always the possibility that the participants in treatment 1 have different characteristics than those in treatment 2. For example, the three participants in treatment 1 may be more intelligent than those in treatment 2 and their higher intelligence caused them to have higher scores. Note that this problem disappears with the repeated-measures design. Specifically, with repeated measures there is no possibility that the participants in one treatment are different from those in another treatment because the same participants are used in all the treatments.

2. Although the two sets of data contain exactly the same scores and have exactly the same 5-point mean difference, you should realize that they are very different in terms of the variance used to compute standard error. For the independent-measures study, you calculate the *SS* or variance for the scores in each of the two separate samples. Note that in each sample there are big differences between participants. In treatment 1, for example, Bill has a score of 33 and John's score is only 18. These individual differences produce a relatively large sample variance and a large standard error. For the independent-measures study, the standard error is 5.77, which produces a *t* statistic of $t = 0.87$. For these data, the hypothesis test concludes that there is no significant difference between treatments.

   In the repeated-measures study, the *SS* and variance are computed for the difference scores. If you examine the repeated-measures data in Table 11.3, you will see that the big differences between John and Bill that exist in treatment 1 and in treatment 2 are eliminated when you get to the difference scores. Because the individual differences are eliminated, the variance and standard error are dramatically reduced. For the repeated-measures study, the standard error is 1.15 and the *t* statistic is $t = -4.35$. With the repeated-measures *t*, the data show a significant difference between treatments. Thus, one big advantage of a repeated-measures study is that it reduces variance by removing individual differences, which increases the chances of finding a significant result.

TIME-RELATED FACTORS AND ORDER EFFECTS

The primary disadvantage of a repeated-measures design is that the structure of the design allows for factors other than the treatment effect to cause a participant's score to change from one treatment to the next. Specifically, in a repeated-measures design, each individual is measured in two different treatment conditions *at two different times.* In this situation, outside factors that change over time may be responsible for changes in the participants' scores. For example, a participant's health or mood may change over time and cause a difference in the participant's scores. Outside factors such as the weather can also change and may have an influence on participants' scores. In general, a repeated-measures study must take place over time, and it is always possible that time-related factors (other than the two treatments) are responsible for causing changes in the participants' scores.

Also, it is possible that participation in the first treatment influences the individual's score in the second treatment. For example, participants may become tired or bored during the first treatment and, therefore, do not perform well in the second treatment. In this situation, the researcher would find a mean difference between the two treatments; however, the difference is not caused by the treatments, instead it is caused by fatigue. Changes in scores that are caused by participation in an earlier treatment are called *order effects* and can distort the mean differences found in repeated-measures research studies.

Counterbalancing   One way to deal with time-related factors and order effects is to counterbalance the order of presentation of treatments. That is, the participants are randomly divided into two groups, with one group receiving treatment 1 followed by treatment 2, and the other group receiving treatment 2 followed by treatment 1. The goal of counterbalancing is to distribute any outside effects evenly over the two treatments. For example, if fatigue is a problem, then half of the participants (one of the two groups) will experience fatigue in treatment 1 and the other half will experience fatigue in treatment 2. Thus, fatigue influences the two treatments equally.

Finally, if there is reason to expect strong time-related effects or strong order effects, your best strategy is not to use a repeated-measures design. Instead, use independent-measures (or a matched-subjects design) so that each individual participates in only one treatment and is measured only one time.

ASSUMPTIONS OF THE RELATED-SAMPLES *t* TEST

The related-samples *t* statistic requires two basic assumptions:

1. The observations within each treatment condition must be independent (see page 253). Notice that the assumption of independence refers to the scores *within* each treatment. Inside each treatment, the scores are obtained from different individuals and should be independent of one another.

2. The population distribution of difference scores (*D* values) must be normal.

As before, the normality assumption is not a cause for concern unless the sample size is relatively small. In the case of severe departures from normality, the validity of the *t* test may be compromised with small samples. However, with relatively large samples ($n > 30$), this assumption can be ignored.

LEARNING CHECK

1. What assumptions must be satisfied for repeated-measures *t* tests to be valid?

2. Describe some situations for which a repeated-measures design is well suited.

3. How is a matched-subjects design similar to a repeated-measures design? How do they differ?

4. The data from a research study consist of 10 scores in each of two different treatment conditions. How many individual subjects would be needed to produce these data

   a. For an independent-measures design?

   b. For a repeated-measures design?

   c. For a matched-subjects design?

ANSWERS   1. The observations within a treatment are independent. The population distribution of *D* scores is assumed to be normal.

2. The repeated-measures design is suited to situations in which a particular type of subject is not readily available for study. This design is helpful because it uses fewer subjects (only one sample is needed). Certain questions are addressed more adequately by a repeated-measures design—for example, anytime one would like to study changes across time in the same individuals. Also, when individual differences are large, a repeated-measures design is helpful because it reduces the amount of this type of error in the statistical analysis.

3. They are similar in that the role of individual differences in the experiment is reduced. They differ in that there are two samples in a matched-subjects design and only one in a repeated-measures study.

4. **a.** The independent-measures design would require 20 subjects (two separate samples with $n = 10$ in each).

   b. The repeated-measures design would require 10 subjects (the same 10 individuals are measured in both treatments).

   c. The matched-subjects design would require 20 subjects (10 matched pairs).

SUMMARY

1. In a related-samples research study, the individuals in one treatment condition are directly related, one-to-one, with the individuals in the other treatment condition(s). The most common related-samples study is a repeated-measures design, in which the same sample of individuals is tested in all of the treatment conditions. This design literally repeats measurements on the same subjects. An alternative is a matched-subjects design, in which the individuals in one sample are matched one-to-one with individuals in another sample. The matching is based on a variable relevant to the study.

2. The repeated-measures *t* test begins by computing a difference between the first and second measurements for each subject (or the difference for each matched pair). The difference scores, or *D* scores, are obtained by

   $$D = X_2 - X_1$$

   The sample mean, M_D, and sample variance, s^2, are used to summarize and describe the set of difference scores.

3. The formula for the repeated-measures *t* statistic is

   $$t = \frac{M_D - \mu_D}{s_{M_D}}$$

   In the formula, the null hypothesis specifies $\mu_D = 0$, and the estimated standard error is computed by

   $$s_{M_D} = \sqrt{\frac{s^2}{n}}$$

4. A repeated-measures design may be preferred to an independent-measures study when one wants to observe changes in behavior in the same subjects, as in learning or developmental studies. An important advantage of the repeated-measures design is that it removes or reduces individual differences, which in turn lowers sample variability and tends to increase the chances for obtaining a significant result.

5. For a repeated-measures design, effect size can be measured using either r^2 (the percentage of variance

accounted for) or Cohen's d (the standardized mean difference). The value of r^2 is computed the same for both independent- and repeated-measures designs,

$$r^2 = \frac{t^2}{t^2 + df}$$

Cohen's d is defined as the sample mean difference divided by standard deviation for both repeated- and independent-measures designs. For repeated-measures studies, Cohen's d is estimated as

$$\text{estimated } d = \frac{M_D}{s}$$

KEY TERMS

repeated-measures design	difference scores	repeated-measures	order effects
within-subjects design	estimated standard	t statistic	
matched-subjects design	error for M_D	individual differences	
related-samples design			

RESOURCES

There are a tutorial quiz and other learning exercises for Chapter 11 at the Wadsworth website, www.cengage.com/psychology/gravetter. The site also provides access to a workshop entitled *Independent vs. Repeated t-tests* that compares the t test presented in this chapter with the independent-measures test that was presented in Chapter 10. See page 29 for more information about accessing items on the website.

ENHANCED WebAssign

Guided interactive tutorials, end-of-chapter problems, and related testbank items may be assigned online at WebAssign.

WebTUTOR

For those using WebTutor along with this book, there is a WebTutor section corresponding to this chapter. The WebTutor contains a brief summary of Chapter 11, hints for learning the concepts and the formulas for the repeated-measures t test, cautions about common errors, and sample exam items including solutions.

SPSS

General instructions for using SPSS are presented in Appendix D. Following are detailed instructions for using SPSS to perform **The Repeated-Measures t Test** presented in this chapter.

Data Entry

1. Enter the data into two columns (VAR0001 and VAR0002) in the data editor with the first score for each participant in the first column and the second score in the second column. The two scores for each participant must be in the same row.

Data Analysis

1. Click **Analyze** on the tool bar, select **Compare Means,** and click on **Paired-Samples T Test.**
2. Highlight both of the column labels for the two data columns (click on one, then click on the second) and click the arrow to move them into the **Paired Variables** box.
3. Click **OK.**

SPSS Output

SPSS produces a summary table showing descriptive statistics for each of the two sets of scores (the mean, the number of scores, the standard deviation, and the standard error for the mean). The second table shows the correlation between the two sets of scores. Correlations are presented in Chapter 16. The final table presents the results of the repeated-measures *t* test. The output shows the mean and the standard deviation for the difference scores, and the standard error for the mean difference. The table includes a 95% confidence interval that estimates the size of the mean difference (confidence intervals are presented in Chapter 12). Finally, the table reports the value for *t,* the value for *df,* and the level of significance (the *p* value or alpha level for the test).

FOCUS ON PROBLEM SOLVING

1. Once data have been collected, we must then select the appropriate statistical analysis. How can you tell whether the data call for a repeated-measures *t* test? Look at the experiment carefully. Is there only one sample of subjects? Are the same subjects tested a second time? If your answers are yes to both of these questions, then a repeated-measures *t* test should be done. There is only one situation in which the repeated-measures *t* can be used for data from two samples, and that is for *matched-subjects* studies (page 341).

2. The repeated-measures *t* test is based on difference scores. In finding difference scores, be sure you are consistent with your method. That is, you may use either $X_2 - X_1$ or $X_1 - X_2$ to find *D* scores, but you must use the same method for all subjects.

DEMONSTRATION 11.1

A REPEATED-MEASURES *t* TEST

A major oil company would like to improve its tarnished image following a large oil spill. Its marketing department develops a short television commercial and tests it on a sample of $n = 7$ participants. People's attitudes about the company are measured with a short questionnaire, both before and after viewing the commercial. The data are as follows:

Person	X_1 (Before)	X_2 (After)	D (Difference)
A	15	15	0
B	11	13	+2
C	10	18	+8
D	11	12	+1
E	14	16	+2
F	10	10	0
G	11	19	+8

$\Sigma D = 21$

$M_D = 21/7 = 3.00$

$SS = 74$

Was there a significant change? Note that participants are being tested twice—once before and once after viewing the commercial. Therefore, we have a repeated-measures design.

STEP 1   State the hypotheses, and select an alpha level.

The null hypothesis states that the commercial has no effect on people's attitude, or in symbols,

$$H_0: \quad \mu_D = 0 \quad \text{(The mean difference is zero.)}$$

The alternative hypothesis states that the commercial does alter attitudes about the company, or

$$H_1: \quad \mu_D \neq 0 \quad \text{(There is a mean change in attitudes.)}$$

For this demonstration, we will use an alpha level of .05 for a two-tailed test.

STEP 2   Locate the critical region.

Degrees of freedom for the repeated-measures t test are obtained by the formula

$$df = n - 1$$

For these data, degrees of freedom equal

$$df = 7 - 1 = 6$$

The t distribution table is consulted for a two-tailed test with $\alpha = .05$ for $df = 6$. The critical t values for the critical region are $t = \pm 2.447$.

STEP 3   Compute the test statistic.

Once again, we suggest that the calculation of the t statistic be divided into a three-part process.

*Variance for the **D** scores*   The variance for the sample of D scores is

$$s^2 = \frac{SS}{n - 1} = \frac{74}{6} = 12.33$$

Estimated standard error for M_D   The estimated standard error for the sample mean difference is computed as follows:

$$s_{M_D} = \sqrt{\frac{s^2}{n}} = \sqrt{\frac{12.33}{7}} = \sqrt{1.76} = 1.33$$

The repeated-measures t statistic   Now we have the information required to calculate the *t* statistic.

$$t = \frac{M_D - \mu_D}{s_{M_D}} = \frac{3 - 0}{1.33} = 2.26$$

STEP 4   Make a decision about H_0, and state the conclusion.

The obtained *t* value is not extreme enough to fall in the critical region. Therefore, we fail to reject the null hypothesis. We conclude that there is no evidence that the commercial will change people's attitudes, $t(6) = 2.26$, $p > .05$, two-tailed. (Note that we state that *p* is *greater than* .05 because we failed to reject H_0.)

DEMONSTRATION 11.2

EFFECT SIZE FOR THE REPEATED-MEASURES *t*

We will estimate Cohen's *d* and calculate r^2 for the data in Demonstration 11.1. The data produced a sample mean difference of $M_D = 3.00$ with a sample variance of $s^2 = 12.33$. Based on these values, Cohen's *d* is

$$\text{estimated } d = \frac{\text{mean difference}}{\text{standard deviation}} = \frac{M_D}{s} = \frac{3.00}{\sqrt{12.33}} = \frac{3.00}{3.51} = 0.86$$

The hypothesis test produced $t = 2.26$ with $df = 6$. Based on these values,

$$r^2 = \frac{t^2}{t^2 + df} = \frac{(2.26)^2}{(2.26)^2 + 6} = \frac{5.11}{11.11} = 0.46 \quad (\text{or } 46\%)$$

PROBLEMS

1. For the following studies, indicate whether a repeated-measures *t* test is the appropriate analysis. Explain your answers.
   a. A researcher is comparing academic performance for college students who participate in organized sports and those who do not.
   b. A researcher is testing the effectiveness of a new blood-pressure medication by comparing blood-pressure readings before and after each participant takes the medication.
   c. A researcher is examining how light intensity influences color perception. A sample of 50 college students is obtained, and each student is asked to judge the differences between color patches under conditions of bright light and dim light.

2. Participants enter a research study with unique characteristics that produce different scores from one person to another. For an independent-measures study, these individual differences can cause problems. Briefly

explain how these problems are eliminated or reduced with a repeated-measures study.

3. Explain the difference between a matched-subjects design and a repeated-measures design.

4. A researcher conducts an experiment comparing two treatment conditions and obtains data with 10 scores for each treatment condition.
   a. If the researcher used an independent-measures design, how many subjects participated in the experiment?
   b. If the researcher used a repeated-measures design, how many subjects participated in the experiment?
   c. If the researcher used a matched-subjects design, how many subjects participated in the experiment?

5. A sample of $n = 16$ individuals participates in a repeated-measures study that produces a sample mean difference of $M_D = 9$ with $SS = 960$ for the difference scores.

a. Calculate the standard deviation for the sample of difference scores. Briefly explain what is measured by the standard deviation.

b. Calculate the estimated standard error for the sample mean difference. Briefly explain what is measured by the estimated standard error.

6. a. A repeated-measures study with a sample of $n = 9$ participants produces a mean difference of $M_D = 4$ with a standard deviation of $s = 12$. Based on the mean and standard deviation you should be able to visualize (or sketch) the sample distribution. Use a two-tailed hypothesis test with $\alpha = .05$ to determine whether it is likely that this sample came from a population with $\mu_D = 0$.

b. Now assume that the sample standard deviation is $s = 3$, and once again visualize the sample distribution. Use a two-tailed hypothesis test with $\alpha = .05$ to determine whether it is likely that this sample came from a population with $\mu_D = 0$.

c. Explain how the size of the sample standard deviation influences the likelihood of finding a significant mean difference.

7. a. A repeated-measures study with a sample of $n = 9$ participants produces a mean difference of $M_D = 3$ with a standard deviation of $s = 6$. Based on the mean and standard deviation you should be able to visualize (or sketch) the sample distribution. Use a two-tailed hypothesis test with $\alpha = .05$ to determine whether it is likely that this sample came from a population with $\mu_D = 0$.

b. Now assume that the sample mean difference is $M_D = 12$, and once again visualize the sample distribution. Use a two-tailed hypothesis test with $\alpha = .05$ to determine whether it is likely that this sample came from a population with $\mu_D = 0$.

c. Explain how the size of the sample mean difference influences the likelihood of finding a significant mean difference.

8. A sample of difference scores from a repeated-measures experiment has a mean of $M_D = 5$ with a variance of $s^2 = 64$.

a. If $n = 4$, is this sample sufficient to reject the null hypothesis using a two-tailed test with $\alpha = .05$?

b. Would you reject H_0 if $n = 16$? Again, assume a two-tailed test with $\alpha = .05$.

c. Explain how the size of the sample influences the likelihood of finding a significant mean difference.

9. In the Preview section of Chapter 2, we discussed a study that examined the effect of humor on memory (Schmidt, 1994). In one part of the study, participants were presented with a list containing a mix of humorous and nonhumorous sentences, and were then asked to recall as many sentences as possible. Schmidt recorded the number of humorous and the number of nonhumorous sentences recalled by each individual. Notice that the data consist of two memory scores for each participant. Suppose that a difference score is computed for each individual in a sample of $n = 16$ and the resulting data show that participants recalled an average of $M_D = 3.25$ more humorous sentences than nonhumorous, with $SS = 135$. Are these results sufficient to conclude that humor has a significant effect on memory? Use a two-tailed test with $\alpha = .05$.

10. Research has shown that losing even one night's sleep can have a significant effect on performance of complex tasks such as problem solving (Linde & Bergstroem, 1992). To demonstrate this phenomenon, a sample of $n = 25$ college students was given a problem-solving task at noon on one day and again at noon on the following day. The students were not permitted any sleep between the two tests. For each student, the difference between the first and second score was recorded. For this sample, the students averaged $M_D = 4.7$ points better on the first test with a variance of $s^2 = 64$ for the difference scores.

a. Do the data indicate a significant change in problem-solving ability? Use a two-tailed test with $\alpha = .05$.

b. Compute an estimated Cohen's d to measure the size of the effect.

11. Strack, Martin, and Stepper (1988) reported that people rate cartoons as funnier when holding a pen in their teeth (which forced them to smile) than when holding a pen in their lips (which forced them to frown). A researcher attempted to replicate this result using a sample of $n = 16$ adults between the ages of 40 and 45. For each person, the researcher recorded the difference between the rating obtained while smiling and the rating obtained while frowning. On average the cartoons were rated as funnier when the participants were smiling, with an average difference of $M_D = 3.6$ with $SS = 960$.

a. Do the data indicate that the cartoons are rated significantly funnier when the participants are smiling? Use a one-tailed test with $\alpha = .05$.

b. Compute r^2 to measure the size of the treatment effect.

12. The stimulant Ritalin has been shown to increase attention span and improve academic performance in children with ADHD (Evans, Pelham, Smith, et al., 2001). To demonstrate the effectiveness of the drug, a researcher selects a sample of $n = 25$ children diagnosed with the disorder and measures each child's attention span before and after taking the drug. The data show an average increase of attention span of $M_D = 6.8$ minutes with a standard deviation of $s = 5.5$.

a. Is this result sufficient to conclude that Ritalin has a significant effect on attention span? Use a two-tailed test with $\alpha = .01$.

b. Compute r^2 and estimate Cohen's *d* to measure the effect size of the drug.

13. A recent report indicates that physically attractive people are also perceived as being more intelligent (Eagly, Ashmore, Makhijani, & Longo, 1991). As a demonstration of this phenomenon, a researcher obtained a set of 10 photographs, 5 showing men who were judged to be attractive and 5 showing men who were judged as unattractive. The photographs were shown to a sample of $n = 25$ college students and the students were asked to rate the intelligence of the person in the photo on a scale from 1 to 10. For each student, the researcher determined the average rating for the 5 attractive photos and the average for the 5 unattractive photos, and then computed the difference between the two scores. For the entire sample, the average difference was $M_D = 2.7$ (attractive photos rated higher) with $s = 2.00$. Are the data sufficient to conclude that there was a significant difference in perceived intelligence for the two sets of photos? Use a two-tailed test at the .05 level of significance.

14. In Chapter 8 (page 233) we presented an example of a research study examining how an antioxidant supplement (such as blueberries) affects the cognitive functioning of elderly adults. In the example, we compared one sample of adults who received the blueberry supplement with the general population. An alternative approach to the research question is to use a repeated-measures design. Suppose a researcher obtains a sample of $n = 16$ adults who are between the ages of 65 and 75. The researcher uses a standardized test to measure cognitive performance for each individual. The participants then begin a 2-month program in which they receive daily doses of the blueberry supplement. At the end of the 2-month period, the researcher again measures cognitive performance for each participant. The results show an average increase in performance of $M_D = 7.4$ with $SS = 1215$. Does this result support the conclusion that the antioxidant supplement has a significant effect on cognitive performance? Use a two-tailed test with $\alpha = .05$.

15. The following data are from a repeated-measures study examining the effect of a treatment by measuring a group of $n = 4$ participants before and after they receive the treatment.
a. Calculate the difference scores and M_D.
b. Compute SS, sample variance, and estimated standard error.

c. Is there a significant treatment effect? Use $\alpha = .05$, two tails.

Participant	Before Treatment	After Treatment
A	7	10
B	6	13
C	9	12
D	5	8

16. A researcher for a cereal company wanted to demonstrate the health benefits of eating oatmeal. A sample of 9 volunteers was obtained and each participant ate a fixed diet without any oatmeal for 30 days. At the end of the 30-day period, cholesterol was measured for each individual. Then the participants began a second 30-day period in which they repeated exactly the same diet except that they added 2 cups of oatmeal each day. After the second 30-day period, cholesterol levels were measured again and the researcher recorded the difference between the two scores for each participant. For this sample, cholesterol scores averaged $M_D = 16$ points lower with the oatmeal diet with $SS = 538$ for the difference scores. Are the data sufficient to indicate a significant change in cholesterol level? Use a two-tailed test with $\alpha = .01$

17. A variety of research results suggest that visual images interfere with visual perception. In one study, Segal and Fusella (1970) had participants watch a screen, looking for brief presentations of a small blue arrow. On some trials, the participants were also asked to form a mental image (for example, imagine a volcano). The results for a sample of $n = 6$, show that participants made an average of $M_D = 4.3$ more errors while forming images than while not forming images. The difference scores had $SS = 63$. Do the data indicate a significant difference between the two conditions? Use a two-tailed test with $\alpha = .05$.

18. One of the primary advantages of a repeated-measures design compared to independent-measures, is that it reduces the overall variability by removing variance caused by individual differences. The following data are from a research study comparing two treatment conditions.
a. Assume that the data are from an independent-measures study using two separate samples, each with $n = 6$ participants. Compute the pooled variance and the estimated standard error for the mean difference.
b. Now assume that the data are from a repeated-measures study using the same sample of $n = 6$ participants in both treatment conditions. Compute

the variance for the sample of difference scores and the estimated standard error for the mean difference. (You should find that the repeated-measures design substantially reduces the variance and the standard error.)

Treatment 1	Treatment 2	Difference
10	13	3
12	12	0
8	10	2
6	10	4
5	6	1
7	9	2
$M = 8$	$M = 10$	$M_D = 2$
$SS = 34$	$SS = 30$	$SS = 10$

19. The previous problem demonstrates that removing individual differences can substantially reduce variance and lower the standard error. However, this benefit only occurs if the individual differences are consistent across treatment conditions. In problem 18 for example, the first two participants (top two rows) consistently had the highest scores in both treatment conditions. Similarly, the last two participants consistently had the lowest scores in both treatments. To construct the following data, we started with the scores in Problem 18 and eliminated the consistency of the individual differences. For example, the first participant now has the lowest score in treatment 1 but the highest score in treatment 2.
   a. Assume that the data are from an independent-measures study using two separate samples, each with $n = 6$ participants. Compute the pooled variance and the estimated standard error for the mean difference.
   b. Now assume that the data are from a repeated-measures study using the same sample of $n = 6$ participants in both treatment conditions. Compute the variance for the sample of difference scores and the estimated standard error for the mean difference. (This time you should find that removing the individual differences does not reduce the variance or the standard error.)

Treatment 1	Treatment 2	Difference
5	13	8
7	12	5
8	10	4
6	10	2
12	6	−6
10	9	−1
$M = 8$	$M = 10$	$M_D = 2$
$SS = 34$	$SS = 30$	$SS = 120$

20. A researcher uses a matched-samples design to investigate whether single people who own pets are generally happier than singles without pets. A mood inventory questionnaire is administered to a group of 20- to 29-year-old non–pet owners and a similar age group of pet owners. The pet owners are matched one to one with the non–pet owners for income, number of close friendships, and general health. The data are as follows:

Matched Pair	Non-pet Owner	Pet Owner
A	10	15
B	9	9
C	11	10
D	11	19
E	5	17
F	9	15

Is there a significant difference in the mood scores for non–pet owners versus pet owners? Test with $\alpha = .05$ for two tails.

21. Placing children in time out can be an effective form of punishment. Research suggests that this same technique can also be effective for adults (Chelonis, Bastilla, Brown, & Gardner, 2007). In the study, participants were shown a series of colored shapes and had to classify each stimulus into one of two categories. Each correct response was rewarded with a nickel and each incorrect response was punished by a time-out period of 5 or 10 seconds. During the time out, no stimulus was presented and the participants had no opportunity to earn more nickels. The dependent variable was the overall level of accuracy during a series of trials. Assume that each participant completed a series of trials using a 5-second time-out period and another series using a 10-second time-out period. Data for a sample of $n = 9$ participants are as follows. Do the data indicate a significant difference between the two conditions? Use a two-tailed test with $\alpha = .05$.

5-Second Time Out	10-Second Time Out
71	86
68	80
91	88
65	74
73	82
81	89
85	85
86	88
65	76

22. In the Preview section of this chapter, we discussed a report describing how the accuracy of Olympic marksmen can be affected by heartbeats. Pelton (1983) reported that Olympic-level marksmen shoot much better if they fire between heartbeats rather than squeezing the trigger during a heartbeat. The small vibration caused by a heartbeat seems to be sufficient to affect the marksmen's aim. The following hypothetical data demonstrate this phenomenon. A sample of $n = 8$ Olympic marksmen fire a series of rounds while a researcher records heartbeats. For each marksman, the total score is recorded for shots fired during heartbeats and for shots fired between heartbeats. Do these data indicate a significant difference? Test with $\alpha = .05$.

Participant	During Heartbeats	Between Heartbeats
A	93	98
B	90	94
C	95	96
D	92	91
E	95	97
F	91	97
G	92	95
H	93	97

23. A researcher is examining the effect of fatigue on performance of a perceptual-speed task (see Figure 5.1, p 138). A sample of $n = 12$ college students is obtained and each student completes the speed task at the beginning of the study. Each student then spends 1 hour reading a boring 25-page report and answering a series of questions about the report. The students are then tested again on the perceptual-speed task and the difference between the first and second score is computed for each student. The difference scores for this sample are as follows:

Difference Scores: $-3, -7, -8, -2, +1, -5,$
$-3, -7, -10, -6, -8, -4$

A negative score indicates that performance declined for the second test. Can the researcher conclude that the boring task had a significant effect on performance? Use a two-tailed test with $\alpha = .05$.

24. A researcher studies the effect of a drug (MAO inhibitor) on the number of nightmares occurring in veterans with post-traumatic stress disorder (PTSD). A sample of PTSD clients records each incident of a nightmare for 1 month before treatment. Participants are then given the medication for 1 month, and they continue to report each occurrence of a nightmare. For the following hypothetical data, determine whether the MAO inhibitor significantly reduces nightmares. Use the .05 level of significance and a one-tailed test.

Number of Nightmares	
1 Month Before Treatment	1 Month During Treatment
6	1
10	2
3	0
5	5
7	2

CHAPTER

12

Estimation

Tools You Will Need

The following items are considered essential background material for this chapter. If you doubt your knowledge of any of these items, you should review the appropriate chapter or section before proceeding.

- Single-sample t statistic (Chapter 9)

- Independent-measures t statistic (Chapter 10)

- Related-samples t statistic (Chapter 11)

Preview

12.1   An Overview of Estimation

12.2   Estimation with the t Statistic

12.3   A Final Look at Estimation

Summary

Focus on Problem Solving

Demonstrations 12.1 and 12.2

Problems

Preview

In the preview for Chapter 9, we discussed an experiment by Scaife (1976) to determine whether the eye-spot patterns on the wings of moths are effective deterrents for warding off predators such as moth-eating birds. In the study, birds were placed in a box with two separate chambers, one with plain walls and one with eye-spot patterns on the walls. If the spots have no influence, the birds should wander randomly between the two chambers, spending an average of $\mu = 30$ minutes per hour in the plain-wall chamber. The goal of the study was to demonstrate that the birds tend to avoid the eye-spot patterns and spend the majority of their time on the plain side, $\mu_{\text{plain side}} \neq 30$ minutes.

In the preview for Chapter 10 we discussed a study by Katona (1940) that compared the effectiveness of two methods of instruction. Two groups of participants learned to solve a problem involving a pattern of squares made of matchsticks. One group simply memorized the solution and the other group had to study the problem and discover the solution on their own. Later both groups were tested again on a similar problem and Katona compared their scores. The goal of the study was to demonstrate that learning by understanding produces bettter performance than learning by memorization, $\mu_{\text{understanding}} \neq \mu_{\text{memorization}}$.

Finally, in the preview for Chapter 11 we discussed a study by Pelton (1983) that compared the scores for Olympic marksmen for shots fired during hearbeats with the scores for shots fired between heartbeats. The goal was to demonstrate that the small vibrations caused by a heartbeat are enough to disrupt the marksmens' aim and affect their scores, $\mu_{\text{during}} \neq \mu_{\text{between}}$.

The Problem: Notice that in each of the three studies the goal of the hypothesis test is to demonstrate that the population mean (or mean difference) is *not equal* to some specific value. In Scaife's study, the goal was to show that the birds do not spend an equal amount of time in both chambers. Katona's goal was to demonstrate that the two methods of learning are not equivalent, there is a real, nonzero difference between them. Pelton also intended to show a real, nonzero difference by demonstrating that the marksmen's scores during heartbeats are not equal to their socres between heartbeats. In each case, the hypothesis test avoids the question of how much. If the birds don't spend 30 minutes on the plain side, how much time do they spend? If performance is not the same for the two methods of learning, how much difference is there? If marksmen's scores are different for shots fired between heartbeats compared with shots fired during heartbeats, how much difference is there?

The Solution: In this chapter we introduce a new inferential procedure know as *estimation*. Like hypothesis testing, estimation uses sample data to answer questions about populations. However, estimation uses sample statistics as the basis for estimating the corresponding population parameters. Specifically, we will use sample means to estimate population means, and use the difference between two sample means to estimate the difference between two population means. A hypothesis test can tell us that the difference between two means is not equal to zero but estimation can tell us how big the difference is. In general, hypothesis tests are used to determine whether a treatment effect exists and estimation is used to determine how much.

12.1 AN OVERVIEW OF ESTIMATION

In Chapter 8, we introduced hypothesis testing as a statistical procedure that allows researchers to use sample data to draw inferences about populations. Hypothesis testing is probably the most frequently used inferential technique, but it is not the only one. In this chapter, we examine the process of estimation, which provides researchers with an additional method for using samples as the basis for drawing general conclusions about populations.

The basic principle underlying all of inferential statistics is that samples are representative of the populations from which they come. The most direct application of this principle is the use of sample values as estimators of the corresponding population values—that is, using statistics to estimate parameters. This process is called *estimation*.

DEFINITION

The inferential process of using sample statistics to estimate population parameters is called **estimation**.

The use of samples to estimate populations is quite common. For example, you often hear news reports such as "Sixty percent of the general public approves of the president's new budget plan." Clearly, the percentage that is reported was obtained from a sample (they don't ask everyone's opinion), and this sample statistic is being used as an estimate of the population parameter.

We already have encountered estimation in earlier sections of this book. For example, the formula for sample variance (Chapter 4) was developed so that the sample value would give an accurate and unbiased estimate of the population variance. Now we examine the process of using sample means as the basis for estimating population means.

PRECISION AND CONFIDENCE IN ESTIMATION

Before we begin the actual process of estimation, a few general points should be kept in mind. First, a sample will not give a perfect picture of the whole population. A sample is expected to be representative of the population, but there always will be some differences between the sample and the entire population. These differences are referred to as *sampling error*. Second, there are two distinct ways of making estimates. Suppose, for example, you are asked to estimate the age of the authors of this book. If you look in the frontmatter of the book, just before the Contents, you will find pictures of Gravetter and Wallnau. We are roughly the same age, so pick either one of us and estimate how old we are. Note that you could make your estimate using a single value (for example, Gravetter appears to be 55 years old) or you could use a range of values (for example, Wallnau seems to be between 50 and 60 years old). The first estimate, using a single number, is called a *point estimate*.

DEFINITION

For a **point estimate**, you use a single number as your estimate of an unknown quantity.

Point estimates have the advantage of being very precise; they specify a particular value. On the other hand, you generally do not have much confidence that a point estimate is correct. For example, most of you would not be willing to bet that Gravetter is exactly 55 years old.

The second type of estimate, using a range of values, is called an *interval estimate*. Interval estimates do not have the precision of point estimates, but they do give you more confidence. For example, it would be reasonably safe for you to bet that Wallnau is between 40 and 60 years old. At the extreme, you would be very confident betting that Wallnau is between 20 and 70 years old. Note that there is a trade-off between precision and confidence. As the interval gets wider and wider, your confidence grows. At the same time, however, the precision of the estimate gets worse. We will be using samples to make both point and interval estimates of a population mean. Because the interval estimates are associated with confidence, they usually are called *confidence intervals*.

DEFINITIONS

For an **interval estimate**, you use a range of values as your estimate of an unknown quantity.

When an interval estimate is accompanied by a specific level of confidence (or probability), it is called a **confidence interval**.

Estimation is used in the same general situations in which we have already used hypothesis testing. In fact, there is an estimation procedure that accompanies each of the hypothesis tests we presented in the preceding chapters. Figure 12.1 shows an example of a research situation in which either hypothesis testing or estimation could be used. The figure shows a population with an unknown mean (the population after treatment). A sample is selected from the unknown population. The goal of estimation is to use the sample data to obtain an estimate of the unknown population mean.

COMPARISON OF HYPOTHESIS TESTS AND ESTIMATION

The situation shown in Figure 12.1 has been used to demonstrate hypothesis tests in the past. The research question concerns how the treatment affects the individuals in the population. Because it usually is impossible to administer the treatment to the entire population, a sample is selected and the treatment is administered to the sample. The treated sample then serves as a representative of the unknown treated population. In many ways, hypothesis testing and estimation are similar. They both use sample data and either z-scores or t statistics to answer questions about unknown populations. However, these two inferential procedures are designed to answer different questions. Using the situation in Figure 12.1 as an example, a hypothesis test would determine whether the treatment has any effect. Notice that this is a yes-no question. The test would differentiate between two alternatives:

1. No. The treatment does not have an effect (H_0).

2. Yes. The treatment does have an effect (H_1).

The purpose of estimation, on the other hand, is to determine the value of the unknown population mean after treatment. Essentially, estimation will determine *how much* effect the treatment has (Box 12.1). For example, a sample mean of $M = 38$ would produce a point estimate of $\mu = 38$ for the unknown population mean. In this case, we are estimating that the effect of the treatment is to increase scores by 8 points from the original mean of $\mu = 30$ to the estimated post-treatment mean of $\mu = 38$.

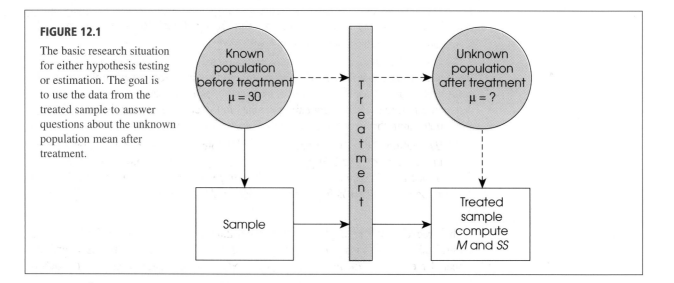

FIGURE 12.1

The basic research situation for either hypothesis testing or estimation. The goal is to use the data from the treated sample to answer questions about the unknown population mean after treatment.

| BOX 12.1 | HYPOTHESIS TESTING VERSUS ESTIMATION: STATISTICAL SIGNIFICANCE VERSUS PRACTICAL SIGNIFICANCE |

As we already noted, hypothesis tests tend to involve a yes-no decision. Either we decide to reject H_0, or we fail to reject H_0. The language of hypothesis testing reflects this process. The outcome of the hypothesis test is one of two conclusions:

There is no evidence for a treatment effect (fail to reject H_0)

or

There is a statistically significant effect (H_0 is rejected)

For example, a researcher studies the effect of a new drug on people with high cholesterol. In hypothesis testing, the question is whether the drug has a significant effect on cholesterol levels. Suppose the hypothesis test revealed that the drug did produce a significant decrease in cholesterol. The next question might be, How much of a reduction occurs? This question calls for estimation, in which the size of a treatment effect for the population is estimated.

Estimation can be of great practical importance because the presence of a "statistically significant" effect does not necessarily mean the results are large enough for use in practical applications. Consider the following possibility: Before drug treatment, the sample of patients had a mean cholesterol level of 225. After drug treatment, their cholesterol reading was 210. When analyzed, this 15-point change reached statistical significance (H_0 was rejected). Although the hypothesis test revealed that the drug produced a *statistically significant* change, it may not be *clinically significant*. That is, a cholesterol level of 210 is still quite high. In estimation, we would estimate the population mean cholesterol level for patients who are treated with the drug. This estimated value may reveal that even though the drug does in fact reduce cholesterol levels, it does not produce a large enough change (notice we are looking at a "how much" question) to make it of any practical value. Thus, the hypothesis test might reveal that an effect occurred, but estimation indicates it is small and of little *practical significance* in real-world applications.

WHEN TO USE ESTIMATION

There are three situations in which estimation commonly is used:

1. Estimation is used after a hypothesis test when H_0 is rejected. Remember that when H_0 is rejected, the conclusion is that the treatment does have an effect. The next logical question would be, How much effect? This is exactly the question that estimation is designed to answer.

2. Estimation is used when you already know that there is an effect and simply want to find out how much. For example, the city school board probably knows that a special reading program will help students. However, they want to be sure that the effect is big enough to justify the cost. Estimation is used to determine the size of the treatment effect.

3. Estimation is used when you simply want some basic information about an unknown population. Suppose, for example, you want to know about the political attitudes of students at your college. You could use a sample of students as the basis for estimating the population mean.

THE LOGIC OF ESTIMATION

In the preceding three chapters we presented three different situations for hypothesis testing using t statistics: the single sample t, the independent-measures t, and the repeated-measures t. In this chapter, we introduce estimation as it applies to each of these three situations. Recall that the general form of the *t* statistic that is used for hypothesis testing and for estimation is as follows:

$$t = \frac{\text{sample mean (or mean difference)} - \text{population mean (or mean difference)}}{\text{estimated standard error}}$$

In the formula, the population mean (or mean difference) is the unknown value. However, we have a sample mean (or mean difference) that should be representative of the population value. Also, we can compute a standard error that measures how much discrepancy is expected, on average, between the sample statistic and the corresponding population parameter. Finally, we know about the distribution of *t* statistics. In particular, we know that high-probability values for *t* are located in the center of the distribution, around *t* = 0, and we know that values in the extreme tails of the distribution are low-probability outcomes (Figure 12.2).

As we have noted, estimation and hypothesis testing are both inferential procedures that use sample data as the basis for drawing conclusions about an unknown population. A hypothesis test uses the sample to evaluate a hypothesis about the unknown population mean (or mean difference). Estimation uses the sample to estimate the value for the unknown population mean (or mean difference). Because hypothesis testing and estimation have different goals, they will follow different logical paths. These different paths are outlined in Table 12.1.

Because the goal of the estimation process is to compute a value for the unknown population mean or mean difference, it usually is easier to regroup the terms in the *t* formula so that the population value is isolated on one side of the equation. In algebraic terms, we are solving the equation for the unknown population parameter. The result takes the following form:

$$\text{population mean (or mean difference)} = \text{sample mean (or mean difference)} \pm t(\text{estimated standard error}) \qquad (12.1)$$

This is the general equation that we will use for estimation. Consider the following two points about Equation 12.1:

1. On the right-hand side of the equation, the values for the sample mean and the estimated standard error can be computed directly from the sample data. Thus, only the value of *t* is unknown. If we can determine this missing value, then we can use the equation to calculate the unknown population mean.

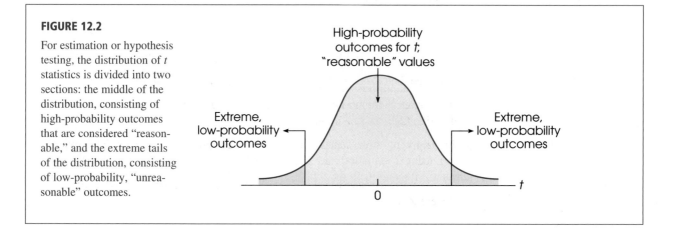

FIGURE 12.2

For estimation or hypothesis testing, the distribution of *t* statistics is divided into two sections: the middle of the distribution, consisting of high-probability outcomes that are considered "reasonable," and the extreme tails of the distribution, consisting of low-probability, "unreasonable" outcomes.

High-probability outcomes for *t*; "reasonable" values

Extreme, low-probability outcomes

Extreme, low-probability outcomes

t

0

TABLE 12.1

The logic behind hypothesis tests and estimation.

Hypothesis Test	Estimation
Goal: To test a hypothesis about an unknown parameter, usually the null hypothesis, which states that the treatment has no effect.	*Goal:* To estimate the value of an unknown parameter.
A. Begin by hypothesizing a value for the unknown parameter.	A. Do not attempt to calculate a *t* statistic. Instead, begin by estimating what the *t* value ought to be. The strategy for making this estimate is to pick a reasonable, high-probability value for *t* (near zero).
B. The hypothesized value is substituted into the formula and the value for *t* is calculated.	B. The reasonable value for *t* is substituted into the formula and the value for the unknown parameter is calculated.
C. If the hypothesized value produces a reasonable outcome for *t* (near zero), we conclude that the hypothesis was reasonable and we fail to reject H_0. If the outcome is an extreme, low-probability value for t, we reject H_0.	C. Because we used a reasonable value for *t*, it is assumed that the calculations will produce a reasonable estimate of the population parameter.

2. Although the specific value for the *t* statistic cannot be determined, we do know what the entire distribution of *t* statistics looks like. We can use the distribution to *estimate* what the *t* statistic ought to be.

   a. For a point estimate, the best bet is to use $t = 0$, the exact center of the distribution. There is no reason to suspect that the sample data are biased (either above average or below average), so $t = 0$ is a sensible value. Also, $t = 0$ is the most likely value, with probabilities decreasing steadily as you move away from zero toward the tails of the distribution.

   b. For an interval estimate, we will use a range of *t* values around zero. For example, to be 90% confident that our estimation is correct, we simply use the range of *t* values that forms the middle 90% of the distribution. Note that we are estimating that the sample corresponds to a *t* statistic located somewhere in the middle 90% of the distribution, so we are 90% confident that the estimate is correct.

Once we have estimated a value for *t*, then we have all the numbers on the right-hand side of the equation and we can calculate a value for the unknown population mean. Because one of the numbers on the right-hand side is an estimated value, the population mean that we calculate is also an estimated value.

LEARNING CHECK

1. Describe the difference between a point estimate and an interval estimate.

2. In a research situation using a sample to evaluate the effect of a treatment, describe the goal of a hypothesis and the goal of estimation.

3. Explain why estimation is often used after a hypothesis for which the decision is to reject the null hypothesis.

4. What *t* value is used for a point estimate of a population mean or mean difference?

1. A point estimate uses a single value to estimate an unknown parameter. An interval estimate uses a range of values. A point estimate is more precise but an interval estimate provides more confidence.

2. The goal of a hypothesis test is to determine whether the treatment has an effect on the individuals in the population. The goal of estimation is to determine how much effect.

3. If the null hypothesis is rejected, the conclusion is that the treatment does have an effect. The next logical question is "How much effect?" which is exactly what estimation seeks to determine.

4. A value of $t = 0$ is used for point estimates of population means or mean differences.

12.2   ESTIMATION WITH THE t STATISTIC

The three t statistics that were introduced in the preceding three chapters can all be adapted to be used for estimation. As we saw in the previous section, the general form of the t equation for estimation is as follows:

$$\begin{matrix} \text{population mean} \\ \text{(or mean difference)} \end{matrix} = \begin{matrix} \text{sample mean} \\ \text{(or mean difference)} \end{matrix} \pm t(\text{estimated standard error})$$

With the single-sample t (Chapter 9), we use a sample mean, M, to estimate an unknown population mean, μ. The formula for estimation with the single-sample t is

$$\mu = M \pm ts_M \tag{12.2}$$

With the independent-measures t (Chapter 10), we use the difference between two sample means, $M_1 - M_2$, to estimate the size of the difference between two population means, $\mu_1 - \mu_2$. The formula for estimation with the independent-measures t is

$$\mu_1 - \mu_2 = M_1 - M_2 \pm ts_{(M_1 - M_2)} \tag{12.3}$$

Finally, the repeated-measures t (Chapter 11) uses the mean difference for a sample, M_D, to estimate the corresponding mean difference for the population, μ_D. The formula for estimation with the repeated-measures t is

$$\mu_D = M_D \pm ts_{M_D} \tag{12.4}$$

To use the t statistic formulas for estimation, we must determine all of the values on the right-hand side of the equation (including an estimated value for t) and then use these numbers to compute an estimated value for the population mean or mean difference. Specifically, you first compute the sample mean (or mean difference) and the estimated standard error from the sample data. Next, you estimate a value, or a range of values, for t. More precisely, you are estimating where the sample data are located in the t distribution. These values complete the right-hand side of the equation and allow you to compute an estimated value for the mean (or the mean difference). The following examples demonstrate the estimation procedure with each of the three t statistics.

ESTIMATION OF μ FOR SINGLE-SAMPLE STUDIES

In Chapter 9, we introduced single-sample studies and hypothesis testing with the *t* statistic. Now we will use a single-sample study to estimate the value for μ, using point and interval estimates.

EXAMPLE 12.1

For several years researchers have noticed that there appears to be a regular, year-by-year increase in the average IQ for the general population. This phenomenon is called the Flynn effect after the researcher who first reported it (Flynn, 1984, 1999), and it means that psychologists must continuously update IQ tests to keep the population mean at μ = 100. To evaluate the size of the effect, a researcher obtained a 10-year-old IQ test that was standardized to produce a mean IQ of μ = 100 for the population 10 years ago. The test was then given to a sample of *n* = 64 of today's 20-year-old adults. The average score for the sample was *M* = 107 with a standard deviation of *s* = 12. The researcher would like to use the data to estimate how much IQ scores have changed during the past 10 years. Specifically, the researcher would like to make a point estimate and an 80% confidence interval estimate of the population mean for people taking a 10-year-old IQ test. The structure for this research study is shown in Figure 12.3.

In this example, we are using a single sample to estimate the mean for a single population. In this case, the estimation formula for the single-sample *t* is

$$\mu = M \pm t s_M$$

To use the equation, we must first compute the estimated standard error and then determine the estimated value(s) to be used for the *t* statistic.

Compute the estimated standard error, s_M   To compute the estimated standard error, it is first necessary to calculate the sample variance. For this example, we are given a sample standard deviation of *s* = 12, so the sample variance is $s^2 = (12)^2 = 144$. The estimated standard error is

$$s_M = \sqrt{\frac{s^2}{n}} = \sqrt{\frac{144}{64}} = \frac{12}{8} = 1.50$$

FIGURE 12.3

The structure of the research study described in Example 12.1. The goal is to use the sample to estimate the population mean IQ for people taking a 10-year-old IQ test. We can then estimate how much IQ scores have changed during the past 10 years.

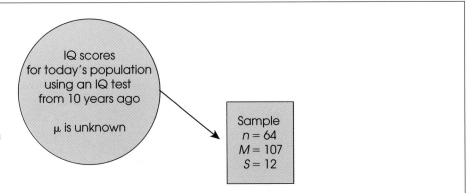

IQ scores for today's population using an IQ test from 10 years ago

μ is unknown

Sample
n = 64
M = 107
S = 12

The point estimate   As noted earlier, a point estimate involves selecting a single value for t. Because the t distribution is always symmetrically distributed with a mean of zero, we will always use $t = 0$ as the best choice for a point estimate. Using the sample data and the estimate of $t = 0$, we obtain

$$\mu = M \pm ts_M$$
$$= 107 \pm 0(1.50)$$
$$= 107$$

This is our point estimate of the population mean. Note that we simply have used the sample mean, M, to estimate the population mean, μ. The sample is the only information that we have about the population, and it provides an unbiased estimate of the population mean (Chapter 7, page 205). That is, on average, the sample mean provides an accurate representation of the population mean. Based on this point estimate, our conclusion is that today's population would have a mean IQ of $\mu = 107$ on an IQ test from 10 years ago. Thus, we are estimating that there has been a 7-point increase in IQ scores (from $\mu = 100$ to $\mu = 107$) during the past decade.

To have 80% in the middle there must be 20% (or .20) in the tails. To find the t values, look under two tails, .20 in the t table.

The interval estimate   For an interval estimate, select a range of t values that is determined by the level of confidence. In this example, we want 80% confidence in our estimate of μ. Therefore, we estimate that the t statistic for this sample is located somewhere in the middle 80% of the t distribution. With $df = n - 1 = 63$, the middle 80% of the distribution is bounded by t values of $+1.296$ and -1.296 (using $df = 60$ from the table). These values are shown in Figure 12.4. Using the sample data and this estimated range of t values, we obtain

$$\mu = M \pm t(s_M) = 107 \pm 1.296(1.50) = 107 \pm 1.944$$

At one end of the interval we obtain 108.944 ($107 + 1.944$), and at the other end of the interval we obtain 105.056 ($107 - 1.944$). Our conclusion is that today's population would have a mean IQ between 105.056 and 108.944 if they used an IQ test from 10 years ago. In other words, we are concluding that the mean IQ has increased over the past 10 years, and we are estimating with 80% confidence that the size of the increase is between 5.056 and 8.944 points. The confidence comes from the fact that the calculation was based on only one assumption. Specifically, we assumed that the t statistic was located between $+1.296$ and -1.296, and we are 80%

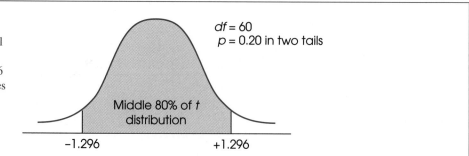

FIGURE 12.4

The 80% confidence interval with $df = 60$ is constructed using t values of $t = -1.296$ and $t = +1.296$. The t values are obtained from the table using 20% (0.20) as the proportion remaining in the two tails of the distribution.

$df = 60$
$p = 0.20$ in two tails

Middle 80% of t distribution

-1.296 $+1.296$

confident that this assumption is correct because 80% of all the possible *t* values are located in this interval. Finally, note that the confidence interval is constructed around the sample mean. As a result, the sample mean, $M = 107$, is located exactly in the center of the interval.

Figure 12.5 provides a visual presentation of the results from Example 12.1. The original population mean from 10 years ago is shown along with the two estimates (point and interval) of today's mean. The estimates clearly indicate that there has been an increase in IQ scores over the past 10 years, and they provide an indication of how large the increase is.

Interpretation of the confidence interval   In the preceding example, we computed an 80% confidence interval to estimate an unknown population mean. We obtained an interval ranging from 105.056 to 108.944, and we are 80% confident that the unknown population mean is located within this interval. You should note, however, that the 80% confidence applies to the *process* of computing the interval rather than the specific end points for the interval. For example, if we repeated the process over and over, we could eventually obtain hundreds of different samples (each with $n = 64$ scores) and we could calculate hundreds of different confidence intervals. However, each interval is computed using the same procedure. Specifically, each interval is centered around its own sample mean, and each interval extends from a value corresponding to $t = -1.296$ at one end to $t = +1.296$ at the other end.

Figure 12.6 shows the distribution of *t* values with the middle 80% highlighted. Note that the value of zero in the middle of the distribution corresponds to the population mean, μ, because $t = 0$ only when the sample mean is exactly equal to the population mean. Also note that all of the *t* values in the middle 80% are within 1.296 points of $t = 0$ (the population mean). Finally, the figure shows two possible samples and their corresponding *t* values.

1. Sample 1 is an example of a sample with a *t* value located inside the 80% boundaries. The 80% confidence interval for sample 1 is centered at the sample mean and extends for a distance corresponding to $t = 1.296$ in both directions. Thus, the confidence interval for sample 1 overlaps $t = 0$, which means that the unknown population mean μ is contained within the interval.

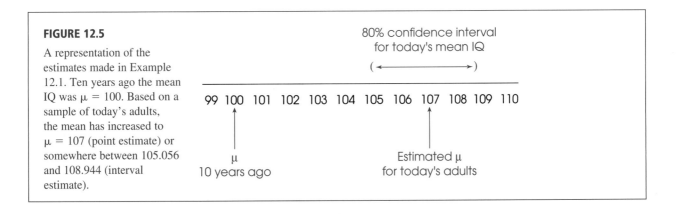

FIGURE 12.5

A representation of the estimates made in Example 12.1. Ten years ago the mean IQ was $\mu = 100$. Based on a sample of today's adults, the mean has increased to $\mu = 107$ (point estimate) or somewhere between 105.056 and 108.944 (interval estimate).

FIGURE 12.6

Interpretation of the 80% confidence for an 80% confidence interval. Of all the possible samples, 80% will have *t* scores located in the middle 80% of the distribution and will produce confidence intervals that overlap and contain the population mean. Thus, 80% of all the possible confidence intervals will contain the true value for μ.

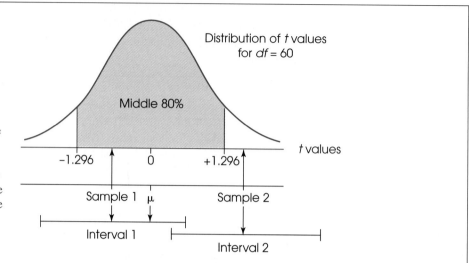

2. Sample 2, on the other hand, corresponds to a *t* value located outside the 80% boundaries. The 80% confidence for sample 2 extends for a distance corresponding to *t* = 1.296 in both directions from the sample mean. This time, however, the interval does not extend far enough to include *t* = 0, which means that the value of μ is not included in the interval.

Out of all the possible samples, 80% will be similar to sample 1. That is, they will correspond to *t* values between ±1.296 and will produce confidence intervals that contain the population mean. The other 20% of the possible samples will be similar to sample 2. They will correspond to *t* values outside the ±1.296 boundaries, and they will produce confidence intervals that do not contain μ. Thus, out of all the different confidence intervals that we could calculate, 80% will actually contain the population mean and 20% will not.

Note that the population mean, μ, is a constant value. Because the mean does not change, it is incorrect to think that sometimes μ is in a specific interval and sometimes it is not. Instead, the intervals change from one sample to another, so that μ is in some of the intervals and is not in others. The probability that any individual confidence interval actually contains the mean is determined by the level of confidence (the percentage) used to construct the interval.

ESTIMATION OF $\mu_1 - \mu_2$ FOR INDEPENDENT-MEASURES STUDIES

The independent-measures *t* statistic uses the data from two separate samples to evaluate the mean difference between two populations. In Chapter 10, we used this statistic to answer a yes-no question: Is there any difference between the two population means? With estimation, we ask, *How much* difference? In this case, the independent-measures *t* statistic is used to estimate the value of $\mu_1 - \mu_2$. The following example demonstrates the process of estimation with the independent-measures *t* statistic.

EXAMPLE 12.2   Recent studies have allowed psychologists to establish definite links between specific foods and specific brain functions. For example, lecithin (found in soybeans, eggs, and liver) has been shown to increase the concentration of certain brain chemicals that help regulate memory and motor coordination. This experiment is designed to demonstrate the importance of this particular food substance.

The experiment involves two separate samples of newborn rats (an independent-measures experiment). The 10 rats in the first sample are given a normal diet containing standard amounts of lecithin. The 5 rats in the other sample are fed a special diet, which contains almost no lecithin. After 6 months, each of the rats is tested on a specially designed learning problem that requires both memory and motor coordination. The purpose of the experiment is to demonstrate the deficit in performance that results from lecithin deprivation. The score for each animal is the number of errors it makes before it solves the learning problem. The data from this experiment are as follows:

Regular Diet	No-Lecithin Diet
$n = 10$	$n = 5$
$M = 25$	$M = 33$
$SS = 61.0$	$SS = 36.5$

Because we fully expect that there will be a significant difference between these two treatments, we will not do the hypothesis test (although you should be able to do it). We want to use these data to obtain an estimate of the size of the difference between the two population means; that is, how much does lecithin affect learning performance? We will use a point estimate and a 95% confidence interval.

The basic equation for estimation with an independent-measures experiment is

$$\mu_1 - \mu_2 = (M_1 - M_2) \pm ts_{(M_1 - M_2)}$$

The first step is to obtain the known values from the sample data. The sample mean difference is easy; one group averaged $M = 25$, and the other averaged $M = 33$, so there is an 8-point difference. Note that it is not important whether we call this a $+8$ or a -8 difference. In either case, the size of the difference is 8 points, and the regular diet group scored lower. Because it is easier to do arithmetic with positive numbers, we will use

$$M_1 - M_2 = 8$$

Compute the standard error   To find the standard error, we first must pool the two variances:

$$s_p^2 = \frac{SS_1 + SS_2}{df_1 + df_2} = \frac{61.0 + 36.5}{9 + 4} = \frac{97.5}{13} = 7.50$$

Next, the pooled variance is used to compute the standard error:

$$s_{(M_1 - M_2)} = \sqrt{\frac{s_p^2}{n_1} + \frac{s_p^2}{n_2}} = \sqrt{\frac{7.50}{10} + \frac{7.50}{5}} = \sqrt{0.75 + 1.50} = \sqrt{2.25} = 1.50$$

Recall that this standard error combines the error from the first sample and the error from the second sample. Because the first sample is much larger, $n = 10$, it should have less error. This difference shows up in the formula. The larger sample contributes an error of 0.75 points and the smaller sample contributes 1.50 points, which combine for a total error of 2.25 points under the square root.

Sample 1 has $df = 9$, and sample 2 has $df = 4$. The t statistic has $df = 9 + 4 = 13$.

Estimate the value(s) for t   The final value needed on the right-hand side of the equation is t. The data from this experiment would produce a t statistic with $df = 13$. With 13 degrees of freedom, we can sketch the distribution of all the possible t values. This distribution is shown in Figure 12.7. The t statistic for our data is somewhere in this distribution. The problem is to estimate where. For a point estimate, the best bet is to use $t = 0$. This is the most likely value, located exactly in the middle of the distribution. To gain more confidence in the estimate, you can select a range of t values. For 95% confidence, for example, you would estimate that the t statistic is somewhere in the middle 95% of the distribution. Checking the table, you find that the middle 95% is bounded by values of $t = +2.160$ and $t = -2.160$.

Using these t values and the sample values computed earlier, we now can estimate the magnitude of the performance deficit caused by lecithin deprivation.

Compute the point estimate   For a point estimate, use the single-value (point) estimate of $t = 0$:

$$\mu_1 - \mu_2 = (M_1 - M_2) \pm ts_{(M_1 - M_2)}$$
$$= 8 + 0(1.50)$$
$$= 8$$

Note that the result simply uses the sample mean difference to estimate the population mean difference. The conclusion is that lecithin deprivation produces an average of 8 more errors on the learning task. (Based on the fact that the nondeprived animals averaged around 25 errors, an 8-point increase would mean a performance deficit of approximately 30%.)

Construct the interval estimate   For an interval estimate, or confidence interval, use the range of t values. With 95% confidence, at one extreme,

$$\mu_1 - \mu_2 = (M_1 - M_2) + ts_{(M_1 - M_2)}$$
$$= 8 + 2.160(1.50)$$
$$= 8 + 3.24$$
$$= 11.24$$

and at the other extreme,

$$\mu_1 - \mu_2 = (M_1 - M_2) - ts_{(M_1 - M_2)}$$
$$= 8 - 2.160(1.50)$$
$$= 8 - 3.24$$
$$= 4.76$$

FIGURE 12.7

The distribution of t values with $df = 13$. Note that t values pile up around zero and that 95% of the values are located between -2.160 and $+2.160$.

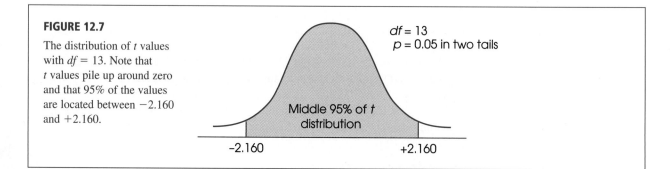

$df = 13$
$p = 0.05$ in two tails

Middle 95% of t distribution

-2.160 $+2.160$

This time we conclude that the effect of lecithin deprivation is to increase errors, with an average increase somewhere between 4.76 and 11.24 errors. We are 95% confident of this estimate because our only estimation was the location of the *t* statistic, and we used the middle 95% of all the possible *t* values.

Note that the result of the point estimate is to say that lecithin deprivation will increase errors by *exactly* 8 points. To gain confidence, you must lose precision and say that errors will increase by *around* 8 points (for 95% confidence, we say that the average increase will be 8 ± 3.24).

ESTIMATION OF μ_D FOR REPEATED-MEASURES STUDIES

Finally, we turn our attention to the repeated-measures study. Remember that this type of study has a single group of participants, with the same individuals measured in two different treatment conditions. By finding the difference between the score for treatment 1 and the score for treatment 2, we calculate a difference score for each participant.

$$D = X_2 - X_1$$

The mean for the sample of *D* scores, M_D, is used to estimate the population mean μ_D, which is the mean for the entire population of difference scores.

EXAMPLE 12.3

Research indicates that even the suggestion of hypnosis can be an effective treatment for pain (Hylands-White & Derbyshire, 2007). In their study, college students experienced a painful stimulus before and after listening to a recording of relaxation instructions. However, the students were told that the instructions were "a hypnotic induction to help you become hypnotized." The painful experience consisted of immersing their right hands in an ice water bath for 5 minutes and reporting the level of pain. The results indicate that the reported level of pain was significantly lower after listening to the hypnosis instructions. By the way, in another part of the study, a different group of students received exactly the same treatment but the word "hypnosis" was never used. For the second group, who were told that they were listening to relaxation instructions, there was no significant decrease in the level of pain.

Suppose a similar study is conducted with a sample of $n = 16$ college students and the results show that pain reports are $M_D = 2.2$ points lower after the "hypnosis" instructions with $SS = 84$ for the difference scores. We will use these results to estimate how much pain relief the hypnosis instructions would have for the entire population. Specifically, we will make a point estimate and a 90% confidence interval estimate for the population mean difference, μ_D.

You should recognize that this study requires a repeated-measures *t* statistic. For estimation, the repeated-measures *t* equation is as follows:

$$\mu_D = M_D \pm t s_{M_D}$$

The sample mean is $M_D = 2.2$, so all that remains is to compute the estimated standard error and estimate the appropriate value(s) for *t*.

Compute the standard error   To find the standard error, we first must compute the sample variance:

$$s^2 = \frac{SS}{n-1} = \frac{84}{15} = 5.60$$

Now the estimated standard error is

$$s_{M_D} = \sqrt{\frac{s^2}{n}} = \sqrt{\frac{5.60}{16}} = \sqrt{0.35} = 0.59$$

To complete the estimate of μ_D, we must identify the value of t. We will consider the point estimate and the interval estimate separately.

Compute the point estimate   To obtain a point estimate, a single value of t is selected to approximate the location of M_D. Remember that the t distribution is symmetrical and bell-shaped with a mean of zero (see Figure 12.8). Because $t = 0$ is the most frequently occurring value in the distribution, this is the t value used for the point estimate. Using this value in the estimation formula gives

$$\mu_D = M_D \pm t s_{M_D} = 2.2 \pm 0(0.59) = 2.2$$

For this example, our best estimate is that the hypnosis instructions reduce pain reports in the general population by an average of $\mu_D = 2.2$ points. As noted several times before, the sample mean $M_D = 2.2$, provides the best point estimate of the population mean.

Construct the interval estimate   We also want to make an interval estimate to be 90% confident that we have accurately estimated the value of μ_D. To get the interval, it is necessary to determine what t values form the boundaries of the middle 90% of the t distribution. With 90% in the middle, the remaining area in the two tails must total 10% or $p = .10$. Using $df = n - 1 = 15$ and $p = .10$ for two tails, you should find the values $+1.753$ and -1.753 in the t table. These values form the boundaries for the middle 90% of the t distribution (Figure 12.8). We are confident that the t value for our sample is in this range because 90% of all the possible t values are there. Using these values in the estimation formula, we obtain the following: On one end of the interval,

$$\mu_D = M_D - t s_{M_D}$$
$$= 2.2 - 1.753(0.59)$$
$$= 2.2 - 1.03$$
$$= 1.17$$

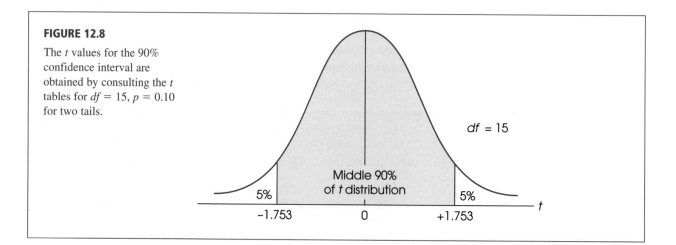

FIGURE 12.8

The t values for the 90% confidence interval are obtained by consulting the t tables for $df = 15, p = 0.10$ for two tails.

and on the other end of the interval,

$$\mu_D = 2.2 + 1.753(0.59)$$

$$= 2.2 + 1.03$$

$$= 3.23$$

Based on these results, the researchers estimate that the hypnosis instructions would reduce pain in the general population by an average of 2.2 points, and they can be 90% confident that the true population mean is in an interval from 1.17 to 3.23 points.

LEARNING CHECK

1. A researcher would like to determine the average reading ability for third-grade students in the local school district. A sample of $n = 25$ students is selected and each student takes a standardized reading achievement test. The average score for the sample is $M = 72$ with $SS = 2400$. Use the sample results to construct a 99% confidence interval for the population mean.

2. A researcher is testing the effectiveness of a new antidepressant medication. A sample of $n = 9$ clinical depressed patients is obtained and the current level of depression is measured for each individual. The patients are then given the medication daily for three weeks and depression is measured again. On average, the patients showed a decrease in depression of $M_D = 18$ points with $SS = 152$. Determine the 95% confidence interval for the population mean difference, μ_D.

3. In families with several children, the first-born tend to be more reserved and serious, whereas the last-born tend to be more outgoing and happy-go-lucky. A psychologist is using a standardized personality inventory to measure the magnitude of this difference. Two samples are used: 8 first-born children and 8 last-born children. Each child is given the personality test. The results are as follows:

First-born	Last-born
$M = 11.4$	$M = 13.9$
$SS = 26$	$SS = 30$

   a. Use these sample statistics to make a point estimate of the population mean difference in personality for first-born versus last-born children.

   b. Make an interval estimate of the population mean difference so that you are 80% confident that the true mean difference is in your interval.

ANSWERS

1. The sample variance is $s^2 = 100$ and the estimated standard error is $s_M = 2$ points. With $df = 24$ and 99% confidence, the t statistic should be between $+2.797$ and -2.797. With 99% confidence, we estimate that the population mean is between 66.406 and 77.594.

2. $s^2 = 19$, $s_{M_D} = 1.45$, $df = 8$, $t = \pm 2.306$; estimate that μ_D is between 14.66 and 21.34.

3. a. For a point estimate, use the sample mean difference: $M_1 - M_2 = 2.5$ points.

   b. Pooled variance $= 4$, estimated standard error $= 1$, $df = 14$, $t = \pm 1.345$. The 80% confidence interval is 1.16 to 3.85.

12.3   A FINAL LOOK AT ESTIMATION

**FACTORS AFFECTING
THE WIDTH OF A
CONFIDENCE INTERVAL**

Two characteristics of the confidence interval should be noted. First, notice what happens to the width of the interval when you change the level of confidence (the percent confidence). To gain more confidence in your estimate, you must increase the width of the interval. Conversely, to have a smaller interval, you must give up confidence. This is the basic trade-off between precision and confidence that was discussed earlier. In the estimation formula, the percentage of confidence influences the width of the interval by way of the t value. The larger the level of confidence (the percentage), the larger the t value and the larger the interval. This relationship can be seen in Figure 12.8. In the figure, we identified the middle 90% of the t distribution in order to find a 90% confidence interval. It should be obvious that if we were to increase the confidence level to 95%, it would be necessary to increase the range of t values and thereby increase the width of the interval.

Second, note what would happen to the interval width if you had a different sample size. This time the basic rule is as follows: The bigger the sample (n), the smaller the interval. This relationship is straightforward if you consider the sample size as a measure of the amount of information. A bigger sample gives you more information about the population and allows you to make a more precise estimate (a narrower interval). The sample size controls the magnitude of the standard error in the estimation formula. As the sample size increases, the standard error decreases, and the interval gets smaller.

With t statistics, the sample size has an additional effect on the width of a confidence interval. Remember that the exact shape of the t distribution depends on degrees of freedom. As the sample size gets larger, df also gets larger, and the t values associated with any specific percentage of confidence get smaller. This fact simply enhances the general relationship that the larger the sample, the smaller the confidence interval.

**ESTIMATION, EFFECT SIZE,
AND HYPOTHESIS TESTS**

The process of estimation, especially the estimation of mean differences, provides a relatively simple and direct method for evaluating effect size. For example, the outcome of the study in Example 12.3 indicates that the hyponosis/relaxation instruction reduce pain by an estimated 2.2 points. Given that the students' average pain rating was around 12 points, this is a decrease of almost 20%. In this case, the estimation produces a clear and understandable measure of how large the treatment effect actually is.

In addition to describing the size of a treatment effect, estimation can be used to get an indication of the *significance* of the effect. Example 12.2 presented an independent-measures research study examining the effect of lecithin on problem-solving performance for rats. Based on the results of this study, it was estimated that the mean difference in performance produced by lecithin was $\mu_1 - \mu_2 = 8$ points. The 95% confidence interval estimated the mean difference to be between 4.76 points and 11.24 points. The confidence interval estimate is shown in Figure 12.9. In addition to the confidence interval for $\mu_1 - \mu_2$, we have marked the spot where the mean difference is equal to zero. You should recognize that a mean difference of zero is exactly what would be predicted by the null hypothesis if we were doing a hypothesis test. You also should realize that a zero difference ($\mu_1 - \mu_2 = 0$) is *outside* the 95% confidence interval. In other words, $\mu_1 - \mu_2 = 0$ is not an acceptable value if we want 95% confidence in our estimate. To conclude that a value of zero is *not acceptable* with 95% confidence is equivalent to concluding that a value of zero is *rejected* with 95% confidence. This conclusion is equivalent to rejecting H_0 with $\alpha = .05$. On the other hand, if a mean difference of zero was included within the 95% confidence interval, then we would have to conclude that $\mu_1 - \mu_2 = 0$ is an acceptable value, which is the same as failing to reject H_0.

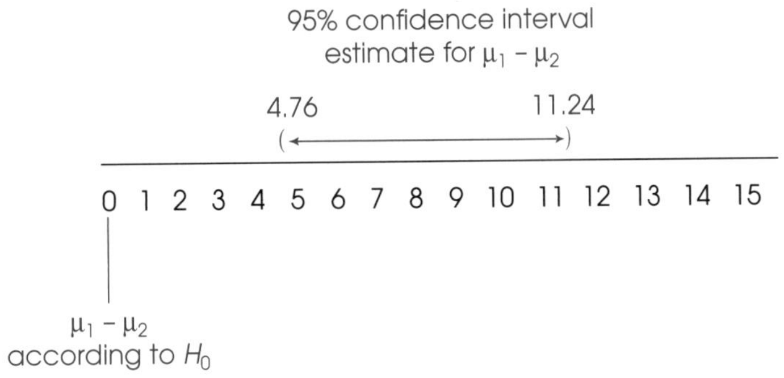

FIGURE 12.9

The 95% confidence interval for the population mean difference ($\mu_1 - \mu_2$) from Example 12.2. Note that $\mu_1 - \mu_2 = 0$ is excluded from the confidence interval, indicating that a zero difference is not an acceptable value (H_0 would be rejected in a hypothesis test).

LEARNING CHECK

1. If all other factors are held constant, an 80% confidence interval will be wider than a 90% confidence interval. (True or false?)

2. If all other factors are held constant, a confidence interval computed from a sample of $n = 25$ will be wider than a confidence interval from a sample of $n = 100$. (True or false?)

3. A 99% confidence interval for a population mean difference (μ_D) extends from -1.50 to $+3.50$. If a repeated-measures hypothesis test with two tails and $\alpha = .01$ were conducted using the same data, the decision would be to fail to reject the null hypothesis. (True or false?)

ANSWERS

1. False. Greater confidence requires a wider interval.

2. True. The smaller sample will produce a wider interval.

3. True. The value $\mu_D = 0$ is included within the 99% confidence interval, which means that it is an acceptable value with $\alpha = .01$ and would not be rejected.

SUMMARY

1. Estimation is a procedure that uses sample data to obtain an estimate of a population mean or mean difference. The estimate can be either a point estimate (single value) or an interval estimate (range of values). Point estimates have the advantage of precision, but they do not give much confidence. Interval estimates provide confidence, but you lose precision as the interval grows wider.

2. Estimation and hypothesis testing are similar processes: Both use sample data to answer questions about populations. However, these two procedures are designed to answer different questions. Hypothesis testing will tell you whether a treatment effect exists (yes or no). Estimation will tell you how much treatment effect there is.

3. The estimation process begins by solving the t-statistic equation for the unknown population mean (or mean difference).

$$\begin{matrix} \text{population mean} \\ \text{(or mean difference)} \end{matrix} = \begin{matrix} \text{sample mean} \\ \text{(or mean difference)} \\ \pm\, t(\text{estimated standard error}) \end{matrix}$$

Except for the value of t, the numbers on the right-hand side of the equation are all obtained from the sample data. By using an estimated value for t, you can then compute an estimated value for the population mean (or mean difference). For a point estimate, use $t = 0$. For an interval estimate, first select a level of confidence and then look up the corresponding range of t values from the t-distribution table. For example, for 90% confidence, use the range of t values that determine the middle 90% of the distribution.

4. For a single-sample study, the mean from one sample is used to estimate the mean for the corresponding population.

$$\mu = M \pm ts_M$$

For an independent-measures study, the means from two separate samples are used to estimate the mean difference between two populations.

$$(\mu_1 - \mu_2) = (M_1 - M_2) \pm ts_{(M_1 - M_2)}$$

For a repeated-measures study, the mean from a sample of difference scores (D values) is used to estimate the mean difference for the general population.

$$\mu_D = M_D \pm ts_{M_D}$$

5. The width of a confidence interval is an indication of its precision: A narrow interval is more precise than a wide interval. The interval width is influenced by the sample size and the level of confidence.
   a. As sample size (n) gets larger, the interval width gets smaller (greater precision).
   b. As the percentage confidence increases, the interval width gets larger (less precision).

KEY TERMS

estimation   point estimate   interval estimate   confidence interval

RESOURCES

There are a tutorial quiz and other learning exercises for Chapter 12 at the Wadsworth website, www.cengage.com/psychology/gravetter. See page 29 for more information about accessing items on the website.

ENHANCED
WebAssign

Guided interactive tutorials, end-of-chapter problems, and related testbank items may be assigned online at WebAssign.

WebTUTOR

For those using WebTutor along with this book, there is a WebTutor section corresponding to this chapter. The WebTutor contains a brief summary of Chapter 12, hints for learning the concepts and the formulas for estimation, cautions about common errors, and sample exam items including solutions.

SPSS

General instructions for using SPSS are presented in Appendix D. The SPSS program can be used to compute **The Confidence Intervals** presented in this chapter. When SPSS is used to perform any of the three t tests presented in this book (single-sample, independent-measures, and repeated-measures), the output automatically includes a 95% confidence interval. If you want a different level of confidence (for example, 80%), you can click on the **Options** box and enter your own percentage before clicking the final **OK** for the t test. Detailed instructions for each of the three t tests are presented in the SPSS section at the end of the appropriate chapter (Chapter 9 for the single-sample test, Chapter 10 for the independent-measures test, and Chapter 11 for the repeated-measures test).

FOCUS ON PROBLEM SOLVING

1. Although hypothesis tests and estimation are similar in some respects, remember that they are separate statistical techniques. A hypothesis test is used to determine whether there is evidence for a treatment effect. Estimation is used to determine how much effect a treatment has.

2. When students perform a hypothesis test and estimation with the same set of data, a common error is to take the t statistic from the hypothesis test and use it in the estimation formula. For estimation, the t value is determined by the level of confidence and must be looked up in the appropriate table.

3. Now that you are familiar with several different formulas for hypothesis tests and estimation, one problem will be determining which formula is appropriate for each set of data. When the data consist of a single sample selected from a single population, the appropriate statistic is the single-sample t. For an independent-measures design, you will always have two separate samples. In a repeated-measures design, there is only one sample, but each individual is measured twice so that difference scores can be computed.

DEMONSTRATION 12.1

ESTIMATION WITH A SINGLE-SAMPLE t STATISTIC

A sample of $n = 16$ is randomly selected from a population with unknown parameters. The sample has a mean of $M = 11$ with $SS = 112$. We will use the sample data to estimate the value of the population mean using a point estimate and a 90% confidence interval. Because we are using a single sample to estimate the mean for a single population, the formula is

$$\mu = M \pm ts_M$$

STEP 1   Compute the sample variance and the estimated standard error, s_M.

Variance. The sample variance is computed for these data.

$$s^2 = \frac{SS}{n-1} = \frac{112}{16-1} = \frac{112}{15} = 7.47$$

Estimated standard error. We can now determine the estimated standard error.

$$s_M = \sqrt{\frac{s^2}{n}} = \sqrt{\frac{7.47}{16}} = \sqrt{0.467} = 0.68$$

STEP 2   Compute the point estimate for μ.

For a point estimate, we use $t = 0$. Using the estimation formula, we obtain

$$\mu = M \pm ts_M$$

$$= 11 \pm 0(0.68)$$

$$= 11 \pm 0 = 11$$

The point estimate for the population mean is $\mu = 11$.

STEP 3   Determine the confidence interval for μ.

For these data, we want the 90% confidence interval. Therefore, we will use a range of t values that form the middle 90% of the distribution. For this demonstration, degrees of freedom are

$$df = n - 1 = 16 - 1 = 15$$

If we are looking for the middle 90% of the distribution, then 10% ($p = 0.10$) would lie in both tails outside of the interval. To find the t values, we look up $p = 0.10$, two tails, for $df = 15$ in the t-distribution table. The t values for the 90% confidence interval are $t = \pm 1.753$. Using the estimation formula, one end of the confidence interval is

$$\mu = M - ts_M$$

$$= 11 - 1.753(0.68)$$

$$= 11 - 1.19 = 9.81$$

For the other end of the confidence interval, we obtain

$$\mu = M + ts_M$$

$$= 11 + 1.753(0.68)$$

$$= 11 + 1.19 = 12.19$$

Thus, the 90% confidence interval for μ is from 9.81 to 12.19.

DEMONSTRATION 12.2

ESTIMATION WITH THE INDEPENDENT-MEASURES t STATISTIC

Samples are taken from two school districts, and knowledge of American history is tested with a short questionnaire. For the following sample data, estimate the amount of mean difference between the students of these two districts. Specifically, provide a point estimate and a 95% confidence interval for $\mu_1 - \mu_2$.

District A	District B
$n = 4$	$n = 4$
$M = 18$	$M = 10$
$SS = 54$	$SS = 30$

STEP 1   Compute the pooled variance and the estimated standard error, $s_{(M_1 - M_2)}$.

Pooled variance. For pooled variance, we use the SS and df values from both samples. For district A, $df_1 = n_1 - 1 = 3$. For district B, $df_2 = n_2 - 1 = 3$. Pooled variance is

$$s_p^2 = \frac{SS_1 + SS_2}{df_1 + df_2} = \frac{54 + 30}{3 + 3} = \frac{84}{6} = 14$$

Estimated standard error. The estimated standard error for mean difference can now be calculated.

$$s_{(M_1 - M_2)} = \sqrt{\frac{s_p^2}{n_1} + \frac{s_p^2}{n_2}} = \sqrt{\frac{14}{4} + \frac{14}{4}} = \sqrt{3.5 + 3.5}$$
$$= \sqrt{7} = 2.65$$

STEP 2   Compute the point estimate for $\mu_1 - \mu_2$.

For the point estimate, we use a t value of zero. Using the sample means and estimated standard error from previous steps, we obtain

$$\mu_1 - \mu_2 = (M_1 - M_2) \pm ts_{(M_1 - M_2)}$$
$$= (18 - 10) \pm 0(2.65)$$
$$= 8 \pm 0 = 8$$

STEP 3   Determine the confidence interval for $\mu_1 - \mu_2$.

For the independent-measures t statistic, degrees of freedom are determined by

$$df = n_1 + n_2 - 2$$

For these data, df is

$$df = 4 + 4 - 2 = 6$$

With a 95% level of confidence, 5% of the distribution falls in the tails outside the interval. Therefore, we consult the t-distribution table for $p = 0.05$, two tails, with $df = 6$. The t values from the table are $t = \pm 2.447$. On one end of the confidence interval, the population mean difference is

$$\mu_1 - \mu_2 = (M_1 - M_2) - ts_{(M_1 - M_2)}$$
$$= (18 - 10) - 2.447(2.65)$$
$$= 8 - 6.48$$
$$= 1.52$$

On the other end of the confidence interval, the population mean difference is

$$\mu_1 - \mu_2 = (M_1 - M_2) + ts_{(M_1 - M_2)}$$
$$= (18 - 10) + 2.447(2.65)$$
$$= 8 + 6.48$$
$$= 14.48$$

Thus, the 95% confidence interval for population mean difference is from 1.52 to 14.48.

PROBLEMS

1. Explain how the purpose of estimation differs from the purpose of a hypothesis test.

2. Explain why it would *not* be reasonable to use estimation after a hypothesis test for which the decision was "fail to reject H_0."

3. Explain the trade off between precision and confidence for interval estimates.

4. A researcher has constructed an 80% confidence interval of $\mu = 45 \pm 8$, using a sample of $n = 25$ scores.
   a. What would happen to the width of the interval if the researcher had used a larger sample size? (Assume other factors are held constant.)
   b. What would happen to the width of the interval if the researcher had used 90% confidence instead of 80%?
   c. What would happen to the width of the interval if the sample variance increased? (Assume other factors are held constant.)

5. For the following studies, state whether estimation or hypothesis testing is required. Also, is an independent- or a repeated-measures *t* statistic appropriate?
   a. A researcher would like to determine whether there is any difference in the amount of homework between senior year of high school and freshman year of college. A sample of high school seniors is surveyed and, 1 year later, the students are surveyed again during their first year of college.
   b. A doctor would like to determine how much a new medication affects sleeping for insomniacs. A sample of people suffering from insomnia is obtained and half are given the new medication and half are given a placebo. Each person records the number of hours of sleep each night for a week.
   c. A teacher would like to know whether a new computer-based education program improves performance for high school math students. Two high school math classes are selected. One class gets the computer program and the other receives regular classroom instruction. At the end of the year, both classes take the same exam.
   d. A researcher would like to determine how much people's moods are affected by seasonal changes. A sample of 100 adults is obtained and each individual is given a mood-analysis questionnaire in the summer and again in the winter.

6. An elementary school principal would like to know how many hours the students spend watching TV each day. A sample of $n = 25$ children is selected, and a survey is sent to each child's parents. The results indicate an average of $M = 3.1$ hours per day with a standard deviation of $s = 3.0$.
   a. Make a point estimate of the mean number of hours of TV per day for the population of elementary school children
   b. Make an interval estimate of the mean so that you are 90% confident that the true mean is in your interval.

7. A sample of $n = 9$ scores is obtained from an unknown population. The sample mean is $M = 46$ with a standard deviation of $s = 6$.
   a. Use the sample data to make an 80% confidence interval estimate of the unknown population mean.
   b. Make a 90% confidence interval estimate of μ.
   c. Make a 95% confidence interval estimate of μ.
   d. In general, how is the width of a confidence interval related to the percentage of confidence?

8. Problem 16 in Chapter 9 described a study by Harlow (1959) in which infant monkeys were placed in cages with two artificial "mothers." One mother was made of wire mesh and had a bottle from which the infant could feed, and the second mother was made of soft terry cloth but did not provide any food. Data for a sample of $n = 9$ monkeys showed that the infants spent an average of $M = 15.3$ hours per day with $SS = 216$ with the terry cloth mother. Use the data to estimate how many hours per day would be spent with the terry cloth mother for the entire population of infant monkeys. Make a point estimate and an 80% confidence interval estimate of the population mean.

9. Problem 18 in Chapter 9 described a study showing that children who attended a high-quality daycare program before they enter school perform better on math tests than children who do not have daycare experience. The data showed that a sample of $n = 25$ children who attended day care before starting school had an average score of $M = 87$ with $SS = 1536$ on a standardized math test for which the population mean is $\mu = 81$.
   a. Use the sample data to make a point estimate of the average math test score for the population of children with a history of high-quality daycare. Based on your estimate, how much higher than average do these children score?
   b. Make an interval estimate the population mean score for the daycare children so that you are 95% confident that the true mean is in your interval.

10. Standardized measures seem to indicate that the average level of anxiety has increased gradually over the past 50 years (Twenge, 2000). In the 1950s, the

average score on the Child Manifest Anxiety Scale was $\mu = 15.1$. A sample of $n = 16$ of today's children produces a mean score of $M = 23.3$ with $SS = 240$.
  a. Based on the sample, make a point estimate of the population mean anxiety score for today's children.
  b. Make a 90% confidence interval estimate of today's population mean.

11. A common test of short-term memory requires participants to repeat a random string of digits that was presented a few seconds earlier. The number of digits is increased on each trial until the person begins to make mistakes. The longest string that can be reported accurately determines the participant's score. The following data were obtained from a sample of $n = 9$ participants. The scores are 4, 3, 12, 5, 7, 8, 10, 5, 9.
  a. Compute the mean and variance for the sample.
  b. Use the data to make a point estimate of the population mean.
  c. Make an 80% confidence interval estimate of μ.

12. A developmental psychologist would like to determine how much fine motor skill improves for children from age 3 to age 4. A random sample of $n = 15$ 3-year-old children and a second sample of $n = 15$ 4-year-olds are obtained. Each child is given a manual dexterity test that measures fine motor skills. The average score for the older children was $M = 40.6$ with $SS = 430$ and the average for the younger children was $M = 35.4$ with $SS = 410$. Using these data,
  a. Make a point estimate of the population mean difference in fine motor skills.
  b. Make an interval estimate so you are 95% confident that the real mean difference is in your interval.
  c. Make an interval estimate so you are 99% confident that the real mean difference is in your interval.
  d. Based on your answers from b and c, do these data indicate a significant change using a two-tailed test with $\alpha = .05$? Is the difference significant with $\alpha = .01$?

13. Example 10.1 (page 318) presented a research study demonstrating that high school students who regularly watched Sesame Street as children had higher grades than their peers who did not watch the program (Anderson, Huston, Wright, & Collins, 1998). One sample of $n = 10$ students who watched the program had an average grade of $M = 93$ with $SS = 200$. A second sample of $n = 10$ students who did not watch Sesame Street had an average grade of $M_D = 85$ with $SS = 160$. Use the data to estimate the mean difference in high school grades between the two populations of students represented by the two samples. Make a point estimate and a 90% confidence interval estimate.

14. There is some evidence suggesting that you are likely to improve your test score if you rethink and change

answers on a multiple-choice exam (Johnston, 1975). To examine this phenomenon, a teacher gave the same final exam to two sections of a psychology course. The students in one section were told to turn in their exams immediately after finishing, without changing any of their answers. In the other section, students were encouraged to reconsider each question and to change answers whenever they felt it was appropriate. The average score for the $n = 20$ students in the no-change section was $M = 74.2$ with $SS = 460.6$. The average for the $n = 20$ students in the change section was $M = 78.6$ with $SS = 512.2$.
  a. Make a point estimate of the population mean difference between the two test conditions.
  b. Make a 95% confidence interval estimate of the mean difference between the two conditions.
  c. Based on your answer to part b, would a two-tailed hypothesis test with $\alpha = .05$ conclude that there is a significant difference between the two conditions? (Is $\mu_1 - \mu_2 = 0$ an acceptable hypothesis?)

15. Research results suggest that children who show high levels of stimulation seeking as preschoolers, will also show higher IQs as young adolescents (Raine, Reynolds, Verables, and Mednick, 2002). In an attempt to evaluate the size of this effect, a researcher obtained measures of stimulation seeking for a sample of 30 3-year-old children. The children were separated into two groups consisting of the $n = 15$ highest and $n = 15$ lowest stimulation seekers. At age 11 each child was given an IQ test. The IQ scores for the high-stimulation-seeking group had a mean of $M = 113.6$ with $SS = 2780$ and the mean for the low group was $M = 102.4$ with $SS = 2470$. Use the sample data to estimate the population mean difference in IQ at age 11 between children who exhibited high stimulation seeking and children who showed low stimulation seeking at age 3. Make a point estimate and a 90% confidence interval estimate.

16. People's handshakes seem to be related to other aspects of their personalities (Chaplin, Phillips, Brown, Clanton, & Stein, 2000). For example, a firm handshake is negatively related to shyness. To measure the magnitude of this effect, a researcher obtained a sample of $n = 12$ people who were rated as having firm handshakes and $n = 8$ who were rated as having weak handshakes. Both groups had 50% males and 50% females. Each individual was given a personality test measuring shyness. The average score for the firm handshake group was $M = 43$ with $SS = 1170$ and the average for the weak handshake group was $M = 61$ with $SS = 990$. Use the sample data to estimate how much difference there is in shyness between people with firm handshakes and people with weak handshakes. Make a point estimate and a 95% confidence interval estimate.

17. Problem 14 in Chapter 10 described a study by Rozin, Bauer, and Catanese (2003) comparing attitudes toward eating for male and female college students. The results showed that females are much more concerned about weight gain and other negative aspects of eating than are men. The following data represent the results from one measure of concern about weight gain.

Males	Females
$n = 6$	$n = 9$
$M = 31$	$M = 45$
$SS = 490$	$SS = 680$

a. Use the data to make a point estimate of the population mean difference between males' and females' level of concern about weight gain.
b. Make a 90% confidence interval estimate of the population mean difference.

18. Problem 15 in Chapter 10 described a study by Laughlin, Zander, Knievel, and Tan (2003) showing that groups of three people consistently outperformed the best of the individuals working alone on a letters-to-numbers substitution problem. The following data describe the number of trials needed to solve each problem for a sample of $n = 15$ groups and a sample of $n = 15$ individuals.

Groups	Individuals
$n = 15$	$n = 15$
$M = 3.8$	$M = 4.7$
$SS = 15.4$	$SS = 18.2$

a. Make a point estimate of how much better groups perform than individuals for the general population. (Estimate the mean difference in the number of trials needed for a solution.)
b. Make an interval estimate of the mean difference so that you are 95% confident that the true mean difference is in your interval.
c. Based on your answer to part b, what would be the decision from a hypothesis test evaluating the significance of the mean difference using a two-tailed test with $\alpha = .05$. Explain your answer.

19. Problem 23 in Chapter 10 described a study demonstrating that elderly people who own dogs are less likely to pay visits to their doctors than are those who do not own pets. The data are presented again here.

Number of Doctor Visits in Past Year	
Control Group	Dog Owners
10	7
8	4
7	9
9	3
13	7
7	
6	
12	

Use these data to estimate how much difference there is in the number of doctor visits for people with dogs compared to those without.
a. Make a point estimate of the mean difference between the two populations.
b. Make a 90% confidence interval estimate of the population mean difference.

20. A researcher would like to estimate how much reaction time is affected by a common over-the-counter cold medication. The researcher measures reaction time for a sample of $n = 36$ particiants. Each is then given a dose of the cold medication and reaction time is measured again. For this sample, reaction time increased after the medication by an average of $M_D = 24$ milliseconds with $s = 8$.
a. Make a point estimate of the mean difference in reaction time caused by the medicine.
b. Make a 95% confidence interval estimate of the population mean difference.

21. A researcher would like to determine how physical endurance is affected by a common herbal supplement. The researcher measures endurance for a sample of $n = 9$ participants. Each individual is then given a 30-day supply of the herbs and, 1 month later, endurance is measured again. For this sample, endurance increased by an average of $M_D = 6$ points with $SS = 216$.
a. Make a point estimate of the population mean difference in endurance after taking the herbs.
b. Make a 95% confidence interval estimate of the population mean difference.

22. Problem 12 in Chapter 11 presented data from a study testing the effectiveness of Ritalin for increasing attention span for children with ADHD. Results from a sample of $n = 25$ children showed an average increase

of attention span of $M_D = 6.8$ minutes with a standard deviation of $s = 5.5$ after the children were administered the drug.

a. Make a point estimate of the population mean difference in attention span after children with ADHD take Ritalin.

b. Make an interval estimate of the population mean difference so that you are 99% confident that the true mean difference is in your interval.

23. A research study has demonstrated that self-hypnosis can be an effective treatment for allergies (Langewitz, Izakovic, & Wyler, 2005). The researchers recruited a sample of patients with moderate to severe allergic reactions. The patients were trained to focus their minds on a specific place, such as a ski slope in winter, where allergies did not bother them. The participants were then tested for allergic reactions to pollen under two circumstances: once without using self-hypnosis and once after using self-hypnosis. This sample of $n = 16$ averaged $M_D = 21$ points lower when they were using self-hypnosis with $SS = 1215$ for the difference scores. Use the data to estimate how much effect self-hypnosis would have on the general population of individuals with moderate to severe allergies. Make a point estimate and an 80% confidence interval estimate.

24. Problem 16 in Chapter 11 described a study examining the effect of eating oatmeal regularly on cholesterol. Cholesterol was measured before and after adding oatmeal to the diet of a sample of $n = 9$ participants. For this sample, cholesterol scores averaged $M_D = 16$ points lower with the oatmeal diet with $SS = 538$ for the difference scores. Use the data estimate how much effect oatmeal would have on the cholesterol level for the general population. Make a point estimate and a 95% confidence interval estimate of the population mean difference.

25. In Chapter 11, problem 22 presented data comparing the accuracy of Olympic marksmen for shots fired during heartbeat compared to shots fired between heartbeat. The data are presented again here.

Participant	During Heartbeats	Between Heartbeats
A	93	98
B	90	94
C	95	96
D	92	91
E	95	97
F	91	97
G	92	95
H	93	97

Use the data to estimate how much effect the vibration from a heartbeat affects a marksman's score. Make a point estimate and an 80% cofidence interval estimate of the population mean difference.

C H A P T E R

13

Introduction to Analysis of Variance

Tools You Will Need

The following items are considered essential background material for this chapter. If you doubt your knowledge of any of these items, you should review the appropriate chapter or section before proceeding.

- Variability (Chapter 4)
  - Sum of squares
  - Sample variance
  - Degrees of freedom
- Introduction to hypothesis testing (Chapter 8)
  - The logic of hypothesis testing
- Independent-measures t statistic (Chapter 10)

Preview

13.1   Introduction

13.2   The Logic of Analysis of Variance

13.3   ANOVA Notation and Formulas

13.4   The Distribution of F-Ratios

13.5   Examples of Hypothesis Testing and Effect Size with ANOVA

13.6   Post Hoc Tests

13.7   The Relationship Between ANOVA and t Tests

Summary

Focus on Problem Solving

Demonstrations 13.1 and 13.2

Problems

Preview

"But I read the chapter four times! How could I possibly have failed the exam!"

Most of you probably have had the experience of reading a textbook and suddenly realizing that you have no idea of what was said on the past few pages. Although you have been reading the words, your mind has wandered off, and the meaning of the words has never reached memory. In an influential paper on human memory, Craik and Lockhart (1972) proposed a *levels of processing* theory of memory that can account for this phenomenon. In general terms, this theory says that all perceptual and mental processing leaves behind a memory trace. However, the quality of the memory trace depends on the level or the depth of the processing. If you superficially skim the words in a book, your memory also will be superficial. On the other hand, when you think about the meaning of the words and try to understand what you are reading, the result will be a good, substantial memory that should serve you well on exams. In general, deeper processing results in better memory.

Rogers, Kuiper, and Kirker (1977) conducted an experiment demonstrating the effect of levels of processing. Participants in this experiment were shown lists of words and asked to answer questions about each word. The questions were designed to require different levels of processing, from superficial to deep. In one experimental condition, participants were simply asked to judge the physical characteristics of each printed word ("Is it printed in capital letters or small letters?"). A second condition asked about the sound of each word ("Does it rhyme with 'boat'?"). In a third condition, participants were required to process the meaning of each word ("Does it have the same meaning as 'attractive'?"). The final condition required participants to understand each word and relate its meaning to themselves ("Does this word describe you?"). After going through the complete list, all participants were given a surprise memory test. As you can see in Figure 13.1, deeper processing resulted in better memory. Remember that the participants were not trying to memorize the words; they were simply reading through the list answering questions. However, the more they processed and understood the words, the better they recalled the words on the test.

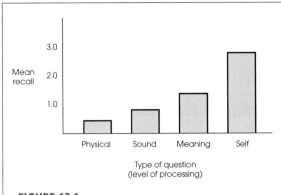

FIGURE 13.1

Mean recall as a function of the level of processing.

Rogers, T. B., Kuiper, N. A., & Kirker, W. S. (1977). Self-reference and the encoding of personal information. *Journal of Personality and Social Psychology, 35*, 677–688. Copyright (1977) by the American Psychological Association. Adapted by permission of the author.

The Problem: In terms of human memory, the Rogers, Kuiper, and Kirker experiment is notable because it demonstrates the importance of "self" in memory. You are most likely to remember material that is directly related to you. In terms of statistics, however, this study is notable because it compares four different treatment conditions in a single experiment. We now have four different means and need a hypothesis test to evaluate the mean differences. Unfortunately, the *t* tests introduce in Chapter 10 and 11 are limited to comparing only two treatments. A new hypothesis test is needed for this kind of data.

The Solution: In this chapter we introduce a new hypothesis test known as *analysis of variance* that is designed to evaluate the results from research studies producing two or more mean difference. Although "two or more" may seem like a small step from "two," this new hypothesis testing procedure provides researchers with a tremendous gain in experimental sophistication. In this chapter, and the two that follow, we examine some of the many applications of analysis of variance.

13.1   INTRODUCTION

Analysis of variance (ANOVA) is a hypothesis-testing procedure that is used to evaluate mean differences between two or more treatments (or populations). As with all inferential procedures, ANOVA uses sample data as the basis for drawing general conclusions about populations. It may appear that analysis of variance and *t* tests are simply two different ways of doing exactly the same job: testing for mean differences. In some respects, this is true—both tests use sample data to test hypotheses about population means. However, ANOVA has a tremendous advantage over *t* tests. Specifically, *t* tests are limited to situations in which there are only two treatments to compare. The major advantage of ANOVA is that it can be used to compare *two or more treatments*. Thus, ANOVA provides researchers with much greater flexibility in designing experiments and interpreting results.

Figure 13.2 shows a typical research situation for which analysis of variance would be used. Note that the study involves three samples representing three populations. The goal of the analysis is to determine whether the mean differences observed among the samples provide enough evidence to conclude that there are mean differences among the three populations. Specifically, we must decide between two interpretations:

1. There really are no differences between the populations (or treatments). The observed differences between the sample means are caused by random, unsystematic factors (sampling error) that differentiate one sample from another.

2. The populations (or treatments) really do have different means, and these population mean differences are responsible for causing systematic differences between the sample means.

You should recognize that these two interpretations correspond to the two hypotheses (null and alternative) that are part of the general hypothesis-testing procedure.

TERMINOLOGY IN ANALYSIS OF VARIANCE

Before we continue, it is necessary to introduce some of the terminology that is used to describe the research situation shown in Figure 13.2. Recall (from Chapter 1) that when a researcher manipulates a variable to create the treatment conditions in an

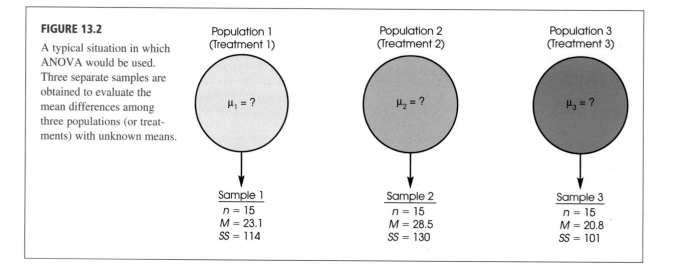

FIGURE 13.2

A typical situation in which ANOVA would be used. Three separate samples are obtained to evaluate the mean differences among three populations (or treatments) with unknown means.

experiment, the variable is called an independent variable. For example, Figure 13.2 could represent a study examining behavior in three different temperature conditions that have been created by the researcher. On the other hand, when a researcher uses a nonmanipulated variable to designate groups, the variable is called a *quasi-independent variable*. For example, the three groups in Figure 13.2 could represent 6-year-old, 8-year-old, and 10-year-old children. In the context of analysis of variance, an independent variable or a quasi-independent variable is called a *factor*. Thus, Figure 13.2 could represent a study in which temperature is the factor being evaluated or it could represent a study in which age is the factor being examined.

DEFINITION

In analysis of variance, the variable (independent or quasi-independent) that designates the groups being compared is called a **factor**.

In addition, the individual groups or treatment conditions that are used to make up a factor are called the *levels* of the factor. For example, a study that examined performance under three different temperature conditions would have three levels of temperature.

DEFINITION

The individual conditions or values that make up a factor are called the **levels** of the factor.

Like the *t* tests presented in Chapters 10 and 11, ANOVA can be used with either an independent-measures or a repeated-measures design. Recall that an independent-measures design means that there is a separate sample for each of the treatments (or populations) being compared. In a repeated-measures design, on the other hand, the same sample is tested in all of the different treatment conditions. In addition, ANOVA can be used to evaluate the results from a research study that involves more than one factor. For example, a researcher may want to compare two different therapy techniques, examining their immediate effectiveness as well as the persistence of their effectiveness over time. In this situation, the research study could involve two different groups of participants, one for each therapy, and measure each group at several different points in time. The structure of this design is shown in Figure 13.3. Notice that the study uses two factors, one independent-measures factor and one repeated-measures factor:

1. Therapy technique: A separate group is used for each technique (independent measures).

2. Time: Each group is tested at three different times (repeated measures).

In this case, the ANOVA would evaluate mean differences between the two therapies as well as mean differences between the scores obtained at different times. A study that combines two factors, like the one in Figure 13.3, is called a *two-factor design* or a *factorial design*.

The ability to combine different factors and to mix different designs within one study provides researchers with the flexibility to develop studies that address scientific questions that could not be answered by a single design using a single factor.

Although ANOVA can be used in a wide variety of research situations, this chapter introduces ANOVA in its simplest form. Specifically, we consider only *single-factor* designs. That is, we examine studies that have only one independent variable (or only one quasi-independent variable). Second, we consider only *independent-measures* designs; that is, studies that use a separate sample for each treatment condition. The basic logic and procedures that are presented in this chapter

FIGURE 13.3

A research design with two factors. The research study uses two factors: One factor uses two levels of therapy technique (I versus II), and the second factor uses three levels of time (before, after, and 6 months after). Also notice that the therapy factor uses two separate groups (independent measures) and the time factor uses the same group for all three levels (repeated measures).

THERAPY TECHNIQUE	TIME		
	Before Therapy	After Therapy	6 Months After Therapy
Therapy I (Group 1)	Scores for group 1 measured before Therapy I	Scores for group 1 measured after Therapy I	Scores for group 1 measured 6 months after Therapy I
Therapy II (Group 2)	Scores for group 2 measured before Therapy II	Scores for group 2 measured after Therapy II	Scores for group 2 measured 6 months after Therapy II

form the foundation for more complex applications of ANOVA. For example, in Chapter 14 we extend the analysis to single-factor, repeated-measures designs, and in Chapter 15 we consider two-factor designs. But for now, in this chapter, we will limit our discussion of ANOVA to *single-factor, independent-measures* research studies.

STATISTICAL HYPOTHESES FOR ANOVA

The following example introduces the statistical hypotheses for ANOVA. Suppose that a psychologist examined learning performance under three temperature conditions: 50°, 70°, and 90°. Three samples of subjects are selected, one sample for each treatment condition. The purpose of the study is to determine whether room temperature affects learning performance. In statistical terms, we want to decide between two hypotheses: the null hypothesis (H_0), which says that temperature has no effect, and the alternative hypothesis (H_1), which states that temperature does affect learning. In symbols, the null hypothesis states

$$H_0: \quad \mu_1 = \mu_2 = \mu_3$$

In words, the null hypothesis states that temperature has no effect on performance. That is, the population means for the three temperature conditions are all the same. In general, H_0 states that there is no *treatment effect*. Once again, notice that the hypotheses are always stated in terms of population parameters, even though we use sample data to test them.

For the alternative hypothesis, we may state that

$$H_1: \quad \text{At least one population mean is different from another.}$$

In general, H_1 states that the treatment conditions are not all the same; that is, there is a real treatment effect.

Notice that we have not given any specific alternative hypothesis. This is because many different alternatives are possible, and it would be tedious to list them all. One alternative, for example, would be that the first two populations are identical, but that the third is different. Another alternative states that the last two means are the same, but that the first is different. Other alternatives might be

$$H_1: \quad \mu_1 \neq \mu_2 \neq \mu_3 \qquad \text{(All three means are different.)}$$

$$H_1: \quad \mu_1 = \mu_3, \text{ but } \mu_2 \text{ is different}$$

We should point out that a researcher typically entertains only one (or at most a few) of these alternative hypotheses. Usually a theory or the outcomes of previous studies will dictate a specific prediction concerning the treatment effect. For the sake of simplicity, we will state a general alternative hypothesis rather than try to list all possible specific alternatives.

| THE TEST STATISTIC FOR ANOVA | The test statistic for ANOVA is very similar to the t statistics used in earlier chapters. For the t statistic, we first computed the standard error, which measures the standard distance that is reasonable to expect between two sample means if there is no treatment effect (that is, if H_0 is true). Then we computed the t statistic with the following structure |

$$t = \frac{\text{obtained difference between two sample means}}{\text{standard error (the difference expected with no treatment effect)}}$$

For ANOVA, the test statistic is called an F-*ratio* and has the following structure:

$$F = \frac{\text{variance (differences) between sample means}}{\text{variance (differences) expected with no treatment effect}}$$

Note that the F-ratio is based on *variance* instead of sample mean *difference*. The reason for this change is that it becomes impossible to compute a sample mean difference when there are more than two samples involved. For example, if there are only two samples and they have means of $M = 20$ and $M = 30$, then there is a 10-point difference between the sample means. Suppose, however, that we add a third sample with a mean of $M = 35$. Now how much difference is there between the sample means? It should be clear that the concept of *a mean difference* becomes difficult to define and impossible to calculate when there are more than two means. The solution to this problem is to use variance to define and measure the size of the differences among the sample means. Consider the following two sets of sample means:

Set 1	Set 2
$M_1 = 20$	$M_1 = 28$
$M_2 = 30$	$M_2 = 30$
$M_3 = 35$	$M_3 = 31$

If you compute the variance for the three numbers in each set, then the variance you obtain for set 1 is $s^2 = 58.33$ and the variance for set 2 is $s^2 = 2.33$. Notice that the two variances provide an accurate representation of the size of the differences. In set 1 there are relatively large differences between sample means and the variance is relatively large. In set 2 the mean differences are small and the variance is small. Thus, the variance in the numerator of the F-ratio provides a single number that describes how big the differences are among all of the sample means.

In much the same way, the variance in the denominator of the F-ratio and the standard error in the denominator of the t statistic are both measuring the mean differences that would be expected if there is no treatment effect. Remember that two samples are not expected to be identical even if there is no treatment effect whatsoever. In the independent-measures t statistic we computed an estimated standard error to measure how much difference is reasonable to expect between two sample means. In ANOVA, we will compute a variance to measure how big the mean differences should be if there is no treatment effect.

Finally, you should realize that the t statistic and the F-ratio provide the same basic information. In each case, the numerator of the ratio measures the actual difference obtained from the sample data, and the denominator measures the difference that would be expected if there were no treatment effect. With either the F-ratio or the t statistic, a large value provides evidence that the sample mean difference is more than would be expected by chance alone (Box 13.1).

BOX 13.1   TYPE I ERRORS AND MULTIPLE-HYPOTHESIS TESTS

If we already have t tests for comparing mean differences, you might wonder why ANOVA is necessary. Why create a whole new hypothesis-testing procedure that simply duplicates what the t tests can already do? The answer to this question is based in a concern about Type I errors.

Remember that each time you do a hypothesis test, you select an alpha level that determines the risk of a Type I error. With $\alpha = .05$, for example, there is a 5%, or a 1-in-20, risk of a Type I error. Thus, for every hypothesis test you do, there is a risk of a Type I error. The more tests you do, the more risk there is of a Type I error. For this reason, researchers often make a distinction between the *testwise* alpha level and the *experimentwise* alpha level. The testwise alpha level is simply the alpha level you select for each individual hypothesis test. The experimentwise alpha level is the total probability of a Type I error accumulated from all of the separate tests in the experiment. As the number of separate tests increases, so does the experimentwise alpha level.

For an experiment involving three treatments, you would need three separate t tests to compare all of the mean differences:

Test 1 compares treatment I versus treatment II.

Test 2 compares treatment I versus treatment III.

Test 3 compares treatment II versus treatment III.

If all tests use $\alpha = .05$, then there is a 5% risk of a Type I error for the first test, a 5% risk for the second test, and another 5% risk for the third test. The three separate tests accumulate to produce a relatively large experimentwise alpha level. The advantage of ANOVA is that it performs all three comparisons simultaneously in the same hypothesis test. Thus, no matter how many different means are being compared, ANOVA uses one test with one alpha level to evaluate the mean differences and thereby avoids the problem of an inflated experimentwise alpha level.

13.2   THE LOGIC OF ANALYSIS OF VARIANCE

The formulas and calculations required in ANOVA are somewhat complicated, but the logic that underlies the whole procedure is fairly straightforward. Therefore, this section gives a general picture of analysis of variance before we start looking at the details. We will introduce the logic of ANOVA with the help of the hypothetical data in Table 13.1.

TABLE 13.1

Hypothetical data from an experiment examining learning performance under three temperature conditions.*

Treatment 1 50° (Sample 1)	Treatment 2 70° (Sample 2)	Treatment 3 90° (Sample 3)
0	4	1
1	3	2
3	6	2
1	3	0
0	4	0
$M = 1$	$M = 4$	$M = 1$

*Note that there are three separate samples, with $n = 5$ in each sample. The dependent variable is the number of problems solved correctly.

These data represent the results of an independent-measures experiment comparing learning performance under three temperature conditions.

One obvious characteristic of the data in Table 13.1 is that the scores are not all the same. In everyday language, the scores are different; in statistical terms, the scores are variable. Our goal is to measure the amount of variability (the size of the differences) and to explain why the scores are different.

The first step is to determine the total variability for the entire set of data. To compute the total variability, we will combine all the scores from all the separate samples to obtain one general measure of variability for the complete experiment. Once we have measured the total variability, we can begin to break it apart into separate components. The word *analysis* means dividing into smaller parts. Because we are going to analyze variability, the process is called *analysis of variance*. This analysis process divides the total variability into two basic components.

1. Between-Treatments Variance. Looking at the data in Table 13.1, we clearly see that much of the variability in the scores results from general differences between treatment conditions. For example, the scores in the 70° condition tend to be much higher ($M = 4$) than the scores in the 50° condition ($M = 1$). We will calculate the variance between treatments to provide a measure of the overall differences between treatment conditions. Notice that the variance between treatments is really measuring the differences between sample means.

2. Within-Treatment Variance. In addition to the general differences between treatment conditions, there is variability within each sample. Looking again at Table 13.1, we see that the scores in the 70° condition are not all the same; they are variable. The within-treatments variance provides a measure of the variability inside each treatment condition.

Analyzing the total variability into these two components is the heart of analysis of variance. We will now examine each of the components in more detail.

BETWEEN-TREATMENTS VARIANCE

Remember that calculating variance is simply a method for measuring how big the differences are for a set of numbers. When you see the term *variance*, you can automatically translate it into the term *differences*. Thus, the *between-treatments variance* simply measures how much difference exists between the treatment conditions.

In addition to measuring the differences between treatments, the overall goal of ANOVA is to interpret the differences between treatments. Specifically, the purpose for the analysis is to distinguish between two alternative explanations:

1. The differences between treatments are not caused by any treatment effect but are simply the naturally occurring differences that exist between one sample and another. That is, the differences are the result of sampling error.

2. The differences between treatments are significantly greater than can be explained by sampling error; that is, the differences have been caused by the treatment effects.

Thus, there are always two possible explanations for the difference (or variance) that exists between treatments:

1. Systematic Differences Caused by the Treatments For the data in Table 13.1, the scores in sample 1 were obtained in a 50° room and the scores in sample 2 were obtained in a 70° room. It is possible that the difference between sample means is caused by the different temperatures.

2. Random, Unsystematic Differences If there is no treatment effect at all, you would still expect some differences between samples. The samples consist of different individuals with different scores and it should not be surprising that they have different means. In general these differences are random and unsystematic, and they cannot be explained by any action on the part of the researcher. Two primary sources are usually identified for these unpredictable differences.

a. *Individual differences:* Because each treatment condition has a different sample of participants, you would expect the individuals in one treatment to have different scores than the individuals in another treatment. Although it is reasonable to expect different samples to produce different scores, it is impossible to predict exactly what the differences will be.

b. *Experimental error:* Whenever you make a measurement, there is potential for some degree of error. Even if you measure the same individual under the same conditions, it is possible that you will obtain two different measurements. Again, these differences are unexplained and unpredictable.

Thus, when we compute the between-treatments variance, we are measuring differences that could be caused by a systematic treatment effect or could simply be random and unsystematic mean differences caused by sampling error. To demonstrate that there really is a treatment effect, we must establish that the differences between treatments are bigger than would be expected by sampling error alone. To accomplish this goal, we will determine how big the differences are when there is no systematic treatment effect; that is, we will measure how much difference (or variance) can be explained by random and unsystematic factors. To measure these differences, we compute the variance within treatments.

WITHIN-TREATMENTS VARIANCE

Inside each treatment condition, we have a set of individuals who all receive exactly the same treatment; that is, the researcher does not do anything that would cause these individuals to have different scores. In Table 13.1, for example, the data show that five individuals were tested in a 70° room (treatment 2). Although these five individuals were all treated exactly the same, their scores are different. Why are the scores different? The answer is that the differences within a treatment represent random and unsystematic differences.

Thus, the *within-treatments variance* provides a measure of how much difference is reasonable to expect from random and unsystematic factors. In particular, the within-treatments variance measures the naturally occurring differences that exist when there is no treatment effect; that is, how big the differences are when H_0 is true.

Figure 13.4 shows the overall ANOVA and identifies the sources of variability that are measured by each of the two basic components.

THE *F*-RATIO: THE TEST STATISTIC FOR ANOVA

Once we have analyzed the total variability into two basic components (between treatments and within treatments), we simply compare them. The comparison is made by computing a statistic called an *F-ratio*. For the independent-measures ANOVA, the *F*-ratio has the following structure:

$$F = \frac{\text{variance between treatments}}{\text{variance within treatments}} = \frac{\text{differences including any treatment effects}}{\text{differences with no treatment effects}} \quad (13.1)$$

When we express each component of variability in terms of its sources (see Figure 13.4), the structure of the *F*-ratio is

$$F = \frac{\text{systematic treatment effects} + \text{random, unsystematic differences}}{\text{random, unsystematic differences}} \quad (13.2)$$

The value obtained for the *F*-ratio helps determine whether any treatment effects exist. Consider the following two possibilities:

1. When there are no systematic treatment effects, the differences between treatments (numerator) are entirely caused by random, unsystematic factors. In this case, the numerator and the denominator of the *F*-ratio are both measuring random differences and should be roughly the same size. With the numerator and denominator roughly equal, the *F*-ratio should have a value around 1.00. In terms of the formula, when the treatment effect is zero, we obtain

$$F = \frac{0 + \text{random, unsystematic differences}}{\text{random, unsystematic differences}}$$

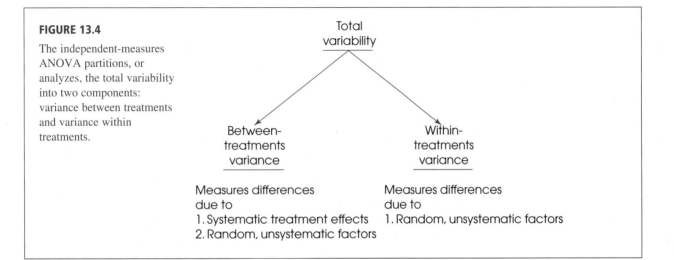

FIGURE 13.4

The independent-measures ANOVA partitions, or analyzes, the total variability into two components: variance between treatments and variance within treatments.

Total variability

Between-treatments variance

Within-treatments variance

Measures differences due to
1. Systematic treatment effects
2. Random, unsystematic factors

Measures differences due to
1. Random, unsystematic factors

Thus, an F-ratio near 1.00 indicates that the differences between treatments (numerator) are random and unsystematic, just like the differences in the denominator. With an F-ratio near 1.00, we conclude that there is no evidence to suggest that the treatment has any effect.

2. When the treatment does have an effect, causing systematic differences between samples, then the combination of systematic and random differences in the numerator should be larger than the random differences alone in the denominator. In this case, the numerator of the F-ratio should be noticeably larger than the denominator, and we should obtain an F-ratio noticeably larger than 1.00. Thus, a large F-ratio is evidence for the existence of systematic treatment effects; that is, there are significant differences between treatments.

In more general terms, the denominator of the F-ratio measures only random and unsystematic variability. For this reason, the denominator of the F-ratio is called the *error term*. The numerator of the F-ratio always includes the same unsystematic variability as in the error term, but it also includes any systematic differences caused by the treatment effect. The goal of ANOVA is to find out whether a treatment effect exists.

DEFINITION

For ANOVA, the denominator of the F-ratio is called the **error term**. The error term provides a measure of the variance due to random, unsystematic differences. When the treatment effect is zero (H_0 is true), the error term measures the same sources of variance as the numerator of the F-ratio, so the value of the F-ratio is expected to be nearly equal to 1.00.

LEARNING CHECK

1. ANOVA is a statistical procedure that compares two or more treatment conditions for differences in variance. (True or false?)

2. In ANOVA, what value is expected, on the average, for the F-ratio when the null hypothesis is true?

3. What happens to the value of the F-ratio if differences between treatments are increased? What happens to the F-ratio if variability inside the treatments is increased?

4. In ANOVA, the total variability is partitioned into two parts. What are these two variability components called, and how are they used in the F-ratio?

ANSWERS

1. False. Although ANOVA uses variance in the computations, the purpose of the test is to evaluate differences in *means* between treatments.

2. When H_0 is true, the expected value for the F-ratio is 1.00 because the top and bottom of the ratio are both measuring the same variance.

3. As differences between treatments increase, the F-ratio will increase. As variability within treatments increases, the F-ratio will decrease.

4. The two components are between-treatments variability and within-treatments variability. Between-treatments variance is the numerator of the F-ratio, and within-treatments variance is the denominator.

13.3 ANOVA NOTATION AND FORMULAS

Because ANOVA most often is used to examine data from more than two treatment conditions (and more than two samples), we need a notational system to help keep track of all the individual scores and totals. To help introduce this notational system, we use the hypothetical data from Table 13.1 again. The data are reproduced in Table 13.2 along with some of the notation and statistics that will be described.

1. The letter k is used to identify the number of treatment conditions—that is, the number of levels of the factor. For an independent-measures study, k also specifies the number of separate samples. For the data in Table 13.2, there are three treatments, so $k = 3$.

2. The number of scores in each treatment is identified by a lowercase letter n. For the example in Table 13.2, $n = 5$ for all the treatments. If the samples are of different sizes, you can identify a specific sample by using a subscript. For example, n_2 is the number of scores in treatment 2.

3. The total number of scores in the entire study is specified by a capital letter N. When all the samples are the same size (n is constant), $N = kn$. For the data in Table 13.2, there are $n = 5$ scores in each of the $k = 3$ treatments, so we have a total of $N = 3(5) = 15$ scores in the entire study.

Because ANOVA formulas require ΣX for each treatment and ΣX for the entire set of scores, we have introduced new notation (T and G) to help identify which ΣX is being used. Remember: T stands for *treatment total*, and G stands for *grand total*.

4. The sum of the scores (ΣX) for each treatment condition is identified by the capital letter T (for treatment total). The total for a specific treatment can be identified by adding a numerical subscript to the T. For example, the total for the second treatment in Table 13.2 is $T_2 = 20$.

5. The sum of all the scores in the research study (the grand total) is identified by G. You can compute G by adding up all N scores or by adding up the treatment totals: $G = \Sigma T$.

6. Although there is no new notation involved, we also have computed SS and M for each sample, and we have calculated ΣX^2 for the entire set of $N = 15$ scores in the study. These values are given in Table 13.2 and will be important in the formulas and calculations for ANOVA.

TABLE 13.2

Hypothetical data from an experiment examining learning performance under three temperature conditions.*

Temperature Conditions			
1 50°	2 70°	3 90°	
0	4	1	$\Sigma X^2 = 106$
1	3	2	$G = 30$
3	6	2	$N = 15$
1	3	0	$k = 3$
0	4	0	
$T_1 = 5$	$T_2 = 20$	$T_3 = 5$	
$SS_1 = 6$	$SS_2 = 6$	$SS_3 = 4$	
$n_1 = 5$	$n_2 = 5$	$n_3 = 5$	
$M_1 = 1$	$M_2 = 4$	$M_3 = 1$	

*Summary values and notation for an ANOVA are also presented.

Finally, we should note that there is no universally accepted notation for ANOVA. Although we are using Gs and Ts, for example, you may find that other sources use other symbols.

ANOVA FORMULAS

Because ANOVA requires extensive calculations and many formulas, one common problem for students is simply keeping track of the different formulas and numbers. Therefore, we will examine the general structure of the procedure and look at the organization of the calculations before we introduce the individual formulas.

1. The final calculation for ANOVA is the F-ratio, which is composed of two variances:

$$F = \frac{\text{variance between treatments}}{\text{variance within treatments}}$$

2. Recall that variance for sample data has been defined as

$$\text{sample variance} = s^2 = \frac{SS}{df}$$

Therefore, we need to compute an SS and a df for the variance between treatments (numerator of F), and we need another SS and df for the variance within treatments (denominator of F). To obtain these SS and df values, we must go through two separate analyses: First, compute SS for the total study, and analyze it into two components (between and within). Then compute df for the total study, and analyze it into two components (between and within).

Thus, the entire process of ANOVA will require nine calculations: three values for SS, three values for df, two variances (between and within), and a final F-ratio. However, these nine calculations are all logically related and are all directed toward finding the final F-ratio. Figure 13.5 shows the logical structure of ANOVA calculations.

FIGURE 13.5

The structure and sequence of calculations for the ANOVA.

The final goal for the ANOVA is an *F*-ratio	$F = \dfrac{\text{Variance between treatments}}{\text{Variance within treatments}}$
Each variance in the *F*-ratio is computed as *SS/df*	$\text{Variance between treatments} = \dfrac{SS\text{ between}}{df\text{ between}}$ $\text{Variance within treatments} = \dfrac{SS\text{ within}}{df\text{ within}}$
To obtain each of the *SS* and *df* values, the total variability is analyzed into the two components	SS total $\diagup \diagdown$ SS between SS within df total $\diagup \diagdown$ df between df within

ANALYSIS OF SUM OF SQUARES (*SS*)

The ANOVA requires that we first compute a total sum of squares and then partition this value into two components: between treatments and within treatments. This analysis is outlined in Figure 13.6. We will examine each of the three components separately.

1. Total Sum of Squares, SS_{total}. As the name implies, SS_{total} is the sum of squares for the entire set of *N* scores. We calculate this value by using the computational formula for *SS*:

$$SS = \Sigma X^2 - \frac{(\Sigma X)^2}{N}$$

To make this formula consistent with the ANOVA notation, we substitute the letter *G* in place of ΣX and obtain

$$SS_{total} = \Sigma X^2 - \frac{G^2}{N} \qquad (13.3)$$

Applying this formula to the set of data in Table 13.2, we obtain

$$SS_{total} = 106 - \frac{30^2}{15}$$
$$= 106 - 60$$
$$= 46$$

2. Within-Treatments Sum of Squares, $SS_{within\ treatments}$. Now we are looking at the variability inside each of the treatment conditions. We already have computed the *SS* within each of the three treatment conditions (Table 13.2): $SS_1 = 6$, $SS_2 = 6$, and $SS_3 = 4$. To find the overall within-treatment sum of squares, we simply add these values together:

$$SS_{within\ treatments} = \Sigma SS_{inside\ each\ treatment} \qquad (13.4)$$

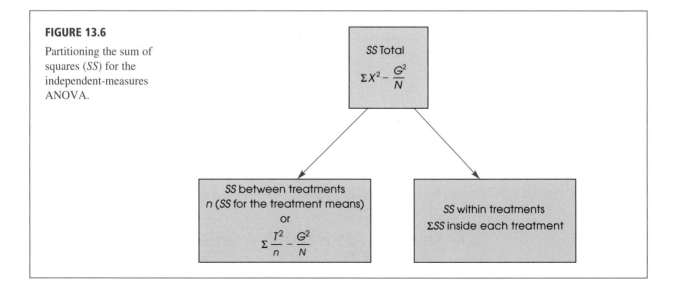

FIGURE 13.6

Partitioning the sum of squares (*SS*) for the independent-measures ANOVA.

For the data in Table 13.2, this formula gives

$$SS_{\text{within treatments}} = 6 + 6 + 4$$

$$= 16$$

3. Between-Treatments Sum of Squares, $SS_{\text{between treatments}}$. Before we introduce any equations for $SS_{\text{between treatments}}$, consider what we have found so far. The total variability for the data in Table 13.2 is $SS_{\text{total}} = 46$. We intend to partition this total into two parts (see Figure 13.6). One part, $SS_{\text{within treatments}}$, has been found to be equal to 16. This means that $SS_{\text{between treatments}}$ must be equal to 30 so that the two parts (16 and 30) to add up to the total (46). Thus, the value for $SS_{\text{between treatments}}$ can be found simply by subtraction:

The simplify the notation we will use the subscripts *between* and *within* in place of *between treatments* and *within treatments*.

$$SS_{\text{between}} = SS_{\text{total}} - SS_{\text{within}} \tag{13.5}$$

However, it is also possible to compute SS_{between} directly, then check your calculations by ensuring that the two components, between and within, add up to the total. Therefore, we present two different formulas for calculating SS_{between}.

Recall that the variability between treatments is measuring the differences between treatment means. Conceptually, the most direct way of measuring the amount of variability among the treatment means is to compute the sum of squares for the set of sample means. Specifically,

An alternative explanation for Equation 13.6 is presented in Box 13.2.

$$SS_{\text{between}} = n(SS_{\text{means}}) \tag{13.6}$$

For the data in Table 13.2, the sample means are 1, 4, and 1, and $n = 5$. The calculation of SS_{means} is shown as follows:

Mean (X)	X^2
1	1
4	16
1	1
6	18

$$SS = \Sigma X^2 - \frac{(\Sigma X)^2}{n}$$

$$= 18 - \frac{(6)^2}{3}$$

$$= 18 - 12 = 6$$

Thus, the sum of squares for the three treatment means in Table 13.2 is $SS_{\text{means}} = 6$, and each treatment contains $n = 5$ scores. Therefore,

$$SS_{\text{between}} = n(SS_{\text{means}}) = 5(6) = 30$$

The intent of Equation 13.6 is to emphasize that SS_{between} measures the mean differences. In this case, we are measuring the differences among three treatment means. However, Equation 13.6 can only be used when all of the samples are exactly the same size (equal ns), and the equation can be very awkward, especially when the treatment means are not whole numbers. Therefore, we also present a computational formula for SS_{between} that uses the treatment totals (T) instead of the treatment means.

$$SS_{\text{between}} = \Sigma \frac{T^2}{n} - \frac{G^2}{N} \tag{13.7}$$

In the formula, each treatment total (T) is squared and then divided by the number of scores in the treatment. These values are added to produce the first term in the formula. Next, the grand total (G) is squared and divided by the total number of scores in the entire study to produce the second term in the formula. Finally, the second term is subtracted from the first. The formula is demonstrated using the data from Table 13.2 as follows:

$$SS_{between} = \frac{5^2}{5} + \frac{20^2}{5} + \frac{5^2}{5} - \frac{30^2}{15}$$

$$= 5 + 80 + 5 - 60$$

$$= 90 - 60$$

$$= 30$$

BOX 13.2	AN ALTERNATIVE VIEW OF EQUATION 13.6

Equation 13.6 states that $SS_{between} = n(SS_{means})$. To demonstrate an alternative explanation for this equation, we begin by modifying the original set of scores in Table 13.2. The modification involves eliminating all of the within-treatments variability but maintaining the between-treatments differences. To accomplish this, each of the scores within each treatment is set equal to the treatment mean. For example, the mean for the 50° group is $M = 1$, so everyone in the group is assigned a score of $X = 1$. Note that the modified scores have exactly the same differences between treatments as the original scores; that is, the treatment totals and the treatment means are the same as in the original data.

Temperature Conditions			
1	2	3	
50°	70°	90°	
1	4	1	$N = 15$
1	4	1	$G = 30$
1	4	1	$\Sigma X^2 = 90$
1	4	1	
1	4	1	
$T = 5$	$T = 20$	$T = 5$	
$M = 1$	$M = 4$	$M = 1$	
$SS = 0$	$SS = 0$	$SS = 0$	

These scores have two important characteristics:
1. Because the data have no variability within treatments, all of the variability is now accounted for by differences between treatments. Therefore,

calculating the total variability is the same as calculating the between-treatments variability:

$$SS_{total} = SS_{between}$$

2. The set of scores now consists of the set of sample means, with each mean appearing $n = 5$ times. Therefore, the SS for this set of scores is equal to n times the SS for the set of sample means:

$$SS_{total} = n(SS_{means})$$

Combining these two observations produces Equation 13.6.

$$SS_{between} = n(SS_{means})$$

To demonstrate that this reasoning is sound, we will calculate SS_{total} for these scores and demonstrate that it is equal to $SS_{between}$ from the original analysis.

$$SS_{total} = \Sigma X^2 - \frac{G^2}{N}$$

$$= 90 - \frac{30^2}{15}$$

$$= 90 - 60$$

$$= 30$$

Again, the total variability for these scores is exactly the same as the between-treatments variability for the original set of scores.

Note that all three techniques (Equations 13.5, 13.6, and 13.7) produce the same result, $SS_{between} = 30$. Also note that the two components, between and within, add up to the total. For the data in Table 13.2,

$$SS_{total} = SS_{within} + SS_{between}$$

$$46 = 16 + 30$$

THE ANALYSIS OF DEGREES OF FREEDOM (*df*)

The analysis of degrees of freedom (*df*) follows the same pattern as the analysis of *SS*. First, we find *df* for the total set of *N* scores, and then we partition this value into two components: degrees of freedom between treatments and degrees of freedom within treatments. In computing degrees of freedom, there are two important considerations to keep in mind:

1. Each *df* value is associated with a specific *SS* value.

2. Normally, the value of *df* is obtained by counting the number of items that were used to calculate *SS* and then subtracting 1. For example, if you compute *SS* for a set of *n* scores, then $df = n - 1$.

With this in mind, we will examine the degrees of freedom for each part of the analysis.

1. **Total Degrees of Freedom, df_{total}.** To find the *df* associated with SS_{total}, you must first recall that this *SS* value measures variability for the entire set of *N* scores. Therefore, the *df* value is

$$df_{total} = N - 1 \tag{13.8}$$

For the data in Table 13.2, the total number of scores is $N = 15$, so the total degrees of freedom are

$$df_{total} = 15 - 1$$

$$= 14$$

2. **Within-Treatments Degrees of Freedom, df_{within}.** To find the *df* associated with SS_{within}, we must look at how this *SS* value is computed. Remember, we first find *SS* inside of each of the treatments and then add these values together. Each of the treatment *SS* values measures variability for the *n* scores in the treatment, so each *SS* will have $df = n - 1$. When all these individual treatment values are added together, we obtain

$$df_{within} = \Sigma(n - 1) = \Sigma df_{in\ each\ treatment} \tag{13.9}$$

For the experiment we have been considering, each treatment has $n = 5$ scores. This means there are $n - 1 = 4$ degrees of freedom inside each treatment. Because there are three different treatment conditions, this gives a total of 12 for the within-treatments degrees of freedom. Notice that this formula for *df* simply adds up the number of scores in each treatment (the *n* values) and subtracts 1 for each treatment. If these two stages are done separately, you obtain

$$df_{within} = N - k \tag{13.10}$$

(Adding up all the n values gives N. If you subtract 1 for each treatment, then altogether you have subtracted k because there are k treatments.) For the data in Table 13.2, $N = 15$ and $k = 3$, so

$$df_{within} = 15 - 3$$
$$= 12$$

3. Between-Treatments Degrees of Freedom, $df_{between}$. The df associated with $SS_{between}$ can be found by considering the SS formula. This SS formula measures the variability for the set of treatment totals (or means). To find $df_{between}$, simply count the number of T values and subtract 1. Because the number of treatments is specified by the letter k, the formula for df is

$$df_{between} = k - 1 \tag{13.11}$$

For the data in Table 13.2, there are three different treatment conditions (three T values), so the between-treatments degrees of freedom are computed as follows:

$$df_{between} = 3 - 1$$
$$= 2$$

Notice that the two parts we obtained from this analysis of degrees of freedom add up to equal the total degrees of freedom:

$$df_{total} = df_{within} + df_{between}$$
$$14 = 12 + 2$$

The complete analysis of degrees of freedom is shown in Figure 13.7.

As you are computing the SS and df values for ANOVA, keep in mind that the labels that are used for each value can help you to understand the formulas. Specifically,

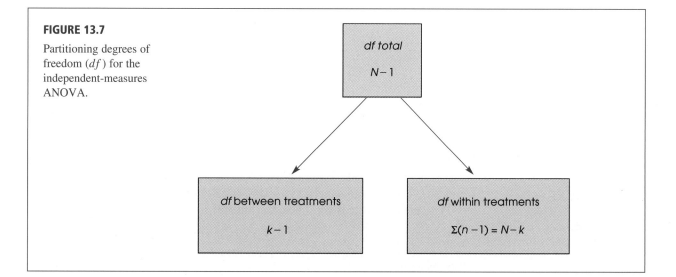

FIGURE 13.7

Partitioning degrees of freedom (df) for the independent-measures ANOVA.

1. The term **total** refers to the entire set of scores. We compute SS for the whole set of N scores, and the df value is simply $N - 1$.

2. The term **within treatments** refers to differences that exist inside the individual treatment conditions. Thus, we compute SS and df inside each of the separate treatments.

3. The term **between treatments** refers to differences from one treatment to another. With three treatments, for example, we are comparing three different means (or totals) and have $df = 3 - 1 = 2$.

CALCULATION OF VARIANCES
(*MS*) AND THE *F*-RATIO

The next step in the analysis of variance procedure is to compute the variance between treatments and the variance within treatments to calculate the F-ratio (see Figure 13.5).

In ANOVA, it is customary to use the term *mean square,* or simply *MS,* in place of the term *variance.* Recall (from Chapter 4) that variance is defined as the mean of the squared deviations. In the same way that we use SS to stand for the sum of the squared deviations, we now will use MS to stand for the mean of the squared deviations. For the final F-ratio we will need an MS (variance) between treatments for the numerator and an MS (variance) within treatments for the denominator. In each case

$$MS \text{ (variance)} = s^2 = \frac{SS}{df} \tag{13.12}$$

For the data we have been considering,

$$MS_{\text{between}} = s^2_{\text{between}} = \frac{SS_{\text{between}}}{df_{\text{between}}} = \frac{30}{2} = 15$$

and

$$MS_{\text{within}} = s^2_{\text{within}} = \frac{SS_{\text{within}}}{df_{\text{within}}} = \frac{16}{12} = 1.33$$

We now have a measure of the variance (or differences) between the treatments and a measure of the variance within the treatments. The F-ratio simply compares these two variances:

$$F = \frac{S^2_{\text{between}}}{S^2_{\text{within}}} = \frac{MS_{\text{between}}}{MS_{\text{within}}} \tag{13.13}$$

For the experiment we have been examining, the data give an F-ratio of

$$F = \frac{15}{1.33} = 11.28$$

It is useful to organize the results of the analysis in one table called an *ANOVA summary table.* The table shows the source of variability (between treatments, within treatments, and total variability), SS, df, MS, and F. For the previous computations, the ANOVA summary table is constructed as follows:

Source	SS	df	MS	
Between treatments	30	2	15	$F = 11.28$
Within treatments	16	12	1.33	
Total	46	14		

Although these tables are no longer used in published reports, they are a common part of computer printouts, and they do provide a concise method for presenting the results of an analysis. (Note that you can conveniently check your work: Adding the first two entries in the SS column $(30 + 16)$ yields the total SS. The same applies to the df column.) When using ANOVA, you might start with a blank ANOVA summary table and then fill in the values as they are calculated. With this method, you are less likely to "get lost" in the analysis, wondering what to do next.

For this example, the obtained value of $F = 11.28$ indicates that the numerator of the F-ratio is substantially bigger than the denominator. If you recall the conceptual structure of the F-ratio as presented in Equations 13.1 and 13.2, the F value we obtained indicates that the differences between treatments are more than 11 times bigger than what would be expected if there is no treatment effect. Stated in terms of the experimental variables, it appears that temperature does have an effect on learning performance. However, to properly evaluate the F-ratio, we must examine the F distribution.

LEARNING CHECK

1. Calculate SS_{total}, $SS_{between}$, and SS_{within} for the following set of data:

Treatment 1	Treatment 2	Treatment 3	
$n = 10$	$n = 10$	$n = 10$	$N = 30$
$T = 10$	$T = 20$	$T = 30$	$G = 60$
$SS = 27$	$SS = 16$	$SS = 23$	$\Sigma X^2 = 206$

2. A researcher uses an ANOVA to compare three treatment conditions with a sample of $n = 8$ in each treatment. For this analysis, find df_{total}, $df_{between}$, and df_{within}.

3. A researcher reports an F-ratio with $df_{between} = 2$ and $df_{within} = 30$ for an independent-measures ANOVA. How many treatment conditions were compared in the experiment? How many subjects participated in the experiment?

4. A researcher conducts an experiment comparing four treatment conditions with a separate sample of $n = 6$ in each treatment. An ANOVA is used to evaluate the data, and the results of the ANOVA are presented in the following table. Complete all missing values in the table. *Hint:* Begin with the values in the df column.

Source	SS	df	MS	
Between treatments	__	__	__	$F = $ ____
Within treatments	__	__	2	
Total	58	__		

ANSWERS

1. $SS_{total} = 86$; $SS_{between} = 20$; $SS_{within} = 66$

2. $df_{total} = 23$; $df_{between} = 2$; $df_{within} = 21$

3. There were 3 treatment conditions ($df_{between} = k - 1 = 2$). A total of $N = 33$ individuals participated ($df_{within} = 30 = N - k$).

4.

Source	SS	df	MS	
Between treatments	18	3	6	$F = 3.00$
Within treatments	40	20	2	
Total	58	23		

13.4 THE DISTRIBUTION OF *F*-RATIOS

In analysis of variance, the *F*-ratio is constructed so that the numerator and denominator of the ratio are measuring exactly the same variance when the null hypothesis is true (see Equation 13.2). In this situation, we expect the value of *F* to be around 1.00. The problem now is to define precisely what we mean by "around 1.00." What values are considered to be close to 1.00, and what values are far away? To answer this question, we need to look at all the possible *F* values—that is, the *distribution of F-ratios*.

Before we examine this distribution in detail, you should note two obvious characteristics:

1. Because *F*-ratios are computed from two variances (the numerator and denominator of the ratio), *F* values always are positive numbers. Remember that variance is always positive.

2. When H_0 is true, the numerator and denominator of the *F*-ratio are measuring the same variance. In this case, the two sample variances should be about the same size, so the ratio should be near 1. In other words, the distribution of *F*-ratios should pile up around 1.00.

With these two factors in mind, we can sketch the distribution of *F*-ratios. The distribution is cut off at zero (all positive values), piles up around 1.00, and then tapers off to the right (Figure 13.8). The exact shape of the *F* distribution depends on the degrees of freedom for the two variances in the *F*-ratio. You should recall that the precision of a sample variance depends on the number of scores or the degrees of freedom. In general, the variance for a large sample (large *df*) provides a more accurate estimate of the population variance. Because the precision of the *MS* values depends on *df*, the shape of the *F* distribution also will depend on the *df* values for the numerator and denominator of the *F*-ratio. With very large *df* values, nearly all the *F*-ratios will be clustered very near to 1.00. With the smaller *df* values, the *F* distribution is more spread out.

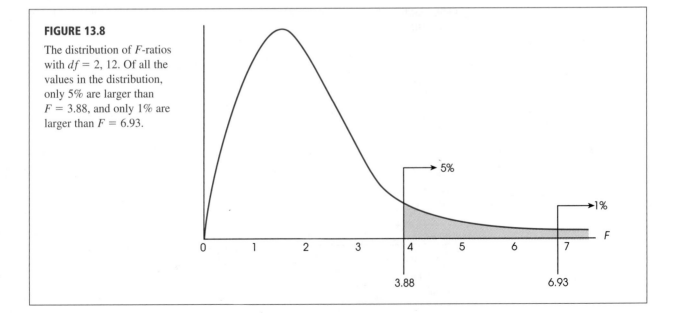

FIGURE 13.8

The distribution of *F*-ratios with $df = 2, 12$. Of all the values in the distribution, only 5% are larger than $F = 3.88$, and only 1% are larger than $F = 6.93$.

THE *F* DISTRIBUTION TABLE

For ANOVA, we expect F near 1.00 if H_0 is true, and we expect a large value for F if H_0 is not true. In the F distribution, we need to separate those values that are reasonably near 1.00 from the values that are significantly greater than 1.00. These critical values are presented in an F distribution table in Appendix B, page 731. A portion of the F distribution table is shown in Table 13.3. To use the table, you must know the df values for the F-ratio (numerator and denominator), and you must know the alpha level for the hypothesis test. It is customary for an F table to have the df values for the numerator of the F-ratio printed across the top of the table. The df values for the denominator of F are printed in a column on the left-hand side. For the temperature experiment we have been considering, the numerator of the F-ratio (between treatments) has $df = 2$, and the denominator of the F-ratio (within treatments) has $df = 12$. This F-ratio is said to have "degrees of freedom equal to 2 and 12." The degrees of freedom would be written as $df = 2, 12$. To use the table, you would first find $df = 2$ across the top of the table and $df = 12$ in the first column. When you line up these two values, they point to a pair of numbers in the middle of the table. These numbers give the critical cutoffs for $\alpha = .05$ and $\alpha = .01$. With $df = 2, 12$, for example, the numbers in the table are 3.88 and 6.93. These values indicate that the most unlikely 5% of the distribution ($\alpha = .05$) begins at a value of 3.88. The most extreme 1% of the distribution begins at a value of 6.93 (see Figure 13.8).

In the temperature experiment, we obtained an F-ratio of 11.28. According to the critical cutoffs in Figure 13.8, this value is extremely unlikely (it is in the most extreme 1%). Therefore, we would reject H_0 with α set at either .05 or .01 and conclude that temperature does have a significant effect on learning performance.

TABLE 13.3

A portion of the F distribution table. Entries in roman type are critical values for the .05 level of significance, and bold type values are for the .01 level of significance. The critical values for $df = 2, 12$ have been highlighted (see text).

Degrees of Freedom: Denominator	Degrees of Freedom: Numerator					
	1	2	3	4	5	6
10	4.96	4.10	3.71	3.48	3.33	3.22
	10.04	**7.56**	**6.55**	**5.99**	**5.64**	**5.39**
11	4.84	3.98	3.59	3.36	3.20	3.09
	9.65	**7.20**	**6.22**	**5.67**	**5.32**	**5.07**
12	4.75	3.88	3.49	3.26	3.11	3.00
	9.33	**6.93**	**5.95**	**5.41**	**5.06**	**4.82**
13	4.67	3.80	3.41	3.18	3.02	2.92
	9.07	**6.70**	**5.74**	**5.20**	**4.86**	**4.62**
14	4.60	3.74	3.34	3.11	2.96	2.85
	8.86	**6.51**	**5.56**	**5.03**	**4.69**	**4.46**

LEARNING CHECK

1. A researcher obtains $F = 4.18$ with $df = 2, 15$. Is this value sufficient to reject H_0 with $\alpha = .05$? Is it big enough to reject H_0 if $\alpha = .01$?

2. With $\alpha = .05$, what value forms the boundary for the critical region in the distribution of F-ratios with $df = 2, 24$?

ANSWERS

1. For $\alpha = .05$, the critical value is 3.68 and you should reject H_0. For $\alpha = .01$, the critical value is 6.36 and you should fail to reject H_0.

2. The critical value is 3.40.

13.5 EXAMPLES OF HYPOTHESIS TESTING AND EFFECT SIZE WITH ANOVA

Although we have now seen all the individual components of ANOVA, the following example demonstrates the complete ANOVA process using the standard four-step procedure for hypothesis testing.

EXAMPLE 13.1   The data depicted in Table 13.4 were obtained from an independent-measures experiment designed to measure the effectiveness of three pain relievers (A, B, and C). A fourth group that received a placebo (sugar pill) also was tested.

TABLE 13.4

The effect of drug treatment on the amount of time (in seconds) a stimulus is endured.

Placebo	Drug A	Drug B	Drug C	
3	4	6	7	$N = 20$
0	3	3	6	$G = 60$
2	1	4	5	$\Sigma X^2 = 262$
0	1	3	4	
0	1	4	3	
$T = 5$	$T = 10$	$T = 20$	$T = 25$	
$SS = 8$	$SS = 8$	$SS = 6$	$SS = 10$	

The purpose of the analysis is to determine whether these sample data provide evidence of any significant differences among the four drugs. The dependent variable is the amount of time (in seconds) that subjects can withstand a painfully hot stimulus.

Before we begin the hypothesis test, note that we have already computed several summary statistics for the data in Table 13.4. Specifically, the treatment totals (T) and SS values are shown for each sample, and the grand total (G) as well as N and ΣX^2 are shown for the entire set of data. Having these summary values simplifies the computations in the hypothesis test, and we suggest that you always compute these summary statistics before you begin an ANOVA.

STEP 1   The first step is to state the hypotheses and select an alpha level:

$$H_0: \quad \mu_1 = \mu_2 = \mu_3 = \mu_4 \quad \text{(There is no treatment effect.)}$$

$$H_1: \quad \text{At least one of the treatment means is different.}$$

We will use $\alpha = .05$.

STEP 2   To locate the critical region for the F-ratio, we first must determine degrees of freedom for $MS_{\text{between treatments}}$ and $MS_{\text{within treatments}}$ (the numerator and denominator of F). For these data, the total degrees of freedom are

Often it is easier to postpone finding the critical region until after step 3, where you compute the df values as part of the calculations for the F-ratio.

$$df_{\text{total}} = N - 1$$

$$= 20 - 1$$

$$= 19$$

Analyzing this total into two components, we obtain

$$df_{between} = k - 1$$
$$= 4 - 1$$
$$= 3$$

$$df_{within} = \Sigma df_{inside\ each\ treatment} = 4 + 4 + 4 + 4 = 16$$

The F-ratio for these data have $df = 3, 16$. The distribution of all the possible F-ratios with $df = 3, 16$ is presented in Figure 13.9. Note that F-ratios larger than 3.24 are extremely rare ($p < .05$) if H_0 is true and, therefore, will form the critical region for the test.

STEP 3   To compute the F-ratio for these data, you must go through the series of calculations outlined in Figure 13.5. The calculations can be summarized as follows:

   a. Analyze the SS to obtain $SS_{between}$ and SS_{within}.

   b. Use the SS values and the df values (from step 2) to calculate the two variances, $MS_{between}$ and MS_{within}.

   c. Finally, use the two MS values (variances) to compute the F-ratio.

Analysis of SS. First, we compute the total SS and then the two components, as indicated in Figure 13.6.

SS_{total} is simply the SS for the total set of $N = 20$ scores.

$$SS_{total} = \Sigma X^2 - \frac{G^2}{N}$$
$$= 262 - \frac{60^2}{20}$$
$$= 262 - 180$$
$$= 82$$

SS_{within} combines the SS values from inside each of the treatment conditions.

$$SS_{within} = \Sigma SS_{inside\ each\ treatment} = 8 + 8 + 6 + 10 = 32$$

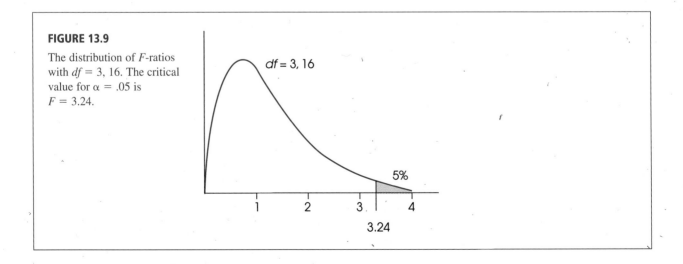

FIGURE 13.9

The distribution of F-ratios with $df = 3, 16$. The critical value for $\alpha = .05$ is $F = 3.24$.

$SS_{between}$ measures the differences between the four treatment means (or treatment totals).

Because we have already calculated SS_{total} and SS_{within}, the simplest way to obtain $SS_{between}$ is by subtraction (Equation 13.5).

$$SS_{between} = SS_{total} - SS_{within}$$
$$= 82 - 32$$
$$= 50$$

However we will also compute the value using the computational formula (Equation 13.7).

$$SS_{between} = \Sigma \frac{T^2}{n} - \frac{G^2}{N}$$
$$= \frac{5^2}{5} + \frac{10^2}{5} + \frac{20^2}{5} + \frac{25^2}{5} - \frac{60^2}{20}$$
$$= 5 + 20 + 80 + 125 - 180$$
$$= 50$$

Calculation of mean squares. Because we already found the *df* values (Step 2), we now can compute the variance or *MS* value for each of the two components.

The *df* values ($df_{between} = 3$ and $df_{within} = 16$) were computed in step 2 when we located the critical region.

$$MS_{between} = \frac{SS_{between}}{df_{between}} = \frac{50}{3} = 16.67$$

$$MS_{within} = \frac{SS_{within}}{df_{within}} = \frac{32}{16} = 2.00$$

Calculation of F. We compute the *F*-ratio:

$$F = \frac{MS_{between}}{MS_{within}} = \frac{16.67}{2.00} = 8.33$$

STEP 4   Finally, we make the statistical decision. The *F* value we obtained, $F = 8.33$, is in the critical region (see Figure 13.9). It is very unlikely ($p < .05$) that we will obtain a value this large if H_0 is true. Therefore, we reject H_0 and conclude that there is a significant treatment effect.

Example 13.1 demonstrated the complete, step-by-step application of the ANOVA procedure. There are two additional points that can be made using this example.

First, you should look carefully at the statistical decision. We have rejected H_0 and concluded that not all the treatments are the same. But we have not determined which ones are different. Is drug A different from the placebo? Is drug A different from drug B? Unfortunately, these questions remain unanswered. We do know that at least one difference exists (we rejected H_0), but additional analysis is necessary to find out exactly where this difference is. We address this problem in Section 13.6.

Second, as noted earlier, all of the components of the analysis (the *SS*, *df*, *MS*, and *F*) can be presented together in one summary table. The summary table for the analysis in Example 13.1 is as follows:

Source	SS	df	MS	
Between treatments	50	3	16.67	$F = 8.33$
Within treatments	32	16	2.00	
Total	82	19		

Although these tables are very useful for organizing the components of an ANOVA, they are not commonly used in published reports. The current method for reporting the results from an ANOVA is presented on page 419.

MEASURING EFFECT SIZE FOR ANOVA

As we noted previously, a *significant* mean difference simply indicates that the difference observed in the sample data is very unlikely to have occurred just by chance. Thus, the term significant does not necessarily mean *large*, it simply means larger than expected by chance. To provide an indication of how large the effect actually is, it is recommended that researchers report a measure of effect size in addition to the measure of significance.

For ANOVA, the simplest and most direct way to measure effect size is to compute r^2, the percentage of variance accounted for. In simpler terms, r^2 measures how much of the differences between scores is accounted for by the differences between treatments. For ANOVA, the calculation and the concept of r^2 is extremely straightforward.

$SS_{between}$ measures the variability accounted for by the treatment differences, and SS_{total} measures the total variability. Therefore,

$$r^2 = \frac{SS_{between}}{SS_{total}} \tag{13.14}$$

For the data from Example 13.1, we obtain

$$r^2 = \frac{50}{82} = 0.61 \quad (\text{or } 61\%)$$

In published reports of ANOVA results, the percentage of variance accounted for by the treatment effect is usually called η^2 (the Greek letter *eta squared*) instead of using r^2. Thus, for the study in Example 13.1, $\eta^2 = 0.61$.

MORE ABOUT r^2 AND η^2

The r^2 value that is used to measure effect size for t tests and the η^2 that is used for ANOVA are both described as measuring a percentage of variance. However, as we saw with the t tests, they actually measure a percentage of the sum of squared deviations (SS). This concept is demonstrated using the data from Example 13.1.

Figure 13.10(a) shows the four samples from Example 13.1 as one distribution of scores. If the null hypothesis is true and there are no differences among the four treatments, then the four samples are all representing the same population mean ($\mu_1 = \mu_2 = \mu_3 = \mu_4$) and can be combined to form a single set of $n = 20$ scores with a mean of $M = 3$.

For this example, however, we did find a treatment effect that causes scores to be higher for participants receiving drugs B and C, and causes scores to be lower for those receiving drug A and the placebo. Notice that the treatment effects tend to push the scores away from $M = 3$ and, therefore, increase the variability. To measure how much the variability is influenced by the treatment effects, we calculate SS for the

FIGURE 13.10

(a) The four samples from Example 13.1 combined as a single distribution of scores with a mean of $M = 3$. Note that the treatment effects tend to move the samples away from $M = 3$ which increases the variability. For these scores, $SS = 82$. (b) The same four samples after the treatment effects have been removed so that all four samples now have $M = 3$. For these scores, $SS = 32$. Removing the treatment effects reduces the variability by 50 points (from $SS = 82$ to $SS = 32$).

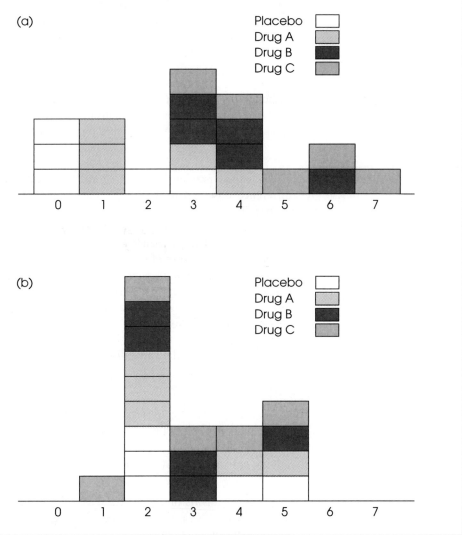

original scores, then remove the treatment effects and compute SS again. The difference between the two SS values describes how much the treatment effects contribute to the variability of the scores.

We have already computed SS for the original set of scores, and obtained $SS_{total} = 82$. This total includes differences caused by the treatment effects. To eliminate the treatment effects, we adjust the scores so that all four samples end up with the same mean. To accomplish this, we add 2 points to each score in the placebo condition, add 1 point to each individual in the drug A condition, subtract 1 point for each of the drug B scores, and subtract 2 points for each of the drug C scores. In the resulting data, all four samples have $M = 3$, so there are no treatment differences. The distribution of scores with the treatment effects removed is shown in Figure 13.10(b). These scores have $SS_{total} = 32$. (Because the mean is a whole number, you should be able to verify the calculation easily.) You should also recognize that $SS = 32$ is exactly the value we obtained for SS_{within} in the ANOVA. After removing the variability between treatments, all of the remaining variability is within treatments.

Finally, we compare the SS values obtained for the two sets of scores. For the distribution in Figure 13.10(a), which includes the treatment effects we obtained $SS_{total} = 82$. After the treatment effects were removed [Figure 13.10(b)], we obtained $SS_{total} = 32$. The difference between these values, $82 - 32 = 50$, is the amount of variability contributed by the treatments. Expressed as a percentage of the total variability, the amount contributed by the treatments is

$$\text{percentage from the treatments} = \frac{SS \text{ from the treatments}}{\text{total } SS} = \frac{50}{82} = 0.61 \text{ or } 61\%$$

You should recognize this as exactly the same value (and exactly the same calculation) that we used to compute η^2 earlier.

IN THE LITERATURE:
REPORTING THE RESULTS OF ANALYSIS OF VARIANCE

The APA format for reporting the results of ANOVA begins with a presentation of the treatment means and standard deviations in the narrative of the article, a table or a graph. These descriptive statistics are not needed in the calculations of the actual ANOVA, but you can easily determine the treatment means from n and $T(M = T/n)$ and the standard deviations from the SS of each treatment [$s = \sqrt{SS/(n - 1)}$]. Next, report the results of the ANOVA. For the study described in Example 13.1, the report might state the following:

The means and standard deviations are presented in Table 1. The analysis of variance revealed a significant difference, $F(3, 16) = 8.33$, $p < .05$, $\eta^2 = 0.61$.

TABLE 1

Amount of time (seconds) the stimulus was endured

	Placebo	Drug A	Drug B	Drug C
M	1.00	2.00	4.00	5.00
SD	1.41	1.41	1.22	1.58

Note how the F-ratio is reported. In this example, degrees of freedom for between and within treatments are $df = 3, 16$, respectively. These values are placed in parentheses immediately following the symbol F. Next, the calculated value for F is reported, followed by the probability of committing a Type I error (the alpha level) and the measure of effect size.

When an ANOVA is done using a computer program, the F-ratio is usually accompanied by an exact value for p. The data from Example 13.1 were analyzed using the SPSS program (see Resources at the end of this chapter) and the computer output included a significance level of $p = .001$. Using the exact p value from the computer output, the research report would conclude, "The analysis of variance revealed a significant difference, $F(3, 16) = 8.33$, $p = .001$, $\eta^2 = 0.61$." ❑

A CONCEPTUAL VIEW OF ANOVA

Because analysis of variance requires relatively complex calculations, students encountering this statistical technique for the first time often tend to be overwhelmed by the formulas and arithmetic and lose sight of the general purpose for the analysis. The following two examples are intended to minimize the role of the formulas and shift attention back to the conceptual goal of the ANOVA process.

EXAMPLE 13.2

The following data represent the outcome of an experiment using two separate samples to evaluate the mean difference between two treatment conditions. Take a minute to look at the data and, without doing any calculations, try to predict the outcome of an ANOVA for these values. Specifically, predict what values should be obtained for $SS_{between}$, $MS_{between}$, and the F-ratio. If you do not "see" the answer after 20 or 30 seconds, try reading the hints that follow the data.

Treatment I	Treatment II	
4	2	$N = 8$
0	1	$G = 16$
1	0	$\Sigma X^2 = 56$
3	5	
$T = 8$	$T = 8$	
$SS = 10$	$SS = 14$	

If you are having trouble predicting the outcome of the ANOVA, read the following hints, and then go back and look at the data.

Hint 1: Remember: $SS_{between}$ and $MS_{between}$ provide a measure of how much difference there is *between* treatment conditions.

Hint 2: Find the mean or total (T) for each treatment, and determine how much difference there is between the two treatments.

You should realize by now that the data have been constructed so that there is zero difference between treatments. The two sample means (and totals) are identical, so $SS_{between} = 0$, $MS_{between} = 0$, and the F-ratio is zero.

Conceptually, the numerator of the F-ratio always measures how much difference exists between treatments. In Example 13.2, we constructed an extreme set of scores with zero difference. However, you should be able to look at any set of data and quickly compare the means (or totals) to determine whether there are big differences between treatments or small differences between treatments.

Being able to estimate the magnitude of between-treatment differences is a good first step in understanding ANOVA and should help you to predict the outcome of an analysis of variance. However, the *between-treatment* differences are only one part of the analysis. You must also understand the *within-treatment* differences that form the denominator of the F-ratio. The following example is intended to demonstrate the concepts underlying SS_{within} and MS_{within}. In addition, the example should give you a better understanding of how the between-treatment differences and the within-treatment differences act together within the ANOVA.

EXAMPLE 13.3

The purpose of this example is to present a visual image for the concepts of between-treatments variability and within-treatments variability. In this example, we compare two hypothetical outcomes for the same experiment. In each case, the experiment uses two separate samples to evaluate the mean difference between two treatments. The following data represent the two outcomes, which we call experiment A and experiment B.

	Experiment A		Experiment B	
	Treatment		Treatment	
	I	II	I	II
	8	12	4	12
	8	13	11	9
	7	12	2	20
	9	11	17	6
	8	13	0	16
	9	12	8	18
	7	11	14	3
	$M = 8$	$M = 12$	$M = 8$	$M = 12$
	$s = 0.82$	$s = 0.82$	$s = 6.35$	$s = 6.35$

The data from experiment A are displayed in a frequency distribution graph in Figure 13.11(a). Notice that there is a 4-point difference between the treatment means

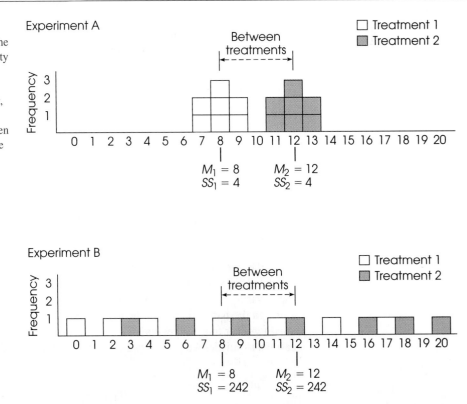

FIGURE 13.11

A visual representation of the between-treatments variability and the within-treatments variability that form the numerator and denominator, respectively, of the F-ratio. In (a), the difference between treatments is relatively large and easy to see. In (b), the same 4-point difference between treatments is relatively small and is overwhelmed by the within-treatments variability.

($M_1 = 8$ and $M_2 = 12$). This is the *between-treatments* difference that contributes to the numerator of the *F*-ratio. Also notice that the scores in each treatment are clustered close around the mean, indicating that the variance inside each treatment is relatively small. This is the *within-treatments* variance that contributes to the denominator of the *F*-ratio. Finally, you should realize that it is easy to see the mean difference between the two samples. The fact that there is a clear mean difference between the two treatments is confirmed by computing the *F*-ratio for experiment A.

$$F = \frac{\text{between-treatments difference}}{\text{within-treatments differences}} = \frac{MS_{\text{between}}}{MS_{\text{within}}} = \frac{56}{0.667} = 83.96$$

An *F*-ratio of $F = 83.96$ is sufficient to reject the null hypothesis with $\alpha = .05$, so we conclude that there is a significant difference between the two treatments.

Now consider the data from experiment B, which are shown in Figure 13.11(b) and present a very different picture. This experiment has the same 4-point difference between treatment means that we found in experiment A ($M_1 = 8$ and $M_2 = 12$). However, for these data the scores in each treatment are scattered across the entire scale, indicating relatively large variance inside each treatment. Notice that the large variance within treatments makes it almost impossible to see the mean difference between treatments. The *F*-ratio for these data confirms that there is no clear mean difference between treatments.

$$F = \frac{\text{between-treatments difference}}{\text{within-treatments differences}} = \frac{MS_{\text{between}}}{MS_{\text{within}}} = \frac{56}{40.33} = 1.39$$

For experiment B, the *F*-ratio is not large enough to reject the null hypothesis, so we conclude that there is no significant difference between the two treatments. Once again, the statistical conclusion is consistent with the appearance of the data in Figure 13.11(b). Looking at the figure, we see that the scores from the two samples appear to be intermixed randomly with no clear distinction between treatments.

As a final point, note that the denominator of the *F*-ratio, MS_{within}, is a measure of the variability (or variance) within each of the separate samples. As we have noted in previous chapters, high variability makes it difficult to see any patterns in the data. In Figure 13.11(a), the 4-point mean difference between treatments is easy to see because the sample variability is small. In Figure 13.11(b), the 4-point difference gets lost because the sample variability is large.

MS_{within} AND POOLED VARIANCE

You may have recognized that the two research outcomes presented in Example 13.3 are similar to those presented earlier in Example 10.4 in Chapter 10. Both examples are intended to demonstrate the role of variance in a hypothesis test. Both examples show that large values for sample variance can obscure any patterns in the data and reduce the potential for finding significant differences between means.

For the independent-measures *t* statistic in Chapter 10, the sample variance contributed directly to the standard error in the bottom of the *t* formula. Now, the sample variance contributes directly to the value of MS_{within} in the bottom of the *F*-ratio. In the *t*-statistic and in the *F*-ratio the variances from the separate samples are pooled together to create one average value for sample variance. For the independent-measures *t* statistic, we pooled two samples together to compute

$$\text{pooled variance} = s_p^2 = \frac{SS_1 + SS_2}{df_1 + df_2}$$

Now, in ANOVA, we are combining two or more samples to calculate

$$MS_{within} = \frac{SS_{within}}{df_{within}} = \frac{\Sigma SS}{\Sigma df} = \frac{SS_1 + SS_2 + SS_3 + \cdots}{df_1 + df_2 + df_3 + \cdots}$$

Notice that the concept of pooled variance is the same whether you have exactly two samples or more than two samples. In either case, you simply add the SS values and divide by the sum of the df values. The result is an average of all the different sample variances. As always, when variance is large, it means that the scores are scattered far and wide and it is difficult to identify sample means or mean differences (see Figure 13.11). In general, you can think of variance as measuring the amount of "noise" or "confusion" in the data. With large variance there is a lot of noise and confusion and it is difficult to see any clear patterns.

Although Examples 13.2 and 13.3 present somewhat simplified demonstrations with exaggerated data, the general point of the examples is to help you *see* what happens when you perform an ANOVA. Specifically:

1. The numerator of the F-ratio ($MS_{between}$) measures how much difference exists between the treatment means. The bigger the mean differences, the bigger is the F-ratio.

2. The denominator of the F-ratio (MS_{within}) measures the variance of the scores inside each treatment; that is, the variance for each of the separate samples. As seen in Example 13.3, large sample variances can make it difficult to see a mean difference. In general, the larger the sample variances, the smaller is the F-ratio.

AN EXAMPLE WITH UNEQUAL SAMPLE SIZES

In the previous examples, all the samples were exactly the same size (equal ns). However, the formulas for ANOVA can be used when the sample size varies within an experiment. With unequal sample sizes, you must take care to be sure that each value of n is matched with the proper T value in the equations. You also should note that the general ANOVA procedure is most accurate when used to examine experimental data with equal sample sizes. Therefore, researchers generally try to plan experiments with equal ns. However, there are circumstances in which it is impossible or impractical to have an equal number of subjects in every treatment condition. In these situations, ANOVA still provides a valid test, especially when the samples are relatively large and when the discrepancy between sample sizes is not extreme.

The following example demonstrates an ANOVA with samples of different sizes.

EXAMPLE 13.4

A psychologist conducts a research study to compare learning performance for three species of monkeys. The animals are tested individually on a delayed-response task. A raisin is hidden in one of three containers while the animal watches from its cage window. A shade is then pulled over the window for 1 minute to block the view. After this delay period, the monkey is allowed to respond by tipping over one container. If its response is correct, the monkey is rewarded with the raisin. The number of trials it takes before the animal makes five consecutive correct responses is recorded. The researcher used all of the available animals from each species, which resulted in unequal sample sizes (n). The data are summarized in Table 13.5.

TABLE 13.5

The performance of different species of monkeys on a delayed-response task.

Vervet	Rhesus	Baboon	
$n = 4$	$n = 10$	$n = 6$	$N = 20$
$M = 9$	$M = 14$	$M = 4$	$G = 200$
$T = 36$	$T = 140$	$T = 24$	$\Sigma X^2 = 3400$
$SS = 200$	$SS = 500$	$SS = 320$	

STEP 1 State the hypotheses, and select the alpha level.

$$H_0: \quad \mu_1 = \mu_2 = \mu_3$$

$$H_1: \quad \text{At least one population is different.}$$

$$\alpha = .05$$

STEP 2 Locate the critical region. To find the critical region, we first must determine the df values for the F-ratio:

$$df_{total} = N - 1 = 20 - 1 = 19$$

$$df_{between} = k - 1 = 3 - 1 = 2$$

$$df_{within} = N - k = 20 - 3 = 17$$

The F-ratio for these data will have $df = 2, 17$. With $\alpha = .05$, the critical value for the F-ratio is 3.59.

STEP 3 Compute the F-ratio. First, compute the three SS values. As usual, SS_{total} is the SS for the total set of $N = 20$ scores, and SS_{within} combines the SS values from inside each of the treatment conditions.

$$SS_{total} = \Sigma X^2 - \frac{G^2}{N} \qquad\qquad SS_{within} = \Sigma SS_{\text{inside each treatment}}$$

$$= 3400 - \frac{200^2}{20} \qquad\qquad\qquad = 200 + 500 + 320$$

$$= 3400 - 2000 \qquad\qquad\qquad\quad = 1020$$

$$= 1400$$

$SS_{between}$ can be found by subtraction.

$$SS_{between} = SS_{total} - SS_{within}$$

$$= 1400 - 1020$$

$$= 380$$

Or, $SS_{\text{between treatments}}$ can be calculated using the computational formula (Equation 13.7). If you use the computational formula, be careful to match each treatment total with the appropriate sample size.

Finally, compute the MS values and the F-ratio:

$$MS_{between} = \frac{SS}{df} = \frac{380}{2} = 190$$

$$MS_{within} = \frac{SS}{df} = \frac{1020}{17} = 60$$

$$F = \frac{MS_{between}}{MS_{within}} = \frac{190}{60} = 3.17$$

STEP 4   Make a decision. Because the obtained F-ratio is not in the critical region, we fail to reject H_0 and conclude that these data do not provide evidence of significant differences among the three populations of monkeys in terms of average learning performance.

LEARNING CHECK

1. A researcher used ANOVA and computed $F = 4.25$ for the following data.

	Treatments	
I	II	III
$n = 10$	$n = 10$	$n = 10$
$M = 20$	$M = 28$	$M = 35$
$SS = 1005$	$SS = 1391$	$SS = 1180$

   a. If the mean for treatment III were changed to $M = 25$, what would happen to the size of the F-ratio (increase or decrease)? Explain your answer.

   b. If the SS for treatment I were changed to $SS = 1400$, what would happen to the size of the F-ratio (increase or decrease)? Explain your answer.

2. A research study comparing three treatment conditions produces $T = 20$ with $n = 4$ for the first treatment, $T = 10$ with $n = 5$ for the second treatment, and $T = 30$ with $n = 6$ for the third treatment. Calculate $SS_{between\ treatments}$ for these data.

ANSWERS

1. **a.** If the mean for treatment III were changed to $M = 25$, it would reduce the size of the mean differences (the three means would be closer together). This would reduce the size of $MS_{between}$ and would reduce the size of the F-ratio.

   b. If the SS in treatment I were increased to $SS = 1400$, it would increase the size of the variability within treatments. This would increase MS_{within} and would reduce the size of the F-ratio.

2. With $G = 60$ and $N = 15$, $SS_{between} = 30$.

13.6   POST HOC TESTS

As noted earlier, the primary advantage of ANOVA (compared to t tests) is it allows researchers to test for significant mean differences when there are *more than two* treatment conditions. ANOVA accomplishes this feat by comparing all the individual mean differences simultaneously within a single test. Unfortunately, the process of combining several mean differences into a single test statistic creates some difficulty when it is time to interpret the outcome of the test. Specifically, when you obtain a significant F-ratio (reject H_0), it simply indicates that somewhere among the entire set of mean differences there is at least one that is statistically significant. In other words, the overall F-ratio only tells you that a significant difference exists; it does not tell exactly which means are significantly different and which are not.

Consider, for example, a research study that uses three samples to compare three treatment conditions. Suppose that the three sample means are $M_1 = 3$, $M_2 = 5$, and $M_3 = 10$. In this hypothetical study there are three mean differences:

1. There is a 2-point difference between M_1 and M_2.
2. There is a 5-point difference between M_2 and M_3.
3. There is a 7-point difference between M_1 and M_3.

If an ANOVA were used to evaluate these data, a significant F-ratio would indicate that at least one of the sample mean differences is large enough to satisfy the criterion of statistical significance. In this example, the 7-point difference is the biggest of the three and, therefore, it must indicate a significant difference between the first treatment and the third treatment ($\mu_1 \neq \mu_3$). But what about the 5-point difference? Is it also large enough to be significant? And what about the 2-point difference between M_1 and M_2? Is it also significant? The purpose of *post hoc tests* is to answer these questions.

DEFINITION

Post hoc tests (or **posttests**) are additional hypothesis tests that are done after an ANOVA to determine exactly which mean differences are significant and which are not.

As the name implies, post hoc tests are done after an ANOVA. More specifically, these tests are done after ANOVA when

1. You reject H_0 and
2. There are three or more treatments ($k \geq 3$).

Rejecting H_0 indicates that at least one difference exists among the treatments. With $k = 3$ or more, the problem is to find where the differences are. Note that when you have two treatments, rejecting H_0 indicates that the two means are not equal ($\mu_1 \neq \mu_2$). In this case there is no question about *which* means are different, and there is no need to do posttests.

POSTTESTS AND TYPE I ERRORS

In general, a post hoc test enables you to go back through the data and compare the individual treatments two at a time. In statistical terms, this is called making *pairwise comparisons*. For example, with $k = 3$, we would compare μ_1 versus μ_2, then μ_2 versus μ_3, and then μ_1 versus μ_3. In each case, we are looking for a significant mean difference. The process of conducting pairwise comparisons involves performing a series of separate hypothesis tests, and each of these tests includes the risk of a Type I error. As you do more and more separate tests, the risk of a Type I error accumulates and is called the *experimentwise alpha level* (see Box 13.1).

DEFINITION

The **experimentwise alpha level** is the overall probability of a Type I error that accumulates over a series of separate hypothesis tests. Typically, the experiment-wise alpha level is substantially greater than the value of alpha used for any one of the individual tests.

We have seen, for example, that a research study with three treatment conditions produces three separate mean differences, each of which could be evaluated using a post hoc test. If each test uses $\alpha = .05$, then there is a 5% risk of a Type I error for the first posttest, another 5% risk for the second test, and one more 5% risk for the third test. Although the probability of error is not simply the sum across the three tests, it should be clear that increasing the number of separate tests will definitely increase the total, experimentwise probability of a Type I error.

Whenever you are conducting posttests, you must be concerned about the experimentwise alpha level. Statisticians have worked with this problem and have developed several methods for trying to control Type I errors in the context of post hoc tests.

PLANNED VERSUS UNPLANNED COMPARISONS

Statisticians often distinguish between what are called *planned* and *unplanned* comparisons. As the name implies, *planned comparisons* refer to specific mean differences that are relevant to specific hypotheses the researcher had in mind before the study was conducted. For example, suppose a researcher suspects that a 70° room is the best environment for human problem solving. Any warmer or colder temperature will interfere with performance. The researcher conducts a study comparing three different temperature conditions: 60°, 70°, and 80°. The researcher is predicting that performance in the 70° condition will be significantly better than in the 60° condition (comparison 1) and that 70° will be significantly better than 80° (comparison 2). Because these specific comparisons were planned before the data were collected, many statisticians would argue that the two specific posttests could be conducted without concern about inflating the risk of a Type I error. When the primary interest is in a few specific comparisons, the overall *F*-ratio for all the groups may be viewed as relatively unimportant and the overall ANOVA may not even be conducted. Instead, individual tests (*t* tests or two-group ANOVAs) could be done to evaluate the specific, planned comparisons. Although the planned comparisons could all be done with a standard alpha level, it is often recommended that researchers protect against an inflated experimentwise alpha level by dividing α equally among the planned comparisons. For example, with two comparisons, an alpha level of .05 would be divided equally so that $\alpha = .025$ is used for each comparison and the overall risk of a Type I error is limited to .05 (.025 for the first test and .025 for the second). This technique for controlling the experimentwise alpha level is referred to as the Dunn test.

Unplanned comparisons, on the other hand, typically involve sifting through the data (a "fishing expedition") by conducting a large number of comparisons in the hope that a significant difference will turn up. In this situation it is necessary that a researcher take deliberate steps to limit the risk of a Type I error. Fortunately, many different procedures for conducting post hoc tests have been developed, each using its own technique to control the experimentwise alpha level. We will examine two commonly used procedures: Tukey's HSD test and the Scheffé test. Incidentally, the safest plan (and the authors' recommendation) for posttests is to use one of the special procedures (such as Tukey or Scheffé) for *all* posttests whether they are planned or unplanned.

TUKEY'S HONESTLY SIGNIFICANT DIFFERENCE (HSD) TEST

The first post hoc test we consider is *Tukey's HSD test*. We selected Tukey's HSD test because it is a commonly used test in psychological research. Tukey's test allows you to compute a single value that determines the minimum difference between treatment means that is necessary for significance. This value, called the *honestly significant difference,* or HSD, is then used to compare any two treatment conditions. If the mean difference exceeds Tukey's HSD, you conclude that there is a significant difference between the treatments. Otherwise, you cannot conclude that the treatments are significantly different. The formula for Tukey's HSD is

$$HSD = q\sqrt{\frac{MS_{within}}{n}} \qquad\qquad (13.15)$$

The q value used in Tukey's HSD test is called a Studentized range statistic.

where the value of q is found in Table B.5 (Appendix B, page 734), $MS_{within\ treatments}$ is the within-treatments variance from the ANOVA, and n is the number of scores in each treatment. Tukey's test requires that the sample size, n, be the same for all treatments. To locate the appropriate value of q, you must know the number of treatments in the overall experiment (k) and the degrees of freedom for $MS_{within\ treatments}$ (the error term in the F-ratio) and select an alpha level (generally the same α used for the ANOVA).

EXAMPLE 13.5

To demonstrate the procedure for conducting post hoc tests with Tukey's HSD, we use the hypothetical data shown in Table 13.6. The data represent the results of a study comparing scores in three different treatment conditions. Note that the table displays summary statistics for each sample and the results from the overall ANOVA. With $\alpha = .05$ and $k = 3$ treatments, the value of q for the test is $q = 3.53$ (check the table). Therefore, Tukey's HSD is

$$HSD = q\sqrt{\frac{MS_{within}}{n}} = 3.53\sqrt{\frac{4.00}{9}} = 2.36$$

Thus, the mean difference between any two samples must be at least 2.36 to be significant. Using this value, we can make the following conclusions:

1. Treatment A is significantly different from treatment B ($M_A - M_B = 2.44$).
2. Treatment A is also significantly different from treatment C ($M_A - M_C = 4.00$).
3. Treatment B is not significantly different from treatment C ($M_B - M_C = 1.56$).

THE SCHEFFÉ TEST

Because it uses an extremely cautious method for reducing the risk of a Type I error, the *Scheffé test* has the distinction of being one of the safest of all possible post hoc tests (smallest risk of a Type I error). The Scheffé test uses an F-ratio to evaluate the significance of the difference between any two treatment conditions. The numerator of the F-ratio is an MS between treatments that is calculated using *only the two treatments you want to compare*. The denominator is the same MS within treatments that was used

TABLE 13.6

Hypothetical results from a research study comparing three treatment conditions. Summary statistics are presented for each treatment along with the outcome from the analysis of variance.

Treatment A	Treatment B	Treatment C		Source	SS	df	MS
				Between	73.19	2	36.60
$n = 9$	$n = 9$	$n = 9$		Within	96.00	24	4.00
$T = 27$	$T = 49$	$T = 63$		Total	169.19	26	
$M = 3.00$	$M = 5.44$	$M = 7.00$		Overall $F(2, 24) = 9.15$			

for the overall ANOVA. The "safety factor" for the Scheffé test comes from the following two considerations:

1. Although you are comparing only two treatments, the Scheffé test uses the value of k from the original experiment to compute df between treatments. Thus, df for the numerator of the F-ratio is $k - 1$.

2. The critical value for the Scheffé F-ratio is the same as was used to evaluate the F-ratio from the overall ANOVA. Thus, Scheffé requires that every posttest satisfy the same criterion that was used for the complete ANOVA. The following example uses the data from Table 13.6 to demonstrate the Scheffé posttest procedure.

EXAMPLE 13.6 Remember that the Scheffé procedure requires a separate $SS_{between}$, $MS_{between}$, and F-ratio for each comparison being made. Although Scheffé computes $SS_{between}$ using the regular computational formula (Equation 13.7), you must remember that all the numbers in the formula are entirely determined by the two treatment conditions being compared. To demonstrate the Scheffé test, we will use the data in Table 13.6. We begin by comparing treatment A (with $T = 27$ and $n = 9$) and treatment B (with $T = 49$ and $n = 9$). The first step is to compute $SS_{between}$ for these two groups. In the formula for SS, notice that the grand total for the two groups is $G = 27 + 49 = 76$, and the total number of scores for the two groups is $N = 9 + 9 = 18$.

$$SS_{between} = \Sigma \frac{T^2}{n} - \frac{G^2}{N}$$

$$= \frac{27^2}{9} + \frac{49^2}{9} - \frac{76^2}{18}$$

$$= 81 + 266.78 - 320.89$$

$$= 26.89$$

Although we are comparing only two groups, these two were selected from a study consisting of $k = 3$ samples. The Scheffé test uses the overall study to determine the degrees of freedom between treatments. Therefore, $df_{between} = 3 - 1 = 2$, and the MS between treatments is

$$MS_{between} = \frac{SS_{between}}{df_{between}} = \frac{26.89}{2} = 13.45$$

Finally, the Scheffé procedure uses the error term from the overall ANOVA to compute the F-ratio. In this case, $MS_{within} = 4.00$ with $df_{within} = 24$. Thus, the Scheffé test produces an F-ratio of

$$F_{A \text{ verus } B} = \frac{MS_{between}}{MS_{within}} = \frac{13.45}{4.00} = 3.36$$

With $df = 2, 24$ and $\alpha = .05$, the critical value for F is 3.40 (see Table B.4). Therefore, our obtained F-ratio is not in the critical region, and we must conclude that these data show no significant difference between treatment A and treatment B.

The second comparison involves treatment B ($T = 49$) versus treatment C ($T = 63$). This time the data produce $SS_{between} = 10.89$, $MS_{between} = 5.45$, and $F(2, 24) = 1.36$ (check the calculations for yourself). Once again the critical value for F is 3.40, so we must conclude that the data show no significant difference between treatment B and treatment C.

The final comparison is treatment A ($T = 27$) versus treatment C ($T = 63$). This time the data produce $SS_{between} = 72$, $MS_{between} = 36$, and $F(2, 24) = 9.00$ (check the calculations for yourself). Once again the critical value for F is 3.40, and this time we conclude that the data show a significant difference.

Thus, the Scheffé posttest indicates that the only significant difference is between treatment A and treatment C.

There are two interesting points to be made from the posttest outcomes presented in the preceding two examples. First, the Scheffé test was introduced as being one of the safest of the posttest techniques because it provides the greatest protection from Type I errors. To provide this protection, the Scheffé test simply requires a larger sample mean difference before you may conclude that the difference is significant. In Example 13.5 we found that the difference between treatment A and treatment B was large enough to be significant according to Tukey's test, but this same difference failed to reach significance according to Scheffé (Example 13.6). The discrepancy between the results is an example of Scheffé's extra demands: The Scheffé test simply requires more evidence and, therefore, its use is less likely to lead to a Type I error.

The second point concerns the pattern of results from the three Scheffé tests in Example 13.6. You may have noticed that the posttests produce what are apparently contradictory results. Specifically, the tests show no significant difference between A and B and they show no significant difference between B and C. This combination of outcomes might lead you to suspect that there is no significant difference between A and C. However, the test did show a significant difference. The answer to this apparent contradiction lies in the criterion of statistical significance. The differences between A and B and between B and C are too small to satisfy the criterion of significance. However, when these differences are combined, the total difference between A and C is large enough to meet the criterion for significance.

LEARNING CHECK

1. With $k = 2$ treatments, are post hoc tests necessary when the null hypothesis is rejected? Explain why or why not.

2. An ANOVA comparing three treatments produces an overall F-ratio with $df = 2, 27$. If the Scheffé test was used to compare two of the three treatments, then the Scheffé F-ratio would also have $df = 2, 27$. (True or false?)

3. Using the data and the results from Example 13.1,
   a. Use Tukey's HSD test to determine whether there is a significant mean difference between drug B and the placebo. Use $\alpha = .05$.
   b. Use the Scheffé test to determine whether there is a significant mean difference between drug B and the placebo. Use $\alpha = .05$.

ANSWERS

1. No. Post hoc tests are used to determine which treatments are different. With only two treatment conditions, there is no uncertainty as to which two treatments are different.

2. True

3. a. For this test, $q = 4.05$ and HSD = 2.55. There is a 3-point mean difference between drug B and the placebo, which is large enough to be significant.
   b. The Scheffé $F = 3.75$, which is greater than the critical value of 3.24. Conclude that the mean difference between drug B and the placebo is significant.

13.7 THE RELATIONSHIP BETWEEN ANOVA AND *t* TESTS

When you are evaluating the mean difference from an independent-measures study comparing only two treatments (two separate samples), you can use either an independent-measures *t* test (Chapter 10) or the ANOVA presented in this chapter. In practical terms, it makes no difference which you choose. These two statistical techniques always will result in the same statistical decision. In fact the two methods use many of the same calculations and are very closely related in several other respects. The basic relationship between *t* statistics and *F*-ratios can be stated in an equation:

$$F = t^2$$

This relationship can be explained by first looking at the structure of the formulas for *F* and *t*.

The structure of the *t* statistic compares the obtained difference between sample means (numerator) with the standard error or standard difference that is reasonable to expect if the two samples come from populations with equal means (H_0 is true).

$$t = \frac{\text{obtained difference between sample means}}{\text{standard difference between means if } H_0 \text{ is true}}$$

The *F*-ratio in ANOVA measures the same values in terms of variance.

$$F = \frac{\text{variance (differences) between sample means}}{\text{variance (differences) expected if } H_0 \text{ is true}}$$

However, variances in the *F*-ratio are simply measures of squared distance. Therefore, the *F*-ratio can be expressed as follows

$$F = \frac{(\text{difference between sample means})^2}{(\text{standard difference if } H_0 \text{ is true})^2}$$

The fact that the *t* statistic is based on differences and the *F*-ratio is based on *squared* differences leads to the basic relationship $F = t^2$.

There are several other points to consider in comparing the *t* statistic to the *F*-ratio.

1. It should be obvious that you will be testing the same hypotheses whether you choose a *t* test or an ANOVA. With only two treatments, the hypotheses for either test are

    $H_0: \quad \mu_1 = \mu_2$

    $H_1: \quad \mu_1 \neq \mu_2$

2. The degrees of freedom for the *t* statistic and the *df* for the denominator of the *F*-ratio (df_{within}) are identical. For example, if you have two samples, each with six scores, the independent-measures *t* statistic will have $df = 10$, and the *F*-ratio will have $df = 1, 10$. In each case, you are adding the *df* from the first sample $(n - 1)$ and the *df* from the second sample $(n - 1)$.

3. The distribution of *t* and the distribution of *F*-ratios match perfectly if you take into consideration the relationship $F = t^2$. Consider the *t* distribution with $df = 18$ and the corresponding *F* distribution with $df = 1, 18$ that are presented in Figure 13.12. Notice the following relationships:

    a. If each of the *t* values is squared, then all of the negative values become positive. As a result, the whole left-hand side of the *t* distribution (below

FIGURE 13.12

The distribution of t statistics with $df = 18$ and the corresponding distribution of F-ratios with $df = 1, 18$. Notice that the critical values for $\alpha = .05$ are $t = \pm2.101$ and that $F = 2.101^2 = 4.41$.

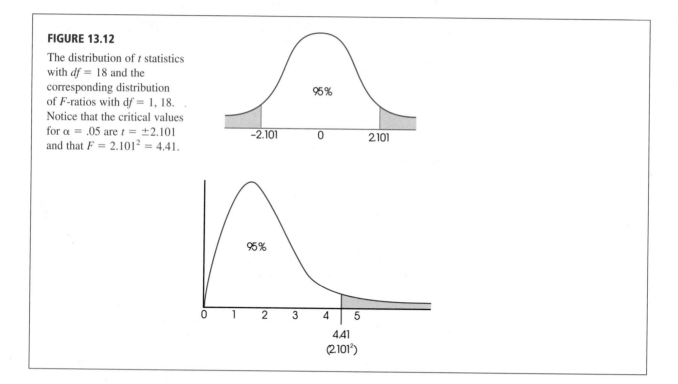

zero) will be flipped over to the positive side. This creates an asymmetrical, positively skewed distribution—that is, the F distribution.

b. For $\alpha = .05$, the critical region for t is determined by values greater than $+2.101$ or less than -2.101. When these boundaries are squared, you get

$$\pm2.101^2 = 4.41$$

Notice that 4.41 is the critical value for $\alpha = .05$ in the F distribution. Any value that is in the critical region for t will end up in the critical region for F-ratios after it is squared.

ASSUMPTIONS FOR THE INDEPENDENT-MEASURES ANOVA

The independent-measures ANOVA requires the same three assumptions that were necessary for the independent-measures t hypothesis test:

1. The observations within each sample must be independent (see page 253).

2. The populations from which the samples are selected must be normal.

3. The populations from which the samples are selected must have equal variances (homogeneity of variance).

Ordinarily, researchers are not overly concerned with the assumption of normality, especially when large samples are used, unless there are strong reasons to suspect the assumption has not been satisfied. The assumption of homogeneity of variance is an important one. If a researcher suspects it has been violated, it can be tested by Hartley's F-max test for homogeneity of variance (Chapter 10, page 328).

1. A researcher uses an independent-measures t test to evaluate the mean difference obtained in a research study, and obtains a t statistic of $t = 3.00$. If the researcher had used an ANOVA to evaluate the results, the F-ratio would be $F = 9.00$. (True or false?)

2. An ANOVA produces an F-ratio with $df = 1, 34$. Could the data have been analyzed with a t test? What would be the degrees of freedom for the t statistic?

ANSWERS

1. True. $F = t^2$

2. If the F-ratio has $df = 1, 34$, then the experiment compared only two treatments, and you could use a t statistic to evaluate the data. The t statistic would have $df = 34$.

SUMMARY

1. Analysis of variance (ANOVA) is a statistical technique that is used to test for mean differences among two or more treatment conditions. The null hypothesis for this test states that in the general population there are no mean differences among the treatments. The alternative states that at least one mean is different from another.

2. The test statistic for ANOVA is a ratio of two variances called an F-ratio. The variances in the F-ratio are called mean squares, or MS values. Each MS is computed by

$$MS = \frac{SS}{df}$$

3. For the independent-measures ANOVA, the F-ratio is

$$F = \frac{MS_{between}}{MS_{within}}$$

The $MS_{between}$ measures differences between the treatments by computing the variability of the treatment means or totals. These differences are assumed to be produced by
a. Treatment effects (if they exist)
b. Differences due to chance

The MS_{within} measures variability inside each of the treatment conditions. Because individuals inside a treatment condition are all treated exactly the same, any differences within treatments cannot be caused by treatment effects. Thus, the within-treatments MS is produced only by differences due to chance. With these factors in mind, the F-ratio has the following structure:

$$F = \frac{\text{treatment effect} + \text{differences due to chance}}{\text{differences due to chance}}$$

When there is no treatment effect (H_0 is true), the numerator and the denominator of the F-ratio are measuring the same variance, and the obtained ratio should be near 1.00. If there is a significant treatment effect, the numerator of the ratio should be larger than the denominator, and the obtained F value should be much greater than 1.00.

4. The formulas for computing each SS, df, and MS value are presented in Figure 13.13, which also shows the general structure for the ANOVA.

5. The F-ratio has two values for degrees of freedom, one associated with the MS in the numerator and one associated with the MS in the denominator. These df values are used to find the critical value for the F-ratio in the F distribution table.

6. Effect size for the independent-measures ANOVA is measured by computing eta squared, the percentage of variance accounted for by the treatment effect.

$$\eta^2 = \frac{SS_{between}}{SS_{between} + SS_{within}}$$
$$= \frac{SS_{between}}{SS_{total}}$$

7. When the decision from an ANOVA is to reject the null hypothesis and when the experiment contained more than two treatment conditions, it is necessary to continue the analysis with a post hoc test, such as Tukey's HSD test or the Scheffé test. The purpose of these tests is to determine exactly which treatments are significantly different and which are not.

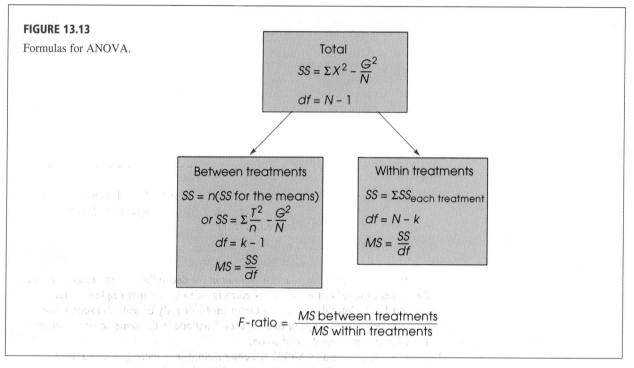

FIGURE 13.13

Formulas for ANOVA.

KEY TERMS

analysis of variance (ANOVA)	treatment effect	error term	post hoc tests
factor	F-ratio	mean square (MS)	pairwise comparisons
levels	individual differences	ANOVA summary table	experimentwise alpha level
between-treatments variance	experimental error	distribution of F-ratios	Tukey's HSD test
	within-treatments variance	eta squared (η^2)	Scheffé test

RESOURCES

There are a tutorial quiz and other learning exercises for Chapter 13 at the Wadsworth website, www.cengage.com/psychology/gravetter. See page 29 for more information about accessing items on the website.

ENHANCED
WebAssign

Guided interactive tutorials, end-of-chapter problems, and related testbank items may be assigned online at WebAssign.

WebTUTOR

If you are using WebTutor along with this book, there is a WebTutor section corresponding to this chapter. The WebTutor contains a brief summary of Chapter 13, hints for learning the concepts and the formulas for analysis of variance, cautions about common errors, and sample exam items including solutions.

SPSS

General instructions for using SPSS are presented in Appendix D. Following are detailed instructions for using SPSS to perform **The Single-Factor, Independent-Measures Analysis of Variance (ANOVA)** presented in this chapter.

Data Entry

1. The scores are entered in a *stacked format* in the data editor, which means that all the scores from all of the different treatments are entered in a single column (VAR00001). Enter the scores for treatment #2 directly beneath the scores from treatment #1 with no gaps or extra spaces. Continue in the same column with the scores from treatment #3, and so on.
2. In the second column (VAR00002), enter a number to identify the treatment condition for each score. For example, enter a 1 beside each score from the first treatment, enter a 2 beside each score from the second treatment, and so on.

Data Analysis

1. Click **Analyze** on the tool bar, select **Compare Means,** and click on **One-Way ANOVA.**
2. Highlight the column label for the set of scores (VAR0001) in the left box and click the arrow to move it into the **Dependent List** box.
3. Highlight the label for the column containing the treatment numbers (VAR0002) in the left box and click the arrow to move it into the **Factor** box.
4. If you want descriptive statistics for each treatment, click on the **Options** box, select **Descriptives,** and click **Continue.**
5. Click **OK.**

SPSS Output

If you selected the Descriptives Option, SPSS will produce a table showing descriptive statistics for each of the samples. This table includes the number of scores, the mean, the standard deviation, and the standard error for the mean for each of the individual samples as well as the same statistics for the entire group of participants. This table also includes a 95% confidence interval for each mean. The second part of the output presents a summary table showing the results from the analysis of variance including all three SS and df values, the two MS values (variances), the F-ratio, and the level of significance (the p value or alpha level for the test).

FOCUS ON PROBLEM SOLVING

1. It can be helpful to compute all three SS values separately, then check to verify that the two components (between and within) add up to the total. However, you can greatly

simply the calculations if you simply find SS_{total} and $SS_{within\ treatments}$, then obtain $SS_{between\ treatments}$ by subtraction.

2. Remember that an F-ratio has two separate values for df: a value for the numerator and one for the denominator. Properly reported, the $df_{between}$ value is stated first. You will need both df values when consulting the F distribution table for the critical F value. You should recognize immediately that an error has been made if you see an F-ratio reported with a single value for df.

3. When you encounter an F-ratio and its df values reported in the literature, you should be able to reconstruct much of the original experiment. For example, if you see "$F(2, 36) = 4.80$," you should realize that the experiment compared $k = 3$ treatment groups (because $df_{between} = k - 1 = 2$), with a total of $N = 39$ subjects participating in the experiment (because $df_{within} = N - k = 36$).

DEMONSTRATION 13.1

ANALYSIS OF VARIANCE

A human factors psychologist studied three computer keyboard designs. Three samples of individuals were given material to type on a particular keyboard, and the number of errors committed by each participant was recorded. The data are as follows:

Keyboard A	Keyboard B	Keyboard C	
0	6	6	$N = 15$
4	8	5	$G = 60$
0	5	9	$\Sigma X^2 = 356$
1	4	4	
0	2	6	
$T = 5$	$T = 25$	$T = 30$	
$SS = 12$	$SS = 20$	$SS = 14$	

Are these data sufficient to conclude that there are significant differences in typing performance among the three keyboard designs?

STEP 1   State the hypotheses, and specify the alpha level.

The null hypothesis states that there is no difference among the keyboards in terms of number of errors committed. In symbols, we would state

$$H_0: \quad \mu_1 = \mu_2 = \mu_3 \qquad \text{(Type of keyboard used has no effect.)}$$

As noted previously in this chapter, there are a number of possible statements for the alternative hypothesis. Here we state the general alternative hypothesis:

$$H_1: \quad \text{At least one of the treatment means is different.}$$

We will set alpha at $\alpha = .05$.

STEP 2   Locate the critical region.

To locate the critical region, we must obtain the values for $df_{between}$ and df_{within}.

$$df_{between} = k - 1 = 3 - 1 = 2$$

$$df_{within} = N - k = 15 - 3 = 12$$

The F-ratio for this problem will have $df = 2, 12$. Consult the F-distribution table for $df = 2$ in the numerator and $df = 12$ in the denominator. The critical F value for $\alpha = .05$ is $F = 3.88$. The obtained F-ratio must exceed this value to reject H_0.

STEP 3   Perform the analysis.

The analysis involves the following steps:

1. Perform the analysis of SS.
2. Perform the analysis of df.
3. Calculate mean squares.
4. Calculate the F-ratio.

Perform the analysis of SS. We will compute SS_{total} followed by its two components.

$$SS_{total} = \Sigma X^2 - \frac{G^2}{N} = 356 - \frac{60^2}{15} = 356 - \frac{3600}{15}$$

$$= 356 - 240 = 116$$

$$SS_{within} = \Sigma SS_{inside each treatment}$$

$$= 12 + 20 + 14$$

$$= 46$$

$$SS_{between} = \Sigma \frac{T^2}{n} - \frac{G^2}{N}$$

$$= \frac{5^2}{5} + \frac{25^2}{5} + \frac{30^2}{5} - \frac{60^2}{15}$$

$$= \frac{25}{5} + \frac{625}{5} + \frac{900}{5} - \frac{3600}{15}$$

$$= 5 + 125 + 180 - 240$$

$$= 70$$

Analyze degrees of freedom. We will compute df_{total}. Its components, $df_{between}$ and df_{within}, were previously calculated (step 2).

$$df_{total} = N - 1 = 15 - 1 = 14$$

$$df_{between} = 2$$

$$df_{within} = 12$$

Calculate the MS values. The values for $MS_{between}$ and MS_{within} are determined.

$$MS_{between} = \frac{SS_{between}}{df_{between}} = \frac{70}{2} = 35$$

$$MS_{within} = \frac{SS_{within}}{df_{within}} = \frac{46}{12} = 3.83$$

Compute the F-ratio. Finally, we can compute F.

$$F = \frac{MS_{between}}{MS_{within}} = \frac{35}{3.83} = 9.14$$

STEP 4 Make a decision about H_0, and state a conclusion.

The obtained F of 9.14 exceeds the critical value of 3.88. Therefore, we can reject the null hypothesis. The type of keyboard used has a significant effect on the number of errors committed, $F(2, 12) = 9.14, p < .05$. The following table summarizes the results of the analysis:

Source	SS	df	MS	
Between treatments	70	2	35	$F = 9.14$
Within treatments	46	12	3.83	
Total	116	14		

DEMONSTRATION 13.2

COMPUTING EFFECT SIZE FOR ANALYSIS OF VARIANCE

We will compute eta squared (η^2), the percentage of variance explained, for the data that were analyzed in Demonstration 13.1. The data produced a between-treatments SS of 70 and a total SS of 116. Thus,

$$\eta^2 = \frac{SS_{between}}{SS_{total}} = \frac{70}{116} = 0.60 \quad (or\ 60\%)$$

PROBLEMS

1. Explain why the expected value for an F-ratio is equal to 1.00 when there is no treatment effect.

2. Describe the similarities between an F-ratio and a t statistic.

3. Several factors influence the size of the F-ratio. For each of the following, indicate whether it would influence the numerator or the denominator of the F-ratio, and indicate whether the size of the F-ratio would increase or decrease.
   a. Increase the differences between the sample means.
   b. Increase the size of the sample variances.

4. Explain why you should use ANOVA instead of several t tests to evaluate mean differences when an experiment consists of three or more treatment conditions.

5. Posttests are done after an ANOVA.
   a. What is the purpose for posttests?
   b. Explain why you would not do posttests if the analysis is comparing only two treatments.

   c. Explain why you would not do posttests if the decision from the ANOVA was to fail to reject the null hypothesis.

6. A research study comparing three treatment conditions produced means of $M_1 = 2$, $M_2 = 4$, and $M_3 = 6$.
   a. Compute the variance for the set of three means. (Treat the means as a sample of $n = 3$ values and compute the sample variance.)
   b. Now we will change the third mean from $M_3 = 6$ to $M_3 = 15$. Notice that we have substantially increased the differences among the three means. Compute the variance for the new set of $n = 3$ means. You should find that the variance is much larger than the value obtained in part a. Note: The variance provides a measure of the size of the mean differences.

7. The following data represent the results from an independent-measures study comparing three treatments.

a. Compute SS for the set of 3 treatment means.
   (Use the three means as a set of $n = 3$ scores and
   compute SS.)
b. Using the result from part a, compute $n(SS_{means})$.
   Note that this value is equal to $SS_{between}$ (see
   Equation 13.6).
c. Now, compute $SS_{between}$ with the computational
   formula using the T values (Equation 13.7). You
   should obtain the same result as in part b.

Treatment		
I	II	III
$n = 10$	$n = 10$	$n = 10$
$M = 2$	$M = 3$	$M = 7$
$T = 20$	$T = 30$	$T = 70$

8. The following data represent the results from an
   independent-measures experiment comparing three
   treatment conditions. Use an analysis of variance with
   $\alpha = .05$ to determine whether these data are sufficient
   to conclude that there are significant differences
   between the treatments.

Treatment			
I	II	III	
0	2	5	$N = 15$
0	4	6	
2	0	4	$G = 45$
1	1	7	
2	3	8	$\Sigma X^2 = 229$
$M = 1$	$M = 2$	$M = 6$	
$T = 5$	$T = 10$	$T = 30$	
$SS = 4$	$SS = 10$	$SS = 10$	

9. In the previous problem the treatments were significantly
   different primarily because the mean for sample III is
   noticeably different from the other two samples
   ($M_1 = 1$, $M_2 = 2$, and $M_3 = 6$). For the following data
   we have taken the scores from problem 8 and reduced
   the differences between treatments by moving sample III
   closer to the other two samples. Specifically, we have
   subtracted 3 points from each score in the third sample
   to produce sample means of $M_1 = 1$, $M_2 = 2$, and
   $M_3 = 3$.
   a. Before you begin any calculations, predict how the
      changes in the data should influence the outcome of
      the analysis. That is, how will the F-ratio for these
      data compare with the F-ratio from problem 8?

b. Use an ANOVA with $\alpha = .05$ to determine
   whether there are any significant differences
   among the three treatments. (Does your answer
   agree with your prediction in part a?)

Treatment			
I	II	III	
0	2	2	$N = 15$
0	4	3	
2	0	1	$G = 30$
1	1	4	
2	3	5	$\Sigma X^2 = 94$
$M = 1$	$M = 2$	$M = 3$	
$T = 5$	$T = 10$	$T = 15$	
$SS = 4$	$SS = 10$	$SS = 10$	

10. The following data are from an experiment comparing
    three treatment conditions with a separate sample of
    $n = 4$ in each treatment.
    a. Use an ANOVA with $\alpha = .05$ to determine whether
       there are any significant differences among the
       three treatments.
    b. Compute η^2 for these data.

Treatment			
I	II	III	
2	3	7	$N = 12$
6	7	5	$G = 60$
2	6	4	$\Sigma X^2 = 344$
6	4	8	
$M = 4$	$M = 5$	$M = 6$	
$T = 16$	$T = 20$	$T = 24$	
$SS = 16$	$SS = 10$	$SS = 10$	

11. In the preceding problem, the three sample means
    are close together and there are no significant mean
    differences. To construct the following data, we started
    with the scores in problem 10, then increased the size
    of the mean differences. Specifically, we lowered the
    smallest mean by subtracting 2 points from each score
    in treatment I, and we increased the largest mean by
    adding 2 points to each score in treatment III. As a
    result, the three sample means are now much more
    spread out.
    a. Use an ANOVA with $\alpha = .05$ to determine whether
       there are any significant differences among the

three treatments. Compare the outcome with the results in problem 10.

b. Compute η^2 for these data. Compare the outcome with the result from problem 10.

Treatment			
I	II	III	
0	3	9	$N = 12$
4	7	7	$G = 60$
0	6	6	$\Sigma X^2 = 408$
4	4	10	
$M = 2$	$M = 5$	$M = 8$	
$T = 8$	$T = 20$	$T = 32$	
$SS = 16$	$SS = 10$	$SS = 10$	

12. The following data were obtained in a study using three separate samples to compare three different treatments.

a. Use an analysis of variance with $\alpha = .05$ to determine whether there are any significant differences among the treatments.

b. Compute the value for η^2 for these data.

Treatment			
I	II	III	
4	3	8	$N = 12$
3	1	4	$G = 48$
5	3	6	$\Sigma X^2 = 238$
4	1	6	
$M = 4$	$M = 2$	$M = 6$	
$T = 16$	$T = 8$	$T = 24$	
$SS = 2$	$SS = 4$	$SS = 8$	

13. The data from problem 12 show significant differences between treatments. To construct the following data, we started with the numbers in problem 12 but we increased the variance within each sample. Specifically, we kept the same means as in problem 12, but we spread out the scores inside each of the treatments.

a. Before you begin any calculations, predict how the changes in the data should influence the outcome of the analysis. That is, how will the F-ratio for these data compare with the F-ratio from problem 12?

b. Use an ANOVA with $\alpha = .05$ to determine whether there are any significant differences among the three treatments. (Does your answer agree with your prediction in part a?)

Treatment			
I	II	III	
4	4	9	$N = 12$
2	0	3	$G = 48$
6	3	6	$\Sigma X^2 = 260$
4	1	6	
$M = 4$	$M = 2$	$M = 6$	
$T = 16$	$T = 8$	$T = 24$	
$SS = 8$	$SS = 10$	$SS = 18$	

14. A researcher reports an F-ratio with $df = 2, 24$ for an independent-measures research study.

a. How many treatment conditions were compared in the study?

b. How many subjects participated in the entire study?

15. The results from an independent-measures research study indicate that there are significant differences between treatments, $F(3, 36) = 3.28, p < .05$.

a. How many treatment conditions were compared in the study?

b. How many individuals participated in the entire study?

16. Recent research indicates that the effectiveness of antidepressant medication is directly related to the severity of the depression (Khan, Brodhead, Kolts, & Brown, 2005). Based on pre-treatment depression scores, patients were divided into four groups based on their level of depression. After receiving the antidepressant medication, depression scores were measured again and the amount of improvement was recorded for each patient. The following data are similar to the results of the study.

a. Do the data indicate significant differences among the four levels of severity? Test with $\alpha = .05$.

b. Compute η^2, the percentage of variance explained by the group differences.

Low Moderate	High Moderate	Moderately Severe	Severe	
0	1	4	5	$N = 16$
2	3	6	6	$G = 48$
2	2	2	6	$\Sigma X^2 = 204$
0	2	4	3	
$M = 1$	$M = 2$	$M = 4$	$M = 5$	
$T = 4$	$T = 8$	$T = 16$	$T = 20$	
$SS = 4$	$SS = 2$	$SS = 8$	$SS = 6$	

17. In the previous problem, the total SS was partitioned into two components, $SS_{between}$ and SS_{within}. To create the following data, we started with the scores in problem 16 but eliminated all of the within-treatments variability. (Note that all the scores in each treatment are equal to the treatment mean.) As a result, all of the variability in the scores (SS_{total}) comes from differences between treatments ($SS_{between}$). Calculate SS_{total} for the following scores and verify that the result you get is equal to $SS_{between}$ from problem 16.

Low Moderate	High Moderate	Moderately Severe	Severe	
1	2	4	5	$N = 16$
1	2	4	5	$G = 48$
1	2	4	5	$\Sigma X^2 = 184$
1	2	4	5	
$T = 4$	$T = 8$	$T = 16$	$T = 20$	
$SS = 0$	$SS = 0$	$SS = 0$	$SS = 0$	

18. The following data are constructed so that all three treatments have exactly the same mean, $M = 4$.
  a. Before you begin any calculations, predict what value should be obtained for $SS_{between}$. (*Hint:* How big are the differences between treatments?)
  b. Calculate SS_{total} and SS_{within}. You should find that all of the variability in the data is within-treatments variability ($SS_{total} = SS_{within}$).
  c. Complete the analysis of variance with $\alpha = .05$ to determine whether there are any significant differences among the three treatments. (Does your answer agree with your prediction in part a?)

Treatment			
I	II	III	
0	2	3	$N = 12$
6	6	2	$G = 48$
4	4	5	$\Sigma X^2 = 234$
6	4	6	
$M = 4$	$M = 4$	$M = 4$	
$SS = 24$	$SS = 8$	$SS = 10$	

19. There is some evidence that high school students justify cheating in class on the basis of poor teacher skills or low levels of teacher caring (Murdock, Miller, and Kohlhardt, 2004). Students appear to rationalize their illicit behavior based on perceptions of how their teachers view cheating. Poor teachers are thought not to know or care whether students cheat, so cheating in their classes is okay. Good teachers, on the other hand, do care and are alert to cheating, so students tend not to cheat in their classes. Following are hypothetical data similar to the actual research results. The scores represent judgments of the acceptability of cheating for the students in each sample. Use an ANOVA with $\alpha = .05$ to determine whether there are significant differences in student judgments depending on how they see their teachers.

Poor Teacher	Average Teacher	Good Teacher	
$n = 6$	$n = 8$	$n = 10$	$N = 24$
$M = 6$	$M = 2$	$M = 2$	$G = 72$
$SS = 30$	$SS = 33$	$SS = 42$	$\Sigma X^2 = 393$

20. The following summary table presents the results from an ANOVA comparing three treatment conditions with $n = 12$ participants in each condition. Complete all missing values. (*Hint:* Start with the df column.)

Source	SS	df	MS	
Between Treatments	____	____	9	$F =$ ____
Within Treatments	____	____	____	
Total	117	____		

21. A pharmaceutical company has developed a drug that is expected to reduce hunger. To test the drug, three samples of rats are selected with $n = 10$ in each sample. The first sample receives the drug every day. The second sample is given the drug once a week, and the third sample receives no drug at all. The dependent variable is the amount of food eaten by each rat over a 1-month period. These data are analyzed by an ANOVA, and the results are reported in the following summary table. Fill in all missing values in the table. (*Hint:* Start with the df column.)

Source	SS	df	MS	
Between Treatments	____	____	15	$F = 7.50$
Within Treatments	____	____	____	
Total	____	____		

22. A developmental psychologist is examining the development of language skills from age 2 to age 5. Four different groups of children are obtained, one for each age, with $n = 15$ children in each group. Each child is given a language skills assessment test. The resulting data were analyzed with an ANOVA to test for mean differences between age groups. The results of the ANOVA are presented in the following table. Fill in all missing values.

Source	SS	df	MS	
Between Treatments	81	___	___	$F =$ ___
Within Treatments	___	___	___	
Total	249	___		

23. The following data were obtained from an independent-measures research study comparing three treatment conditions. Use an ANOVA with $\alpha = .05$ to determine whether there are any significant mean differences among the treatments.

Treatment			
I	II	III	
2	5	7	$N = 14$
5	2	3	$G = 42$
0	1	6	$\Sigma X^2 = 182$
1	2	4	
2			
2			
$T = 12$	$T = 10$	$T = 20$	
$SS = 14$	$SS = 9$	$SS = 10$	

24. The following values summarize the results from an independent-measures study comparing two treatment conditions.
    a. Use an independent-measures t test with $\alpha = .05$ to determine whether there is a significant mean difference between the two treatments.
    b. Use an ANOVA with $\alpha = .05$ to determine whether there is a significant mean difference between the two treatments.

Treatment		
I	II	
$n = 8$	$n = 4$	
$M = 4$	$M = 10$	$N = 12$
$T = 32$	$T = 40$	$G = 72$
$SS = 45$	$SS = 15$	$\Sigma X^2 = 588$

25. The following data represent the results from an independent-measures study comparing two treatment conditions.
    a. Use an independent-measures t test with $\alpha = .05$ to determine whether there is a significant mean difference between the two treatments.
    b. Use an ANOVA with $\alpha = .05$ to determine whether there is a significant mean difference between the two treatments.

Treatment		
I	II	
8	2	$N = 10$
7	3	$G = 50$
6	3	$\Sigma X^2 = 306$
5	5	
9	2	
$M = 7$	$M = 3$	
$T = 35$	$T = 15$	
$SS = 10$	$SS = 6$	

26. One possible explanation for why some birds migrate and others maintain year round residency in a single location is intelligence. Specifically, birds with small brains, relative to their body size, are simply not smart enough to find food during the winter and must migrate to warmer climates where food is easily available (Sol, Lefebvre, & Rodriguez-Teijeiro, 2005). Birds with bigger brains, on the other hand, are more creative and can find food even when the weather turns harsh. Following are hypothetical data similar to the actual research results. The numbers represent relative brain size for the individual birds in each sample.

Non-Migrating	Short-Distance Migrants	Long-Distance Migrants	
18	6	4	$N = 18$
13	11	9	$G = 180$
19	7	5	$\Sigma X^2 = 2150$
12	9	6	
16	8	5	
12	13	7	
$M = 15$	$M = 9$	$M = 6$	
$T = 90$	$T = 54$	$T = 36$	
$SS = 48$	$SS = 34$	$SS = 16$	

a. Use an ANOVA with $\alpha = .05$ to determine whether there are any significant mean differences among the three groups of birds.

b. Compute η^2, the percentage of variance explained by the group differences, for these data.

c. Use the Tukey HSD posttest to determine which groups are significantly different.

27. First-born children tend to develop language skills faster than their younger siblings. One possible explanation for this phenomenon is that first-borns have undivided attention from their parents. If this explanation is correct, then it is also reasonable that twins should show slower language development than single children and that triplets should be even slower. Davis (1937) found exactly this result. The following hypothetical data demonstrate the relationship. The dependent variable is a measure of language skill at age 3 for each child. Do the data indicate any significant differences? Test with $\alpha = .05$.

Single Child	Twin	Triplet
8	6	5
7	4	5
10	6	8
6	7	3
9	4	5
8	9	4

a. Use an ANOVA with $\alpha = .05$ to determine whether there are any significant mean differences among the three groups of children.

b. Use the Scheffé posttest to determine which groups are significantly different.

C H A P T E R

14

Repeated-Measures Analysis of Variance (ANOVA)

Tools You Will Need

The following items are considered essential background material for this chapter. If you doubt your knowledge of any of these items, you should review the appropriate chapter or section before proceeding.

- Introduction to analysis of variance (Chapter 13)

- The logic of analysis of variance
  - ANOVA notation and formulas
  - Distribution of F-ratios

- Repeated-measures design (Chapter 11)

Preview

14.1   Overview

14.2   Testing Hypotheses with the Repeated-Measures ANOVA

14.3   An Alternative View of the Repeated-Measures ANOVA

14.4   Advantages and Disadvantages of the Repeated-Measures Design

14.5   Individual Differences and the Consistency of the Treatment Effects

Summary

Focus on Problem Solving

Demonstrations 14.1 and 14.2

Problems

Preview

Suppose that you were offered a choice between receiving $1000 in 5 years or a smaller amount today. How much would you be willing to accept today in exchange for a $1000 payment in 5 years?

The general result for this kind of decision is that the longer the $1000 payment is delayed, the smaller the amount that people will accept today. For example, you may be willing to take $300 today rather than waiting 5 years for the $1000. However, you might be willing to settle for $100 today if you have to wait 10 years for the $1000. This phenomenon is known as delayed discounting because people discount the value of a future reward depending on how long it is delayed (Green, Fry, & Myerson, 1994). In a typical study examining delayed discounting, people are asked to place a value on a future reward for several different delay periods. For example, how much would you accept today instead of waiting for a future reward of $1000 if you had to wait 1 month before receiving the payment? How about waiting 6 months? 12 months? 24 months? 60 months?

Typical results for a sample of college students are shown in Figure 14.1. Note that the average value declines regularly as the delay period increases. The statistical question is whether the mean differences from one delay period to another are significant.

The Problem: You should recognize that evaluating mean differences for more than two sample means is a job for analysis of variance. However, the discounting study is a repeated-measures design with five scores for each individual, and ANOVA introduced in Chapter 13 is intended for independent-measures studies. Once again, a new hypothesis test is needed.

The Solution: In this chapter we introduce the *repeated-measures ANOVA*. As the name implies, this new procedure is used to evaluate the differences between two or more sample means obtained from a repeated-measures research study. As you will see, many of the notational symbols and computations are the same as those used for the independent-measures ANOVA. In fact, your best preparation for this chapter is a good understanding of the basic ANOVA procedure presented in Chapter 13.

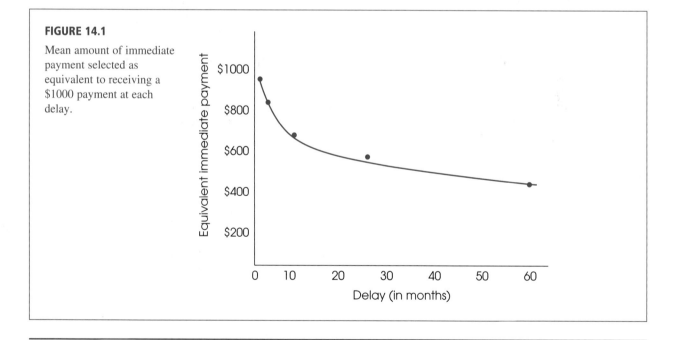

FIGURE 14.1

Mean amount of immediate payment selected as equivalent to receiving a $1000 payment at each delay.

14.1 OVERVIEW

In the preceding chapter, we introduced analysis of variance (ANOVA) as a hypothesis-testing procedure that is used to evaluate mean differences among two or more treatment conditions. The primary advantage of analysis of variance (compared to *t* tests) is that it allows researchers to test for significant mean differences in situations in which there are *more than two* treatment conditions. The *F*-ratio in the ANOVA compares all the different sample means in a single test statistic using a single alpha level. If the *t* statistic were used in a situation with more than two sample means, it would require multiple *t* tests to evaluate all the mean differences. Because each *t* test involves a risk of a Type I error, doing multiple tests would inflate the overall, experimentwise alpha level and produce an unacceptably high risk of a Type I error.

In this chapter, we present ANOVA as it is used for *repeated-measures* designs. You should recall that the defining characteristic of a repeated-measures study is that the same group of individuals participates in all of the different treatment conditions. Occasionally, researchers will try to duplicate some of the characteristics of a repeated-measures study by using a *matched-subjects* design. A matched-subjects design requires that each individual is matched, one-to-one, with an "equivalent" individual in each of the other treatments (see page 341 in Chapter 11). Although matched-subjects designs are relatively rare in current research, you should realize that the ANOVA presented in this chapter applies equally well to either a repeated-measures or a matched-subjects design.

THE SINGLE-FACTOR, REPEATED-MEASURES DESIGN

An independent variable is a manipulated variable in an experiment. A quasi-independent variable is not manipulated but defines the groups of scores in a nonexperimental design.

Chapter 13 introduced the general logic underlying ANOVA and presented the equations used for analyzing data from a single-factor, independent-measures research study. The term *independent-measures* indicates that the study uses a separate sample for each treatment condition being compared. In this chapter we extend the ANOVA procedure to research situations using single-factor, *repeated-measures* designs. Recall that the term "factor" is used in analysis of variance to refer to an independent (or quasi-independent) variable. Also recall that a repeated-measures design uses a single sample, with the same set of individuals participating in all of the different treatment conditions. Table 14.1 shows two sets of data representing common examples of single-factor, repeated-measures designs. Table 14.1(a), for example, shows data from a research study examining visual perception. The participants' task is to detect a very faint visual stimulus under three different conditions: (1) When there is no distraction, (2) when there is visual distraction from flashing lights, and (3) when there is auditory distraction from banging noises. You should notice that this study is an experiment because the researcher is manipulating an independent variable (the type of distraction). You should also realize that this is a repeated-measures study because the same sample is being measured in all three treatment conditions.

The single-factor, repeated-measures design can also be used with nonexperimental research in which the different conditions are not created by manipulating an independent variable. Table 14.1(b) shows results from a study examining the effectiveness of a clinical therapy for treating depression. In this example, measurements of depression were obtained for a single sample of patients when they first entered therapy, when therapy is completed, and again 6 months later to evaluate the long-term effectiveness of the therapy. The goal of the study is to evaluate mean differences in the level of depression as a function of the therapy. Notice that the three sets of data are all obtained from the same sample of participants; this is the essence of a repeated-measures design. In this case, a nonmanipulated variable (time) is the single factor in the study, and the three measurement times define the three levels of the factor. Another common example of this type of design is found in developmental psychology, when the participants' age is the quasi-independent variable. For example, a psychologist could examine the development of vocabulary skill by measuring vocabulary for a sample of

TABLE 14.1

Two sets of data representing typical examples of single-factor, repeated-measures research designs.

(a) Data from an experimental study evaluating the effects of different types of distraction on the performance of a visual detection task.

	Visual Detection Scores		
Participant	No Distraction	Visual Distraction	Auditory Distraction
A	47	22	41
B	57	31	52
C	38	18	40
D	45	32	43

(b) Data from a nonexperimental design evaluating the effectiveness of a clinical therapy for treating depression.

	Depression Scores		
Participant	Before Therapy	After Therapy	6-Month Follow Up
A	71	53	55
B	62	45	44
C	82	56	61
D	77	50	46
E	81	54	55

3-year-old children, then measuring the same children again at age 4 and at age 5. Again, this is a nonexperimental study (age is not manipulated), with the individuals measured at three different times.

HYPOTHESES FOR THE REPEATED-MEASURES ANOVA

The hypotheses for the repeated-measures ANOVA are exactly the same as those for the independent-measures ANOVA presented in Chapter 13. Specifically, the null hypothesis states that for the general population there are no mean differences among the treatment conditions being compared. In symbols,

$$H_0: \quad \mu_1 = \mu_2 = \mu_3 = \cdots$$

The null hypothesis states that, on average, all of the treatments have exactly the same effect. According to the null hypothesis, any differences that may exist among the sample means are not caused by systematic treatment effects but rather are the result of random and unsystematic factors.

The alternative hypothesis states that there are mean differences among the treatment conditions. Rather than specifying exactly which treatments are different, we use a generic version of H_1 which simply states that differences exist:

$$H_1: \quad \text{At least one treatment mean } (\mu) \text{ is different from another.}$$

Notice that the alternative says that, on average, the treatments do have different effects. Thus, the treatment conditions may be responsible for causing mean differences among the samples. As always, the goal of the ANOVA is to use the sample data to determine which of the two hypotheses is more likely to be correct.

INDIVIDUAL DIFFERENCES AND THE REPEATED-MEASURES *F*-RATIO

The test statistic for the repeated-measures ANOVA has the same structure that was used for the independent-measures ANOVA in Chapter 13. In each case, the F-*ratio* compares the actual mean differences between treatments with the amount of difference that would be expected just by chance. The numerator of the *F*-ratio measures the mean differences between treatments. The denominator measures how much difference is expected just by chance—that is, how big the differences should be if there is no treatment effect. As always, the *F*-ratio uses variance to measure the size of the differences. Thus, the *F*-ratio for the repeated-measures ANOVA has the general structure

$$F = \frac{\text{variance (differences) between treatments}}{\text{variance (differences) expected if there is no treatment effect}}$$

A large value for the *F*-ratio indicates that the differences between treatments are greater than would be expected by chance or error alone. If the *F*-ratio is larger than the critical value in the *F* distribution table, we can conclude that the differences between treatments are *significantly* larger than would be due to chance.

Although the structure of the *F*-ratio is the same for independent-measures and repeated-measures designs, there is a fundamental difference between the two designs that produces a corresponding difference in the two *F*-ratios. Specifically, one of the characteristics of a repeated-measures design is that it eliminates or removes the variance caused by individual differences. This point was first made when we introduced the repeated-measures design in Chapter 11 (page 354), but we will repeat it briefly now.

First, recall that the term *individual differences* refers to participant characteristics such as age, personality, and gender that vary from one person to another and may influence the measurements that you obtain for each person. In some research designs it is possible that the participants assigned to one treatment will have higher scores simply because they have different characteristics than the participants assigned to another treatment. For example, the participants in one treatment may be smarter, or older, or taller than those in another treatment. In this case, the mean difference between treatments is not necessarily caused by the treatments; instead, it may be caused by individual differences. With a repeated-measures design, however, you never need to worry about this problem. In a repeated-measures study, the participants in one treatment are exactly the same as the participants in every other treatment. Thus, individual differences cannot be responsible for causing differences between treatments. In terms of the *F*-ratio for a repeated-measures design, the variance between treatments (the numerator) does not contain any individual differences.

A repeated-measures design also allows you to remove individual differences from the variance in the denominator of the *F*-ratio. Because the same individuals are measured in every treatment condition, it is possible to measure the size of the individual differences. In Table 14.1(a), for example, participant A has scores that are consistently 10 points lower than the scores for participant B. Because the individual differences are systematic and predictable, they can be measured and separated from the random, unsystematic differences in the denominator of the *F*-ratio.

Thus, individual differences are automatically eliminated from the numerator of the repeated-measures *F*-ratio. In addition, they can be measured and removed from the denominator. As a result, the structure of the final *F*-ratio is as follows:

$$F = \frac{\begin{array}{c}\text{variance/differences between treatments}\\ \text{(without individual differences)}\end{array}}{\begin{array}{c}\text{variance/differences expected by chance}\\ \text{(with individual differences removed)}\end{array}}$$

The process of removing individual differences is an important part of the procedure for a repeated-measures analysis of variance.

The general purpose of a repeated-measures ANOVA is to determine whether the differences that are found between treatment conditions are significantly greater than would be expected if there is no treatment effect. In the numerator of the F-ratio, the *between-treatments variance* measures the actual mean differences between the treatment conditions. The variance in the denominator is intended to measure how much difference is reasonable to expect if there are no systematic treatment effects and no systematic individual differences. In other words, the denominator measures variability caused entirely by random and unsystematic factors. For this reason, the variance in the denominator is called the *error variance*. In this section we examine the elements that make up the two variances in the repeated-measures F-ratio.

The numerator of the F-ratio: between-treatments variance Logically, any differences that are found between treatments can be explained by only two factors:

1. Systematic Differences Caused by the Treatments. It is possible that the different treatment conditions really do have different effects and, therefore, cause the individuals' scores in one condition to be higher (or lower) than in another. Remember that the purpose for the research study is to determine whether or not a *treatment effect* exists.

2. Random, Unsystematic Differences. Even if there is no treatment effect, it is possible for the scores in one treatment condition to be different from the scores in another. For example, suppose that I measure your IQ score on a Monday morning. A week later I come back and measure your IQ again under exactly the same conditions. Will you get exactly the same IQ score both times? In fact, minor differences between the two measurement situations will probably cause you to end up with two different scores. For example, for one of the IQ tests you might be more tired, or hungry, or worried, or distracted than you were on the other test. These differences can cause your scores to vary. The same thing can happen in a repeated-measures research study. The same individuals are being measured two different times and, even though there may be no difference between the two treatment conditions, you can still end up with different scores. However, these differences are random and unsystematic and are classified as error variance.

Thus, it is possible that any differences (or variance) found between treatments could be caused by treatment effects, and it is possible that the differences could simply be due to chance. On the other hand, it is *impossible* that the differences between treatments are caused by individual differences. Because the repeated-measures design uses exactly the same individuals in every treatment condition, individual differences are *automatically eliminated* from the variance between treatments in the numerator of the F-ratio.

The denominator of the F-ratio: error variance The goal of the ANOVA is to determine whether the differences that are observed in the data are greater than would be expected without any systematic treatment effects. To accomplish this goal, the denominator of the F-ratio is intended to measure how much difference (or variance) is reasonable to expect from random and unsystematic factors. This means that we must measure the variance that exists when there are no treatment effects or any other systematic differences.

We begin exactly as we did with the independent-measures F-ratio; specifically, we calculate the variance that exists within treatments. Recall from Chapter 13 that within each treatment all of the individuals are treated exactly the same, so any differences that exist within treatments cannot be caused by treatment effects.

In a repeated-measures design, however, it is also possible that individual differences can cause systematic differences between the scores within treatments. For example, one individual may score consistently higher than another. To eliminate the individual differences from the denominator of the F-ratio, we measure the individual differences and then subtract them from the rest of the variability. The variance that remains is a measure of pure *error* without any systematic differences that can be explained by treatment effects or by individual differences.

In summary, the F-ratio for a repeated-measures ANOVA has the same basic structure as the F-ratio for independent measures (Chapter 13) except that it includes no variability due to individual differences. The individual differences are automatically eliminated from the variance between treatments (numerator) because the repeated-measures design uses the same subjects in all treatments. In the denominator, the individual differences are subtracted out during the analysis. As a result, the repeated-measures F-ratio has the following structure:

$$F = \frac{\text{between-treatments variance}}{\text{error variance}}$$

$$= \frac{\text{treatment effects} + \text{random, unsystematic differences}}{\text{random, unsystematic differences}} \tag{14.1}$$

Note that this F-ratio is structured so that there are no individual differences contributing to either the numerator or the denominator. When there is no treatment effect, the F-ratio is balanced because the numerator and denominator are both measuring exactly the same variance. In this case, the F-ratio should have a value near 1.00. When research results produce an F-ratio near 1.00, we conclude that there is no evidence of a treament effect and we fail to reject the null hypothesis. On the other hand, when a treatment effect does exist, it will contribute only to the numerator and should produce a large value for the F-ratio. Thus, a large value for F indicates that there is a real treatment effect and therefore we should reject the null hypothesis.

LEARNING CHECK

1. Explain why individual differences do not contribute to the between-treatments variability in a repeated-measures study.

2. What sources of variability contribute to the within-treatment variability for a repeated-measures study.

3. Describe the structure of the F-ratio for the repeated-measures ANOVA.

ANSWERS

1. Because the individuals in one treatment are exactly the same as the individuals in every other treatment, there are no individual differences from one treatment to another.

2. Variability (differences) within treatments is caused by individual differences and random, unsystematic differences.

3. The numerator of the F-ratio measures between treatments variability, which consists of treatment effects and random, unsystematic differences. The denominator measures variability that is exclusively caused by random, unsystematic differences.

14.2   TESTING HYPOTHESES WITH THE REPEATED-MEASURES ANOVA

The overall structure of the repeated-measures ANOVA is shown in Figure 14.2. Note that the ANOVA can be viewed as a two-stage process. In the first stage, the total variance is partitioned into two components: *between-treatments variance* and *within-treatments variance.* This stage is identical to the analysis that we conducted for an independent-measures design in Chapter 13.

The second stage of the analysis is intended to remove the individual differences from the denominator of the *F*-ratio. In the second stage, we begin with the variance within treatments and then measure and subtract out the *between-subject variance,* which measures the size of the individual differences. The remaining variance, often called the *residual variance,* or *error variance,* provides a measure of how much variance is reasonable to expect after the treatment effects and individual differences have been removed. The second stage of the analysis is what differentiates the repeated-measures ANOVA from the independent-measures ANOVA. Specifically, the repeated-measures design requires that the individual differences be removed.

DEFINITION

In a repeated-measures ANOVA, the denominator of the *F*-ratio is called the **residual variance,** or the **error variance,** and measures how much variance is expected if there are no systematic treatment effects and no individual differences contributing to the variability of the scores.

FIGURE 14.2

The partitioning of variance for a repeated-measures experiment.

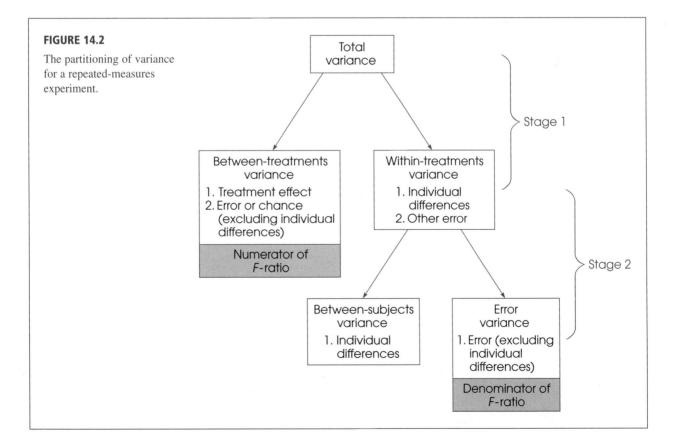

NOTATION FOR THE REPEATED-MEASURES ANOVA

We will use the data in Table 14.2 to introduce the notation for the repeated-measures ANOVA. The data represent the results of a study comparing three brands of pain reliever as well as a fourth condition in which the participants receive a placebo (sugar pill). You may have noticed that this research study and the numerical values in the table are identical to those used to demonstrate the independent-measures ANOVA in the previous chapter (Example 13.1, page 414). In this case, however, the data represent a repeated-measures study in which the same group of $n = 5$ individuals is tested in all four treatment conditions.

You should recognize that most of the notation in Table 14.2 is identical to the notation used in an independent-measures analysis (Chapter 13). For example, there are $n = 5$ participants who are tested in $k = 4$ treatment conditions, producing a total of $N = 20$ scores that add up to a grand total of $G = 60$. Note, however, that $N = 20$ now refers to the total number of scores in the study, not the number of participants.

The repeated-measures ANOVA introduces only one new notational symbol. The letter P is used to represent the total of all the scores for each individual in the study. You can think of the P values as "Person totals" or "Participant totals." In Table 14.2, for example, participant A had scores of 3, 4, 6, and 7 for a total of $P = 20$. The P values will be used to define and measure the magnitude of the individual differences in the second stage of the analysis.

EXAMPLE 14.1

We use the data in Table 14.2 to demonstrate the repeated-measures ANOVA. Again, the goal of the test is to determine whether there are any significant differences among the four drugs being compared. Specifically, are any of the sample mean differences greater than would be expected by chance alone, without any treatment effect?

STAGE 1 OF THE REPEATED-MEASURES ANALYSIS

The first stage of the repeated-measures analysis is identical to the independent-measures ANOVA that was presented in Chapter 13. Specially, the SS and df for the total set of scores are analyzed into within-treatments and between-treatments components.

Because the numerical values in Table 14.2 are the same as the values used in Example 13.1 (page 414), the computations for the first stage of the repeated-measures analysis are identical to those in Example 13.1. Rather than repeating the same arithmetic, the results of the first stage of the repeated-measures analysis can be summarized as follows:

TABLE 14.2

The effect of drug treatment on the amount of time (in seconds) a stimulus is endured.

Person	Placebo	Drug A	Drug B	Drug C	Person Totals	
A	3	4	6	7	$P = 20$	$n = 5$
B	0	3	3	6	$P = 12$	$k = 4$
C	2	1	4	5	$P = 12$	$N = 20$
D	0	1	3	4	$P = 8$	$G = 60$
E	0	1	4	3	$P = 8$	$\Sigma X^2 = 262$
	$T = 5$	$T = 10$	$T = 20$	$T = 25$		
	$SS = 8$	$SS = 8$	$SS = 6$	$SS = 10$		

Total:

For more details on the formulas
and calculations see Example
13.1, pages 414–416.

$$SS_{\text{total}} = \Sigma X^2 - \frac{G^2}{N} = 262 - \frac{(60)^2}{20} = 262 - 180 = 82$$

$$df_{\text{total}} = N - 1 = 19$$

Within Treatments:

$$SS_{\text{within treatments}} = \Sigma SS_{\text{inside each treatment}} = 8 + 8 + 6 + 10 = 32$$

$$df_{\text{within treatments}} = \Sigma df_{\text{inside each treatment}} = 4 + 4 + 4 + 4 = 16$$

Between Treatments:

$$SS_{\text{between treatments}} = \Sigma\frac{T^2}{n} - \frac{G^2}{N} = \frac{5^2}{5} + \frac{10^2}{5} + \frac{20^2}{5} + \frac{25^2}{5} - \frac{60^2}{20} = 50$$

$$df_{\text{between treatments}} = k - 1 = 3$$

This completes the first stage of the repeated-measures analysis. Note that the two components, between and within, add up to the total for the SS values and for the df values. Also note that the between-treatments SS and df values provide a measure of the mean differences between treatments and will be used to compute the variance in the numerator of the final F-ratio.

STAGE 2 OF THE REPEATED-MEASURES ANALYSIS

The second stage of the analysis involves removing the individual differences from the denominator of the F-ratio. Because the same individuals are used in every treatment, it is possible to measure the size of the individual differences. For the data in Table 14.2, for example, person A tends to have the highest scores and participants D and E tend to have the lowest scores. These individual differences are reflected in the P values, or person totals in the right-hand column. We will use these P values to calculate an SS between subjects in much the same way that we used the treatment totals, the T values, to compute the SS between treatments. Specifically, the formula for SS between subjects is

$$SS_{\text{between subjects}} = \Sigma\frac{P^2}{k} - \frac{G^2}{N} \tag{14.2}$$

Notice that the formula for the between-subjects SS has exactly the same structure as the formula for the between-treatments SS. In this case we use the person totals (P values) instead of the treatment totals (T values). Each P value is squared and divided by the number of scores that were added to obtain the total. In this case, each person has k scores, one for each treatment. Box 14.1 presents another demonstration of the similarity of the formulas for SS between subjects and SS between treatments. For the data in Table 14.2,

$$SS_{\text{between subjects}} = \frac{20^2}{4} + \frac{12^2}{4} + \frac{12^2}{4} + \frac{8^2}{4} + \frac{8^2}{4} - \frac{60^2}{20}$$

$$= 100 + 36 + 36 + 16 + 16 - 180$$

$$= 24$$

The value of $SS_{\text{between subjects}}$ provides a measure of the size of the individual differences—that is, the differences between subjects. In the second stage of the analysis, we simply subtract out the individual differences to obtain the measure of

BOX 14.1 $SS_{\text{between subjects}}$ AND $SS_{\text{between treatments}}$

The data for a repeated-measures study are normally presented in a matrix, with the treatment conditions determining the columns and the participants defining the rows. The data in Table 14.2 demonstrate this normal presentation. The calculation of $SS_{\text{between treatments}}$ provides a measure of the differences between treatment conditions—that is, a measure of the mean differences between the *columns* in the data matrix. For the data in Table 14.2, the column totals are 5, 10, 20, and 25. These values are variable, and $SS_{\text{between treatments}}$ measures the amount of variability.

The following table reproduces the data from Table 14.2, but now we have turned the data matrix on its side so that the people define the columns and the treatment conditions define the rows.

In this new format, the differences between the columns represent the between-subjects variability. The column totals are now P values (instead of T values) and the number of scores in each column is now identified by k (instead of n). With these changes in notation, the formula for $SS_{\text{between subjects}}$ has exactly the same structure as the formula for $SS_{\text{between treatments}}$. If you examine the two equations, the similarity should be clear.

	A	B	Person C	D	E	
Placebo	3	0	2	0	0	T = 5
Drug A	4	3	1	1	1	T = 10
Drug B	6	3	4	3	4	T = 20
Drug C	7	6	5	4	3	T = 25
	P = 20	P = 12	P = 12	P = 8	P = 8	

error that will form the denominator of the F-ratio. Thus, the final step in the analysis of SS is

$$SS_{\text{error}} = SS_{\text{within treatments}} - SS_{\text{between subjects}} \qquad (14.3)$$

For the data in Table 14.2,

$$SS_{\text{error}} = 32 - 24 = 8$$

The analysis of degrees of freedom follows exactly the same pattern that was used to analyze SS. Remember that we are using the P values to measure the magnitude of the individual differences. The number of P values corresponds to the number of subjects, n, so the corresponding df is

$$df_{\text{between subjects}} = n - 1 \qquad (14.4)$$

For the data in Table 14.2, there are $n = 5$ subjects and

$$df_{\text{between subjects}} = 5 - 1 = 4$$

Next, we subtract the individual differences from the within-subjects component to obtain a measure of error. In terms of degrees of freedom,

$$df_{\text{error}} = df_{\text{within treatments}} - df_{\text{between subjects}} \qquad (14.5)$$

For the data in Table 14.2,

$$df_{\text{error}} = 16 - 4 = 12$$

Remember: The purpose for the second stage of the analysis is to measure the individual differences and then remove the individual differences from the denominator of the F-ratio. This goal is accomplished by computing SS and df between subjects (the individual differences) and then subtracting these values from the within-treatments values. The result is a measure of variability due to error with the individual differences removed. This error variance (SS and df) will be used in the denominator of the F-ratio.

CALCULATION OF THE VARIANCES (MS VALUES) AND THE F-RATIO

The final calculation in the analysis is the F-ratio, which is a ratio of two variances. Each variance is called a *mean square,* or *MS,* and is obtained by dividing the appropriate SS by its corresponding df value. The MS in the numerator of the F-ratio measures the size of the differences between treatments and is calculated as

$$MS_{\text{between treatments}} = \frac{SS_{\text{between treatments}}}{df_{\text{between treatments}}}$$

(14.6)

For the data in Table 14.2,

$$MS_{\text{between treatments}} = \frac{50}{3} = 16.67$$

The denominator of the F-ratio measures how much difference is reasonable to expect just by chance, after the individual differences have been removed. This is the error variance or the residual obtained in stage 2 of the analysis.

$$MS_{\text{error}} = \frac{SS_{\text{error}}}{df_{\text{error}}}$$

(14.7)

For the data in Table 14.2,

$$MS_{\text{error}} = \frac{8}{12} = 0.67$$

Finally, the F-ratio is computed as

$$F = \frac{MS_{\text{between treatments}}}{MS_{\text{error}}}$$

(14.8)

For the data in Table 14.2,

$$F = \frac{16.67}{0.67} = 24.88$$

Once again, notice that the repeated-measures ANOVA uses MS_{error} in the denominator of the F-ratio. This MS value is obtained in the second stage of the analysis, after the individual differences have been removed. As a result, individual differences are completely eliminated from the repeated-measures F-ratio, so that the general structure is

$$F = \frac{\text{treatment effects} + \text{unsystematic differences (without individual diff's)}}{\text{unsystematic differences (without individual diff's)}}$$

For the data we have been examining, the F-ratio is $F = 24.88$, indicating that the differences between treatments (numerator) are almost 25 times bigger than you would expect without any treatment effects (denominator). A ratio this large provides clear evidence that there is a real treatment effect. To verify this conclusion you must consult the F distribution table to determine the appropriate critical value for the test. The degrees of freedom for the F-ratio are determined by the two variances that form the numerator and the denominator. For a repeated-measures ANOVA, the df values for the F-ratio are reported as

$$df = df_{\text{between treatments}}, df_{\text{error}}$$

For the example we are considering, the F-ratio has $df = 2, 12$ ("degrees of freedom equal two and twelve"). Using the F distribution table (page 731) with $\alpha = .05$, the critical value is $F = 3.88$, and with $\alpha = .01$ the critical value is $F = 6.93$. Our obtained F-ratio, $F = 24.88$, is well beyond either of the critical values, so we can conclude that the differences between treatments are *significantly* greater than expected by chance using either $\alpha = .05$ or $\alpha = .01$.

The summary table for the repeated-measures ANOVA from Example 14.1 is presented in Table 14.3. Although these tables are no longer commonly used in research reports, they provide a concise format for displaying all of the elements of the analysis.

MEASURING EFFECT SIZE FOR THE REPEATED-MEASURES ANOVA

The most common method for measuring effect size with ANOVA is to compute the percentage of variance that is explained by the treatment differences. The percentage of explained variance is usually identified by r^2, but in the context of ANOVA it is commonly identified as η^2 (eta squared). In Chapter 13, for the independent-measures design, we computed η^2 as

$$\eta^2 = \frac{SS_{\text{between treatments}}}{SS_{\text{between treatments}} + SS_{\text{within treatments}}} = \frac{SS_{\text{between treatments}}}{SS_{\text{total}}}$$

The intent is to measure how much of the total variability is explained by the differences between treatments. With a repeated-measures design, however, there is another component that can explain some of the variability in the data. Specifically, part of the variability is caused by differences between individuals. In Table 14.2, for example, person A consistently scored higher than person B. This consistent difference explains some of the variability in the data. When computing the size of the treatment effect, it is customary to remove any variability that can be explained by other factors, and then compute the percentage of the remaining variability that can be explained by the treatment effects. Thus, for a repeated-measures ANOVA, the variability from the individual differences is removed before computing η^2. As a result, η^2 is computed as

$$\eta^2 = \frac{SS_{\text{between treatments}}}{SS_{\text{total}} - SS_{\text{between subjects}}} \tag{14.9}$$

TABLE 14.3

A summary table for the repeated-measures ANOVA for the data from Example 14.1.

Source	SS	df	MS	F
Between treatments	50	3	16.67	$F(3,12) = 24.88$
Within treatments	32	16		
Between subjects	24	4		
Error	8	12	0.67	
Total	82	19		

Because Equation 14.9 computes a percentage that is not based on the total variability of the scores (one part, $SS_{\text{between subjects}}$, is removed), the result is often called a *partial eta squared.*

The general goal of Equation 14.9 is to calculate a percentage of the variability that has not already been explained by other factors. Thus, the denominator of Equation 14.9 is limited to variability from the treatment differences and variability that is exclusively from random, unsystematic factors. With this in mind, an equivalent version of the η^2 formula is

$$\eta^2 = \frac{SS_{\text{between treatments}}}{SS_{\text{between treatments}} + SS_{\text{error}}} \tag{14.10}$$

In this new version of the eta-squared formula, the denominator consists of the variability that is explained by the treatment differences plus the other *unexplained* variability. Using either formula, the data from Example 14.1 produce

$$\eta^2 = \frac{50}{58} = 0.862 \quad (\text{or } 86.2\%)$$

This result means that 86.2% of the variability in the data (except for the individual differences) is accounted for by the differences between treatments.

IN THE LITERATURE
REPORTING THE RESULTS OF A REPEATED-MEASURES ANOVA

As described in Chapter 13 (page 419), the format for reporting ANOVA results in journal articles consists of

1. A summary of descriptive statistics (at least treatment means and standard deviations, and tables or graphs as needed)

2. A concise statement of the outcome of the ANOVA

For the study in Example 14.1, the report could state:

> The means and standard deviations for the four drug conditions are shown in Table 1. A repeated-measures analysis of variance indicated significant differences in pain tolerance among the four drugs, $F(3, 12) = 24.88$, $p < .01$, $\eta^2 = 0.862$.
>
> **TABLE 1**
>
> Amount of pain tolerated
>
> | | Placebo | Drug A | Drug B | Drug C |
> | ---- | ------- | ------ | ------ | ------ |
> | *M* | 1.00 | 2.00 | 4.00 | 5.00 |
> | *SD* | 1.41 | 1.41 | 1.22 | 1.58 |

POST HOC TESTS WITH REPEATED MEASURES   Recall that ANOVA provides an overall test of significance for the mean differences between treatments. When the null hypothesis is rejected, it indicates only that there is a difference between at least two of the treatment means. If $k = 2$, it is obvious where the difference lies in the experiment. However, when k is greater than 2, the situation becomes more complex. To determine exactly where significant differences exist, the researcher must follow the ANOVA with post hoc tests. In Chapter 13, we used Tukey's HSD and the Scheffé test to make these multiple comparisons among treatment means. These two procedures attempt to control the overall alpha level by making adjustments for the number of potential comparisons.

For a repeated-measures ANOVA, Tukey's HSD and the Scheffé test can be used in the exact same manner as was done for the independent-measures ANOVA, *provided* that you substitute MS_{error} in place of $MS_{within\ treatments}$ in the formulas and use df_{error} in place of $df_{within\ treatments}$ when locating the critical value in a statistical table. Note that statisticians are not in complete agreement about the appropriate error term in post hoc tests for repeated-measures designs (for a discussion, see Keppel, 1973, or Keppel & Zedeck, 1989).

ASSUMPTIONS OF THE REPEATED-MEASURES ANOVA

The basic assumptions for the repeated-measures ANOVA are identical to those required for the independent-measures ANOVA.

1. The observations within each treatment condition must be independent (see page 253).
2. The population distribution within each treatment must be normal. (As before, the assumption of normality is important only with small samples.)
3. The variances of the population distributions for each treatment should be equivalent.

For the repeated-measures ANOVA, there is an additional assumption, called homogeneity of covariance. Basically, it refers to the requirement that the relative standing of each subject be maintained in each treatment condition. This assumption is violated if the effect of the treatment is not consistent for all of the subjects or if order effects exist for some, but not other, subjects. This issue is very complex and is beyond the scope of this book. However, methods do exist for dealing with violations of this assumption (for a discussion, see Keppel, 1973).

LEARNING CHECK

1. Explain how SS_{error} is computed in the repeated-measures ANOVA.

2. A repeated-measures study is used to evaluate the mean differences among three treatment conditions using a sample of $n = 8$ participants. What are the df values for the F-ratio?

3. For the following data, compute $SS_{between\ treatments}$ and $SS_{between\ subjects}$.

	Treatment				
Subject	1	2	3	4	
A	2	2	2	2	$G = 32$
B	4	0	0	4	$\Sigma X^2 = 96$
C	2	0	2	0	
D	4	2	2	4	
	$T = 12$	$T = 4$	$T = 6$	$T = 10$	
	$SS = 4$	$SS = 4$	$SS = 3$	$SS = 11$	

ANSWERS   1. $SS_{error} = SS_{within\ treatments} - SS_{between\ subjects}$ (Variability from individual differences is subtracted from the within-treatments variability.)

2. $df = 2, 14$

3. $SS_{between\ treatments} = 10$, $SS_{between\ subjects} = 8$

14.3   AN ALTERNATIVE VIEW OF THE REPEATED-MEASURES ANOVA

As we noted earlier in the chapter, the repeated-measures ANOVA can be viewed as a two-stage process. In the first stage, the variability from systematic treatment effects is measured and removed. In the second stage, the variability from systematic individual differences is measured and removed. Although we have presented formulas for computing these different components of variability, it is also possible to simply remove the treatment effects and individual differences from the data and then compute the sum of squared deviations, SS, for the resulting scores. In this section, we demonstrate this process.

We begin with the original scores from Example 14.1, which are reproduced in Table 14.4(a). To measure the total variability for these scores, we computed $SS_{total} = 82$. Now we will modify the scores by first removing the treatment effects and then removing the individual differences.

TABLE 14.4

The step-by-step process of removing treatment effects and individual differences from repeated-measures data. (a) The original data. (b) The data after the treatment effects have been removed, reducing SS by 50 points. (c) The data after the individual differences are also removed, reducing SS by an additional 24 points.

(a) The original data, including treatment effects and individual differences. For these scores, $SS_{total} = 82$.

Person	Placebo	Drug A	Drug B	Drug C	Person Totals	
A	3	4	6	7	$P = 20$	$n = 5$
B	0	3	3	6	$P = 12$	$k = 4$
C	2	1	4	5	$P = 12$	$N = 20$
D	0	1	3	4	$P = 8$	$G = 60$
E	0	1	4	3	$P = 8$	$\Sigma X^2 = 262$
	$M = 1$	$M = 2$	$M = 4$	$M = 5$		
	$T = 8$	$T = 10$	$T = 20$	$T = 25$		

(b) The data after the treatment effects have been removed. For these scores $SS_{total} = 32$. Removing the treatment effects reduces the SS by 50 points.

Person	Placebo	Drug A	Drug B	Drug C	Person Totals	
A	5	5	5	5	$P = 20$	$n = 5$
B	2	4	2	4	$P = 12$	$k = 4$
C	4	2	3	3	$P = 12$	$N = 20$
D	2	2	2	2	$P = 8$	$G = 60$
E	2	2	3	1	$P = 8$	$\Sigma X^2 = 212$
	$M = 3$	$M = 3$	$M = 3$	$M = 3$		
	$T = 15$	$T = 15$	$T = 15$	$T = 15$		

TABLE 14.4 (Continued)

(c) The data after the individual differences (as well as the treatment effects) have been removed. For these scores, $SS_{total} = 8$. Removing the individual differences reduces the SS by an additional 24 points.

Person	Placebo	Drug A	Drug B	Drug C	Person Totals	
A	3	3	3	3	$P = 12$	$n = 5$
B	2	4	2	4	$P = 12$	$k = 4$
C	4	2	3	3	$P = 12$	$N = 20$
D	3	3	3	3	$P = 12$	$G = 60$
E	3	3	4	2	$P = 12$	$\Sigma X^2 = 188$
	$M = 3$	$M = 3$	$M = 3$	$M = 3$		
	$T = 15$	$T = 15$	$T = 15$	$T = 15$		

REMOVING AND MEASURING TREATMENT EFFECTS

First, notice that the four treatment conditions all have different means. These mean differences are the treatment effect. We can remove the treatment effect by simply eliminating the mean differences between treatments. To accomplish this,

1. Add 2 points to the score for each individual in the placebo condition.
2. Add 1 point to each score for individuals who received drug A.
3. Subtract 1 point from score for individuals receiving drug B.
4. Subtract 2 points from each score for individuals receiving drug C.

The resulting scores are shown in Table 14.4(b). Notice that all four treatment conditions now have exactly the same mean, $M = 3$, so there are no differences between treatments. Once again, we can measure the total variability for these scores by computing SS_{total}. For the adjusted scores, without any treatment effect,

$$SS_{total} = \Sigma X^2 - \frac{G^2}{N}$$

$$= 212 - \frac{(60)^2}{20}$$

$$= 212 - 180$$

$$= 32$$

Thus, the original scores, including the treatment effects, have $SS_{total} = 82$. After removing the treatment effect, the scores have $SS_{total} = 32$. Therefore, the treatment effects contributed $SS = 50$ points to the variability in the original scores. This is exactly the same value that we obtained for the between-treatments component in the analysis of variance, $SS_{between\ treatments} = 50$.

REMOVING AND MEASURING INDIVIDUAL DIFFERENCES

In the same way that we removed the treatment effects by eliminating the mean differences between treatments, we can remove the individual differences by eliminating the mean differences between people. To accomplish this, we begin with the modified scores (without treatment effects) in Table 14.4(b) and then,

1. Subtract 2 points from each score for person A.
2. Add 1 point to each score for person D.
3. Add 1 point to each score for person E.

The resulting scores are shown in Table 14.4(c). Notice that all five people now have exactly the same total, $P = 12$, so there are no differences between individuals. Once again, we can measure the total variability for these scores by computing SS_{total}. For the scores in Table 14.4(c), without any treatment effect and without individual differences,

$$SS_{total} = \Sigma X^2 - \frac{G^2}{N}$$

$$= 188 - \frac{(60)^2}{20}$$

$$= 188 - 180$$

$$= 8$$

There are two points to be made from this final calculation. First, when the treatment effect was removed [(Table 14.4(b)], the scores had $SS_{total} = 32$. Now that we have also removed the individual differences, the variability has been reduced to $SS_{total} = 8$. The difference between these two values, $32 - 8 = 24$ points, is the variability caused by the individual differences. Again, this is exactly the same value that we obtained for the between-subjects variability in the analysis of variance, $SS_{between subjects} = 24$. Second, the scores that remain after the treatment effects and the individual differences have been removed [Table 14.4(c)] have no systematic variability. The random, unsystematic variability that exists in these scores is exactly what we want for the denominator of the repeated-measures F-ratio. We calculated the total variability for these scores and found $SS_{total} = 8$, which is exactly the same value that we obtained for the error variability in the analysis of variance, $SS_{error} = 8$.

By removing the systematic sources of variability, one at a time, we are able to measure the amount of variability from each source, and we are able to determine how much variability remains when the systematic portion in eliminated. For this example, we first removed the treatment effects to determine how much variability was caused by differences between treatments. Then we removed the individual differences to determine how much variability was caused by differences between people. Finally, the variability that remained after treatment effects and individual differences were both removed, is the random, unsystematic variability that appears in the denominator of the F-ratio.

14.4 | ADVANTAGES AND DISADVANTAGES OF THE REPEATED-MEASURES DESIGN

When we first encountered the repeated-measure design (Chapter 11), we noted that this type of research study has certain advantages and disadvantages (pages 353–355). On the bright side, a repeated-measures study may be desirable if the supply of participants is limited. A repeated-measures study is economical in that the research requires relatively few participants. Also, a repeated-measures design eliminates or minimizes most of the problems associated with individual differences. However, disadvantages also exist. These take the form of order effects, such as fatigue, that can make the interpretation of the data difficult.

Now that we have introduced the repeated-measures ANOVA we can examine one of the primary advantages of this design—namely, the elimination of variability caused by individual differences. Consider the structure of the F-ratio for both the independent- and the repeated-measures designs.

$$F = \frac{\text{treatment effects} + \text{random, unsystematic differences}}{\text{random, unsystematic differences}}$$

In each case, the goal of the analysis is to determine whether the data provide evidence for a treatment effect. If there is no treatment effect, the numerator and denominator are both measuring the same random, unsystematic variance and the F-ratio should produce a value near 1.00. On the other hand, the existence of a treatment effect should make the numerator substantially larger than the denominator and result in a large value for the F-ratio.

For the independent-measures design, the unsystematic differences include individual differences as well as other random sources of error. Thus, for the independent-measures ANOVA, the F-ratio has the following structure:

$$F = \frac{\text{treatment effect} + \text{individual differences and other error}}{\text{individual differences and other error}}$$

For the repeated-measures design, the individual differences are eliminated or subtracted out, and the resulting F-ratio is structured as follows:

$$F = \frac{\text{treatment effects} + \text{error (excluding individual differences}}{\text{error (excluding individual differences)}}$$

The removal of individual differences from the analysis becomes an advantage in situations in which very large individual differences exist among the participants being studied.

When individual differences are extraordinarily large, the presence of a treatment effect may be masked if an independent-measures study is performed. In this case, a repeated-measures design would be more sensitive in detecting a treatment effect because individual differences do not influence the value of the F-ratio.

This point will become evident in the following example. Suppose that an experiment is performed in two ways, with an independent-measures design and as a repeated-measures experiment. Also, let's suppose that we know how much variability is accounted for by the different sources of variance. For example,

$$\text{treatment effect} = 10 \text{ units of variance}$$

$$\text{individual differences} = 1000 \text{ units of variance}$$

$$\text{other error} = 1 \text{ unit of variance}$$

Notice that a very large amount of the variability in the experiment is due to individual differences. By comparing the F-ratios of both types of experiments, we are able to see a fundamental difference between the two types of experimental designs. For the independent-measures experiment, we obtain

$$F = \frac{\text{treatment effect} + \text{individual differences} + \text{error}}{\text{individual differences} + \text{error}}$$

$$= \frac{10 + 1000 + 1}{1000 + 1}$$

$$= \frac{1011}{1001}$$

$$= 1.01$$

Thus, the independent-measures ANOVA produces an F-ratio of $F = 1.01$. Recall that the F-ratio is structured to produce $F = 1.00$ if there is no treatment effect whatsoever. In this case, the F-ratio is almost identical to 1.00 and strongly suggests that there is little or no treatment effect. In fact, the 10-point treatment effect that does exist has been overwhelmed by all the other variance so that it is almost invisible.

Now consider what happens with a repeated-measures analysis. With the individual differences removed, the F-ratio becomes:

$$F = \frac{\text{treatment effect} + \text{error}}{\text{error}}$$

$$= \frac{10 + 1}{1}$$

$$= \frac{11}{1}$$

$$= 11$$

For the repeated-measures analysis, the numerator of the F-ratio (which includes the treatment effect) is 11 times larger than the denominator (which has no treatment effect). This result strongly indicates that there is a substantial treatment effect. In this example, the F-ratio is much larger for the repeated-measures study because the individual differences, which are extremely large, have been removed. In the independent-measures ANOVA, the presence of a treatment effect is obscured by the influence of individual differences. This problem is remedied by the repeated-measures design, in which variability due to individual differences has been partitioned out of the analysis. When the individual differences are large, a repeated-measures experiment may provide a more sensitive test for a treatment effect. In statistical terms, a repeated-measures test has more *power* than an independent-measures test; that is, it is more likely to detect a real treatment effect.

14.5 INDIVIDUAL DIFFERENCES AND THE CONSISTENCY OF THE TREATMENT EFFECTS

As we have demonstrated, one major advantage of a repeated-measures design is that it removes individual differences from the denominator of the F-ratio, which usually increases the likelihood of obtaining a significant result. However, removing individual differences is an advantage only when the treatment effects are reasonably consistent for all of the participants. If the treatment effects are not consistent across participants, the individual differences tend to disappear and the denominator is not noticeably reduced by removing them. This phenomenon is demonstrated in the following example.

TABLE 14.5

Data from a repeated-measures study comparing three treatments. The data show consistent treatment effects from one participant to another which produce consistent and relatively large differences in the individual P totals.

		Treatment		
Person	I	II	III	
A	0	1	2	$P = 3$
B	1	2	3	$P = 6$
C	2	4	6	$P = 12$
D	3	5	7	$P = 15$
	$T = 6$	$T = 12$	$T = 18$	
	$SS = 5$	$SS = 10$	$SS = 17$	

EXAMPLE 14.2

Table 14.5 presents hypothetical data from a repeated-measures research study. We constructed the data specifically to demonstrate the relationship between consistent treatment effects and large individual differences.

First, notice the consistency of the treatment effects. Treatment II has the same effect on every participant, increasing everyone's score by 1 or 2 points compared to treatment I. Also, treatment III produces a consistent increase of 1 or 2 points compared to treatment II. One consequence of the consistent treatment effects is that the individual differences are maintained in all of the treatment conditions. For example, participant A has the lowest score in all three treatments, and participant D always has the highest score. The participant totals (P values) reflect the consistent differences. For example, participant D has the largest score in every treatment and, therefore, has the largest P value. Also notice that there are big differences between the P totals from one individual to the next. For these data, $SS_{\text{between subjects}} = 30$ points.

Now consider the data in Table 14.6. To construct these data we started with the same numbers within each treatment that were used in Table 14.5. However, we scrambled the numbers to eliminate the consistency of the treatment effects. In Table 14.6, for example, two participants show an increase in scores as they go from treatment I to treatment II, and two show a decrease. The data also show an inconsistent treatment effect as the participants go from treatment II to treatment III. One consequence of the inconsistent treatment effects is that there are no consistent

TABLE 14.6

Data from a repeated-measures study comparing three treatments. The data show treatment effects that are not consistent from one participant to another and, as a result, produce relatively small differences in the individual P totals. Note that the data have exactly the same scores within each treatment as the data in Table 14.5, however, the scores have been scrambled to eliminate the consistency of the treatment effects.

		Treatment		
Person	I	II	III	
A	0	4	3	$P = 7$
B	1	5	2	$P = 8$
C	2	1	7	$P = 10$
D	3	2	6	$P = 11$
	$T = 6$	$T = 12$	$T = 18$	
	$SS = 5$	$SS = 10$	$SS = 17$	

individual differences between participants. Participant C, for example, has the lowest score in treatment II and the highest score in treatment III. As a result, there are no longer consistent differences between the individual participants. All of the P totals are about the same. For these data, $SS_{\text{between subjects}} = 3.33$ points. Because the two sets of data (Tables 14.5 and 14.6) have the same treatment totals (T values) and SS values, they have the same $SS_{\text{between treatments}}$ and $SS_{\text{within treatments}}$. For both sets of data,

$$SS_{\text{between treatments}} = 18 \text{ and } SS_{\text{within treatments}} = 32$$

However, there is a huge difference between the two sets of data when you compute SS_{error} for the denominator of the F-ratio. For the data in Table 14.5, with consistent treatment effects and large individual differences,

$$SS_{\text{error}} = SS_{\text{within treatments}} - SS_{\text{between subjects}}$$
$$= 32 - 30$$
$$= 2$$

For the data in Table 14.6, with no consistent treatment effects and relatively small differences between the individual P totals,

$$SS_{\text{error}} = SS_{\text{within treatments}} - SS_{\text{between subjects}}$$
$$= 32 - 3.33$$
$$= 28.67$$

Thus, consistent treatment effects tend to produce a relatively small error term for the F-ratio. As a result, consistent treatment effects are more likely to be statistically significant (reject the null hypothesis). For the examples we have been considering, the data in Table 14.5 produce an F-ratio of $F = 27.0$. With $df = 2, 6$, this F-ratio is well into the critical region for $\alpha = .05$ or $.01$ and we conclude that there are significant differences among the three treatments. On the other hand, the same mean differences in Table 14.6 produce $F = 1.88$. With $df = 2, 6$, this value is not in the critical region for $\alpha = .05$ or $.01$ and we conclude that there are no significant differences.

In summary, when treatment effects are consistent from one individual to another, the individual differences also tend to be consistent and relatively large. The large individual differences get subtracted from the denominator of the F-ratio producing a larger value for F and increasing the likelihood that the F-ratio will be in the critical region.

SUMMARY

1. The repeated-measures ANOVA is used to evaluate the mean differences obtained in a research study comparing two or more treatment conditions using the same sample of individuals in each condition. The test statistic is an F-ratio, in which the numerator measures the variance (differences) between treatments and the denominator measures the variance (differences) due to chance or error.

$$F = \frac{MS_{\text{between treatments}}}{MS_{\text{error}}}$$

2. The first stage of the repeated-measures ANOVA is identical to the independent-measures analysis and separates the total variability into two components: between-treatments and within-treatments. Because a repeated-measures design uses the same subjects in

every treatment condition, the differences between treatments cannot be caused by individual differences. Thus, individual differences are automatically eliminated from the between-treatments variance in the numerator of the F-ratio.

3. In the second stage of the repeated-measures analysis, individual differences are computed and removed from the denominator of the F-ratio. To remove the individual differences, you first compute the variability between subjects (SS and df) and then subtract these values from the corresponding within-treatments values. The residual provides a measure of error excluding individual differences, which is the appropriate denominator for the repeated-measures F-ratio. The complete set of equations for analyzing SS and df for the repeated-measures analysis of variance are presented in Figure 14.3.

4. Effect size for the repeated-measures ANOVA is measured by computing eta squared, the percentage of variance accounted for by the treatment effect. For the repeated-measures ANOVA

$$\eta^2 = \frac{SS_{\text{between treatments}}}{SS_{\text{total}} - SS_{\text{between subjects}}}$$

$$= \frac{SS_{\text{between treatments}}}{SS_{\text{between treatments}} + SS_{\text{error}}}$$

Because part of the variability (the SS due to individual differences) is removed before computing η^2, this measure of effect size is often called a partial eta squared.

5. When the obtained F-ratio is significant (that is, H_0 is rejected), it indicates that a significant difference lies between at least two of the treatment conditions. To determine exactly where the difference lies, post hoc comparisons may be made. Post hoc tests, such as Tukey's HSD, use MS_{error} rather than $MS_{\text{within treatments}}$ and df_{error} instead of $df_{\text{within treatments}}$.

6. A repeated-measures ANOVA eliminates the influence of individual differences from the analysis. If individual differences are extremely large, a treatment effect might be masked in an independent-measures experiment. In this case, a repeated-measures design might be a more sensitive test for a treatment effect.

KEY TERMS

individual differences	treatment effect	between-subjects variance	error variance
F-ratio	error	within-treatments variance	mean squares
between-treatments variance			

RESOURCES

There are a tutorial quiz and other learning exercises for Chapter 14 at the Wadsworth website, www.cengage.com/psychology/gravetter. See page 29 for more information about accessing the website.

ENHANCED
Web**Assign**

Guided interactive tutorials, end-of-chapter problems, and related testbank items may be assigned online at WebAssign.

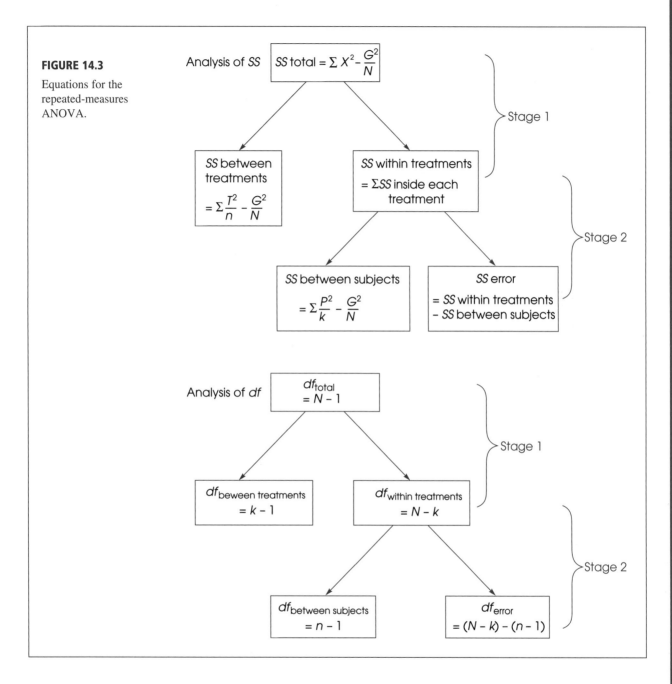

FIGURE 14.3

Equations for the repeated-measures ANOVA.

Analysis of SS

$$SS\text{ total} = \Sigma X^2 - \frac{G^2}{N}$$

SS between treatments
$$= \Sigma \frac{T^2}{n} - \frac{G^2}{N}$$

SS within treatments
$$= \Sigma SS \text{ inside each treatment}$$

SS between subjects
$$= \Sigma \frac{P^2}{k} - \frac{G^2}{N}$$

SS error
$$= SS \text{ within treatments} - SS \text{ between subjects}$$

Stage 1

Stage 2

Analysis of df

$$df_{\text{total}} = N - 1$$

$$df_{\text{beween treatments}} = k - 1$$

$$df_{\text{within treatments}} = N - k$$

$$df_{\text{between subjects}} = n - 1$$

$$df_{\text{error}} = (N - k) - (n - 1)$$

Stage 1

Stage 2

WebTUTOR

If you are using WebTutor along with this book, there is a WebTutor section corresponding to this chapter. The WebTutor contains a brief summary of Chapter 14, hints for learning the concepts and formulas for the repeated-measures ANOVA, cautions about common errors, and sample exam items including solutions.

General instructions for using SPSS are presented in Appendix D. Following are detailed instructions for using SPSS to perform the **Single-Factor, Repeated-Measures Analysis of Variance (ANOVA)** presented in this chapter.

Data Entry

1. Enter the scores for each treatment condition in a separate column, with the scores for each individual in the same row. All the scores for the first treatment go in the VAR00001 column, the second treatment scores in the VAR00002 column, and so on.

Data Analysis

1. Click **Analyze** on the tool bar, select **General Linear Model,** and click on **Repeated-Measures.**
2. SPSS will present a box titled **Repeated-Measures Define Factors.** Within the box, the Within-Subjects Factor Name should already contain **Factor 1.** If not, type in Factor 1.
3. Enter the **Number of levels** (number of different treatment conditions) in the next box.
4. Click on **Add.**
5. Click **Define.**
6. One by one, move the column labels for your treatment conditions into the **Within Subjects Variables** box. (Highlight the column label on the left and click the arrow to move it into the box.)
7. If you want descriptive statistics for each treatment, click on the **Options** box, select **Descriptives,** and click **Continue.**
8. Click **OK.**

SPSS Output

If you selected the Descriptives Option, SPSS will produce a table showing the mean and standard deviation for each treatment condition. The rest of the SPSS output is relatively complex and includes a lot of statistical information that goes well beyond the scope of this book. Therefore, direct your attention to the table titled **Test of Within-Subjects Effects.** The top line of the FACTOR1 box (sphericity assumed) shows the between-treatments sum of squares, degrees of freedom, and mean square that form the numerator of the *F*-ratio. The same line reports the value of the *F*-ratio and the level of significance (the *p* value or alpha level). Similarly, the top line of the Error(FACTOR1) box shows the sum of squares, the degrees of freedom, and the mean square for the error term (the denominator of the *F*-ratio).

FOCUS ON PROBLEM SOLVING

1. Before you begin a repeated-measures ANOVA, complete all the preliminary calculations needed for the ANOVA formulas. This requires that you find the total for each treatment (*T*s), the total for each person (*P*s), the grand total (*G*), the *SS* for each

treatment condition, and ΣX^2 for the entire set of N scores. As a partial check on these calculations, be sure that the T values add up to G and that the P values have a sum of G.

2. To help remember the structure of repeated-measures ANOVA, keep in mind that a repeated-measures experiment eliminates the contribution of individual differences. There are no individual differences contributing to the numerator of the F-ratio ($MS_{\text{between treatments}}$) because the same individuals are used for all treatments. Therefore, you must also eliminate individual differences in the denominator. This is accomplished by partitioning within-treatments variability into two components: between-subjects variability and error variability. It is the MS value for error variability that is used in the denominator of the F-ratio.

3. As with any ANOVA, it helps to organize your work so you do not get lost. Compute the SS values first, followed by the df values. You can then calculate the MS values and the F-ratio. Filling in the columns on an ANOVA summary table (for example, page 456) as you do the computations will help guide your way through the problem.

4. Be careful when using df values to find the critical F value. Remember that df_{error} (*not* $df_{\text{within treatments}}$) is used for the denominator.

DEMONSTRATION 14.1

REPEATED-MEASURES ANOVA

The following data were obtained from a research study examining the effect of sleep deprivation on motor-skills performance. A sample of five participants was tested on a motor-skills task after 24 hours of sleep deprivation, tested again after 36 hours, and tested once more after 48 hours. The dependent variable is the number of errors made on the motor-skills task. Do these data indicate that the number of hours of sleep deprivation has a significant effect on motor skills performance?

Participant	24 Hours	36 Hours	48 Hours	P totals	
A	0	0	6	6	$N = 15$
B	1	3	5	9	$G = 45$
C	0	1	5	6	$\Sigma X^2 = 245$
D	4	5	9	18	
E	0	1	5	6	
	$T = 5$	$T = 10$	$T = 30$		
	$SS = 12$	$SS = 16$	$SS = 12$		

STEP 1 State the hypotheses, and specify alpha.
The null hypothesis states that for the general population there are no differences among the three deprivation conditions. Any differences that exist among the samples are simply due to chance or error. In symbols,

$$H_0: \quad \mu_1 = \mu_2 = \mu_3$$

The alternative hypothesis states that there are differences among the conditions.

H_1:   At least one of the treatment means is different.

We will use $\alpha = .05$.

STEP 2 The repeated-measures analysis.
Rather than compute the df values and look for a critical value for F at this time, we will proceed directly to the ANOVA.

STAGE 1 The first stage of the analysis is identical to the independent-measures ANOVA presented in Chapter 13.

$$SS_{total} = \Sigma X^2 - \frac{G^2}{N}$$

$$= 245 - \frac{45^2}{15} = 110$$

$$SS_{within} = \Sigma SS_{inside\ each\ treatment} = 12 + 16 + 12 = 40$$

$$SS_{between} = \Sigma \frac{T^2}{n} - \frac{G^2}{N}$$

$$= \frac{5^2}{5} + \frac{10^2}{5} + \frac{30^2}{5} - \frac{45^2}{15}$$

$$= 70$$

and the corresponding degrees of freedom are

$$df_{total} = N - 1 = 14$$

$$df_{within} = \Sigma df = 4 + 4 + 4 = 12$$

$$df_{between} = k - 1 = 2$$

STAGE 2 The second stage of the repeated-measures analysis measures and removes the individual differences from the denominator of the F-ratio.

$$SS_{between\ subjects} = \Sigma \frac{P^2}{k} - \frac{G^2}{N}$$

$$= \frac{6^2}{3} + \frac{9^2}{3} + \frac{6^2}{3} + \frac{18^2}{3} + \frac{6^2}{3} - \frac{45^2}{15}$$

$$= 36$$

$$SS_{error} = SS_{within} - SS_{between\ subjects}$$

$$= 40 - 36$$

$$= 4$$

and the corresponding df values are

$$df_{between\ subjects} = n - 1 = 4$$

$$df_{error} = df_{within} - df_{between\ subjects}$$

$$= 12 - 4$$

$$= 8$$

The mean square values that form the F-ratio are as follows:

$$MS_{between} = \frac{SS_{between}}{df_{between}} = \frac{70}{2} = 35$$

$$MS_{error} = \frac{SS_{error}}{df_{error}} = \frac{4}{8} = 0.50$$

Finally, the F-ratio is

$$F = \frac{MS_{between}}{MS_{error}} = \frac{35}{0.50} = 70.00$$

STEP 3 Make a decision and state a conclusion.

With $df = 2, 8$ and $\alpha = .05$, the critical value is $F = 4.46$. Our obtained F-ratio ($F = 70.00$) is well into the critical region, so our decision is to reject the null hypothesis and conclude that there are significant differences among the three levels of sleep deprivation.

DEMONSTRATION 14.2

EFFECT SIZE FOR THE REPEATED-MEASURES ANOVA

We will compute η^2, the percentage of variance explained by the treatment differences, for the data in Demonstration 14.1. Using Equation 14.9, we obtain

$$\eta^2 = \frac{SS_{between\ treatments}}{SS_{total} - SS_{between\ subjects}} = \frac{70}{110 - 36} = \frac{70}{74} = 0.95 \quad (or\ 95\%)$$

Equivalently, Equation 14.10 produces

$$\eta^2 = \frac{SS_{between\ treatments}}{SS_{between\ treatments} + SS_{error}} = \frac{70}{70 + 4} = \frac{70}{74} = 0.95 \quad (or\ 95\%)$$

PROBLEMS

1. How does the denominator of the F-ratio (the error term) differ for a repeated-measures ANOVA compared to an independent-measures ANOVA?

2. The repeated-measures ANOVA can be viewed as a two-stage process. What is the purpose for the second stage?

3. A researcher conducts an experiment comparing three treatment conditions. The data consist of $n = 20$ scores for each treatment condition.

   a. If the researcher used an independent-measures design, then how many individuals were needed for the study and what are the df values for the F-ratio?

   b. If the researcher used a repeated-measures design, then how many individuals were needed for the study and what are the df values for the F-ratio?

4. A researcher conducts a repeated-measures experiment using a sample of $n = 10$ subjects to evaluate the differences among three treatment conditions. If the

results are examined with an ANOVA, what are the *df* values for the *F*-ratio?

5. A researcher uses a repeated-measures ANOVA to evaluate the results from a research study and reports an *F*-ratio with *df* = 3, 36.
   a. How many treatment conditions were compared in the study?
   b. How many individuals participated in the study?

6. A published report of a repeated-measures research study includes the following description of the statistical analysis. "The results show significant differences among the treatment conditions, $F(2, 20)$ = 6.10, $p < .01$."
   a. How many treatment conditions were compared in the study?
   b. How many individuals participated in the study?

7. The following data were obtained from a repeated-measures study comparing three treatment conditions. Use a repeated-measures ANOVA with $\alpha = .05$ to determine whether there are significant mean differences among the three treatments.

Person	Treatments I	II	III	Person Totals
A	0	4	2	$P = 6$
B	1	5	6	$P = 12$ $N = 18$
C	3	3	3	$P = 9$ $G = 48$
D	0	1	5	$P = 6$ $\Sigma X^2 = 184$
E	0	2	4	$P = 6$
F	2	3	4	$P = 9$

$$M = 1 \quad M = 3 \quad M = 4$$
$$T = 6 \quad T = 18 \quad T = 24$$
$$SS = 8 \quad SS = 10 \quad SS = 10$$

8. The following data were obtained from a repeated-measures study comparing two treatment conditions. Use a repeated-measures ANOVA with $\alpha = .05$ to determine whether there are significant mean differences between the two treatments.

Person	Treatments I	II	Person Totals
A	3	5	$P = 8$
B	5	9	$P = 14$ $N = 16$
C	1	5	$P = 6$ $G = 80$
D	1	7	$P = 8$ $\Sigma X^2 = 500$
E	5	9	$P = 14$
F	3	7	$P = 10$
G	2	6	$P = 8$
H	4	8	$P = 12$

$$M = 3 \quad M = 7$$
$$T = 24 \quad T = 56$$
$$SS = 18 \quad SS = 18$$

9. For the data in problem 8,
   a. Compute SS_{total} and $SS_{between\ treatments}$.
   b. Eliminate the mean differences between treatments by adding 2 points to each score in treatment I and subtracting 2 points from each score in treatment II. (Both treatments should end up with $M = 5$ and $T = 40$.)
   c. Calculate SS_{total} for the modified scores. (Caution: You first must find the new value for ΣX^2.)
   d. Because the treatment effects were eliminated in part b, you should find that SS_{total} for the modified scores is smaller than SS_{total} for the original scores. The difference between the two SS values should be exactly equal to the value of $SS_{between\ treatments}$ for the original scores.

10. The following data were obtained from a repeated-measures study comparing three treatment conditions.
   a. Use a repeated-measures ANOVA with $\alpha = .05$ to determine whether there are significant mean differences among the three treatments.
   b. Compute η^2, the percentage of variance accounted for by the mean differences, to measures the size of the treatment effects.

| | Treatments | | | |
Person	I	II	III	Person Totals
A	1	1	4	$P = 6$
B	3	4	8	$P = 15$
C	0	2	7	$P = 9$
D	0	0	6	$P = 6$
E	1	3	5	$P = 9$

$N = 15$
$G = 45$
$\Sigma X^2 = 231$

$$M = 1 \quad M = 2 \quad M = 6$$
$$T = 5 \quad T = 10 \quad T = 30$$
$$SS = 6 \quad SS = 10 \quad SS = 10$$

11. For the data in problem 10,
  a. Compute SS_{total} and $SS_{between\ subjects}$.
  b. Eliminate the mean differences between people by adding 1 point to each score for person A, subtracting 2 points from each score for person B, and adding 1 point to each score for person D. (All five people should end up with $P = 9$.)
  c. Calculate SS_{total} for the modified scores. (Caution: You first must find the new value for ΣX^2.)
  d. Because the individual differences were eliminated in part b, you should find that SS_{total} for the modified scores is smaller than SS_{total} for the original scores. The difference between the two SS values should be exactly equal to the value of $SS_{between\ subjects}$ for the original scores.

12. The following data were obtained from a repeated-measures study comparing three treatment conditions.

| | Treatment | | | |
Subject	I	II	III	P
A	6	8	10	24
B	5	5	5	15
C	1	2	3	6
D	0	1	2	3

$G = 48$
$\Sigma X^2 = 294$

$$T = 12 \quad T = 16 \quad T = 20$$
$$SS = 26 \quad SS = 30 \quad SS = 38$$

Use a repeated-measures ANOVA with $\alpha = .05$ to determine whether these data are sufficient to demonstrate significant differences between the treatments.

13. In Problem 12 the data show large and consistent differences between subjects. For example, subject A has the largest score in every treatment and subject D always has the smallest score. In the second stage of the ANOVA, the large individual differences get subtracted out of the denominator of the F-ratio, which results in a larger value for F.

The following data were created by using the same numbers that appeared in Problem 12. However, we eliminated the consistent individual differences by scrambling the scores within each treatment.

| | Treatment | | | |
Subject	I	II	III	P
A	6	2	3	11
B	5	1	5	11
C	0	5	10	15
D	1	8	2	11

$G = 48$
$\Sigma X^2 = 294$

$$T = 12 \quad T = 16 \quad T = 20$$
$$SS = 26 \quad SS = 30 \quad SS = 38$$

  a. Use a repeated-measures ANOVA with $\alpha = .05$ to determine whether these data are sufficient to demonstrate significant differences between the treatments.
  b. Explain how the results of this analysis compare with the results from Problem 12.

14. The following data are from an experiment comparing three different treatment conditions:

A	B	C	
0	1	2	$N = 15$
2	5	5	$\Sigma X^2 = 354$
1	2	6	
5	4	9	
2	8	8	

| $T = 10$ | $T = 20$ | $T = 30$ |
| $SS = 14$ | $SS = 30$ | $SS = 30$ |

  a. If the experiment uses an *independent-measures design,* can the researcher conclude that the treatments are significantly different? Test at the .05 level of significance.
  b. If the experiment is done with a *repeated-measures design,* should the researcher conclude that the treatments are significantly different? Set alpha at .05 again.

c. Explain why the results are different in the analyses of parts a and b.

15. Loss of hearing can be a significant problem for older adults. Although hearing aids can correct the physical problem, people who have lived with hearing impairment often develop poor communication strategies and social skills. To address this problem, a home education program has been developed to help people who are receiving hearing aids for the first time. The program emphasizes communication skills. To evaluate the program, overall quality of life and satisfaction were measured before treatment, again at the end of the training program, and once more at a 6-month follow-up (Kramer, Allessie, Dondorp, Zekveld, & Kapteyn, 2005). Data similar to the results obtained in the study are shown below. Do the data indicate a significant improvement in quality of life following the training program? Test at the .05 level of significance.

	Quality-of-Life Scores			
Person	Before	After	6-Months	
A	3	7	8	$N = 12$
B	0	5	7	$G = 60$
C	4	9	5	$\Sigma X^2 = 384$
D	1	7	4	
	$T = 8$	$T = 28$	$T = 24$	
	$SS = 10$	$SS = 8$	$SS = 10$	

16. A researcher is evaluating the effectiveness of a speed-reading course. A standardized reading test, measuring both speed and comprehension, was given to a sample of $n = 15$ students before the course started and again at the end of the course. The following table presents the results from the repeated-measures ANOVA. Fill in the missing values in the table. (*Hint:* Start with the *df* values.)

K = 2
n = 15
N = 30

Source	SS	df	MS	
Between treatments	12	1	12	$F =$ 4
Within treatments	74	28		
Between subjects	32	14		
Error	42	14	3	
Total	86	29		

17. The following summary table presents the results from a repeated-measures ANOVA comparing four treatment conditions, each with a sample of $n = 10$ subjects. Fill in the missing values in the table. (*Hint:* Start with the *df* values.)

Source	SS	df	MS	
Between treatments	___	___	20	$F = 5.00$
Within treatments	158	___		
Between subjects	___	___		
Error	108	___	___	
Total	___	___		

18. The following summary table presents the results from a repeated-measures ANOVA comparing three treatment conditions, each with a sample of $n = 12$ participants. Fill in the missing values in the table. (*Hint:* Start with the *df* values.)

Source	SS	df	MS	
Between treatments	30	___	___	$F = 6.00$
Within treatments	___	___		
Between subjects	___	___		
Error	___	___	2.50	
Total	155	___		

19. The following data represent idealized results from an experiment comparing two treatments. Notice that the mean for treatment II is 4 points higher than the mean for treatment I. Also notice that this 4-point treatment effect is perfectly consistent for all participants.

	Treatment		
Person	I	II	P
A	1	5	6
B	4	8	12
C	7	11	18
D	4	8	12
	$T = 16$	$T = 32$	
	$M = 4$	$M = 8$	
	$SS = 18$	$SS = 18$	

a. Calculate the within-treatments *SS* for these data.
b. Calculate the between-subjects *SS* for these data. You should find that all of the within-treatments variability is accounted for by variability between subjects; that is, $SS_{\text{within treatments}} = SS_{\text{between subjects}}$. For these data, the treatment effect is perfectly consistent across subjects, and there is no error variability ($SS_{\text{error}} = 0$).

20. A repeated-measures experiment comparing only two treatments can be evaluated with either a *t* statistic or an ANOVA. As we found with the independent-measures design, the *t* test and the ANOVA produce equivalent conclusions, and the two test statistics are related by the equation $F = t^2$.

The following data are from a repeated-measures study:

Subject	Treatment 1	Treatment 2	Difference
1	2	4	+2
2	1	3	+2
3	0	10	+10
4	1	3	+2

a. Use a repeated-measures t statistic with $\alpha = .05$ to determine whether the data provide evidence of a significant difference between the two treatments. (*Caution:* ANOVA calculations are done with the X values, but for t you use the difference scores.)

b. Use a repeated-measures ANOVA with $\alpha = .05$ to evaluate the data. (You should find $F = t^2$.)

21. For either independent-measures or repeated-measures designs comparing two treatments, the mean difference can be evaluated with either a t test or an ANOVA. The two tests are related by the equation $F = t^2$. For the following data,

a. Use a repeated-measures t test with $\alpha = .05$ to determine whether the mean difference between treatments is statistically significant.

b. Use a repeated-measures ANOVA with $\alpha = .05$ to determine whether the mean difference between treatments is statistically significant. (You should find that $F = t^2$.)

	Treatments		
Person	I	II	Difference Scores
A	4	7	3
B	2	11	9
C	3	6	3
D	7	10	3
	$M = 4$	$M = 8.5$	$M_D = 4.5$
	$T = 16$	$T = 34$	
	$SS = 14$	$SS = 17$	$SS = 27$

22. An educational psychologist is studying student motivation in elementary school. A sample of $n = 5$ students is followed over 3 years from fourth grade to sixth grade. Each year the students complete a questionnaire measuring their motivation and enthusiasm for school. The psychologist would like to know whether there are significant changes in motivation across the three grade levels. The data from this study are as follows:

Student	Fourth Grade	Fifth Grade	Sixth Grade
A	4	3	1
B	8	6	4
C	5	3	3
D	7	4	2
E	6	4	0

a. Compute the mean motivation score for each grade level.

b. Use an ANOVA to determine whether there are any significant differences in motivation among the three grade levels. Use the .05 level of significance.

23. In a study examining cardiovascular responses to embarrassment, Harris (2001) asked participants to sing the Star Spangled Banner in front of a video camera while she recorded their heart rate and blood pressure. The data indicate that heart rate increases quickly during the first minute of an embarrassing episode, then drops quickly during the second minute. Blood pressure, on the other hand, increases steadily for a full 2 minutes before gradually returning to normal. Following are heart rate data similar to the research results. Do the data show significant changes in heart rate over time? Use a repeated-measures ANOVA with $\alpha = .05$.

Person	Baseline	1 Minute	2 Minutes	P	
A	74	88	75	237	$N = 12$
B	76	90	77	243	$G = 960$
C	78	89	76	243	$\Sigma X^2 = 77{,}208$
D	76	85	76	237	
	$M = 76$	$M = 88$	$M = 76$		
	$SS = 8$	$SS = 14$	$SS = 2$		

24. A habituation task is often used to test memory for infants. In the habituation procedure, a stimulus is shown to the infant for a brief period, and the researcher records how much time the infant spends looking at the stimulus. This same process is repeated again and again. If the infant begins to lose interest in the stimulus (decreases the time looking at it), the researcher can conclude that the infant "remembers" the earlier presentations and is demonstrating habituation to an old familiar stimulus. Hypothetical data from a habituation experiment are as follows:

Infant	Amount of Time (In Sec.) Attending to the Stimulus		
	First Presentation	Second Presentation	Third Presentation
A	112	81	20
B	97	35	42
C	82	58	27
D	104	70	39
E	78	51	46

a. Use an ANOVA with $\alpha = .01$ to determine whether there is a significant change in behavior for successive presentations.

b. Compute η^2 to measure the effect size for these data.

25. The endorphins released by the brain act as natural painkillers. For example, Gintzler (1970) monitored endorphin activity and pain thresholds in pregnant rats during the days before they gave birth. The data showed an increase in pain threshold as the pregnancy progressed. The change was gradual until 1 or 2 days before birth, at which point there was an abrupt increase in pain threshold. Apparently a natural painkilling mechanism was preparing the animals for the stress of giving birth. The following data represent pain-threshold scores similar to the results obtained by Gintzler. Do these data indicate a significant change in pain threshold? Use a repeated-measures ANOVA with $\alpha = .05$.

Subject	Days Before Giving Birth				
	9	7	5	3	1
A	35	39	41	49	52
B	33	38	40	44	55
C	39	41	44	47	57
D	37	40	45	46	56
E	36	37	40	44	55

C H A P T E R

15

Two-Factor Analysis of Variance (Independent Measures)

Tools You Will Need

The following items are considered essential background material for this chapter. If you doubt your knowledge of any of these items, you should review the appropriate chapter or section before proceeding.

- Introduction to analysis of variance (Chapter 13)
  - The logic of analysis of variance
  - ANOVA notation and formulas
  - Distribution of *F*-ratios

Preview

15.1   Overview

15.2   Main Effects and Interactions

15.3   Notation and Formulas

15.4   Interpreting the Results from a Two-Factor ANOVA

15.5   An Alternative View of the Two-Factor ANOVA

15.6   Assumptions for the Two-Factor ANOVA

Summary

Focus on Problem Solving

Demonstrations 15.1 and 15.2

Problems

Preview

Imagine that you are seated at your desk, ready to take the final exam in statistics. Just before the exams are handed out, a television crew appears and sets up a camera and lights aimed directly at you. They explain they are filming students during exams for a television special. You are told to ignore the camera and go ahead with your exam.

Would the presence of a TV camera affect your performance on an exam? For some of you, the answer to this question is "definitely yes" and for others, "probably not." In fact, both answers are right; whether or not the TV camera affects performance depends on your personality. Some of you would become terribly distressed and self-conscious, while others really could ignore the camera and go on as if everything were normal.

In an experiment that duplicates the situation we have described, Shrauger (1972) tested participants on a concept formation task. Half the participants worked alone (no audience), and half worked with an audience of people who claimed to be interested in observing the experiment. Shrauger also divided the participants into two groups on the basis of personality: those high in self-esteem and those low in self-esteem. The dependent variable for this experiment was the number of errors on the concept formation task. Data similar to those obtained by Shrauger are shown in Figure 15.1. Notice that the audience had no effect on the high-self-esteem participants. However, the low-self-esteem participants made nearly twice as many errors with an audience as when working alone.

The Problem: Shrauger's study is an example of research that involves two independent variables in the same study. The independent variables are:

1. Audience (present or absent)
2. Self-esteem (high or low)

The results of the study indicate that the effect of one variable (audience) *depends on* another variable (self-esteem).

You should realize that it is quite common to have two variables that interact in this way. For example, a drug may have profound effect on some patients and have no effect whatsoever on others. Some children survive abusive

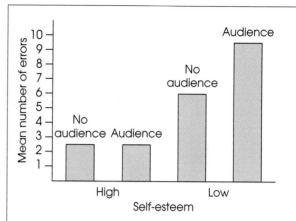

FIGURE 15.1

Results of an experiment examining the effect of an audience on the number of errors made on a concept formation task for participants who are rated either high or low in self-esteem. Notice that the effect of the audience depends on the self-esteem of the participants.

Shrauger, J. S. (1972). Self-esteem and reactions to being observed by others. *Journal of Personality and Social Psychology, 23,* 192–200. Copyright 1972 by the American Psychological Association. Adapted by permission of the author.

environments and live normal, productive lives, while others show serious difficulties. To observe how one variable interacts with another, it is necessary to study both variables simultaneously in one study. However, the analysis of variance (ANOVA) procedures introduced in Chapters 13 and 14 are limited to evaluating mean differences produced by one independent variable and are not appropriate for mean differences involving two (or more) independent variables.

The Solution: ANOVA is a very flexible hypothesis testing procedure and can be modified again to evaluate the mean differences produced in a research study with two (or more) independent variables. In this chapter we intorduce the *two-factor* ANOVA, which tests the significance of each independent variable acting alone as well as the interaction between variables.

15.1 OVERVIEW

In most research situations, the goal is to examine the relationship between two variables. Typically, the research study attempts to isolate the two variables to eliminate or reduce the influence of any outside variables that may distort the relationship being studied. A typical experiment, for example, focuses on one independent variable (which is expected to influence behavior) and one dependent variable (which is a measure of the behavior). In real life, however, variables rarely exist in isolation. That is, behavior usually is influenced by a variety of different variables acting and interacting simultaneously. To examine these more complex, real-life situations, researchers often design research studies that include more than one independent variable. Thus, researchers systematically change two (or more) variables and then observe how the changes influence another (dependent) variable.

An independent variable is a manipulated variable in an experiment. A quasi-independent variable is not manipulated but defines the groups of scores in a nonexperimental study.

In the preceding two chapters, we examined ANOVA for *single-factor* research designs—that is, designs that included only one independent variable or only one quasi-independent variable. When a research study involves more than one factor, it is called a *factorial design*. In this chapter, we consider the simplest version of a factorial design. Specifically, we examine ANOVA as it applies to research studies with exactly two factors. In addition, we limit our discussion to studies that use a separate sample for each treatment condition—that is, independent-measures designs. Finally, we consider only research designs for which the sample size (n) is the same for all treatment conditions. In the terminology of ANOVA, this chapter examines *two-factor, independent-measures, equal* n *designs*. The following example demonstrates the general structure of this kind of research study.

EXAMPLE 15.1

Most of us find it difficult to think clearly or to work efficiently on hot summer days. If you listen to people discussing this problem, you will occasionally hear comments like "It's not the heat; it's the humidity." To evaluate this claim scientifically, you would need to design a study in which both heat and humidity are manipulated within the same experiment and then observe behavior under a variety of different heat and humidity combinations. Table 15.1 shows the structure of a hypothetical study intended to examine the heat-versus-humidity question. Note that the study involves two independent variables: The temperature (heat) is varied from 70° to 80° to 90°, and the humidity is varied from low to high. The two independent variables are used to create a *matrix* with the different values of temperature defining the columns and the different levels of humidity defining the rows. The resulting two-by-three matrix shows six different combinations of the variables, producing six treatment conditions. Thus, the research study would require six separate samples, one for each of the *cells*, or boxes, in the matrix. The dependent variable for our study would be a measure of

TABLE 15.1

The structure of a two-factor experiment presented as a matrix. The factors are humidity and temperature. There are two levels for the humidity factor (low and high), and there are three levels for the temperature factor (70°, 80°, and 90°).

		Factor B: Temperature		
		70° Room	80° Room	90° Room
Factor A: Humidity	Low Humidity	Scores for $n = 15$ participants tested in a 70° room with low humidity	Scores for $n = 15$ participants tested in an 80° room with low humidity	Scores for $n = 15$ participants tested in a 90° room with low humidity
	High Humidity	Scores for $n = 15$ participants tested in a 70° room with high humidity	Scores for $n = 15$ participants tested in an 80° room with high humidity	Scores for $n = 15$ participants tested in a 90° room with high humidity

thinking/working proficiency (such as performance on a problem-solving task) for people observed in each of the six conditions.

The two-factor ANOVA tests for mean differences in research studies that are structured like the heat-and-humidity example in Table 15.1. For this example, the two-factor ANOVA evaluates three separate sets of mean differences:

1. What happens to the mean level of performance as the temperature is changed from 70° to 80° to 90°?

2. What happens to the mean level of performance as the humidity is changed form low to high?

3. How do specific combinations of temperature and humidity affect performance? (For example, increasing humidity may have no effect on performance when the temperature is low but could be very disruptive when the temperature is high.)

Thus, the two-factor ANOVA allows us to examine three types of mean differences within one analysis. In particular, we conduct three separate hypotheses tests for the same data, with a separate F-ratio for each test. The three F-ratios have the same basic structure:

$$F = \frac{\text{variance (differences) between treatments}}{\text{variance (differences) expected if there is no treatment effect}}$$

In each case, the numerator of the F-ratio measures the actual mean differences in the data, and the denominator measures the differences that would be expected if there was no treatment effect. As always in ANOVA, a large value for the F-ratio indicates that the sample mean differences are greater than would be expected by chance alone, and therefore provide evidence of a treatment effect. To determine whether the obtained F-radios are *significant*, we need to compare each F-ratio with the critical values found in the F-distribution table in Appendix B.

15.2 MAIN EFFECTS AND INTERACTIONS

As noted in the previous section, a two-factor ANOVA actually involves three distinct hypothesis tests. In this section, we examine these three tests in more detail.

Traditionally, the two independent variables in a two-factor experiment are identified as factor A and factor B. For the experiment presented in Table 15.1, humidity is factor A, and temperature is factor B. The goal of the experiment is to evaluate the mean differences that may be produced by either of these factors independently or by the two factors acting together.

MAIN EFFECTS One purpose of the experiment is to determine whether differences in humidity (factor A) result in differences in performance. To answer this question, we compare the mean score for all participants tested with low humidity versus the mean score for those tested with high humidity. Note that this process evaluates the mean difference between the top row and the bottom row in Table 15.1.

To make this process more concrete, we have presented a set of hypothetical data in Table 15.2. The table shows the mean score for each of the treatment conditions (cells)

TABLE 15.2

Hypothetical data from an experiment examining two different levels of humidity (factor A) and three different levels of temperature (factor B).

	70°	80°	90°	
Low humidity	$M = 85$	$M = 80$	$M = 75$	$M = 80$
High humidity	$M = 75$	$M = 70$	$M = 65$	$M = 70$
	$M = 80$	$M = 75$	$M = 70$	

as well as the overall mean for each column (each temperature) and the overall mean for each row (humidity level). These data indicate that participants in the low-humidity condition (the top row) obtained an average score of $M = 80$. This overall mean was obtained by computing the average of the three means in the top row. In contrast, high humidity resulted in a mean score of $M = 70$ (the overall mean for the bottom row). The difference between these means constitutes what is called the *main effect* for humidity, or the *main effect for factor A*.

Similarly, the main effect for factor B (temperature) is defined by the mean differences among columns of the matrix. For the data in Table 15.2, the two groups of participants tested with a temperature of 70° obtained an overall mean score of $M = 80$. Participants tested with the temperature at 80° averaged only $M = 75$, and those tested at 90° achieved a mean score of $M = 70$. The differences among these means constitute the *main effect* for temperature, or the *main effect for factor* B.

DEFINITION

The mean differences among the levels of one factor are referred to as the **main effect** of that factor. When the design of the research study is represented as a matrix with one factor determining the rows and the second factor determining the columns, then the mean differences among the rows describe the main effect of one factor, and the mean differences among the columns describe the main effect for the second factor.

The mean differrences between columns or rows simply *describe* the main effects for a two-factor study. As we have observed in earlier chapters, the existence of sample mean differences does not necessarily imply that the differences are *statistically significant*. In general, two samples are not expected to have exactly the same means. There will always be small differences from one sample to another, and you should not automatically assume that these differences are an indication of a systematic treatment effect. In the case of a two-factor study, any main effects that are observed in the data must be evaluated with a hypothesis test to determine whether they are statistically significant effects. Unless the hypothesis test demonstrates that the main effects are significant, you must conclude that the observed mean differences are simply the result of sampling error.

The evaluation of main effects accounts for two of the three hypothesis tests in a two-factor ANOVA. We state hypotheses concerning the main effect of factor A and the main effect of factor B and then calculate two separate F-ratios to evaluate the hypotheses.

For the example we are considering, factor A involves the comparison of two different levels of humidity. The null hypothesis would state that there is no difference between the two levels; that is, humidity has no effect on performance. In symbols,

$$H_0: \quad \mu_{A_1} = \mu_{A_2}$$

The alternative hypothesis is that the two different levels of humidity do produce different scores:

$$H_1: \quad \mu_{A_1} \neq \mu_{A_2}$$

To evaluate these hypotheses, we will compute an F-ratio that compares the actual mean differences between the two humidity levels versus the amount of difference that would be expected without any systematic treatment effects.

$$F = \frac{\text{variance (differences) between the means for factor } A}{\text{variance (differences) expected if there is no treatment effect}}$$

$$F = \frac{\text{variance (differences) between the row means}}{\text{variance (differences) expected if there is no treatment effect}}$$

Similarly, factor B involves the comparison of the three different temperature conditions. The null hypothesis states that overall there are no differences in mean performance among the three temperatures. In symbols,

$$H_0: \quad \mu_{B_1} = \mu_{B_2} = \mu_{B_3}$$

As always, the alternative hypothesis states that there are differences:

$$H_1: \quad \text{At least one mean is different from another.}$$

Again, the F-ratio compares, the obtained mean differences among the three temperature conditions versus the amount of difference that would be expected if there are no systematic treatment effects.

$$F = \frac{\text{variance (differences) between the means for factor } B}{\text{variance (differences) expected if there is no treatment effect}}$$

$$F = \frac{\text{variance (differences) between the column means}}{\text{variance (differences) expected if there is no treatment effect}}$$

INTERACTIONS   In addition to evaluating the main effect of each factor individually, the two-factor ANOVA allows you to evaluate other mean differences that may result from unique combinations of the two factors. For example, specific combinations of heat and humidity may have effects that are different from the overall effects of heat or humidity acting alone. Any "extra" mean differences that are not explained by the main effects are called an *interaction,* or an *interaction between factors.* The real advantage of combining two factors within the same study is the ability to examine the unique effects caused by an interaction.

D E F I N I T I O N

An **interaction** between two factors occurs whenever the mean differences between individual treatment conditions, or cells, are different from what would be predicted from the overall main effects of the factors.

To make the concept of an interaction more concrete, we reexamine the data shown in Table 15.2. For these data, there is no interaction; that is, there are no extra mean differences that are not explained by the main effects. For example, within each temperature condition (each column of the matrix) the participants scored 10 points higher in the low-humidity condition than in the high-humidity condition. This 10-point mean difference is exactly what is predicted by the overall main effect for humidity.

Now consider the data shown in Table 15.3. These new data show exactly the same main effects that existed in Table 15.2 (the column means and the row means have not been changed). But now there is an interaction between the two factors. For example, when the temperature is 90° (third column), there is a 20-point difference between the low-humidity and high-humidity conditions. This 20-point difference cannot be explained by the 10-point main effect for humidity. Also, when the temperature is 70° (first column), the data show no difference between the two humidity conditions. Again, this zero difference is not what would be expected based on the 10-point main effect for humidity. Mean differences that are not explained by the main effects are an indication of an interaction between the two factors.

To evaluate the interaction, the two-factor ANOVA first identifies mean differences that are not explained by the main effects. The extra mean differences are then evaluated by an F-ratio with the following structure:

$$F = \frac{\text{variance (mean differences) not explained by main effects}}{\text{variance (differences) expected if there is no treatment effects}}$$

The null hypothesis for this F-ratio simply states that there is no interaction:

H_0:   There is no interaction between factors A and B. All the mean differences between treatment conditions are explained by the main effects of the two factors.

The alternative hypothesis is that there is an interaction between the two factors:

H_1:   There is an interaction between factors. The mean differences between treatment conditions are not what would be predicted from the overall main effects of the two factors.

MORE ABOUT INTERACTIONS

In the previous section, we introduced the concept of an interaction as the unique effect produced by two factors working together. This section presents two alternative definitions of an interaction. These alternatives are intended to help you understand the concept of an interaction and to help you identify an interaction when you encounter one in a set of data. You should realize that the new definitions are equivalent to the original and simply present slightly different perspectives on the same concept.

TABLE 15.3

Hypothetical data from an experiment examining two different levels of humidity (factor A) and three different temperature conditions (factor B). (These data show the same main effects as the data in Table 15.2, but the individual treatment means have been modified to produce an interaction.)

	70°	80°	90°	
Low humidity	$M = 80$	$M = 80$	$M = 80$	$M = 80$
High humidity	$M = 80$	$M = 70$	$M = 60$	$M = 70$
	$M = 80$	$M = 75$	$M = 70$	

The first new perspective on the concept of an interaction focuses on the notion of independence for the two factors. More specifically, if the two factors are independent, so that one factor does not influence the effect of the other, then there is no interaction. On the other hand, when the two factors are not independent, so that the effect of one factor *depends on* the other, then there is an interaction. The notion of dependence between factors is consistent with our earlier discussion of interactions. If one factor influences the effect of the other, then unique combinations of the factors produce unique effects.

DEFINITION

When the effect of one factor depends on the different levels of a second factor, then there is an **interaction** between the factors.

This definition of an interaction should be familiar in the context of a "drug interaction." Your doctor and pharmacist are always concerned that the effect of one medication may be altered or distorted if a second medication is being taken at the same time. Thus, the effect of one drug (factor A) depends on a second drug (factor B), and you have an interaction between the two drugs.

Returning to the data in Table 15.2, you will notice that the size of the humidity effect (top row versus bottom row) *does not depend on* the temperature. For these data, the change in humidity shows the same 10-point effect for all three levels of temperature. Thus, the humidity effect does not depend on temperature, and there is no interaction. Now consider the data in Table 15.3. This time, the effect of changing humidity *depends on* the temperature, and there is an interaction. At 70°, for example, there is no difference between the high and low humidity conditions. However, there is a 10-point difference between high and low humidity when the temperature is 80°, and there is a 20-point effect at 90°. Thus, the effect of humidity depends on the temperature, which means that there is an interaction between the two factors.

The second alternative definition of an interaction is obtained when the results of a two-factor study are presented in a graph. In this case, the concept of an interaction can be defined in terms of the pattern displayed in the graph. Figure 15.2 shows the two sets of data we have been considering. The original data from Table 15.2, where there is no interaction, are presented in Figure 15.2(a). To construct this figure, we selected one of the factors to be displayed on the horizontal axis; in this case, the different levels of temperature are displayed. The dependent variable, mean level of performance, is shown on the vertical axis. Note that the figure actually contains two separate graphs: The top line shows the relationship between temperature and mean performance when the humidity is low, and the bottom line shows the relationship when the humidity is high. In general, the picture in the graph matches the structure of the data matrix; the columns of the matrix appear as values along the *X*-axis, and the rows of the matrix appear as separate lines in the graph.

For the original set of data, Figure 15.2(a), note that the two lines are parallel; that is, the distance between lines is constant. In this case, the distance between lines reflects the 10-point difference in mean performance between high and low humidity, and this 10-point difference is the same for all three temperature conditions.

Now look at a graph that is obtained when there is an interaction in the data. Figure 15.2(b) shows the data from Table 15.3. This time, note that the lines in the graph are not parallel. The distance between the lines changes as you scan from left to right. For these data, the distance between the lines corresponds to the humidity effect—that is, the mean difference in performance for low humidity versus high humidity. The fact that this difference depends on temperature indicates an interaction between factors.

FIGURE 15.2

(a) Graph showing the data from Table 15.2, where there is no interaction. (b) Graph showing the data from Table 15.3, where there is an interaction.

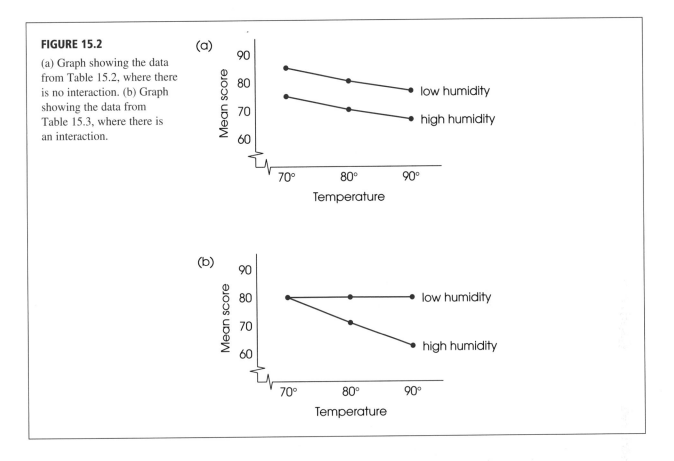

D E F I N I T I O N

When the results of a two-factor study are presented in a graph, the existence of nonparallel lines (lines that cross or converge) indicates an **interaction** between the two factors.

The $A \times B$ interaction typically is called "A by B" interaction. If there is an interaction of temperature and humidity, it may be called the "temperature by humidity" interaction.

For many students, the concept of an interaction is easiest to understand using the perspective of interdependency; that is, an interaction exists when the effects of one variable *depend* on another factor. However, the easiest way to identify an interaction within a set of data is to draw a graph showing the treatment means. The presence of nonparallel lines is an easy way to spot an interaction.

INDEPENDENCE OF MAIN EFFECTS AND INTERACTIONS

The two-factor ANOVA consists of three hypothesis tests, each evaluating specific mean differences: the A effect, the B effect, and the $A \times B$ interaction. As we have noted, these are three *separate* tests, but you should also realize that the three tests are *independent*. That is, the outcome for any one of the three tests is totally unrelated to the outcome for either of the other two. Thus, it is possible for data from a two-factor study to display any possible combination of significant and/or nonsignificant main effects and interactions. The data sets in Table 15.4 show several possibilities.

Table 15.4(a) shows data with mean differences between levels of factor A (an A effect) but no mean differences for factor B or for an interaction. To identify the A effect, notice that the overall mean for A_1 (the top row) is 10 points higher than the overall mean for A_2 (the bottom row). This 10-point difference is the main effect for

TABLE 15.4

Three sets of data showing different combinations of main effects and interaction for a two-factor study. (The numerical value in each cell of the matrices represents the mean value obtained for the sample in that treatment condition.)

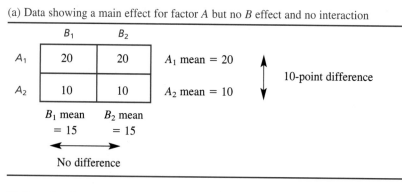

(a) Data showing a main effect for factor A but no B effect and no interaction

	B_1	B_2
A_1	20	20
A_2	10	10

A_1 mean = 20
A_2 mean = 10

B_1 mean = 15 B_2 mean = 15

No difference

10-point difference

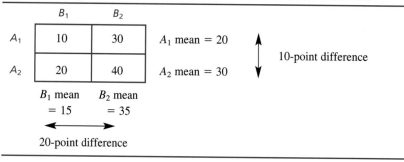

(b) Data showing main effects for both factor A and factor B but no interaction

	B_1	B_2
A_1	10	30
A_2	20	40

A_1 mean = 20
A_2 mean = 30

B_1 mean = 15 B_2 mean = 35

20-point difference

10-point difference

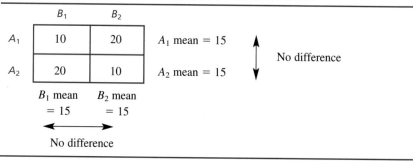

(c) Data showing no main effect for either factor but an interaction

	B_1	B_2
A_1	10	20
A_2	20	10

A_1 mean = 15
A_2 mean = 15

B_1 mean = 15 B_2 mean = 15

No difference

No difference

factor A. To evaluate the B effect, notice that both columns have exactly the same overall mean, indicating no difference between levels of factor B; hence, there is no B effect. Finally, the absence of an interaction is indicated by the fact that the overall A effect (the 10-point difference) is constant within each column; that is, the A effect *does not depend* on the levels of factor B. (Alternatively, the data indicate that the overall B effect is constant within each row.)

Table 15.4(b) shows data with an A effect and a B effect but no interaction. For these data, the A effect is indicated by the 10-point mean difference between rows, and the B effect is indicated by the 20-point mean difference between columns. The fact that the 10-point A effect is constant within each column indicates no interaction.

Finally, Table 15.4(c) shows data that display an interaction but no main effect for factor A or for factor B. For these data, note that there is no mean difference between rows (no A effect) and no mean difference between columns (no B effect). However, within each row (or within each column), there are mean differences. The "extra" mean differences within the rows and columns cannot be explained by the overall main effects and therefore indicate an interaction.

BOX 15.1   GRAPHING RESULTS FROM A TWO-FACTOR DESIGN

One of the best ways to get a quick overview of the results from a two-factor study is to present the data in a graph. Because the graph must display the means obtained for *two* independent variables (two factors), constructing the graph can be a bit more complicated than constructing the single-factor graphs we presented in Chapter 3 (pages 94–96).

Figure 15.3 shows two possible graphs presenting the results from a two-factor study with 2 levels of factor A and 3 levels of factor B. With a 2x3 design, there are a total of 6 different treatment means that are shown in the following matrix.

		Factor B		
		B_1	B_2	B_3
	A_1	10	40	20
Factor A	A_2	30	50	30

In the two graphs, note that values for the dependent variable (the treatment means) are shown on the vertical axis. Also note that the levels for one factor (we selected factor B) are displayed on the horizontal axis. Either factor can be used, but it is better to select one for which the different levels are measured on an interval or ratio scale. The reason for this suggestion is that an interval or ratio scale will permit you to construct a line graph as in Figure 15.3(a). If neither of the two factors is measured on an interval or ratio scale, you must use a bar graph as in Figure 15.3(b).

Line graphs: In Figure 15.3(a), we have assumed that factor B is an interval or a ratio variable, and the three levels for this factor are listed on the horizontal axis. Directly above the B_1 value on the horizontal axis, we have placed two dots corresponding to the two means in the B_1 column of the data matrix. Similarly, we have placed two dots above B_2 and another two dots above B_3. Finally, we have drawn a line connecting the three dots corresponding to level 1 of factor A (the three means in the top row of the data matrix). We have also drawn a second line that connects the three dots corresponding to level 2 of factor A. These lines are labeled A_1 and A_2 in the figure.

Bar graphs: Figure 15.3(b) also shows the three levels of factor B displayed on the horizontal axis. This time, however, we assume that factor B is measured on a nominal or ordinal scale, and the result is a bar graph. Directly above the B_1 value, we have drawn two bars so that the height of the bars correspond to the two means in the B_1 column of the data matrix. Similarly, we have drawn two bars above B_2 and two more bars above B_3. Finally, the three bars corresponding to level 1 of factor A (the top row of the data matrix) are all colored (or shaded) to differentiate them from the three bars for level 2 of factor A.

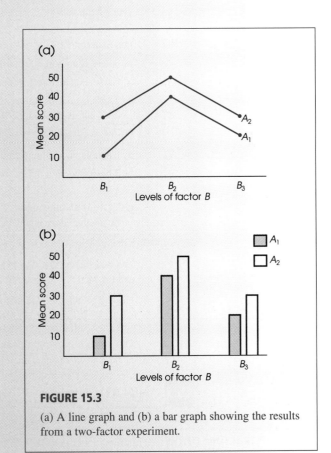

FIGURE 15.3

(a) A line graph and (b) a bar graph showing the results from a two-factor experiment.

LEARNING CHECK

1. Each of the following matrices represents a possible outcome of a two-factor experiment. For each experiment:

   a. Describe the main effect for factor A.

   b. Describe the main effect for factor B.

   c. Does there appear to be an interaction between the two factors?

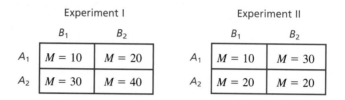

Experiment I

	B_1	B_2
A_1	$M = 10$	$M = 20$
A_2	$M = 30$	$M = 40$

Experiment II

	B_1	B_2
A_1	$M = 10$	$M = 30$
A_2	$M = 20$	$M = 20$

2. In a graph showing the means from a two-factor experiment, parallel lines indicate that there is no interaction. (True or false?)

3. A two-factor ANOVA consists of three hypothesis tests. What are they?

4. It is impossible to have an interaction unless you also have main effects for at least one of the two factors. (True or false?)

ANSWERS

1. For Experiment I:

   a. There is a main effect for factor A; the scores in A_2 average 20 points higher than in A_1.

   b. There is a main effect for factor B; the scores in B_2 average 10 points higher than in B_1.

   c. There is no interaction; there is a constant 20-point difference between A_1 and A_2 that does not depend on the levels of factor B.

   For Experiment II:

   a. There is no main effect for factor A; the scores in A_1 and in A_2 both average 20.

   b. There is a main effect for factor B; on average, the scores in B_2 are 10 points higher than in B_1.

   c. There is an interaction. The difference between A_1 and A_2 depends on the level of factor B. (There is a $+10$ difference in B_1 and a -10 difference in B_2.)

2. True.

3. The two-factor ANOVA evaluates the main effect for factor A, the main effect for factor B, and the interaction between the two factors.

4. False. The existence of main effects and interactions is completely independent.

15.3   NOTATION AND FORMULAS

STRUCTURE OF THE TWO-FACTOR ANALYSIS

The two-factor ANOVA is composed of three distinct hypothesis tests:

1. The main effect of factor A (often called the A-effect). Assuming that factor A is used to define the rows of the matrix, the main effect of factor A evaluates the mean differences between rows.

2. The main effect of factor B (called the B-effect). Assuming that factor B is used to define the columns of the matrix, the main effect of factor B evaluates the mean differences between columns.

3. The interaction (called the $A \times B$ interaction). The interaction evaluates mean differences between treatment conditions that are not predicted from the overall main effects from factor A or factor B.

For each of these three tests, we are looking for mean differences between treatments that are larger than would be expected by chance. In each case, the magnitude of the treatment effect is evaluated by an F-ratio. All three F-ratios have the same basic structure:

$$F = \frac{\text{variance (mean differences) between treatments}}{\text{variance (mean differences) expected if there are no treatment effects}} \qquad (15.1)$$

The general structure of the two-factor ANOVA is shown in Figure 15.4. Note that the overall analysis is divided into two stages. In the first stage, the total variability is separated into two components: between-treatments variability and within-treatments variability. This first stage is identical to the single-factor ANOVA introduced in Chapter 13 with each cell in the two-factor matrix viewed as a separate treatment condition. The within-treatments variability that is obtained in stage 1 of the analysis is used to compute the denominator for the F-ratios. As we noted in Chapter 13, within each treatment, all of the participants are treated exactly the same. Thus, any differences that exist within the treatments cannot be caused by treatment effects. As a result, the within-treatments variability provides a measure of the differences that exist when there are no systematic treatment effects influencing the scores (see Equation 15.1).

The between-treatments variability obtained in stage 1 of the analysis combines all the mean differences produced by factor A, factor B, and the interaction. The purpose of the second stage is to partition the differences into three separate components: differences attributed to factor A, differences attributed to factor B, and any remaining mean differences that define the interaction. These three components form the numerators for the three F-ratios in the analysis.

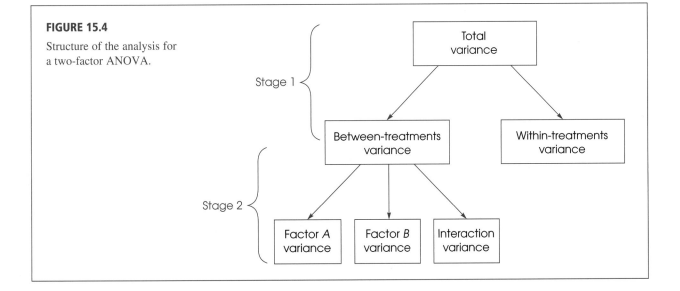

FIGURE 15.4

Structure of the analysis for a two-factor ANOVA.

The goal of this analysis is to compute the variance values needed for the three F-ratios. We will need three between-treatments variances (one for factor A, one for factor B, and one for the interaction), and we will need a within-treatments variance. Each of these variances (or mean squares) will be determined by a sum of squares value (SS) and a degrees of freedom value (df):

Remember that in ANOVA a variance is called a mean square, or MS.

$$\text{mean square} = MS = \frac{SS}{df}$$

EXAMPLE 15.2 The hypothetical data shown in Table 15.5 will be used to demonstrate the two-factor ANOVA. The data are representative of many studies examining the relationship between arousal and performance. The general result of these studies is that increasing the level of arousal (or motivation) tends to improve the level of performance. For very difficult tasks, however, increasing arousal beyond a certain point tends to lower the level of performance. (This relationship is generally known as the Yerkes–Dodson law). The data are displayed in a matrix with the two levels of task difficulty (factor A) making up the rows and the three levels of arousal (factor B) making up the columns. For the easy task, note that performance scores increase consistently as arousal increases. For the difficult task, on the other hand, performance peaks at a medium level of arousal and drops when arousal is increased to a high level. Note that the data matrix has a total of six *cells* or treatment conditions with a separate sample of $n = 5$ subjects in each condition. Most of the notation should be familiar from the single-factor ANOVA presented in Chapter 13. Specifically, the treatment totals are identified by T values, the total number of scores in the entire study is $N = 30$, and the

TABLE 15.5

Hypothetical data for a two-factor research study comparing two levels of task difficulty (easy and hard) and three levels of arousal (low, medium, and high). The study involves a total of six different treatment conditions with $n = 5$ participants in each condition.

		Factor B Arousal Level				
		Low	Medium	High		
		3	2	9		
		1	5	9		
		1	9	13		
		6	7	6	$T_{ROW1} = 90$	
	Easy	4	7	8		
		$M = 3$	$M = 6$	$M = 9$		
		$T = 15$	$T = 30$	$T = 45$		
		$SS = 18$	$SS = 28$	$SS = 26$		$N = 30$
Factor A						$G = 120$
Task Difficulty		0	3	0		$\Sigma X^2 = 840$
		2	8	0		
		0	3	0		
		0	3	5	$T_{ROW2} = 30$	
	Difficult	3	3	0		
		$M = 1$	$M = 4$	$M = 1$		
		$T = 5$	$T = 20$	$T = 5$		
		$SS = 8$	$SS = 20$	$SS = 20$		

$$T_{COL1} = 20 \quad T_{COL2} = 50 \quad T_{COL3} = 50$$

grand total (sum) of all 30 scores is $G = 120$. In addition to these familiar values, we have included the totals for each row and for each column in the matrix. The goal of the ANOVA is to determine whether the mean differences observed in the data are significantly greater than would be expected if there are no treatment effects.

STAGE 1 OF THE TWO-FACTOR ANALYSIS

The first stage of the two-factor analysis separates the total variability into two components: between-treatments and within-treatments. The formulas for this stage are identical to the formulas used in the single-factor ANOVA in Chapter 13 with the provision that each cell in the two-factor matrix is treated as a separate treatment condition. The formulas and the calculations for the data in Table 15.5 are as follows:

Total variability

$$SS_{total} = \Sigma X^2 - \frac{G^2}{N} \tag{15.2}$$

For these data,

$$SS_{total} = 840 - \frac{120^2}{30}$$

$$= 840 - 480$$

$$= 360$$

This SS value measures the variability for all $N = 30$ scores and has degrees of freedom given by

$$df_{total} = N - 1 \tag{15.3}$$

For the data in Table 15.5, $df_{total} = 29$.

Between-treatments variability   Remember that each treatment condition corresponds to a cell in the matrix. With this in mind, the between-treatments SS is computed as

$$SS_{between\ treatments} = \Sigma \frac{T^2}{n} - \frac{G^2}{N} \tag{15.4}$$

For the data in Table 15.5, there are six treatments (six T values), each with $n = 5$ scores, and the between-treatments SS is

$$SS_{between\ treatments} = \frac{15^2}{5} + \frac{30^2}{5} + \frac{45^2}{5} + \frac{5^2}{5} + \frac{20^2}{5} + \frac{5^2}{5} - \frac{120^2}{30}$$

$$= 45 + 180 + 405 + 5 + 80 + 5 - 480$$

$$= 240$$

The between-treatments df value is determined by the number of treatments (or the number of T values) minus one. For a two-factor study, the number of treatments is equal to the number of cells in the matrix. Thus,

$$df_{between\ treatments} = number\ of\ cells - 1 \tag{15.5}$$

For these data, $df_{\text{between treatments}} = 5$.

Within-treatments variability   To compute the variance within treatments, we first compute SS and $df = n - 1$ for each of the individual treatment conditions. Then the within-treatments SS is defined as

$$SS_{\text{within treatments}} = \Sigma SS_{\text{each treatment}} \tag{15.6}$$

And the within-treatments df is defined as

$$df_{\text{within treatments}} = \Sigma df_{\text{each treatment}} \tag{15.7}$$

For the data in Table 15.5,

$$SS_{\text{within treatments}} = 18 + 28 + 26 + 8 + 20 + 20$$

$$= 120$$

$$df_{\text{within treatments}} = 4 + 4 + 4 + 4 + 4 + 4$$

$$= 24$$

This completes the first stage of the analysis. Note that the two components add to equal the total for both SS values and df values.

$$SS_{\text{between treatments}} + SS_{\text{within treatments}} = SS_{\text{total}}$$

$$240 + 120 = 360$$

$$df_{\text{between treatments}} + df_{\text{within treatments}} = df_{\text{total}}$$

$$5 + 24 = 29$$

STAGE 2 OF THE TWO-FACTOR ANALYSIS   The second stage of the analysis will determine the numerators for the three F-ratios. Specifically, this stage will determine the between-treatments variance for factor A, factor B, and the interaction.

1. **Factor A.**   The main effect for factor A evaluates the mean differences between the levels of factor A. For this example, factor A defines the rows of the matrix, so we are evaluating the mean differences between rows. To compute the SS for factor A, we calculate a between-treatment SS using the row totals exactly the same as we computed $SS_{\text{between treatments}}$ using the treatment totals (T values) earlier. For factor A, the row totals are 90 and 30, and each total was obtained by adding 15 scores. Therefore,

$$SS_A = \Sigma \frac{T^2_{\text{ROW}}}{n_{\text{ROW}}} - \frac{G^2}{N} \tag{15.8}$$

For our data,

$$SS_A = \frac{90^2}{15} + \frac{30^2}{15} - \frac{120^2}{30}$$

$$= 540 + 60 - 480$$

$$= 120$$

Factor A involves two treatments (or two rows), easy and difficult, so the df value is

$$df_A = \text{number of rows} - 1 \tag{15.9}$$

$$= 2 - 1$$

$$= 1$$

2. Factor B. The calculations for factor B follow exactly the same pattern that was used for factor A, except for substituting columns in place of rows. The main effect for factor B evaluates the mean differences between the levels of factor B, which define the columns of the matrix.

$$SS_B = \Sigma \frac{T^2_{\text{COL}}}{n_{\text{COL}}} - \frac{G^2}{N}$$

(15.10)

For our data, the column totals are 20, 50, and 50, and each total was obtained by adding 10 scores. Thus,

$$SS_B = \frac{20^2}{10} + \frac{50^2}{10} + \frac{50^2}{10} - \frac{120^2}{30}$$

$$= 40 + 250 + 250 - 480$$

$$= 60$$

$$df_B = \text{number of columns} - 1$$

(15.11)

$$= 3 - 1$$

$$= 2$$

3. The $A \times B$ Interaction. The $A \times B$ interaction is defined as the "extra" mean differences not accounted for by the main effects of the two factors. We use this definition to find the SS and df values for the interaction by simple subtraction. Specifically, the between-treatments variability is partitioned into three parts: the A effect, the B effect, and the interaction (see Figure 15.4). We have already computed the SS and df values for A and B, so we can find the interaction values by subtracting to find out how much is left. Thus,

$$SS_{A \times B} = SS_{\text{between treatments}} - SS_A - SS_B$$

(15.12)

For our hypothetical data,

$$SS_{A \times B} = 240 - 120 - 60$$

$$= 60$$

Similarly,

$$df_{A \times B} = df_{\text{between treatments}} - df_A - df_B$$

(15.13)

$$= 5 - 1 - 2$$

$$= 2$$

MEAN SQUARES AND F-RATIOS FOR THE TWO-FACTOR ANALYSIS

The two-factor ANOVA consists of three separate hypothesis tests with three separate F-ratios. The denominator for each F-ratio is intended to measure the variance (differences) that would be expected if there are no treatment effects. As we saw in Chapter 13, the within-treatments variance is the appropriate denominator for an independent-measures design. Remember that inside each treatment all of the individuals are treated exactly the same, which means that the differences that exist were not caused by any systematic treatment effects (see Chapter 13, page 400). The within-treatments variance is called a *mean square*, or *MS*, and its computed as follows:

$$MS_{\text{within treatments}} = \frac{SS_{\text{within treatments}}}{df_{\text{within treatments}}}$$

For the data in Table 15.5,

$$MS_{\text{within treatments}} = \frac{120}{24} = 5.00$$

This value forms the denominator for all three F-ratios.

The numerators of the three F-ratios all measured variance or differences between treatments: differences between levels of factor A, differences between levels of factor B, and extra differences that are attributed to the $A \times B$ interaction. These three variances are computed as follows:

$$MS_A = \frac{SS_A}{df_A} \qquad MS_B = \frac{SS_B}{df_B} \qquad MS_{A \times B} = \frac{SS_{A \times B}}{df_{A \times B}}$$

For the data in Table 15.5, the three MS values are

$$MS_A = \frac{SS_A}{df_A} = \frac{120}{1} = 120 \qquad MS_B = \frac{SS_B}{df_B} = \frac{60}{2} = 30$$

$$MS_{A \times B} = \frac{SS_{A \times B}}{df_{A \times B}} = \frac{60}{2} = 30$$

Finally, the three F-ratios are

$$F_A = \frac{MS_A}{MS_{\text{within treatments}}} = \frac{120}{5} = 24.00$$

$$F_B = \frac{MS_B}{MS_{\text{within treatments}}} = \frac{30}{5} = 6.00$$

$$F_{A \times B} = \frac{MS_{A \times B}}{MS_{\text{within treatments}}} = \frac{30}{5} = 6.00$$

To determine the significance of each F-ratio, we must consult the F distribution table using the df values for each of the individual F-ratios. For this example, the F-ratio for factor A has $df = 1$ for the numerator and $df = 24$ for the denominator. Checking the table with $df = 1, 24$, we find a critical value of 4.26 for $\alpha = .05$ and a critical value of 7.82 for $\alpha = .01$. Our obtained F-ratio, $F = 24.00$ exceeds both of these values, so we conclude that there is a significant difference between the levels of factor A. That is, performance on the easy task (top row) is significantly different from performance on the difficult task (bottom row).

Similarly, the F-ratio for factor B has $df = 2, 24$. The critical values obtained from the table are 3.40 for $\alpha = .05$ and 5.61 for $\alpha = .01$. Again, our obtained F-ratio, $F = 6.00$, exceeds both values, so we can conclude that there are significant differences among the levels of factor B. For this study, the three levels of arousal result in significantly different levels of performance.

Finally, the F-ratio for the $A \times B$ interaction has $df = 2, 24$ (the same as factor B). With critical values of 3.40 for $\alpha = .05$ and 5.61 for $\alpha = .01$, our obtained F-ratio of $F = 6.00$ is sufficient to conclude that there is a significant interaction between task difficulty and level of arousal.

Table 15.6 is a summary table for the complete two-factor ANOVA from Example 15.2. Although these tables are no longer commonly used in research reports, they provide a concise format for displaying all of the elements of the analysis.

TABLE 15.6

A summary table for the repeated-measures ANOVA for the data from Example 15.2.

Source	SS	df	MS	F
Between treatments	240	5		
Factor A (difficulty)	120	1	120.0	$F(1, 24) = 24.00$
Factor B (arousal)	60	2	30.0	$F(2, 24) = 6.00$
$A \times B$ interaction	60	2	30.0	$F(2, 24) = 6.00$
Within treatments	120	24	5.0	
Total	360	29		

LEARNING CHECK

1. Explain why the within-treatment variability is the appropriate denominator for the two-factor independent-measures F-ratios.

2. The following data summarize the results from a two-factor independent-measures experiment:

Factor B

		B_1	B_2	B_3
Factor A	A_1	$n = 10$ $T = 0$ $SS = 30$	$n = 10$ $T = 10$ $SS = 40$	$n = 10$ $T = 20$ $SS = 50$
	A_2	$n = 10$ $T = 40$ $SS = 60$	$n = 10$ $T = 30$ $SS = 50$	$n = 10$ $T = 20$ $SS = 40$

a. Calculate the totals for each level of factor A, and compute SS for factor A.

b. Calculate the totals for factor B, and compute SS for this factor. (*Note:* You should find that the totals for B are all the same, so there is no variability for this factor.)

c. Given that the between-treatments (or between-cells) SS is equal to 100, what is the SS for the interaction?

ANSWERS

1. Within each treatment condition, all individuals are treated exactly the same. Therefore, the within-treatment variability measures the differences that exist between one score and another when there is no treatment effect causing the scores to be different. This is exactly the variance that is needed for the denominator of the F-ratios.

2. **a.** The totals for factor A are 30 and 90, and each total is obtained by adding 30 scores. $SS_A = 60$.

   b. All three totals for factor B are equal to 40. Because they are all the same, there is no variability, and $SS_B = 0$.

   c. The interaction is determined by differences that remain after the main effects have been accounted for. For these data,

$$SS_{A \times B} = SS_{\text{between treatments}} - SS_A - SS_B$$
$$= 100 - 60 - 0$$
$$= 40$$

MEASURING EFFECT SIZE FOR THE TWO-FACTOR ANOVA

The general technique for measuring effect size with an ANOVA is to compute a value for η^2, the percentage of variance that is explained by the treatment effects. For a two-factor ANOVA, we compute three separate values for eta squared: one measuring how much of the variance is explained by the main effect for factor A, one for factor B, and a third for the interaction. As we did with the repeated-measures ANOVA (page 456) we remove any variability that can be explained by other sources before we calculate the percentage for each of the three specific effects. Thus, for example, before we compute the η^2 for factor A, we remove the variability that is explained by factor B and the variability explained by the interaction. The resulting equation is,

$$\text{for factor } A, \ \eta^2 = \frac{SS_A}{SS_{\text{total}} - SS_B - SS_{A \times B}} \tag{15.14}$$

Note that the denominator of Equation 15.14 consists of the variability that is explained by factor A and the other *unexplained* variability. Thus, an equivalent version of the equation is,

$$\text{for factor } A, \ \eta^2 = \frac{SS_A}{SS_A + SS_{\text{within treatments}}} \tag{15.15}$$

Similarly, the η^2 formulas for factor B and for the interaction are as follows:

$$\text{for factor } B, \ \eta^2 = \frac{SS_B}{SS_{\text{total}} - SS_A - SS_{A \times B}} = \frac{SS_B}{SS_B + SS_{\text{within treatments}}} \tag{15.16}$$

$$\text{for } A \times B, \ \eta^2 = \frac{SS_{A \times B}}{SS_{\text{total}} - SS_A - SS_B} = \frac{SS_{A \times B}}{SS_{A \times B} + SS_{\text{within treatments}}} \tag{15.17}$$

Because each of the η^2 equations computes a percentage that is not based on the total variability of the scores, the results are often called *partial* eta squares. For the data in Example 15.2, the equations produce the following values:

$$\eta^2 \text{ for factor } A \text{ (difficulty)} = \frac{120}{360 - 60 - 60} = \frac{120}{240} = 0.50 \quad \text{(or 50\%)}$$

$$\eta^2 \text{ for factor } B \text{ (arousal)} = \frac{60}{360 - 120 - 60} = \frac{60}{180} = 0.33 \quad \text{(or 33\%)}$$

$$\eta^2 \text{ for the interaction} = \frac{60}{360 - 120 - 60} = \frac{60}{180} = 0.33 \quad \text{(or 33\%)}$$

IN THE LITERATURE
REPORTING THE RESULTS OF A TWO-FACTOR ANOVA

The APA format for reporting the results of a two-factor ANOVA follows the same basic guidelines as the single-factor report. First, the means and standard deviations are reported. Because a two-factor design typically involves several treatment conditions, these descriptive statistics usually are presented in a table or a graph. Next, the results of all three hypothesis tests (F-ratios) are reported. For the research study in Example 15.2, the report would have the following form:

The means and standard deviations for all treatment conditions are shown in Table 1. The two-factor analysis of variance showed a significant main effect for task difficulty, $F(1, 24) = 24.00$, $p < .01$, $\eta^2 = 0.50$; a significant main effect for level of arousal, $F(2, 24) = 6.00$, $p < .01$, $\eta^2 = 0.33$; and a significant interaction between difficulty and arousal, $F(2, 24) = 6.00$, $p < .01$, $\eta^2 = 0.33$.

TABLE 1

Mean performance score for each treatment condition

		Level of Arousal		
		Low	Medium	High
Difficulty	Easy	$M = 3.0$ $SD = 2.12$	$M = 6.00$ $SD = 2.65$	$M = 9.00$ $SD = 2.55$
	Hard	$M = 1.00$ $SD = 1.41$	$M = 4.00$ $SD = 2.24$	$M = 1.00$ $SD = 2.24$

15.4   INTERPRETING THE RESULTS FROM A TWO-FACTOR ANOVA

Because the two-factor ANOVA involves three separate tests, you must consider the overall pattern of results rather than focusing on the individual main effects or the interaction. In particular, whenever there is a significant interaction, you should be cautious about accepting the main effects at face value (whether they are significant or not). Remember, an interaction means that the effect of one factor *depends on* the level of the second factor. Because the effect changes from one level to the next, there is no consistent "main effect."

Figure 15.5 shows the sample means obtained from the task difficulty and arousal study. Recall that the analysis showed that both main effects and the interaction were

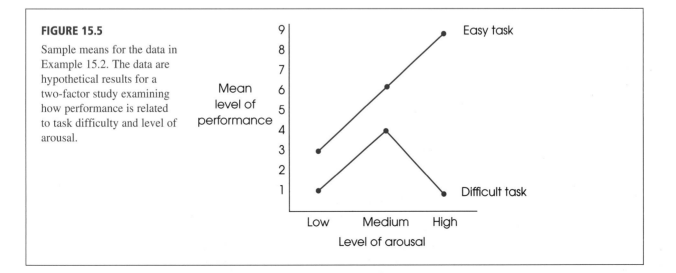

FIGURE 15.5

Sample means for the data in Example 15.2. The data are hypothetical results for a two-factor study examining how performance is related to task difficulty and level of arousal.

significant. The main effect for factor A (task difficulty) can be seen by the fact that the scores on the easy task are generally higher than scores on the difficult task.

On the other hand, the main effect for factor B (arousal) is not that easy to see. Although there is a tendency for scores to increase as the level of arousal increases, this is not a completely consistent trend. In fact, the scores on the difficult task show a sharp *decrease* when arousal is increased from moderate to high. This is an example of the complications that can occur when you have a significant interaction. Remember that an interaction means that a factor does not have a uniformly consistent effect. Instead, the effect of one factor *depends on* the other factor. For the data in Figure 15.5, the effect of increasing arousal depends on the task difficulty. For the easy task, increasing arousal produces increased performance. For the difficult task, however, increasing arousal beyond a moderate level produces decreased performance. Thus, the consequences of increasing arousal *depend on* the difficulty of the task. This interdependence between factors is the source of the significant interaction.

TESTING SIMPLE MAIN EFFECTS

The existence of a significant interaction indicates that the effect (mean differences) for one factor depends on the levels of the second factor. When the data are presented in a matrix showing treatment means, a significant interaction indicates that the mean differences within one column (or row) show a different pattern than the mean differences within another column (or row). In this case, a researcher may want to perform a separate analysis for each of the individual columns (or rows). In effect, the researcher is separating the two-factor experiment into a series of separate single-factor experiments. The process of testing the significance of mean differences within one column (or one row) of a two-factor design is called testing *simple main effects*. To demonstrate this process, we once again use the data from the task-difficulty and arousal study (Example 15.2), which are summarized in Figure 15.5.

EXAMPLE 15.3

For this demonstration we test for significant mean differences within each column of the two-factor data matrix. That is, we test for significant mean differences between the two levels of task difficulty for the low level of arousal, then repeat the test for the medium level of arousal, and once more for the high level. In terms of the two-factor notational system, we test the simple main effect of factor A for each level of factor B.

For the low level of arousal We begin by considering only the low level of arousal. Because we are restricting the data to the first column of the data matrix, the data effectively have been reduced to a single-factor study comparing only two treatment conditions. Therefore, the analysis is essentially a single-factor ANOVA duplicating the procedure presented in Chapter 13. To facilitate the change from a two-factor to a single-factor analysis, the data for the low level of arousal (first column of the matrix) are reproduced as follows using the notation for a single-factor study.

Low Level of Arousal		
Easy Task	Difficult Task	
$n = 5$	$n = 5$	$N = 10$
$M = 3$	$M = 1$	$G = 20$
$T = 15$	$T = 5$	

STEP 1   State the hypothesis. For this restricted set of the data, the null hypothesis would state that there is no difference between the mean for the easy task condition and the mean for the difficult task condition. In symbols,

$$H_0: \quad \mu_{\text{easy}} = \mu_{\text{difficult}} \qquad \text{for the low level of arousal}$$

STEP 2   To evaluate this hypothesis, we use an F-ratio for which the numerator, $MS_{\text{between treatments}}$, is determined by the mean differences between these two groups and the denominator consists of $MS_{\text{within treatments}}$ from the original ANOVA. Thus, the F-ratio has the structure

$$F = \frac{\text{variance (differences) for the means in column 1}}{\text{variance (differences) expected if there are no treatment effects}}$$

$$= \frac{MS_{\text{between treatments}} \text{ for the two treatments in column 1}}{MS_{\text{within treatments}} \text{ from the original ANOVA}}$$

To compute the $MS_{\text{between treatments}}$, we begin with the two treatment totals $T = 15$ and $T = 5$. Each of these totals is based on $n = 5$ scores, and the two totals add up to a grand total of $G = 20$. The $SS_{\text{between treatments}}$ for the two treatments is

$$SS_{\text{between treatments}} = \Sigma \frac{T^2}{n} - \frac{G^2}{N}$$

$$= \frac{15^2}{5} + \frac{5^2}{5} - \frac{20^2}{10}$$

$$= 45 + 5 - 40$$

$$= 10$$

Because this SS value is based on only two treatments, it has $df = 1$. Therefore,

Remember that the F-ratio uses $MS_{\text{within treatments}}$ from the original ANOVA. This $MS = 5$ with $df = 24$.

$$MS_{\text{between treatments}} = \frac{SS}{df} = \frac{10}{1} = 10$$

Using $MS_{\text{within treatments}} = 5$ from the original two-factor analysis, the final F-ratio is

$$F = \frac{MS_{\text{between treatments}}}{MS_{\text{within treatments}}} = \frac{10}{5} = 2.00$$

Note that this F-ratio has the same df values (1, 24) as the test for factor A main effects (easy versus difficult) in the original ANOVA. Therefore, the critical value for the F-ratio is the same as that in the original ANOVA. With $df = 1, 24$ the critical value is 4.26. In this case, our F-ratio fails to reach the critical value, so we conclude that there is no significant difference between the two tasks, easy and difficult, at a low level of arousal.

For the medium level of arousal   The test for the medium level of arousal follows the same process. The data for the medium level are as follows:

Medium Level of Arousal		
Easy Task	Difficult Task	
$n = 5$	$n = 5$	$N = 10$
$M = 6$	$M = 4$	$G = 50$
$T = 30$	$T = 20$	

Note that these data show a 2-point mean difference between the two conditions ($M = 6$ and $M = 4$), which is exactly the same as the 2-point difference that we evaluated for the low level of arousal ($M = 3$ and $M = 1$). Because the mean difference is the same for these two levels of arousal, the F-ratios will also be identical. For the low level of arousal, we obtained $F(1, 24) = 2.00$ which was not significant. This test also produces $F(1, 24) = 2.00$ and again we conclude that there is no significant difference. (*Note:* You should be able to complete the test to verify this decision.)

For the high level of arousal The data for the high level are as follows:

High Level of Arousal		
Easy Task	Difficult Task	
$n = 5$	$n = 5$	$N = 10$
$M = 9$	$M = 1$	$G = 50$
$T = 45$	$T = 5$	

For these data,

$$SS_{\text{between treatments}} = \Sigma \frac{T^2}{n} - \frac{G^2}{N}$$

$$= \frac{45^2}{5} + \frac{5^2}{5} - \frac{50^2}{10}$$

$$= 405 + 5 - 250$$

$$= 160$$

Again, we are comparing only two treatment conditions, so $df = 1$ and

$$MS_{\text{between treatments}} = \frac{SS}{df} = \frac{160}{1} = 160$$

Thus, for the high level of arousal, the final F-ratio is

$$F = \frac{MS_{\text{between treatments}}}{MS_{\text{within treatments}}} = \frac{160}{5} = 32.00$$

As before, this F-ratio has $df = 1, 24$ and is compared with the critical value $F = 4.26$. This time the F-ratio is far into the critical region and we conclude that there is a significant difference between the easy task and the difficult task for the high level of arousal.

As a final note, we should point out that the evaluation of simple main effects is used to account for the interaction as well as the overall main effect for one factor. In Example 15.2, the significant interaction indicates that the effect of arousal level (factor B) depends on the difficulty of the task (factor A). The evaluation of the simple main effects demonstrates this dependency. Specifically, task difficulty has no significant effect on performance when arousal level is low or medium, but does have a significant effect when arousal level is high. Thus, the analysis of simple main effects provides a detailed evaluation of the effects of one factor *including its interaction with a second factor.*

The fact that the simple main effects for one factor encompass both the interaction and the overall main effect of the factor can be seen if you consider the SS values. For this demonstration,

Simple Main Effects for Arousal	Interaction and Main Effect for Arousal
$SS_{\text{low arousal}} = 10$	$SS_{A \times B} = 60$
$SS_{\text{medium arousal}} = 10$	$SS_A = 120$
$SS_{\text{high arousal}} = 160$	
Total $SS = 180$	Total $SS = 180$

Notice that the total variability from the simple main effects of difficulty (factor A) completely accounts for the total variability of factor A and the $A \times B$ interaction.

15.5 AN ALTERNATIVE VIEW OF THE TWO-FACTOR ANOVA

The two-factor ANOVA involves three separate hypothesis tests that evaluate

1. The mean differences between levels of factor A.
2. The mean differences between levels of factor B.
3. Remaining mean differences that make up the interaction.

Although we have presented formulas for computing the SS values corresponding to each of these three components, it also is possible to measure the three SS values by simply removing the mean differences and observing what happens to the SS value for the total set of scores. In this section, we demonstrate this process.

ELIMINATING THE VARIABILITY FROM FACTOR A

We will begin with the original set of data that was used to demonstrate the two-factor ANOVA (see Table 15.5 on page 490). These data produced $SS_{\text{total}} = 360$. We also found that factor A, the difficulty factor, produced $SS_A = 120$. The SS for factor A measures the variability contributed by the mean differences between rows in the original data. We can also find this SS value by simply eliminating the mean differences between rows and observing what happens to the overall SS for the resulting scores. To eliminate the mean differences between rows,

1. Subtract 2 points from each individual in the first row.
2. Add 2 points to each individual in the second row.

The resulting data are shown in Table 15.7. Note that both rows now have a total of $T = 60$ and a row mean of $M = 4$. Thus, there are no mean differences between the levels of task difficulty and we have eliminated the variability contributed by factor A. To determine how much variability has been removed, we will compute SS for the scores in Table 15.7.

$$SS_{total} = \Sigma X^2 - \frac{G^2}{N}$$

$$= 720 - \frac{(120)^2}{30}$$

$$= 720 - 480$$

$$= 240$$

Eliminating the mean difference between rows has reduced the total variability from $SS_{total} = 360$ for the original data, to $SS_{total} = 240$ for the modified scores. Thus, eliminating the main effect for factor A reduces the total variability by 120 points, which is exactly the value we obtained for factor A in the original data, $SS_A = 120$.

ELIMINATING THE VARIABILITY FROM FACTOR B

First, note that the mean differences between columns for the modified data in Table 15.7 are the same as in the original data in Table 15.5. Because the totals and means for factor B (arousal level) are the same for both sets of data, SS_B will also be the same. For the original scores, we found $SS_B = 60$. This SS value measures the

TABLE 15.7

Data from a two-factor study for which the mean difference between rows has been eliminated. Starting with the data for Example 15.1 (Table 15.5), 2 points were subtracted from each score in the top row and 2 points were added to each score in the bottom row. In the resulting data, shown here, both rows have $T = 60$ and $M = 4$. For these scores, $SS_{total} = 240$. For the original scores, $SS_{total} = 360$. Thus, eliminating the main effect for factor A reduced the SS value by 120 points.

		Factor B Arousal Level			
		Low	Medium	High	
Factor A Task Difficulty	Easy	1 −1 −1 4 2 $M = 1$ $T = 5$ $SS = 18$	0 3 7 5 5 $M = 4$ $T = 20$ $SS = 28$	7 7 11 4 6 $M = 7$ $T = 35$ $SS = 26$	$T_{ROW1} = 60$
	Difficult	2 4 2 2 5 $M = 3$ $T = 15$ $SS = 8$	5 10 5 5 5 $M = 6$ $T = 30$ $SS = 20$	2 2 2 7 2 $M = 3$ $T = 15$ $SS = 20$	$T_{ROW2} = 60$

$N = 30$
$G = 120$
$\Sigma X^2 = 720$

$T_{COL1} = 20$ $T_{COL2} = 50$ $T_{COL3} = 50$

variability contributed by mean differences between columns. The same SS value can also be found by eliminating the mean differences between columns and observing what happens to the overall variability for the resulting scores. We begin with the scores in Table 15.7. To eliminate the mean differences between columns,

1. Add 2 points to each individual in the first column.
2. Subtract 1 point from each individual in the second column.
3. Subtract 1 point from each individual in the third column.

The resulting data are shown in Table 15.8. Note that all three columns now have a total of $T = 40$ and a column mean of $M = 4$. Thus, there are no mean differences between the levels of task arousal and we have eliminated the variability contributed by factor B. To determine how much variability has been removed, we will compute SS for the scores in Table 15.8.

$$SS_{total} = \Sigma X^2 - \frac{G^2}{N}$$

$$= 660 - \frac{(120)^2}{30}$$

$$= 660 - 480$$

$$= 180$$

TABLE 15.8

Data from a two-factor study for which the mean differences between columns have been eliminated. Starting with the data Table 15.7, 2 points were added to each score in the first column and 1 point was subtracted from each score in the second and third columns. In the resulting data, shown here, all three columns have $T = 40$ and $M = 4$. For these scores, $SS_{total} = 180$. For the original scores in Table 15.7, $SS_{total} = 240$. Thus, eliminating the main effect for factor B reduced the SS value by 60 points.

		Factor B Arousal Level			
		Low	Medium	High	
	Easy	3 1 1 6 4 $M = 3$ $T = 15$ $SS = 18$	−1 2 6 4 4 $M = 3$ $T = 15$ $SS = 28$	6 6 10 3 5 $M = 6$ $T = 30$ $SS = 26$	$T_{ROW1} = 60$
Factor A Task Difficulty	Difficult	4 6 4 4 7 $M = 5$ $T = 25$ $SS = 8$	4 9 4 4 4 $M = 5$ $T = 25$ $SS = 20$	1 1 1 6 1 $M = 2$ $T = 10$ $SS = 20$	$T_{ROW2} = 60$

$N = 30$
$G = 120$
$\Sigma X^2 = 660$

$T_{COL1} = 40 \quad T_{COL2} = 40 \quad T_{COL3} = 40$

Eliminating the mean difference between columns has reduced the total variability from $SS_{total} = 240$ for the scores in Table 15.7, to $SS_{total} = 180$ for the modified scores in Table 15.8. Thus, we reduced the total variability by 60 points, which is exactly the value we obtained for factor B in the original data, $SS_A = 60$.

ELIMINATING THE INTERACTION

For the data in Table 15.8, we have systematically eliminated the mean differences between rows and the mean differences between columns. Therefore, the data no longer have any main effects for factor A or for factor B. However, there are still mean differences remaining between the individual treatment conditions (the individual cells in the matrix). These remaining mean differences are the interaction. To measure the SS value for the interaction, we eliminate the remaining mean differences and then determine how this reduces the SS value for the resulting scores. To remove the mean differences, we begin with the data in Table 15.8 and,

1. Add 1 point to each individual in the easy low-arousal condition.
2. Add 1 point to each individual in the easy medium-arousal condition.
3. Subtract 2 points from each individual in the easy high-arousal condition.
4. Subtract 1 point from each individual in the difficult low-arousal condition.
5. Subtract 1 point from each individual in the difficult medium-arousal condition.
6. Add 2 points to each individual in the difficult high-arousal condition.

The resulting data are shown in Table 15.9. Note that there are no mean differences between any of the treatment conditions; all cells in the matrix now have $T = 20$ and $M = 4$. To determine how much variability was removed by eliminating the final mean differences, we will calculate SS for the scores in Table 15.9.

TABLE 15.9

Data from a two-factor study for which all the mean differences beween treatment conditions (cells) have been eliminated. We start with the data Table 15.8, for which the main effects for both factors had already been eliminated. Then, a constant amount was added to (or subtracted from) the scores in each cell so that the resulting data, shown here, have $T = 20$ and $M = 4$ for every treatment condition. For these scores, $SS_{total} = 120$. For the original scores in Table 15.8, $SS_{total} = 180$. Thus, eliminating the remaining mean differences (the interaction) reduced the SS value by 60 points.

		Factor B Arousal Level			
		Low	Medium	High	
	Easy	4	0	4	
		2	3	4	
		2	7	8	
		7	5	1	$T_{ROW1} = 60$
		5	5	3	
		$M = 4$	$M = 4$	$M = 4$	
		$T = 20$	$T = 20$	$T = 20$	
		$SS = 18$	$SS = 28$	$SS = 26$	$N = 30$
Factor A Task Difficulty					$G = 120$
	Difficult	3	3	3	$\Sigma X^2 = 600$
		5	8	3	
		3	3	3	
		3	3	8	
		6	3	3	$T_{ROW2} = 60$
		$M = 4$	$M = 4$	$M = 4$	
		$T = 20$	$T = 20$	$T = 20$	
		$SS = 8$	$SS = 20$	$SS = 20$	

$T_{COL1} = 40$ $T_{COL2} = 40$ $T_{COL3} = 40$

$$SS_{\text{total}} = \Sigma X^2 - \frac{G^2}{N}$$

$$= 600 - \frac{(120)^2}{30}$$

$$= 600 - 480$$

$$= 120$$

This final calculation introduces two points of interest. First, removing the last of the mean differences reduced the variability by 60 points, from $SS_{\text{total}} = 180$ for the scores in Table 15.8 to $SS_{\text{total}} = 120$ for the scores in Table 15.9. Thus, eliminating the final mean differences completely accounts for the interaction in the original data, $SS_{A \times B} = 60$. Second, we have eliminated both main effects and the interaction from the data in Table 15.9. There are no remaining mean differences in the data. The only variability that now exists is within treatments, and the total variability for these scores, $SS_{\text{total}} = 120$, is exactly the same as the within-treatments variability for the original data, $SS_{\text{within treatments}} = 120$.

By systematically eliminating the mean differences, first for factor A, then for factor B, and finally for the interaction, we are able to measure the amount of variability contributed by each source and we are able to determine how much variability remains when the systematic sources have been eliminated.

15.6   ASSUMPTIONS FOR THE TWO-FACTOR ANOVA

The validity of the ANOVA presented in this chapter depends on the same three assumptions we have encountered with other hypothesis tests for independent-measures designs (the t test in Chapter 10 and the single-factor ANOVA in Chapter 13):

1. The observations within each sample must be independent (see page 328).
2. The populations from which the samples are selected must be normal.
3. The populations from which the samples are selected must have equal variances (homogeneity of variance).

As before, the assumption of normality generally is not a cause for concern, especially when the sample size is relatively large. The homogeneity of variance assumption is more important, and if it appears that your data fail to satisfy this requirement, you should conduct a test for homogeneity before you attempt the ANOVA. Hartley's F-max test (see page 328) allows you to use the sample variances from your data to determine whether there is evidence for any differences among the population variances.

HIGHER-ORDER FACTORIAL DESIGNS

The basic concepts of the two-factor ANOVA can be extended to more complex designs involving three or more factors. A three-factor design, for example, might look at academic performance scores for two different teaching methods (factor A), for boys versus girls (factor B), and for first-grade versus second-grade classes (factor C). The logic of the analysis and many of the formulas from the two-factor ANOVA are simply extended to the situation with three (or more) factors. In a three-factor experiment, for example, you would evaluate the main effects for each of the three factors, and you would evaluate a set of two-way interactions: $A \times B$, $B \times C$, and $A \times C$. In addition, however, the extra factor introduces the potential for a three-way interaction: $A \times B \times C$.

The general logic for defining and interpreting higher-order interactions follows the pattern set by two-way interactions. For example, a two-way interaction, $A \times B$, means that the effect of factor A depends on the levels of factor B. Extending this definition, a three-way interaction, $A \times B \times C$, indicates that the two-way interaction between A and B depends on the levels of factor C. Although you may have a good understanding of two-way interactions and you may grasp the general idea of a three-way interaction, most people have great difficulty comprehending or interpreting a four-way (or more) interaction. For this reason, factorial experiments involving three or more factors can produce very complex results that are difficult to understand and thus often have limited practical value.

SUMMARY

1. A research study with two independent variables is called a two-factor design. Such a design can be diagramed as a matrix with the levels of one factor defining the row and the levels of the other factor defining the columns. Each cell in the matrix corresponds to a specific combination of the two factors.

2. Traditionally, the two factors are identified as factor A and factor B. The purpose of the ANOVA is to determine whether there are any significant mean differences among the treatment conditions or cells in the experimental matrix. These treatment effects are classified as follows:
   a. The A-effect: Overall mean differences among the levels of factor A.
   b. The B-effect: Overall mean differences among the different levels of factor B.
   c. The $A \times B$ interaction: Extra mean differences that are not accounted for by the main effects.

3. The two-factor ANOVA produces three F-ratios: one for factor A, one for factor B, and one for the $A \times B$ interaction. Each F-ratio has the same basic structure:

$$F = \frac{MS_{\text{treatment effect}}(\text{either } A \text{ or } B \text{ or } A \times B)}{MS_{\text{within treatments}}}$$

The formulas for the SS, df, and MS values for the two-factor ANOVA are presented in Figure 15.6.

4. Effect size for each main effect and for the interaction is computed as a percentage of variance explained by the specific main effect or interaction. In each case, the variability that is explained by other sources is removed before the percentage is computed.

$$\text{For factor } A, \eta^2 = \frac{SS_A}{SS_{\text{total}} - SS_B - SS_{A \times B}}$$

$$= \frac{SS_A}{SS_A + SS_{\text{within treatments}}}$$

$$\text{For factor } B, \eta^2 = \frac{SS_B}{SS_{\text{total}} - SS_A - SS_{A \times B}}$$

$$= \frac{SS_B}{SS_B + SS_{\text{within treatments}}}$$

$$\text{For } A \times B, \eta^2 = \frac{SS_{A \times B}}{SS_{\text{total}} - SS_A - SS_B}$$

$$= \frac{SS_{A \times B}}{SS_{A \times B} + SS_{\text{within treatments}}}$$

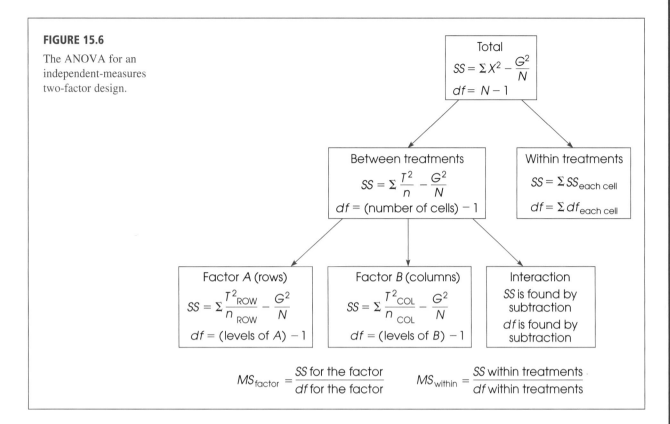

FIGURE 15.6

The ANOVA for an independent-measures two-factor design.

Total

$$SS = \Sigma X^2 - \frac{G^2}{N}$$

$$df = N - 1$$

Between treatments

$$SS = \Sigma \frac{T^2}{n} - \frac{G^2}{N}$$

$$df = (\text{number of cells}) - 1$$

Within treatments

$$SS = \Sigma SS_{\text{each cell}}$$

$$df = \Sigma df_{\text{each cell}}$$

Factor A (rows)

$$SS = \Sigma \frac{T^2_{\text{ROW}}}{n_{\text{ROW}}} - \frac{G^2}{N}$$

$$df = (\text{levels of } A) - 1$$

Factor B (columns)

$$SS = \Sigma \frac{T^2_{\text{COL}}}{n_{\text{COL}}} - \frac{G^2}{N}$$

$$df = (\text{levels of } B) - 1$$

Interaction

SS is found by subtraction

df is found by subtraction

$$MS_{\text{factor}} = \frac{SS \text{ for the factor}}{df \text{ for the factor}} \qquad MS_{\text{within}} = \frac{SS \text{ within treatments}}{df \text{ within treatments}}$$

KEY TERMS

factorial design   matrix   main effect   simple main effects

two-factor design   cells   interaction

RESOURCES

There are a tutorial quiz and other learning exercises for Chapter 15 at the Wadsworth website, www.cengage.com/psychology/gravetter. See page 29 for more information about accessing the website.

ENHANCED

WebAssign

Guided interactive tutorials, end-of-chapter problems, and related testbank items may be assigned online at WebAssign.

WebTUTOR

If you are using WebTutor along with this book, there is a WebTutor section corresponding to this chapter. The WebTutor contains a brief summary of Chapter 15, hints for learning the concepts and formulas for the two-factor ANOVA, cautions about common errors, and sample exam items including solutions.

SPSS

General instructions for using SPSS are presented in Appendix D. Following are detailed instructions for using SPSS to perform **the Two-Factor, Independent-Measures Analysis of Variance (ANOVA)** presented in this chapter.

Data Entry

1. The scores are entered into the SPSS data editor in a *stacked format,* which means that all the scores from all the different treatment conditions are entered in a single column (VAR00001).
2. In a second column (VAR00002) enter a code number to identify the level of factor *A* for each score. If factor *A* defines the rows of the data matrix, enter a 1 beside each score from the first row, enter a 2 beside each score from the second row, and so on.
3. In a third column (VAR00003) enter a code number to identify the level of factor *B* for each score. If factor *B* defines the columns of the data matrix, enter a 1 beside each score from the first column, enter a 2 beside each score from the second column, and so on.

Thus, each row of the SPSS data editor will have one score and two code numbers, with the score in the first column, the code for factor *A* in the second column, and the code for factor *B* in the third column.

Data Analysis

1. Click **Analyze** on the tool bar, select **General Linear Model,** and click on **Univariant.**
2. Highlight the column label for the set of scores (VAR0001) in the left box and click the arrow to move it into the **Dependent Variable** box.
3. One by one, highlight the column labels for the two factor codes and click the arrow to move them into the **Fixed Factors** box.
4. If you want descriptive statistics for each treatment, click on the **Options** box, select **Descriptives,** and click **Continue.**
5. Click **OK.**

SPSS Output

If you selected the Descriptives Options, SPSS will produce a table showing the means and standard deviations for each treatment condition (each cell) as well as the mean and standard deviation for each level of both factors. The results of the ANOVA are shown in the table labeled **Tests of Between-Subjects Effects.** The top row (*Corrected Model*) presents the between-treatments *SS* and *df* values. The second row (*Intercept*) is not relevant for our purposes. The next three rows present the two main effects and the interaction (the *SS, df,* and *MS* values, as well as the *F*-ratio and the level of significance), with each factor identified by its column number from the SPSS data editor. The next row (*Error*) describes the error term (denominator of the *F*-ratio), and the final row (*Corrected Total*) describes the total variability for the entire set of scores. (Ignore the row labeled *Total.*)

FOCUS ON PROBLEM SOLVING

1. Before you begin a two-factor ANOVA, take time to organize and summarize the data. It is best if you summarize the data in a matrix with rows corresponding to the levels of one factor and columns corresponding to the levels of the other factor. In each cell of the matrix, show the number of scores (*n*), the total and mean for the cell, and the *SS* within the cell. Also compute the row totals and column totals that will be needed to calculate main effects.

2. To draw a graph of the result from a two-factor experiment, first prepare a matrix with each cell containing the mean for that treatment condition.

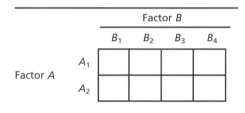

Next, list the levels of factor *B* on the *X*-axis (B_1, B_2, etc.), and put the scores (dependent variable) on the *Y*-axis. Starting with the first row of the matrix, place a dot above each *B* level so that the height of the dot corresponds to the cell mean. Connect the dots with a line, and label the line A_1. In the same way, construct a separate line for each level of factor *A* (that is, a separate line for each row of the matrix). In general, your graph will be easier to draw and easier to understand if the factor with the larger number of levels is placed on the *X*-axis (factor *B* in this example).

3. The concept of an interaction is easier to grasp if you sketch a graph showing the means for each treatment. Remember that parallel lines indicate no interaction. Crossing or converging lines indicate that the effect of one treatment depends on the levels of the other treatment you are examining. This indicates that an interaction exists between the two treatments.

4. For a two-factor ANOVA, there are three separate *F*-ratios. These three *F*-ratios use the same error term in the denominator (MS_{within}). On the other hand, these *F*-ratios will have different numerators and may have different *df* values associated with each of these numerators. Therefore, be careful when you look up the critical *F* values in the table. The two factors and the interaction may have different critical *F* values.

5. As we have mentioned in previous ANOVA chapters, it helps tremendously to organize your computations; start with *SS* and *df* values, and then compute *MS* values and *F*-ratios. Once again, using an ANOVA summary table (see page 495) can be of great assistance.

DEMONSTRATION 15.1

TWO-FACTOR ANOVA

The following data are representative of the results obtained in a research study examining the relationship between eating behavior and body weight (Schachter, 1968). The two factors in this study were:

1. The participant's weight (normal or obese)
2. The participant's state of hunger (full stomach or empty stomach)

All participants were led to believe that they were taking part in a taste test for several types of crackers, and they were allowed to eat as many crackers as they wanted. The dependent variable was the number of crackers eaten by each participant. There were two specific predictions for this study. First, it was predicted that normal participants' eating behavior would be determined by their state of hunger. That is, people with empty stomachs would eat more and people with full stomachs would eat less. Second, it was predicted that eating behavior for obese participants would not be related to their state of hunger. Specifically, it was predicted that obese participants would eat the same amount whether their stomachs were full or empty. Note that the researchers are predicting an interaction: The effect of hunger will be different for the normal participants and the obese participants. The data are as follows:

		Factor B: Hunger		
		Empty stomach	Full stomach	
Fector A: Weight	Normal	$n = 20$ $M = 22$ $T = 440$ $SS = 1540$	$n = 20$ $M = 15$ $T = 300$ $SS = 1270$	$T_{normal} = 740$
	Obese	$n = 20$ $M = 17$ $T = 340$ $SS = 1320$	$n = 20$ $M = 18$ $T = 360$ $SS = 1266$	$T_{normal} = 700$

$G = 1440$
$N = 80$
$\Sigma X^2 = 31{,}836$

$T_{empty} = 780 \quad T_{full} = 660$

STEP 1   State the hypotheses, and select alpha.
For a two-factor study, there are three separate hypotheses, the two main effects and the interaction.

For factor A, the null hypothesis states that there is no difference in the amount eaten for normal participants versus obese participants. In symbols,

$$H_0: \quad \mu_{normal} = \mu_{obese}$$

For factor B, the null hypothesis states that there is no difference in the amount eaten for full-stomach versus empty-stomach conditions. In symbols,

$$H_0: \quad \mu_{full} = \mu_{empty}$$

For the $A \times B$ interaction, the null hypothesis can be stated two different ways. First, the difference in eating between the full-stomach and empty-stomach conditions will be the same for normal and obese participants. Second, the difference in eating between the normal and obese participants will be the same for the full-stomach and empty-stomach conditions. In more general terms,

H_0:   The effect of factor A does not depend on the levels of factor B (and B does not depend on A).

We will use $\alpha = .05$ for all tests.

STEP 2   The two-factor analysis.

Rather than compute the df values and look up critical values for F at this time, we will proceed directly to the ANOVA.

STAGE 1   The first stage of the analysis is identical to the independent-measures ANOVA presented in Chapter 13, where each cell in the data matrix is considered a separate treatment condition.

$$SS_{total} = \Sigma X^2 - \frac{G^2}{N}$$

$$= 31{,}836 - \frac{1440^2}{80} = 5916$$

$$SS_{within\ treatments} = \Sigma SS_{inside\ each\ treatment} = 1540 + 1270 + 1320 + 1266 = 5396$$

$$SS_{between\ treatments} = \Sigma \frac{T^2}{n} - \frac{G^2}{N}$$

$$= \frac{440^2}{20} + \frac{300^2}{20} + \frac{340^2}{20} + \frac{360^2}{20} - \frac{1440^2}{80}$$

$$= 520$$

The corresponding degrees of freedom are

$$df_{total} = N - 1 = 79$$

$$df_{within\ treatments} = \Sigma df = 19 + 19 + 19 + 19 = 76$$

$$df_{between\ treatments} = \text{number of treatments} - 1 = 3$$

STAGE 2   The second stage of the analysis partitions the between-treatments variability into three components: the main effect for factor A, the main effect for factor B, and the $A \times B$ interaction.

For factor A (normal/obese),

$$SS_A = \Sigma \frac{T^2_{ROWS}}{n_{ROWS}} - \frac{G^2}{N}$$

$$= \frac{740^2}{40} + \frac{700^2}{40} + \frac{1440^2}{80}$$

$$= 20$$

For factor B (full/empty),

$$SS_B = \Sigma \frac{T^2_{\text{COLS}}}{n_{\text{COLS}}} - \frac{G^2}{N}$$

$$= \frac{780^2}{40} + \frac{660^2}{40} + \frac{1440^2}{80}$$

$$= 180$$

For the $A \times B$ interaction,

$$SS_{A \times B} = SS_{\text{between treatments}} - SS_A \times SS_B$$

$$= 520 - 20 - 180$$

$$= 320$$

The corresponding degrees of freedom are

$$df_A = \text{number of rows} - 1 = 1$$

$$df_B = \text{number of columns} - 1 = 1$$

$$df_{A \times B} = df_{\text{between treatments}} - df_A - df_B$$

$$= 3 - 1 - 1$$

$$= 1$$

The MS values needed for the F-ratios are

$$MS_A = \frac{SS_A}{df_A} = \frac{20}{1} = 20$$

$$MS_B = \frac{SS_B}{df_B} = \frac{180}{1} = 180$$

$$MS_{A \times B} = \frac{SS_{A \times B}}{df_{A \times B}} = \frac{320}{1} = 320$$

$$MS_{\text{within treatments}} = \frac{SS_{\text{within treatments}}}{df_{\text{within treatments}}} = \frac{5396}{76} = 71$$

Finally, the F-ratios are

$$F_A = \frac{MS_A}{MS_{\text{within treatments}}} = \frac{20}{71} = 0.28$$

$$F_B = \frac{MS_B}{MS_{\text{within treatments}}} = \frac{180}{71} = 2.54$$

$$F_{A \times B} = \frac{MS_{A \times B}}{MS_{\text{within treatments}}} = \frac{320}{71} = 4.51$$

STEP 3 Make a decision and state a conclusion.

All three F-ratios have $df = 1, 76$. With $\alpha = .05$, the critical F value is 3.98 for all three tests.

For these data, factor A (weight) has no significant effect; $F(1, 76) = 0.28$. Statistically, there is no difference in the number of crackers eaten by normal versus obese participants.

Similarly, factor B (fullness) has no significant effect; $F(1, 76) = 2.54$. Statistically, the number of crackers eaten by full participants is no different from the number eaten by hungry participants. (*Note:* This conclusion concerns the combined group of normal and obese participants. The interaction concerns these two groups separately.)

These data produce a significant interaction; $F(1, 76) = 4.51, p < .05$. This means that the effect of fullness does depend on weight. A closer look at the original data shows that the degree of fullness did affect the normal participants, but it had no effect on the obese participants.

DEMONSTRATION 15.2

MEASURING EFFECT SIZE FOR THE TWO-FACTOR ANOVA

Effect size for each main effect and for the interaction is measured by eta squared (η^2), the percentage of variance explained by the specific main effect or interaction. In each case, the variability that is explained by other sources is removed before the percentage is computed. For the two-factor ANOVA in Demonstration 15.1,

$$\text{For factor } A, \ \eta^2 = \frac{SS_A}{SS_{\text{total}} - SS_B - SS_{A \times B}} = \frac{20}{5916 - 180 - 320} = 0.004 \text{ (or 0.4\%)}$$

$$\text{For factor } B, \ \eta^2 = \frac{SS_B}{SS_{\text{total}} - SS_A - SS_{A \times B}} = \frac{180}{5916 - 20 - 320} = 0.032 \text{ (or 3.2\%)}$$

$$\text{For } A \times B, \ \eta^2 = \frac{SS_{A \times B}}{SS_{\text{total}} - SS_A - SS_B} = \frac{320}{5916 - 20 - 180} = 0.056 \text{ (or 5.6\%)}$$

PROBLEMS

1. Define each of the following terms:
   a. Factor
   b. Level
   c. Two-factor study

2. The structure of a two-factor study can be presented as a matrix with the levels of one factor determining the rows and the levels of the second factor determining the columns. With this structure in mind, describe the mean differences that are evaluated by each of the three hypothesis tests that make up a two-factor ANOVA.

3. Briefly explain what happens during the second stage of the two-factor ANOVA.

4. What additional information is obtained by using a two-factor design instead of two separate single-factor designs (one for each factor)?

5. For the data in the following matrix:

	No Treatment	Treatment	
Male	$M = 5$	$M = 3$	Overall $M = 4$
Female	$M = 9$	$M = 13$	Overall $M = 11$
	overall $M = 7$	overall $M = 8$	

   a. Describe the mean difference that is the main effect for the treatment.
   b. Describe the mean difference that is the main effect for gender.

c. Is there an interaction between gender and treatment? Explain your answer.

6. The following matrix presents the results from an independent-measures, two-factor study with a sample of $n = 10$ participants in each treatment condition. Note that one treatment mean is missing.

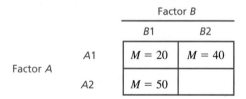

Factor B

	B1	B2
A1	$M = 20$	$M = 40$
A2	$M = 50$	

Factor A

a. What value for the missing mean would result in no main effect for factor A?
b. What value for the missing mean would result in no main effect for factor B?
c. What value for the missing mean would result in no interaction?

7. The following matrix presents the results of a two-factor study with $n = 10$ scores in each of the six treatment conditions. Note that one of the treatment means is missing.

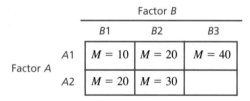

Factor B

	B1	B2	B3
A1	$M = 10$	$M = 20$	$M = 40$
A2	$M = 20$	$M = 30$	

Factor A

a. What value for the missing mean would result in no main effect for factor A?
b. What value for the missing mean would result in no interaction?

8. For the data in the following graph:

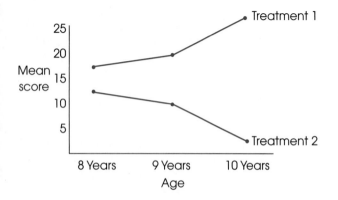

a. Is there a main effect for the treatment factor?
b. Is there a main effect for the age factor?
c. Is there an interaction between age and treatment?

9. Each of the following sets of data represent the outcome from a two-factor study evaluating males and females (factor A) and two different ages (factor B). For each set of data, determine whether the data indicate a main effect for factor A (overall mean difference between males and females); a main effect for factor B (overall mean difference between the two ages); and an interaction (mean differences not explained by the main effects).

a.

		Factor B	
		Age 5	Age 6
Factor A	Male	$M = 10$	$M = 10$
	Female	$M = 20$	$M = 20$

b.

		Factor B	
		Age 5	Age 6
Factor A	Male	$M = 10$	$M = 20$
	Female	$M = 20$	$M = 30$

c.

		Factor B	
		Age 5	Age 6
Factor A	Male	$M = 10$	$M = 20$
	Female	$M = 10$	$M = 40$

10. A researcher conducts an independent-measures, two-factor study with two levels of factor A and four levels of factor B, using a separate sample of $n = 10$ participants in each treatment condition.
a. What are the df values for the F-ratio evaluating the main effect of factor A?
b. What are the df values for the F-ratio evaluating the main effect of factor B?
c. What are the df values for the F-ratio evaluating the interaction?

11. A researcher conducts an independent-measures, two-factor study using a separate sample of $n = 8$ participants in each treatment condition. The results are evaluated using an ANOVA and the researcher reports an F-ratio with $df = 1, 42$ for factor A, and an F-ratio with $df = 2, 42$ for factor B.

 a. How many levels of factor A were used in the study?

 b. How many levels of factor B were used in the study?

 c. What are the df values for the F-ratio evaluating the interaction?

12. The following results are from an independent-measures, two-factor study with $n = 10$ participants in each treatment condition.

		Factor B	
		B1	B2
Factor A	A1	$T = 40$ $M = 4$ $SS = 50$	$T = 10$ $M = 1$ $SS = 30$
	A2	$T = 50$ $M = 5$ $SS = 60$	$T = 20$ $M = 2$ $SS = 40$

$N = 40 \quad G = 120 \quad \Sigma X^2 = 640$

 a. Use a two-factor ANOVA with $\alpha = .05$ to evaluate the main effects and the interaction.

 b. Compute η^2 to measure the effect size for each of the main effects and the interaction.

13. The following results are from an independent-measures, two-factor study with $n = 5$ participants in each treatment condition.

		Factor B		
		B1	B2	B3
Factor A	A1	$T = 25$ $M = 5$ $SS = 30$	$T = 40$ $M = 8$ $SS = 38$	$T = 70$ $M = 14$ $SS = 46$
	A2	$T = 15$ $M = 3$ $SS = 22$	$T = 20$ $M = 4$ $SS = 26$	$T = 40$ $M = 8$ $SS = 30$

$N = 30 \quad G = 210 \quad \Sigma X^2 = 2062$

 a. Use a two-factor ANOVA with $\alpha = .05$ to evaluate the main effects and the interaction.

 b. Test the simple main effects using $\alpha = .05$ to evaluate the mean difference between treatment $A1$ and $A2$ for each level of factor B.

14. Most sports injuries are immediate and obvious, like a broken leg. However, some can be more subtle, like the neurological damage that may occur when soccer players repeatedly head a soccer ball. To examine long-term effects of repeated heading, Downs and Abwender (2002) examined two different age groups of soccer players and swimmers. The dependent variable was performance on a conceptual thinking task. Following are hypothetical data, similar to the research results.

 a. Use a two-factor ANOVA with $\alpha = .05$ to evaluate the main effects and interaction.

 b. Calculate the effects size (η^2) for the main effects and the interaction.

 c. Briefly describe the outcome of the study.

		Factor B: Age	
		College	Older
Factor A: Sport	Soccer	$n = 20$ $M = 9$ $T = 180$ $SS = 380$	$n = 20$ $M = 4$ $T = 80$ $SS = 390$
	Swimming	$n = 20$ $M = 9$ $T = 180$ $SS = 350$	$n = 20$ $M = 8$ $T = 160$ $SS = 400$

$\Sigma X^2 = 6360$

15. A recent study of driving behavior suggests that self-reported measures of high driving skills and low ratings of safety skills create a dangerous combination (Šumer, Özkan, & Lajunen, 2006). *Note:* Those who rate themselves as highly skilled drivers are probably overly confident. Drivers were classified as high or low in self-rated skill based on responses to a driver skill inventory, then classified as high or low in safety skill based on responses to a driver-aggression scale. An overall measre of driving risk was obtained by combining several variables such as number of accidents, tickets, tendency to speed, and tendency to pass other cars. The following data represent results similar to those obtained in the study. Use a two-factor ANOVA with $\alpha = .05$ to evaluate the results.

		Driving Skill		
		Low	High	
Driving Safety	Low	$n = 8$ $M = 5$ $T = 40$ $SS = 52$	$n = 8$ $M = 8.5$ $T = 68$ $SS = 71$	$N = 32$ $G = 160$ $\Sigma X^2 = 1151$
	High	$n = 8$ $M = 3$ $T = 24$ $SS = 34$	$n = 8$ $M = 3.5$ $T = 28$ $SS = 46$	

16. The following table summarizes the results from a two-factor study with 2 levels of factor A and 3 levels of factor B using a separate sample of $n = 8$ participants in each treatment condition. Fill in the missing values. (*Hint:* Start with the *df* values.)

Source	SS	df	MS	
Between treatments	60	——		
Factor A	——	——	5	$F =$ ——
Factor B	——	——	——	$F =$ ——
$A \times B$ Interaction	25	——	——	$F =$ ——
Within treatments	——	——	2.5	
Total	——	——		

17. The following table summarizes the results from a two-factor study with 3 levels of factor A and 3 levels of factor B using a separate sample of $n = 9$ participants in each treatment condition. Fill in the missing values. (*Hint:* Start with the *df* values.)

Source	SS	df	MS	
Between treatments	144	——		
Factor A	——	——	18	$F =$ ——
Factor B	——	——	——	$F =$ ——
$A \times B$ Interaction	——	——	——	$F = 7.0$
Within treatments	——	——	——	
Total	360	——		

18. The following table summarizes the results from a two-factor study with 2 levels of factor A and 2 levels of factor B using a separate sample of $n = 15$ participants in each treatment condition. Fill in the missing values. (*Hint:* Start with the *df* values.)

Source	SS	df	MS	
Between treatments	——	——		
Factor A	——	——	——	$F = 7$
Factor B	——	——	——	$F = 8$
$A \times B$ Interaction	——	——	——	$F = 3$
Within treatments	280	——	——	
Total	——	——		

19. The following data are from a two-factor study examining the effects of two treatment conditions on males and females.
  a. Use an ANOVA with $\alpha = .05$ for all tests to evaluate the significance of the main effects and the interaction.
  b. Compute η^2 to measure the size of the effect for each main effect and the interaction.

		Treatments		
		I	II	
Factor A: Gender	Male	3 8 9 4 $M = 6$ $T = 24$ $SS = 26$	2 8 7 7 $M = 6$ $T = 24$ $SS = 22$	$T_{male} = 48$
	Female	0 0 2 6 $M = 2$ $T = 8$ $SS = 24$	12 6 9 13 $M = 10$ $T = 40$ $SS = 30$	$T_{female} = 48$
		$T_I = 32$	$T_{II} = 64$	

$N = 16$
$G = 96$
$\Sigma X^2 = 806$

20. For the data in the previous problem
  a. Compute SS_{total} and the SS value for factor B (treatments).
  b. The SS for factor B measures the variability in the scores that results from the mean difference between the two treatments. Eliminate this mean difference by adding 2 points to each score in treatment I and subtracting 2 points from each score in treatment II. For the resulting data, both treatments should have $T = 48$ and $M = 6$.

c. Eliminating the mean difference between treatments in part b should also reduce the overall variability of the scores. Compute SS_{total} for the new data and compare the value you obtain with SS_B from the original data. You should find that the difference is equal to SS_B, which was removed when the mean difference was eliminated. (*Caution:* You must find the new ΣX^2 value for the new scores.)

21. The following data are from a two-factor study examining the effects of three treatment conditions on males and females.
  a. Use an ANOVA with $\alpha = .05$ for all tests to evaluate the significance of the main effects and the interaction.
  b. Test the simple main effects using $\alpha = .05$ to evaluate the mean difference between males and females for each of the three treatments.

Treatments

	I	II	III
Male	1	7	9
	2	2	11
	6	9	7
	$M = 3$	$M = 6$	$M = 9$
	$T = 9$	$T = 18$	$T = 27$
	$SS = 14$	$SS = 26$	$SS = 8$

$T_{male} = 54$

Factor A: Gender

$N = 18$
$G = 144$
$\Sigma X^2 = 1608$

Female	3	10	16
	1	11	18
	5	15	11
	$M = 3$	$M = 12$	$M = 15$
	$T = 9$	$T = 36$	$T = 45$
	$SS = 8$	$SS = 14$	$SS = 26$

$T_{female} = 90$

22. For the data in the previous problem
  a. Compute SS_{total} and the SS value for factor A (gender).
  b. The SS for factor A measures the variability in the scores that results from the mean difference between the males and the females. Eliminate this mean difference by adding 2 points to each score for the males and subtracting 2 points from each score for the females. For the resulting data, both males and females should have $T = 72$ and $M = 8$.
  c. Eliminating the mean difference between genders in part b should also reduce the overall variability of the scores. Compute SS_{total} for the new data and compare the value you obtain with SS_A from the original data. You should find that the difference is equal to SS_A, which was removed when the mean difference was eliminated. (*Caution:* You must find the new ΣX^2 value for the new scores.)

23. Hyperactivity in children usually is treated by counseling, by drugs, or by both. The following data are from an experiment designed to evaluate the effectiveness of these different treatments. The dependent variable is a measure of attention span (how long each child was able to concentrate on a specific task).

	Drug	No Drug
Counseling	$n = 10$	$n = 10$
	$T = 140$	$T = 80$
	$SS = 40$	$SS = 36$
No Counseling	$n = 10$	$n = 10$
	$T = 120$	$T = 100$
	$SS = 45$	$SS = 59$

$\Sigma X^2 = 5220$

  a. Use an ANOVA with $\alpha = .05$ to evaluate these data.
  b. Do the data indicate that the drug has a significant effect? Does the counseling have an effect? Describe these results in terms of the effectiveness of the drug and counseling and their interaction.

24. In the Preview section of this chapter, we presented an experiment that examined the effect of an audience on the performance of two different personality types. Data from this experiment are as follows. The dependent variable is the number of errors made by each participant.

Self-esteem	Alone	Audience
High	3	9
	6	4
	2	5
	2	8
	4	4
	7	6
Low	7	10
	7	14
	2	11
	6	15
	8	11
	6	11

  a. Use an ANOVA with $\alpha = .05$ to evaluate the data. Describe the effect of the audience and the effect of self-esteem on performance.
  b. Calculate the effect size (η^2) for each main effect and for the interaction.

25. The following data are from a study examining the influence of a specific hormone on eating behavior. Three different drug doses were used, including a control condition (no drug), and the study measured eating behavior for males and females. The dependent variable was the amount of food consumed over a 48-hour period.

	No Drug	Small Dose	Large Dose
Males	1	7	3
	6	7	1
	1	11	1
	1	4	6
	1	6	4
Females	0	0	0
	3	0	2
	7	0	0
	5	5	0
	5	0	3

Use an ANOVA with $\alpha = .05$ to evaluate these data, and describe the results (i.e., the drug effect, the sex difference, and the interaction). Test the simple main effects at the .05 level of significance.

C H A P T E R

16

Correlation

Tools You Will Need

The following items are considered essential background material for this chapter. If you doubt your knowledge of any of these items, you should review the appropriate chapter or section before proceeding.

- Sum of squares (*SS*) (Chapter 4)
  - Computational formula
  - Definitional formula
- *z*-scores (Chapter 5)
- Hypothesis testing (Chapter 8)

Preview

16.1   Overview

16.2   The Pearson Correlation

16.3   Understanding and Interpreting the Pearson Correlation

16.4   Hypothesis Tests with the Pearson Correlation

16.5   The Spearman Correlation

16.6   Other Measures of Relationship

Summary

Focus on Problem Solving

Demonstrations 16.1 and 16.2

Problems

Preview

Having been a student and taken exams for much of your life, you probably have noticed a curious phenomenon. In every class, there are some students who zip through exams and turn in their papers while everyone else is still on page 1. Other students cling to their exams and are still working frantically when the instructor announces that time is up and demands that all papers be turned in. Have you wondered what grades these students receive? Are the students who finish first the best in the class, or are they completely unprepared and simply accepting their failure? Are the *A* students the last to finish because they are compulsively checking and rechecking their answers? To help answer these questions, we carefully observed a recent exam and recorded the amount of time each student spent on the exam and the grade each student received. These data are shown in Figure 16.1. Note that we have listed time along the *X*-axis and grade on the *Y*-axis. Each student is identified by a point on the graph that is located directly above the student's time and directly across from the student's grade. Also note that we have drawn a line through the middle of the data points in Figure 16.1. The line helps make the relationship between time and grade more obvious. The graph shows that the highest grades tend to go to the students who finished their exams early. Students who held their papers to the bitter end tended to have low grades.

The Problem: Although the data in Figure 16.1 appear to show a clear relationship, we need some procedure to measure the relationship and a hypothesis test to determine whether it is significant. In the preceding chapter, we described relationships between variables in terms of mean differences between two or more groups of scores, and we used hypothesis tests that evaluate the significance of mean differences. For the data in Figure 16.1, there is only one group of scores, and calculating a mean is not going to help describe the relationship. To evaluate these data, a completely different approach is needed for both descriptive and inferential statistics.

The Solution: The data in Figure 16.1 are an example of the results from a correlational research study. In Chapter 1, the correlational design was introduced as a method for examining the relationship between two variables by measuring two different variables for each individual in one group of participants. The relationship obtained in a correlational study is typically described and evaluated with a statistical measure known as a *correlation*. Just as a sample mean provides a concise description of an entire sample, a correlation provides a description of a relationship. We will look at how correlations are used and interpreted. For example, now that you have seen the relationship between time and grades, don't you think it might be a good idea to start turning in your exam papers a little sooner? Wait and see.

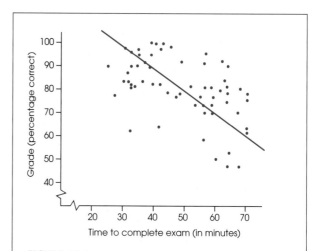

FIGURE 16.1

The relationship between exam grade and time needed to complete the exam. Notice the general trend in these data: Students who finish the exam early tend to have better grades.

16.1 OVERVIEW

Correlation is a statistical technique that is used to measure and describe a relationship between two variables. Usually the two variables are simply observed as they exist naturally in the environment—there is no attempt to control or manipulate the variables. For example, a researcher could check high school records (with permission) to obtain a measure of each student's academic performance, and then survey each family to obtain a measure of income. The resulting data could be used to determine whether there is a relationship between high school grades and family income. Notice that the researcher is not manipulating any student's grade or any family's incomes, but is simply observing what occurs naturally. You also should notice that a correlation requires two scores for each individual (one score from each of the two variables). These scores normally are identified as X and Y. The pairs of scores can be listed in a table, or they can be presented graphically in a scatter plot (Figure 16.2). In the scatter plot, the values for the X variable are listed on the horizontal axis and the Y values are listed on the vertical axis. Each individual is then represented by a single point in the graph so that the horizontal position corresponds to the individual's X value and the vertical position corresponds to the Y value. The value of a scatter plot is that it allows you to see any patterns or trends that exist in the data. The scores in Figure 16.2, for example, show a clear relationship between family income and student grades; as income increases, grades also increase.

THE CHARACTERISTICS OF A RELATIONSHIP

A correlation is a numerical value that describes and measures three characteristics of the relationship between X and Y. These three characteristics are as follows:

1. The Direction of the Relationship. The sign of the correlation, positive or negative, describes the direction of the relationship.

DEFINITIONS

In a **positive correlation**, the two variables tend to move in the same direction: As the value of the X variable increases from one individual to another, the Y variable also tends to increase; when the X variable decreases, the Y variable also decreases.

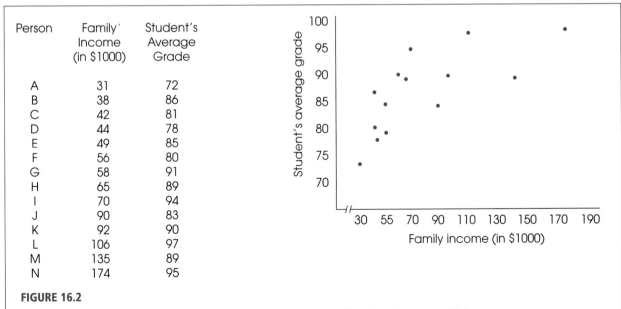

Person	Family Income (in $1000)	Student's Average Grade
A	31	72
B	38	86
C	42	81
D	44	78
E	49	85
F	56	80
G	58	91
H	65	89
I	70	94
J	90	83
K	92	90
L	106	97
M	135	89
N	174	95

FIGURE 16.2

Correlational data showing the relationship between family income (X) and student grades (Y) for a sample of n = 14 high school students. The scores are listed in order from lowest to highest family income and are shown in a scatter plot.

In a **negative correlation**, the two variables tend to go in opposite directions. As the *X* variable increases, the *Y* variable decreases. That is, it is an inverse relationship.

The following examples illusrate positive and negative relationships.

EXAMPLE 16.1

Remember that the actual data appear as *points* in the figures. The dashed lines have been added as visual aids to help make the direction of the relationship easier to see.

Suppose that you run the drink concession at the football stadium. After several seasons, you begin to notice a relationship between the temperature at game time and the beverages you sell. Specifically, you have noted that when the temperature is high, you tend to sell a lot of beer. When the temperature is low, you sell relatively little beer [Figure 16.3(a)]. This is an example of a positive correlation. At the same time, you have noted a relationship between temperature and coffee sales: On cold days, you sell much more coffee than on hot days [see Figure 16.3(b)]. This is an example of a negative relationship.

2. The Form of the Relationship. In the preceding coffee and beer examples, the relationships tend to have a linear form; that is, the points in the scatter plot tend to form a straight line. Notice that we have drawn a line through the middle of the data points in each figure to help show the relationship. The most common use of correlation is to measure straight-line relationships. However, you should note that other forms of relationship do exist and that there are special correlations used to measure them. (We will examine alternatives in Sections 16.5 and 16.6.) Figure 16.4(a) shows the relationship between reaction time and age. In this scatter plot, there is a curved relationship. Reaction time improves with age until the late teens, when it reaches a peak; after that, reaction time starts to get worse. Figure 16.4(b) shows the

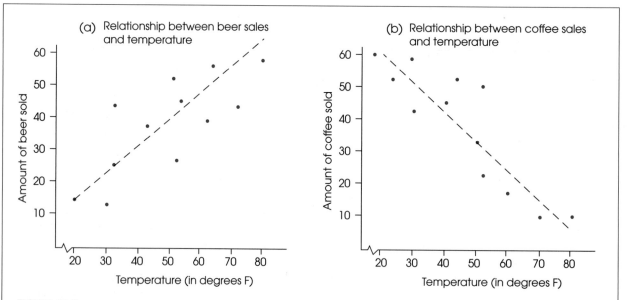

FIGURE 16.3

Examples of positive and negative relationships. (a) Beer sales are positively related to temperature. (b) Coffee sales are negatively related to temperature.

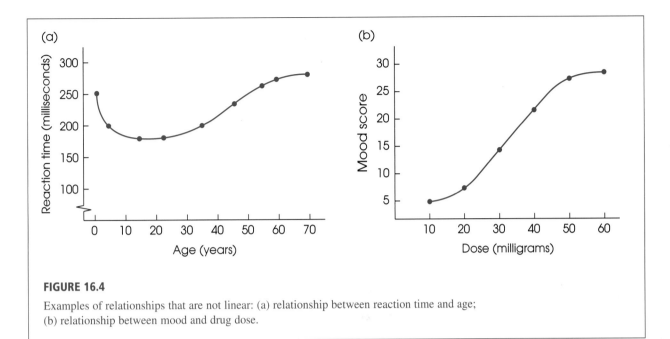

FIGURE 16.4

Examples of relationships that are not linear: (a) relationship between reaction time and age;
(b) relationship between mood and drug dose.

typical dose-response relationship. Again, this is not a straight-line relationship. In this graph, the elevation in mood increases rather rapidly with dose increases. However, beyond a certain dose, the amount of improvement in mood levels off. Many types of measurements exist for correlation. In general, each type is designed to evaluate a specific form of relationship. In this text, we concentrate on the correlation that measures linear relationships.

3. The Strength or Consistency of the Relationship. Finally, the correlation measures the consistency of the relationship. For a linear relationship, for example, the data points could fit perfectly on a straight line. Every time *X* increases by 1 point, the value of *Y* also changes by a consistent and predictable amount. Figure 16.5(a) shows an example of a perfect linear relationship. However, relationships are usually not perfect. Although there may be a tendency for the value of *Y* to increase whenever *X* increases, the amount that *Y* changes is not always the same, and, occasionally, *Y* decreases when *X* increases. In this situation, the data points will not fall perfectly on a straight line. The consistency of the relationship is measured by the numerical value of the correlation. A *perfect correlation* always is identified by a correlation of 1.00 and indicates a perfectly consistent erelationship. For a correlation of 1.00 (or −1.00), each change in *X* is accompanied by a perfectly prredictable change in *Y*. At the other extreme, a correlation of zero indicates no consistency at all. For a correlation of zero, the data points are scattered randomly with no clear trend [see Figure 16.5(b)]. Intermediate values between 0 and 1 indicate the degree of consistency.

Examples of different values for linear correlations are shown in Figure 16.5. In each example, we have sketched a line around the data points. This line, called an *envelope* because it encloses the data, often helps you to see the overall trend in the data. As a rule of thumb, when the envelope is shaped roughly like a football, the correlation is around 0.7. Envelopes that are fatter than a football indicate correlations closer to zero, and narrower shapes indicate correlations closer to 1.00.

FIGURE 16.5

Examples of different values for linear correlations: (a) a perfect negative correlation, −1.00; (b) no linear trend, 0.00; (c) a strong positive relationship, approximately +0.90; (d) a relatively weak negative correlation, approximately −0.40.

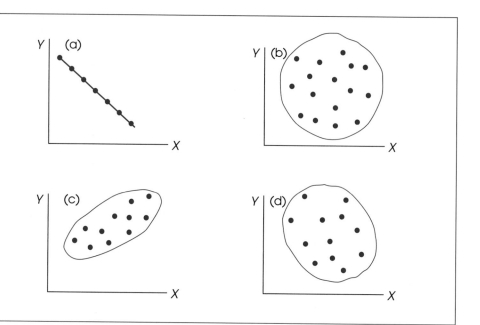

WHERE AND WHY CORRELATIONS ARE USED	Although correlations have a number of different applications, a few specific examples are presented next to give an indication of the value of this statistical measure.

    1. Prediction. If two variables are known to be related in some systematic way, it is possible to use one of the variables to make accurate predictions about the other. For example, when you applied for admission to college, you were required to submit a great deal of personal information, including your scores on the Scholastic Achievement Test (SAT). College officials want this information so they can predict your chances of success in college. It has been demonstrated over several years that SAT scores and college grade point averages are correlated. Students who do well on the SAT tend to do well in college; students who have difficulty with the SAT tend to have difficulty in college. Based on this relationship, the college admissions office can make a prediction about the potential success of each applicant. You should note that this prediction is not perfectly accurate. Not everyone who does poorly on the SAT will have trouble in college. That is why you also submit letters of recommendation, high school grades, and other information with your application.

    2. Validity. Suppose that a psychologist develops a new test for measuring intelligence. How could you show that this test truly is measuring what it claims; that is, how could you demonstrate the validity of the test? One common technique for demonstrating validity is to use a correlation. If the test actually is measuring intelligence, then the scores on the test should be related to other measures of intelligence—for example, standardized IQ tests, performance on learning tasks, problem-solving ability, and so on. The psychologist could measure the correlation between the new test and each of these other measures of intelligence to demonstrate that the new test is valid.

    3. Reliability. In addition to evaluating the validity of a measurement procedure, correlations are used to determine reliability. A measurement procedure is considered reliable to the extent that it produces stable, consistent measurements. That is, a reliable measurement procedure produces the same (or nearly the same) scores when the same individuals are measured under the same conditions. For example, if your IQ were measured as 113 last week, you would expect to obtain nearly the same score if your IQ were measured again this week. One way to evaluate reliability is to use correlations to determine the relationship between two sets of measurements. When reliability is high, the correlation between two measurements should be strong and positive.

4. Theory Verification. Many psychological theories make specific predictions about the relationship between two variables. For example, a theory may predict a relationship between brain size and learning ability; a developmental theory may predict a relationship between the parents' IQs and the child's IQ; a social psychologist may have a theory predicting a relationship between personality type and behavior in a social situation. In each case, the prediction of the theory could be tested by determining the correlation between the two variables.

1. For each of the following, indicate whether you would expect a positive or a negative correlation.
   a. Height and weight for a group of adults
   b. Daily high temperature and daily energy consumption for 30 days in the summer
   c. Daily high temperature and daily energy consumption for 30 days in the winter

2. The data points would be clustered more closely around a straight line for a correlation of $-.80$ than for a correlation of $+.05$. (True or false?)

3. If the data points are tightly clustered around a line that slopes down from left to right, then a good estimate of the correlation would be $+.90$. (True or false?)

4. A correlation can never be greater than $+1.00$. (True or false?)

ANSWERS

1. **a.** Positive: Taller people tend to weigh more.
   b. Positive: Higher temperature tends to increase the use of air-conditioning.
   c. Negative: Higher temperature tends to decrease the need for heating.

2. True. The numerical value indicates the strength of the relationship. The sign only indicates direction.

3. False. The sign of the correlation would be negative.

4. True.

16.2   THE PEARSON CORRELATION

By far the most common correlation is the Pearson correlation (or the Pearson product-moment correlation) which measures the degree of straight-line relationship.

DEFINITION

The **Pearson correlation** measures the degree and direction of linear relationship between two variables.

The Pearson correlation is identified by the letter r. Conceptually, this correlation is computed by

$$r = \frac{\text{degree to which } X \text{ and } Y \text{ vary together}}{\text{degree to which } X \text{ and } Y \text{ vary separately}}$$

$$= \frac{\text{covariability of } X \text{ and } Y}{\text{variability of } X \text{ and } Y \text{ separately}}$$

When there is a perfect linear relationship, every change in the X variable is accompanied by a corresponding change in the Y variable. In Figure 16.5(a), for example, every time the value of X increases, there is a perfectly predictable decrease in the value of Y. The result is a perfect linear relationship, with X and Y always varying together. In this case, the covariability (X and Y together) is identical to the variability of X and Y separately, and the formula produces a correlation of 1.00 or -1.00. At the other extreme, when there is no linear relationship, a change in the X variable does not correspond to any predictable change in the Y variable. In this case, there is no covariability, and the resulting correlation is zero.

THE SUM OF PRODUCTS OF DEVIATIONS

To calculate the Pearson correlation, it is necessary to introduce one new concept: the *sum of products* of deviations, or *SP*. This new value is similar to *SS* (the sum of squared deviations) which is used to measure variability for a single variable. Now, we will use *SP* to measure the amount of covariability between two variables. The value for *SP* can be calculated with either a definitional formula or a computational formula.

The *definitional formula* for the sum of products is

$$SP = \Sigma(X - M_X)(Y - M_Y) \tag{16.1}$$

where M_X is the mean for the X scores and M_Y is the mean for the Ys.

This formula instructs you to find the X deviation and the Y deviation for each individual, then multiply the deviations to obtain the product for each individual, then add up the products. Notice that the terms in the formula define the value being calculated: the sum of the products of the deviations.

The *computational formula* for the sum of products is

$$SP = \Sigma XY - \frac{\Sigma X \Sigma Y}{n} \tag{16.2}$$

Caution: The n in this formula refers to the number of pairs of scores.

Because the computational formula uses the original scores (X and Y values), it usually results in easier calculations than those required with the definitional formula. However, both formulas will always produce the same value for *SP*.

You may have noted that the formulas for *SP* are similar to the formulas you have learned for *SS* (sum of squares). The relationship between the two sets of formulas is described in Box 16.1. The following example demonstrates the calculation of *SP* with both formulas.

EXAMPLE 16.2

The same set of $n = 4$ pairs of scores will be used to calculate *SP*, first using the definitional formula and then using the computational formula.

For the definitional formula, you need deviation scores for each of the X values and each of the Y values. Note that the mean for the Xs is $M_X = 3$ and that the mean

BOX 16.1   COMPARING THE *SP* AND *SS* FORMULAS

It will help you to learn the formulas for *SP* if you note the similarity between the two *SP* formulas and the corresponding formulas for *SS* that were presented in Chapter 4. The definitional formula for *SS* is

$$SS = \Sigma(X - M)^2$$

In this formula, you must square each deviation, which is equivalent to multiplying it by itself. With this in mind, the formula can be rewritten as

$$SS = \Sigma(X - M)(X - M)$$

The similarity between the *SS* formula and the *SP* formula should be obvious—the *SS* formula uses squares and the *SP* formula uses products. This same relationship exists for the computational formulas. For *SS*, the computational formula is

$$SS = \Sigma X^2 - \frac{(\Sigma X)^2}{n}$$

As before, each squared value can be rewritten so that the formula becomes

$$SS = \Sigma XX - \frac{\Sigma X \Sigma X}{n}$$

Again, note the similarity in structure between the *SS* formula and the *SP* formula. If you remember that *SS* uses squares and *SP* uses products, the two new formulas for the sum of products should be easy to learn.

---

for the *Y*s is $M_Y = 5$. The deviations and the products of deviations are shown in the following table:

Caution: The signs ($+$ and $-$) are critical in determining the sum of products, *SP*.

Scores		Deviations		Products
X	Y	$X - M_X$	$Y - M_Y$	$(X - M_X)(Y - M_Y)$
1	3	-2	-2	$+4$
2	6	-1	$+1$	-1
4	4	$+1$	-1	-1
5	7	$+2$	$+2$	$+4$
				$+6 = SP$

For these scores, the sum of the products of the deviations is $SP = +6$.

For the computational formula, you need the *X* value, the *Y* value, and the *XY* product for each individual. Then you find the sum of the *X*s, the sum of the *Y*s, and the sum of the *XY* products. These values are as follows:

X	Y	XY	
1	3	3	
2	6	12	
4	4	16	
5	7	35	
12	20	66	Totals

Substituting the sums in the formula gives

$$SP = \Sigma XY - \frac{\Sigma X \Sigma Y}{n}$$

$$= 66 - \frac{12(20)}{4}$$

$$= 66 - 60$$

$$= 6$$

Note that both formulas produce the same result, $SP = 6$.

CALCULATION OF THE
PEARSON CORRELATION

As noted earlier, the Pearson correlation consists of a ratio comparing the covariability of X and Y (the numerator) with the variability of X and Y separately (the denominator). In the formula for the Pearson r, we use SP to measure the covariability of X and Y. The variability of X and Y is measured by computing SS for the X scores and SS for the Y scores separately. With these definitions, the formula for the Pearson correlation becomes

Note that you *multiply* SS for X by SS for Y in the denominator of the Pearson formula.

$$r = \frac{SP}{\sqrt{SS_X SS_Y}} \tag{16.3}$$

The following example demonstrates the use of this formula with a simple set of scores.

EXAMPLE 16.3

The Pearson correlation is computed for the set of $n = 5$ pairs of scores shown in the margin on page 529.

Before starting any calculations, it is useful to put the data in a scatter plot and make a preliminary estimate of the correlation. These data have been graphed in Figure 16.6. Looking at the scatter plot, it appears that there is a very good (but not

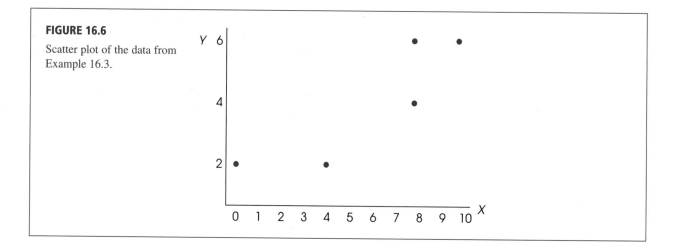

FIGURE 16.6

Scatter plot of the data from Example 16.3.

X	Y
0	2
10	6
4	2
8	4
8	6

perfect) positive correlation. You should expect an approximate value of $r = +.8$ or $+.9$. To find the Pearson correlation, we will need SP, SS for X, and SS for Y. The following table presents the calculations for each of these values using the definitional formulas. (Note that the mean for the X values is $M_X = 6$ and the mean for the Y scores is $M_Y = 4$.)

Scores		Deviations		Squared Deviations		Products
X	Y	$X - M_X$	$Y - M_Y$	$(X - M_X)^2$	$(Y - M_Y)^2$	$(X - M_X)(Y - M_Y)$
0	2	−6	−2	36	4	+12
10	6	+4	+2	16	4	+8
4	2	−2	−2	4	4	+4
8	4	+2	0	4	0	0
8	6	+2	+2	4	4	+4
				$SS_X = 64$	$SS_Y = 16$	$SP = +28$

Using these values, the Pearson correlation is

$$r = \frac{SP}{\sqrt{(SS_X)(SS_Y)}} = \frac{28}{\sqrt{(64)(16)}} = \frac{28}{32} = +0.875$$

Note that the value we obtained for the correlation is perfectly consistent with the pattern shown in Figure 16.6. First, the positive value of the correlation indicates that the points are clustered around a line that slopes up to the right. Second, the high value for the correlation (near 1.00) indicates that the points are very tightly clustered close to the line. Thus, the value of the correlation describes the relationship that exists in the data.

THE PEARSON CORRELATION AND z-SCORES

The Pearson correlation measures the relationship between an individual's location in the X distribution and his or her location in the Y distribution. For example, a positive correlation means that individuals who score high on X also tend to score high on Y. Similarly, a negative correlation indicates that individuals with high X scores tend to have low Y scores.

Recall from Chapter 5 that z-scores identify the exact location of each individual score within a distribution. With this in mind, each X value can be transformed into a z-score, z_X, using the mean and standard deviation for the set of Xs. Similarly, each Y score can be transformed into z_Y. If the X and Y values are viewed as a sample, the transformation is completed using the sample formula for z (Equation 5.3). If the X and Y values form a complete population, the z-scores are computed using Equation 5.1. After the transformation, the formula for the Pearson correlation can be expressed entirely in terms of z-scores.

Note that the population value is identified with a Greek letter, in this case the letter rho (ρ), which is the Greek equivalent of the letter r.

For a sample, $r = \dfrac{\Sigma z_X z_Y}{(n - 1)}$ (16.4)

For a population, $\rho = \dfrac{\Sigma z_X z_Y}{N}$ (16.5)

LEARNING CHECK

1. Describe what is measured by a Pearson correlation.

2. Can *SP* ever have a value less than zero?

3. Calculate the sum of products of deviations (*SP*) for the following set of scores. Use the definitional formula and then the computational formula. Verify that you get the same answer with both formulas.

X	Y
1	0
3	1
7	6
5	2
4	1

4. Compute the Pearson correlation for the following data:

X	Y
2	9
1	10
3	6
0	8
4	2

ANSWERS

Remember: It is useful to sketch a scatterplot and make an estimate of the correlation before you begin calculations.

1. The Pearson correlation measures the degree and direction of linear relationship between two variables.

2. Yes. *SP* can be positive, negative, or zero depending on the relationship between X and Y.

3. $SP = 19$

4. $r = -\dfrac{16}{20} = -0.80$

16.3   UNDERSTANDING AND INTERPRETING THE PEARSON CORRELATION

When you encounter correlations, there are four additional considerations that you should bear in mind:

1. Correlation simply describes a relationship between two variables. It does not explain why the two variables are related. Specifically, a correlation should not and cannot be interpreted as proof of a cause-and-effect relationship between the two variables.

2. The value of a correlation can be affected greatly by the range of scores represented in the data.

3. One or two extreme data points, often called *outliers,* can have a dramatic effect on the value of a correlation.

4. When judging how "good" a relationship is, it is tempting to focus on the numerical value of the correlation. For example, a correlation of $+.5$ is halfway between 0 and 1.00 and therefore appears to represent a moderate degree of relationship. However, a correlation should not be interpreted as a proportion. Although a correlation of 1.00 does mean that there is a 100% perfectly predictable relationship between X and Y, a correlation of .5 does not mean that you can make predictions with 50% accuracy. To describe how accurately one variable predicts the other, you must square the correlation. Thus, a correlation of $r = .5$ means that one variable *partially* predicts the other, but the predictable portion is only $r^2 = .5^2 = 0.25$ (or 25%) of the total variability.

We now discuss each of these four points in detail.

CORRELATION AND CAUSATION	One of the most common errors in interpreting correlations is to assume that a correlation necessarily implies a cause-and-effect relationship between the two variables. (Even Pearson blundered by asserting causation from correlational data [Blum, 1978].) We constantly are bombarded with reports of relationships: Cigarette smoking is related to heart disease; alcohol consumption is related to birth defects; carrot consumption is related to good eyesight. Do these relationships mean that cigarettes cause heart disease or carrots cause good eyesight? The answer is *no*. Although there may be a causal relationship, the simple existence of a correlation does not prove it. Earlier, for example, we discussed a study showing a relationship between high school grades and family income. However, this result does not mean that having a higher family income *causes* students to get better grades. To establish a cause-and-effect relationship, it would be necessary to conduct a true experiment (see page 12) in which family income was manipulated by a researcher and other variables were rigorously controlled. The fact that a correlation does not establish causation is demonstrated in the following example.
EXAMPLE 16.4	Suppose that we select a variety of different cities and towns throughout the United States and measure the number of serious crimes (Y variable) and the number of churches (X variable) for each. A scatter plot showing hypothetical data for this study is presented in Figure 16.7. Notice that this scatter plot shows a strong, positive correlation between churches and crime. You also should note that these are realistic data. It is reasonable that the smaller towns would have less crime and fewer churches and that the large cities would have large values for both variables. Does this relationship mean that churches cause crime? Does it mean that crime causes churches? It should be clear that the answer to both is no. Although a strong correlation exists between churches and crime, the real cause of the relationship is the size of the population.
CORRELATION AND RESTRICTED RANGE	Whenever a correlation is computed from scores that do not represent the full range of possible values, you should be cautious in interpreting the correlation. Suppose, for example, that you are interested in the relationship between IQ and creativity. If you select a sample of your fellow college students, your data probably will represent only a

FIGURE 16.7

Hypothetical data showing the logical relationship between the number of churches and the number of serious crimes for a sample of U.S. cities.

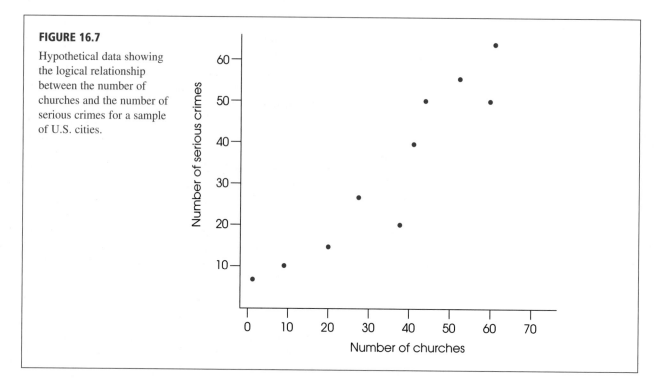

limited range of IQ scores (most likely from 110 to 130). The correlation within this *restricted range* could be completely different from the correlation that would be obtained from a full range of IQ scores. For example, Figure 16.8 shows a strong positive relationship between X and Y when the entire range of scores is considered. However, this relationship is obscured when the data are limited to a restricted range.

To be safe, you should not generalize any correlation beyond the range of data represented in the sample. For a correlation to provide an accurate description for the general population, there should be a wide range of X and Y values in the data.

FIGURE 16.8

In this example, the full range of X and Y values shows a strong, positive correlation, but the restricted range of scores produces a correlation near zero.

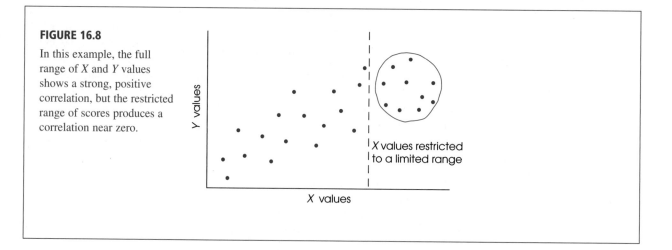

OUTLIERS   An outlier is an individual with X and/or Y values that are substantially different (larger or smaller) from the values obtained for the other individuals in the data set. The data point of a single outlier can have a dramatic influence on the value obtained for the correlation. This effect is illustrated in Figure 16.9. Figure 16.9(a) shows a set of $n = 5$ data points for which the correlation between X and Y variables is nearly zero (actually $r = -0.08$). In Figure 16.9(b), one extreme data point (14, 12) has been added to the original data set. When this outlier is included in the analysis, a strong, positive correlation emerges (now $r = +0.85$). Note that the single outlier drastically alters the value for the correlation and thereby can affect one's interpretation of the relationship between variables X and Y. Without the outlier, one would conclude there is no relationship between the two variables. With the extreme data point, $r = +0.85$ implies that as X increases, Y will increase—and do so consistently. The problem of outliers is a good reason why you should always look at a scatter plot, instead of simply basing your interpretation on the numerical value of the correlation. If you only "go by the numbers," you might overlook the fact that one extreme data point inflated the size of the correlation.

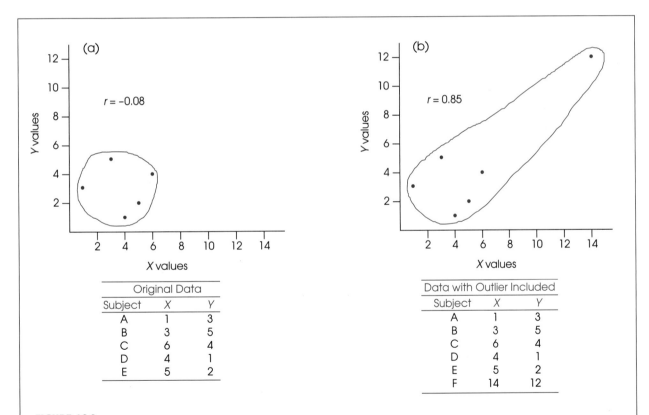

FIGURE 16.9

A demonstration of how one extreme data point (an outlier) can influence the value of a correlation.

CORRELATION AND THE STRENGTH OF THE RELATIONSHIP

A correlation measures the degree of relationship between two variables on a scale from 0 to 1.00. Although this number provides a measure of the degree of relationship, many researchers prefer to square the correlation and use the resulting value to measure the strength of the relationship.

One of the common uses of correlation is for prediction. If two variables are correlated, you can use the value of one variable to predict the other. For example, college admissions officers do not just guess which applicants are likely to do well; they use other variables (SAT scores, high school grades, and so on) to predict which students are most likely to be successful. These predictions are based on correlations. By using correlations, the admissions officers expect to make more accurate predictions than would be obtained by chance. In general, the squared correlation (r^2) measures the gain in accuracy that is obtained from using the correlation for prediction. The squared correlation measures the proportion of variability in the data that is explained by the relationship between X and Y. It is sometimes called the *coefficient of determination*.

DEFINITION

The value r^2 is called the **coefficient of determination** because it measures the proportion of variability in one variable that can be determined from the relationship with the other variable. A correlation of $r = 0.80$ (or -0.80), for example, means that $r^2 = 0.64$ (or 64%) of the variability in the Y scores can be predicted from the relationship with X.

In earlier chapters (see pages 292, 321, and 349) we introduced r^2 as a method for measuring effect size for research studies where mean differences were used to compare treatments. Specifically, we measured how much of the variance in the scores was accounted for by the differences between treatments. In experimental terminology, r^2 measures how much of the variance in the dependent variable is accounted for by the independent variable. Now we are doing the same thing, except that there is no independent or dependent variable. Instead, we simply have two variables, X and Y, and we use r^2 to measure how much of the variance in one variable can be determined from its relationship with the other variable.

Just as r^2 was used to evaluate effect size for mean differences in Chapters 9, 10, and 11, r^2 can now be used to evaluate the size or strength of the correlation. The same standards that were introduced in Table 9.3 (page 294), apply to both uses of the r^2 measure. Specifically, an r^2 value of 0.01 indicates a small effect or a small correlation, r^2 of 0.09 indicates a medium correlation, and r^2 of 0.25 or larger indicates a large correlation.

A more detailed discussion of the coefficient of determination is presented in Section 16.6 and in Chapter 17. For now, you simply should realize that whenever two variables are consistently related, it is possible to use one variable to predict the values of the second variable. The following example demonstrates this concept.

EXAMPLE 16.5

Figure 16.10 shows three sets of data representing different degrees of linear relationship. The first set of data [Figure 16.10(a)] shows the relationship between IQ and shoe size. In this case, the correlation is $r = 0$ (and $r^2 = 0$), and you have no ability to predict a person's IQ based on his or her shoe size. Knowing a person's shoe size provides no information (0%) about the person's IQ. In this case, shoe size provides no information (0%) to help explain why different people have different IQs.

Now consider the data in Figure 16.10(b). These data show a moderate, positive correlation, $r = +0.60$, between IQ scores and college grade point averages (GPA). Students with high IQs tend to have higher grades than students with low IQs. From this relationship, it is possible to predict a student's GPA based on his or her IQ. However, you should realize that the prediction is not perfect. Although students with high IQs *tend* to have high GPAs, this is not always true. Thus, knowing a student's IQ provides some information about the student's grades, or knowing a student's grades provides some information about the student's IQ. In this case, IQ scores help explain the fact that different students have different GPAs. Specifically, you can say that *part* of the differences in GPA are accounted for by IQ. With a correlation of $r = +0.60$, we obtain $r^2 = 0.36$, which means that 36% of the differences in GPA can be explained by IQ.

Finally, consider the data in Figure 16.10(c). This time we show a perfect linear relationship ($r = +1.00$) between monthly salary and yearly salary for a group of college employees. With $r = 1.00$ and $r^2 = 1.00$, there is 100% predictability. If you know a person's monthly salary, you can predict perfectly the person's annual salary. If two people have different annual salaries, the difference can be completely explained (100%) by the difference in their monthly salaries.

One final comment concerning the interpretation of a correlation is presented in Box 16.2.

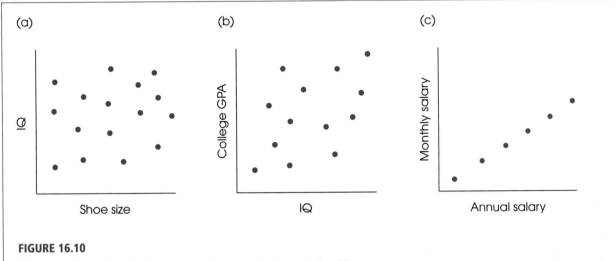

FIGURE 16.10

Three sets of data showing three different degrees of linear relationship.

| BOX 16.2 | **REGRESSION TOWARD THE MEAN** |

Consider the following problem.

Explain why the rookie of the year in major-league baseball usually does not perform as well in his second season.

Notice that this question does not appear to be statistical or mathematical in nature. However, the answer to the question is directly related to the statistical concepts of correlation and regression (Chapter 17). Specifically, there is a simple observation about correlations known as *regression toward the mean.*

D E F I N I T I O N   When there is a less-than-perfect correlation between two variables, extreme scores (high or low) on one variable tend to be paired with the less extreme scores (more toward the mean) on the second variable. This fact is called **regression toward the mean.**

Figure 16.11 shows a scatter plot with a less-than-perfect correlation between two variables. The data points in this figure might represent batting averages for baseball rookies in 2007 (variable 1) and batting averages for the same players in 2008 (variable 2). Because the correlation is less than perfect, the highest scores on variable 1 are generally *not* the highest scores on variable 2. In baseball terms, the rookies with the highest averages in 2007 do not have the highest averages in 2008.

Remember that a correlation does not explain *why* one variable is related to the other; it simply says that there is a relationship. The correlation cannot explain why the best rookie does not perform as well in his second year. But, because the correlation is not perfect, it is a statistical fact that extremely high scores in one year generally will *not* be paired with extremely high scores in the next year.

Regression toward the mean often poses a problem for interpreting experimental results. Suppose, for example, that you want to evaluate the effects of a special preschool program for disadvantaged children. You select a sample of children who score extremely low on an academic performance test. After participating in your preschool program, these children score significantly higher on the test. Why did their scores improve? One answer is that the special program helped. But an alternative answer is regression toward the mean. If there is a less-than-perfect correlation between scores on the first test and scores on the second test (which is usually the case), individuals with extremely low scores on test 1 will tend to have higher scores on test 2. It is a statistical fact of life, not necessarily the result of any special program.

Now try using the concept of regression toward the mean to explain the following phenomena:

1. You have a truly outstanding meal at a restaurant. However, when you go back with a group of friends, you find that the food is disappointing.

2. You have the highest score on exam I in your statistics class, but score only a little above average on exam II.

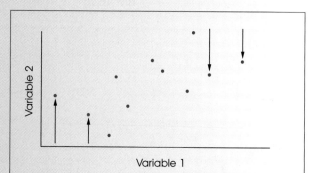

FIGURE 16.11

A demonstration of regression toward the mean. The figure shows a scatterplot for a set of data with a less-than-perfect correlation. Notice that the highest scores on variable 1 (extreme right-hand points) are not the highest scores on variable 2, but are displaced downward toward the mean. Also, the lowest scores on variable 1 (extreme left-hand points) are not the lowest scores on variable 2, but are displaced upward toward the mean.

16.4 HYPOTHESIS TESTS WITH THE PEARSON CORRELATION

The Pearson correlation is generally computed for sample data. As with most sample statistics, however, a sample correlation often is used to answer questions about the coresponding population correlation. For example, a psychologist would like to know whether there is a relationship between IQ and creativity. This is a general question concerning a population. To answer the question, a sample would be selected, and the sample data would be used to compute the correlation value. You should recognize this process as an example of inferential statistics: using samples to draw inferences about populations. In the past, we have been concerned primarily with using sample means as the basis for answering questions about population means. In this section, we examine the procedures for using a sample correlation as the basis for testing hypotheses about the corresponding population correlation.

THE HYPOTHESES

The basic question for this hypothesis test is whether a correlation exists in the population. The null hypothesis is "No, there is no correlation in the population" or "The population correlation is zero." The alternative hypothesis is "Yes, there is a real, nonzero correlation in the population." Because the population correlation is traditionally represented by ρ (the Greek letter rho), these hypotheses would be stated in symbols as

H_0: $\rho = 0$ (There is no population correlation.)

H_1: $\rho \neq 0$ (There is a real correlation.)

When there is a specific prediction about the direction of the correlation, it is possible to do a directional, or one-tailed test. For example, if a researcher is predicting a positive relationship, the hypotheses would be

H_0: $\rho \leq 0$ (The population correlation is not positive.)

H_1: $\rho > 0$ (The population correlation is positive.)

The correlation from the sample data is used to evaluate these hypotheses. For the regular, nondirectional test, a sample correlation near zero provides support for H_0 and a sample value far from zero tends to refute H_0. For a directional test, a positive value for the sample correlation would tend to refute a null hypothesis stating that the population correlation is not positive.

Although sample correlations are used to test hypotheses about population correlations, you should keep in mind that samples are not expected to be identical to the populations from which they come; there will be some discrepancy (sampling error) between a sample statistic and the corresponding population parameter. Specifically, you should always expect some error between a sample correlation and the population correlation it represents. One implication of this fact is that even when there is no correlation in the population ($\rho = 0$), you are still likely to obtain a nonzero value for the sample correlation. This is particularly true for small samples. Figure 16.12 illustrates how a small sample from a population with a near-zero correlation could result in a correlation that deviates from zero. The colored dots in the figure represent the entire population and the three circled dots represent a random sample. Note that the three sample points show a relatively good, positive correlation even though there is no linear trend ($\rho = 0$) for the population.

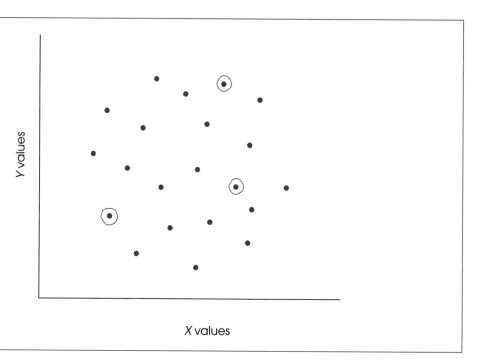

FIGURE 16.12

Scatterplot of a population of X and Y values with a near-zero correlation. However, a small sample of $n = 3$ data points from this population shows a relatively strong, positive correlation. Data points in the sample are circled.

Y values

X values

The purpose of the hypothesis test is to decide between the following two alternatives:

1. There is no correlation in the population ($\rho = 0$), and the sample value is the result of sampling error. Remember, a sample is not expected to be identical to the population. There always will be some error between a sample statistic and the corresponding population parameter. This is the situation specified by H_0.

2. The nonzero sample correlation accurately represents a real, nonzero correlation in the population. This is the alternative stated in H_1.

DEGREES OF FREEDOM FOR THE CORRELATION TEST

The hypothesis test for the Pearson correlation has degrees of freedom defined by $df = n - 2$. An intuitive explanation for this value is that a sample with only $n = 2$ data points has no degrees of freedom. Specifically, if there are only two points, they will fit perfectly on a straight line, and the sample will produce a perfect correlation of $r = +1.00$ or $r = -1.00$. Because the first two points always produce a perfect correlation, the sample correlation is free to vary only when the data set contains more than two points. Thus, $df = n - 2$.

THE HYPOTHESIS TEST

The table lists critical values in terms of degrees of freedom: $df = n - 2$. Remember to subtract 2 when using this table.

Although it is possible to conduct the hypothesis test by computing either a t statistic or an F-ratio, the computations have been completed and are summarized in Table B.6 in Appendix B. The table is based on the concept that a sample is expected to be representative of the population from which it was obtained. In particular, a sample correlation should be similar to the population correlation. If the population correlation is zero, as specified in the null hypothesis, then the sample correlation should be near to zero. Thus, a sample correlation that is close to zero provides support for H_0 and a sample correlation that is far from zero contradicts the null hypothesis. Table B6 identifies exactly which sample correlations are likely to be obtained from a population with $\rho = 0$ and which values are very unlikely. To use the table, you need to know the sample size (n) and the alpha level. With a sample size of $n = 20$ and an alpha level of .05, for example, you locate $df = n - 2 = 18$ in the left-hand column and the value .05 for either one tail or two tails across the top of the table. For $df = 18$ and $\alpha = .05$ for a two-tailed test, the table shows a value of 0.444. Thus, if the null hypothesis is true and there is no correlation in the population, then the sample correlation should be near zero. According to the table, the sample correlation should have a value between $+0.444$ and -0.444. If H_0 is true, it is very unlikely ($\alpha = .05$) to obtain a sample correlation outside this range. Therefore, a sample correlation beyond ± 0.444 will lead to rejecting the null hypothesis. The following examples demonstrate the use of the table.

EXAMPLE 16.6

A researcher is using a regular, two-tailed test with $\alpha = .05$ to determine whether a nonzero correlation exists in the population. A sample of $n = 30$ individuals is obtained. With $\alpha = .05$ and $n = 30$, the table lists a value of 0.361. Thus, the sample correlation (independent of sign) must have a value greater than or equal to 0.361 to reject H_0 and conclude that there is a significant correlation in the population. Any sample correlation between 0.361 and -0.361 is considered within the realm of sampling error and, therefore, not significant.

EXAMPLE 16.7

This time the researcher is using a directional, one-tailed test to determine whether there is a positive correlation in the population.

H_0: $\rho \leq 0$ (There is not a positive correlation.)

H_1: $\rho > 0$ (There is a positive correlation.)

With $\alpha = .05$ and a sample of $n = 30$, the table lists a value of 0.306 for a one-tailed test. Thus, the researcher must obtain a sample correlation that is positive (as predicted) and has a value greater than or equal to 0.306 to reject H_0 and conclude that there is a significant positive correlation in the population.

IN THE LITERATURE
REPORTING CORRELATIONS

When correlations are computed, the results are reported using APA format. The statement should include the sample size, the calculated value for the correlation, whether it is a statistically significant relationship, the probability level, and the type of test used (one- or two-tailed). For example, a correlation might be reported as follows:

> A correlation for the data revealed that amount of education and annual income were significantly related, $r = +.65$, $n = 30$, $p < .01$, two tails.

Sometimes a study might look at several variables, and correlations between all possible variable pairings are computed. Suppose, for example, that a study measured people's annual income, amount of education, age, and intelligence. With four variables, there are six possible pairings leading to six different correlations. The results from multiple correlations are most easily reported in a table called a correlation matrix, using footnotes to indicate which correlations are significant. For example, the report might state:

The analysis examined the relationships between income, amount of education, age, and intelligence for $n = 30$ participants. The correlations between pairs of variables are reported in Table 1. Significant correlations are noted in the table.

TABLE 1

Correlation matrix for income, amount of education, age, and intelligence

	Education	Age	IQ
Income	+.65*	+.41**	+.27
Education		+.11	+.38**
Age			−.02

$n = 30$
*$p < .01$, two tails
**$p < .05$, two tails

LEARNING CHECK

1. A researcher obtains a correlation of $r = -.41$ for a sample of $n = 25$ individuals. Does this sample provide sufficient evidence to conclude that there is a significant, nonzero correlation in the population? Assume a nondirectional test with $\alpha = .05$.

2. For a sample of $n = 10$, how large a correlation is needed to conclude at the .05 level that there is a nonzero correlation in the population? Assume a nondirectional test.

3. As sample size gets smaller, what happens to the magnitude of the correlation necessary for significance? Explain why this occurs.

ANSWERS

1. Yes. For $n = 25$, the critical value is $r = .396$. The sample value is in the critical region.

2. For $n = 10$, the critical value is $r = .632$.

3. As the sample size gets smaller, the magnitude of the correlation needed for significance gets larger. With a small sample, it is easy to get a relatively good correlation just by chance. Therefore, a small sample requires a very large correlation before you can be confident that there is a real (nonzero) relationship in the population.

16.5   THE SPEARMAN CORRELATION

The Pearson correlation specifically measures the degree of linear relationship between two variables when the data (X and Y values) consist of numerical scores from an interval or ratio scale of measurement. However, other correlations have been developed for nonlinear relationships and for other types of data. One of these useful measures is called the *Spearman correlation*. The Spearman correlation is used in two situations.

First, the Spearman correlation is used to measure the relationship between X and Y when both variables are measured on an ordinal scale of measurement. Recall from Chapter 1 that an ordinal scale typically involves ranking individuals rather than obtaining numerical scores. Rank-order data are fairly common because they are often easier to obtain than interval or ratio scale data. For example, a teacher may feel confident about rank-ordering students' leadership abilities but would find it difficult to measure leadership on some other scale.

In addition to measuring relationships for ordinal data, the Spearman correlation can be used as a valuable alternative to the Pearson correlation, even when the original raw scores are on an interval or a ratio scale. As we have noted, the Pearson correlation measures the degree of *linear relationship* between two variables—that is, how well the data points fit on a straight line. However, a researcher often expects the data to show a consistently one-directional relationship but not necessarily a linear relationship. For example, Figure 16.13 shows the typical relationship between practice and performance. For nearly any skill, increasing amounts of practice tend to be associated with improvements in performance (the more you practice, the better you get). However, it is not a straight-line relationship. When you are first learning a new skill, practice produces large improvement in performance. After you have been performing a skill for several years, however, additional practice produces only minor changes in performance. Although there is a consistent relationship between the amount of practice and the quality of performance, it clearly is not linear. If the Pearson correlation were computed for these data, it would not produce a correlation of 1.00 because the data do not fit perfectly on a straight line. In a situation like this, the Spearman corrrelation can be used to measure the consistency of the relationship, independent of its form.

The reason that the Spearman correlation measures consistency, rather than form, comes from a simple observation: When two variables are consistently related, their ranks will be linearly related. For example, a perfectly consistent positive relationship means that every time the X variable increases, the Y variable also increases. Thus, the smallest value of X is paired with the smallest value of Y, the second-smallest value of X is paired with the second-smallest value of Y, and so on. Every time the rank for X goes up by 1 point, the rank for Y also goes up by 1 point. As a result, the ranks fit perfectly on a straight line. This phenomenon is demonstrated in the following example.

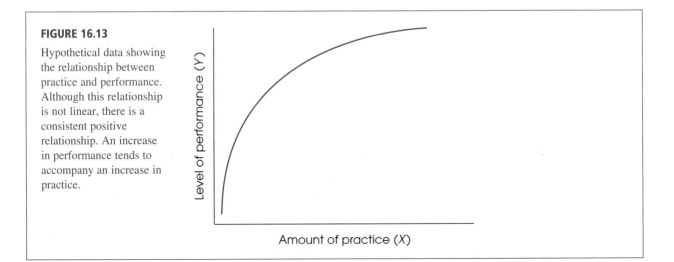

FIGURE 16.13

Hypothetical data showing the relationship between practice and performance. Although this relationship is not linear, there is a consistent positive relationship. An increase in performance tends to accompany an increase in practice.

EXAMPLE 16.8

TABLE 16.1

Scores for Example 16.8.

Person	X	Y
A	4	9
B	2	2
C	10	10
D	3	8

TABLE 16.2

Ranks for Example 16.8.

Person	X Rank	Y Rank
A	3	3
B	1	1
C	4	4
D	2	2

The data in Table 16.1 represent X and Y scores for a sample of $n = 4$ people. Note that the data show a perfectly consistent relationship. Each increase in X is accompanied by an increase in Y. However the relationship is not linear, as can be seen in the graph of the data in Figure 16.14(a).

Next, we convert the scores to ranks. The lowest X is assigned a rank of 1, the next lowest a rank of 2, and so on. The ranking procedure is then repeated with the Y scores. The ranks are listed in Table 16.2(b) and shown in Figure 16.14(b). Note that the perfect consistency for the scores produces a perfect linear relationship for the ranks.

The preceding example has demonstrated that a consistent relationship among scores produces a linear relationship when the scores are converted to ranks. Thus, if you want to measure the consistency of a relationship for a set of scores, you can simply convert the scores to ranks and then use the Pearson correlation formula to measure the linear relationship for the ranked data. The degree of linear relationship for the ranks provides a measure of the degree of consistency for the original scores.

To summarize, the Spearman correlation measures the relationship between two variables when both are measured on ordinal scales (ranks). There are two general situations in which the Spearman correlation is used:

1. Spearman is used when the original data are ordinal; that is, when the X and Y values are ranks. In this case, you simply apply the Pearson correlation formula to the set of ranks.

2. Spearman is used when a researcher wants to measure the consistency of a relationship between X and Y, independent of the specific form of the relationship. In this case, the original scores are first converted to ranks;

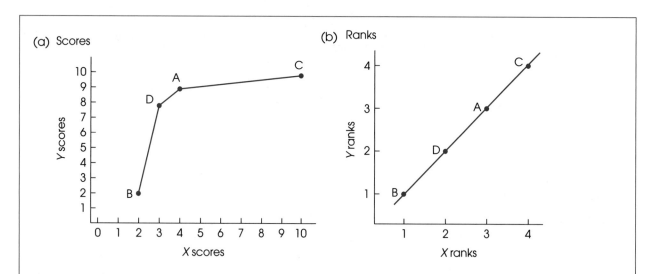

FIGURE 16.14

Scatter plots showing (a) the scores and (b) the ranks for the data in Example 16.8. Notice that there is a consistent, positive relationship between the X and Y scores, although it is not a linear relationship. Also notice that the scatter plot of the ranks shows a perfect linear relationship.

then the Pearson correlation formula is used with the ranks. Because the Pearson formula measures the degree to which the ranks fit on a straight line, it also measures the degree of consistency in the relationship for the original scores. Incidentally, when there is a consistently one-directional relationship between two variables, the relationship is said to be *monotonic*. Thus, the Spearman correlation can be used to measure the degree of monotonic relationship between two variables.

The word *monotonic* describes a sequence that is consistently increasing (or decreasing). Like the word *monotonous*, it means constant and unchanging.

In either case, the Spearman correlation is identified by the symbol r_s to differentiate it from the Pearson correlation. The complete process of computing the Spearman correlation, including ranking scores, is demonstrated in Example 16.9.

EXAMPLE 16.9

The following data show a nearly perfect monotonic relationship between X and Y. When X increases, Y tends to decrease, and there is only one reversal in this general trend. To compute the Spearman correlation, we first rank the X and Y values, and we then compute the Pearson correlation for the ranks.

We have listed the X values in order so that the trend is easier to recognize.

Original Data		Ranks		
X	Y	X	Y	XY
3	12	1	5	5
4	10	2	3	6
8	11	3	4	12
10	9	4	2	8
13	3	5	1	5
				$36 = \Sigma XY$

The scatter plots for the original data and the ranks are shown in Figure 16.15. To compute the correlation, we will need SS for X, SS for Y, and SP. Remember that all these values are computed with the ranks, not the original scores.

The X ranks are simply the integers 1, 2, 3, 4, and 5. These values have $\Sigma X = 15$ and $\Sigma X^2 = 55$. The SS for the X ranks is

FIGURE 16.15

Scatter plots showing (a) the scores and (b) the ranks for the data in Example 16.9.

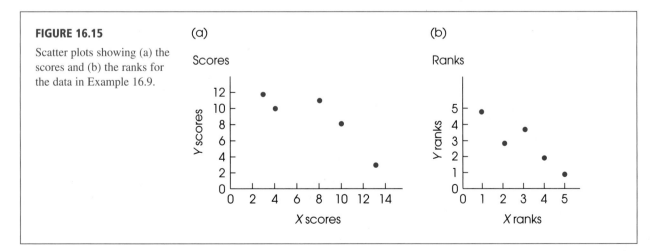

$$SS_X = \Sigma X^2 - \frac{(\Sigma X)^2}{n}$$

$$= 55 - \frac{(15)^2}{5}$$

$$= 55 - 45$$

$$= 10$$

Note that the ranks for Y are identical to the ranks for X; that is, they are the integers 1, 2, 3, 4, and 5. Therefore, the SS for Y will be identical to the SS for X:

$$SS_Y = 10$$

To compute the SP value, we need ΣX, ΣY, and ΣXY for the ranks. The XY values are listed in the table with the ranks, and we already have found that both the Xs and the Ys have a sum of 15. Using these values, we obtain

$$SP = \Sigma XY - \frac{(\Sigma X)(\Sigma Y)}{n}$$

$$= 36 - \frac{(15)(15)}{5}$$

$$= 36 - 45$$

$$= -9$$

The final Spearman correlation is

$$r_S = \frac{SP}{\sqrt{(SS_X)(SS_Y)}}$$

$$= \frac{-9}{\sqrt{10(10)}}$$

$$= -0.9$$

The Spearman correlation indicates that the data show a strong (nearly perfect) negative trend.

RANKING TIED SCORES   When you are converting scores into ranks for the Spearman correlation, you may encounter two (or more) identical scores. Whenever two scores have exactly the same value, their ranks should also be the same. This is accomplished by the following procedure:

1. List the scores in order from smallest to largest. Include tied values in the list.

2. Assign a rank (first, second, etc.) to each position in the ordered list.

3. When two (or more) scores are tied, compute the mean of their ranked positions, and assign this mean value as the final rank for each score.

The process of finding ranks for tied scores is demonstrated here. These scores have been listed in order from smallest to largest.

Scores	Rank Position	Final Rank	
3	1	1.5	} Mean of 1 and 2
3	2	1.5	
5	3	3	
6	4	5	
6	5	5	} Mean of 4, 5, and 6
6	6	5	
12	7	7	

Note that this example has seven scores and uses all seven ranks. For $X = 12$, the largest score, the appropriate rank is 7. It cannot be given a rank of 6 because that rank has been used for the tied scores.

SPECIAL FORMULA FOR THE SPEARMAN CORRELATION

After the original X values and Y values have been ranked, the calculations necessary for SS and SP can be greatly simplified. First, you should note that the X ranks and the Y ranks are really just a set of integers: 1, 2, 3, 4, ..., n. To compute the mean for these integers, you can locate the midpoint of the series by $M = (n + 1)/2$. Similarly, the SS for this series of integers can be computed by

$$SS = \frac{n(n^2 - 1)}{12} \quad \text{(Try it out.)}$$

Also, because the X ranks and the Y ranks are the same values, the SS for X will be identical to the SS for Y.

Because calculations with ranks can be simplified and because the Spearman correlation uses ranked data, these simplifications can be incorporated into the final calculations for the Spearman correlation. Instead of using the Pearson formula after ranking the data, you can put the ranks directly into a simplified formula:

Caution: In this formula, you compute the value of the fraction and then subtract from 1. The first 1 is not part of the fraction.

$$r_S = 1 - \frac{6\Sigma D^2}{n(n^2 - 1)} \quad (16.5)$$

where D is the difference between the X rank and the Y rank for each individual. This special formula will produce the same result that would be obtained from the Pearson formula. However, you should note that this special formula can be used only after the scores have been converted to ranks and only when there are no ties among the ranks. If there are relatively few tied ranks, the formula still may be used, but it loses accuracy as the number of ties increases. The application of this formula is demonstrated in the following example.

EXAMPLE 16.10

To demonstrate the special formula for the Spearman correlation, we use the same data that were presented in Example 16.9. The ranks for these data are shown again in the following table along with the difference between ranks and the squared difference for each individual. Note that the sign of the difference scores is not important because they will all be squared. However, the signed values can help you avoid mistakes because you can check that the signed difference scores add up to zero.

Ranks		Difference	
X	Y	D	D^2
1	5	4	16
2	3	1	1
3	4	1	1
4	2	−2	4
5	1	−4	16
			$\overline{38} = \Sigma D^2$

Using the special formula for the Spearman correlation, we obtain

$$r_S = 1 - \frac{6\Sigma D^2}{n(n^2 - 1)}$$

$$= 1 - \frac{6(38)}{5(25 - 1)}$$

$$= 1 - \frac{228}{120}$$

$$= 1 - 1.90$$

$$= -0.90$$

Notice that this is exactly the same answer that we obtained in Example 16.9, using the Pearson formula on the ranks.

TESTING THE SIGNIFICANCE OF THE SPEARMAN CORRELATION

Testing a hypothesis for the Spearman correlation is similar to the procedure used for the Pearson r. The basic question is whether a correlation exists in the population. The sample correlation could be due to chance, or perhaps it reflects an actual relationship between the variables in the population. For the Pearson correlation, the Greek letter rho (ρ) was used for the population correlation. For the Spearman, ρ_S is used for the population parameter. Note that this symbol is consistent with the sample statistic, r_S. The null hypothesis states that there is no correlation (no monotonic relationship) between the variables for the population, or in symbols:

$$H_0: \quad \rho_S = 0 \quad \text{(The population correlation is zero.)}$$

The alternative hypothesis predicts that a nonzero correlation exists in the population, which can be stated in symbols as

$$H_1: \quad \rho_S \neq 0 \quad \text{(There is a real correlation.)}$$

To determine whether the Spearman correlation is statistically significant (that is, H_0 should be rejected), consult Table B.7. This table is similar to the one used to determine the significance of Pearson's r (Table B.6); however, the first column is sample size (n) rather than degrees of freedom. To use the table, line up the sample size in the first column with the alpha level at the top. The values in the body of the table identify the magnitude of the Spearman correlation that is necessary to be significant. The table is built on the concept that a sample correlation should be representative of the corresponding population value. In particular, if the population correlation is $\rho_S = 0$ (as specified in H_0), then the sample correlation should be near zero. For each sample size and alpha level, the table identifies the minimum sample correlation that is significantly different from zero. The following example demonstrates the use of the table.

EXAMPLE 16.11   An industrial psychologist selects a sample of $n = 15$ employees. These employees are ranked in order of work productivity by their manager. They also are ranked by a peer. The Spearman correlation computed for these data revealed a correlation of $r_S = .60$. Using Table B.7 with $n = 15$ and $\alpha = .05$, a correlation of $\pm.45$ is needed to reject H_0. The observed correlation for the sample easily surpasses this critical value. The correlation between manager and peer ratings is statistically significant.

LEARNING CHECK

1. Describe what is measured by a Spearman correlation, and explain how this correlation is different from the Pearson correlation.

2. Identify the two procedures that can be used to compute the Spearman correlation.

3. Compute the Spearman correlation for the following set of scores:

X	Y
2	7
12	38
9	6
10	19

ANSWERS

1. The Spearman correlation measures the consistency of the direction of the relationship between two variables. The Spearman correlation does not depend on the form of the relationship, whereas the Pearson correlation measures how well the data fit a linear form.

2. After the X and Y values have been ranked, you can compute the Spearman correlation by using either the special formula or the Pearson formula.

3. $r_S = 0.80$

16.6   OTHER MEASURES OF RELATIONSHIP

Although the Pearson correlation is the most commonly used method for evaluating the relationship between two variables, there are many other measures of relationship that exist. We already examined one such alternative, the Spearman correlation. Like the Spearman correlation, many of the alternative measures are simply special applications of the Pearson formula. In this section, we examine two more correlational methods. Notice that each alternative simply applies the Pearson formula to data with special characteristics.

THE POINT-BISERIAL
CORRELATION
AND MEASURING
EFFECT SIZE WITH r^2

A *dichotomous* variable is also called a *binomial* variable.

In Chapters 9, 10, and 11 we introduced r^2 as a measure of effect size that often accompanies a hypothesis test using the *t* statistic. The r^2 used to measure effect size and the *r* used to measure a correlation are directly related, and we now have an opportunity to demonstrate the relationship. Specifically, we compare the independent-measures *t* test (Chapter 10) and a special version of the Pearson correlation known as the *point-biserial correlation*.

The point-biserial correlation is used to measure the relationship between two variables in situations where one variable consists of regular, numerical scores, but the second variable has only two values. A variable with only two values is called a *dichotomous variable*. Some examples of dichotomous variables are

1. Male versus female

2. College graduate versus not a college graduate

3. First-born child versus later-born child

4. Success versus failure on a particular task

5. Older than 30 years versus younger than 30 years

It is customary to use the numerical values 0 and 1, but any two different numbers would work equally well and would not affect the value of the correlation.

To compute the point-biserial correlation, the dichotomous variable is first converted to numerical values by assigning a value of zero (0) to one category and a value of one (1) to the other category. Then the regular Pearson correlation formula is used with the converted data.

To demonstrate the point-biserial correlation and its association with the r^2 measure of effect size, we will use the data from Example 10.1 (page 318). The original examle compared high school grades for two groups of students: one group who regularly watched Sesame Street as 5-year-old children and one who did not watch the program. The data from the independent-measures study are presented on the left side of Table 16.3. Notice that the data consist of two separate samples and the independent-measures *t* was used to determine whether there was a significant mean difference between the two populations represented by the samples.

On the right-hand side of Table 16.3 we have reorganized the data into a form that is suitable for a point-biserial correlation. Specifically, we used each student's high school grade as the *X* value and we have created a new variable, *Y*, to represent the group or condition for each student. In this case, we have used $Y = 1$ for students who watched Sesame Street and $Y = 0$ for students who did not watch the program.

When the data in Table 16.3 were originally presented in Chapter 10, we conducted an independent-measures *t* hypothesis test and obtained $t = 4.00$ with $df = 18$. We measured the size of the treatment effect by calculating r^2, the percentage of variance accounted for, and obtained $r^2 = 0.47$.

Calculating the point-biserial correlation for these data also produces a value for *r*. Specifically, the *X* scores produce $SS = 680$; the *Y* values produce $SS = 5.00$, and the sum of the products of the *X* and *Y* deviations produces $SP = 40$. The point-biserial correlation is

$$r = \frac{SP}{\sqrt{(SS_X)(SS_Y)}} = \frac{40}{\sqrt{(680)(5)}} = \frac{40}{58.31} = 0.686$$

Notice that squaring the value of the point-biserial correlation produces $r^2 = (0.686)^2 = 0.47$, which is exactly the value of r^2 we obtained measuring effect size.

In some respects, the point-biserial correlation and the independent-measures hypothesis test are evaluating the same thing. Specifically, both are examining the

TABLE 16.3

The same data are organized in two different formats. On the left-hand side, the data appear as two separate samples appropriate for an independent-measures t hypothesis test. On the right-hand side, the same data are shown as a single sample, with two scores for each individual: the original high school grade and a dichotomous score (Y) that identifies the condition (Seasame Street or not) in which the participant is located. The data on the right are appropriate for a point-biserial correlation.

Data for the Independent-Measures t. Two separate samples, each with $n = 10$ scores.

Average High School Grade			
Watched Seasame Street		Did Not Watch Seasame Street	
86	99	90	79
87	97	89	83
91	94	82	86
97	89	83	81
98	92	85	92
$n = 10$		$n = 10$	
$M = 93$		$M = 85$	
$SS = 200$		$SS = 160$	

Data for the Point-Biserial Correlation. Two scores, X and Y for each of the $n = 20$ participants.

Participant	Grade X	Condition Y
A	86	1
B	87	1
C	91	1
D	97	1
E	98	1
F	99	1
G	97	1
H	94	1
I	89	1
J	92	1
K	90	0
L	89	0
M	82	0
N	83	0
O	85	0
P	79	0
Q	83	0
R	86	0
S	81	0
T	92	0

relationship between the TV viewing habits of 5-year-old children and their future academic performance in high school.

1. The correlation is measuring the *strength* of the relationship between the two variables. A large correlation (near 1.00 or −1.00) would indicate that there is a consistent, predictable relationship between high school grades and watching Sesame Street as a 5-year-old child. In particular, the value of r^2 measures how much of the variability in grades can be predicted by knowing whether the participants watched Sesame Street

2. The t test evaluates the *significance* of the relationship. The hypothesis test determines whether the mean difference in grades between the two groups is greater than can be reasonably explained by chance alone.

As we noted in Chapter 10 (pages 320–321), the outcome of the hypothesis test and the value of r^2 are often reported together. The t value measures statistical significance and r^2 measures the effect size. Also, as we noted in Chapter 10, the values for t and r^2 are directly related. In fact, either can be calculated from the other by the equations

$$r^2 = \frac{t^2}{t^2 + df} \quad \text{and} \quad t^2 = \frac{r^2}{(1 - r^2)/df}$$

where df is the degrees of freedom for the t statistic.

However, you should note that r^2 is determined entirely by the size of the correlation, whereas t is influenced by the size of the correlation and the size of the sample. For example, a correlation of $r = 0.30$ produces $r^2 = 0.09$ (9%) no matter how large the sample may be. On the other hand, a point-biserial correlation of $r = 0.30$ for a total sample of 10 people ($n = 5$ in each group) produces a nonsignificant value of $t = 0.791$. If the sample is increased to 50 people ($n = 25$ in each group), the same correlation produces a significant t value of $t = 4.75$. Although t and r^2 are related, they are measuring different things.

THE PHI-COEFFICIENT

When both variables (X and Y) measured for each individual are dichotomous, the correlation between the two variables is called the *phi-coefficient*. To compute phi (ϕ), you follow a two-step procedure:

1. Convert each of the dichotomous variables to numerical values by assigning a 0 to one category and a 1 to the other category for each of the variables.

2. Use the regular Pearson formula with the converted scores.

This process is demonstrated in the following example.

EXAMPLE 16.12

A researcher is interested in examining the relationship between birth-order position and personality. A random sample of $n = 8$ individuals is obtained, and each individual is classified in terms of birth-order position as first-born or only child versus later-born. Then each individual's personality is classified as either introvert or extrovert.

The original measurements are then converted to numerical values by the following assignments:

Birth Order	Personality
1st or only child = 0	Introvert = 0
Later-born child = 1	Extrovert = 1

The original data and the converted scores are as follows:

Original Data		Converted Scores	
Birth Order X	Personality Y	Birth Order X	Personality Y
1st	Introvert	0	0
3rd	Extrovert	1	1
Only	Extrovert	0	1
2nd	Extrovert	1	1
4th	Extrovert	1	1
2nd	Introvert	1	0
Only	Introvert	0	0
3rd	Extrovert	1	1

The Pearson correlation formula would then be used with the converted data to compute the phi-coefficient.

Because the assignment of numerical values is arbitrary (either category could be designated 0 or 1), the sign of the resulting correlation is meaningless. As with most correlations, the *strength* of the relationship is best described by the value of r^2, the coefficient of determination, which measures how much of the variability in one variable is predicted or determined by the association with the second variable.

We also should note that although the phi-coefficient can be used to assess the relationship between two dichotomous variables, the more common statistical procedure is a chi-square statistic, which is examined in Chapter 18.

LEARNING CHECK

1. Define a *dichotomous* variable.

2. The following data represent job-related stress scores for a sample of $n = 8$ individuals. These people also are classified by salary level.
   a. Convert the data into a form suitable for the point-biserial correlation.
   b. Compute the point-biserial correlation for these data.

Salary More Than $40,000	Salary Less Than $40,000
8	4
6	2
5	1
3	3

ANSWERS

1. A dichotomous variable has only two possible values.

2. a. Salary level is a dichotomous variable and can be coded as $Y = 1$ for individuals with salary more than $40,000 and $Y = 0$ for salary less than $40,000. The stress scores produce $SS_X = 36$, the salary codes produce $SS_Y = 2$, and $SP = 6$.   b. The point-biserial correlation is 0.71.

SUMMARY

1. A correlation measures the relationship between two variables, X and Y. The relationship is described by three characteristics:
   a. *Direction.* A relationship can be either positive or negative. A positive relationship means that X and Y vary in the same direction. A negative relationship means that X and Y vary in opposite directions. The sign of the correlation ($+$ or $-$) specifies the direction.
   b. *Form.* The most common form for a relationship is a straight line. However, special correlations exist for measuring other forms. The form is specified by the type of correlation used. For example, the Pearson correlation measures linear form.
   c. *Strength or consistency.* The numerical value of the correlation measures the strength or consistency of the relationship. A correlation of 1.00 indicates a perfectly consistent relationship and 0.00 indicates no relationship at all. For the Pearson correlation, $r = 1.00$ (or -1.00) means that the data points fit perfectly on a straight line.

2. The most commonly used correlation is the Pearson correlation, which measures the degree of linear relationship. The Pearson correlation is identified by the letter r and is computed by

$$r = \frac{SP}{\sqrt{SS_X SS_Y}}$$

In this formula, *SP* is the sum of products of deviations and can be calculated with either a definitional formula or a computational formula:

definitional formula: $SP = \Sigma(X - M_X)(Y - M_Y)$

computational formula: $SP = \Sigma XY - \dfrac{\Sigma X \Sigma Y}{n}$

3. The Pearson correlation and z-scores are closely related because both are concerned with the location of individuals within a distribution. When *X* and *Y* scores are transformed into z-scores, the Pearson correlation can be computed.

For a sample, $r = \dfrac{\Sigma z_X z_Y}{(n - 1)}$

For a population, $\rho = \dfrac{\Sigma z_X z_Y}{N}$

4. A correlation between two variables should not be interpreted as implying a causal relationship. Simply because *X* and *Y* are related does not mean that *X* causes *Y* or that *Y* causes *X*.

5. When the *X* or *Y* values used to compute a correlation are limited to a relatively small portion of the potential range, you should exercise caution in generalizing the value of the correlation. Specifically, a limited range of values can either obscure a strong relationship or exaggerate a poor relationship.

6. To evaluate the strength of a relationship, you should square the value of the correlation. The resulting value, r^2, is called the *coefficient of determination* because it measures the portion of the variability in one variable that can be predicted using the relationship with the second variable.

7. The Spearman correlation (r_S) measures the consistency of direction in the relationship between *X* and *Y*—that is, the degree to which the relationship is one-directional, or monotonic. The Spearman correlation is appropriate when both variables are measured on an ordinal scale. If the data consist of numerical scores (interval or ratio scale), the Spearman correlation can be used to measure the degree of monotonic relationship. In this case, the Spearman correlation is computed by a two-stage process:
   a. Rank the *X* scores and the *Y* scores separately.
   b. Compute the Pearson correlation using the ranks.

Note: After the *X* and *Y* values are ranked, you may use a special formula to determine the Spearman correlation:

$$r_S = 1 - \dfrac{6\Sigma D^2}{n(n^2 - 1)}$$

where *D* is the difference between the *X* rank and the *Y* rank for each individual. The formula is accurate only when there are no tied scores in the data.

8. Special correlations are used to measure relationships between variables with specific characteristics. The Spearman correlation is used when both variables are measured on ordinal scales. The point-biserial correlation is used to measure the strength of the relationship when one of the two variables is dichotomous. The dichotomous variable is coded using values of 0 and 1, and the regular Pearson formula is applied. Squaring the point-biserial correlation produces the same r^2 value that is obtained to measure effect size for the independent-measures *t* test. The phi-coefficient is used when both variables are dichotomous. Each dichotomous variable is coded using 0 and 1, and the regular Pearson formula is applied to the coded data.

KEY TERMS

correlation	Pearson correlation	regression toward the mean
positive correlation	sum of products (*SP*)	Spearman correlation
negative correlation	restricted range	point-biserial correlation
perfect correlation	coefficient of determination	phi-coefficient

RESOURCES

There are a tutorial quiz and other learning exercises for Chapter 16 at the Wadsworth website, www.cengage.com/psychology/gravetter. The site also provides access to two workshops titled *Correlation* and *Bivariate Scatter Plots* that both review the concept and calculation of correlation presented in this chapter. See page 29 for more information about accessing the website.

ENHANCED WebAssign

Guided interactive tutorials, end-of-chapter problems, and related testbank items may be assigned online at WebAssign.

WebTUTOR

For those using WebTutor along with this book, there is a WebTutor section corresponding to this chapter. The WebTutor contains a brief summary of Chapter 16, hints for learning the concepts and the formulas for correlation, cautions about common errors, and sample exam items including solutions.

SPSS

General instructions for using SPSS are presented in Appendix D. Following are detailed instructions for using SPSS to perform the **Pearson, Spearman,** and the **point-biserial correlations.** Note: We will focus on the Pearson correlation and then describe how slight modifications to this procedure can be made to compute the Spearman and point-biserial correlations. Separate instructions for the **phi-coefficient** are presented at the end of this section.

Data Entry

1. The data are entered into two columns in the data editor, one for the *X* values (VAR00001) and one for the *Y* values (VAR00002), with the two scores for each individual in the same row.

Data Analysis

1. Click **Analyze** on the tool bar, select **Correlate,** and click on **Bivariate.**
2. One by one move the labels for the two data columns into the **Variables** box. (Hightlight each label and click the arrow to move it into the box.)
3. The **Pearson** box should be checked, but, at this point, you can switch to the Spearman correlation by clicking the appropriate box.
4. Click **OK.**

SPSS Output

The program will produce a correlation matrix showing all the possible correlations, including the correlation of *X* with *X* and the correlation of *Y* with *Y* (both are perfect correlations). You want the correlation of *X* and *Y,* which is contained in the upper right corner (or the lower left). The output includes the significance level (*p* value or alpha level) for the correlation.

To compute the **Spearman** correlation, enter either the X and Y ranks or the X and Y scores into the first two columns. Then follow the same Data Analysis instructions that were presented for the Pearson correlation. At step 5 in the instructions, click on the **Spearman** box before the final OK. (*Note:* If you enter X and Y scores into the data editor, SPSS will convert the scores to ranks before computing the Spearman correlation.)

To compute the **point-biserial** correlation, enter the scores (X values) in the first column and enter the numerical values (usually 0 and 1) for the dichotomous variable in the second column. Then, follow the same Data Analysis instructions that were presented for the Pearson correlation.

The **phi-coefficient** can also be computed by entering the complete string of 0s and 1s into two columns of the SPSS data editor, then following the same Data Analysis instructions that were presented for the Pearson correlation. However, this can be tedious, especially with a large set of scores. The following is an alternative procedure for computing the phi-coefficient with large data sets.

Data Entry

1. Enter the values, 0, 0, 1, 1 (in order) into the first column of the SPSS data editor.
2. Enter the values 0, 1, 0, 1 (in order) into the second column.
3. Count the number of individuals in the sample who are classified with $X = 0$ and $Y = 0$. Enter this frequency in the top box in the third column of the data editor. Then, count how many have $X = 0$ and $Y = 1$ and enter the frequency in the second box of the third column. Continue with the number who have $X = 1$ and $Y = 0$, and finally the number who have $X = 1$ and $Y = 1$. You should end up with 4 values in column three.
4. Click **Data** on the Tool Bar at the top of the SPSS Data Editor page and select **Weight Cases** at the bottom of the list.
5. Click the **weight cases by** circle, then highlight the label for the column containing your frequencies (VAR00003) on the left and move it into the **Frequency Variable** box by clicking on the arrow.
6. Click **OK.**
7. Click **Analyze** on the tool bar, select **Correlate,** and click on **Bivariate.**
8. One by one move the labels for the two data columns containing the 0s and 1s (probably VAR00001 and VAR00002) into the **Variables** box. (Highlight each label and click the arrow to move it into the box.)
9. Verify that the **Pearson** box is checked.
10. Click **OK.**

SPSS Output

The program will produce the same correlation matrix that was described for the Pearson correlation. Again, you want the correlation between X and Y, which is in the upper right corner (or lower left). Remember, with the phi-coefficient, the sign of the correlation is meaningless.

FOCUS ON PROBLEM SOLVING

1. A correlation always has a value from $+1.00$ to -1.00. If you obtain a correlation outside this range, then you have made a computational error.

2. When interpreting a correlation, do not confuse the sign ($+$ or $-$) with its numerical value. The sign and numerical value must be considered separately. Remember that the sign indicates the direction of the relationship between X and Y. On the other hand,

the numerical value reflects the strength of the relationship or how well the points approximate a linear (straight-line) relationship. Therefore, a correlation of −0.90 is just as strong as a correlation of +0.90. The signs tell us that the first correlation is an inverse relationship.

3. Before you begin to calculate a correlation, you should sketch a scatterplot of the data and make an estimate of the correlation. (Is it positive or negative? Is it near 1 or near 0?) After computing the correlation, compare your final answer with your original estimate.

4. The definitional formula for the sum of products (SP) should be used only when you have a small set (n) of scores and the means for X and Y are both whole numbers. Otherwise, the computational formula will produce quicker, easier, and more accurate results.

5. For computing a correlation, n is the number of individuals (and, therefore, the number of *pairs* of X and Y values).

6. When using the special formula for the Spearman correlation, remember that the fraction is computed separately and then subtracted from 1. Students often include the 1 as a part of the numerator, or they get so absorbed in computing the fractional part of the equation that they forget to subtract it from 1. Be careful when using this formula.

7. When computing a Spearman correlation, be sure that both X and Y values have been ranked. Sometimes the data will consist of one variable already ranked with the other variable on an interval or a ratio scale. If one variable is ranked, do not forget to rank the other. When interpreting a Spearman correlation, remember that it measures how monotonic (consistent) the relationship is between X and Y.

DEMONSTRATION 16.1

THE PEARSON CORRELATION

For the following data, calculate the Pearson correlation:

Person	X	Y	XY	
A	0	4	0	
B	2	1	2	$SS_X = 40$
C	8	10	80	$SS_Y = 54$
D	6	9	54	
E	4	6	24	
	20	30	160	Column totals

STEP 1   *Sketch a scatter plot.* We have constructed a scatter plot for the data (Figure 16.16) and placed an envelope around the data points to make a preliminary estimate of the correlation. Note that the envelope is narrow and elongated. This indicates that the correlation is large— perhaps 0.80 to 0.90. Also, the correlation is positive because increases in X are generally accompanied by increases in Y.

STEP 2   *Compute the value for SP.* For this sample of n = 5 individuals, $\Sigma X = 20$, $\Sigma Y = 30$, and $\Sigma XY = 160$.

FIGURE 16.16

The scatter plot for the data of Demonstration 16.1. An envelope is drawn around the points to estimate the magnitude of the correlation. A line is drawn through the middle of the envelope.

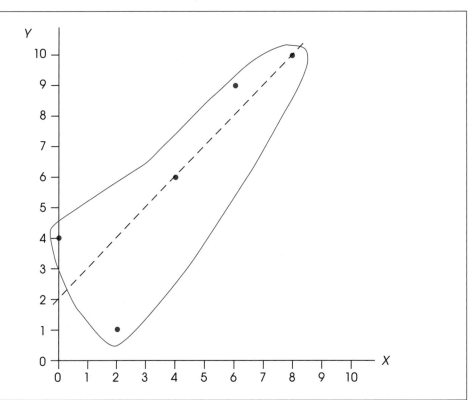

$$SP = \Sigma XY - \frac{(\Sigma X)(\Sigma Y)}{n}$$

$$= 160 - \frac{(20)(30)}{5}$$

$$= 160 - 120$$

$$= 40$$

STEP 3   *Compute the Pearson correlation.* For these data, the Pearson correlation is

$$r = \frac{SP}{\sqrt{SS_X SS_Y}} = \frac{40}{\sqrt{40(54)}} = \frac{40}{\sqrt{2160}} = \frac{40}{46.48}$$

$$= 0.861$$

In step 1, our preliminary estimate for the correlation was between $+0.80$ and $+0.90$. The calculated correlation is consistent with this estimate.

DEMONSTRATION 16.2

THE SPEARMAN CORRELATION

The following data will be used to demonstrate the calculation of the Spearman correlation. The X and Y values are both measurements on interval scales so the scores have been ranked and the difference between the ranks has been computed for each individual.

X Score	Y Score	X Rank	Y Rank	D
5	12	2	2	0
7	18	3	5	2
2	9	1	1	0
15	14	5	4	−1
10	13	4	3	−1

STEP 1   Use the special Spearman formula to compute the correlation. For these scores, the D^2 values are 0, 4, 0, 1, and 1. The sum is $\Sigma D^2 = 6$. Using this value in the Spearman formula, we obtain

$$r_S = 1 - \frac{6\Sigma D^2}{n(n^2 - 1)}$$

$$= 1 - \frac{6(6)}{5(24)}$$

$$= 1 - \frac{36}{120}$$

$$= 1 - 0.30$$

$$= 0.70$$

There is a positive relationship between X and Y for these data. The Spearman correlation is fairly high, which indicates a very consistent positive relationship.

PROBLEMS

1. What information is provided by the sign ($+$ or $-$) of the Pearson correlation?

2. What information is provided by the numerical value of the Pearson correlation?

3. Calculate SP (the sum of products of deviations) for the following scores. *Note:* Both means are

whole numbers, so the definitional formula works well.

X	Y
2	4
6	3
1	8
3	5

4. Calculate SP (the sum of products of deviations) for the following scores. *Note:* Both means are decimal values, so the computational formula works well.

X	Y
1	4
4	3
5	9
0	2

5. For the following scores,

X	Y
7	6
9	6
6	3
12	5
9	6
5	4

   a. Sketch a scatter plot showing the six data points

   b. Just looking at the scatter plot, estimate the value of the Pearson correlation.

   c. Compute the Pearson correlation.

6. For the following scores,

X	Y
1	5
2	9
4	3
5	1
3	2

   a. Sketch a scatter plot showing the five data points.

   b. Just looking at the scatter plot, estimate the value of the Pearson correlation.

   c. Compute the Pearson correlation.

7. To create the following data, we used the same X and Y values that appeared in problem 6, but we have changed the X and Y pairings.

X	Y
1	1
2	3
4	9
5	5
3	2

   a. Sketch a scatter plot showing the reorganized data points.

b. Just looking at the scatter plot, estimate the value of the Pearson correlation.

c. Compute the Pearson correlation. (*Note:* Because the X values and the Y values are the same as in problem 5, the values for SS_X and SS_Y will also be the same.)

8. For the following scores,

X	Y
3	12
6	7
3	9
5	7
3	10

   a. Sketch a scatter plot showing the five data points.

   b. Just looking at the scatter plot, estimate the value of the Pearson correlation.

   c. Compute the Pearson correlation.

9. With a small sample, a single point can have a large effect on the magnitude of a correlation. To create the following data, we started with the scores in problem 8 and changed $X = 5$ to $X = 0$. This single change has a dramatic effect on the correlation.

X	Y
3	12
6	7
3	9
0	7
3	10

   a. Sketch a scatter plot showing the five data points.

   b. Just looking at the scatter plot, estimate the value of the Pearson correlation.

   c. Compute the Pearson correlation.

10. In the following data, there are three scores (X, Y, and Z) for each of the $n = 5$ individuals:

X	Y	Z
3	5	5
4	3	2
2	4	6
1	1	3
0	2	4

   a. Sketch a graph showing the relationship between X and Y. Compute the Pearson correlation between X and Y.

b. Sketch a graph showing the relationship between Y and Z. Compute the Pearson correlation between Y and Z.

c. Given the results of parts a and b, what would you predict for the correlation between X and Z?

d. Sketch a graph showing the relationship between X and Z. Compute the Pearson correlation for these data.

e. What general conclusion can you make concerning relationships among correlations? If X is related to Y and Y is related to Z, does this necessarily mean that X is related to Z?

11. For the following set of scores,

X	Y
3	8
8	1
5	6
6	3
6	6
8	6

a. Compute the Pearson correlation.

b. Add 2 points to each X value and compute the Pearson correlation for the modified scores.

c. Comparing the values obtained in parts a and b, how does adding a constant to every score affect the value of the correlation?

12. For the following set of scores,

X	Y
0	3
1	1
2	5
3	9
4	7

a. Compute the Pearson correlation.

b. Multiply each X value by 2 and compute the Pearson correlation for the modified scores.

c. Comparing the values obtained in parts a and b, how does multiplying each score by a constant affect the value of the correlation?

13. Correlation studies are often used to help determine whether certain characteristics are controlled more by genetic influences or by environmental influences. These studies often examine adopted children and compare their behaviors with the behaviors of their birth parents and their adoptive parents. One study examined how much time individuals spend watching TV (Plomin, Corley, DeFries, & Fulker, 1990). The

following data are similar to the results obtained in the study.

Amount of Time Spent Watching TV		
Adopted Children	Birth Parents	Adoptive Parents
2	0	1
3	3	4
6	4	2
1	1	0
3	1	0
0	2	3
5	3	2
2	1	3
5	3	3

a. Compute the correlation between the children and their birth parents.

b. Compute the correlation between the children and their adoptive parents.

c. Based on the two correlations, does TV watching appear to be inherited from the birth parents or is it learned from the adoptive parents?

14. It is well known that similarity in attitudes, beliefs, and interests plays an important role in interpersonal attraction (see Byrne, 1971, for example). Thus, correlations for attitudes between married couples should be strong. Suppose a researcher developed a questionnaire that measures how liberal or conservative one's attitudes are. Low scores indicate that the person has liberal attitudes, whereas high scores indicate conservatism. The following hypothetical data are scores for married couples.

Couple	Wife	Husband
A	11	14
B	6	7
C	16	15
D	4	7
E	1	3
F	10	9
G	5	9
H	3	8

Compute the Pearson correlation for these data, and determine whether there is a significant correlation between attitudes for husbands and wives. Set alpha at .05, two tails.

15. Identifying individuals with a high risk of Alzheimer's disease usually involves a long series of cognitive tests. However, researchers have developed a 7-Minute

Screen, which is a quick and easy way to accomplish the same goal. The question is whether the 7-Minute Screen is as effective as the complete series of tests. To address this question, Ijuin et al. (2008) administered both tests to a group of patients and compared the results. The following data represent results similar to those obtained in the study.

Patient	7-Minute Screen	Cognitive Series
A	3	11
B	8	19
C	10	22
D	8	20
E	4	14
F	7	13
G	4	9
H	5	20
I	14	25

a. Compute the Pearson correlation to measure the degree of relationship between the two test scores.
b. Is the correlation statistically significant? Use a two-tailed test with $\alpha = .01$.
c. What percentage of variance for the cognitive scores is predicted from the 7-Minute Screen scores? (Compute the value of r^2.)

16. Assuming a two-tailed test with $\alpha = .05$, how large a correlation is needed to be statistically significant for each of the following samples?
a. A sample of $n = 10$
b. A sample of $n = 20$
c. A sample of $n = 30$

17. For each of the following, does the sample provide enough evidence to conclude that there is a significant, nonzero correlation in the population. In each case, assume a two-tailed test.
a. $r = 0.40$ for a sample of $n = 20$ with $\alpha = .05$
b. $r = 0.40$ for a sample of $n = 30$ with $\alpha = .05$
c. $r = -0.60$ for a sample of $n = 10$ with $\alpha = .01$
d. $r = -0.60$ for a sample of $n = 20$ with $\alpha = .01$

18. As we have noted in previous chapters, even a very small effect can be significant if the sample is large enough. For each of the following, determine how large a sample is necessary for the correlation to be significant. Assume a two-tailed test with $\alpha = .05$.
   (*Note:* Because the table does not list every possible *df* value, you cannot determine every possible sample size. In each case, use the sample size corresponding to the appropriate *df* value in the table.)

a. A correlation of $r = 0.40$.
b. A correlation of $r = 0.30$.
c. A correlation of $r = 0.20$.

19. What is the major distinction between the Pearson and Spearman correlations?

20. The following data consist of numerical scores measured on an interval scale of measurement. Convert the scores to ranks and compute the Spearman correlation.

X	Y
0	15
2	10
6	5
10	2
3	12

21. A common concern for students (and teachers) is the assignment of grades for essays or term papers. Because there are no absolute right or wrong answers, these grades must be based on a judgment of quality. To demonstrate that these judgments actually are reliable, an English instructor asked a colleague to rank-order a set of term papers. The ranks and the instructor's grades for these papers are as follows:

Rank	Grade
1	A
2	B
3	A
4	B
5	B
6	C
7	D
8	C
9	C
10	D
11	E

a. Compute the Spearman correlation for these data. (*Note:* You must convert the letter grades to ranks, using tied ranks to represent tied grades.)
b. Is there a significant correlation between the two instructors' judgments of the papers. Use a two-tailed test with $\alpha = .05$.

22. It appears that there is a significant relationship between cognitive ability and social rank, at least for birds. Boogert, Reader, and Laland (2006) measured social rank and individual learning ability for a group of starlings. The following data represent results similar to those obtained in the study.

Subject	Social Rank	Learning Score
A	1	3
B	3	10
C	2	7
D	3	11
E	5	19
F	4	17
G	5	17
H	2	4
I	4	12
J	2	3

a. Because social rank is an ordinal variable, the Spearman correlation is appropriate for these data. The social rank categories and the learning scores should both be converted to ranks before the correlation is computed.

b. Based on the correlation from part a, is there a significant relationship between social rank and learning ability? Use a two-tailed test with $\alpha = .05$.

23. The point-biserial correlation and the independent-measures t test can both be used to evaluate the same set of data.

a. What is measured by the correlation? What is measured by the t test?

b. What factor can influence the outcome of the t test, but has no influence on the size of the correlation?

24. To test the effectiveness of a new studying strategy, a psychologist randomly divides a sample of 8 students into two groups, with $n = 4$ in each group. The students in one group receive training in the new studying strategy. Then all students are given 30 minutes to study a chapter from a history textbook before they take a quiz on the chapter. The quiz scores for the two groups are as follows:

Training	No Training
9	4
7	7
6	3
10	6

a. Convert these data into a form suitable for the point-biserial correlation. (Use $X = 1$ for training, $X = 0$ for no training, and the quiz score for Y.)

b. Calculate the point-biserial correlation for these data.

25. Problem 14 in Chapter 10 described a study by Rozin, Bauer, and Catanese (2003) comparing attitudes toward eating for male and female college students.

The results showed that females are much more concerned about weight gain and other negative aspects of eating than are males. The following data represent the results from one measure of concern about weight gain.

Males	Females
22	54
44	57
39	32
27	53
35	49
19	41
	35
	36
	48

Convert the data into a form suitable for the point-biserial correlation and compute the correlation.

26. Problem 23 in Chapter 10 presented hypothetical data showing that elderly people who own dogs are significantly less likely to pay visits to their doctors than those who do not own pets. The independent-measures t test produced $t = 2.11$ with $df = 11$ and a value of $r = 0.288$ (28.8%).

a. Convert the data from this problem into a form suitable for the point-biserial correlation (use 1 for the control group and 0 for the dog owners), and then compute the correlation.

b. Square the value of the point-biserial correlation to verify that you obtain the same r^2 value that was computed in Chapter 10.

27. A researcher would like to evaluate the relationship between a person's age and his or her preference between two leading brands of cola. In a sample of 12 people, the researcher found that 5 out of 8 people 30 years or older preferred brand A and only 1 out of 4 people younger than 30 years old preferred brand A.

a. Convert the data to a form suitable for computing the phi-coefficient. (Code the two age categories as 0 and 1 for the X variable, and code the preferred brand of soft drink as 0 and 1 for the Y variable.)

b. Compute the phi-coefficient for the data.

28. A researcher is comparing two sets of instructions for assembling a bicycle. Of the 10 people using the first set of instructions, only 2 were able to assemble the bike completely in 30 minutes. Of the 10 people using the second set of instructions, 7 completed assembly in 30 minutes. Convert the data into a form suitable for the phi-coefficient and compute the phi-correlation for these data.

C H A P T E R

17

Introduction to Regression

Tools You Will Need

The following items are considered essential background material for this chapter. If you doubt your knowledge of any of these items, you should review the appropriate chapter or section before proceeding.

- Sum of squares (*SS*) (Chapter 4)
  - Computational formula
  - Definitional formula
- *z*-scores (Chapter 5)
- Analysis of variance (Chapter 13)
  - *MS* values and *F*-ratios
- Pearson correlation (Chapter 16)
  - Sum of products (*SP*)

Preview

17.1   Introduction to Linear Equations and Regression

17.2   Testing the Significance of the Regression Equation: Analysis of Regression

17.3   Introduction to Multiple Regression with Two Predictor Variables

17.4   Evaluating the Contribution of Each Predictor Variable

Summary

Focus on Problem Solving

Demonstrations 17.1 and 17.2

Problems

Preview

In Chapter 16, we noted that one common application of correlations is for purposes of prediction. Whenever there is a consistent relationship between two variables, it is possible to use the value of one variable to predict the value of another. Managers at the electric company, for example, can use the weather forecast to predict power demands for upcoming days. If exceptionally hot summer weather is forecast, they can anticipate an exceptionally high demand for electricity. In the field of psychology, a known relationship between certain personality characteristics and eating disorders can allow clinicians to predict that individuals who show specific characteristics are more likely to develop disorders. A common prediction that is especially relevant for college students (and potential college students) is based on the relationship between scores on aptitude tests (such as the SAT) and future grade point averages in college. Each year, SAT scores from thousands of high school students are used to help college admissions officers decide who should be admitted and who should not.

The Problem: The correlations introduced in Chapter 16 allow researchers to measure and describe correlations, and the hypothesis tests allow researchers to evaluate the significance of correlations. However, we now want to go one step further and actually use a correlation to make predictions.

The Solution: In this chapter we introduce some of the statistical techniques that are used to make predictions based on correlations. Whenever there is a linear relationship (Pearson correlation) between two variables, it is possible to compute an equation that provides a precise, mathematical description of the relationship. With the equation, it is possible to plug in the known value for one variable (for example, your SAT score), then calculate a predicted value for the second variable (for example, your college grade point average). The general statistical process of finding and using a prediction equation is known as *regression*.

Beyond finding a prediction equation, however, it is reasonable to ask how good are the predictions it makes. For example, I can make predictions about the outcome of a coin toss by simply guessing. However, my predictions are correct only about 50% of the time. In statistical terms, my predictions are not significantly better than chance. In the same way, it is appropriate to challenge the significance of any prediction equation. In this chapter we introduce the techniques that are used to find prediction equations, as well as the techniques that are used to determine whether their predictions are statistically significant. Incidentally, although there is some controversy about the practice of using SAT scores to predict college performance, there is a great deal of research showing that SAT scores really are valid and significant predictors (Camera & Echternacht, 2000; Geiser & Studley, 2002).

17.1 INTRODUCTION TO LINEAR EQUATIONS AND REGRESSION

In Chapter 16, we introduced the Pearson correlation as a technique for describing and measuring the linear relationship between two variables. Figure 17.1 presents hypothetical data showing the relationship between SAT scores and college grade point average (GPA). Note that the figure shows a good, but not perfect, positive relationship. Also, note that we have drawn a line through the middle of the data points. This line serves several purposes:

a. The line makes the relationship between SAT and GPA easier to see.

b. The line identifies the center, or "central tendency," of the relationship, just as the mean describes central tendency for a set of scores. Thus, the line provides a simplified description of the relationship. For example, if the data points were removed, the straight line would still give a general picture of the relationship between SAT and GPA.

c. Finally, the line can be used for prediction. The line establishes a precise relationship between each X value (SAT score) and a corresponding Y value (GPA). For example, an SAT score of 620 corresponds to a GPA of 3.25 (see Figure 17.1). Thus, the college admissions office could use the straight-line

FIGURE 17.1

Hypothetical data showing the relationship between SAT scores and GPA with a regression line drawn through the data points. The regression line defines a precise, one-to-one relationship between each X value (SAT score) and its corresponding Y value (GPA).

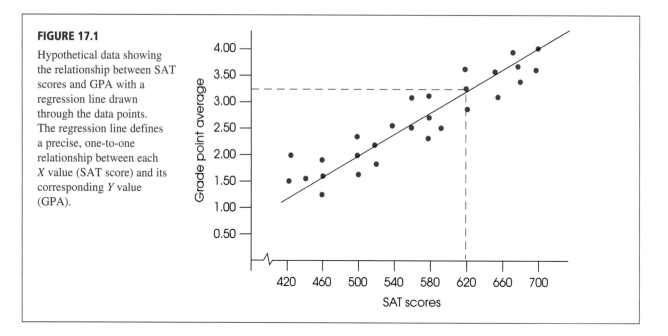

relationship to predict that a student entering college with an SAT score of 620 should achieve a college GPA of approximately 3.25.

Our goal in this section is to develop a procedure that identifies and defines the straight line that provides the best fit for any specific set of data. You should realize that this straight line does not have to be drawn on a graph; it can be presented in a simple equation. Thus, our goal is to find the equation for the line that best describes the relationship for a set of X and Y data.

LINEAR EQUATIONS

In general, a *linear relationship* between two variables X and Y can be expressed by the equation

$$Y = bX + a \qquad\qquad (17.1)$$

where a and b are fixed constants.

For example, a local tennis club charges a fee of $5 per hour plus an annual membership fee of $25. With this information, the total cost of playing tennis can be computed using a *linear equation* that describes the relationship between the total cost (Y) and the number of hours (X).

$$Y = 5X + 25$$

Note that a positive slope means that Y increases when X increases, and a negative slope indicates that Y decreases when X increases.

In the general linear equation, the value of b is called the *slope*. The slope determines how much the Y variable will change when X is increased by one point. For the tennis club example, the slope is $b = \$5$ and indicates that your total cost will increase by $5 for each hour you play. The value of a in the general equation is called the *Y-intercept* because it determines the value of Y when $X = 0$. (On a graph, the a value identifies the point where the line intercepts the Y-axis.) For the tennis club example, $a = \$25$; there is a $25 charge even if you never play tennis.

Figure 17.2 shows the general relationship between cost and number of hours for the tennis club example. Notice that the relationship results in a straight line. To obtain this graph, we picked any two values of X and then used the equation to compute the corresponding values for Y. For example,

when $X = 10$:	when $X = 30$:
$Y = bX + a$	$Y = bX + a$
$= \$5(10) + \$25$	$= \$5(30) + \$25$
$= \$50 + \$25$	$= \$150 + \$25$
$= \$75$	$= \$175$

When drawing a graph of a linear equation, it is wise to compute and plot at least three points to be certain you have not made a mistake.

Next, these two points are plotted on the graph: one point at $X = 10$ and $Y = 75$, the other point at $X = 30$ and $Y = 175$. Because two points completely determine a straight line, we simply drew the line so that it passed through these two points.

FIGURE 17.2

Relationship between total cost and number of hours playing tennis. The tennis club charges a $25 membership fee plus $5 per hour. The relationship is described by a linear equation:

total cost =
  $5 (number of hours) + $25

$Y = bX + a$

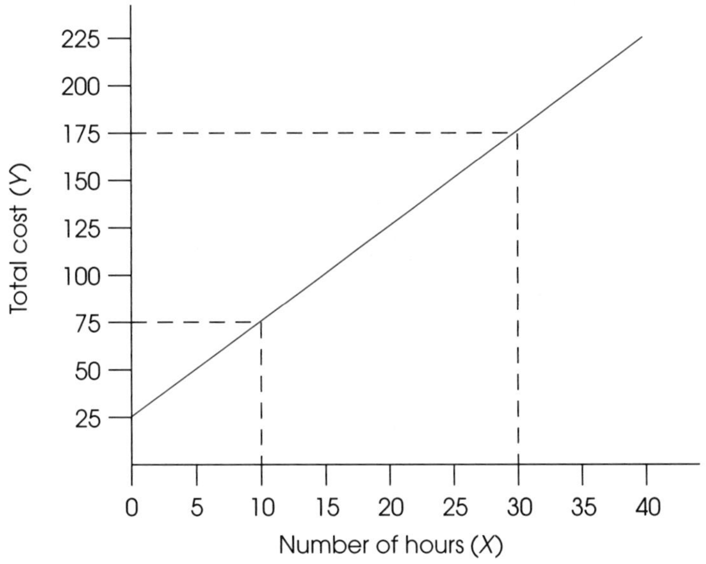

LEARNING CHECK

1. Identify the slope and Y-intercept for the following linear equation:

   $Y = -3X + 7$

2. Use the linear equation $Y = 2X - 7$ to determine the value of Y for each of the following values of X: 1, 3, 5, 10.

3. If the slope constant (b) in a linear equation is positive, then a graph of the equation will be a line tilted from lower left to upper right. (True or false?)

ANSWERS   **1.** Slope $= -3$ and Y-intercept $= +7$.

2.

X	Y
1	−5
3	−1
5	3
10	13

3. True. A positive slope indicates that Y increases (goes up in the graph) when X increases (goes to the right in the graph).

REGRESSION

Because a straight line can be extremely useful for describing a relationship between two variables, a statistical technique has been developed that provides a standardized method for determining the best-fitting straight line for any set of data. The statistical procedure is *regression,* and the resulting straight line is called the *regression line.*

DEFINITION

The statistical technique for finding the best-fitting straight line for a set of data is called **regression**, and the resulting straight line is called the **regression line**.

The goal for regression is to find the best-fitting straight line for a set of data. To accomplish this goal, however, it is first necessary to define precisely what is meant by "best fit." For any particular set of data, it is possible to draw lots of different straight lines that all appear to pass through the center of the data points. Each of these lines can be defined by a linear equation of the form $Y = bX + a$, where b and a are constants that determine the slope and Y-intercept of the line, respectively. Each individual line has its own unique values for b and a. The problem is to find the specific line that provides the best fit to the actual data points.

THE LEAST-SQUARED-ERROR SOLUTION AND THE REGRESSION EQUATION

To determine how well a line fits the data points, the first step is to define mathematically the distance between the line and each data point. For every X value in the data, the linear equation will determine a Y value on the line. This value is the predicted Y and is called $\hat{Y}$ ("Y hat"). The distance between this predicted value and the actual Y value in the data is determined by

$$\text{distance} = Y - \hat{Y}$$

Notice that we simply are measuring the vertical distance between the actual data point (Y) and the predicted point on the line. This distance measures the error between the line and the actual data (Figure 17.3).

Because some of these distances will be positive and some will be negative, the next step is to square each distance to obtain a uniformly positive measure of error. Finally, to determine the total error between the line and the data, we add the squared errors for all of the data points. The result is a measure of overall squared error between the line and the data:

$$\text{total squared error} = \Sigma(Y - \hat{Y})^2$$

FIGURE 17.3

The distance between the actual data point (Y) and the predicted point on the line ($\hat{Y}$) is defined as $Y - \hat{Y}$. The goal of regression is to find the equation for the line that minimizes these distances.

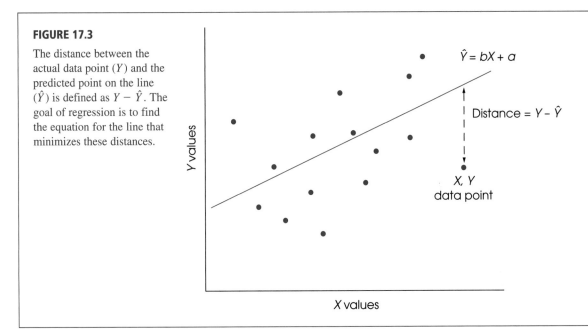

Now we can define the *best-fitting* line as the one that has the smallest total squared error. For obvious reasons, the resulting line is commonly called the *least-squared-error solution.*

In symbols, we are looking for a linear equation of the form

$$\hat{Y} = bX + a$$

For each value of X in the data, this equation will determine the point on the line ($\hat{Y}$) *that gives the best prediction of Y.* The problem is to find the specific values for a and b that will make this the best-fitting line.

The calculations that are needed to find this equation require calculus and some sophisticated algebra, so we will not present the details of the solution. The results, however, are relatively straightforward, and the solutions for b and a are as follows:

$$b = \frac{SP}{SS_X} \tag{17.2}$$

where SP is the sum of products and SS_X is the sum of squares for the X scores.

A commonly used alternative formula for the slope is based on the standard deviations for X and Y. The alternative formula is

$$b = r\frac{s_Y}{s_X} \tag{17.3}$$

where s_Y is the standard deviation for the Y scores, s_X is the standard deviation for the X scores, and r is the Pearson correlation between X and Y. The value of the constant a in the equation is determined by

$$a = M_Y - bM_X \tag{17.4}$$

Note that these formulas determine the linear equation that provides the best prediction of Y values. This equation is called the *regression equation for* Y.

DEFINITION

The **regression equation for Y** is the linear equation

$$\hat{Y} = bX + a \tag{17.5}$$

where the constant b is determined by Equation 17.2 or 17.3, and the constant a is determined by Equation 17.4. This equation results in the least squared error between the data points and the line.

The slope and Y-intercept values in the regression equation introduce some simple and very predictable facts about the regression line. First, the calculation of the Y-intercept (Equation 17.4) ensures that the regression line passes through the point defined by the mean for X and the mean for Y. That is, the point identified by the coordinates M_X, M_Y will always be on the line. Second, the sign of the correlation ($+$ or $-$) is the same as the sign of the slope of the regression line. Specifically, if the correlation is positive, then the slope will also be positive and the regression line will slope up to the right. On the other hand, if the correlation is negative, the slope will be negative and the line will slope down to the right. Finally, if the correlation is zero, the slope will also be zero and the regression equation will produce a horizontal line that passes through the data at a level equal to the mean for the Y values.

EXAMPLE 17.1

The scores in the following table will be used to demonstrate the calculation and use of the regression equation for predicting Y.

X	Y	$X - M_X$	$Y - M_Y$	$(X - M_X)^2$	$(X - M_X)(Y - M_Y)$
2	3	-2	-5	4	10
6	11	2	3	4	6
0	6	-4	-2	16	8
4	6	0	-2	0	0
5	7	1	-1	1	-1
7	12	3	4	9	12
5	10	1	2	1	2
3	9	-1	1	1	-1

$$SS_X = 36 \qquad SP = 36$$

For these data, $\Sigma X = 32$, so $M_X = 4$. Also, $\Sigma Y = 64$, so $M_Y = 8$. These values have been used to compute the deviation scores for each X and Y value. The final two columns show the squared deviations for X and the products of the deviation scores. Based on these values,

$$SP = \Sigma(X - M_X)(Y - M_Y) = 36$$
$$SS_X = \Sigma(X - M_X)^2 = 36$$

Our goal is to find the values for b and a in the regression equation. Using Equations 17.2 and 17.4, the solutions for b and a are

$$b = \frac{SP}{SS_X} = \frac{36}{36} = 1.00$$

$$a = M_Y - bM_X$$
$$= 8 - 1(4)$$
$$= 4.00$$

The resulting equation is

$$\hat{Y} = X + 4$$

The original data and the regression line are shown in Figure 17.4.

As we noted at the beginning of this section, one common use of regression equations is for prediction. For any given value of X, we can use the equation to compute a predicted value for Y. Using the equation from Example 17.1, an individual with a score of $X = 1$ would be predicted to have a Y score of

$$\hat{Y} = X + 4$$
$$= 1 + 4$$
$$= 5$$

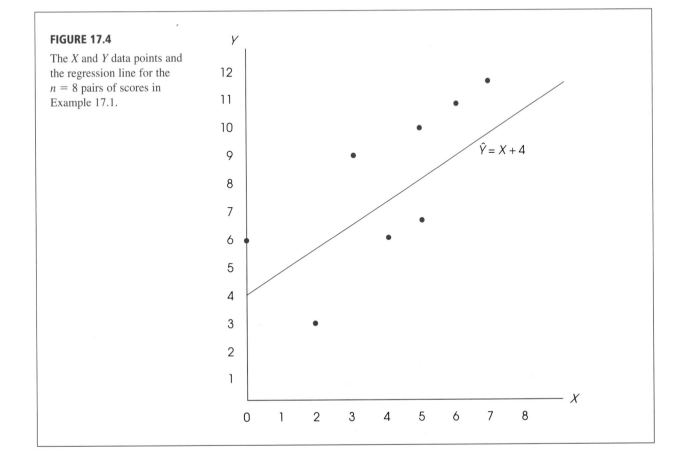

FIGURE 17.4

The X and Y data points and the regression line for the $n = 8$ pairs of scores in Example 17.1.

Although regression equations can be used for prediction, there are two cautions that should be considered whenever you are interpreting the predicted values:

1. The predicted value is not perfect (unless $r = +1.00$ or -1.00). If you examine Figure 17.4, it should be clear that the data points do not fit perfectly on the line. In general, there will be some error between the predicted Y values (on the line) and the actual data. Although the amount of error varies from point to point, on average the errors are directly related to the magnitude of the correlation. With a correlation near 1.00 (or -1.00), the data points are generally close to the line (small error), but as the correlation gets nearer to zero, the magnitude of the error increases. The statistical procedure for measuring this error is described in the following section.

2. The regression equation should not be used to make predictions for X values that fall outside the range of values covered by the original data. For Example 17.1, the X values ranged from $X = 0$ to $X = 7$, and the regression equation was calculated as the best-fitting line within this range. Because you have no information about the X-Y relationship outside this range, the equation should not be used to predict Y for any X value lower than 0 or greater than 7.

LEARNING CHECK

1. Sketch a scatterplot for the following data—that is, a graph showing the X, Y data points:

X	Y
2	4
3	9
4	8

a. Find the regression equation for predicting Y and X. Draw this line on your graph. Does it look like the best-fitting line?

b. Use the regression equation to find the predicted Y value corresponding to each X in the data.

ANSWER

1. a. $SS_X = 2$, $SP = 4$, $b = 2$, $a = 1$.
   The equation is

   $$\hat{Y} = 2X + 1$$

   b. The predicted Y values are 5, 7, and 9.

STANDARDIZED FORM OF THE REGRESSION EQUATION

So far we have presented the regression equation in terms of the original values, or raw scores, for X and Y. Occasionally, however, researchers standardize the scores by transforming the X and Y values into z-scores before finding the regression equation. The resulting equation is often called the standardized form of the regression equation and is greatly simplified compared to the raw-score version. The simplification comes from the fact that z-scores have standardized characteristics. Specifically, the mean for a set of z-scores is always zero and the standard deviation is always 1. As a result, the standardized form of the regression equation becomes

$$\hat{z}_Y = (\text{beta})z_X \tag{17.6}$$

First notice that we are now using the z-score for each X value to predict the z-score for the corresponding Y value. Also, note that the slope constant that was identified as b in the raw-score formula is now identified as beta. Because both sets of z-scores have a mean of zero, the constant a disappears from the regression equation. Finally, when one variable, X, is being used to predict a second variable, Y, the value of beta is equal to the Pearson correlation for X and Y. Thus, the standardized form of the regression equation can also be written as

$$\hat{z}_Y = rz_X \tag{17.7}$$

Because the process of transforming all of the original scores into z-scores can be tedious, researchers usually compute the raw-score version of the regression equation (Equation 17.5) instead of the standardized form. However, most computer programs report the value of beta as part of the output from linear regression, and you should understand what this value represents.

THE STANDARD ERROR OF ESTIMATE

It is possible to determine a best-fitting regression equation for any set of data by simply using the formulas already presented. The linear equation you obtain is then used to generate predicted Y values for any known value of X. However, it should be clear that the accuracy of this prediction depends on how well the points on the line correspond to the actual data points—that is, the amount of error between the predicted values, $\hat{Y}$, and the actual scores, Y values. Figure 17.5 shows two different sets of data that have exactly the same regression equation. In one case, there is a perfect correlation ($r = +1$) between X and Y, so the linear equation fits the data perfectly. For the second set of data, the predicted Y values on the line only approximate the real data points.

A regression equation, by itself, allows you to make predictions, but it does not provide any information about the accuracy of the predictions. To measure the precision of the regression, it is customary to compute a *standard error of estimate.*

DEFINITION

The **standard error of estimate** gives a measure of the standard distance between a regression line and the actual data points.

Conceptually, the standard error of estimate is very much like a standard deviation: Both provide a measure of standard distance. Also note that the calculation of the standard error of estimate is very similar to the calculation of standard deviation.

To calculate the standard error of estimate, we first will find a sum of squared deviations (SS). Each deviation will measure the distance between the actual Y value (from the data) and the predicted Y value (from the regression line). This sum of squares is commonly called SS_{residual} because it is based on the remaining distance between the actual Y scores and the predicted values.

$$SS_{\text{residual}} = \Sigma(Y - \hat{Y})^2 \tag{17.8}$$

The obtained SS value is then divided by its degrees of freedom to obtain a measure of variance. This procedure should be very familiar:

$$\text{Variance} = \frac{SS}{df}$$

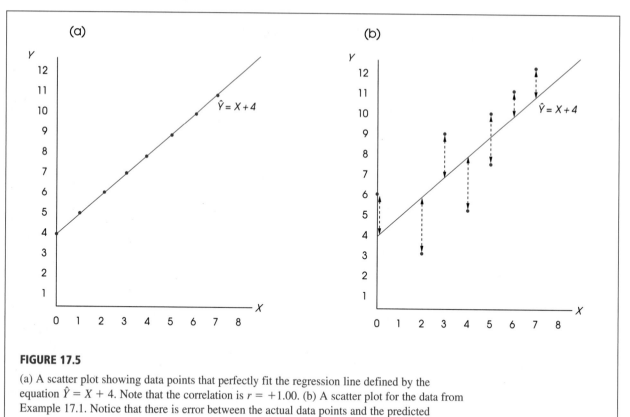

FIGURE 17.5

(a) A scatter plot showing data points that perfectly fit the regression line defined by the equation $\hat{Y} = X + 4$. Note that the correlation is $r = +1.00$. (b) A scatter plot for the data from Example 17.1. Notice that there is error between the actual data points and the predicted Y values of the regression line.

The degrees of freedom for the standard error of estimate are $df = n - 2$. The reason for having $n - 2$ degrees of freedom, rather than the customary $n - 1$, is that we now are measuring deviations from a line rather than deviations from a mean. Recall that it is necessary to know SP to find the slope of the regression line (the value of b in the equation). To calculate SP, you must know the means for both the X and the Y scores. Specifying these two means places two restrictions on the variability of the data, with the result that the scores have only $n - 2$ degrees of freedom. (A more intuitive explanation for the fact that the SS_{residual} has $df = n - 2$ comes from the simple observation that it takes exactly two points to determine a straight line. If there are only two data points, they always fit perfectly on a straight line, so there will be no error. It is only when you have more than two points that there is some freedom in determining the best-fitting line.)

Recall that variance measures the average squared distance.

The final step in the calculation of the standard error of estimate is to take the square root of the variance to obtain a measure of standard distance. The final equation is

$$\text{standard error of estimate} = \sqrt{\frac{SS_{\text{residual}}}{df}} = \sqrt{\frac{\Sigma(Y - \hat{Y})^2}{n - 2}} \qquad (17.9)$$

The following example demonstrates the calculation of this standard error.

EXAMPLE 17.2 The data in Example 17.1 are used to demonstrate the calculation of the standard error of estimate. These data have the regression equation

$$\hat{Y} = X + 4$$

Using this regression equation, we have computed the predicted Y value, the residual, and the squared residual for each individual, using the data from Example 17.1.

Data		Predicted Y Values	Residual	Squared Residual
X	Y	$\hat{Y} = X + 4$	$Y - \hat{Y}$	$(Y - \hat{Y})^2$
2	3	6	−3	9
6	11	10	1	1
0	6	4	2	4
4	6	8	−2	4
5	7	9	−2	4
7	12	11	1	1
5	10	9	1	1
3	9	7	2	4
			$SS_{\text{residual}} = 28$	

For these data, the sum of the squared residuals is $SS_{\text{residual}} = 28$. With $n = 8$, the data have $df = n - 2 = 6$, so the standard error of estimate is

$$\text{standard error of estimate} = \sqrt{\frac{SS_{\text{residual}}}{df}} = \sqrt{\frac{28}{6}} = 2.16$$

Remember: The standard error of estimate provides a measure of how accurately the regression equation predicts the Y values. In this case, the standard distance between the actual data points and the regression line is measured by standard error of estimate = 2.16.

RELATIONSHIP BETWEEN THE STANDARD ERROR AND THE CORRELATION

It should be clear from Example 17.2 that the standard error of estimate is directly related to the magnitude of the correlation between X and Y. If the correlation is near 1.00 (or −1.00), the data points will be clustered close to the line, and the standard error of estimate will be small. As the correlation gets nearer to zero, the line will provide less accurate predictions, and the standard error of estimate will grow larger. In Chapter 16 (page 534) we observed that squaring the correlation provides a measure of the accuracy of prediction: r^2 is called the coefficient of determination because it determines what proportion of the variability in Y is predicted by the relationship with X.

Because r^2 measures the portion of the variability in the Y scores that is predicted by the regression equation, we can use the expression $(1 - r^2)$ to measure the unpredicted portion. Thus,

$$\text{Predicted variability} = SS_{\text{regression}} = r^2 SS_Y \tag{17.10}$$

$$\text{Unpredicted variability} = SS_{\text{residual}} = (1 - r^2)SS_Y \tag{17.11}$$

For example, if $r = 0.80$, then $r^2 = 0.64$ (or 64%) of the variability for the Y scores is predicted by the relationship with X and the remaining 36% $(1 - r^2)$ is the unpredicted portion. Note that when $r = 1.00$, the prediction is perfect and there are no residuals. As the correlation approaches zero, the data points move farther off the line and the residuals grow larger. Using Equation 17.11 to compute $SS_{residual}$, the standard error of estimate can be computed as

$$\text{standard error of estimate} = \sqrt{\frac{SS_{residual}}{df}} = \sqrt{\frac{(1 - r^2)SS_Y}{n - 2}}$$

The following example demonstrates this new formula.

EXAMPLE 17.3   The same data used in Examples 17.1 and 17.2 are reproduced in the following table:

X	Y	$X - M_X$	$Y - M_Y$	$(X - M_X)^2$	$(Y - M_X)^2$	$(X - M_X)(Y - M_Y)$
2	3	−2	−5	4	25	10
6	11	2	3	4	9	6
0	6	−4	−2	16	4	8
4	6	0	−2	0	4	0
5	7	1	−1	1	1	−1
7	12	3	4	9	16	12
5	10	1	2	1	4	2
3	9	−1	1	1	1	−1
				$SS_X = 36$	$SS_Y = 64$	$SP = 36$

For these data, the Pearson correlation is

$$r = \frac{36}{\sqrt{36(64)}} = \frac{36}{48} = 0.75$$

With $SS_Y = 64$ and a correlation of $r = 0.75$, the predicted variability from the regression equation is

$$SS_{regression} = r^2 SS_Y = (0.75^2)(64) = 0.5625(64) = 36.00$$

Similarly, the unpredicted variability is

$$SS_{residual} = (1 - r^2)SS_Y = (1 - 0.75^2)(64) = 0.4375(64) = 28.00$$

Notice that the new formula for $SS_{residual}$ produces exactly the same value that we obtained by adding the squared residuals in Example 17.2. Also note that this new formula is generally much easier to use because it requires only the correlation value (r) and the SS for Y. The primary point of this example, however, is that the standard error of estimate is closely related to the value of the correlation. With a large correlation (near $+1.00$ or -1.00), the data points will be close to the regression line, and the standard error of estimate will be small. As a correlation gets smaller (near zero), the data points move away from the regression line, and the standard error of estimate gets larger.

Because it is possible to have the same regression line for sets of data that have different correlations, it is important to examine r^2 and the standard error of estimate. The regression equation simply describes the best-fitting line and is used for making predictions. However, r^2 and the standard error of estimate indicate how accurate these predictions will be.

PREDICTED AND UNPREDICTED DEVIATIONS

Although a regression equation is typically used to predict scores, it also can be viewed as predicting deviations from the mean. Recall that the regression line always passes through the point corresponding to the mean for X and the mean for Y. Figure 17.6 shows an example of a regression line with a positive slope. Notice that all of the points on the line that are above the mean for X are also above the mean for Y. Similarly, all of the points below the mean for X are also below the mean for Y. Thus, every individual with a positive deviation for X is predicted to have a positive deviation for Y, and everyone with a negative deviation for X is predicted to have a negative deviation for Y.

However, the predictions from the regression line are typically not perfect, so the line predicts a portion of each individual's deviation, but also fails to predict a portion. These two portions, predicted and unpredicted, produce the deviations that are used to compute $SS_{\text{regression}}$ and SS_{residual}.

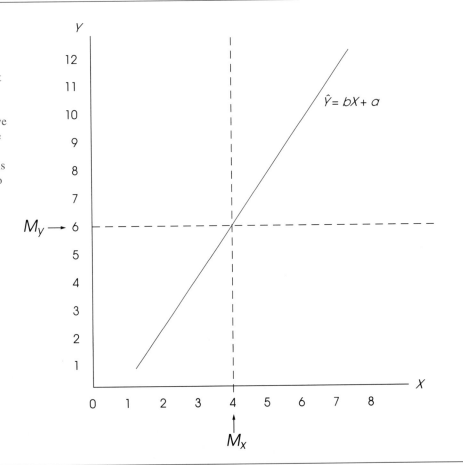

FIGURE 17.6

A regression line with a positive slope. Because the line passes through the point defined by the mean for X and the mean for Y, every point on the line that is above the mean for X is also above the mean for Y. Similarly, every point on the line that is below the mean for X is also below the mean for Y.

For example, the scores in Examples 17.1, 17.2, and 17.3 have a mean of $M_Y = 8$ for the Y values. The first individual in these examples has scores of $X = 2$ and $Y = 3$. Note that the Y value is below the mean by 5 points (a deviation of -5). When the X score is plugged into the regression equation, the result is a predicted value of $Y = 6$, which is below the mean by 2 points (a deviation of -2). Thus, the equation predicts 2 points of this individual's deviation and the remaining 3 points is residual.

By repeating this process, the predicted and unpredicted deviation can be computed for each individual. The resulting values are shown in Table 17.1. If you square all of the predicted deviations and add the squared values, you will obtain $SS_{\text{regression}} = 36$. Similarly, if you square and add all of the residual deviations, you will obtain $SS_{\text{residual}} = 28$. These are exactly the same values we obtained earlier using r^2 and $(1 - r^2)$ proportions of SS_Y.

TABLE 17.1

The predicted and unpredicted deviations for the scores from Examples 17.1, 17.2, and 17.3.

Actual Y Score	Predicted Y Score	Actual Deviation from $M_Y = 8$	Predicted Deviation from $M_Y = 8$	Residual Deviation (Actual − Predicted)
3	6	−5	−2	−3
11	10	3	2	1
6	4	−2	−4	2
6	8	−2	0	−2
7	9	−1	1	−2
12	11	4	3	1
10	9	2	1	1
9	7	1	−1	2

LEARNING CHECK

1. Use the following set of data:

X	Y	
1	4	$SS_X = 10$
2	1	$SS_Y = 90$
3	7	$SP = 24$
4	13	
5	10	

a. Find the regression equation for predicting Y from X.

b. Use the regression equation to find the predicted Y value for each X in the data.

c. Find the residual $(Y - \hat{Y})$ for each data point. Square each residual value, and add the results to find SS_{residual}.

d. Calculate the Pearson correlation for these data.

e. Use the correlation and SS_Y to compute SS_{residual}.

2. Assuming that all other factors are held constant, what happens to the standard error of estimate as the correlation between X and Y moves toward zero?

ANSWERS   **1. a.** $\hat{Y} = 2.4X - 0.2$

      b. c.

Y	$\hat{Y}$	Residual	Residual^2
4	2.2	1.8	3.24
1	4.6	−3.6	12.96
7	7.0	0	0
13	9.4	3.6	12.96
10	11.8	−1.8	3.24
			$32.40 = SS_{residual}$

    d. $r = 0.80$

    e. $(1 - r^2)SS_Y = 0.36(90) = 32.40 = SS_{residual}$

  2. The standard error of estimate would get larger.

17.2 TESTING THE SIGNIFICANCE OF THE REGRESSION EQUATION: ANALYSIS OF REGRESSION

As we noted in Chapter 16, a sample correlation is expected to be representative of its population correlation. For example, if the population correlation is zero, the sample correlation is expected to be near zero. Note that we do not expect the sample correlation to be exactly equal to zero. This is the general concept of *sampling error* that was introduced in Chapter 1 (page 6). The principle of sampling error is that there will typically be some discrepancy or error between the value obtained for a sample statistic and the corresponding population parameter. Thus, when there is no relationship whatsoever in the population ($\rho = 0$), you are still likely to obtain a nonzero value for the sample correlation. In this situation, however, the sample correlation is meaningless and a hypothesis test will usually demonstrate that the correlation is not significant.

Whenever you obtain a nonzero value for a sample correlation, you will also obtain real, numerical values for the regression equation. However, if there is no real relationship in the population, both the sample correlation and the regression equation are meaningless—they are simply the result of sampling error and should not be viewed as an indication of any relationship between X and Y. In the same way that we tested the significance of a Pearson correlation, we can test the significance of the regression equation. In fact, when a single variable X is being used to predict a single variable Y, the two tests are equivalent. In each case, the purpose for the test is to determine whether the sample correlation represents a real relationship or is simply the result of sampling error. For both tests, the null hypothesis states that there is no relationship between the two variables in the population. In the context of regression, there are two more specific alternatives for the null hypothesis. The first version states that the true, population value for the slope constant in the regression equation is zero. Remember that, whenever the correlation is zero, the slope constant also is zero. According to this version of H_0, any nonzero value for the slope (b or beta) in the regression equation is meaningless and is simply the result of sampling error. The second version of H_0 states that the regression equation does not account for a significant proportion of the variance in the scores. According to this version, the constants in the regression equation are meaningless, so the equation is not capable of producing predictions that are significantly better than chance.

The process of testing the significance of a regression equation is called analysis of regression and is very similar to the analysis of variance (ANOVA) presented in Chapter 13. As with ANOVA, the regression analysis uses an F-ratio to determine whether the amount of variance accounted for by the regression equation is significantly greater than would be expected by chance alone. The F-ratio is a ratio of two variances, or mean square (MS) values, and each variance is obtained by dividing an SS value by its corresponding degrees of freedom. The numerator of the F-ratio is $MS_{\text{regression}}$, which measures the variance that is predicted by the regression equation. The denominator is MS_{residual}, which measures unpredicted variance. The two MS values are defined as

$$MS_{\text{regression}} = \frac{SS_{\text{regression}}}{df_{\text{regression}}} \text{ with } df = 1 \quad \text{and} \quad MS_{\text{residual}} = \frac{SS_{\text{residual}}}{df_{\text{residual}}} \text{ with } df = n - 2$$

The F-ratio is

$$F = \frac{MS_{\text{regression}}}{MS_{\text{residual}}} \quad \text{with } df = 1, n - 2 \tag{17.12}$$

The complete analysis of SS and degrees of freedom is diagrammed in Figure 17.7. The analysis of regression procedure is demonstrated in the following example, using the same data that we used in Examples 17.1, 17.2, and 17.3.

FIGURE 17.7

The partitioning of SS and df for analysis of regression. The variability in the original Y scores (both SS_Y and df_Y) is partitioned into two components: (1) the variability that is explained by the regression equation, and (2) the residual variability.

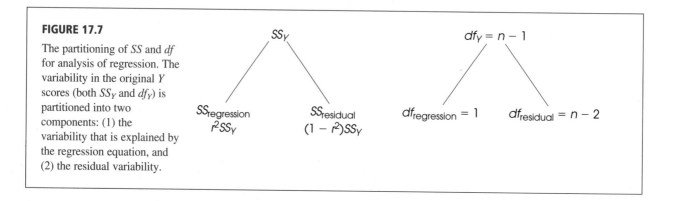

EXAMPLE 17.4

STEP 1   State the hypothesis and select an; alpha level.

The null hypothesis can be phrased either in terms of the slope constant or in terms of the predicted variance:

H_0: For the population, b, or beta, is equal to zero, or

H_0: The regression equation does not account for a significant portion of the variance in the Y scores.

The corresponding alternative hypothesis states that the slope constant is not zero or that the equation does account for a significant portion of the variance.

We will use $\alpha = .05$.

STEP 2   Locate the critical region.

With $n = 8$, the F-ratio will have $df = 1, 6$. For $\alpha = .05$, the critical value is 5.99.

STEP 3   Compute the test statistic.

As noted in the previous section, the SS value for the Y scores can be separated into two components: the predicted portion corresponding to r^2 and the unpredicted, or residual, portion corresponding to $1 - r^2$. For the data in the three previous examples, we obtained $SS_Y = 64$ with a correlation of $r = 0.75$, producing $r^2 = 0.5625$. Thus, we obtained

$$\text{predicted variability} = SS_{\text{regression}} = r^2 SS_Y = 0.5625(64) = 36.00$$

$$\text{unpredicted, residual variability} = SS_{\text{residual}} = (1 - r^2)SS_Y = 0.4375(64) = 28.00$$

Using these SS values and the corresponding df values, we calculate a variance or MS for each component:

$$MS_{\text{regression}} = \frac{SS_{\text{regression}}}{df_{\text{regression}}} = \frac{36}{1} = 36.00$$

$$MS_{\text{residual}} = \frac{SS_{\text{residual}}}{df_{\text{residual}}} = \frac{28}{6} = 4.67$$

Finally, the F-ratio for evaluating the significance of the regression equation is:

$$F = \frac{MS_{\text{regression}}}{MS_{\text{residual}}} = \frac{36.00}{4.67} = 7.71$$

STEP 4   Make a decision.

With a critical value of 5.99, we reject the null hypothesis and conclude that the regression equation does account for a significant portion of the variance for the Y scores. Note that this conclusion is perfectly consistent with the corresponding test for the significance of the Pearson correlation. For these data, the Pearson correlation is $r = 0.75$ with $n = 8$. Checking Table B6 in Appendix B, you can verify that the correlation is also significant. Thus, the data we have been evaluating provide sufficient evidence to conclude that there is a significant relationship between X and Y either in terms of a significant correlation or in terms of a significant regression equation. The complete analysis of regression is summarized in Table 17.2, which is a common format for computer printouts of regression analysis.

TABLE 17.2

A summary table showing the results of the analysis of regression in Example 17.4.

Source	SS	df	MS	F
Regression	36	1	36.60	7.71
Residual	28	6	4.67	
Total	64	7		

LEARNING CHECK

1. A set of $n = 12$ pairs of scores produces a Pearson correlation of $r = 0.40$ with $SS_Y = 120$. Find $SS_{regression}$ and $SS_{residual}$ and compute the F-ratio to evaluate the significance of the regression equation for predicting Y.

ANSWER

1. $SS_{regression} = 19.2$ with $df = 1$. $SS_{residual} = 100.8$ with $df = 10$. $F = 1.90$. With $df = 1, 10$ the F-ratio is not significant.

17.3 INTRODUCTION TO MULTIPLE REGRESSION WITH TWO PREDICTOR VARIABLES

Thus far, we have looked at regression in situations in which one variable is being used to predict a second variable. For example, IQ scores can be used to predict academic performance for a group of college students. However, a variable such as academic performance is usually related to a variety of other factors. For example, college GPA is probably related to motivation, self-esteem, SAT score, rank in high school graduating class, parents' highest level of education, and many other variables. In this case, it is possible to combine several predictor variables to obtain a more accurate prediction. For example, IQ predicts some of academic performance, but you can probably get a better prediction if you use IQ and SAT scores together. The process of using several predictor variables to help obtain more accurate predictions is called *multiple regression.*

Although it is possible to combine a large number of predictor variables in a single multiple-regression equation, we limit our discussion to the two-predictor case. There are two practical reasons for this limitation.

1. Multiple regression, even limited to two predictors, can be relatively complex. Although we present equations for the two-predictor case, the calculations are usually performed by a computer, so there is not much point in developing a set of complex equations when people are going to use a computer instead.

2. Usually, different predictor variables are related to each other, which means that they are often measuring and predicting the same thing. Because the variables may overlap with each other, adding another predictor variable to a regression equation does not always add to the accuracy of prediction. This situation is shown in Figure 17.8. In the figure, IQ overlaps with academic performance, which means that part of academic performance can be predicted by IQ. The figure also shows that SAT scores overlap with academic performance, which means that part of academic performance can be predicted by knowing SAT

FIGURE 17.8

Predicting the variance in academic performance from IQ and SAT scores. The overlap between IQ and academic performance indicates that 40% of the variance in academic performance can be predicted from IQ scores. Similarly, 30% of the variance in academic performance can be predicted from SAT scores. However, IQ and SAT also overlap, so that SAT scores contribute an additional predication of only 10% beyond what is already predicted by IQ.

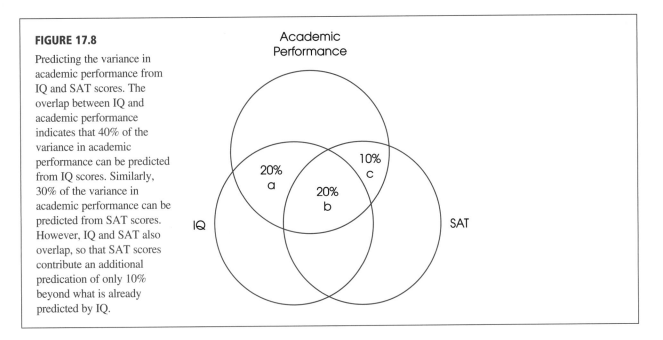

scores. In this example, IQ overlaps (predicts) 40% of the variance in academic performance (combine sections a and b in the figure). SAT scores overlap or predict 30% of the variance (combine sections b and c). However, there is also a lot of overlap between SAT scores and IQ. In particular, much of the prediction from SAT scores overlaps with the prediction from IQ (section b). As a result, adding SAT scores as a second predictor only adds 10% to the amount already predicted by IQ (section c). Because variables tend to overlap in this way, adding new variables beyond the first one or two predictors often does not add significantly to the quality of the prediction.

REGRESSION EQUATIONS WITH TWO PREDICTORS

We identify the two predictor variables as X_1 and X_2. The variable we are trying to predict is identified as Y. Using this notation, the general form of the multiple regression equation with two predictors is

$$\hat{Y} = b_1X_1 + b_2X_2 + a \tag{17.13}$$

If all three variables, X_1, X_2, and Y, have been standardized by transformation into z-scores, then the standardized form of the multiple regression equation predicts the z-score for each Y value. The standardized form is

$$\hat{z}_Y = (\text{beta}_1)z_{X1} + (\text{beta}_2)z_{X2} \tag{17.14}$$

Researchers rarely transform raw X and Y scores into z-scores before finding a regression equation, however, the beta values are meaningful and are usually reported by computer programs conducting multiple regression. We return to the discussion of beta values later in this section.

The goal of the multiple-regression equation is to produce the most accurate estimated values for Y. As with the single predictor regression, this goal is accomplished with a least-squared solution. First, we define "error" as the difference between the predicted Y value and the actual Y value for each individual. Each error is then squared to produce uniformly positive values, and then we add the squared errors. Finally, we calculate values for b_1, b_2, and a that produce the smallest possible sum of squared errors. The derivation of the final values is beyond the scope of this text, but the final equations are as follows:

$$b_1 = \frac{(SP_{X1Y})(SS_{X2}) - (SP_{X1X2})(SP_{X2Y})}{(SS_{X1})(SS_{X2}) - (SP_{X1X2})^2} \qquad (17.15)$$

$$b_2 = \frac{(SP_{X2Y})(SS_{X1}) - (SP_{X1X2})(SP_{X1Y})}{(SS_{X1})(SS_{X2}) - (SP_{X1X2})^2} \qquad (17.16)$$

$$a = M_Y - b_1 M_{X1} - b_2 M_{X2} \qquad (17.17)$$

In these equations, you should recognize the following SS and SP values:

SS_{X1} is the sum of squared deviations for X_1

SS_{X2} is the sum of squared deviations for X_2

SP_{X1Y} is the sum of products of deviations for X_1 and Y

SP_{X2Y} is the sum of products of deviations for X_2 and Y

SP_{X1X2} is the sum of products of deviations for X_1 and X_2

Note: More detailed information about the calculation of SS is presented in Chapter 4 (pages 113–115) and information concerning SP is in Chapter 16 (pages 526–528). The following example demonstrates multiple regression with two predictor variables.

EXAMPLE 17.5   We use the data in Table 17.3 to demonstrate multiple regression. Note that each individual has a Y score and two X scores that are used as predictor variables. Also

TABLE 17.3

Hypothetical data consisting of three scores for each person. Two of the scores, X_1 and X_2, are used to predict the Y score for each individual.

Person	Y	X_1	X_2	
A	11	4	10	$SP_{X1Y} = 52$
B	5	5	6	$SP_{X2Y} = 47$
C	7	3	7	$SP_{X1X2} = 35$
D	3	2	4	
E	4	1	3	
F	12	7	5	
G	10	8	8	
H	4	2	4	
I	8	6	10	
J	6	2	3	

$M_Y = 7$ $M_{X1} = 4$ $M_{X2} = 6$
$SS_Y = 90$ $SS_{X1} = 52$ $SS_{X2} = 64$

note that we have already computed the SS values for Y and for both of the X scores, as well as all of the SP values. These values are used to compute the coefficients, b_1 and b_2, and the constant, a, for the regression equation.

$$\hat{Y} = b_1X_1 + b_2X_2 + a$$

$$b_1 = \frac{(SP_{X1Y})(SS_{X2}) - (SP_{X1X2})(SP_{X2Y})}{(SS_{X1})(SS_{X2}) - (SP_{X1X2})^2} = \frac{(52)(64) - (35)(47)}{(52)(64) - (35)^2} = 0.800$$

$$b_2 = \frac{(SP_{X2Y})(SS_{X1}) - (SP_{X1X2})(SP_{X1Y})}{(SS_{X1})(SS_{X2}) - (SP_{X1X2})^2} = \frac{(47)(52) - (35)(52)}{(52)(64) - (35)^2} = 0.297$$

$$a = M_Y - b_1M_{X1} - b_2M_{X2} = 7 - 0.800(4) - 0.297(6) = 7 - 3.2 - 1.782 = 2.018$$

Thus, the final regression equation is,

$$\hat{Y} = 0.800X_1 + 0.297X_2 + 2.018$$

LEARNING CHECK

1. A researcher computes a multiple-regression equation for predicting annual income for 40-year-old men based on their level of education (X_1 = number of years after high school) and their social skills (X_2 = score from a self-report questionnaire). The regression equation is $\hat{Y} = 8.3X_1 + 2.1X_2 + 3.5$ and predicts income in thousands of dollars. Two individuals are selected from the sample. One has $X_1 = 0$ and $X_2 = 16$; the other has $X_1 = 3$ and $X_2 = 12$. Compute the predicted income for each.

ANSWER

1. The first man has a predicted income of $\hat{Y} = 37.1$ thousand dollars and the second has $\hat{Y} = 53.6$ thousand dollars.

PERCENTAGE OF VARIANCE ACCOUNTED FOR AND RESIDUAL VARIANCE

In the same way that we computed an r^2 value to measure the percentage of variance accounted for with the single-predictor regression, it is possible to compute a corresponding percentage for multiple regression. For a multiple-regression equation, this percentage is identified by the symbol R^2. The value of R^2 describes the proportion of the total variability of the Y scores that is accounted for by the regression equation. In symbols,

$$R^2 = \frac{SS_{\text{regression}}}{SS_Y} \quad \text{or} \quad SS_{\text{regression}} = R^2SS_Y$$

For a regression equation with two predictor variables, R^2 can be computed directly using the following equation

$$R^2 = \frac{b_1 SP_{X1Y} + b_2 SP_{X2Y}}{SS_Y} \qquad (17.18)$$

For the data in Table 17.2, we obtain a value of

$$R^2 = \frac{0.800(52) + 0.297(47)}{90} = \frac{55.559}{90} = 0.617 \text{ (or 61.7\%)}$$

COMPUTING R^2 AND $1 - R^2$ FROM THE RESIDUALS   The value of R^2 can also be obtained indirectly, by computing the residual, or difference between the predicted Y and the actual Y for each individual, then computing the sum of the squared residuals. The resulting value is $SS_{residual}$ and measures the unpredicted portion of the variability of Y, which is equal to $(1 - R^2)SS_Y$. For the data in Table 17.3, we first use the multiple-regression equation to compute the predicted value of Y for each individual. The process of finding and squaring each residual is shown in Table 17.4.

Note that the sum of the squared residuals, the unpredicted portion of SS_Y, is 34.44. This value corresponds to 38.3% of the variability for the Y scores:

$$\frac{SS_{residual}}{SS_Y} = \frac{34.44}{90} = 0.383 \text{ or } 38.3\%$$

Because the unpredicted portion of the variability is $1 - R^2 = 38.3\%$, we conclude once again that the predicted portion is $R^2 = 61.7\%$.

THE STANDARD ERROR OF ESTIMATE   On page 571 we defined the standard error of estimate for a linear regression equation as the standard distance between the regression line and the actual data points. In more general terms, the standard error of estimate can be defined as the standard distance between the predicted Y values (from the regression equation) and the actual Y values (in the data). The more general definition applies equally well to both linear and multiple regression.

TABLE 17.4

The predicted Y values and the residuals for the data in Table 17.3. The predicted Y values were obtained using the values of X_1 and X_2 in the multiple-regression equation for each individual.

Actual Y	Predicted Y $\hat{Y}$	Residual $(Y - \hat{Y})$	Squared Residual $(Y - \hat{Y})^2$
11	8.18	2.81	7.91
5	7.80	−2.80	7.84
7	6.50	.50	0.25
3	4.81	−1.81	3.26
4	3.71	0.29	0.08
12	9.10	2.90	8.39
10	10.79	−0.79	0.63
4	4.81	−0.81	0.65
8	9.79	−1.79	3.20
6	4.51	1.49	2.22

$$34.44 = SS_{residuals}$$

For either linear regression or multiple regression, part of the computation involves finding $SS_{residual}$. For linear regression with one predictor, $SS_{residual} = (1 - r^2)SS_Y$ and has $df = n - 2$. For multiple regression with two predictors, $SS_{residual} = (1 - R^2)SS_Y$ and has $df = n - 3$. In each case, we can use the SS and df values to compute a variance or $MS_{residual}$.

$$MS_{residual} = \frac{SS_{residual}}{df}$$

The variance or MS value is a measure of the average squared distance between the actual Y values and the predicted Y values. By simply taking the square root, we obtain a measure of standard deviation or standard distance. This standard distance for the residuals is exactly what is measured by the standard error of estimate. Thus, for both linear regression and multiple regression

$$\text{the standard error of estimate} = \sqrt{MS_{residual}}$$

For either type of regression, you do not expect the predictions from the regression equation to be perfect. In general, there will be some discrepancy between the predicted values of Y and the actual values. The standard error of estimate provides a measure of how much discrepancy, on average, there is between the $\hat{Y}$ values and the actual Y values.

TESTING THE SIGNIFICANCE OF THE MULTIPLE REGRESSION EQUATION: ANALYSIS OF REGRESSION

Just as we did with the single predictor equation, we can evaluate the significance of a multiple-regression equation by computing an F-ratio to determine whether the equation predicts a significant portion of the variance for the Y scores. The total variability of the Y scores is partitioned into two components, $SS_{regression}$ and $SS_{residual}$. With two predictor variables, $SS_{regression}$ has $df = 2$, and $SS_{residual}$ has $df = n - 3$. Therefore, the two MS values for the F-ratio are

$$MS_{regression} = \frac{SS_{regression}}{2} \qquad (17.19)$$

and

$$MS_{residual} = \frac{SS_{residual}}{n - 3} \qquad (17.20)$$

The data for the $n = 10$ people in Table 17.2 have produced $R^2 = 0.617$ (or 61.7%) and $SS_Y = 90$. Thus,

$$SS_{regression} = R^2 SS_Y = 0.617(90) = 55.53$$

$$SS_{residual} = (1 - R^2)SS_Y = 0.383(90) = 34.47$$

Because of rounding error, the value we obtain for $SS_{residual}$ is slightly different from the value in Table 17.4.

Therefore, $\quad MS_{regession} = \dfrac{55.53}{2} = 27.77 \quad$ and $\quad MS_{residual} = \dfrac{34.47}{7} = 4.92$

and $\quad F = \dfrac{MS_{regression}}{MS_{residual}} = \dfrac{27.77}{4.92} = 5.64$

With $df = 2, 7$, this F-ratio is significant with $\alpha = .05$, so we can conclude that the regression equation accounts for a significant portion of the variance for the Y scores.

The analysis of regression is summarized in the following table, which is a common component of the output from most computer versions of multiple regression.

Source	SS	df	MS	F
Regression	55.53	2	27.77	5.64
Residual	34.47	7	4.92	
Total	90.00	9		

LEARNING CHECK

1. Data from a sample of $n = 25$ individuals are used to compute a multiple-regression equation with two predictor variables. If the equation has $R^2 = 0.27$ and $SS_Y = 384$, find $SS_{regression}$ and $SS_{residual}$ and compute the F-ratio to evaluate the significance of the regression equation.

ANSWER

1. $SS_{regression} = 103.68$ with $df = 2$. $SS_{residual} = 280.32$ with $df = 22$. $F = 4.07$. With $df = 2$, 22 the F-ratio is significant with $\alpha = .05$.

17.4 EVALUATING THE CONTRIBUTION OF EACH PREDICTOR VARIABLE

In addition to evaluating the overall significance of the multiple-regression equation, researchers are often interested in the relative contribution of each of the two predictor variables. Is one of the predictors responsible for more of the prediction than the other? Unfortunately, the b values in the regression equation are influenced by a variety of other factors and do not address this issue. If b_1 is larger than b_2, it does not necessarily mean that X_1 is a better predictor than X_2. However, in the standardized form of the regression equation, the relative size of the beta values is an indication of the relative contribution of the two variables. For the data in Table 17.3, the standardized regression equation is

$$\hat{z}_Y = (beta_1)z_{X1} + (beta_2)z_{X2}$$

$$= 0.608z_{X1} + 0.250z_{X2} \qquad (17.21)$$

In this case, the larger beta value for the X_1 predictor indicates that X_1 predicts more of the variance than does X_2. The signs of the beta values are also meaningful. In this example, both betas are positive indicating the both X_1 and X_2 are positively related to Y.

Beyond judging the relative contribution for each of the predictor variables, it also is possible to evaluate the significance of each contribution. For example, does variable X_2 make a significant contribution beyond what is already predicted by variable X_1? The null hypothesis states that the multiple-regression equation (using X_2 in addition to X_1) is not any better than the simple regression equation using X_1 as a single predictor variable. An alternative view of the null hypothesis is that the b_2 (or $beta_2$) value in the equation is not significantly different from zero.

To test this hypothesis, we first determine how much more variance is predicted using X_1 and X_2 together than is predicted using X_1 alone.

For the data in Table 17.2, the correlation between X_1 and Y can be computed as,

$$r = \frac{SP_{X1Y}}{\sqrt{(SS_{X1})(SS_Y)}} = \frac{52}{\sqrt{(52)(90)}} = \frac{52}{68.41} = 0.760$$

Thus, $r^2 = (0.760)^2 = 0.578$ or 57.8%. This means that the relationship with X_1 predicts 57.8% of the variance for the Y scores. The total variability for the Y scores is $SS_Y = 90$, so the portion predicted by X_1 is 57.8% of 90, which is $SS_{\text{regression with }X1\text{ alone}}$ = 52.02. Earlier we found that the predicted variance using the multiple-regression equation was $SS_{\text{regression with }X1\text{ and }X2} = 55.56$. Therefore, the contribution made by adding X_2 to the regression equation can be computed as

$$SS \text{ contributed by } X_2 = SS_{\text{additional}}$$

$$= SS_{\text{regression with }X1\text{ and }X2} - SS_{\text{regression with }X1\text{ alone}}$$

$$= 55.56 - 52.02$$

$$= 3.54$$

This SS value has $df = 1$, and can be used to compute an F-ratio evaluating the significance of the contribution of X_2. First,

$$MS_{\text{additional}} = \frac{SS_{\text{additional}}}{1} = \frac{3.54}{1} = 3.54$$

This MS value is evaluated by computing an F-ratio with the MS_{residual} value from the multiple regression as the denominator. (*Note:* This is the same denominator that was used in the F-ratio to evaluate the significance of the multiple-regression equation.) For these data, we obtain

In computer printouts, the test for the contribution of each variable is often reported as a t statistic, where $t = \sqrt{F}$.

$$F = \frac{MS_{\text{additional}}}{MS_{\text{residual}}} = \frac{3.54}{4.92} = 0.720$$

With $df = 1, 7$, this F-ratio is not significant. Therefore, we conclude that adding X_2 to the regression equation does not significantly improve the prediction compared to the regression equation using X_1 as a single predictor.

USING MULTIPLE REGRESSION TO CONTROL A THIRD VARIABLE

As we demonstrated in the previous section, it is possible to add a second predictor variable to a regression equation and see how much the new variable contributes to the prediction after the influence of the first predictor has been considered. One interesting application of this process of adding predictor variables one at a time is to examine the relationship between two variables while eliminating the influence of a third variable that might otherwise interfere with the relationship. As an example, consider research examining the relationship between exposure to sexual content on TV and sexual behavior among adolescents (Collins, et al., 2004). The study consisted of a survey of 1792 adolescents, 12 to 17 years old, who reported their TV viewing habits and their sexual behaviors. The results showed a clear relationship between TV viewing and behaviors. Specifically, the more sexual content the adolescents saw on TV, the more likely they were to engage in sexual behaviors. One concern for the researchers was that the observed relationship may be influenced by the age of the participants. For example, as the adolescents mature from age 12 to age 17, they increasingly watch TV programs with sexual content, and they increase their own sexual behaviors. Thus, the viewing of sexual content on TV and the participants' sexual behaviors are increasing together; however, the observed relationship may simply be the result of age differences. To address this problem, the researcher used multiple regression. First, age was used as a single variable to predict sexual behaviors, then TV sexual content was added as a second predictor variable. The results clearly showed that TV sexual content was still a significant predictor of sexual behavior even after the influence of the participants' ages was accounted for.

MULTIPLE REGRESSION AND PARTIAL CORRELATIONS

Although multiple regression provides researchers with a method for controlling a third variable, a more direct procedure is to calculate a partial correlation. A *partial correlation* allows researchers to measure the relationship between two variables while eliminating or holding constant the influence of a third variable. This can be a valuable tool for researchers who suspect that the relationship between two variables is being influenced or distorted by a third variable.

A **partial correlation** measures the relationship between two variables while controlling the influence of a third variable by holding it constant.

In the previous section, we discussed a research study examining the relationship between sexual content observed on TV and the sexual behavior of adolescents. However, the researchers were concerned that both variables might be influenced by age. As the age of the adolescents increases from 12 to 17 years, the amount of sexual content watched on TV and the amount of sexual behavior also increase. A partial correlation would allow researchers to determine the correlation between TV content and behavior while eliminating the influence of age by effectively holding the age variable constant.

In a situation with three variables, X, Y, and Z, it is possible to compute three individual Pearson correlations:

1. r_{XY} measuring the correlation between X and Y
2. r_{XZ} measuring the correlation between X and Z
3. r_{YZ} measuring the correlation between Y and Z

These three individual correlations can then be used to compute a partial correlation. For example, the partial correlation between X and Y, holding Z constant, is determined by the formula

$$r_{XY \cdot Z} = \frac{r_{XY} - (r_{XY}\, r_{YZ})}{\sqrt{(1 - r^2_{XZ})\,(1 - r^2_{YZ})}}$$

The following examples demonstrate the calculation and interpretation of partial correlations. In each case, a third variable influences both X and Y to produce what appears to be a strong positive correlation between X and Y. In one case, however, there is no relationship between X and Y when the third variable is held constant (the partial correlation is zero). In another case, the partial correlation between X and Y is still positive when the third variable is controlled, and in the final case, the partial correlation between X and Y is negative.

EXAMPLE 17.6   We begin with the data shown in Figure 17.9. For these scores, the individual Pearson correlations are all strong and positive: $r_{XY} = 0.943$, $r_{XZ} = 0.971$, and $r_{YZ} = 0.971$. However, the figure reveals that the third variable, Z, separates the scores into three distinct groups: when $Z = 1$, the X and Y scores are low; when $Z = 2$, the X and Y scores are moderate; and when $Z = 3$ the X and Y scores are large. Thus, as the X values increase, the Y values also increase, and the result is a strong positive correlation between X and Y. Within each of the three groups, however, there is no linear relationship between X and Y. If the group differences are eliminated by holding variable Z constant, the partial correlation between X and Y is

$$r_{XY \cdot Z} = \frac{0.943 - 0.971(0.971)}{\sqrt{(1 - 0.971^2)(1 - 0.971^2)}}$$

$$= \frac{0}{0.00325}$$

$$= 0$$

When the group differences are eliminated, the remaining correlation between X and Y is $r = 0$. The group differences, which correspond to the different values of the Z variable, are eliminated mathematically in the calculation of the partial correlation. However, it is possible to visualize how these differences are eliminated in the actual data. For the scores in Figure 17.9, the overall mean for both X and Y is $M = 10$. For the four individuals with $Z = 1$, however, the mean for both X and Y is only $M = 5$. Thus, these four individuals average 5 points below the mean. We can correct for this group difference by simply adding 5 points to each individual's X score and Y score. After the 5 points are added, the scores for this group are identical to the scores for the four individuals who have $Z = 2$. Looking at Figure 17.9, imagine that the four points in the bottom left corner are moved up and right so that they overlap the four points in the center of the figure.

Now consider the four individuals with $Z = 3$. For this group, the mean for both X and Y is $M = 15$, which is 5 points above the overall mean. To correct for this group difference, we subtract 5 points from each individual's X score and Y score. After the subtraction, the scores for this group are also identical to the scores for the four individuals with $Z = 2$. This time, imagine the four points in the upper right corner being shifted down and left so that they also overlap with the four points in the center of the graph.

After completing the addition and subtraction, there no longer are any group differences. All three groups are exactly the same and correspond to the four points in the center of Figure 17.9. It should be clear that the correlation for these four points is $r = 0$.

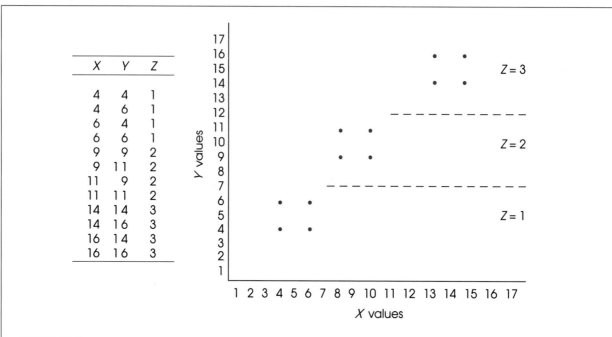

FIGURE 17.9

Data showing a strong positive correlation between X and Y. However, the relationship between X and Y is entirely explained by the third variable Z. When the effect of Z is eliminated, the partial correlation between X and Y is $r = 0$.

The fact that the partial correlation between X and Y is equal to zero is confirmed by the multiple-regression equation using X and Z to predict Y. The general form of the equation is

$$\hat{Y} = b_1 X + b_2 Z + a$$

For these data $b_1 = 0$, $b_2 = 5$, and $a = 0$. The fact that $b_1 = 0$ (and the corresponding beta equals zero) indicates that the X variable makes no contribution to predicting Y beyond the prediction already made by variable Z. In other words, there is no relationship between X and Y except for their common association with Z.

EXAMPLE 17.7   This time we will use the scores presented in Figure 17.10. Once again, the individual Pearson correlations are all strong and positive: $r_{XY} = 0.974$, $r_{XZ} = 0.933$, and $r_{YZ} = 0.933$. As before, the variable Z separates the X and Y scores into three groups: the lowest four scores for both X and Y occur when $Z = 1$, the middle four scores occur with $Z = 2$, and the highest X and Y scores occur with $Z = 3$. The fact that X and Y are both systematically related to Z is responsible for part of the relationship between X and Y. This time, however, there is still a strong, positive relationship between X and Y within each of the three groups. For these data, if the group differences are eliminated by holding variable Z constant, the partial correlation between X and Y is

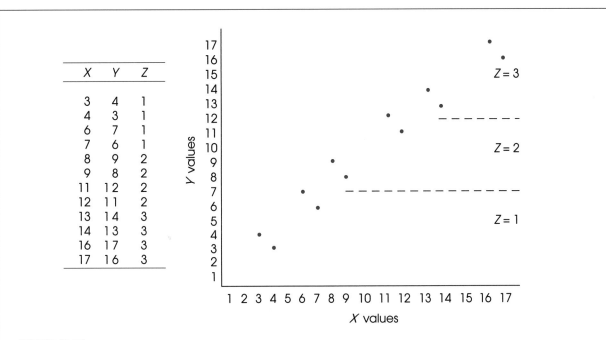

FIGURE 17.10

Data showing a strong positive correlation between X and Y. Part of the relationship is explained by the third variable, Z. However, when the effect of Z is eliminated, there is still a strong positive correlation between X and Y. The partial correlation between X and Y, holding Z constant, is $r = +0.80$.

$$r_{XY \cdot Z} = \frac{0.974 - 0.933(0.933)}{\sqrt{(1 - 0.933^2)(1 - 0.933^2)}}$$

$$= \frac{0.104}{0.130} = 0.800$$

When the group differences are eliminated, the remaining correlation between X and Y is $r = 0.800$. Again, you can visualize the elimination of the group differences by looking at Figure 17.10. First, imagine that the four points in the bottom left corner are shifted up and to the right so that they overlap the four points in the center of the figure. Then, imagine that the four points in the top right corner are shifted down and left so that they also overlap the four points in the center. When the group differences are removed, it should be clear that the resulting data produce a positive correlation between X and Y.

The fact that the partial correlation between X and Y is large and positive is confirmed by the multiple-regression equation using X and Z to predict Y. The general form of the equation is

$$\hat{Y} = b_1 X + b_2 Z + a$$

For these data $b_1 = 0.8$, $b_2 = 1.00$, and $a = 0$. The fact that $b_1 = 0.8$ indicates that the X variable makes a contribution to predicting Y beyond the prediction already made by variable Z. More importantly, the beta value for the X variable is also 0.80, which indicates a strong, positive relationship between X and Y when the variable Z is controlled. (*Note:* The beta value gives an indication of the direction and the relative size of the partial correlation, but usually is not equal to the partial correlation. The two values happen to be identical for this example because the data were designed to provide an extreme demonstration.)

EXAMPLE 17.8   Finally, consider the data shown in Figure 17.11. As before, the individual Pearson correlations are all strong and positive: $r_{XY} = 0.765$, $r_{XZ} = 0.933$, and $r_{YZ} = 0.933$. Again, the variable Z separates the X and Y scores into three groups: the lowest four scores for both X and Y occur when $Z = 1$; the middle four scores occur with $Z = 2$; and the highest X and Y scores occur when $Z = 3$. The fact that X and Y are both systematically related to Z accounts for the positive relationship between X and Y. Within each group, however, there is a strong, negative relationship between X and Y. For these data, the partial correlation between X and Y is

$$r_{XY \cdot Z} = \frac{0.765 - 0.933(0.933)}{\sqrt{(1 - 0.933^2)(1 - 0.933^2)}}$$

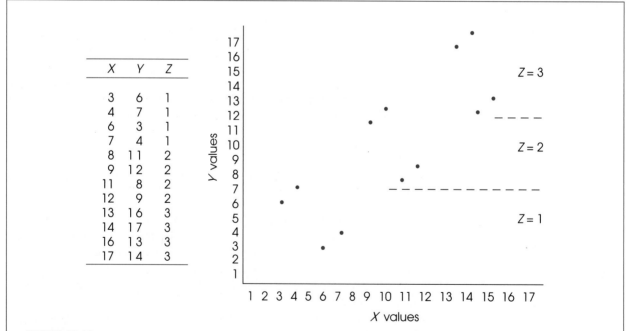

FIGURE 17.11

Data showing a strong positive correlation between X and Y. However, the positive relationship is entirely explained by the third variable, Z. When the effect of Z is eliminated, there is a strong negative correlation between X and Y. The partial correlation between X and Y, holding Z constant, is $r = -0.80$.

$$= \frac{-0.105}{0.130} = -0.808$$

The true value for this partial correlation is $r = -0.800$. The computed value is the result of rounding error from the multiple calculations required in the formula.

Eliminating, the group differences for these data produces a partial correlation between X and Y of $r = -0.808$. Again, you can visualize the elimination of the group differences by imagining that the four points in the lower left corner of Figure 17.11 are moved up and right, and the four points in the upper right corner are moved down and left so that all three groups end up being identical to the four points in the center of the figure. It should be clear that these four points will produce a negative correlation.

The fact that the partial correlation between X and Y is large and negative is confirmed by the multiple-regression equation using X and Z to predict Y. The general form of the equation is

$$\hat{Y} = b_1X + b_2Z + a$$

For these data $b_1 = -0.8$, $b_2 = 9.00$, and $a = 0$. The fact that $b_1 = -0.8$ indicates that the X variable makes a contribution to predicting Y beyond the prediction already made by variable Z. More importantly, the beta value for the X variable is -0.80, which indicates a strong, negative relationship between X and Y when the variable Z is controlled.

LEARNING CHECK

1. Sales figures show a positive relationship between temperature and ice cream consumption; as temperature increases, ice cream consumption also increases. Other research shows a positive relationship between temperature and crime rate (Cohn & Rotton, 2000). As a result, when the temperature increases, both ice cream consumption and crime rates tend to increase. As a result, there is a positive correlation between ice cream consumption and crime rate. However, what do you think is the true relationship between ice cream consumption and crime rate? Specifically, what value would you predict for the partial correlation between the two variables if temperature were held constant?

ANSWERS

1. There should be no systematic relationship between ice cream consumption and crime rate. The partial correlation should be near zero.

SUMMARY

1. When there is a general linear relationship between two variables, X and Y, it is possible to construct a linear equation that allows you to predict the Y value corresponding to any known value of X:

predicted Y value $= \hat{Y} = bX + a$

The technique for determining this equation is called regression. By using a *least-squares* method to minimize the error between the predicted Y values and the actual Y values, the best-fitting line is achieved when the linear equation has

$$b = \frac{SP}{SS_X} = r\frac{s_Y}{s_X} \quad \text{and} \quad a = M_Y - bM_X$$

2. The linear equation generated by regression (called the regression equation) can be used to compute a predicted Y value for any value of X. However, the prediction is not perfect, so for each Y value, there is a predicted portion and an unpredicted, or residual, portion. Overall, the predicted portion of the Y score variability is measured by r^2, and the residual portion is measured by $1 - r^2$.

Predicted variability $= SS_{\text{regression}} = r^2SS_Y$

Unpredicted variability $= SS_{\text{residual}} = (1 - r^2)SS_Y$

3. The residual variability can be used to compute the standard error of estimate, which provides a measure of the standard distance (or error) between the predicted Y values on the line and the actual data points. The standard error of estimate is computed by

$$\text{standard error of estimate} = \sqrt{\frac{SS_{\text{residual}}}{n-2}}$$

$$= \sqrt{MS_{\text{residual}}}$$

4. It is also possible to compute an F-ratio to evaluate the significance of the regression equation. The process is called analysis of regression and determines whether the equation predicts a significant portion of the variance for the Y scores. First a variance, or MS, value is computed for the predicted variability and the residual variability,

$$MS_{\text{regression}} = \frac{SS_{\text{regression}}}{df_{\text{regression}}} \qquad MS_{\text{residual}} = \frac{SS_{\text{residual}}}{df_{\text{residual}}}$$

where $df_{\text{regression}} = 1$ and $df_{\text{residual}} = n - 2$. Next, an F-ratio is computed to evaluate the significance of the regression equation.

$$F = \frac{MS_{\text{regression}}}{MS_{\text{residual}}} \qquad \text{with } df = 1, n - 2$$

5. Multiple regression involves finding a regression equation with more than one predictor variable. With two predictors (X_1 and X_2), the equation becomes

$$\hat{Y} = b_1X_1 + b_2X_2 + a$$

with the values for b_1, b_2, and a computed using equations 17.15, 17.16, and 17.17.

6. For multiple regression, the value of R^2 describes the proportion of the total variability of the Y scores that is accounted for by the regression equation. With two predictor variables,

$$R^2 = \frac{b_1SP_{X1Y} + b_2SP_{X2Y}}{SS_Y}$$

Predicted variability $= SS_{\text{regression}} = R^2SS_Y$

Unpredicted variability $= SS_{\text{residual}} =$ $(1 - R^2)SS_Y$

7. The residual variability for the multiple-regression equation can be used to compute a standard error of estimate, which provides a measure of the standard distance (or error) between the predicted Y values from the equation and the actual data points. For multiple regression with two predictors, the standard error of estimate is computed by

$$\text{standard error of estimate} = \sqrt{\frac{SS_{\text{residual}}}{n-3}}$$

$$= \sqrt{MS_{\text{residual}}}$$

8. Evaluating the significance of the two-predictor multiple-regression equation involves computing an F-ratio that divides the $MS_{\text{regression}}$ (with $df = 2$) by the MS_{residual} (with $df = n - 3$). A significant F-ratio indicates that the regression equation accounts for a significant portion of the variance for the Y scores.

9. An F-ratio can also be used to determine whether a second predictor variable (X_2) significantly improves the prediction beyond what was already predicted by X_1. The numerator of the F-ratio measures the additional SS that is predicted by adding X_2 as a second predictor.

$$SS_{\text{additional}} = SS_{\text{regression with X1 and X2}} - SS_{\text{regression with X1 alone}}$$

This SS value has $df = 1$. The denominator of the F-ratio is the MS_{residual} from the two-predictor regression equation.

10. A partial correlation measures the relationship that exists between two variables after a third variable has been controlled or held constant (see Equation 17.21).

KEY TERMS

linear relationship	regression	standard error of estimate	partial correlation
linear equation	regression line	predicted variability ($SS_{regression}$)	
slope	least-squared-error solution	unpredicted variability ($SS_{residual}$)	
Y-intercept	regression equation for Y	multiple regression	

RESOURCES

There are a tutorial quiz and other learning exercises for Chapter 17 at the Wadsworth website, www.cengage.com/psychology/gravetter. The site also provides access to a workshop entitled Correlation that includes information on regression. See page 29 for more information about accessing items on the website.

ENHANCED WebAssign

Guided interactive tutorials, end-of-chapter problems, and related testbank items may be assigned online at WebAssign.

WebTUTOR

For those using WebTutor along with this book, there is a WebTutor section corresponding to this chapter. The WebTutor contains a brief summary of Chapter 17, hints for learning the concepts of and the formulas for regresion, cautions about common errors, and sample exam items including solutions.

SPSS

General instructions for using SPSS are presented in Appendix D. Following are detailed instructions for using SPSS to perform the **Linear Regression, Multiple Regression,** and **Partial Correlation** presented in this chapter.

Linear and Multiple Regression

Data Entry

1. With one predictor variable (X), you enter the X values in one column and the Y values in a second column of the SPSS data editor. With two predictors (X_1 and X_2), enter the X_1 values in one column, X_2 in a second column, and Y in a third column.

Data Analysis

1. Click **Analyze** on the tool bar, select **Regression,** and click on **Linear.**
2. In the left-hand box, highlight the column label for the Y values, then click the arrow to move the column label into the **Dependent Variable** box.
3. For one predictor variable, highlight the column label for the X values and click the arrow to move it into the **Independent Variable(s)** box. For two predictor variables, highlight the X_1 and X_2 column labels, one at a time, and click the arrow to move them into the **Independent Variable(s)** box.
4. Click **OK.**

SPSS Output

The first table in the output simply lists the predictor variables that were entered into the regression equation. The second table (Model Summary) presents the values for R, R^2, and the standard error of estimate. (*Note:* For a single predictor, R is simply the Pearson correlation between X and Y.) The third table (ANOVA) presents the analysis of regression evaluating the significance of the regression equation, including the F-ratio and the level of significance (the p value or alpha level for the test). The final table, summarizes the unstandardized and the standardized coefficients for the regression equation. For one predictor, the table shows the values for the constant (a) and the coefficient (b). For two predictors, the table shows the constant (a) and the two coefficients (b_1 and b_2). The standardized coefficients are the beta values. For one predictor, beta is simply the Pearson correlation between X and Y. Finally, the table uses a t statistic to evaluate the significance of each predictor variable. For one predictor variable, this is identical to the significance of the regression equation and you should find that t is equal to the square root of the F-ratio from the analysis of regression. For two predictor variables, the t values measure the significance of the contribution of each variable beyond what is already predicted by the other variable.

Partial Correlation

Data Entry

1. Enter the X values in one column, the Y values in a second column, and the Z values in a third column of the SPSS data editor. The three scores for each participant should all be in the same row.

Data Analysis

1. Click **Analyze** on the tool bar, select **Correlate**, and click on **Partial**.
2. One at a time, highlight the column labels for X and Y in the left-hand column and move them into the **Variables** box.
3. Highlight the column label for Z and move it into the **Controlling for** box.
4. Click **OK**.

SPSS Output

The program produces a correlation matrix showing all the possible correlations, including the correlation of X with X and the correlation of Y with Y. You want the partial correlation of X and Y which is in the upper right-hand corner of the matrix. Beneath the correlation is the significance level. A significance level less than .05 indicates a significant correlation.

FOCUS ON PROBLEM SOLVING

1. A basic understanding of the Pearson correlation, including the calculation of SP and SS values, is critical for understanding and computing regression equations.

2. You can calculate $SS_{residual}$ directly by finding the residual (the difference between the actual Y and the predicted Y for each individual), squaring the residuals, and adding the squared values. However, it usually is much easier to compute r^2 (or R^2) and then find $SS_{residual} = (1 - r^2)SS_Y$.

3. The F-ratio for analysis of regression is usually calculated using the actual $SS_{regression}$ and $SS_{residual}$. However, you can simply use r^2 (or R^2) in place of $SS_{regression}$ and you can use $1 - r^2$ or $(1 - R^2)$ in place of $SS_{residual}$. *Note:* You must still use the correct df value for the numerator and the denominator.

DEMONSTRATION 17.1

LINEAR REGRESSION

The same data that were used to demonstrate the Pearson correlation will be used to demonstrate the process of linear regression (see Demonstration 16.1 on page 555). The data and summary statistics are as follows:

Person	X	Y
A	0	4
B	2	1
C	8	10
D	6	9
E	4	6

$M_X = 4$ with $SS_X = 40$
$M_Y = 6$ with $SS_Y = 54$
$SP = 40$

These data produced a Pearson correlation of $r = 0.861$.

STEP 1 *Compute the values for the regression equation.* The general form of the regression equation is

$$\hat{Y} = bX + a \quad \text{where } b = \frac{SP}{SS_X} \quad \text{and} \quad a = M_Y - bM_X$$

For these data, $b = \dfrac{40}{40} = 1.00$ and $a = 6 - 1(4) = +2.00$

Thus, the regession equation is $\hat{Y} = (1)X + 2.00$ or simply, $\hat{Y} = X + 2$

STEP 2 *Evaluate the significance of the regression equation.* The null hypothesis states that the regression equation does not predict a significant portion of the variance for the Y scores. To conduct the test, the total variability for the Y scores, $SS_Y = 54$, is partitioned into the portion predicted by the regression equation and the residual portion.

$$SS_{\text{regression}} = r^2(SS_Y) = 0.741(54) = 40.01 \qquad \text{with } df = 1$$

$$SS_{\text{residual}} = (1 - r^2)(SS_Y) = 0.259(54) = 13.99 \qquad \text{with } df = n - 2 = 3$$

The two *MS* values (variances) for the *F*-ratio are

$$MS_{\text{regression}} = \frac{SS_{\text{regression}}}{df} = \frac{40.01}{1} = 40.01$$

$$MS_{\text{residual}} = \frac{SS_{\text{residual}}}{df} = \frac{13.99}{3} = 4.66$$

And the *F*-ratio is

$$F = \frac{MS_{\text{regression}}}{MS_{\text{residual}}} = \frac{40.01}{4.66} = 8.59$$

With $df = 1, 3$ and $\alpha = .05$, the critical value for the *F*-ratio is 10.13. Therefore, we fail to reject the null hypothesis and conclude that the regression equation does not predict a significant portion of the variance for the *Y* scores.

DEMONSTRATION 17.2

MULTIPLE REGRESSION

The following data will be used to demonstrate the process of multiple regression. Note that there are two predictor variables, X_1 and X_2, that will be used to compute a predicted *Y* score for each individual.

Person	X_1	X_2	Y
A	0	5	2
B	3	1	4
C	5	2	7
D	6	0	9
E	8	4	5
F	2	6	3

STEP 1 *Compute the values for the multiple regression equation.* The general form of the multiple-regression equation is

$$\hat{Y} = b_1X_1 + b_2X_2 + a$$

To compute the values for b_1, b_2, and *a*, we will need to compute *SS* for X_1 and X_2 as well as several sum-of-products (*SP*) values. The necessary sums for these values are shown in the following table

X_1	X_2	Y	Y^2	X_1^2	X_2^2	X_1Y	X_2Y	X_1X_2
0	5	2	4	0	25	0	10	0
3	1	4	16	9	1	12	4	3
5	2	7	49	25	4	35	14	10
6	0	9	81	36	0	54	0	0
8	4	5	25	64	16	40	20	32
2	6	3	9	4	36	6	18	12
Sums 24	18	30	184	138	82	147	66	57

Using these sums, we obtain

$$SS_{X1} = \Sigma X_1^2 - \frac{(\Sigma X_1)^2}{n} = 138 - \frac{24^2}{6} = 42$$

$$SS_{X2} = \Sigma X_2^2 - \frac{(\Sigma X_2)^2}{n} = 82 - \frac{18^2}{6} = 28$$

$$SP_{X1Y} = \Sigma X_1 Y - \frac{(\Sigma X_1)(\Sigma Y)}{n} = 147 - \frac{24(30)}{6} = 27$$

$$SP_{X2Y} = \Sigma X_1 Y - \frac{(\Sigma X_2)(\Sigma Y)}{n} = 66 - \frac{18(30)}{6} = -24$$

$$SS_Y = \Sigma Y^2 - \frac{(\Sigma Y)^2}{n} = 184 - \frac{30^2}{6} = 34$$

$$SP_{X2X2} = \Sigma X_1 X_2 - \frac{(\Sigma X_1)(\Sigma X_2)}{n} = 57 - \frac{24(18)}{6} = -15$$

The values for the multiple regression equation are

$$b_1 = \frac{(SP_{X1Y})(SS_{X2}) - (SP_{X1X2})(SP_{X2Y})}{(SS_{X1})(SS_{X2}) - (SP_{X1X2})^2} = \frac{(27)(28) - (-15)(-24)}{(42)(28) - (-15)^2} = 0.416$$

$$b_2 = \frac{(SP_{X2Y})(SS_{X1}) - (SP_{X1X2})(SP_{X1Y})}{(SS_{X1})(SS_{X2}) - (SP_{X1X2})^2} = \frac{(-24)(42) - (-15)(27)}{(42)(28) - (-15)^2} = -0.634$$

$$a = M_Y - b_1 M_{X1} - b_2 M_{X2} = 5 - 0.416(4) - -0.634(3) = 5 - 1.664 + 1.902 = 5.238$$

The multiple-regression equation is

$$\hat{Y} = 0.416X_1 - 0.634X_2 + 5.238$$

STEP 2   *Evaluate the significance of the regression equation.* The null hypothesis states that the regression equation does not predict a significant portion of the variance for the Y scores. To conduct the test, the total variability for the Y scores, $SS_Y = 34$, is partitioned into the portion predicted by the regression equation and the residual portion. To find each portion, we must first compute the value of R^2.

$$R^2 = \frac{b_1 SP_{X1Y} + b_2 SP_{X2Y}}{SS_Y} = \frac{0.416(27) + -0.634(-24)}{34} = \frac{26.448}{34} = 0.778 \text{ or } 77.8\%$$

Then, the two components for the F-ratio are

$$SS_{regression} = R^2(SS_Y) = 0.778(34) = 26.45 \qquad \text{with } df = 2$$

$$SS_{residual} = (1 - R^2)(SS_Y) = 0.222(34) = 7.55 \qquad \text{with } df = n - 3 = 3$$

The two MS values (variances) and the F-ratio are

$$MS_{regression} = \frac{SS_{regression}}{df} = \frac{26.45}{2} = 13.23$$

$$MS_{residual} = \frac{SS_{residual}}{df} = \frac{7.55}{3} = 2.52$$

$$F = \frac{MS_{regression}}{MS_{residual}} = \frac{13.23}{2.52} = 5.25$$

with $df = 2, 3$, the F-ratio is not significant.

PROBLEMS

1. Sketch a graph showing the line for the equation $Y = -2X + 4$. On the same graph, show the line for $Y = X - 4$.

2. The regression equation is intended to be the "best fitting" straight line for a set of data. What is the criterion for "best fitting"?

3. A set of $n = 20$ pairs of scores (X and Y values) has $SS_X = 25$, $SS_Y = 16$, and $SP = 12.5$. If the mean for the X values is $M = 6$ and the mean for the Y values is $M = 4$.
   a. Calculate the Pearson correlation for the scores.
   b. Find the regression equation for predicting Y from the X values.

4. A set of $n = 25$ pairs of scores (X and Y values) produce a regression equation of $\hat{Y} = 2X - 7$. Find the predicted Y value for each of the following X scores: 0, 1, 3, −2.

5. Briefly explain what is measured by the standard error of estimate.

6. In general, how is the magnitude of the standard error of estimate related to the value of the correlation?

7. For the following set of data, find the linear regression equation for predicting Y from X:

X	Y
0	9
2	9
4	7
6	3

8. For the following data:
   a. Find the regression equation for predicting Y from X.
   b. Use the regression equation to find a predicted Y for each X.
   c. Find the difference between the actual Y value and the predicted Y value for each individual, square the differences, and add the squared values to obtain $SS_{residual}$.
   d. Calculate the Pearson correlation for these data. Use r^2 and SS_Y to compute $SS_{residual}$ with Equation 17.11. You should obtain the same value as in part c.

X	Y
1	2
4	7
3	5
2	1
5	14
3	7

9. Does the regression equation from problem 8 account for a significant portion of the variance in the Y scores? Use $\alpha = .05$ to evaluate the F-ratio.

10. For the following scores,

X	Y
3	0
8	10
7	8
5	3
7	7
6	8

a. Find the regression equation for predicting Y from X.

b. Calculate the predicted Y value for each X.

11. A researcher finds a correlation of $r = 0.60$ between IQ and creativity scores for a sample of $n = 20$ college students. Using these data, the researcher computes the regression equation for predicting creativity from IQ.

a. What percentage of the variance in creativity scores is predicted from IQ, and what percentage is not predicted?

b. If $SS_Y = 10$, does the regression equation predict a significant proportion of the variance in creativity?

c. If $SS_Y = 100$, does the regression equation predict a significant proportion of the variance in creativity?

Note: Comparing your answers for part b and part c, you should find that the value of SS_Y has no influence on the F-ratio for analysis of regression. In fact, SS_Y appears in both the numerator and the denominator of the F-ratio and simply cancels out.

12. A professor obtains SAT scores and freshman grade point averages (GPAs) for a group of $n = 15$ college students. The SAT scores have a mean of $M = 580$ with $SS = 22,400$, and the GPAs have a mean of 3.10 with $SS = 1.26$, and $SP = 84$.

a. Find the regression equation for predicting GPA from SAT scores.

b. What percentage of the variance in GPAs is accounted for by the regression equation? (Compute the correlation, r, then find r^2.)

c. Does the regression equation account for a significant portion of the variance in GPA? Use $\alpha = .05$ to evaluate the F-ratio.

13. Problem 14 in Chapter 16 examined the relationship between political attitudes for husbands and wives for a sample of $n = 8$ married couples. The wives had a mean score of $M = 7$ with $SS = 172$, the husbands had a mean score of $M = 9$ with $SS = 106$, and $SP = 122$.

a. Find the regression equation for predicting a husband's political attitude from his wife's score. (Identify the wives' scores as X values and the husbands' scores as Y values.)

b. What percentage of the variance in the husbands' scores is accounted for by the regression equation? (Compute the correlation, r, then find r^2.)

c. Does the regression equation account for a significant portion of the variance in the husbands' scores? Use $\alpha = .05$ to evaluate the F-ratio.

14. Problem 15 in Chapter 16 described a study examining the effectiveness of a 7-Minute Screen test for Alzheimer's disease. The study evaluated the relationship between scores from the 7-Minute Screen and scores for the same patients from a set of cognitive exams that are typically used to test for Alzheimer's disease. For a sample of $n = 9$ patients, the scores for the 7-Minute Screen averaged $M = 7$ with $SS = 92$. The cognitive test scores averaged $M = 17$ with $SS = 236$. For these data, $SP = 127$.

a. Find the regression equation for predicting the cognitive scores from the 7-Minute Screen score.

b. What percentage of variance in the cognitive scores is accounted for by the regression equation?

c. Does the regression equation account for a significant portion of the variance in the cognitive scores? Use $\alpha = .05$ to evaluate the F-ratio.

15. A set of $n = 15$ pairs of X and Y values has a correlation of $r = +.80$ with $SS_Y = 75$, and the regression equation for predicting Y is computed.

a. Find the standard error of estimate for the regression equation.

b. How big would the standard error be if the sample size were $n = 30$?

16. a. One set of 12 pairs of scores, X and Y values, produces a correlation of $r = 0.90$. If $SS_Y = 200$, find the standard error of estimate for the regression line.

b. A second set of 12 pairs of X and Y values produces of correlation of $r = 0.60$. If $SS_Y = 200$, find the standard error of estimate for the regression line.

17. a. A researcher computes the regression equation for a sample of $n = 18$ pairs of scores, X and Y values. If an analysis of regression is used to test the significance of the equation, what are the df values for the F-ratio?

b. A researcher evaluating the significance of a regression equation obtains an F-ratio with $df = 1, 24$. How many pairs of scores, X and Y values, are in the sample?

18. For the following data:

a. Find the regression equation for predicting Y from X.

b. Use the regression equation to find a predicted Y for each X.

c. Find the difference between the actual Y value and the predicted Y value for each individual, square the differences, and add the squared values to obtain $SS_{residual}$.

d. Calculate the Pearson correlation for these data. Use r^2 and SS_Y to compute $SS_{residual}$ with Equation 17.11. You should obtain the same value as in part c.

X	Y
7	16
5	2
6	1
3	2
4	9

19. A multiple-regression equation with two predictor variables accounts for $R^2 = 18.5\%$ of the variance in the Y scores for a sample of $n = 30$ individuals.

a. If $SS_Y = 20$, does the equation predict a significant portion of the variance for the Y scores? Test with $\alpha = .05$.

b. If $SS_Y = 200$, does the equation predict a significant portion of the variance for the Y scores? Test with $\alpha = .05$.

Note: Comparing your answers for part a and part b, you should find that the value of SS_Y has no influence on the F-ratio for analysis of regression. In fact, it appears in both the numerator and the denominator of the F-ratio and simply cancels out.

20. A researcher obtained the following multiple-regression equation using two predictor variables: $\hat{Y} = 2X_1 + 3.5X_2 + 18$. Given that $SS_Y = 205$, the SP value for X_1 and Y is 18, and the SP value for X_2 and Y is 9, find R^2, the percentage of variance accounted for by the equation.

21. At the end of this chapter (page 587) we talked about a study examining the relationship between exposure to sexual content on TV and sexual behavior among adolescents. Because the researchers were concerned that the age of the participants might distort the results, they compared the multiple-regression equation using age and TV exposure as predictors with the simple regression equation using age as the only predictor. Suppose that the researchers obtained $R^2 = 38.5\%$ for the multiple regression and $r^2 = 27.2\%$ for the one-predictor regression. If the researchers used a sample of $n = 30$ adolescents, are the results sufficient to conclude that there is a significant relationship between TV exposure and sexual behaviors even after the participants' ages are accounted for? Use $\alpha = .05$ to evaluate the F-ratio. *Note:* The value

for SS_Y is not necessary to evaluate the difference between the two regression equations. However, if you want to include SS_Y in your calculations, you may select any number, for example $SS_Y = 100$.

22. Problem 13 in Chapter 16 examined the TV-viewing habits of adopted children in relation to their biological parents and their adoptive parents. The data are reproduced as follows. If both the biological and adoptive parents are used to predict the viewing habits of the children in a multiple-regression equation, what percentage of the variance in the children's scores would be accounted for? That is, compute R^2.

Amount of Time Spent Watching TV			
Adopted Children Y	Birth Parents X_1	Adoptive Parents X_2	
2	0	1	$SP_{X1X2} = 8$
3	3	4	$SP_{X1Y} = 15$
6	4	2	$SP_{X2Y} = 3$
1	1	0	
3	1	0	
0	2	3	
5	3	2	
2	1	3	
5	3	3	
$SS_Y = 32$	$SS_{X1} = 14$	$SS_{X2} = 16$	

23. For the data in problem 22, the correlation between the children's scores and the biological parents' scores is $r = 0.709$. Does adding the adoptive parents' scores as a second predictor significantly improve the ability to predict the children's scores? Use $\alpha = .05$ to evaluate the F-ratio.

24. For the following data, find the multiple-regression equation for predicting Y from X_1 and X_2.

X_1	X_2	Y	
1	3	1	$SS_{X1} = 16$
2	4	2	$SS_{X2} = 28$
3	5	6	$SS_Y = 34$
6	9	8	$SP_{X1X2} = 18$
4	8	3	$SP_{X1Y} = 19$
2	7	4	$SP_{X2Y} = 21$
$M = 3$	$M = 6$	$M = 4$	

25. A researcher evaluates the significance of a multiple-regression equation and obtains an F-ratio with $df = 2, 34$. How many participants were in the sample?

26. A researcher measures three variables, X, Y, and Z for each individual in a sample of $n = 25$. The Pearson correlations for this sample are $r_{XY} = 0.8$, $r_{XZ} = 0.6$, and $r_{YZ} = 0.7$.

a. Find the partial correlation between X and Y, holding Z constant.

b. Find the partial correlation between X and Z, holding Y constant. (*Hint*: Simply switch the labels for the variables Y and Z to correspond with the labels in the equation.)

CHAPTER

18

The Chi-Square Statistic: Tests for Goodness of Fit and Independence

Tools You Will Need

The following items are considered essential background material for this chapter. If you doubt your knowledge of any of these items, you should review the appropriate chapter or section before proceeding.

- Proportions (math review, Appendix A)
- Frequency distributions (Chapter 2)

Preview

18.1   Parametric and Nonparametric Statistical Tests

18.2   The Chi-Square Test for Goodness of Fit

18.3   The Chi-Square Test for Independence

18.4   Measuring Effect Size for the Chi-Square Test for Independence

18.5   Assumptions and Restrictions for Chi-Square Tests

18.6   Special Applications of the Chi-Square Tests

Summary

Focus on Problem Solving

Demonstration 18.1

Problems

Preview

Loftus and Palmer (1974) conducted a classic experiment demonstrating how language can influence eyewitness memory. A sample of 150 students watched a film of an automobile accident. After watching the film, the students were separated into three groups. One group was asked, "About how fast were the cars going when they smashed into each other?" Another group received the same question except that the verb was changed to "hit" instead of "smashed into." A third group served as a control and was not asked any question about the speed of the two cars. A week later, the participants returned and were asked if they remembered seeing any broken glass in the accident. (There was no broken glass in the film.) Notice that the researchers are manipulating the form of the initial question and then measuring a yes/no response to a follow-up question 1 week later. Table 18.1 shows the structure of this design represented by a matrix with the independent variable (different groups) determining the rows of the matrix and the two categories for the dependent variable (yes/no) determining the columns. The number in each cell of the matrix is the frequency count showing how many participants are classified in that category. For example, of the 50 students who heard the word *smashed*, there were 16 (32%) who claimed to remember seeing broken glass even though there was none in the film. By comparison, only 7 of the 50 students (14%) who heard the word *hit* said they recalled seeing broken glass. The researchers would like to use these data to support the argument that a witness's "memory" can be influenced by the language used during questioning. If the two cars smashed into each other, there must have been some broken glass.

The Problem: Although the Loftus and Palmer study involves an independent variable (the form of the question) and a dependent variable (memory for broken glass), you should realize that this study is different from any experiment we have considered in the past. Specifically, the Loftus and Palmer study does not produce a numerical score for each participant. Instead, each participant is simply classified into one of two categories (yes or no). The data consist of *frequencies* or *proportions* describing how many individuals are in each category. You should also note that Loftus and Palmer want to use a hypothesis test to evaluate the data. The null hypothesis would state that the form of the question has no effect on the memory of the witness. The hypothesis test would determine whether the sample data provide enough evidence to reject this null hypothesis.

Because there are no numerical scores, it is impossible to compute a mean or a variance for the sample data. Therefore, it is impossible to use any of the familiar hypothesis tests (such as a *t* test or analysis of variance [ANOVA]) to determine whether there is a significant difference between the treatment conditions. What is needed is a new hypothesis testing procedure that can be used with non-numerical data.

The Solution: In this chapter we introduce two hypothesis tests based on the *chi-square* statistic. Unlike earlier tests that require numerical scores (X values), the chi-square tests use sample frequencies and proportions to test hypothesis about the corresponding population values.

TABLE 18.1

A frequency distribution table showing the number of participants who answered either yes or no when asked whether they recalled seeing any broken glass 1 week after witnessing an automobile accident. Immediately after the accident, one group was asked how fast the cars were going when they smashed into each other. A second group was asked how fast the cars were going when they hit each other. A third group served as a control and was not asked about the speed of the cars.

Verb Used to Ask about the Speed of the Cars	Response to the Question: Did You See Any Broken Glass?	
	Yes	No
Smashed into	16	34
Hit	7	43
Control (Not Asked)	6	44

18.1 PARAMETRIC AND NONPARAMETRIC STATISTICAL TESTS

All the statistical tests we have examined thus far are designed to test hypotheses about specific population parameters. For example, we used t tests to assess hypotheses about μ and later about $\mu_1 - \mu_2$. In addition, these tests typically make assumptions about the shape of the population distribution and about other population parameters. Recall that for ANOVA the population distributions are assumed to be normal and homogeneity of variance is required. Because these tests all concern parameters and require assumptions about parameters, they are called *parametric tests*.

Another general characteristic of parametric tests is that they require a numerical score for each individual in the sample. The scores then are added, squared, averaged, and otherwise manipulated using basic arithmetic. In terms of measurement scales, parametric tests require data from an interval or a ratio scale (see Chapter 1).

Often, researchers are confronted with experimental situations that do not conform to the requirements of parametric tests. In these situations, it may not be appropriate to use a parametric test. Remember: When the assumptions of a test are violated, the test may lead to an erroneous interpretation of the data. Fortunately, there are several hypothesis-testing techniques that provide alternatives to parametric tests. These alternatives are called *nonparametric tests*.

In this chapter, we introduce two commonly used examples of nonparametric tests. Both tests are based on a statistic known as chi-square and both tests use sample data to evaluate hypotheses about the proportions or relationships that exist within populations. Note that the two chi-square tests, like most nonparametric tests, do not state hypotheses in terms of a specific parameter, and they make few (if any) assumptions about the population distribution. For the latter reason, nonparametric tests sometimes are called *distribution-free tests*.

One of the most obvious differences between parametric and nonparametric tests is the type of data they use. All of the parametric tests that we have examined so far require numerical scores. For nonparametric tests, on the other hand, the subjects are usually just classified into categories such as Democrat and Republican, or High, Medium, and Low IQ. Notice that these classifications involve measurement on nominal or ordinal scales, and they do not produce numerical values that can be used to calculate means and variances. Instead, the data for many nonparametric tests are simply frequencies; for example, the number of Democrats and the number of Republicans in a sample of $n = 100$ registered voters.

Occasionally, you have a choice between using a parametric and a nonparametric test. Changing to a nonparametric test usually involves transforming the data from numerical scores to nonnumerical categories. For example, you could start with numerical scores measuring self-esteem and create three categories consisting of high, medium, and low self-exteem. In most situations, the parametric test is preferred because it is more likely to detect a real difference or a real relationship. However, there are situations for which transforming scores into categories might be a better choice.

1. Ocassionally, it is simpler to obtain category measurements. For example, it is easier to classify students as high, medium, or low in leadership ability than to obtain a numerical score measuring each student's ability.

2. The original scores may violate some of the basic assumptions that underline certain statistical procedures. For example, the t tests and ANOVA assume that the data come from normal distributions. Also, the independent-measures tests assume that the different populations all have the same variance (the

homogeneity-of-variance assumption). If a researcher suspects that the data do not satisfy these assumptions, it may be safer to transform the scroes into categories and use a nonparametric test to evaluate the data.

3. The original scores may have unusually high variance. Variance is a major component of the standard error in the denominator of *t* statistics and the error term in the denominator of *F*-ratios. Thus, large variance can greatly reduce the likelihood that these parametric tests will find significant differences. Converting the scores to categories essentially eliminates the variance. For example, all individuals fit into three categories (high, medium, and low) no matter how variable the original scores are.

4. Occasionally, an experiment produces an undetermined, or infinite, score. For example, a rat may show no sign of solving a particular maze after hundreds of trials. This animal has an infinite, or undetermined, score. Although there is no absolute score that can be assigned, you can say that this rat is in the highest category, and then classify the other scores according to their numerical values.

18.2   THE CHI-SQUARE TEST FOR GOODNESS OF FIT

Parameters such as the mean and the standard deviation are the most common way to describe a population, but there are situations in which a researcher has questions about the proportions or relative frequencies for a distribution. For example,

How does the number of women lawyers compare with the number of men in the profession?

Of the two leading brands of cola, which is preferred by most Americans?

In the past 10 years, has there been a significant change in the proportion of college students who declare a business major?

The name of the test comes from the Greek letter χ (chi, pronounced "kye"), which is used to identify the test statistic.

Note that each of the preceding examples asks a question about proportions in the population. In particular, we are not measuring a numerical score for each individual. Instead, the individuals are simply classified into categories and we want to know what proportion of the population is in each category. The *chi-square test for goodness of fit* is specifically designed to answer this type of question. In general terms, this chi-square test uses the proportions obtained for sample data to test hypotheses about the corresponding proportions in the population.

DEFINITION

The **chi-square test for goodness of fit** uses sample data to test hypotheses about the shape or proportions of a population distribution. The test determines how well the obtained sample proportions fit the population proportions specified by the null hypothesis.

Recall from Chapter 2 that a frequency distribution is defined as a tabulation of the number of individuals located in each category of the scale of measurement. In a frequency distribution graph, the categories that make up the scale of measurement are listed on the *X*-axis. In a frequency distribution table, the categories are listed in the first column. With chi-square tests, however, it is customary to present the scale of measurement as a series of boxes, with each box corresponding to a separate category on the scale. The frequency corresponding to each category is simply presented as a number written inside the box. Figure 18.1 shows how a distribution of eye colors for a set of $n = 40$ students can be presented as a graph, a table, or a series of boxes. Notice that the scale of measurement for this example consists of four categories of eye color (blue, brown, green, and other).

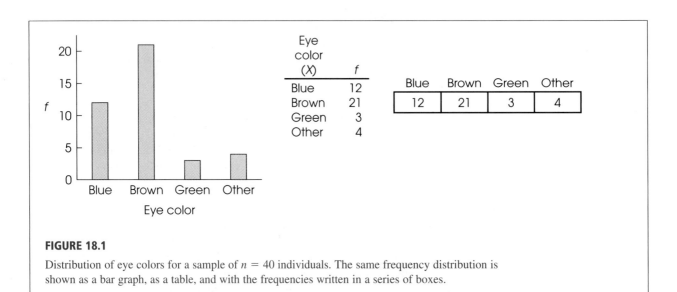

FIGURE 18.1

Distribution of eye colors for a sample of $n = 40$ individuals. The same frequency distribution is shown as a bar graph, as a table, and with the frequencies written in a series of boxes.

THE NULL HYPOTHESIS FOR THE GOODNESS-OF-FIT TEST

For the chi-square test of goodness of fit, the null hypothesis specifies the proportion (or percentage) of the population in each category. For example, a hypothesis might state that 90% of all lawyers are men and only 10% are women. The simplest way of presenting this hypothesis is to put the hypothesized proportions in the series of boxes representing the scale of measurement:

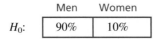

Although it is conceivable that a researcher could choose any proportions for the null hypothesis, there usually is some well-defined rationale for stating a null hypothesis. Generally H_0 falls into one of the following categories:

1. No Preference. The null hypothesis often states that there is no preference among the different categories. In this case, H_0 states that the population is divided equally among the categories. For example, a hypothesis stating that there is no preference among the three leading brands of soft drinks would specify a population distribution as follows:

(Preferences in the population are equally divided among the three soft drinks.)

The no-preference hypothesis is used in situations in which a researcher wants to determine whether there are any preferences among the categories, or whether the proportions differ from one category to another.

2. No Difference from a Known Population. The null hypothesis can state that the proportions for one population are not different from the proportions that are known to exist for another population. For example, suppose it is known that 28% of the licensed drivers in the state are younger that 30 years old and 72% are 30 or older. A researcher might wonder whether this same proportion holds for the distribution of speeding tickets. The null hypothesis would state that tickets are handed out equally across the population of drivers, so there is no difference between the age distribution for drivers and the age distribution for speeding tickets. Specifically, the null hypothesis would be

	Tickets given to drivers younger than 30	Tickets given to drivers 30 or older	
H_0:	28%	72%	(Proportions for the population of tickets are not different from proportions for drivers.)

The no-difference hypothesis is used when a specific population distribution is already known. For example, you may have a known distribution from an earlier time, and the question is whether there has been any change in the proportions. Or, you may have a known distribution for one population (drivers) and the question is whether a second population (speeding tickets) has the same proportions.

Because the null hypothesis for the *goodness-of-fit test* specifies an exact distribution for the population, the alternative hypothesis (H_1) simply states that the population distribution has a different shape from that specified in H_0. If the null hypothesis states that the population is equally divided among three categories, the alternative hypothesis will say that the population is not divided equally.

THE DATA FOR THE GOODNESS-OF-FIT TEST

The data for a chi-square test are remarkably simple. There is no need to calculate a sample mean or *SS;* you just select a sample of n individuals and count how many are in each category. The resulting values are called *observed frequencies*. The symbol for observed frequency is f_o. For example, the following data represent observed frequencies for a sample of $n = 40$ participants. Each person was given a personality questionnaire and classified into one of three personality categories: A, B, or C.

Category A	Category B	Category C	
15	19	6	$n = 40$

Notice that each individual in the sample is classified into one and only one of the categories. Thus, the frequencies in this example represent three completely separate groups of individuals: 15 who were classified as category A, 19 classified as B, and 6 classified as C. Also note that the observed frequencies add up to the total sample size: $\Sigma f_o = n$.

Finally, you should realize that we are not assigning individuals to categories. Instead, we are simply measuring individuals to determine the category in which they belong.

DEFINITION

The **observed frequency** is the number of individuals from the sample who are classified in a particular category. Each individual is counted in one and only one category.

EXPECTED FREQUENCIES

The general goal of the chi-square test for goodness of fit is to compare the data (the observed frequencies) with the null hypothesis. The problem is to determine how well the data fit the distribution specified in H_0—hence the name *goodness of fit*.

The first step in the chi-square test is to construct a hypothetical sample that represents how the sample distribution would look if it were in perfect agreement with the proportions stated in the null hypothesis. Suppose, for example, the null hypothesis states that the population is distributed into three categories with the following proportions:

Category A	Category B	Category C
25%	50%	25%

If this hypothesis is correct, how would you expect a random sample of $n = 40$ individuals to be distributed among the three categories? It should be clear that your best strategy is to predict that 25% of the sample would be in category A, 50% would be in category B, and 25% would be in category C. To find the exact frequency expected for each category, multiply the sample size (n) by the proportion (or percentage) from the null hypothesis. For this example, you would expect

$$25\% \text{ of } 40 = 0.25(40) = 10 \text{ individuals in category A}$$

$$50\% \text{ of } 40 = 0.50(40) = 20 \text{ individuals in category B}$$

$$25\% \text{ of } 40 = 0.25(40) = 10 \text{ individuals in category C}$$

The frequency values predicted from the null hypothesis are called *expected frequencies*. The symbol for expected frequency is f_e, and the expected frequency for each category is computed by

$$\text{expected frequency} = f_e = pn \tag{18.1}$$

where p is the proportion stated in the null hypothesis and n is the sample size.

DEFINITION

The **expected frequency** for each category is the frequency value that is predicted from the null hypothesis and the sample size (n). The expected frequencies define an ideal, *hypothetical* sample distribution that would be obtained if the sample proportions were in perfect agreement with the proportions specified in the null hypothesis.

Note that the no-preference null hypothesis always produces equal f_e values for all categories because the proportions (p) are the same for all categories. On the other hand, the no-difference null hypothesis typically does not produce equal values for the expected frequencies because the hypothesized proportions typically vary from one category to another. You also should note that the expected frequencies are calculated, hypothetical values and the numbers that you obtain may be decimals or fractions. The observed frequencies, on the other hand, always represent real individuals and always are whole numbers.

THE CHI-SQUARE STATISTIC

The general purpose of any hypothesis test is to determine whether the sample data support or refute a hypothesis about the population. In the chi-square test for goodness of fit, the sample is expressed as a set of observed frequencies (f_o values), and the null hypothesis is used to generate a set of expected frequencies (f_e values). The *chi-square statistic* simply measures how well the data (f_o) fit the hypothesis (f_e). The symbol for the chi-square statistic is χ^2. The formula for the chi-square statistic is

$$\text{chi-square} = \chi^2 = \Sigma\frac{(f_o - f_e)^2}{f_e} \tag{18.2}$$

As the formula indicates, the value of chi-square is computed by the following steps:

1. Find the difference between f_o (the data) and f_e (the hypothesis) for each category.

2. Square the difference. This ensures that all values are positive.

3. Next, divide the squared difference by f_e. A justification for this step is given in Box 18.1.

4. Finally, add the values from all the categories.

BOX 18.1	THE CHI-SQUARE FORMULA

The numerator of the chi-square formula is fairly easy to understand. Specifically, the numerator simply measures how much difference there is between the data (the f_o values) and the hypothesis (represented by the f_e values). A large value indicates that the data do not fit the hypothesis, and leads us to reject the hypothesis.

The denominator of the formula, on the other hand, is not so easy to understand. Why must we divide by f_e before we add the category values? The answer to this question is that the obtained discrepancy between f_o and f_e is viewed as *relatively* large or *relatively* small depending on the size of the expected frequency. This point is demonstrated in the following analogy.

Suppose you were going to throw a party and you *expected* 1000 people to show up. However, at the party you counted the number of guests and *observed* that 1040 actually showed up. Forty more guests than expected are no major problem when all along you were planning for 1000. There will still probably be enough beer and potato chips for everyone. On the other hand, suppose you had a party and you expected 10 people to attend, but instead 50 actually showed up. Forty more guests in this case spell big trouble. How "significant" the discrepancy is depends in part on what you were originally expecting. With very large expected frequencies, allowances are made for more error between f_o and f_e. This is accomplished in the chi-square formula by dividing the squared discrepancy for each category, $(f_o - f_e)^2$, by its expected frequency.

THE CHI-SQUARE DISTRIBUTION AND DEGREES OF FREEDOM

It should be clear from the chi-square formula that the numerical value of chi-square is a measure of the discrepancy between the observed frequencies (data) and the expected frequencies (H_0). As usual, the sample data are not expected to provide a perfectly accurate representation of the population. In this case, the proportions or observed frequencies in the sample are not expected to be exactly equal to the proportions in the population. Thus, if there are small discrepancies between the f_o and f_e values, we will obtain a small value for chi-sqaure and we will conclude that there is a good fit between the data and the hypothesis (fail to reject H_0). However, when there are large discrepancies between f_o and f_e, we will obtain a large value for chi-square and conclude that the data do not fit the hypothesis (reject H_0). To decide whether a particular chi-square value is "large" or "small," we must refer to a *chi-square distribution*. This distribution is the set of chi-square values for all the possible random samples when H_0 is true. Much like other distributions we have examined (*t* distribution, *F* distribution), the chi-square distribution is a theoretical distribution with well-defined characteristics. Some of these characteristics are easy to infer from the chi-square formula.

1. The formula for chi-square involves adding squared values, so you can never obtain a negative value. Thus, all chi-square values are zero or larger.

2. When H_0 is true, you expect the data (f_o values) to be close to the hypothesis (f_e values). Thus, we expect chi-square values to be small when H_0 is true.

These two factors suggest that the typical chi-square distribution will be positively skewed (Figure 18.2). Note that small values, near zero, are expected when H_0 is true and large values (in the right-hand tail) are very unlikely. Thus, unusually large values of chi-square form the critical region for the hypothesis test.

Although the typical chi-square distribution is positively skewed, there is one other factor that plays a role in the exact shape of the chi-square distribution—the number of categories. Recall that the chi-square formula requires that you add values from every category. The more categories you have, the more likely it is that you will obtain a large sum for the chi-square value. On average, chi-square will be larger when you are adding more than 10 categories than when you are adding only 3 categories. As a result, there is a whole family of chi-square distributions, with the exact shape of each distribution determined by the number of categories used in the study. Technically, each specific chi-square distribution is identified by degrees of freedom (*df*), rather than the number of categories. For the goodness-of-fit test, the degrees of freedom are determined by

Caution: The *df* for a chi-square test is *not* related to sample size (*n*), as it is in most other tests.

$$df = C - 1 \tag{18.3}$$

where *C* is the number of categories. A brief discussion of this *df* formula is presented in Box 18.2. Figure 18.3 shows the general relationship between *df* and the shape of the chi-square distribution. Note that the chi-square values tend to get larger (shift to the right) as the number of categories and the degrees of freedom increase.

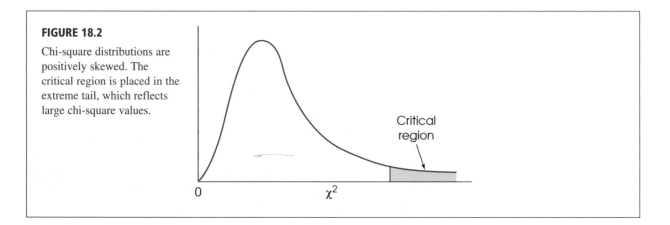

FIGURE 18.2

Chi-square distributions are positively skewed. The critical region is placed in the extreme tail, which reflects large chi-square values.

<table>
<tr><td>BOX
18.2</td><td>A CLOSER LOOK AT DEGREES OF FREEDOM</td></tr>
</table>

Degrees of freedom for the chi-square test literally measure the number of free choices that exist when you are determining the null hypothesis or the expected frequencies. For example, when you are classifying individuals into three categories, you have exactly two free choices in stating the null hypothesis. You may select any two proportions for the first two categories, but then the third proportion is determined. If you hypothesize 25% in the first category and 50% in the second category, then the third category must be 25% in order to account for 100% of the population. In general, you are free to select proportions for all but one of the categories, then the final proportion is determined by the fact that the entire set must total 100%. Thus, you have $C - 1$ free choices, where C is the number of categories: Degrees of freedom, df, equal $C - 1$.

The same restriction holds when you are determining the expected frequencies. Again, suppose you have three categories and a sample of $n = 40$ individuals. If you specify expected frequencies of $f_e = 10$ for the first category and $f_e = 20$ for the second category, then you must use $f_e = 10$ for the final category.

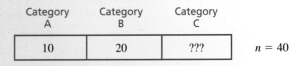

Category A	Category B	Category C
10	20	???

$n = 40$

As before, you may distribute the sample freely among the first $C - 1$ categories, but then the final category is determined by the total number of individuals in the sample.

FIGURE 18.3

The shape of the chi-square distribution for different values of df. As the number of categories increases, the peak (mode) of the distribution has a larger chi-square value.

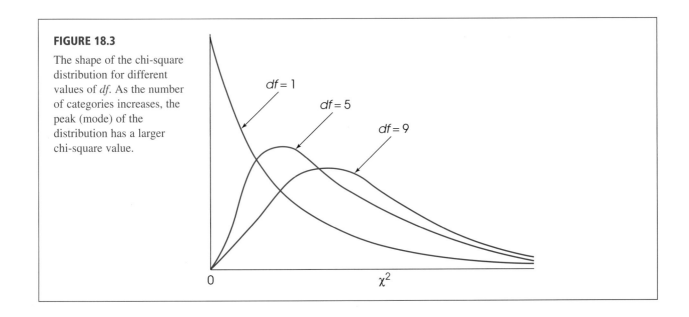

LOCATING THE CRITICAL REGION FOR A CHI-SQUARE TEST   Recall that a large value for the chi-square statistic indicates a big discrepancy between the data and the hypothesis, and suggests that we reject H_0. To determine whether a particular chi-square value is significantly large, you must consult the table entitled The Chi-Square Distribution (Appendix B). A portion of the chi-square table is shown in Table 18.2. The first column lists df values for the chi-square test, and the top row of

TABLE 18.2

A portion of the table of critical values for the chi-square distribution.

df	\multicolumn{5}{c}{Proportion in Critical Region}
	0.10	0.05	0.025	0.01	0.005
1	2.71	3.84	5.02	6.63	7.88
2	4.61	5.99	7.38	9.21	10.60
3	6.25	7.81	9.35	11.34	12.84
4	7.78	9.49	11.14	13.28	14.86
5	9.24	11.07	12.83	15.09	16.75
6	10.64	12.59	14.45	16.81	18.55
7	12.02	14.07	16.01	18.48	20.28
8	13.36	15.51	17.53	20.09	21.96
9	14.68	16.92	19.02	21.67	23.59

the table lists proportions (alpha levels) in the extreme right-hand tail of the distribution. The numbers in the body of the table are the critical values of chi-square. The table shows, for example, that when the null hypothesis is true and $df = 3$, only 5% (.05) of the chi-square value are greater than 7.81, and only 1% (.01) are larger than 11.34.

EXAMPLE OF THE CHI-SQUARE TEST FOR GOODNESS OF FIT

We will use the same step-by-step process for testing hypotheses with chi-square as we used for other hypothesis tests. In general, the steps consist of stating the hypotheses, locating the critical region, computing the test statistic, and making a decision about H_0. The following example demonstrates the complete process of hypothesis testing with the goodness-of-fit test.

EXAMPLE 18.1

A psychologist examining art appreciation selected an abstract painting that had no obvious top or bottom. Hangers were placed on the painting so that it could be hung with any one of the four sides at the top. The painting was shown to a sample of $n = 50$ participants, and each was asked to hang the painting in the orientation that looked correct. The following data indicate how many people chose each of the four sides to be placed at the top:

Top Up (Correct)	Bottom Up	Left Side Up	Right Side Up
18	17	7	8

The question for the hypothesis test is whether there are any preferences among the four possible orientations. Are any of the orientations selected more (or less) often than would be expected simply by chance?

STEP 1   State the hypotheses and select an alpha level. The hypotheses can be stated as follows:

H_0:   In the general population, there is no preference for any specific orientation. Thus, the four possible orientations are selected equally often, and the population distribution has the following proportions:

Top Up (Correct)	Bottom Up	Left Side Up	Right Side Up
25%	25%	25%	25%

H_1:   In the general population, one or more of the orientations is preferred over the others.

We will use $\alpha = .05$.

STEP 2   Locate the critical region. For this example, the value for degrees of freedom is

$$df = C - 1 = 4 - 1 = 3$$

For $df = 3$ and $\alpha = .05$, the table of critical values for chi-square indicates that the critical χ^2 has a value of 7.81. The critical region is sketched in Figure 18.4.

STEP 3   Calculate the chi-square statistic. The calculation of chi-square is actually a two-stage process. First, you must compute the expected frequencies from H_0 and then calculate the value of the chi-square statistic. For this example, the null hypothesis specifies that one-quarter of the population ($p = 25\%$) will be in each of the four categories. According to this hypothesis, we should expect one-quarter of the sample to be in each category. With a sample of $n = 50$ individuals, the expected frequency for each category is

Expected frequencies are computed and may be decimal values. Observed frequencies are always whole numbers.

$$f_e = pn = \tfrac{1}{4}(50) = 12.5$$

The observed frequencies and the expected frequencies are presented in Table 18.3.

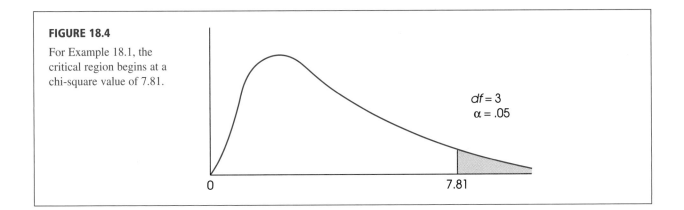

FIGURE 18.4

For Example 18.1, the critical region begins at a chi-square value of 7.81.

$df = 3$
$\alpha = .05$

0   7.81

TABLE 18.3

The observed frequencies and the expected frequencies for the chi-square test in Example 18.1.

Observed Frequencies	Top Up (Correct)	Bottom Up	Left Side Up	Right Side Up
	18	17	7	8

Expected Frequencies	Top Up (Correct)	Bottom Up	Left Side Up	Right Side Up
	12.5	12.5	12.5	12.5

Using these values, the chi-square statistic may now be calculated.

$$\chi^2 = \Sigma \frac{(f_o - f_e)^2}{f_e}$$

$$= \frac{(18 - 12.5)^2}{12.5} + \frac{(17 - 12.5)^2}{12.5} + \frac{(7 - 12.5)^2}{12.5} + \frac{(8 - 12.5)^2}{12.5}$$

$$= \frac{30.25}{12.5} + \frac{20.25}{12.5} + \frac{30.25}{12.5} + \frac{20.25}{12.5}$$

$$= 2.42 + 1.62 + 2.42 + 1.62$$

$$= 8.08$$

STEP 4 State a decision and a conclusion. The obtained chi-square value is in the critical region. Therefore, H_0 is rejected and the researcher may conclude that the four orientations are not equally likely to be preferred. Instead, there are significant differences among the four orientations, with some selected more often and others less often than would be expected by chance.

IN THE LITERATURE
REPORTING THE RESULTS FOR CHI-SQUARE

APA style specifies the format for reporting the chi-square statistic in scientific journals. For the results of Example 18.1, the report might state:

> The participants showed significant preferences among the four orientations for hanging the painting, $\chi^2(3, n = 50) = 8.08, p < .05$.

Note that the form of the report is similar to that of other statistical tests we have examined. Degrees of freedom are indicated in parentheses following the chi-square symbol. Also contained in the parentheses is the sample size (n). This additional information is important because the degrees of freedom value is based on the number of categories (C), not sample size. Next, the calculated value of chi-square is presented, followed by the probability that a Type I error has been committed. Because the null hypothesis was rejected, the probability is *less than* the alpha level.

Additionally, the report should provide the observed frequencies (f_o) for each category. This information may be presented in a simple sentence or in a table (see, for example, Table 18.3). ❑

GOODNESS OF FIT AND THE SINGLE-SAMPLE *t* TEST

We began this chapter with a general discussion of the difference between parametric tests and nonparametric tests. In this context, the chi-square test for goodness of fit is an example of a nonparametric test; that is, it makes no assumptions about the parameters of the population distribution, and it does not require data from an interval or ratio scale. In contrast, the single-sample *t* test introduced in Chapter 9 is an example of a parametric test: It assumes a normal population, it tests hypotheses about the population mean (a parameter), and it requires numerical scores that can be added, squared, divided, and so on.

Although the chi-square test and the single-sample *t* are clearly distinct, they are also very similar. In particular, both tests are intended to use the data from a single sample to test hypotheses about a single population.

The primary factor that determines whether you should use the chi-square test or the *t* test is the type of measurement that is obtained for each participant. If the sample data consist of numerical scores (from an interval or ratio scale), it is appropriate to compute a sample mean and use a *t* test to evaluate a hypothesis about the population mean. For example, a researcher could measure the IQ for each individual in a sample of registered voters. A *t* test could then be used to evaluate a hypothesis about the mean IQ for the entire population of registered voters. On the other hand, if the individuals in the sample are classified into non-numerical categories (on a nominal or ordinal scale), you would use a chi-square test to evaluate a hypothesis about the population proportions. For example, a researcher could classify people according to gender by simply counting the number of males and females in a sample of registered voters. A chi-square test would then be appropriate to evaluate a hypothesis about the population proportions.

LEARNING CHECK

1. A researcher uses a chi-square test for goodness of fit with a sample of $n = 120$ people to determine whether there are any preferences among four different brands of chocolate-chip cookies. What is the *df* value for the chi-square statistic?

2. Is it possible for expected frequencies to be fractions or decimal values?

3. A researcher has developed three different designs for a computer keyboard. A sample of $n = 60$ participants is obtained, and each individual tests all three keyboards and identifies his or her favorite. The frequency distribution of preferences is as follows:

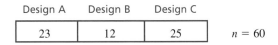

Design A	Design B	Design C	
23	12	25	$n = 60$

Use a chi-square test for goodness of fit with $\alpha = .05$ to determine whether there are any significant preferences among the three designs.

ANSWERS

1. With four categories, $df = 3$.

2. Yes. Expected frequencies are computed and may be fractions or decimal values.

3. The null hypothesis states that there are no preferences; one-third of the population would prefer each design. With $df = 2$, the critical value is 5.99. The expected frequencies are all 20, and the chi-square statistic is 4.90. Fail to reject the null hypothesis. There are no significant preferences.

18.3 THE CHI-SQUARE TEST FOR INDEPENDENCE

The chi-square statistic may also be used to test whether there is a relationship between two variables. In this situation, each individual in the sample is measured or classified on two separate variables. For example, a group of students could be classified in terms of personality (introvert, extrovert) and in terms of color preference (red, yellow, green,

In the context of a chi-square test, the frequency matrix is often called a *contingency table*.

or blue). Usually, the data from this classification are presented in the form of a matrix, where the rows correspond to the categories of one variable and the columns correspond to the categories of the second variable. Table 18.4 presents some hypothetical data for a sample of $n = 200$ students who have been classified by personality and color preference. The number in each box, or cell, of the matrix depicts the frequency of that particular group. In Table 18.4, for example, there are 10 students who were classified as introverted and who selected red as their preferred color. To obtain these data, the researcher first selects a random sample of $n = 200$ students. Each student is then given a personality test, and each student is asked to select a preferred color from among the four choices. Notice that the classification is based on the measurements for each student; the researcher does not assign students to categories. Also, notice that the data consist of frequencies, not scores, from a sample. These sample data are used to test a hypothesis about the corresponding population frequency distribution. Once again, we use the chi-square statistic for the test, but in this case, the test is called the chi-square *test for independence*.

DEFINITION

> The **Chi-square test for independence** uses the frequency data from a sample to evaluate the relationship between two variables in the population.

THE NULL HYPOTHESIS FOR THE TEST FOR INDEPENDENCE

The null hypothesis for the chi-square test for independence states that the two variables being measured are independent; that is, for each individual, the value obtained for one variable is not related to (or influenced by) the value for the second variable. This general hypothesis can be expressed in two different conceptual forms, each viewing the data and the test from slightly different perspectives. The data in Table 18.4 describing color preference and personality are used to present both versions of the null hypothesis.

H_0 **version 1**   For this version of H_0, the data are viewed as a single sample with each individual measured on two variables. The goal of the chi-square test is to evaluate the relationship between the two variables. For the example we are considering, the goal is to determine whether there is a consistent, predictable relationship between personality and color preference. That is, if I know your personality, will it help me to predict your color preference? The null hypothesis states that there is no relationship. The alternative hypothesis, H_1, states that there is a relationship between the two variables.

> H_0:   For the general population of students, there is no relationship between color preference and personality.

This version of H_0 demonstrates the similarity between the chi-square test for independence and a correlation. In each case, the data consist of two measurements (X and Y) for each individual, and the goal is to evaluate the relationship between the

TABLE 18.4

Color preferences according to personality types.

	Red	Yellow	Green	Blue	
Introvert	10	3	15	22	50
Extrovert	90	17	25	18	150
	100	20	40	40	$n = 200$

two variables. The correlation, however, requires numerical scores for X and Y. The chi-square test, on the other hand, simply uses frequencies for individuals classified into categories.

H_0 **version 2**   For this version of H_0, the data are viewed as two (or more) separate samples representing two (or more) separate populations. The goal of the chi-square test is to determine whether there are significant differences between the populations. For the example we are considering, the data in Table 18.4 would be viewed as a sample of $n = 50$ introverts (top row) and a separate sample of $n = 150$ extroverts (bottom row). The chi-square test determines whether the distribution of color preferences for introverts is significantly different from the distribution of color preferences for extroverts. From this perspective, the null hypothesis is stated as follows:

> H_0: In the population of students, there is no difference between the distribution of color preferences for introverts and the distribution of color preferences for extroverts. The two distributions have the same shape (same proportions).

This version of H_0 demonstrates the similarity between the chi-square test and an independent-measures t test (or ANOVA). In each case, the data consist of two (or more) separate samples that are being used to test for differences between two (or more) populations. The t test (or ANOVA) requires numerical scores to compute means and mean differences. However, the chi-square test simply uses frequencies for individuals classified into categories. The null hypothesis for the chi-square test states that the populations have the same proportions (same shape). The alternative hypothesis, H_1, simply states that the populations have different proportions. For the example we are considering, H_1 states that the distribution of color preferences for introverts is different from the distribution of color preferences for extroverts.

Equivalence of H_0 version 1 and H_0 version 2   Although we have presented two different statements of the null hypothesis, you should realize that these two versions are equivalent. The first version of H_0 states that color preference is not related to personality. If this hypothesis is correct, then the distribution of color preferences should not depend on personality. In other words, the distribution of color preferences should be the same for introverts and extroverts, which is the second version of H_0.

For example, if we found that 60% of the introverts preferred red, then H_0 would predict that we also should find that 60% of the extroverts prefer red. In this case, knowing that an individual prefers red does not help you predict his or her personality. Notice that finding the *same proportions* indicates *no relationship*.

On the other hand, if the proportions were different, it would suggest that there is a relationship. For example, if red were preferred by 60% of the extroverts but only 10% of the introverts, then there is a clear, predictable relationship between personality and color preference. (If I know your personality, I can predict your color preference.) Thus, finding *different proportions* means that there is *a relationship* between the two variables.

D E F I N I T I O N   Two variables are **independent** when there is no consistent, predictable relationship between them. In this case, the frequency distribution for one variable is not related to (or dependent on) the categories of the second variable. As a result, when two variables are independent, the frequency distribution for one variable will have the same shape for all categories of the second variable.

Thus, stating that there is no relationship between two variables (version 1 of H_0) is equivalent to stating that the distributions have equal proportions (version 2 of H_0).

OBSERVED AND EXPECTED
FREQUENCIES

The chi-square test for independence uses the same basic logic that was used for the goodness-of-fit test. First, a sample is selected, and each individual is classified or categorized. Because the test for independence considers two variables, every individual is classified on both variables, and the resulting frequency distribution is presented as a two-dimensional matrix (see Table 18.4). As before, the frequencies in the sample distribution are called observed frequencies and are identified by the symbol f_o.

The next step is to find the expected frequencies, or f_e values, for this chi-square test. As before, the expected frequencies define an ideal hypothetical distribution that is in perfect agreement with the null hypothesis. Once the expected frequencies are obtained, we compute a chi-square statistic to determine how well the data (observed frequencies) fit the null hypothesis (expected frequencies).

Although you can use either version of the null hypothesis to find the expected frequencies, the logic of the process is much easier when you use H_0 stated in terms of equal proportions. For the example we are considering, the null hypothesis states

> H_0: The frequency distribution of color preference has the same shape (same proportions) for both categories of personality.

To find the expected frequencies, we first determine the overall distribution of color preferences and then apply this distribution to both categories of personality. Table 18.5 shows an empty matrix corresponding to the data from Table 18.4. Notice that the empty matrix includes all of the row totals and the column totals from the original sample data. The row totals and the column totals are essential for computing the expected frequencies.

TABLE 18.5

An empty frequency distribution matrix showing only the row totals and column totals. (These numbers describe the basic characteristics of the sample from Table 18.4.)

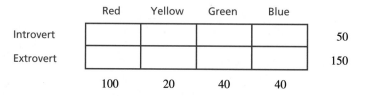

The column totals for the matrix describe the overall distribution of color preferences. For these data, 100 people selected red as their preferred color. Because the total sample consists of 200 people, it is easy to determine that the proportion selecting red is 100 out of 200 or 50%. The complete set of color preference proportions is as follows:

$$100 \text{ out of } 200 = 50\% \text{ prefer red}$$

$$20 \text{ out of } 200 = 10\% \text{ prefer yellow}$$

$$40 \text{ out of } 200 = 20\% \text{ prefer green}$$

$$40 \text{ out of } 200 = 20\% \text{ prefer blue}$$

The row totals in the matrix define the two samples of personality types. For example, the matrix shows a total of 50 introverts (the top row) and a sample of 150 extroverts (the bottom row). According to the null hypothesis, both personality groups should have the same proportions for color preferences. To find the expected frequencies, we simply apply the overall distribution of color preferences to each sample. Beginning with the sample of 50 introverts in the top row, we obtain expected frequencies of

50% prefer red: $f_e = 50\%$ of $50 = 0.50(50) = 25$

10% prefer yellow: $f_e = 10\%$ of $50 = 0.10(50) = 5$

20% prefer green: $f_e = 20\%$ of $50 = 0.20(50) = 10$

20% prefer blue: $f_e = 20\%$ of $50 = 0.20(50) = 10$

Using exactly the same proportions for the sample of $n = 150$ extroverts in the bottom row, we obtain expected frequencies of

50% prefer red: $f_e = 50\%$ of $150 = 0.50(50) = 75$

10% prefer yellow: $f_e = 10\%$ of $150 = 0.10(50) = 15$

20% prefer green: $f_e = 20\%$ of $150 = 0.20(50) = 30$

20% prefer blue: $f_e = 20\%$ of $150 = 0.20(50) = 30$

The complete set of expected frequencies is shown in Table 18.6. Notice that the row totals and the column totals for the expected frequencies are the same as those for the original data (the observed frequencies) in Table 18.4.

TABLE 18.6

Expected frequencies corresponding to the data in Table 18.4. (This is the distribution predicted by the null hypothesis.)

	Red	Yellow	Green	Blue	
Introvert	25	5	10	10	50
Extrovert	75	15	30	30	150
	100	20	40	40	

A SIMPLE FORMULA FOR DETERMINING EXPECTED FREQUENCIES

Although you should understand that expected frequencies are derived directly from the null hypothesis and the sample characteristics, it is not necessary to go through extensive calculations in order to find f_e values. In fact, there is a simple formula that determines f_e for any cell in the frequency distribution table:

$$f_e = \frac{f_c f_r}{n} \tag{18.4}$$

where f_c is the frequency total for the column (column total), f_r is the frequency total for the row (row total), and n is the number of individuals in the entire sample. To demonstrate this formula, we compute the expected frequency for introverts selecting yellow in Table 18.6. First, note that this cell is located in the top row and second column in the table. The column total is $f_c = 20$, the row total is $f_r = 50$, and the sample size is $n = 200$. Using these values in Equation 18.4, we obtain

$$f_e = \frac{f_c f_r}{n} = \frac{20(50)}{200} = 5$$

Notice that this is identical to the expected frequency we obtained using percentages from the overall distribution.

THE CHI-SQUARE STATISTIC AND DEGREES OF FREEDOM

The chi-square test of independence uses exactly the same chi-square formula as the test for goodness of fit:

$$\chi^2 = \Sigma \frac{(f_o - f_e)^2}{f_e}$$

As before, the formula measures the discrepancy between the data (f_o values) and the hypothesis (f_e values). A large discrepancy produces a large value for chi-square and indicates that H_0 should be rejected. To determine whether a particular chi-square statistic is significantly large, you must first determine degrees of freedom (df) for the statistic and then consult the chi-square distribution in the appendix. For the chi-square test of independence, degrees of freedom are based on the number of cells for which you can freely choose expected frequencies. Recall that the f_e values are partially determined by the sample size (n) and by the row totals and column totals from the original data. These various totals restrict your freedom in selecting expected frequencies. This point is illustrated in Table 18.7. Note that the final value in the top row is determined by the row total; the frequency values in the top row must total 50. In the same way, the bottom value in each column is determined by the column total. As a result, once three of the f_e values have been selected, all the other f_e values in the table are also determined. In general, the row totals and the column totals restrict the final choices in each row and column. Thus, we may freely choose all but one f_e in each row and all but one f_e in each column. The total number of f_e values that you can freely choose is $(R - 1)(C - 1)$, where R is the number of rows and C is the number of columns. The degrees of freedom for the chi-square test of independence are given by the formula

$$df = (R - 1)(C - 1) \tag{18.5}$$

AN EXAMPLE OF THE CHI-SQUARE TEST FOR INDEPENDENCE

The steps for the chi-square test of independence should be familiar by now. First, the hypotheses are stated, and an alpha level is selected. Second, the value for degrees of freedom is computed, and the critical region is located. Third, expected frequencies are determined, and the chi-square statistic is computed. Finally, a decision is made

TABLE 18.7

Degrees of freedom and expected frequencies. (Once three values have been selected, all the remaining expected frequencies are determined by the row totals and the column totals. This example has only three free choices, so $df = 3$.)

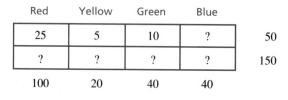

	Red	Yellow	Green	Blue	
	25	5	10	?	50
	?	?	?	?	150
	100	20	40	40	

regarding the null hypothesis. The following example demonstrates the complete hypothesis-testing proedure.

EXAMPLE 18.2   Research has demonstrated strong gender differences in teenagers' approaches to dealing with mental health issues (Chandra & Minkovitz, 2006). In a typical study eighth-grade students are asked to report their willingness to use mental health services in the event they were experiencing emotional or other mental health problems. Typical data for a sample of $n = 150$ students are shown in Table 18.8. Do the data show a significant relationship between gender and willingness to seek mental health assitance?

STEP 1   State the hypotheses, and select a level of significance. According to the null hypothesis, the two variables are independent. This general hypothesis can be stated in two different ways:

Version 1

> H_0: In the general population, there is no relationship between gender and willingness to use mental health services.

This version of H_0 emphasizes the similarity between the chi-square test and a correlation. The corresponding alternative hypothesis would state:

> H_1: In the general population, there is a consistent, predictable relationship between gender and willingness to use mental health services.

Version 2

> H_0: In the general population, the distribution of reported willingness to use mental health services is the same for males and females.

The corresponding alternative hypothesis would state:

> H_1: In the general population, the distribution of reported willingness to use mental health services for males is different from the distribution for females.

The second version of H_0 emphasizes the similarity between the chi-square test and the independent-measures t test.

Remember that the two versions for the hypotheses are equivalent. The choice between them is largely determined by how the researcher wants to describe the outcome. For example, a researcher may want to emphasize the *relationship* between variables or may want to emphasis the *difference* between groups.

TABLE 18.8

A frequency distribution showing willingness to use mental health services according to gender for a sample of $n = 150$ students.

| | Willingness to Use Mental Health Service | | | |
	Probably No	Maybe	Probably Yes	
Males	17	32	11	60
Females	13	43	34	90
	30	75	45	$n = 150$

For this test, we will use $\alpha = .05$.

STEP 2   Determine the degrees of freedom, and locate the critical region. For the chi-square test for independence,

$$df = (R - 1)(C - 1)$$

Therefore, for this study,

$$df = (2 - 1)(3 - 1) = 2$$

With $df = 2$ and $\alpha = .05$, the critical value for chi-square is 5.99 (see Table B.8, page 737).

STEP 3   Determine the expected frequencies, and compute the chi-square statistic. The following table shows an empty matrix with the same row totals and column totals as the original data. The calculation of expected frequencies requires that this table be filled in so that the resulting values provide an ideal frequency distribution that perfectly represents the null hypothesis.

Willingness to Use Mental Health Service

	Probably No	Maybe	Probably Yes	
Males				60
Females				90
	30	75	45	$n = 150$

The column totals indicate that 30 out of 150 students reported that they would probably not use mental health services. This proportion corresponds to $\frac{30}{150}$ or 20% of the total sample. Similarly, $\frac{75}{150} = 50\%$ reported that they may use mental health services, and $\frac{45}{150} = 30\%$ reported that they probably would use the services. The null hypothesis (version 2) states that this distribution is the same for males and females. Therefore, we simply apply the proportions to each group to obtain the expected frequencies. For the group of 60 males, we obtain

20% of 60 = 12 males who would probably not seek services

50% of 60 = 30 males who may seek services

30% of 60 = 18 males who probably would seek services

For the group of 90 females, we would expect

20% of 90 = 18 females who would probably not seek services

50% of 90 = 45 females who may seek services

30% of 90 = 27 females who probably would seek services

These expected frequencies are summarized in Table 18.9.

TABLE 18.9

The expected frequencies (f_e values) that would be expected if willingness to use mental services were compeltely independent of gender.

Willingness to Use Mental Health Service

	Probably No	Maybe	Probably Yes	
Males	12	30	18	60
Females	18	45	27	90
	30	75	45	

The chi-square statistic is now used to measure the discrepancy between the data (the observed frequencies in Table 18.8) and the null hypothesis that was used to generate the expected frequencies in Table 18.9.

$$\chi^2 = \frac{(17-12)^2}{12} + \frac{(32-30)^2}{30} + \frac{(11-18)^2}{18}$$
$$+ \frac{(13-18)^2}{18} + \frac{(43-45)^2}{45} + \frac{(34-27)^2}{27}$$
$$= 2.08 + 0.13 + 2.72 + 1.39 + 0.09 + 1.82$$
$$= 8.23$$

STEP 4   Make a decision regarding the null hypothesis and the outcome of the study. The obtained chi-square value exceeds the critical value (5.99). Therefore, the decision is to reject the null hypothesis. In the literature, this would be reported as a significant result with $\chi^2(2, n = 150) = 8.23$, $p < .05$. According to version 1 of H_0, this means that we have decided that there is a significant relationship between gender and willingness to use mental health services. Expressed in terms of version 2 of H_0, the data show a significant difference between males' and females' attitudes toward using mental health services. To describe the details of the significant result, you must compare the original data (Table 18.8) with the expected frequencies in Table 18.9. Looking at the two tables, it should be clear that males were less willing to use mental health services and females were more willing than would be expected if the two variables were independent.

LEARNING CHECK

1. A researcher suspects that color blindness is inherited by a sex-linked gene. This possibility is examined by looking for a relationship between gender and color vision. A sample of 1000 people is tested for color blindness, and then they are classified according to their sex and color vision status (normal, red-green blind, other color blindness). Is color blindness related to gender? The data are as follows:

Observed Frequencies of Color Vision Status According to Sex

	Normal Color Vision	Red-Green Color Blindness	Other Color Blindness	Totals
Male	320	70	10	400
Female	580	10	10	600
Totals	900	80	20	

   a. State the hypotheses.

   b. Determine the value for *df*, and locate the critical region.

   c. Compute the f_e values and then chi-square.

   d. Make a decision regarding H_0.

ANSWERS   **1. a.** H_0: In the population, there is no relationship between gender and color vision.
          H_1: In the population, gender and color vision are related.

   b. $df = 2$; critical $\chi^2 = 5.99$ for $\alpha = .05$.

   c. f_e values are as follows:

Expected Frequencies

	Normal	Red-Green	Other
Male	360	32	8
Female	540	48	12

Obtained $\chi^2 = 83.44$

   d. Reject H_0.

18.4 MEASURING EFFECT SIZE FOR THE CHI-SQUARE TEST FOR INDEPENDENCE

A hypothesis test, like the chi-square test for independence, evaluates the statistical significance of the results from a research study. Specifically, the intent of the test is to determine whether the patterns or relationships observed in the sample data are likely to have occurred simply by chance; that is, without any corresponding patterns or relationships in the population. Tests of significance are influenced not only by the size or strength of the treatment effects but also by the size of the samples. As a result, even a small effect can be statistically significant if it is observed in a very large sample. Because a significant effect does not necessarily mean a large effect, it is generally recommended that the outcome of a hypothesis test be accompanied by a measure of the effect size. This general recommendation also applies to the chi-square test for independence.

THE PHI-COEFFICIENT
AND CRAMÉR'S *V*

In Chapter 16 (page 550), we introduced the *phi-coefficient* as a measure of correlation for data consisting of two dichotomous variables (both variables have exactly two values). This same situation exists when the data for a chi-square test for independence form a 2×2 matrix (again, each variable has exactly two values). In this case, it is possible to compute the correlation phi (ϕ) in addition to the chi-square hypothesis test for the same set of data. Because phi (ϕ) is a correlation, it measures the strength of the relationship, rather than the significance, and thus provides a measure of effect size. The value for the phi-coefficient can be computed directly from chi-square by the following formula:

Caution: The value of χ^2 is already a squared value. Do not square it again.

$$\phi = \sqrt{\frac{\chi^2}{n}}$$

(18.6)

The value of the phi-coefficient is determined entirely by the *proportions* in the 2×2 data matrix and is completely independent of the absolute size of the frequencies. The chi-square value, however, is influenced by the proportions and by the size of the frequencies. This distinction is demonstrated in the following example.

EXAMPLE 18.3   The following data show a frequency distribution evaluating the relationship between gender and preference between two candidates for student president.

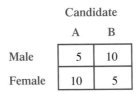

Candidate

	A	B
Male	5	10
Female	10	5

Note that the data show that males prefer candidate B by a 2-to-1 margin and females prefer candidate A by 2 to 1. Also note that the sample includes a total of 15 males and 15 females. We will not perform all the arithmetic here, but these data produce chi-square equal to 3.33 (which is not significant) and a phi-coefficient of 0.3332.

Next we will keep exactly the same proportions in the data, but double all of the frequencies. The resulting data are as follows:

Candidate

	A	B
Male	10	20
Female	20	10

Once again, males prefer candidate B by 2 to 1 and females prefer candidate A by 2 to 1. However, the sample now contains 30 males and 30 females. For these new data, the value of chi-square is 6.66, twice as big as it was before (and now significant with $\alpha = .05$), but the value of the phi-coefficient is still 0.3332.

Because the proportions are the same for the two samples, the value of the phi-coefficient is unchanged. However, the larger sample provides more convincing evidence than the smaller sample, so the larger sample is more likely to produce a significant result.

The interpretation of ϕ follows the same standards used to evaluate a correlation (Table 9.3, page 295 shows the standards for squared correlations): a ϕ value of 0.10 is a small effect, 0.30 is a medium effect, and 0.50 is a large effect. Occasionally, the value of ϕ is squared (ϕ^2) and is reported as a percentage of variance accounted for, exactly the same as r^2.

When the chi-square test involves a matrix larger than 2×2, a modification of the phi-coefficient, known as *Cramér's* V, can be used to measure effect size.

$$V = \sqrt{\frac{\chi^2}{n(df^*)}} \tag{18.7}$$

Note that the formula for Cramér's V (18.7) is identical to the formula for the phi-coefficient (18.6) except for the addition of df^* in the denominator. The df^* value is *not* the same as the degrees of freedom for the chi-square test, but it is related. Recall that the chi-square test for independence has $df = (R - 1)(C - 1)$, where R is the number of rows in the table and C is the number of columns. For Cramér's V, the value of df^* is the smaller of either $(R - 1)$ or $(C - 1)$.

Cohen (1988) has also suggested standards for interpreting Cramér's V that are shown in Table 18.10. Note that when $df^* = 1$, as in a 2×2 matrix, the criteria for interpreting V are exactly the same as the criteria for interpreting a regular correlation or a phi-coefficient.

We use the results from Example 18.2 (page 623) to demonstrate the calculation of Cramér's V. The example evaluated the relationship between gender and willingness to use mental health services. There were two levels of gender and three levels of willingness, producing a 2×3 table with a total of $n = 150$ participants. The data produced $\chi^2 = 8.23$. Using these values we obtain

$$V = \sqrt{\frac{\chi^2}{n(df^*)}} = \sqrt{\frac{8.23}{150(1)}} = \sqrt{0.055} = 0.23$$

According to Cohen's guidelines (Table 18.10), this value indicates a small relationship.

TABLE 18.10

Standards for interpreting Cramér's V as proposed by Cohen (1988).

	Small Effect	Medium Effect	Large Effect
For $df^* = 1$	0.10	0.30	0.50
For $df^* = 2$	0.07	0.21	0.35
For $df^* = 3$	0.06	0.17	0.29

18.5 ASSUMPTIONS AND RESTRICTIONS FOR CHI-SQUARE TESTS

To use a chi-square test for goodness of fit or a test of independence, several conditions must be satisfied. For any statistical test, violation of assumptions and restrictions casts doubt on the results. For example, the probability of committing a Type I error may be distorted when assumptions of statistical tests are not satisfied. Some important assumptions and restrictions for using chi-square tests are the following:

1. Independence of Observations. This is *not* to be confused with the concept of independence between *variables* as seen in the test of independence (Section 18.3). One consequence of independent observations is that each observed frequency is generated by a different subject. A chi-square test would be inappropriate if a person could produce responses that could be classified in more than one category or contribute more than one frequency count to a single category. (See page 253 for more information on independence.)

2. Size of Expected Frequencies. A chi-square test should not be performed when the expected frequency of any cell is less than 5. The chi-square statistic can be distorted when f_e is very small. Consider the chi-square computations for a single cell. Suppose that the cell has values of $f_e = 1$ and $f_o = 5$. The contribution of this cell to the total chi-square value is

$$\text{cell} = \frac{(f_o - f_e)^2}{f_e} = \frac{(5 - 1)^2}{1} = \frac{4^2}{1} = 16$$

Now consider another instance, in which $f_e = 10$ and $f_o = 14$. The difference between the observed and the expected frequencies is still 4, but the contribution of this cell to the total chi-square value differs from that of the first case:

$$\text{cell} = \frac{(f_o - f_e)^2}{f_e} = \frac{(14 - 10)^2}{10} = \frac{4^2}{10} = 1.6$$

It should be clear that a small f_e value can have a great influence on the chi-square value. This problem becomes serious when f_e values are less than 5. When f_e is very small, what would otherwise be a minor discrepancy between f_o and f_e results in large chi-square values. The test is too sensitive when f_e values are extremely small. One way to avoid small expected frequencies is to use large samples.

18.6   SPECIAL APPLICATIONS OF THE CHI-SQUARE TESTS

At the beginning of this chapter, we introduced the chi-square tests as examples of nonparametric tests. Although nonparametric tests serve a function that is uniquely their own, they also can be viewed as alternatives to the common parametric techniques that were examined in earlier chapters. In general, nonparametric tests are used as substitutes for parametric techniques in situations in which one of the following occurs:

1. The data do not meet the assumptions needed for a standard parametric test.
2. The data consist of nominal or ordinal measurements, so that it is impossible to compute standard descriptive statistics such as the mean and standard deviation.

In this section, we examine some of the relationships between chi-square tests and the parametric procedures for which they may substitute.

CHI-SQUARE AND THE PEARSON CORRELATION

The chi-square test for independence and the Pearson correlation are both statistical techniques intended to evaluate the relationship between two variables. The type of data obtained in a research study determines which of these two statistical procedures is appropriate. Suppose, for example, that a researcher is interested in the relationship between self-esteem and academic performance for 10-year-old children. If the researcher obtained numerical scores for both variables, the resulting data would be similar to the values shown in Table 18.11 and the researcher could use a Pearson correlation (Chapter 16) to evaluate the relationship. On the other hand, the researcher

TABLE 18.11

Data appropriate for a Pearson correlation. A numerical score measuring self-esteem and a numerical score measuring academic performance are obtained for each participant.

Participant	Self-Esteem X	Academic Performance Y
A	13	73
B	19	88
C	10	71
D	22	96
E	20	90
F	15	82
.	.	.
.	.	.
.	.	.

may choose simply to classify individuals into categories for both variables. For example, each student could be classified as either high or low for academic performance and classified as high, medium, or low for self-esteem. The resulting data would produce a frequency distribution that could be displayed in a matrix such as the data shown in Table 18.12. Notice that these data do not involve any numerical scores, but rather consist of a set of frequencies appropriate for a chi-square test.

CHI-SQUARE AND THE INDEPENDENT-MEASURES *t* AND ANOVA

Once again, consider a researcher investigating the relationship between self-esteem and academic performance for 10-year-old children. This time, suppose the researcher measured academic performance by simply classifying individuals into two categories, high and low, and then obtained a numerical score for each individual's self-esteem. The resulting data would be similar to the scores in Table 18.13(a), and an independent-measures *t* test would be used to evaluate the mean difference between the two groups of scores. Alternatively, the researcher could measure self-esteem by classifying individuals into three categories: high, medium, and low. If a numerical score is then obtained for each individual's academic performance, the resulting data would look like the scores in Table 18.13(b), and an ANOVA would be used to evaluate the mean differences among the three groups.

The point of these examples is that the chi-square test for independence, the Pearson correlation, and tests for mean differences can all be used to evaluate the relationship between two variables. One main distinction among the different statistical procedures is the form of the data. However, another distinction is the fundamental purpose of these different statistics. The chi-square test and the tests for mean differences (*t* and ANOVA) evaluate the *significance* of the relationship; that is, they determine whether

TABLE 18.12

A frequency distribution showing the level of self-esteem according to the level of academic performance for a sample of $n = 150$ ten-year-old children.

		Level of Self-Esteem			
		High	Medium	Low	
Academic Performance	High	17	32	11	60
	Low	13	43	34	90
		30	75	45	$n = 150$

TABLE 18.13

Data appropriate for an independent-measures *t* test or an ANOVA. In part (a), self-esteem scores are obtained for two groups of students differing in level of academic performance. In part (b), academic performance scores are obtained for three groups of students differing in level of self-esteem.

(a) Self-esteem scores for two groups of students.

Academic Performance	
High	Low
17	13
21	15
16	14
24	20
18	17
15	14
19	12
20	19
18	16

(b) Academic performance scores for three groups of students.

Self-esteem		
High	Medium	Low
94	83	80
90	76	72
85	70	81
84	81	71
89	78	77
96	88	70
91	83	78
85	80	72
88	82	75

the relationship observed in the sample provides enough evidence to conclude that there is a corresponding relationship in the population. You can also evaluate the significance of a Pearson correlation, however, the main purpose of a correlation is to measure the *strength* of the relationship. In particular, squaring the correlation, r^2, provides a measure of effect size, describing the proportion of variance in one variable that is accounted for by its relationship with the other variable.

THE PHI-COEFFICIENT

Earlier in this chapter (page 626), we noted that the relationship between two dichotomous variables (variables that have exactly two values) can be evaluated using either a phi-coefficient correlation or a 2×2 chi-square test for independence. In Example 18.3, we examined the relationship between gender (male/female) and candidate preference in an election (candidate A versus candidate B). The phi-coefficient produces a correlation that measures the strength of the relationship and the chi-square test evaluates the significance of the relationship.

THE MEDIAN TEST FOR INDEPENDENT SAMPLES

The median is the score that divides the population in half, with 50% scoring at or below the median.

The *median test* provides a nonparametric alternative to the independent-measures *t* test (or ANOVA) to determine whether there are significant differences among two or more independent samples. The null hypothesis for the median test states that the different samples come from populations that share a common median (no differences). The alternative hypothesis states that the samples come from populations that are different and do not share a common median.

The logic behind the median test is that whenever several different samples are selected from the same population distribution, roughly half of the scores in each sample should be above the population median and roughly half should be below. That is, all the separate samples should be distributed around the same median. On the other hand, if the samples come from populations with different medians, then the scores in some samples will be consistently higher and the scores in other samples will be consistently lower.

The first step in conducting the median test is to combine all the scores from the separate samples and then find the median for the combined group (see Chapter 3, page 82, for instructions for finding the median). Next, a matrix is constructed with a column for each of the separate samples and two rows: one for individuals above the median and one for individuals below the median. Finally, for each sample, you count how many individuals scored above the combined median and how many scored below. These values are the observed frequencies that are entered in the matrix.

The frequency distribution matrix is evaluated using a chi-square test for independence. The expected frequencies and a value for chi-square are computed exactly as described in Section 18.3. A significant value for chi-square indicates that the discrepancy between the individual sample distributions is greater than would be expected by chance.

The median test is demonstrated in the following example.

EXAMPLE 18.4   The following data represent self-esteem scores obtained from a sample of $n = 40$ children. The children are then separated into three groups based on their level of academic performance (high, medium, low). The median test will evaluate whether there is a significant relationship between self-esteem and level of academic performance.

Self-Esteem Scores for Children
at Three Levels of Academic Performance

High		Medium				Low	
22	14	22	13	24	20	11	19
19	18	18	22	10	16	13	15
12	21	19	15	14	19	20	16
20	18	11	18	11	10	10	18
23	20	12	19	15	12	15	11

The median for the combined group of $n = 40$ scores is $X = 17$ (exactly 20 scores are above this value and 20 are below). For the high performers, 8 out of 10 scores are above the median. For the medium performers, 9 out of 20 are above the median, and for the low performers, only 3 out of 10 are above the median. These observed frequencies are shown in the following matrix:

Academic Performance

	High	Medium	Low
Above Median	8	9	3
Below Median	2	11	7

The expected frequencies for this test are as follows:

Academic Performance

	High	Medium	Low
Above Median	5	10	5
Below Median	5	10	5

The chi-square statistic is

$$\chi^2 = \frac{9}{5} + \frac{9}{5} + \frac{1}{10} + \frac{1}{10} + \frac{4}{5} + \frac{4}{5} = 5.40$$

With $df = 2$ and $\alpha = .05$, the critical value for chi-square is 5.99. The obtained chi-square of 5.40 does not fall in the critical region, so we would fail to reject the null hypothesis. These data do not provide sufficient evidence to conclude that there are significant differences among the self-esteem distributions for these three groups of students.

___

A few words of caution are in order concerning the interpretation of the median test. First, the median test is *not* a test for mean differences. Remember: The mean for a distribution can be strongly affected by a few extreme scores. Therefore, the mean and the median for a distribution are not necessarily the same, and they may not even be related. The results from a median test *cannot* be interpreted as indicating that there is (or is not) a difference between means.

Second, you may have noted that the median test does not directly compare the median from one sample with the median from another. Thus, the median test is not a test for significant differences between medians. Instead, this test compares the distribution of scores for one sample versus the distribution for another sample. If the samples are distributed evenly around a common point (the group median), the test will conclude that there is no significant difference. On the other hand, finding a significant difference simply indicates that the samples are not distributed evenly around the common median. Thus, the best interpretation of a significant result is that there is a *difference in the distributions* of the samples.

SUMMARY

1. Chi-square tests are a type of nonparametric technique that tests hypotheses about the form of the entire frequency distribution. Two types of chi-square tests are the test for goodness of fit and the test for independence. The data for these tests consist of the frequency of observations that fall into various categories of a variable.

2. The test for goodness of fit compares the frequency distribution for a sample to the frequency distribution that is predicted by H_0. The test determines how well the observed frequencies (sample data) fit the expected frequencies (data predicted by H_0).

3. The expected frequencies for the goodness-of-fit test are determined by

$$\text{expected frequency} = f_e = pn$$

where p is the hypothesized proportion (according to H_0) of observations falling into a category and n is the size of the sample.

4. The chi-square statistic is computed by

$$\text{chi-square} = \chi^2 = \Sigma \frac{(f_o - f_e)^2}{f_e}$$

where f_o is the observed frequency for a particular category and f_e is the expected frequency for that category. Large values for χ^2 indicate that there is a large discrepancy between the observed (f_o) and the expected (f_e) frequencies and may warrant rejection of the null hypothesis

5. Degrees of freedom for the test for goodness of fit are

$$df = C - 1$$

where C is the number of categories in the variable. Degrees of freedom measure the number of categories for which f_e values can be freely chosen. As can be seen from the formula, all but the last f_e value to be determined are free to vary.

6. The chi-square distribution is positively skewed and begins at the value of zero. Its exact shape is determined by degrees of freedom.

7. The test for independence is used to assess the relationship between two variables. The null hypothesis states that the two variables in question are independent of each other. That is, the frequency distribution for one variable does not depend on the categories of the second variable. On the other hand, if a relationship does exist, then the form of the distribution for one variable will depend on the categories of the other variable.

8. For the test for independence, the expected frequencies for H_0 can be directly calculated from the marginal frequency totals,

$$f_e = \frac{f_c f_r}{n}$$

where f_c is the total column frequency and f_r is the total row frequency for the cell in question.

9. Degrees of freedom for the test for independence are computed by

$$df = (R - 1)(C - 1)$$

where R is the number of row categories and C is the number of column categories.

10. For the test of independence, a large chi-square value means there is a large discrepancy between the f_o and f_e values. Rejecting H_0 in this test provides support for a relationship between the two variables.

11. Both chi-square tests (for goodness of fit and independence) are based on the assumption that each observation is independent of the others. That is, each observed frequency reflects a different individual, and no individual can produce a response that would be classified in more than one category or more than one frequency in a single category.

12. The chi-square statistic is distorted when f_e values are small. Chi-square tests, therefore, are restricted to situations in which f_e values are 5 or greater. The test should not be performed when the expected frequency of any cell is less than 5.

13. The test for independence also can be used to test whether two sample distributions share a common median. Observed frequencies are obtained by classifying each individual as above or below the median value for the combined samples. A significant value for chi-square indicates a difference between the two sample distributions.

14. The effect size for a chi-square test for independence can be measured computing a phi-coefficient for data that form a 2×2 matrix or computing Cramér's V for a matrix that is larger than 2×2.

$$\text{phi} = \sqrt{\frac{\chi^2}{n}} \qquad \text{Cramér's } V = \sqrt{\frac{\chi^2}{n(df^*)}}$$

where df^* is the smaller of $(R - 1)$ and $(C - 1)$. Both phi and Cramér's V are evaluated using the criteria in Table 18.10.

KEY TERMS

parametric test	observed frequencies	chi-square distribution	Cramér's V
nonparametric test	expected frequencies	test for independence	median test
goodness-of-fit test	chi-square statistic	phi-coefficient	

RESOURCES

There are a tutorial quiz and other learning exercises for Chapter 18 at the Wadsworth website, www.cengage.com/psychology/gravetter. The site also provides access to a workshop entitled *Chi-Square* that reviews the chi-square tests presented in this chapter. See page 29 for more information about accessing items on the website.

WebAssign

Guided interactive tutorials, end-of-chapter problems, and related testbank items may be assigned online at WebAssign.

WebTUTOR

For those using WebTutor along with this book, there is a WebTutor section corresponding to this chapter. The WebTutor contains a brief summary of Chapter 18, hints for learning the concepts and the formulas for the chi-square tests, cautions about common errors, and sample exam items including solutions.

SPSS

General instructions for using SPSS are presented in Appendix D. Following are detailed instructions for using SPSS to perform **The Chi-Square Tests for Goodness of Fit and for Independence** that are presented in this chapter.

The Chi-Square Test for Goodness of Fit

Data Entry

1. Enter the set of observed frequencies in the first column of the SPSS data editor. If there are four categories, for example, enter the four observed frequencies.
2. In the second column, enter the numbers 1, 2, 3, and so on, so there is a number beside each of the observed frequencies in the first column.

Data Analysis

1. Click **Data** on the tool bar at the top of the page and select **weight cases** at the bottom of the list.
2. Click the **weight cases by** circle, then highlight the label for the column containing the observed frequencies (VAR00001) on the left and move it into the **Frequency Variable** box by clicking on the arrow.
3. Click **OK.**
4. Click **Analyze** on the tool bar, select **Nonparametric Tests,** and click on **Chi-Square.**
5. Highlight the label for the column containing the digits 1, 2, 3, and move it into the Test Variables box by clicking on the arrow.
6. To specify the expected frequencies, you can either use the **all categories equal** option, which will automatically compute expected frequencies, or you can enter your own values. To enter your own expected frequencies, click on the **values** option, and one by one enter the expected frequencies into the small box and click **Add** to add each new value to the bottom of the list.
7. Click **OK.**

SPSS Output

The program will produce a table showing the complete set of observed and expected frequencies. A second table provides the value for the chi-square statistic, the degrees of freedom, and the level of significance (the *p* value or alpha level for the test).

The Chi-Square Test for Independence

Data Entry

1. Enter the complete set of observed frequencies in one column of the SPSS data editor (VAR00001).
2. In a second column, enter a number (1, 2, 3, etc.) that identifies the row corresponding to each observed frequency. For example, enter a 1 beside each observed frequency that came from the first row.
3. In a third column, enter a number (1, 2, 3, etc.) that identifies the column corresponding to each observed frequency. Each value from the first column gets a 1, and so on.

Data Analysis

1. Click **Data** on the tool bar at the top of the page and select **weight cases** at the bottom of the list.
2. Click the **weight cases by** circle, then highlight the label for the column containing the observed frequencies (VAR00001) on the left and move it into the **Frequency Variable** box by clicking on the arrow.
3. Click **OK.**
4. Click **Analyze** on the tool bar at the top of the page, select **Descriptive Statistics,** and click on **Crosstabs.**
5. Highlight the label for the column containing the rows (VAR00002) and move it into the **Rows** box by clicking on the arrow.
6. Highlight the label for the column containing the columns (VAR00003) and move it into the **Columns** box by clicking on the arrow.
7. Click on **Statistics,** select **Chi-Square,** and click **Continue.**
8. Click **OK.**

SPSS Output

The program will produce a table presenting valid and missing cases. You should have no missing cases. A second table presents a summary of the observed frequencies. The final table, labeled **Chi-Square Tests,** reports the results. Focus on the top row, the **Pearson Chi-Square,** which reports the calculated chi-square value, the degrees of freedom, and the level of significance (the *p* value or the alpha level for the test).

FOCUS ON PROBLEM SOLVING

1. The expected frequencies that you calculate must satisfy the constraints of the sample. For the goodness-of-fit test, $\Sigma f_e = \Sigma f_o = n$. For the test of independence, the row totals and column totals for the expected frequencies should be identical to the corresponding totals for the observed frequencies.

2. It is entirely possible to have fractional (decimal) values for expected frequencies. Observed frequencies, however, are always whole numbers.

3. Whenever $df = 1$, the difference between observed and expected frequencies $(f_o - f_e)$ will be identical (the same value) for all cells. This makes the calculation of chi-square easier.

4. Although you are advised to compute expected frequencies for all categories (or cells), you should realize that it is not essential to calculate all f_e values separately. Remember that df for chi-square identifies the number of f_e values that are free to vary. Once you have calculated that number of f_e values, the remaining f_e values are determined. You can get these remaining values by subtracting the calculated f_e values from their corresponding row or column totals.

5. Remember that, unlike previous statistical tests, the degrees of freedom (df) for a chi-square test are *not* determined by the sample size (n). Be careful!

DEMONSTRATION 18.1

TEST FOR INDEPENDENCE

A manufacturer of watches would like to examine preferences for digital versus analog watches. A sample of $n = 200$ people is selected, and these individuals are classified by age and preference. The manufacturer would like to know whether there is a relationship between age and watch preference. The observed frequencies (f_o) are as follows:

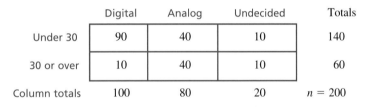

	Digital	Analog	Undecided	Totals
Under 30	90	40	10	140
30 or over	10	40	10	60
Column totals	100	80	20	$n = 200$

STEP 1   State the hypotheses, and select an alpha level.
The null hypothesis states that there is no relationship between the two variables.

H_0: Preference is independent of age. That is, the frequency distribution of preference has the same form for people younger than 30 as for people older than 30.

The alternative hypothesis states that there is a relationship between the two variables.

H_1: Preference is related to age. That is, the type of watch preferred depends on a person's age.

We will set alpha to $\alpha = .05$.

STEP 2   Locate the critical region.

Degrees of freedom for the chi-square test for independence are determined by

$$df = (C - 1)(R - 1)$$

For these data,

$$df = (3 - 1)(2 - 1) = 2(1) = 2$$

For $df = 2$ with $\alpha = .05$, the critical chi-square value is 5.99. Thus, our obtained chi-square must exceed 5.99 to be in the critical region and to reject H_0.

STEP 3   Compute the test statistic. Computing chi-square requires two calculations: finding the expected frequencies and calculating the chi-square statistic.

Expected frequencies, f_e. For the test for independence, the expected frequencies can be found using the column totals (f_c), the row totals (f_r), and the following formula:

$$f_e = \frac{f_c f_r}{n}$$

For people younger than 30, we obtain the following expected frequencies:

$$f_e = \frac{100(140)}{200} = \frac{14{,}000}{200} = 70 \text{ for digital}$$

$$f_e = \frac{80(140)}{200} = \frac{11{,}200}{200} = 56 \text{ for analog}$$

$$f_e = \frac{20(140)}{200} = \frac{2800}{200} = 14 \text{ for undecided}$$

For individuals 30 or older, the expected frequencies are as follows:

$$f_e = \frac{100(60)}{200} = \frac{6000}{200} = 30 \text{ for digital}$$

$$f_e = \frac{80(60)}{200} = \frac{4800}{200} = 24 \text{ for analog}$$

$$f_e = \frac{20(60)}{200} = \frac{1200}{200} = 6 \text{ for undecided}$$

The following table summarizes the expected frequencies:

	Digital	Analog	Undecided
Under 30	70	56	14
30 or over	30	24	6

The chi-square statistic. The chi-square statistic is computed from the formula

$$\chi^2 = \Sigma \frac{(f_o - f_e)^2}{f_e}$$

The following table summarizes the calculations:

Cell	f_o	f_c	$(f_o - f_c)$	$(f_o - f_c)^2$	$(f_o - f_c)^2/f_c$
Under 30—digital	90	70	20	400	5.71
Under 30—analog	40	56	-16	256	4.57
Under 30—undecided	10	14	-4	16	1.14
30 or over—digital	10	30	-20	400	13.33
30 or over—analog	40	24	16	256	10.67
30 or over—undecided	10	6	4	16	2.67

Finally, we can find the sum of the last column to get the chi-square value.

$$\chi^2 = 5.71 + 4.57 + 1.14 + 13.33 + 10.67 + 2.67$$

$$= 38.09$$

STEP 4   Make a decision about H_0, and state the conclusion.
The chi-square value is in the critical region. Therefore, we can reject the null hypothesis. There is a relationship between watch preference and age, $\chi^2(2, n = 200) = 38.09, p < .05$.

PROBLEMS

1. Parametric tests (such as t or ANOVA) differ from nonparametric tests (such as chi-square) primarily in terms of the assumptions they require and the data they use. Explain these differences.

2. The student population at the state college consists of 55% females and 45% males.
   a. The college theater department recently staged a production of a modern musical. A researcher recorded the gender of each student entering the theater and found a total of 385 females and 215 males. Is the gender distribution for theater goers significantly different from the distribution for the general college? Test at the .05 level of significance.
   b. The same researcher also recorded the gender of each student watching a men's basketball game in the college gym and found a total of 83 females and 97 males. Is the gender distribution for basketball fans significantly different from the distribution for the general college? Test at the .05 level of significance.

3. A researcher is investigating the physical characteristics that influence whether a person's face is judged as beautiful. The researcher selects a photograph of a woman and then creates two modifications of the photo by (1) moving the eyes slightly farther apart and (2) moving the eyes slightly closer together. The original photograph and the two modifications are then shown to a sample of $n = 150$ college students, and each student is asked to select the "most beautiful" of the three faces. The distribution of responses was as follows:

Original Photo	Eyes Moved Apart	Eyes Moved Together
51	72	27

Do the data indicate any significant preferences among the three versions of the photograph? Test at the .05 level of significance.

4. Data from the Department of Motor Vehicles indicate that 80% of all licensed drivers are older than age 25.
   a. In a sample of $n = 50$ people who recently received speeding tickets, 32 were older than 25 years and the other 18 were age 25 or younger. Is the age distribution for this sample significantly different from the distribution for the population of licensed drivers? Use $\alpha = .05$.
   b. In sample of $n = 50$ people who recently received parking tickets, 38 were older than 25 years and the other 12 were age 25 or younger. Is the age distribution for this sample significantly different from the distribution for the population of licensed drivers? Use $\alpha = .05$.

5. To investigate the phenomenon of "home team advantage," a researcher recorded the outcomes from 64 college football games on one Saturday in October. Of the 64 games, 42 were won by home teams. Does this result provide enough evidence to conclude that home

teams win significantly more than would be expected by chance? Assume that winning and losing are equally likely events if there is no home team advantage. Use $\alpha = .05$.

6. Langewitz, Izakovic, and Wyler (2005) reported that self-hypnosis can significantly reduce hay fever symptoms. Patients with moderate to severe allergic reactions were trained to focus their minds on specific locations where their allergies do not bother them, such as a beach or a ski resort. In a sample of 64 patients who received this training, suppose that 47 showed reduced allergic reactions and 17 showed an increase in allergic reactions. Are these results sufficient to conclude that the self-hypnosis has a significant effect? Assume that increasing and decreasing allergic reactions are equally likely if the training has no effect. Use $\alpha = .05$.

7. Research has demonstrated that people tend to be attracted to others who are similar to themselves. One study demonstrated that individuals are disproportionately more likely to marry those with surnames that begin with the same last letter as their own (Jones, Pelham, Carvallo, & Mirenberg, 2004). The researchers began by looking at marriage records and recording the surname for each groom and the maiden name of each bride. From these records it is possible to calculate the probability of randomly matching a bride and a groom whose last names begin with the same letter. Suppose that this probability is only 6.5%. Next, a sample of $n = 200$ married couples is selected and the number who shared the same last initial at the time they were married is counted. The resulting observed frequencies are as follows:

Same Initial	Different Initials	
19	181	200

Do these data indicate that the number of couples with the same last initial is significantly different that would be expected if couples were matched randomly? Test with $\alpha = .05$.

8. Suppose that the researcher from the previous problem repeated the study of married couples' initials using twice as many participants and obtaining observed frequencies that exactly double the original values. The resulting data are as follows:

Same Initial	Different Initials	
38	362	400

a. Use a chi-square test to determine whether the number of couples with the same last initial is significantly different that would be expected if couples were matched randomly. Test with $\alpha = .05$.

b. You should find that the data lead to rejecting the null hypothesis. However, in problem 7 the decision was fail to reject. How do you explain the fact that the two samples have the same proportions but lead to different conclusions?

9. A professor in the psychology department would like to determine whether there has been a significant change in grading practices over the years. It is known that the overall grade distribution for the department in 1985 had 14% As, 26% Bs, 31% Cs, 19% Ds, and 10% Fs. A sample of $n = 200$ psychology students from last semester produced the following grade distribution:

A	B	C	D	F
32	61	64	31	12

Do the data indicate a significant change in the grade distribution? Test at the .05 level of significance.

10. Automobile insurance is much more expensive for teenage drivers than for older drivers. To justify this cost difference, insurance companies claim that the younger drivers are much more likely to be involved in costly accidents. To test this claim, a researcher obtains information about registered drivers from the department of motor vehicles and selects a sample of $n = 300$ accident reports from the police department. The motor vehicle department reports the percentage of registered drivers in each age category as follows: 16% are under age 20; 28% are 20 to 29 years old; and 56% are age 30 or older. The number of accident reports for each age group is as follows:

Under age 20	Age 20–29	Age 30 or older
68	92	140

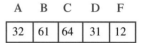

Do the data indicate that the distribution of accidents for the three age groups is significantly different from the distribution of drivers? Test with $\alpha = .05$.

11. The color red is often associated with anger and male dominance. Based on this observation, Hill and Barton (2005) monitored the outcome of four combat sports (boxing, tae kwan do, Greco-Roman wrestling, and freestyle wrestling) during the 2004 Olympic games

and found that participants wearing red outfits won significantly more often than those wearing blue.

a. In 50 wrestling matches involving red versus blue, suppose that the red outfit won 31 times and lost 19 times. Is this result sufficient to conclude that red wins significantly more than would be expected by chance? Test at the .05 level of significance.

b. In 100 matches, suppose red won 62 times and lost 38. Is this sufficient to conclude that red wins significantly more than would be expected by chance? Again, use $\alpha = .05$.

c. Note that the winning percentage for red uniforms in part a is identical to the percentage in part b (31 out of 50 is 62%, and 62 out of 100 is also 62%). Although the two samples have an identical winning percentages, one is significant and the other is not. Explain why the two samples lead to different conclusions.

12. A communications company has developed three new designs for a cell phone. To evaluate consumer response, a sample of 120 college students is selected and each student is given all three phones to use for 1 week. At the end of the week, the students must identify which of the three designs they prefer. The distribution of preference is as follows:

Design 1	Design 2	Design 3
54	38	28

Do the results indicate any significant preferences among the three designs?

13. In problem 12, a researcher asked college students to evaluate three new cell phone designs. However, the researcher suspects that college students may have criteria that are different from those used by older adults. To test this hypothesis, the researcher repeats the study using a sample of $n = 60$ older adults in addition to a sample of $n = 60$ students. The distribution of preference is as follows:

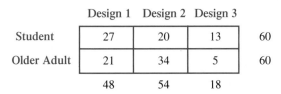

	Design 1	Design 2	Design 3	
Student	27	20	13	60
Older Adult	21	34	5	60
	48	54	18	

Do the data indicate that the distribution of preferences for older adults is significantly different

that the distribution for college students? Test with $\alpha = .05$.

14. Mulvihill, Obuseh, and Caldwell (2008) conducted a survey evaluating healthcare providers' perception of a new state children's insurance program. One question asked the providers whether they viewed the reimbursement from the new insurance as higher, lower, or the same as private insurance. Another question assessed the providers' overall satisfaction with the new insurance. The following table presents observed frequencies similar to the study results.

	Satisfied	Not Satisfied	
Less Reimbursement	46	54	100
Same or More Reimbursement	42	18	60
	88	72	

Do the results indicate that the providers' satisfaction of the new program is related to their perception of the reimbursement rates? Test with $\alpha = .05$.

15. Friedman and Rosenman (1974) have suggested that personality type is related to heart disease. Specifically, type A people, who are competitive, driven, pressured, and impatient, are more prone to heart disease. On the other hand, type B individuals, who are less competitive and more relaxed, are less likely to have heart disease. Suppose that an investigator would like to examine the relationship between personality type and disease. For a random sample of individuals, personality type is assessed with a standardized test. These individuals are then examined and categorized as to whether they have a heart disorder. The observed frequencies are as follows:

	No Heart Disease	Heart Disease	
Type A	32	18	50
Type B	128	22	150
	160	40	

Is there a relationship between personality and disorder? Test at the .05 level of significance.

16. A local county is considering a budget proposal that would allocate extra funding toward the renovation

of city parks. A survey is conducted to measure public opinion concerning the proposal. A total of 150 individuals respond to the survey: 50 who live within the city limits and 100 from the surrounding suburbs. The frequency distribution is as follows:

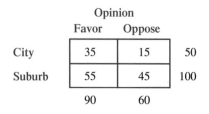

Opinion

	Favor	Oppose	
City	35	15	50
Suburb	55	45	100
	90	60	

a. Is there a significant difference in the distribution of opinions for city residents compared to those in the suburbs? Test at the .05 level of significance.
b. The relationship between home location and opinion can also be evaluated using the phi-coefficient. If the phi-coefficient were computed for these data, what value would be obtained for phi?

17. The data from problem 16 show no significant difference between the opinions for city residents and those who live in the suburbs. To construct the following data, we simply doubled the sample size from problem 15 so that all of the individual frequencies are twice as big. Notice that the sample *proportions* have not changed.

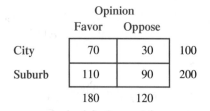

Opinion

	Favor	Oppose	
City	70	30	100
Suburb	110	90	200
	180	120	

a. Test for a significant difference between the city distribution and the suburb distribution using $\alpha = .05$. How does the decision compare with the decision in problem 16? You should find that a larger sample increases the likelihood of a significant result.
b. Compute the phi-coefficient for these data and compare it with the result from problem 16. You should find that the sample size has no effect on the strength of the relationship.

18. Research results suggest that IQ scores for boys are more variable than IQ scores for girls (Arden & Plomin, 2006). A typical study looking at 10-year-old children classifies participants by gender and by low, average, or high IQ. Following are hypothetical data representing the research results. Do the data indicate a significant difference between the frequency

distributions for males and females? Test at the .05 level of significance and describe the difference.

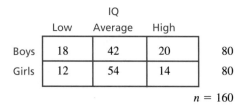

IQ

	Low	Average	High	
Boys	18	42	20	80
Girls	12	54	14	80

$n = 160$

19. Research has demonstrated that involvement in religious activities seems to benefit health and well-being. For example, Reyes-Ortiz, Ayele, Mulligan, Espino, Berges, and Markides (2006) found that church attendance was significantly related to fear of falling in a sample of Mexican-Americans aged 70 and over. The following frequency distribution represents data similar to the results obtained in the study. Do the data indicate a significant relationship between church attendance and fear of falling? Test at the .05 level of significance.

Fear of Falling

	No Fear	Somewhat Afraid	Fairly Afraid	Very Afraid	
Church	51	68	52	29	200
No Church	15	28	29	28	100

$n = 300$

20. Gender differences in dream content are well documented (see Winget & Kramer, 1979). Suppose a researcher studies aggression content in the dreams of men and women. Each participant reports his or her most recent dream. Then each dream is judged by a panel of experts to have low, medium, or high aggression content. The observed frequencies are shown in the following matrix:

Aggression Content

		Low	Medium	High
Gender	Female	18	4	2
	Male	4	17	15

Is there a relationship between gender and the aggression content of dreams? Test with $\alpha = .01$.

21. Recent reports suggest that children who grow up with pets in the home tend to develop resistence to allergies. To test this phenomenon, a researcher interviews a sample of $n = 120$ college students. Each student is asked about pets during childhood and about his or her current allergies.

Number of Dogs or Cats in Childhood Home

	0	1	2 or More	
No Allergies	22	50	18	90
Allergies	18	10	2	30
	40	60	20	

a. Do the data indicate a significant relationship between having pets and developing resistance to allergies? Test at the .05 level of significance.
b. Compute Cramér's *V* to measure the effect size.

22. Research indicates that people who volunteer to participate in research studies tend to have higher intelligence than nonvolunteers. To test this phenomenon, a researcher obtains a sample of 200 high school students. The students are given a description of a psychological research study and asked whether they would volunteer to participate. The researcher also obtains an IQ score for each student and classifies the students into high, medium, and low IQ groups. Do the following data indicate a significant relationship between IQ and volunteering? Test at the .05 level of significance.

IQ

	High	Medium	Low	
Volunteer	43	73	34	150
Not Volunteer	7	27	16	50
	50	100	50	

23. Cialdini, Reno, and Kallgren (1990) examined how people conform to norms concerning littering. The researchers wanted to determine whether a person's tendency to litter depended on the amount of litter already in the area. People were handed a handbill as they entered an amusement park. The entrance area had already been prepared with either no litter, a small amount of litter, or a lot of litter lying on the ground. The people were observed to determine whether they dropped their handbills. The frequency data are as follows:

Amount of Litter

	None	Small Amount	Large Amount
Littering	17	28	49
Not Littering	73	62	41

a. Do the data indicate that people's tendency to litter depends on the amount of litter already on the ground? That is, is there a significant relationship between littering and the amount of existing litter? Test at the .05 level of significance.
b. Compute Cramér's *V* to measure the size of the treatment effect.

24. Although the phenomenon is not well understood, it appears that people born during the winter months are slightly more likely to develop schizophrenia than people born at other times (Bradbury & Miller, 1985). The following hypothetical data represent a sample of 50 individuals diagnosed with schizophrenia and a sample of 100 people with no psychotic diagnosis. Each individual is also classified according to season in which he or she was born. Do the data indicate a significant relationship between schizophrenia and the season of birth? Test at the .05 level of significance.

Season of Birth

	Summer	Fall	Winter	Spring	
No Disorder	26	24	22	28	100
Schizophrenia	9	11	18	12	50
	35	35	40	40	

CHAPTER 19

The Binomial Test

Tools You Will Need

The following items are considered essential background material for this chapter. If you doubt your knowledge of any of these items, you should review the appropriate chapter or section before proceeding.

- Binomial distribution (Chapter 6)
- z-score hypothesis tests (Chapter 8)
- Chi-square test for goodness of fit (Chapter 18)

Preview

19.1   Overview

19.2   The Binomial Test

19.3   The Relationship Between Chi-Square and the Binomial Test

19.4   The Sign Test

Summary

Focus on Problem Solving

Demonstration 19.1

Problems

Preview

In 1960, Gibson and Walk designed a classic piece of apparatus to test depth perception. Their device, called a *visual cliff,* consisted of a wide board with a deep drop (the cliff) to one side and a shallow drop on the other side. An infant was placed on the board and then observed to see whether he or she crawled off the shallow side or crawled off the cliff. (*Note:* Infants who moved to the deep side actually crawled onto a sheet of heavy glass, which prevented them from falling. Thus, the deep side only appeared to be a cliff—hence the name *visual cliff.*)

Gibson and Walk reasoned that if infants are born with the ability to perceive depth, they would recognize the deep side and not crawl off the cliff. On the other hand, if depth perception is a skill that develops over time through learning and experience, then infants should not be able to perceive any difference between the shallow and the deep sides.

Out of 27 infants who moved off the board, only 3 ventured onto the deep side at any time during the experiment. The other 24 infants stayed exclusively on the shallow side. Gibson and Walk interpreted these data as convincing evidence that depth perception is innate. The infants showed a systematic preference for the shallow side. **The Problem:** You should notice immediately that the data from this experiment are different from the data that we usually encounter. There are no scores. Gibson and Walk simply counted the number of infants who went off the deep side and the number who went to the shallow side. Still, we would like to use these data to make statistical decisions. Do these sample data provide sufficient evidence to make a confident conclusion about depth perception in the population? Suppose that 8 of the 27 infants had crawled to the deep side. Would you still be convinced that there is a significant preference for the shallow side? What about 12 out of 27?

The Solution: We are asking a question about statistical significance and will need a hypothesis test to otain an answer. The null hypothesis for the Gibson and Walk study would state that infants have no depth perception and cannot perceive a difference between the shallow and deep sides. In this case, their movment should be random with half going to either side. Notice that the data and the hypothesis both concern frequencies or proportions. This situation is perfect for the chi-square test introduced in Chapter 18, and a chi-square test can be used to evaluate the data. However, when individuals are classfied into exactly two categories (for example, shallow and deep), a special statistical procedure exists. In this chapter, we introduce the *binomial test,* which is used to evaluate and interpret frequency data involving exactly two categories of classification.

19.1 OVERVIEW

Data with exactly two categories are also known as dichotomous data.

In Chapter 6, we introduced the concept of *binomial data.* You should recall that binomial data exist whenever a measurement procedure classifies individuals into exactly two distinct categories. For example, the outcomes from tossing a coin can be classified as heads and tails; people can be classified as male or female; plastic products can be classified as recyclable or nonrecyclable. In general, binomial data exist when

1. The measurement scale consists of exactly two categories.

2. Each individual observation in a sample is classified in only one of the two categories.

3. The sample data consist of the frequency or number of individuals in each category.

The traditional notation system for binomial data identifies the two categories as *A* and *B* and identifies the probability (or proportion) associated with each category as *p* and *q,* respectively. For example, a coin toss results in either heads (*A*) or tails (*B*), with probabilities $p = \frac{1}{2}$ and $q = \frac{1}{2}$.

In this chapter, we examine the statistical process of using binomial data for testing hypotheses about the values of *p* and *q* for the population. This type of hypothesis test is called a *binomial test.*

DEFINITION

> A **binomial test** uses sample data to evaluate hypotheses about the values of p and q for a population consisting of binomial data.

Consider the following two situations:

1. In a sample of $n = 34$ color-blind students, 30 are male, and only 4 are female. Does this sample indicate that color blindness is significantly more common for males in the general population?

2. In 1990, only 10% of American families had incomes below the poverty level. This year, in a sample of 100 families, 19 were below the poverty level. Does this sample indicate that there has been a significant change in the population proportions?

Notice that both of these examples have binomial data (exactly two categories). Although the data are relatively simple, we are asking the same statistical question about significance that is appropriate for a hypothesis test: Do the sample data provide sufficient evidence to make a conclusion about the population?

HYPOTHESES FOR THE BINOMIAL TEST

In the binomial test, the null hypothesis specifies exact values for the population proportions p and q. Theoretically, you could choose any proportions for H_0, but usually there is a clear reason for the values that are selected. The null hypothesis typically falls into one of the following two categories:

1. Just Chance. Often the null hypothesis states that the two outcomes A and B occur in the population with the proportions that would be predicted simply by chance. If you were tossing a coin, for example, the null hypothesis might specify $p(\text{heads}) = \frac{1}{2}$ and $p(\text{tails}) = \frac{1}{2}$. Notice that this hypothesis states the usual, chance proportions for a balanced coin. Also notice that it is not necessary to specify both proportions. Once the value of p is identified, the value of q is determined by $1 - p$. For the coin toss example, the null hypothesis would simply state

$$H_0: \quad p = p(\text{heads}) = \tfrac{1}{2} \quad \text{(The coin is balanced.)}$$

Similarly, if you were selecting cards from a deck and trying to predict the suit on each draw, the probability of predicting correctly would be $p = \frac{1}{4}$ for any given trial. (With four suits, you have a 1-out-of-4 chance of guessing correctly.) In this case, the null hypothesis would state

$$H_0: \quad p = p(\text{guessing correctly}) = \tfrac{1}{4} \quad \text{(The outcome is simply due to chance.)}$$

In each case, the null hypothesis simply states that there is nothing unusual about the population; that is, the outcomes are occurring by chance.

2. No Change or No Difference. Often you may know the proportions for one population and want to determine whether the same proportions apply to a different population. In this case, the null hypothesis would simply specify that there is no difference between the two populations. Suppose that national statistics indicate that 1 out of 12 drivers will be involved in a traffic accident during the next year. Does this same proportion apply to 16-year-olds who are driving for the first time? According to the null hypothesis,

$$H_0: \quad \text{For 16-year-olds, } p = p(\text{accident}) = \tfrac{1}{12} \quad \text{(No different from the general population)}$$

Similarly, suppose that last year, 30% of the freshman class failed the college writing test. This year, the college is requiring all freshmen to take a writing course. Will the course have any effect on the number who fail the test? According to the null hypothesis,

H_0: For this year, $p = p(\text{fail}) = 30\%$ (No different from last year's class)

THE DATA FOR THE BINOMIAL TEST

For the binomial test, a sample of n individuals is obtained and you simply count how many are classified in category A and how many are classified in category B. We focus attention on category A and use the symbol X to stand for the number of individuals classified in category A. Recall from Chapter 6 that X can have any value from 0 to n and that each value of X has a specific probability. The distribution of probabilities for each value of X is called the *binomial distribution*. Figure 19.1 shows an example of a binomial distribution where X is the number of heads obtained in four tosses of a balanced coin.

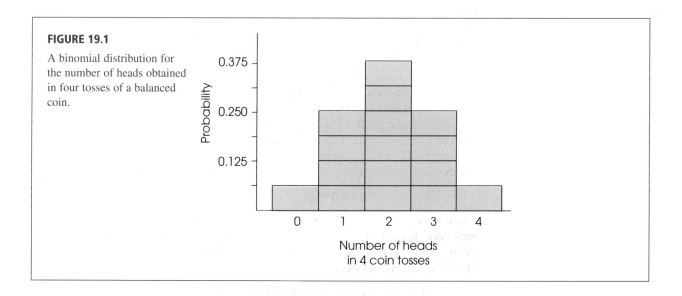

FIGURE 19.1

A binomial distribution for the number of heads obtained in four tosses of a balanced coin.

THE TEST STATISTIC FOR THE BINOMIAL TEST

As we noted in Chapter 6, when the values pn and qn are both equal to or greater than 10, the binomial distribution approximates a normal distribution. This fact is important because it allows us to compute z-scores and use the unit normal table to answer probability questions about binomial events. In particular, when pn and qn are both at least 10, the binomial distribution will have the following properties:

1. The shape of the distribution is approximately normal.

2. The mean of the distribution is $\mu = pn$.

3. The standard deviation of the distribution is

$\sigma = \sqrt{npq}$

With these parameters in mind, it is possible to compute a z-score corresponding to each value of X in the binomial distribution.

$$z = \frac{X - \mu}{\sigma} = \frac{X - pn}{\sqrt{npq}} \qquad \text{(See Equation 6.3.)} \qquad (19.1)$$

This is the basic z-score formula that is used for the binomial test. However, we modify the formula slightly to make it more compatible with the logic of the binomial hypothesis test. The modification consists of dividing both the numerator and the denominator of the z-score by n. (You should realize that dividing both the numerator and the denominator by the same value does not change the value of the z-score.) The resulting equation is

$$z = \frac{X/n - p}{\sqrt{pq/n}} \qquad (19.2)$$

For the binomial test, the values in this formula are defined as follows:

1. X/n is the proportion of individuals in the sample who are classified in category A.

2. p is the hypothesized value (from H_0) for the proportion of individuals in the population who are classified in category A.

3. $\sqrt{pq/n}$ is the standard error for the sampling distribution of X/n and provides a measure of the standard distance between the sample statistic (X/n) and the population parameter (p).

Thus, the structure of the binomial z-score (Equation 19.2) can be expressed as

$$z = \frac{X/n - p}{\sqrt{pq/n}} = \frac{\begin{matrix}\text{sample} \\ \text{proportion} \\ \text{(data)}\end{matrix} - \begin{matrix}\text{hypothesized} \\ \text{population} \\ \text{proportion}\end{matrix}}{\text{standard error}}$$

The logic underlying the binomial test is exactly the same as we encountered with the original z-score hypothesis test in Chapter 8. The hypothesis test involves comparing the sample data with the hypothesis. If the data are consistent with the hypothesis, we conclude that the hypothesis is reasonable. But if there is a big discrepancy between the data and the hypothesis, we reject the hypothesis. The value of the standard error provides a benchmark for determining whether the discrepancy between the data and the hypothesis is more than would be expected by chance. The alpha level for the test provides a criterion for deciding whether the discrepancy is significant. The hypothesis-testing procedure is demonstrated in the following section.

LEARNING CHECK

1. In the Preview we described a research study using a visual cliff. State the null hypothesis for this study in words and as a probability value (p) that an infant will crawl off the deep side.

2. If the visual cliff study had used a sample of $n = 15$ infants, would it be appropriate to use the normal approximation to the binomial distribution? Explain why or why not.

3. If the results from the visual cliff study showed that only 3 out of 36 infants crawled off the deep side, what z-score value would be obtained using Equation 19.1?

ANSWERS

1. The null hypothesis states that the probability of choosing between the deep side and the shallow side is just chance: $p(\text{deep side}) = \frac{1}{2}$.

2. The normal approximation to the binomial distribution requires that both pn and qn are at least 10. With $n = 15$, $pn = qn = 7.5$. The normal approximation should not be used.

3. With $X = 3$, $\mu = \frac{1}{2}(36) = 18$, and $\sigma = \sqrt{\left(\frac{1}{2}\right)\left(\frac{1}{2}\right)(36)} = \sqrt{9} = 3$, the z-score is

   $z = -15/3 = -5.00$.

19.2 THE BINOMIAL TEST

The binomial test follows the same four-step procedure presented earlier with other examples for hypothesis testing. The four steps are summarized as follows.

STEP 1 *State the hypotheses.* In the binomial test, the null hypothesis specifies values for the population proportions p and q. Typically, H_0 specifies a value only for p, the proportion associated with category A. The value of q is directly determined from p by the relationship $q = 1 - p$. Finally, you should realize that the hypothesis, as always, addresses the probabilities or proportions for the *population*. Although we use a sample to test the hypothesis, the hypothesis itself always concerns a population.

STEP 2 *Locate the critical region.* When both values for pn and qn are greater than or equal to 10, the z-scores defined by Equation 19.1 or 19.2 form an approximately normal distribution. Thus, the unit normal table can be used to find the boundaries for the critical region. With $\alpha = .05$, for example, you may recall that the critical region is defined as z-score values greater than $+1.96$ or less than -1.96.

STEP 3 *Compute the test statistic (z-score).* At this time, you obtain a sample of n individuals (or events) and count the number of times category A occurs in the sample. The number of occurrences of A in the sample is the X value for Equation 19.1 or 19.2. Because the two z-score equations are equivalent, you may use either one for the hypothesis test. Usually Equation 19.1 is easier to use because it involves larger numbers (fewer decimals) and it is less likely to be affected by rounding error.

STEP 4 *Make a decision.* If the z-score for the sample data is in the critical region, you reject H_0 and conclude that the discrepancy between the sample proportions and the hypothesized population proportions is significantly greater than chance. That is, the data are not consistent with the null hypothesis, so H_0 must be wrong. On the other hand, if the z-score is not in the critical region, you fail to reject H_0.

The following example demonstrates a complete binomial test.

EXAMPLE 19.1 In the Preview section, we described the *visual cliff* experiment designed to examine depth perception in infants. To summarize briefly, an infant is placed on a wide board that appears to have a deep drop on one side and a relatively shallow drop on the other. An infant who is able to perceive depth should avoid the deep side and move toward the shallow side. Without depth perception, the infant should show no preference between the two sides. Of the 27 infants in the experiment, 24 stayed exclusively on the shallow side and only 3 moved onto the deep side. The purpose of the hypothesis test is to determine whether these data demonstrate that infants have a significant preference for the shallow side.

This is a binomial hypothesis-testing situation. The two categories are

$$A = \text{move onto the deep side}$$

$$B = \text{move onto the shallow side}$$

STEP 1   The null hypothesis states that for the general population of infants, there is no preference between the deep and the shallow sides; the direction of movement is determined by chance. In symbols,

$$H_0: \quad p = p(\text{deep side}) = \tfrac{1}{2} \qquad \left(\text{and } q = \tfrac{1}{2}\right)$$

$$H_1: \quad p \neq \tfrac{1}{2} \qquad \text{(There is a preference.)}$$

We will use $\alpha = .05$.

STEP 2   With a sample of $n = 27$, $pn = 13.5$ and $qn = 13.5$. Both values are greater than 10, so the distribution of z-scores is approximately normal. With $\alpha = .05$, the critical region is determined by boundaries of $z = \pm 1.96$.

STEP 3   For this experiment, the data consist of $X = 3$ out of $n = 27$. Using Equation 19.1, these data produce a z-score value of

$$z = \frac{X - pn}{\sqrt{npq}} = \frac{3 - 13.5}{\sqrt{27\left(\frac{1}{2}\right)\left(\frac{1}{2}\right)}} = \frac{-10.5}{2.60} = -4.04$$

To use Equation 19.2, you first compute the sample proportion, $X/n = 3/27 = 0.111$. The z-score is then

$$z = \frac{X/n - p}{\sqrt{pq/n}} = \frac{0.111 - 0.5}{\sqrt{\frac{1}{2}\left(\frac{1}{2}\right)/27}} = \frac{-0.389}{0.096} = -4.05$$

Within rounding error, the two equations produce the same result.

STEP 4   Because the data are in the critical region, our decision is to reject H_0. These data do provide sufficient evidence to conclude that there is a significant preference for the shallow side. Gibson and Walk (1960) interpreted these data as convincing evidence that depth perception is innate.

REAL LIMITS AND THE BINOMIAL TEST

In Chapter 6, we noted that a binomial distribution forms a discrete histogram (see Figure 19.1), whereas the normal distribution is a continuous curve. The difference between the two distributions was illustrated in Figure 6.17, which is repeated here as Figure 19.2. In the figure, note that each score in the binomial distribution is represented by a bar in the histogram. For example, a score of $X = 6$ actually corresponds to a bar that reaches from a lower real limit of $X = 5.5$ to an upper real limit of 6.5.

When conducting a hypothesis test with the binomial distribution, the basic question is whether a specific score is located in the critical region. However, because each score actually corresponds to an interval, it is possible that part of the score is in the critical region and part is not. Fortunately, this is usually not an

FIGURE 19.2

The relationship between the binomial distribution and the normal distribution. The binomial distribution is always a discrete histogram, and the normal distribution is a continuous, smooth curve. Each X value is represented by a bar in the histogram or a section of the normal distribution.

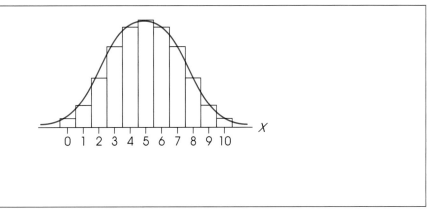

issue. When pn and qn are both equal to or greater than 10 (the criteria for using the normal approximation), each interval in the binomial distribution is extremely small and it is very unlikely that the interval overlaps the critical boundary. For example, the experiment in Example 19.1 produced a score of $X = 3$, and we computed a z-score of $z = -4.04$. Because this value is in the critical region, beyond $z = -1.96$, we rejected H_0. If we had used the real limit boundaries of $X = 2.5$ and $X = 3.5$, instead of $X = 3$, we would have obtained z-scores of

$$z = \frac{2.5 - 13.5}{2.60} \qquad \text{and} \qquad z = \frac{3.5 - 13.5}{2.60}$$

$$= -4.23 \qquad\qquad\qquad = -3.85$$

Thus, a score of $X = 3$ actually corresponds to an interval of z-scores ranging from $z = -3.85$ to $z = -4.23$. However, this entire interval is in the critical region beyond $z = -1.96$, so the decision is still to reject H_0.

In most situations, if the whole-number X value (in this case, $X = 3$) is in the critical region, then the entire interval will also be in the critical region and the correct decision is to reject H_0. The only exception to this general rule occurs when an X value produces a z-score that is barely past the boundary into the critical region. In this situation, you should compute the z-scores corresponding to both real limits to determine whether any part of the z-score interval is not located in the critical region. Suppose, for example, that the researchers in Example 19.1 found that 8 out of 27 infants in the visual cliff experiment moved onto the deep side. A score of $X = 8$ corresponds to

$$z = \frac{8 - 13.5}{2.60}$$

$$z = \frac{-5.5}{2.60}$$

$$= -2.12$$

Because this value is beyond the -1.96 boundary, it appears that we should reject H_0. However, this z-score is only slightly beyond the critical boundary, so it would be wise to check both ends of the interval. For $X = 8$, the real-limit boundaries are 7.5 and 8.5, which correspond to z-scores of

$$z = \frac{7.5 - 13.5}{2.60} \qquad \text{and} \qquad z = \frac{8.5 - 13.5}{2.60}$$

$$= -2.31 \qquad\qquad\qquad\qquad = -1.92$$

Thus, the z-score interval corresponding to $X = 8$ extends from $z = -1.92$ to $z = -2.31$, which means that part of the interval is not in the critical region. You should recall that the critical region is determined by the probability expressed by the alpha level. To reject the null hypothesis with $\alpha = .05$, for example, a score must be located in the extreme 5% of the distribution. For a normal distribution, this means that the score must be beyond the $z = \pm 1.96$ critical boundaries. As we have seen, a score of $X = 8$ is not completely beyond the critical boundary. Therefore, the probability of obtaining $X = 8$ is greater than $\alpha = .05$, and the correct decision is to fail to reject H_0.

In general, it is safe to conduct a binomial test using the whole-number value for X. However, if you obtain a z-score that is only slightly beyond the critical boundary, you also should compute the z-scores for both real limits. If any part of the z-score interval is not in the critical region, the correct decision is to fail to reject H_0.

IN THE LITERATURE
REPORTING THE RESULTS OF A BINOMIAL TEST

Reporting the results of the binomial test typically consists of describing the data and reporting the z-score value and the probability that the results are due to chance. It is also helpful to note that a binomial test was used because z-scores are used in other hypothesis-testing situations (see, for example, Chapter 8). For Example 19.1, the report might state:

> Three out of 27 infants moved to the deep side of the visual cliff. A binomial test revealed that there is a significant preference for the shallow side of the cliff, $z = -4.04$, $p < .05$.

Once again, p is *less than* .05. We have rejected the null hypothesis because it is very unlikely, probability less than 5%, that these results are simply due to chance. ❏

ASSUMPTIONS FOR THE BINOMIAL TEST

The binomial test requires two very simple assumptions:

1. The sample must consist of *independent* observations (see Chapter 8, page 253).

2. The values for pn and qn must both be greater than or equal to 10 to justify using the unit normal table for determining the critical region.

LEARNING CHECK

1. For a binomial test, the null hypothesis always states that $p = \frac{1}{2}$. (True or false?)

2. The makers of brand X beer claim that people like their beer more than the leading brand. The basis for this claim is an experiment in which 64 beer drinkers compared the two brands in a side-by-side taste test. In this sample, 38 preferred brand X, and 26 preferred the leading brand. Do these data support the claim that there is a significant preference? Test at the .05 level.

ANSWERS

1. False.

2. H_0: $p = \frac{1}{2} = q$, $X = 38$, $\mu = 32$, $\sigma = 4$, $z = +1.50$, fail to reject H_0. Conclude that there is no evidence for a significant preference.

19.3 THE RELATIONSHIP BETWEEN CHI-SQUARE AND THE BINOMIAL TEST

You may have noticed that the binomial test evaluates the same basic hypotheses as the chi-square test for goodness of fit; that is, both tests evaluate how well the sample proportions fit a hypothesis about the population proportions. When an experiment produces binomial data, these two tests are equivalent, and either may be used. The relationship between the two tests can be expressed by the equation

$$\chi^2 = z^2$$

where χ^2 is the statistic from the chi-square test for goodness of fit and z is the z-score from the binomial test.

To demonstrate the relationship between the goodness-of-fit test and the binomial test, we reexamine the data from Example 19.1.

STEP 1   *Hypotheses.* In the visual cliff experiment from Example 19.1, the null hypothesis states that there is no preference between the shallow side and the deep side. For the binomial test, the null hypothesis states

$$H_0: \quad p = p(\text{deep side}) = q = p(\text{shallow side}) = \tfrac{1}{2}$$

The chi-square test for goodness of fit would state the same hypothesis, specifying the population proportions as

	Shallow Side	Deep Side
H_0:	$\frac{1}{2}$	$\frac{1}{2}$

STEP 2   *Critical region.* For the binomial test, the critical region is located by using the unit normal table. With $\alpha = .05$, the critical region consists of any z-score value beyond ± 1.96. The chi-square test would have $df = 1$, and with $\alpha = .05$, the critical region consists of chi-square values greater than 3.84. Notice that the basic relationship, $\chi^2 = z^2$, holds:

$$3.84 = (1.96)^2$$

STEP 3   *Test statistic.* For the binomial test (Example 19.1), we obtained a z-score of $z = -4.04$. For the chi-square test, the expected frequencies are

	Shallow Side	Deep Side
f_e	13.5	13.5

With observed frequencies of 24 and 3, respectively, the chi-square statistic is

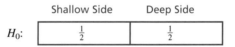

$$\chi^2 = \frac{(24 - 13.5)^2}{13.5} + \frac{(3 - 13.5)^2}{13.5}$$

$$= \frac{(10.5)^2}{13.5} + \frac{(-10.5)^2}{13.5}$$

$$= 8.167 + 8.167$$

$$= 16.33$$

With a little rounding error, the values obtained for the z-score and chi-square are related by the equation

Caution: The χ^2 value is already squared. Do not square it again.

$$\chi^2 = z^2$$

$$16.33 = (-4.04)^2$$

STEP 4 *Decision.* Because the critical values for both tests are related by the equation $\chi^2 = z^2$ and the test statistics are related in the same way, these two tests *always* result in the same statistical conclusion.

19.4 THE SIGN TEST

Although the binomial test can be used in many different situations, there is one specific application that merits special attention. For a repeated-measures study that compares two conditions, it is often possible to use a binomial test to evaluate the results. You should recall that a repeated-measures study involves measuring each individual in two different treatment conditions or at two different points in time. When the measurements produce numerical scores, the researcher can simply subtract to determine the difference between the two scores and then evaluate the data using a repeated-measures t test (see Chapter 11). Occasionally, however, a researcher may record only the *direction* of the difference between the two observations. For example, a clinician may observe patients before therapy and after therapy and simply note whether each patient got better or got worse. Note that there is no measurement of how much change occurred; the clinician is simply recording the direction of change. Also note that the direction of change is a binomial variable; that is, there are only two values. In this situation it is possible to use a binomial test to evaluate the data. Traditionally, the two possible directions of change are coded by signs, with a positive sign indicating an increase and a negative sign indicating a decrease. When the binomial test is applied to signed data, it is called a *sign test*.

An example of signed data is shown in Table 19.1. Notice that the data can be summarized by saying that seven out of eight patients showed a decrease in symptoms after therapy.

TABLE 19.1

Hypothetical data from a research study evaluating the effectiveness of a clinical therapy. For each patient, symptoms are assessed before and after treatment and the data record whether there is an increase or a decrease in symptoms following therapy.

Patient	Direction of Change After Treatment
A	− (decrease)
B	− (decrease)
C	− (decrease)
D	+ (increase)
E	− (decrease)
F	− (decrease)
G	− (decrease)
H	− (decrease)

The null hypothesis for the sign test states that there is no difference between the two treatments. Therefore, any change in a participant's score is due to chance. In terms of probabilities, this means that increases and decreases are equally likely, so

$$p = p(\text{increase}) = \tfrac{1}{2}$$

$$q = p(\text{decrease}) = \tfrac{1}{2}$$

A complete example of a sign test follows.

EXAMPLE 19.2   A researcher testing the effectiveness of acupuncture for treating the symptoms of arthritis obtains a sample of 36 people who have been diagnosed with arthritis. Each person's pain level is measured before treatment starts, and measured again after 4 months of acupuncture treatment. For this sample, 25 people experienced a reduction in pain, and 11 people had more pain after treatment. Do these data indicate a significant treatment effect?

STEP 1   State the hypothesis. The null hypothesis states that acupuncture has no effect. Any change in the level of pain is due to chance, so increases and decreases are equally likely. Expressed as probabilities, the hypotheses are

$$H_0: \quad p = p(\text{increased pain}) = \tfrac{1}{2} \quad \text{and} \quad q = p(\text{decreased pain}) = \tfrac{1}{2}$$

$$H_1: \quad p \neq q \quad \text{(Changes tend to be consistently in one direction.)}$$

Set $\alpha = .05$.

STEP 2   Locate the critical region. With $n = 36$ people, both pn and qn are greater than 10, so the normal approximation to the binomial distribution is appropriate. With $\alpha = .05$, the critical region consists of the most extreme 5% of the normal distribution. This portion consists of z-scores greater than $+1.96$ at one extreme and z-scores less than -1.96 at the other.

STEP 3   Compute the test statistic. For this sample we have $X = 25$ people with decreased pain. This score corresponds to a z-score of

$$z = \frac{X - pn}{\sqrt{npq}} = \frac{25 - 18}{\sqrt{36\left(\frac{1}{2}\right)\left(\frac{1}{2}\right)}} = \frac{7}{3} = 2.33$$

Because the z-score is only slightly beyond the 1.96 critical boundary, we will consider the real limits for $X = 25$ to be certain that the entire interval is beyond the boundary. For $X = 25$, the real limits are 24.5 and 25.5, which correspond to z-scores of

$$z = \frac{24.5 - 18}{3} \qquad \text{and} \qquad z = \frac{25.5 - 18}{3}$$

$$= 2.17 \qquad\qquad\qquad = 2.50$$

Thus, a score of $X = 25$ corresponds to an interval of z-scores ranging from $z = 2.17$ to $z = 2.50$. Note that this entire interval is beyond the 1.96 critical boundary

STEP 4   Make a decision. Because the data are in the critical region, we reject H_0 and conclude that acupuncture treatment has a significant effect on arthritis pain, $z = 2.33, p < .05$.

ZERO DIFFERENCES IN THE SIGN TEST   You should notice that the null hypothesis in the sign test refers only to those individuals who show some difference between treatment 1 versus treatment 2. The null hypothesis states that if there is any change in an individual's score, then the probability of an increase is equal to the probability of a decrease. Stated in this form, the null hypothesis does not consider individuals who show zero difference between the two treatments. As a result, the usual recommendation is that these individuals be discarded from the data and the value of n be reduced accordingly. However, if the null hypothesis is interpreted more generally, it states that there is no difference between the two treatments. Phrased this way, it should be clear that individuals who show no difference actually are supporting the null hypothesis and should not be discarded. Therefore, an alternative approach to the sign test is to divide individuals who show zero differences equally between the positive and negative categories. (With an odd number of zero differences, discard one, and divide the rest evenly.) This alternative results in a more conservative test; that is, the test is more likely to fail to reject the null hypothesis.

EXAMPLE 19.3   It has been demonstrated that stress or exercise causes an increase in the concentration of certain chemicals in the brain called endorphins. Endorphins are similar to morphine and produce a generally relaxed feeling and a sense of well-being. The endorphins may explain the "high" experienced by long-distance runners. To demonstrate this phenomenon, a researcher tested pain tolerance for 40 athletes before and after they completed a mile run. Immediately after running, the ability to tolerate pain increased for 21 of the athletes, decreased for 12, and showed no change for the remaining 7.

Following the standard recommendation for handling zero differences, you would discard the 7 participants who showed no change and conduct a sign test with the remaining $n = 33$ athletes. With the more conservative approach, only 1 of the 7 who showed no difference would be discarded and the other 6 would be divided equally between the two categories. This would result in a total sample of $n = 39$ athletes with $21 + 3 = 24$ in the increased-tolerance category and $12 + 3 = 15$ in the decreased-tolerance category.

WHEN TO USE THE SIGN TEST   In many cases, data from a repeated-measures experiment can be evaluated using either a sign test or a repeated-measures t test. In general, you should use the t test whenever possible. Because the t test uses the actual difference scores (not just the signs), it makes maximum use of the available information and results in a more powerful test. However, there are some cases in which a t test cannot or should not be used, and in these situations, the sign test can be valuable. Four specific cases in which a t test is inappropriate or inconvenient will now be described.

1. When you have infinite or undetermined scores, a t test is impossible, and the sign test is appropriate. Suppose, for example, that you are evaluating the effects of a sedative drug on problem-solving ability. A sample of rats is obtained, and each animal's performance is measured before and after receiving the drug. Hypothetical data are shown in the margin. Note that the third rat in this sample failed to solve the problem after receiving the drug. Because there is no score for this animal, it is impossible to compute a sample mean, an SS, or a t statistic. However, you could do a sign test because you know that the animal made more errors (an increase) after receiving the drug.

Before	After	Difference
20	23	+3
14	39	+25
27	Failed	+??
.	.	.
.	.	.
.	.	.

2. Often it is possible to describe the difference between two treatment conditions without precisely measuring a score in either condition. In a clinical setting, for example, a doctor can say whether a patient is improving, growing worse, or showing no change even though the patient's condition is not precisely measured by a score. In this situation, the data are sufficient for a sign test, but you could not compute a t statistic without individual scores.

3. Often a sign test is done as a preliminary check on an experiment before serious statistical analysis begins. For example, a researcher may predict that scores in treatment 2 should be consistently greater than scores in treatment 1. However, examination of the data after 1 week indicates that only 8 of 15 subjects showed the predicted increase. On the basis of these preliminary results, the researcher may choose to reevaluate the experiment before investing additional time.

4. Occasionally, the difference between treatments is not consistent across participants. This can create a very large variance for the difference scores. As we have noted in the past, large variance decreases the likelihood that a *t* test will produce a significant result. However, the sign test only considers the direction of each difference score and is not influenced by the variance of the scores.

LEARNING CHECK

1. A researcher used a chi-square test for goodness of fit to determine whether people had any preferences among three leading brands of potato chips. Could the researcher have used a binomial test instead of the chi-square test? Explain why or why not.

2. A researcher used a chi-square test to evaluate preferences between two logo designs for a minor-league hockey team. With a sample of $n = 100$ people, the researcher obtained a chi-square of 9.00. If a binomial test had been used instead of chi-square, what value would have been obtained for the z-score?

3. A developmental psychologist is using a behavior-modification program to help control the disruptive behavior of 40 children in a local school. After 1 month, 26 of the children have improved, 10 are worse, and 4 show no change in behavior. On the basis of these data, can the psychologist conclude that the program is working? Test at the .05 level.

ANSWERS

1. No, the binomial test cannot be used when there are three categories.

2. The z-score would be $\sqrt{9} = 3.00$.

3. Discarding the four participants who showed zero difference, $X = 26$ increases out of $n = 36$; $z = 2.67$; reject H_0; the program is working. If the four participants showing no change are divided between the two groups, then $X = 28$ out of $n = 40$; $z = 2.53$ and H_0 is still rejected.

SUMMARY

1. The binomial test is used with dichotomous data—that is, when each individual in the sample is classified in one of two categories. The two categories are identified as *A* and *B*, with probabilities of

$$p(A) = p \quad \text{and} \quad p(B) = q$$

2. The binomial distribution gives the probability for each value of *X*, where *X* equals the number of occurrences of category *A* in a sample of *n* events. For example, *X* equals the number of heads in $n = 10$ tosses of a coin.

3. When *pn* and *qn* are both at least 10, the binomial distribution is closely approximated by a normal distribution with

$$\mu = pn \qquad \sigma = \sqrt{npq}$$

By using this normal approximation, each value of *X* has a corresponding *z*-score:

$$z = \frac{X - \mu}{\sigma} = \frac{X - pn}{\sqrt{npq}} \quad \text{or} \quad z = \frac{X/n - p}{\sqrt{pq/n}}$$

4. The binomial test uses sample data to test hypotheses about the binomial proportions, p and q, for a population. The null hypothesis specifies p and q, and the binomial distribution (or the normal approximation) is used to determine the critical region.

5. Usually the z-score in a binomial test is computed using the whole-number X value from the sample. However, if the z-score is only marginally in the critical region, you should compute the z-scores corresponding to both real limits of the score. If either one of these real-limit z-scores is not in the critical region, the correct decision is to fail to reject the null hypothesis.

6. One common use of the binomial distribution is for the sign test. This test evaluates the difference between two treatments using the data from a repeated measures design. The difference scores are coded as being either increases $(+)$ or decreases $(-)$. Without a consistent treatment effect, the increases and decreases should be mixed randomly, so the null hypothesis states that

$$p(\text{increase}) = \tfrac{1}{2} = p(\text{decrease})$$

With dichotomous data and hypothesized values for p and q, this is a binomial test.

KEY TERMS

binomial data   binomial test   binomial distribution   sign test

RESOURCES

There are a tutorial quiz and other learning exercises for Chapter 19 at the Wadsworth website, www.cengage.com/psychology/gravetter. See page 29 for more information about accessing items on the website.

ENHANCED WebAssign

Guided interactive tutorials, end-of-chapter problems, and related testbank items may be assigned online at WebAssign.

WebTUTOR

If you are using WebTutor along with this book, there is a WebTutor section corresponding to this chapter. The WebTutor contains a brief summary of Chapter 19, hints for learning the concepts and formulas for the binomial test, cautions about common errors, and sample exam items including solutions.

SPSS

General instructions for using SPSS are presented in Appendix D. Following are detailed instructions for using SPSS to perform **The Binomial Test** that is presented in this chapter.

Data Entry

1. Enter the values 0 and 1 in the first column of the SPSS data editor.
2. In the second column, enter the frequencies that were obtained for the two binomial categories. For example, if 21 out of 25 people were classified in category A (and only 4 people in category B), you would enter the values 21 and 4 in the second column.

Data Analysis

1. Click **Data** on the tool bar at the top of the page and select **weight cases** at the bottom of the list.
2. Click the **weight cases by** circle, then highlight the label for the column containing the frequencies for the two categories and move it into the **Frequency Variable** box by clicking on the arrow.
3. Click **OK.**
4. Click **Analyze** on the tool bar, select **Nonparametric Tests,** and click on **Binomial.**
5. Highlight the label for the column containing the digits 0 and 1, and move it into the **Test Variables List** box by clicking on the arrow.
6. Specify the **Test Proportion,** which is the value of p from the null hypothesis. (The box is preset at .50 but you can change it to the value appropriate for your test.)
7. Click **OK.**

SPSS Output

The program will produce a table summarizing the results of the test. The table shows the number of individuals in each category, the proportion of individuals in each category, the test proportion from the null hypothesis, and a level of significance (the p value or alpha level for the test).

FOCUS ON PROBLEM SOLVING

1. For all binomial tests, the values of p and q must add up to 1.00 (or 100%).

2. Remember that both pn and qn must be at least 10 before you can use the normal distribution to determine critical values for a binomial test.

3. Although the binomial test usually specifies the critical region in terms of z-scores, it is possible to identify the X values that determine the critical region. With $\alpha = .05$, the critical region is determined by z-scores greater than 1.96 or less than -1.96. That is, to be significantly different from what is expected by chance, the individual score must be above (or below) the mean by at least 1.96 standard deviations. For example, in an ESP experiment in which an individual is trying to predict the suit of a playing card for a sequence of $n = 64$ trials, chance probabilities are

$$p = p(\text{right}) = \tfrac{1}{4} \qquad q = p(\text{wrong}) = \tfrac{3}{4}$$

For this example, the binomial distribution has a mean of $pn = \left(\tfrac{1}{4}\right)\!\left(64\right) = 16$ right and a standard deviation of $\sqrt{npq} = \sqrt{64\left(\tfrac{1}{4}\right)\left(\tfrac{3}{4}\right)} = \sqrt{12} = 3.46$. To be significantly different from chance, a score must be above (or below) the mean by at least $1.96(3.46) = 6.78$. Thus, with a mean of 16, an individual would need to score above 22.78 $(16 + 6.78)$ or below 9.22 $(16 - 6.78)$ to be significantly different from chance.

DEMONSTRATION 19.1

THE BINOMIAL TEST

The population of students in the psychology department at State College consists of 60% females and 40% males. Last semester, the Psychology of Gender course had a total of 36 students, of whom 26 were female and only 10 were male. Are the proportions of females and males in this class significantly different from what would be expected by chance from a population with 60% females and 40% males? Test at the .05 level of significance.

STEP 1 *State the hypotheses, and specify alpha.* The null hypothesis states that the male/female proportions for the class are not different from what is expected for a population with these proportions. In symbols,

$$H_0: \quad p = p(\text{female}) = 0.60 \quad \text{and} \quad q = p(\text{male}) = 0.40$$

The alternative hypothesis is that the proportions for this class are different from what is expected for these population proportions.

$$H_1: \quad p \neq 0.60 \quad (\text{and } q \neq 0.40)$$

We will set alpha at $\alpha = .05$.

STEP 2 *Locate the critical region.* Because pn and qn are both greater than 10, we can use the normal approximation to the binomial distribution. With $\alpha = .05$, the critical region is defined as any z-score value greater than $+1.96$ or less than -1.96.

STEP 3 *Calculate the test statistic.* The sample has 26 females out of 36 students, so the sample proportion is

$$\frac{X}{n} = \frac{26}{36} = 0.72$$

The corresponding z-score (using Equation 19.2) is

$$z = \frac{X/n - p}{\sqrt{pq/n}} = \frac{0.72 - 0.60}{\sqrt{\dfrac{0.60(0.40)}{36}}} = \frac{0.12}{0.0816} = 1.47$$

STEP 4 *Make a decision about H_0, and state a conclusion.* The obtained z-score is not in the critical region. Therefore, we fail to reject the null hypothesis. On the basis of these data, you conclude that the male/female proportions in the gender class are not significantly different from the proportions in the psychology department as a whole.

PROBLEMS

1. To investigate the phenomenon of "home team advantage," a researcher recorded the outcomes from 64 college football games on one Saturday in October. Of the 64 games, 42 were won by home teams. Does this result provide enough evidence to conclude that home teams win significantly more than would be expected by chance? Use a two-tailed test with $\alpha = .05$.

2. Insurance companies charge young drivers more for automobile insurance because they tend to have more accidents than older drivers. To make this point, an

insurance representative first determines that only 16% of licensed drivers are age 20 or younger. Because this age group makes up only 16% of the drivers, it is reasonable to predict that they should be involved in only 16% of the accidents. In a random sample of 100 accident reports, however, the representative finds 31 accidents that involved drivers who were 20 or younger. Is this sample sufficient to show that younger drivers have significantly more accidents than would be expected from the percentage of young drivers? Use a two-tailed test with $\alpha = .05$.

3. Güven, Elaimis, Binokay, and Tan (2003) studied the distribution of paw preferences in rats using a computerized food-reaching test. For a sample of $n = 144$ rats, they found 104 right-handed animals. Is this significantly more than would be expected if right- and left-handed rats are equally common in the population? Use a two-tailed test with $\alpha = .01$.

4. During the 2004 Olympic Games, Hill and Barton (2005) monitored contests in four combat sports: Greco-Roman wrestling, freestyle wrestling, boxing, and tae kwon do. Half of the competitors were assigned red outfits and half were assigned blue. The results indicate that participants wearing red won significantly more contests. Suppose that a sample of $n = 100$ contests produced 60 winners wearing red and only 40 wearing blue.

  a. Is this enough evidence to justify a conclusion that wearing red produces significantly more wins that would be expected by chance? Use a two-tailed test with $\alpha = .05$.

  b. Because the outcome of the binomial test is a borderline z-score, use the real limits for $X = 60$ to determine if the entire z-score interval is located in the critical region. If any part of the interval is not in the critical region, the correct decision is to fail to reject the null hypothesis.

5. A researcher would like to determine whether people can really tell the difference between bottled water and tap water. Participants are asked to taste two unlabeled glasses of water, one bottled and one tap, and identify the one they think tastes better. Out of 40 people in the sample, 28 picked the bottled water. Was the bottled water selected significantly more often than would be expected by chance? Use a two-tailed test with $\alpha = .05$.

6. In 1985 only 8% of the students in the city school district were classified as being learning disabled. A school psychologist suspects that the proportion of learning-disabled children has changed dramatically over the years. To demonstrate this point, a random sample of $n = 300$ students is selected. In this sample there are 42 students who have been identified as learning-disabled. Is this sample sufficient to indicate that there has been a significant change in the proportion of learning-disabled students since 1985? Use the .05 level of significance.

7. A recent survey of practicing psychotherapists revealed that 25% of the individuals responding agreed with the statement, "Hypnosis can be used to recover accurate memories of past lives" (Yapko, 1994). A researcher would like to determine whether this same level of belief exists in the general population. A sample of 192 adults is surveyed and 65 believe that hypnosis can be used to recover accurate memories of past lives. Based on these data, can you conclude that beliefs held by the general population are significantly different from beliefs held by psychotherapists? Test with $\alpha = .05$.

8. In 2005, Fung et al. published a study reporting that patients prefer technical quality versus interpersonal skills when selecting a primary care physician. Participants were presented with report cards describing pairs of hypothetical physicians and were asked to select the one that they preferred. Suppose that this study is repeated with a sample of $n = 150$ participants, and the results show that physicians with greater technical skill are preferred by 92 participants and physicians with greater interpersonal skills are selected by 58. Are these results sufficient to conclude that there is a significant preference for technical skill?

9. One of the original methods for testing ESP (extrasensory perception) was Zener cards. Each card shows one of five symbols (square, circle, star, wavy lines, cross). The person being tested must predict the symbol before the card is turned over. Chance performance on this task would produce 1 out of 5 (20%) correct predictions. Is 27 correct predictions out of 100 attempts significantly different from chance performance? Use a two-tailed test with $\alpha = .05$.

10. For each of the following, assume that a two-tailed test using the normal approximation to the binomial distribution with $\alpha = .05$ is being used to evaluate the significance of the result.

  a. For a true-false test with 20 questions, how many would you have to get right to do significantly better than chance? That is, what X value is needed to produce a z-score greater than 1.96?

  b. How many would you need to get right on a 40-question true-false test?

  c. How many would you need to get right on a 100-question true-false test?

Remember that each X value corresponds to an interval with real limits. Be sure that the entire interval is in the critical region.

11. On a multiple-choice exam with 100 questions and 4 possible answers for each question, you get a score of $X = 32$. Is your score significantly better than would be expected by chance (by simply guessing for each question)? Use a two tailed test with $\alpha = .05$.

12. For each of the following, assume that a two-tailed test using the normal approximation to the binomial distribution with $\alpha = .05$ is being used to evaluate the significance of the result.
    a. For a multiple-choice test with 48 questions, each with 4 possible answers, how many would you have to get right to do significantly better than chance? That is, what X value is needed to produce a z-score greater than 1.96?
    b. How many would you need to get right on a multiple-choice test with 192 questions to be significantly better than chance?
    Remember that each X value corresponds to an interval with real limits. Be sure that the entire interval is in the critical region.

13. Reed, Vernon, and Johnson (2004) examined the relationship between brain nerve conduction velocity and intelligence in normal adults. Brain nerve conduction velocity was measured three separate ways and nine different measures were used for intelligence. The researchers then correlated each of the three nerve velocity measures with each of the nine intelligence measures for a total of 27 separate correlations. Unfortunately, none of the correlations were significant.
    a. For the 186 males in the study, however, 25 of the 27 correlations were positive. Is this significantly more than would be expected if positive and negative correlations were equally likely? Use a two-tailed test with $\alpha = .05$.
    b. For the 201 females in the study, 20 of the 27 correlations were positive. Is this significantly more than would be expected if positive and negative correlations were equally likely? Use a two-tailed test with $\alpha = .05$.

14. Thirty percent of the students in the local elementary school are classified as only children (no siblings). However, in the special program for talented and gifted children, 43 out of 90 students are only children. Is the proportion of only children in the special program significantly different from the proportion for the school? Test at the .05 level of significance.

15. Stressful or traumatic experiences can often worsen other health-related problems such as asthma or rheumatoid arthritis. However, if patients are instructed to write about their stressful experiences, it can often lead to improvement in health (Smyth, Stone, Hurewitz, & Kaell, 1999). In a typical study, patients with asthma or arthritis are asked to write about the "most stressful event of your life." In a sample of $n = 112$ patients, suppose that 64 showed improvement in their symptoms, 12 showed no change, and 36 showed worsening symptoms.
    a. If the 12 patients showing no change are discarded, are these results sufficient to conclude that the writing had a significant effect? Use a two-tailed test with $\alpha = .05$.
    b. If the 12 patients who showed no change are split between the two groups, are the results sufficient to demonstrate a significant change? Use a two-tailed test with $\alpha = .05$.

16. Langewitz, Izakovic, and Wyler (2005) reported that self-hypnosis can significantly reduce hay-fever symptoms. Patients with moderate to severe allergic reactions were trained to focus their minds on specific locations where their allergies did not bother them, such as a beach or a ski resort. In a sample of 64 patients who received this training, suppose that 47 showed reduced allergic reactions and 17 showed an increase in allergic reactions. Are these results sufficient to conclude that the self-hypnosis has a significant effect? Use a two-tailed test with $\alpha = .05$.

17. Group-housed laying hens appear to prefer having more floor space than height in their cages. Albentosa (2005) tested hens in groups of 10. The birds in each group were given free choice between a cage with a height of 38 cm (low) and a cage with a height of 45 cm (high). The results showed a tendency for the hens in each group to distribute themselves evenly between the two cages, showing no preference for either height. Suppose that a similar study tested a sample of $n = 80$ hens and found that 47 preferred the taller cage. Does this result indicate a significant preference? Use a two-tailed test with $\alpha = .05$

18. The habituation technique is one method that is used to examine memory for infants. The procedure involves presenting a stimulus to an infant (usually projected on the ceiling above the crib) for a fixed time period and recording how long the infant spends looking at the stimulus. After a brief delay, the stimulus is presented again. If the infant spends less time looking at the stimulus during the second presentation, it is interpreted as indicating that the stimulus is remembered and, therefore, is less novel and less interesting than it was on the first presentation. This procedure is used with a sample of $n = 30$ 2-week-old infants. For this sample, 22 infants spent less time looking at the stimulus during the second presentation than during the first. Do these data indicate a significant difference? Test at the .01 level of significance.

19. Most children and adults are able to learn the meaning of new words by listening to sentences in which the words appear. Shulman and Guberman (2007) tested the ability of children to learn word meaning from syntactical cues for three groups: children with autism, children with specific language impairment (SLI), and children with typical language development (TLD). Although the researchers used relatively small samples, their results indicate that the children with TLD and those with autism were able to learn novel words using the syntactical cues in sentences. The children with SLI, on the other hand, experienced significantly more difficulty.

Suppose a similar study is conducted in which each child listens to a set of sentences containing a novel word and then is given a choice of three definitions for the word.

a. If 25 out of 36 autistic children select the correct definition, is this significantly more than would be expected if they were simply guessing? Use a two-tailed test with $\alpha = .05$.

b. If only 16 out of 36 children with SLI select the correct definition, is this significantly more than would be expected if they were simply guessing? Use a two-tailed test with $\alpha = .05$.

20. A researcher is testing the effectiveness of a skills-mastery imagery program for soccer players. A sample of $n = 25$ college varsity players is selected and each player is tested on a ball-handling obstacle course before beginning the imagery program and again after completing the 5-week program. Of the 25 players, 18 showed improved performance on the obstacle course after the imagery program and 7 were worse.

a. Is this result sufficient to conclude that there is a significant change in performance following the imagery program? Use a two-tailed test with $\alpha = .05$.

b. Because the outcome of the binomial test is a borderline z-score, use the real limits for $X = 18$ and verify that the entire z-score interval is located in the critical region.

21. Last year the college counseling center offered a workshop for students who claimed to suffer from

extreme exam anxiety. Of the 45 students who attended the workshop, 31 had higher grade point averages this semester than they did last year. Do these data indicate a significant difference from what would be expected by chance? Test at the .01 level of significance.

22. Trying to fight a drug-resistant bacteria, a researcher tries an experimental drug on infected subjects. Out of 70 monkeys, 42 showed improvement, 22 got worse, and 6 showed no change. Is this researcher working in the right direction? Is there a significant effect of the drug on the infection? Use a two-tailed test at the .05 level of significance.

23. An English professor conducts a special 1-week course in writing skills for freshmen who are poorly prepared for college-level writing. To test the effectiveness of this course, each student in a recent class of $n = 36$ was required to write a short paragraph describing a painting before the course started. At the end of the course, the students were once again required to write a paragraph describing the same painting. The students' paragraphs were then given to another instructor to be evaluated. For 25 students, the writing was judged to be better after the course. For the rest of the class, the first paragraph was judged to be better. Do these results indicate a significant change in performance? Test at the .05 level of significance.

24. Biofeedback training is often used to help people who suffer migraine headaches. A recent study found that 29 out of 50 participants reported a decrease in the frequency and severity of their headaches after receiving biofeedback training. Of the remaining participants in this study, 10 reported that their headaches were worse, and 11 reported no change.

a. Discard the zero-difference participants, and use a sign test with $\alpha = .05$ to determine whether the biofeedback produced a significant difference.

b. Divide the zero-difference participants between the two groups, and use a sign test to evaluate the effect of biofeedback training.

CHAPTER

20

Tools You Will Need

The following items are considered essential background material for this chapter. If you doubt your knowledge of any of these items, you should review the appropriate chapter or section before proceeding.

- Ordinal scales (Chapter 1)

- Probability (Chapter 6)
  - The unit normal table

- Introduction to hypothesis testing (Chapter 8)

- Correlation (Chapter 16)

Statistical Techniques for Ordinal Data: Mann-Whitney, Wilcoxon, Kruskal-Wallis, and Friedman Tests

Preview

20.1   Data from an Ordinal Scale

20.2   The Mann-Whitney *U*-Test: An Alternative to the Independent-Measures *t* Test

20.3   The Wilcoxon Signed-Ranks Test: An Alternative to the Repeated-Measures *t* Test

20.4   The Kruskal-Wallis Test: An Alternative to the Independent-Measures ANOVA

20.5   The Friedman Test: An Alternative to the Repeated-Measures ANOVA

Summary

Focus on Problem Solving

Demonstrations 20.1, 20.2, 20.3, and 20.4

Problems

Preview

People have a passion for ranking things. Everyone wants to know who is number one, and we all tend to describe things in terms of their relationship to others. For example, the Nile is the longest river in the world, the Labrador retriever is the number one registered dog in the United States, the state with the highest average SAT math score is Iowa, *Reader's Digest* is the most popular magazine in the United States, English is the fourth most common native language in the world (after Chinese, Hindi, and Spanish), and Lake Superior is the largest lake in North America and the second largest in the world.

Part of the fascination with ranks is that they are easy to obtain and they are easy to understand. For example, what is your favorite ice-cream flavor? Notice that you do not need to conduct any sophisticated measurement to arrive at an answer. There is no need to rate 31 different flavors on a 100-point scale, and you do not need to worry about how much difference there is between your first and second choices.

You should recall from Chapter 1 that ranking is an example of measurement on an ordinal scale. In general, ordinal scales are less demanding and less sophisticated than the interval or ratio scales that are most commonly used to measure variables. Because ordinal scales are less demanding, they are often easier to use. On the other hand, because they are less sophisticated, they can cause some problems for statistical analysis. Consider the data presented in Table 20.1. The table shows the top 25 cities for singles as ranked by Forbes.com in 2005. The ranking considered the 40 most populated metropolitan areas as defined by the U.S Census, and combined different factors such as culture, nightlife, and the number of other singles to produce an overall rank for each city.

Because the values in Table 20.1 are numbers, it is tempting to try some routine calculations such as means and standard deviations. For example, we could calculate the average score for cities in the East and compare it with the average for cities in the West. However, you should recall that means and standard deviations are measures based on *distance*. The mean for example is a balance point, so the distances above the mean are equal to the distance below. Ranks, on the other hand, do not provide any information about distance. In the table, for example,

San Francisco is ranked #3 and Raleigh-Durham is ranked #4, giving the appearance that these two cities are very close together. However, a detailed examination of how the ranks were obtained shows that San Francisco and Raleigh-Durham have very different profiles. For example, San Francisco is ranked very high in terms of nightlife but Raleigh-Durham is in the bottom half of the list. Raleigh-Durham, on the other hand is ranked very high in terms of job growth, a category in which San Francisco has a poor rank. Although the two cities appear to be very similar in terms of the overall ranking, in many ways they are very different. In general, ranks do not tell you how much difference there is between two individuals.

The Problem: Because ordinal data (ranks) provide limited information they must be used and interpreted carefully. In particular, standard statistical procedures such as computing means or conducting *t* tests should not be used when the data are measurements on an ordinal scale.

The Solution: In this chapter we introduce four *statistical techniques for ordinal data*. The four new hypothesis tests are designed to replace the independent- and repeated-measures *t* tests, and the independent- and repeated-measures analysis of variance (ANOVA) in situations that involve ordinal data.

TABLE 20.1

The best cities for singles as ranked by Forbes.com (2005)

Rank	City	Rank	City
1	Denver-Boulder	14	Phoenix
2	Bostom	15	Sacramento
3	San Francisco	16	Detroit
4	Raleigh–Durham	17	Houston
5	Washington–Baltimore	18	Columbus
6	Atlanta	19	Portland
7	Los Angeles	20	Dallas–Ft. Worth
8	New York	21	San Diego
9	Chicago	22	Nashville
10	Seattle	23	Miami
11	Austin	24	Salt Lake City
12	Philadelphia	25	Las Vegas
13	Minneapolis–St. Paul		

20.1 DATA FROM AN ORDINAL SCALE

Occasionally, a research study generates data that consist of measurements on an ordinal scale. Recall from Chapter 1 that an ordinal scale simply produces a *rank ordering* for the individuals being measured. For example, a kindergarten teacher may rank children in terms of their maturity, or a business manager may rank employee suggestions in terms of creativity. Whenever a set of data consists of ranks, you should be very cautious about choosing a statistical procedure to describe or to interpret the data. Specifically, most of the commonly used statistical methods such as the mean, the standard deviation, hypothesis tests with the *t* statistic, and the Pearson correlation are generally considered to be inappropriate for ordinal data.

The concern about using traditional statistical methods with ranked data stems from the fact that most statistical procedures begin by calculating a sample mean and computing deviations from the mean. When scores are measured on an interval or a ratio scale, the concepts of the *mean* and *distance* are very well defined. On a ruler, for example, a measurement of 2 inches is exactly halfway between a measurement of 1 inch and a measurement of 3 inches. In this case, the mean $M = 2$ provides a precise central location that defines a balance point that is equidistant from the two scores, $X = 1$ and $X = 3$. However, ordinal measurements do not provide this same degree of precision. Specifically, ordinal values (ranks) tell you only the direction from one score to another, but provide no information about the distance between scores. Thus, you know that first place is better than second or third, but you do not know how much better. In a horse race, for example, you know that the second-place horse is somewhere between the first- and third-place horses, but you do not know exactly where. In particular, a rank of second is not necessarily halfway between first and third.

Because the concepts of *mean* and *distance* are not well defined with ordinal data and because most of the traditional statistical methods are based on these concepts, it generally is considered unwise to use traditional statistics with ordinal data. Therefore, statisticians have developed special techniques that are designed specifically for use with ordinal data.

HYPOTHESIS TESTS WITH ORDINAL DATA

In this chapter we introduce four hypothesis-testing procedures that are used with ordinal data. Each of the new tests can be viewed as an alternative for one of the standard, parametric tests that was introduced earlier. The four tests and the situations in which they are used are as follows:

1. The Mann-Whitney test uses data from two separate samples to evaluate the difference between two treatment conditions or two populations. The Mann-Whitney test can be viewed as an alternative to the independent-measures *t* hypothesis test introduced in Chapter 10.

2. The Wilcoxon test uses data from a repeated-measures design to evaluate the difference between two treatment conditions. This test is an alternative to the repeated-measures *t* test from Chapter 11.

3. The Kruskal-Wallis test uses data from three or more separate samples to evaluate the differences between three or more treatment conditions (or populations). The Kruskal-Wallis test is an alternative to the single-factor, independent-measures ANOVA introduced in Chapter 13.

4. The Friedman test uses data from a repeated-measures design to compare the differences between three or more treatment conditions. This test is an alternative to the repeated-measures ANOVA from Chapter 14.

In each case, you should realize that the new, ordinal-data tests are back-up procedures that are available in situations in which the standard, parametric tests cannot be used. In general, if the data are appropriate for an ANOVA or one of the *t* tests, then the standard test is preferred to its ordinal-data alternative.

**OBTAINING ORDINAL
MEASUREMENTS**

Ordinal measurements can be obtained directly from observations of individuals. For example, you could arrange a group of people in order of height simply by comparing them side by side to see who is tallest. Who is next tallest, and so on. This kind of direct ordinal measurement is fairly common because it is easy to do. Notice that you do not need any absolute measurement to complete the ranking: that is, you never need to know anyone's exact height. It is necessary only to make relative judgements—given any two individuals, you must decide who is taller. Because ordinal scales do not require any absolute measurements, they can be used with variables that are not routinely measured. Variables such as beauty or talent can be observed, judged, and ranked, although they may be difficult to define and measure with more-sophisticated scales

In addition to obtaining ranks by direct observation, a researcher may begin with a set of numerical measurements and convert these scores into ranks. For example, if you had a listing of the actual heights for a group of individuals, you could arrange the numbers in order from greatest to least. This process converts data from an interval or a ratio scale into ordinal measurements. In Chapter 18 (page 606), we identify several reasons for converting numerical scores into nominal categories. These same reasons also provide justification for transforming scores into ranks. The following list should give you an idea of why there can be an advantage to using ranks instead of scores.

1. Ranks are simpler. If someone asks you how tall your sister is, you could reply with a specific numerical value, such as 5 feet $7\frac{3}{4}$ inches tall. Or you could answer, "She is a little taller than I am." For many situations, the relative answer would be better.

2. The original scores may violate some of the basic assumptions that underlie certain statistical procedures. For example, the t tests and ANOVA assume that the data come from normal distributions. Also, the independent-measures tests assume that the different populations all have the same variance (the homogeneity-of-variance assumption). If a researcher suspects that the data do not satisfy these assumptions, it may be safer to convert the scores to ranks and use a statistical technique designed for ranks.

3. The original scores may have unusually high variance. Variance is a major component of the standard error in the denominator of t statistics and the error term in the denominator of F-ratios. Thus, large variance can greatly reduce the likelihood that these parametric tests will find significant differences. Converting the scores to ranks essentially eliminates the variance. For example, 10 scores have ranks from 1 to 10 no matter how variable the original scores are.

4. Occasionally, an experiment produces an undetermined, or infinite, score. For example, a rat may show no sign of solving a particular maze after hundreds of trials. This animal has an infinite, or undetermined, score. Although there is no absolute score that can be assigned, you can say that this rat has the highest score for the sample and then rank the rest of the scores by their numerical values.

RANKING TIED SCORES

Whenever you are transforming numerical scores into ranks, you may find two or more scores that have exactly the same value. Because the scores are tied, the transformation process should produce ranks that are also tied. The procedure for converting tied scores into tied ranks was presented in Chapter 16 (page 544) when we introduced the Spearman correlation and is repeated briefly here. First, you list the scores in order, including tied values. Second, you assign each position in the list a rank (1st, 2nd, and so on). Finally, for any scores that are tied, you compute the mean of the tied ranks, and use the mean value as the final rank. The following set of scores demonstrates this process for a set of $n = 8$ scores.

Original scores:	3	4	4	7	9	9	9	12
Position ranks:	1	2	3	4	5	6	7	8
Final ranks:	1	2.5	2.5	4	6	6	6	8

For example, the original scores contain two individuals who are tied with scores of $X = 4$. These two scores are given ranks of 2 and 3. Then, both individuals are assigned a final rank equal to the average of the tied ranks: $(2 + 3)/2 = 2.5$.

20.2 THE MANN-WHITNEY *U*-TEST: AN ALTERNATIVE TO THE INDEPENDENT-MEASURES *t* TEST

Recall that a study using two separate samples is called an *independent-measures* study or a *between-subjects* study.

The Mann-Whitney test is designed to use the data from two separate samples to evaluate the difference between two treatments (or two populations). The calculations for this test require that the individual scores in the two samples be rank-ordered. The mathematics of the Mann-Whitney test are based on the following simple observation:

A real difference between the two treatments should cause the scores in one sample to be generally larger than the scores in the other sample. If the two samples are combined and all the scores placed in rank order on a line, then the scores from one sample should be concentrated at one end of the line, and the scores from the other sample should be concentrated at the other end.

On the other hand, if there is no treatment difference, then large and small scores will be mixed evenly in the two samples because there is no reason for one set of scores to be systematically larger or smaller than the other.

This observation is demonstrated in Figure 20.1.

FIGURE 20.1

In part (a), the scores from the two samples are clustered at opposite ends of the rank ordering. In this case, the data suggest a systematic difference between the two treatments. Part (b) shows the two samples intermixed evenly along the scale, indicating no consistent difference between treatments.

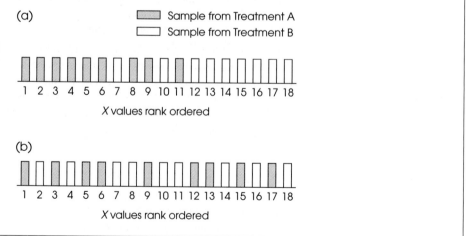

THE NULL HYPOTHESIS FOR THE MANN-WHITNEY TEST

Because the Mann-Whitney test compares two distributions (rather than two means), the hypotheses tend to be somewhat vague. We state the hypotheses in terms of a consistent, systematic difference between the two treatments being compared.

H_0: There is no difference between the two treatments. Therefore, there is no tendency for the ranks in one treatment condition to be systematically higher (or lower) than the ranks in the other treatment condition.

H_1: There is a difference between the two treatments. Therefore, the ranks in one treatment condition are systematically higher (or lower) than the ranks in the other treatment condition.

CALCULATION OF THE MANN-WHITNEY U

The first steps in the calculations for the Mann-Whitney test have already been discussed. To summarize,

1. A separate sample is obtained from each of the two treatments. We use n_A to refer to the number of individuals in sample A and n_B to refer to the number in sample B.

2. These two samples are combined, and the total group of $n_A + n_B$ individuals is rank ordered.

The remaining problem is to decide whether the ranks from the two samples are mixed randomly or are systematically clustered at opposite ends of the scale. This is the familiar question of statistical significance: Are the data simply the result of chance, or has some systematic effect produced these results? We answer this question exactly as we always have answered it. First, look at all the possible results that could have been obtained. Next, separate these outcomes into two groups:

1. Those results that are reasonably likely to occur by chance

2. Those results that are very unlikely to occur by chance (this is the critical region)

For the Mann-Whitney test, the first step is to identify each of the possible outcomes. This is done by assigning a numerical value to every possible set of sample data. This number is called the *Mann-Whitney* U. The value of U is computed as if the two samples were two teams of athletes competing in a sports event. Each individual in sample A (the A team) gets one point whenever he or she is ranked ahead of an individual from sample B. The total number of points accumulated for sample A is called U_A. In the same way, a U value, or team total, is computed for sample B. The final Mann-Whitney U is the smaller of these two values. This process is demonstrated in the following example.

EXAMPLE 20.1

We begin with two separate samples with $n = 6$ scores in each.

Sample A (treatment 1):   27   2   9   48   6   15

Sample B (treatment 2):   71   63   18   68   94   8

Next, the two samples are combined, and all 12 scores are placed in rank order. Each individual in sample A is assigned 1 point for every score in sample B that has a higher rank.

| | Ordered Scores | | Points for |
Rank	Score	Sample	Sample A
1	2	(A)	6 points
2	6	(A)	6 points
3	8	(B)	
4	9	(A)	5 points
5	15	(A)	5 points
6	18	(B)	
7	27	(A)	4 points
8	48	(A)	4 points
9	63	(B)	
10	68	(B)	
11	71	(B)	
12	94	(B)	

Finally, the points from all the individuals in sample A are combined, and the total number of points is computed for the sample. In this example, the total points or the U value for sample A is $U_A = 30$. In the same way, you can compute the U value for sample B. You should obtain $U_B = 6$. As a simple check on your arithmetic, note that

To avoid errors, it is wise to compute both U values and verify that the sum is equal to $n_A n_B$.

$$U_A + U_B = n_A n_B \tag{20.1}$$

For these data,

$$30 + 6 = 6(6)$$

RANKING WITHOUT SCORES

In Example 20.1, we assumed that the researcher had obtained a score (X value) for each individual. However, it is not necessary to have a set of previously obtained scores. The researcher could have started with a total group of 12 individuals (two samples of $n = 6$) and rank-ordered the individuals from 1st to 12th with respect to the variable being measured. For example, a researcher could observe a group of 12 preschool children (6 boys and 6 girls) and rank them in terms of aggressive behavior. Whether or not you have numerical scores, the result is that the two samples are combined into one group and all the individuals are then ranked.

COMPUTING U FOR LARGE SAMPLES

Because the process of counting points to determine the Mann-Whitney U can be tedious, especially with large samples, there is a formula that will generate the U value for each sample. To use this formula, you combine the samples and rank-order all the individuals as before. Then you must find ΣR_A, which is the sum of the ranks for individuals in sample A, and the corresponding ΣR_B for sample B. The U value for each sample is then computed as follows: For sample A,

$$U_A = n_A n_B + \frac{n_A(n_A + 1)}{2} - \Sigma R_A \qquad (20.2)$$

and for sample B,

$$U_B = n_A n_B + \frac{n_B(n_B + 1)}{2} - \Sigma R_B \qquad (20.3)$$

These formulas are demonstrated using the data from Example 20.1. For sample A, the sum of the ranks is

$$\Sigma R_A = 1 + 2 + 4 + 5 + 7 + 8$$
$$= 27$$

For sample B, the sum of the ranks is

$$\Sigma R_B = 3 + 6 + 9 + 10 + 11 + 12$$
$$= 51$$

By using the special formula, for sample A,

$$U_A = n_A n_B + \frac{n_A(n_A + 1)}{2} - \Sigma R_A$$

$$= 6(6) + \frac{6(7)}{2} - 27$$

$$= 36 + 21 - 27$$

$$= 30$$

For sample B,

$$U_B = n_A n_B + \frac{n_B(n_B + 1)}{2} - \Sigma R_B$$

$$= 6(6) + \frac{6(7)}{2} - 51$$

$$= 36 + 21 - 51$$

$$= 6$$

Notice that these are the same U values we obtained in Example 20.1 using the counting method. The Mann-Whitney U value is the smaller of these two,

$$U = 6$$

HYPOTHESIS TESTS WITH THE MANN-WHITNEY *U*

Now that we have developed a method for identifying each rank order with a numerical value, the remaining problem is to decide whether the U value provides evidence for a real difference between the two treatment conditions. We will look at each possibility separately.

A large difference between the two treatments causes all the ranks from sample A to cluster at one end of the scale and all the ranks from sample B to cluster at the other (see Figure 20.1). At the extreme, there is no overlap between the two samples. In this case, the Mann-Whitney U will be zero because one of the samples gets no points at all. In general, a Mann-Whitney U of zero indicates the greatest possible difference between the two samples. As the two samples become more alike, their ranks begin to intermix, and the U becomes larger. If there is no consistent tendency for one treatment to produce larger scores than the other, then the ranks from the two samples should be intermixed evenly.

The null hypothesis for the Mann-Whitney test states that there is no systematic difference between the two treatments being compared; that is, there is no real difference between the two populations from which the samples are selected. In this case, the most likely outcome is that the two samples are similar and that the U value is relatively large. On the other hand, a very small value of U, near zero, is evidence that the two samples are very different. Therefore, a U value near zero would tend to refute the null hypothesis. The distribution of all the possible U values has been constructed, and the critical values for $\alpha = .05$ and $\alpha = .01$ are presented in Table B.9 of Appendix B. When sample data produce a U that is *less than or equal to* the table value, we reject H_0.

> Notice that the null hypothesis does not specify any population parameter.

For example, in Example 20.1 both samples have $n = 6$ scores and the table shows a critical value of $U = 5$ for a two-tailed test with $\alpha = .05$. This means that a value of $U = 5$ or smaller is very unlikely to occur (probability less than .05) if the null hypothesis is true. The data actually produced $U = 6$, which is not in the critical region. Therefore, we would conclude that the data do not provide enough evidence to conclude that there is a significant difference between the two treatments.

IN THE LITERATURE
REPORTING THE RESULTS OF A MANN-WHITNEY U-TEST

Unlike many other statistical results, there are no strict rules for reporting the outcome of a Mann-Whitney U-test. APA guidelines, however, do suggest that the report include a summary of the data (including such information as the sample size and the sum of the ranks) and the obtained statistic and p value. For the study presented in Example 20.1, the results could be reported as follows:

> The original scores were ranked ordered and a Mann-Whitney U-test was used to compare the ranks for the $n = 6$ participants in treatment A and the $n = 6$ participants in treatment B. The results indicate no significant difference between treatments, $U = 6$, $p > .05$, with the sum of the ranks equal to 27 for treatment A and 51 for treatment B.   ❏

NORMAL APPROXIMATION FOR THE MANN-WHITNEY U

When the two samples are both large (about $n = 20$) and the null hypothesis is true, the distribution of the Mann-Whitney U statistic tends to approximate a normal shape. In this case, the Mann-Whitney hypotheses can be evaluated using a z-score statistic and the unit normal distribution. You may have noticed that the table of critical values for the Mann-Whitney test does not list values for samples larger than $n = 20$. This is because the normal approximation typically is used with larger samples. The procedure for this normal approximation is as follows:

1. Find the U values for sample A and sample B as before. The Mann-Whitney U is the smaller of these two values.

2. When both samples are relatively large (around $n = 20$ or more), the distribution of the Mann-Whitney U statistic tends to form a normal distribution with

$$\mu = \frac{n_A n_B}{2} \quad \text{and} \quad \sigma = \sqrt{\frac{n_A n_B (n_A + n_B + 1)}{12}}$$

This approximation is intended for data without tied scores or with very few ties. A special formula has been developed for data with many ties and can be found in most advanced statistics texts such as Hays (1981).

The Mann-Whitney U obtained from the sample data can be located in this distribution using a z-score:

$$z = \frac{X - \mu}{\sigma} = \frac{U - \frac{n_A n_B}{2}}{\sqrt{\dfrac{n_A n_B (n_A + n_B + 1)}{12}}} \tag{20.4}$$

3. Use the unit normal table to establish the critical region for this z-score. For example, with $\alpha = .05$ the critical values would be ± 1.96.

An example of this normal approximation for the Mann-Whitney U follows.

EXAMPLE 20.2

To demonstrate the normal approximation, we will use the same data that were used in Example 20.1. This study compared two treatments, A and B, using a separate sample of $n = 6$ for each treatment. The data produced a value of $U = 6$.

(You should realize that when samples are this small, you normally use the Mann-Whitney table rather than the normal approximation. However, we are using the normal approximation so that we can compare the outcome to the result from the regular Mann-Whitney test.)

Using Equation 20.4, the z-score corresponding to $U = 6$ is

$$z = \frac{U - \frac{n_A n_B}{2}}{\sqrt{\dfrac{n_A n_B (n_A + n_B + 1)}{12}}}$$

$$= \frac{6 - \frac{6(6)}{2}}{\sqrt{\dfrac{6(6)\,(6+6+1)}{12}}}$$

$$= \frac{-12}{\sqrt{\dfrac{468}{12}}}$$

$$= -1.92$$

According to the unit normal table, the extreme 5% of the normal distribution is located in the tails beyond $z = \pm 1.96$ (Figure 20.2). With $\alpha = .05$, a z-score beyond ± 1.96 would lead to rejecting H_0. Our computed z-score, $z = -1.92$, is not in the critical region, so the decision is to fail to reject H_0. Note that we reached the same conclusion for the original test using the critical values in the Mann-Whitney U table.

ASSUMPTIONS AND CAUTIONS FOR THE MANN–WHITNETY *U*

The Mann-Whitney U-test is a very useful alternative to the independent-measures t test. Because the Mann-Whitney test does not require homogeneity of variance or normal distributions, it can be used in situations in which the t test would be inappropriate.

FIGURE 20.2

The normal distribution of z-scores used with the normal approximation to the Mann-Whitney U-test. The critical region for α = .05 has been shaded.

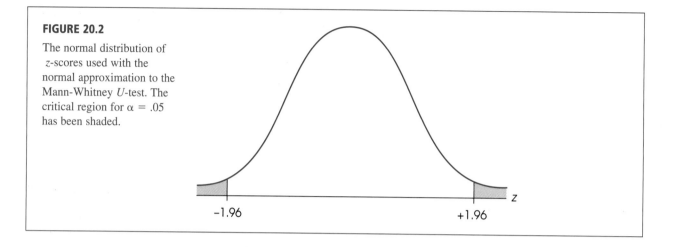

-1.96 $+1.96$

However, the U-test does require independent observations, and it assumes that the dependent variable is continuous. You should recall from Chapter 1 that a continuous scale has an infinite number of distinct points. One consequence of this fact is that it is very unlikely for two individuals to have exactly the same score. This means that there should be few, if any, tied scores in the data. When sample data do have several tied scores, you should suspect that a basic assumption underlying the Mann-Whitney U-test has been violated. In this situation, you should be cautious about using the Mann-Whitney U.

When there are relatively few tied scores in the data, the Mann-Whitney test may be used, but you must follow the standard procedure for ranking tied scores.

LEARNING CHECK

1. Rank the following scores, using tied ranks where necessary: 14, 3, 4, 0, 3, 5, 14, 3.

2. An experiment using $n = 25$ in one sample and $n = 10$ in the other produced a Mann-Whitney U of $U = 50$. Assuming that this is the smaller of the two U values, what was the value of U for the other sample?

3. A developmental psychologist is examining social assertiveness for preschool children. Three- and 4-year-old children are observed for 10 hours in a day-care center. The psychologist records the number of times each child initiates a social interaction with another child. The scores for the sample of 4 boys and 9 girls are as follows:

    Boys' scores: 8   17   14   21

    Girls' scores: 18   25   23   21   34   28   32   30   13

    Use a Mann-Whitney test to determine whether these data provide evidence for a significant difference in social assertiveness between preschool boys and girls. Test at the .05 level.

4. According to the Mann-Whitney table, a value of $U = 30$ is significant (in the critical region) with $\alpha = .05$ when both samples have $n = 11$. If this value is used in the normal approximation, does it produce a z-score in the critical region?

1. Scores: 0 3 3 3 4 5 14   14
    Ranks: 1 3 3 3 5 6 7.5 7.5

2. With $n_A = 25$ and $n_B = 10$, the two U values must total $n_A n_B = 25(10) = 250$. If the smaller value is 50, then the larger value must be 200.

3. For the boys, $U = 31.5$. For the girls, $U = 4.5$. The critical value in the table is 4, so fail to reject H_0.

4. $U = 30$ produces a z-score of $z = -2.00$. This is in the critical region.

20.3 THE WILCOXON SIGNED-RANKS TEST: AN ALTERNATIVE TO THE REPEATED-MEASURES *t* TEST

The Wilcoxon test is designed to evaluate the difference between two treatments, using the data from a repeated-measures experiment. Recall that a repeated-measures study involves only one sample, with each individual in the sample being measured twice: once in the first treatment and once in the second treatment. The difference between the two two measurements for each individual is recorded as the score for that individual. The Wilcoxon test requires that the differences be rank-ordered from smallest to largest in terms of their absolute magnitude, without regard for sign or direction. For example, Table 20.2(a) show nonnumerical differences and ranks for a sample of $n = 4$ participants. If the differences are numerical values, they are also ranked in terms of their absolute magnitude. Table 20.2(b) shows $n = 5$ numerical difference scores and their ranks.

THE HYPOTHESES FOR THE WILCOXON TEST

The null hypothesis for the Wilcoxon test simply states that there is no consistent, systematic difference between the two treatments.

H_0: There is no difference between the two treatments. Therefore, in the general population there is no tendency for the difference scores to be either systematically positive or systematically negative.

H_1: There is a difference between the two treatments. Therefore, in the general population the difference scores are systematically positive or systematically negative.

If the null hypothesis is true, any differences that exist in the sample data must be due to chance. Therefore, we would expect positive and negative differences to be

---

TABLE 20.2

Ranking difference scores. In part (a), differences are measured as relative increases or decreases and the non-numercial values are ranked by magnitude, independent of direction. In part (b), the difference scores are numerical values and the difference are ranked independent of direction.

(a)

Participant	Difference from Treatment 1 to Treatment 2	Rank
A	small increase	2
B	very large decrease	4
C	moderate increase	3
D	very small increase	1

(b)

Participant	Difference Score	Rank
A	+4	2
B	−14	5
C	+9	4
D	−1	1
E	−6	3

intermixed evenly. On the other hand, a consistent difference between the two treatments should cause the scores in one treatment to be consistently larger than the scores in the other. This should produce difference scores that tend to be consistently positive or consistently negative. The Wilcoxon test uses the signs and the ranks of the difference scores to decide whether there is a significant difference between the two treatments.

CALCULATION AND INTERPRETATION OF THE WILCOXON *T*

As with most nonparametric tests, the calculations for the Wilcoxon are quite simple. After ranking the absolute values of the difference scores, you separate the ranks into two groups: those associated with positive differences (increases) and those associated with negative differences (decreases). Next find the sum of the ranks for each group. The smaller of the two sums is the test statistic for the Wilcoxon test and is identified by the letter *T*. For the data in Table 20.2(a) for example, the increases have ranks of 1, 2, and 3 which add up to $\Sigma R = 6$ and there is only one decrease with a rank of 4. The *Wilcoxon T* for this set of data is T = 4. For the difference scores in Table 20.2(b), the positive differences have ranks of 2 and 4, which add to $\Sigma R = 6$ and the negative difference scores have ranks of 1, 3, and 5 which add up to $\Sigma R = 9$. For these scores, $T = 6$

We noted earlier that a strong treatment effect should cause the difference scores to be consistently positive or consistently negative. In the extreme case, all of the differences are in the same direction. This produces a Wilcoxon *T* of zero. For example, when all of the differences are positive, the sum of the negative ranks is zero. On the other hand, when there is no treatment effect, the signs of the difference scores should be intermixed evenly. In this case, the Wilcoxon *T* is relatively large. In general, a small *T* value (near zero) provides evidence for a real difference between the two treatment conditions. The distribution of all the possible *T* values has been constructed, and the critical values for $\alpha = .05$ and $\alpha = .01$ are given in Table B.10 of Appendix B. Whenever sample data produce a *T* that is *less than or equal to* this critical value, we reject H_0. For a repeated-measures study with a sample of $n = 15$, for example, the table shows a critical value of $T = 25$ for a two-tailed test with $\alpha = .05$. This means that a value of $T = 25$ or smaller is very unlikely to occur (probability of .05 or less) if the null hypothesis is true.

TIED SCORES AND ZERO SCORES

Although the Wilcoxon test does not require normal distributions, it does assume that the dependent variable is continuous. As noted with the Mann-Whitney test, this assumption implies that tied scores should be very unlikely. When ties do appear in sample data, you should be concerned that the basic assumption of continuity has been violated, and the Wilcoxon test may not be appropriate. If there are relatively few ties in the data, most researchers assume that the data actually are continuous, but have been crudely measured. In this case, the Wilcoxon test may be used, but the tied values must receive special attention in the calculations.

With the Wilcoxon test, there are two different types of tied scores:

1. A subject may have the same score in treatment 1 and in treatment 2, resulting in a difference score of zero.

2. Two (or more) subjects may have identical difference scores (ignoring the sign of the difference).

When the data include individuals with difference scores of zero, one strategy is to discard these individuals from the analysis and reduce the sample size (*n*). However,

Remember: The null hypothesis says there is no difference between the two treatments. Subjects with difference scores of zero tend to support this hypothesis.

this procedure ignores the fact that a difference score of zero is evidence for retaining the null hypothesis. A better procedure is to divide the zero differences evenly between the positives and negatives. (If you have an odd number of zero differences, discard one and divide the rest evenly.) This second procedure tends to increase ΣR for both the positive and the negative ranks, which increases the final value of T and makes it more likely that H_0 will be retained.

When you have ties among the difference scores, each of the tied scores should be assigned the average of the tied ranks. This procedure was presented in detail in an earlier section of this chapter (see page 667).

HYPOTHESIS TESTS WITH THE WILCOXON *T*

Following is a complete example of the Wilcoxon test using data containing both types of tied scores.

EXAMPLE 20.3

The local Red Cross has conducted an intensive campaign to increase blood donations. This campaign has been concentrated in 10 local businesses. In each company, the goal was to increase the percentage of employees who participate in the blood donation program. Figures showing the percentages of participation from last year (before the campaign) and from this year are as follows. We use the Wilcoxon test to decide whether these data provide evidence that the campaign had a significant impact on blood donations. Note that the 10 companies are listed in rank order according to the absolute value of the difference scores.

Company	Percentage of Participation			Rank Discarding Zeros	Rank Including Zeros
	Before	After	Difference		
A	18	18	0	—	1.5
B	24	24	0	—	1.5
C	31	30	−1	1	3
D	28	24	−4	2	4
E	17	24	+7	3	5
F	16	24	+8	4	6
G	15	26	+11	5.5	7.5
H	18	29	+11	5.5	7.5
I	20	36	+16	7	9
J	9	28	+19	8	10

Discarding zero differences   We first conduct the Wilcoxon test using the recommendation that difference scores of zero be discarded.

STEP 1   The null hypothesis states that the campaign had no effect. Therefore, any differences are due to chance, and there should be no consistent pattern.

STEP 2   The two companies with difference scores of zero are discarded, and *n* is reduced to 8. With $n = 8$ and $\alpha = .05$, the critical value for the Wilcoxon test is $T = 3$. A sample value that is less than or equal to 3 will lead us to reject H_0.

STEP 3   For these data, the positive differences have ranks of 3, 4, 5.5, 5.5, 7, and 8:

$$\Sigma R_+ = 33$$

The negative differences have ranks of 1 and 2:

$$\Sigma R_- = 3$$

The Wilcoxon T is the smaller of these sums, so $T = 3$.

STEP 4   The T value from the data is in the critical region. This value is very unlikely to occur by chance ($p < .05$); therefore, we reject H_0 and conclude that there is a significant change in participation after the Red Cross campaign.

Including zero differences   If we include the two individuals with difference scores of zero, then $n = 10$, and with $\alpha = .05$, the critical value for the Wilcoxon T is 8. Because the difference scores of zero are tied for first and second in the ordering, each is given a rank of 1.5. One of these ranks is assigned to the positive group and one to the negative group. As a result, the sums are

$$\Sigma R_+ = 1.5 + 5 + 6 + 7.5 + 7.5 + 9 + 10$$
$$= 46.5$$

and

$$\Sigma R_- = 1.5 + 3 + 4$$
$$= 8.5$$

The Wilcoxon T is the smaller of these two sums, $T = 8.5$. Because this T value is larger than the critical value, we fail to reject H_0 and conclude that these data do not indicate a significant change in blood donor participation.

By including the difference scores of zero in the test, we have changed the statistical conclusion. Remember: The difference scores of zero are an indication that the null hypothesis is correct. When these scores are considered, the test is more likely to retain H_0.

IN THE LITERATURE
REPORTING THE RESULTS OF A WILCOXON *T*-TEST

As with the Mann-Whitney U-test, there is no specified format for reporting the results of a Wilcoxon T-test. It is suggested, however, that the report include a summary of the data and the value obtained for the test statistic as well as the p value. It also is suggested that the report describe the treatment of zero-difference scores in the analysis. For the study presented in Example 20.5, the report could be as follows:

Two companies showing no change in participation were discarded prior to analysis. The remaining eight companies were rank-ordered by the magnitude of the change in level of participation, and a Wilcoxon T was used to evaluate the data. The results show a significant increase in participation following the campaign, $T = 3, p < .05$, with the ranks for increases totaling 33 and the ranks for decreases totaling 3.   ❏

NORMAL APPROXIMATIONS FOR THE WILCOXON *T*-TEST

When a sample is relatively large, the values for the Wilcoxon T statistic tend to form a normal distribution. In this situation, it is possible to perform the test using a z-score statistic and the normal distribution rather than looking up a T value in the Wilcoxon table. When the sample size is greater than 20, the normal approximation is very accurate and can be used. For samples larger than $n = 50$, the Wilcoxon table typically does not provide any critical values, so the normal approximation must be used. The procedure for the *normal approximation for the Wilcoxon* T is as follows:

1. Find the total for the positive ranks and the total for the negative ranks as before. The Wilcoxon T is the smaller of the two values.

2. With n greater than 20, the Wilcoxon T values form a normal distribution with a mean of

$$\mu = \frac{n(n + 1)}{4}$$

and a standard deviation of

$$\sigma = \sqrt{\frac{n(n + 1)(2n + 1)}{24}}$$

The Wilcoxon T from the sample data corresponds to a z-score in this distribution defined by

$$z = \frac{X - \mu}{\sigma} = \frac{T - \dfrac{n(n + 1)}{4}}{\sqrt{\dfrac{n(n + 1)(2n + 1)}{24}}} \tag{20.5}$$

3. The unit normal table is used to determine the critical region for the z-score. For example, the critical values are ± 1.96 for $\alpha = .05$.

EXAMPLE 20.4

Although the normal approximation is intended for samples with at least $n = 20$ individuals, we will demonstrate the calculations with the same data that were used to introduce the Wilcoxon T in Example 20.3.

Discarding zero differences   When the two companies with zero difference scores were discarded, we obtained $n = 8$ and $T = 3$. Using the normal approximation, these values produce

$$\mu = \frac{n(n + 1)}{4} = \frac{8(9)}{4} = 18$$

$$\sigma = \sqrt{\frac{n(n + 1)(2n + 1)}{24}} = \sqrt{\frac{8(9)(17)}{24}} = \sqrt{51} = 7.14$$

With these values, the obtained $T = 3$ corresponds to a z-score of

$$z = \frac{T - \mu}{\sigma} = \frac{3 - 18}{7.14} = \frac{-15}{7.14} = -2.10$$

With critical boundaries of ± 1.96, the obtained z-score is close to the boundary, but it is enough to be significant at the .05 level. Note that this is exactly the same conclusion we reached using the Wilcoxon T table when the zero differences were discarded in Example 20.3.

Including zero differences When the two zero differences are included, we obtained $n = 10$ and $T = 8.5$. Using the normal approximation, these values produce

$$\mu = \frac{n(n + 1)}{4} = \frac{10(11)}{4} = 27.5$$

$$\sigma = \sqrt{\frac{n(n + 1)(2n + 1)}{24}} = \sqrt{\frac{10(11)(21)}{24}} = \sqrt{96.25} = 9.81$$

With these values, the obtained $T = 8.5$ corresponds to a z-score of

$$z = \frac{T - \mu}{\sigma} = \frac{8.5 - 27.5}{9.81} = \frac{-19}{9.81} = -1.94$$

With critical boundaries of ± 1.96, the obtained z-score is close to the boundary, but it is not quite enough to be significant at the .05 level. Again, this is the same cconclusion we reached using the Wilcoxon T table when the zero differences were included in Example 20.3.

LEARNING CHECK

1. A physician is testing the effectiveness of a new arthritis drug by measuring patients' grip strength before and after they receive the drug. The difference scores for 10 patients are as follows: $+3$, $+46$, $+16$, -2, $+38$, $+14$, 0 (no change), -8, $+25$, and $+41$. Each score is the difference in strength, with a positive value indicating a stronger grip after receiving the drug. Use a Wilcoxon test to determine whether these data provide sufficient evidence to conclude that the drug has a significant effect. Test at the .05 level.

ANSWER

1. The Wilcoxon $T = 4$. Because one patient showed no change, n is reduced to 9, and the critical value is $T = 5$. Therefore, we reject H_0 and conclude that there is a significant effect.

20.4 THE KRUSKAL-WALLIS TEST: AN ALTERNATIVE TO THE INDEPENDENT-MEASURES ANOVA

The *Kruskal-Wallis test* is used to evaluate differences between three or more treatment conditions (or populations) using data from an independent-measures design. You should recognize that this test is an alternative to the single-factor ANOVA introduced

An independent-measures
design simply means that there
is a separate sample for each
treatment condition.

in Chapter 13. However, the ANOVA requires numerical scores that can be used to calculate means and variances. The Kruskal-Wallis test, on the other hand, simply requires that you are able to rank-order the individuals for the variable being measured. You also should recognize that the Kruskal-Wallis test is similar to the Mann-Whitney test introduced earlier in this chapter. However, the Mann-Whitney test is limited to comparing only two treatments, whereas the Kruskal-Wallis test is used to compare three or more treatments.

THE DATA FOR A KRUSKAL-WALLIS TEST

The Kruskal-Wallis test requires three or more separate samples. The samples can represent different treatment conditions or they can represent different preexisting populations. For example, a researcher may want to examine how social input affects creativity. Children are asked to draw pictures under three different conditions: (1) working alone without supervision, (2) working in groups where the children are encouraged to examine and criticize each other's work, and (3) working alone but with frequent supervision and comments from a teacher. Three separate samples are used to represent the three treatment conditions with $n = 6$ children in each condition. At the end of the study, the researcher collects the drawings from all 18 children and rank-orders the complete set of drawings in terms of creativity. The purpose for the study is to determine whether one treatment condition produces drawings that are ranked consistently higher (or lower) than another condition. Notice that the researcher does not need to determine an absolute creativity score for each painting. Instead, the data consist of relative measures; that is, the researcher must decide which painting shows the most creativity, which shows the second most creativity, and so on.

The creativity study that was just described is an example of a research study comparing different treatment conditions. It also is possible that the three groups could be defined by a subject variable so that the three samples represent different populations. For example, a researcher could obtain drawings from a sample of 5-year-old children, a sample of 6-year-old children, and a third sample of 7-year-old children. Again, the Kruskal-Wallis test would begin by rank-ordering all of the drawings to determine whether one age group showed significantly more (or less) creativity than another.

Finally, the Kruskal-Wallis test can be used if the original, numerical data are converted into ordinal values. The following example demonstrates how a set of numerical scores is transformed into ranks to be used in a Kruskal-Wallis analysis.

EXAMPLE 20.5

Table 20.3(a) shows the original data from an independent-measures study comparing three treatment conditions. To prepare the data for a Kruskal-Wallis test, the complete set of original scores is rank-ordered using the standard procedure for ranking tied scores. Each of the original scores is then replaced by its rank to create the transformed data in Table 20.3(b) that are used for the Kruskal-Wallis test.

THE NULL HYPOTHESIS FOR THE KRUSKAL-WALLIS TEST

As with the other tests for ordinal data, the null hypothesis for the Kruskal-Wallis test tends to be somewhat vague. In general, the null hypothesis states that there are no differences among the treatments being compared. Somewhat more specifically, H_0 states that there is no tendency for the ranks in one treatment condition to be systematically higher (or lower) than the ranks in any other condition. Generally, we use the concept of "systematic differences" to phrase the statement of H_0 and H_1. Thus, the hypotheses for the Kruskal-Wallis test are phrased as follows:

TABLE 20.3

Preparing a set of data for analysis using the Kruskal-Wallis test. The original data consisting of numerical scores are shown in table (a). The original scores are listed in order and assigned ranks using the standard procedure for ranking tied scores. The ranks are then substituted for the original scores to create the set of ordinal data shown in table (b).

(a) Original Data Numerical Scores				Find the rank for each score.			(b) Ordinal Data Ranks		
I	II	III		Original Numerical Scores	Ordinal Ranks		I	II	III
14	2	26	$\Rightarrow$	2	1	$\Rightarrow$	9	1	15
3	14	8		3	2		2	9	5
21	9	14		5	3.5		14	6	9
5	12	19		5	3.5		3.5	7	12
16	5	20		8	5		11	3.5	13
				9	6				
				12	7				
				14	9				
				14	9				
				14	9				
				16	11				
				19	12				
				20	13				
				21	14				
				26	15				

H_0: There is no tendency for the ranks in any treatment condition to be systematically higher or lower than the ranks in any other treatment condition. There are no differences between treatments.

H_1: The ranks in at least one treatment condition are systematically higher (or lower) than the ranks in another treatment condition. There are differences between treatments.

FORMULAS AND NOTATION FOR THE KRUSKAL-WALLIS TEST

Table 20.4 presents the transformed data (the ranks) from Example 20.5 along with the notation that is used in the Kruskal-Wallis formula. The notation is relatively simple and involves the following values:

1. The ranks in each treatment are added to obtain a total or T value for that treatment condition. The T values are used in the Kruskal-Wallis formula.

TABLE 20.4

An example of ordinal data including the notation used for a Kruskal-Wallis test. These are the same data that were presented in Table 20.3, and they include some tied measurements that appear as tied ranks.

Ordinal Data Ranks			
I	II	III	
9	1	15	$N = 15$
2	9	5	
14	6	9	
3.5	7	12	
11	3.5	13	
$T_1 = 39.5$	$T_2 = 26.5$	$T_3 = 54$	
$n_1 = 5$	$n_2 = 5$	$n_3 = 5$	

2. The number of subjects in each treatment condition is identified by a lowercase n.

3. The total number of subjects in the entire study is identified by an uppercase N.

The Kruskal-Wallis formula produces a statistic that is usually identified with the letter H and has approximately the same distribution as chi-square, with degrees of freedom defined by the number of treatment conditions minus one. For the data in Table 20.4, there are 3 treatment conditions, so the formula produces a chi-square value with $df = 2$. The formula for the Kruskal-Wallis statistic is

$$H = \frac{12}{N(N + 1)}\left(\Sigma\frac{T^2}{n}\right) - 3(N + 1) \tag{20.6}$$

Using the data in Table 20.4, the Kruskal-Wallis formula produces a chi-square value of

$$H = \frac{12}{15(16)}\left(\frac{39.5^2}{5} + \frac{26.5^2}{5} + \frac{54^2}{5}\right) - 3(16)$$

$$= 0.05(312.05 + 140.45 + 583.2) - 48$$

$$= 0.05(1035.7) - 48$$

$$= 51.785 - 48$$

$$= 3.785$$

With $df = 2$, the chi-square table lists a critical value of 5.99 for $\alpha = .05$. Because the obtained chi-square value (3.785) is not greater than the critical value, our statistical decision is to fail to reject H_0. The data do not provide sufficient evidence to conclude that there are significant differences among the three treatments.

IN THE LITERATURE
REPORTING THE RESULTS OF A KRUSKAL-WALLIS TEST

As with the Mann-Whitney and the Wilcoxon tests, there is no standard format for reporting the outcome of a Kruskal-Wallis test. However, the report should provide a summary of the data, the value obtained for the chi-square statistic, as well as the value of df, N, and the p value. For the Kruskal-Wallis test that we just completed, the results could be reported as follows:

> After ranking the individual scores, a Kruskal-Wallis test was used to evaluate differences among the three treatments. The outcome of the test indicated no significant differences among the treatment conditions, $H = 3.785$ (2, $N = 15$), $p > .05$. ❑

There is one assumption for the Kruskal-Wallis test that is necessary to justify using the chi-square distribution to identify critical values for H. Specifically, each of the treatment conditions must contain at least five scores. If any sample has fewer than five scores, you can still compute the Kruskal-Wallis H statistic but you must consult a special table to evaluate the significance (Siegel, 1956).

LEARNING CHECK

1. Suppose a researcher conducts an independent-measures study comparing three treatment conditions with a separate sample of $n = 5$ participants in each treatment. Consider the two possible extreme outcomes for this study.

   a. First, imagine that there is no overlap between the three treatments. Specifically, the 5 smallest scores are all in treatment 1, the 5 middle scores are all in treatment 2, and the 5 biggest scores are all in treatment 3. In this case, the first treatment has ranks of 1, 2, 3, 4, and 5. The second treatment has ranks of 6, 7, 8, 9, and 10; and the third treatment has ranks of 11, 12, 13, 14, and 15. Predict what outcome should be obtained from a Kruskal-Wallis test for these data, and compute the value for the Kruskal-Wallis H.

   b. Now image that there are no differences at all between the three treatments. Specifically, each treatment has the same value, $T = 40$, for the sum of the ranks. Again, predict what outcome should be obtained for the Kruskal-Wallis test, and compute the value for the Kruskal-Wallis H.

ANSWERS

1. a. This situation represents the largest possible differences between treatments and should produce a significant result. For these data, $T_1 = 15$, $T_2 = 40$, and $T_3 = 65$. The data produce $H = 12.5$. With $df = 2$, the chi-square distribution lists critical values of 5.99 and 9.21 for $\alpha = .05$ and $\alpha = .01$, respectively. The value we obtained is well beyond either of these critical values, so the null hypothesis is rejected and the conclusion is that there are significant differences among the three treatments.

   b. In this situation, there are no differences between treatments and the test should show that there are no significant differences. With all three T values equal to 40, the data produce $H = 0$, and the statistical decision is to fail to reject H_0.

20.5   THE FRIEDMAN TEST: AN ALTERNATIVE TO THE REPEATED-MEASURES ANOVA

A repeated-measures design means that the same group of individuals participates in all of the treatment conditions.

The *Friedman test* is used to evaluate the differences between three or more treatment conditions using data from a repeated-measures design. This test is an alternative to the repeated-measures ANOVA that was introduced in Chapter 14. However, the ANOVA requires numerical scores that can be used to compute means and variances. The Friedman test, however, simply requires that you be able to rank-order the individuals for the variable being measured. The Friedman test is also similar to the Wilcoxon test that was introduced earlier in this chapter. However, the Wilcoxon test is limited to comparing only two treatments, whereas the Friedman test is used to compare three or more treatments.

THE DATA FOR A FRIEDMAN TEST

The Friedman test requires only one sample, with each individual participating in all of the different treatment conditions. The treatment conditions must be rank-ordered for each individual participant. For example, a researcher could observe a group of children diagnosed with ADHD in three different environments: at home, at school, and during unstructured play time. For each child, the researcher observes the degree to which the disorder interferes with normal activity in each environment, then ranks the three environments from most disruptive to least disruptive. In this case, the ranks are obtained by comparing the individual's behavior across the three conditions. It is also possible for each individual to produce his or her own rankings. For example, each individual could be asked to evaluate three different designs for a new computer keyboard. Each person practices with each keyboard and then ranks them, 1st, 2nd, and 3rd in terms of ease of use.

Finally, the Friedman test can be used if the original data consist of numerical scores. However, the scores must be converted to ranks before the Friedman test is used. The following example demonstrates how a set of numerical scores is transformed into ranks for the Friedman test.

EXAMPLE 20.5

To demonstrate the Friedman test, we will use the same data that were used to introduce the repeated-measures ANOVA in Chapter 14 (see page 452). The data are reproduced in Table 20.5(a) and consist of pain-tolerance scores for a sample of $n = 5$ participants tested under four different drug conditions. The original data consist of numerical scores. To convert the data for the Friedman test, the four scores for each participant are replaced with ranks 1, 2, 3, and 4, corresponding to the size of the original scores. For example, the first participant has scores of 3, 4, 6, and 7. For this individual, $X = 3$ is the smallest score and receives a rank of 1; $X = 4$ is the second smallest score and receives a rank of 2; the other two scores are ranked 3 and 4. As usual, tied scores are assigned the mean of the tied ranks. For example, participant B has tied scores of $X = 3$ for drug A and drug B. These two score are tied for 2nd and 3rd and each receives a rank of 2.5. The complete set of ranks is shown in part (b) of the table.

THE NULL HYPOTHESIS FOR THE FRIEDMAN TEST

In general terms, the null hypothesis for the Friedman test states that there are no differences between the treatment conditions being compared. If the null hypothesis is true, the ranks should be distributed randomly and there should be no tendency for

---

TABLE 20.5

Results from a repeated-measures study comparing three drugs and a placebo. Part (a) shows the original scores measuring the degree of pain relief in each of the four treatment conditions. In part (b), the four treatment conditions are rank-ordered according to the degree of pain relief for each participant. The original data appear in Table 14.2 (page 452) and were used to demonsrate the repeated-measures ANOVA.

(a) The original pain relief scores.

Person	Placebo	Drug A	Drug B	Drug C
A	3	4	6	7
B	0	3	3	6
C	2	1	4	5
D	0	1	3	4
E	0	1	4	3

(b) The rank-ordering of the treatment conditions for each participant

Person	Placebo	Drug A	Drug B	Drug C
A	1	2	3	4
B	1	2.5	2.5	4
C	2	1	3	4
D	1	2	3	4
E	1	2	4	3
	$R_1 = 6$	$R_2 = 9.5$	$R_3 = 15.5$	$R_4 = 19$

the ranks in one treatment to be systematically higher (or lower) than the ranks in any other treatment condition. Thus, the hypotheses for the Friedman test can be phrased as follows:

H_0: There is no difference between treatments. Thus, the ranks in one treatment condition should not be systematically higher or lower than the ranks in any other treatment condition.

H_1: There are differences between treatments. Thus, the ranks in at least one treatment condition should be systematically higher or lower than the ranks in another treatment condition.

NOTATION AND CALCULATION FOR THE FRIEDMAN TEST

The first step in the Friedman test is to compute the sum of the ranks for each treatment condition. If the null hypothesis is true, these sums should all be approximately the same size because none of the treatments should have consistently higher or lower ranks than any other treatment. On the other hand, if there is a consistent difference between treatments, then at least one of the sums should be noticeably larger (or smaller) than the sum for another treatment. The sum of the ranks for each treatment is identified by the symbol R, and numerical subscripts can be used to identify individual treatments. In Table 20.5(b) the first treatment (placebo) has ranks that add up to $R_1 = 6$. The second treatment has $R_2 = 9.5$, and third has $R_3 = 15.5$, and the final treatment condition has $R_4 = 19$.

The remaining notation for the Friedman test consists of the symbols n and k, which correspond to the number of individuals in the sample and the number of treatment conditions, respectively. For the data in Table 20.4(b), $n = 5$ and $k = 4$. The Friedman test evaluates the differences between treatments by computing the following test statistic:

$$\chi_r^2 = \frac{12}{nk(k+1)} \Sigma R^2 - 3n(k+1) \tag{20.7}$$

Note that the statistic is identified as chi-square (χ^2) with a subscript r, and corresponds to a chi-square statistic for ranks. This chi-square statistic has degrees of freedom determined by $df = k - 1$, and is evaluated using the critical values in the chi-square distribution shown in Table B8 in Appendix B.

Using the data in Table 20.5(b), the statistic is

$$\chi_r^2 = \frac{12}{5(4)(5)} (6^2 + 9.5^2 + 15.5^2 + 19^2) - 3(5)(5)$$

$$= \frac{12}{100} (36 + 90.25 + 240.25 + 361) - 75$$

$$= 0.12(727.5) - 75$$

$$= 12.3$$

With $df = k - 1 = 3$, the critical value of chi-square is 9.35. Therefore, the decision is to reject the null hypothesis and conclude that there are significant differences among the four treatment conditions. Note that the Friedman test simply concludes that there are significant differences; the ANOVA concluded that there are significant *mean differences*.

IN THE LITERATURE
REPORTING THE RESULTS OF A FRIEDMAN TEST

As with most of the tests for ordinal data, there is no standard format for reporting the results from a Friedman test. However, the report should provide the value obtained for the chi-square statistic as well as the values for *df, n,* and *p*. For the data that were used to demonstrate the Friedman test, the results would be reported as follows:

> After ranking the original scores, a Friedman test was used to evaluate the differences among the four treatment conditions. The outcome indicated that there are significant differences, $\chi_r^2 = 12.3$ (3, $n = 5$), $p < .05$.   ❏

LEARNING CHECK

1. Suppose a researcher conducts a repeated-measures study comparing three treatment conditions using the same sample of $n = 6$ participants in every treatment. Consider the two possible extreme outcomes for this study.

   a. First, imagine that there is no overlap between the three treatments. Specifically, suppose all 6 participants are ranked 1st in treatment 1, 2nd in treatment 2, and 3rd in treatment 3. In this case, the sum of the ranks for the three treatments would be $R_1 = 6$, $R_2 = 12$, and $R_3 = 18$. Predict the outcome for the Friedman test for these data, and compute the value for chi-square.

   b. Now imagine that there are no differences at all between the three treatments. For example, the first three participants have ranks of 1, 2, and 3 across the three treatments, and the last three participants have ranks of 3, 2, and 1. In this case, each treatment has the same value, $R = 12$, for the sum of the ranks. Again, predict the outcome for the Friedman test, and compute the value for chi-square.

ANSWERS

1. a. This situation represents the largest possible differences between treatments and should produce a significant result. With $k = 3$, $n = 6$, and R values of 6, 12, and 18, the data produce a chi-square statistic of 12. With $df = 2$, the chi-square distribution lists critical values of 5.99 and 9.21 for $\alpha = .05$ and $\alpha = .01$, respectively. The value we obtained is well beyond either of these critical values, so the null hypothesis is rejected and the conclusion is that there are significant difference among the three treatments.

   b. In this situation, there are no differences between treatments, and the test should show that there are no significant differences. With all three R values equal to 12, the data produce $\chi_r^2 = 0$, and the statistical decision is to fail to reject H_0.

SUMMARY

1. The Mann-Whitney test, the Wilcoxon test, the Kruskal-Wallis test, and the Friedman test are nonparametric alternatives to the independent-measures t, the repeated-measures t, the single-factor ANOVA, and the repeated-measures ANOVA, respectively. These tests do not require normal distributions or homogeneity of variance. However, the tests do require that the data be rank-ordered, and they assume that the dependent variable is continuously distributed. For all four tests, the null hypothesis states that there are no consistent, systematic differences between the treatments being compared.

2. The Mann-Whitney U can be computed either by a counting process or by a formula. A small value of U (near zero) is evidence of a difference between the two treatments. With the counting procedure, U is determined by the following:
   a. The scores from the two samples are combined and ranked from smallest to largest.
   b. Each individual in sample A is awarded one point for every member of sample B with a larger rank.
   c. U_A equals the total points for sample A. U_B is the total for sample B. The Mann-Whitney U is the smaller of these two values.

   In formula form,

   $$U_A = n_A n_B + \frac{n_A(n_A + 1)}{2} - \Sigma R_A$$

   $$U_B = n_A n_B + \frac{n_B(n_B + 1)}{2} - \Sigma R_B$$

3. For large samples, larger than those normally presented in the Mann-Whitney table, the normal distribution can be used to evaluate the difference between the two treatments. This normal approximation to the Mann-Whitney is used as follows:
   a. Find the value of U as before.
   b. Convert the Mann-Whitney U to a z-score by the formula

   $$z = \frac{U - \frac{n_A n_B}{2}}{\sqrt{\frac{n_A n_B(n_A + n_B + 1)}{12}}}$$

   c. If this z-score is in the critical region of the unit normal distribution, reject the null hypothesis.

4. The test statistic for the Wilcoxon test is called a T score. A small value of T (near zero) provides evidence

of a difference between the two treatments. T is computed as follows:
   a. Compute a difference score (treatment 1 versus treatment 2) for each individual in the sample.
   b. Rank these difference scores from smallest to largest without regard to the signs.
   c. Add the ranks for the positive differences, and add the ranks for the negative differences. T is the smaller of these two sums.

5. For large samples, $n \geq 20$, the Wilcoxon T is converted into a z-score by the formula

   $$z = \frac{X - \mu}{\sigma} = \frac{T - \frac{n(n + 1)}{4}}{\sqrt{\frac{n(n + 1)(2n + 1)}{24}}}$$

   The unit normal table can then be used to find critical values for different alpha levels.

6. The Kruskal-Wallis test is used to evaluate differences among three or more treatment conditions using a separate sample for each treatment. The test statistic, H, is equivalent to a chi-square statistic with degrees of freedom equal to the number of treatment conditions minus one. The calculation of the test statistic involves the following:
   a. The individuals from all the separate samples are combined and ranked.
   b. The sum of the ranks, T, is computed for each treatment.
   c. The test statistic is computed as

   $$H = \frac{12}{N(N + 1)}\left(\Sigma \frac{T^2}{n}\right) - 3(N + 1)$$

   where n is the number of scores in each individual sample and N is the number of scores for all samples combined.

7. The Friedman test is used to evaluate the differences among three or more treatment conditions using the same group of individuals in all of the treatments. The test statistic is equivalent to chi-square, with degrees of freedom determined by the number of treatments minus one. The calculation of the test statistic involves the following steps.
   a. Each individual is ranked across the different treatment conditions. With three treatments, for example, each individual would have ranks of 1st, 2nd, and 3rd, depending on his or her relative performance in the three treatments.

b. The sum of the ranks, R, is obtained for each treatment.

c. With k = the number of treatments and n = the number of participants, the test statistic is computed as follows:

$$\chi_r^2 = \frac{12}{nk(k+1)} \Sigma R^2 - 3n(k+1)$$

The subscript r beside the chi-square symbol simply indicates that the data are ranks.

KEY TERMS

Mann-Whitney U

normal approximation for the Mann-Whitney U

Wilcoxon T

normal approximation for the Wilcoxon T

Kruskal-Wallis test

Friedman test

RESOURCES

There are a tutorial quiz and other learning exercises for Chapter 20 at the Wadsworth website, www.cengage.com/psychology/gravetter. See page 29 for more information about accessing the website.

ENHANCED
WebAssign

Guided interactive tutorials, end-of-chapter problems, and related testbank items may be assigned online at WebAssign.

WebTUTOR

For those using WebTutor along with this book, there is a WebTutor section corresponding to this chapter. The WebTutor contains a brief summary of Chapter 20, hints for learning the concepts and formulas for all four ordinal tests, cautions about common errors, and sample exam items including solutions.

SPSS

General instructions for using SPSS are presented in Appendix D. Following are detailed instructions for using SPSS to perform **The Mann-Whitney Test, The Wilcoxon Test, the Kruskal-Wallis Test, and the Friedman Test** that are presented in this chapter.

The Mann-Whitney Test

Data Entry
Note: SPSS will accept either raw scores (X values before ranking) or ranks. If you enter raw scores, the program will transform the scores into ranks before conducting the Mann-Whitney test.

1. The data are entered in what is called a *stacked format,* which means that all the scores (or ranks) from *both samples* are entered in one column of the data editor (probably VAR00001). Enter the scores for sample #2 directly beneath the scores from sample #1 with no gaps or extra spaces.

2. Values are then entered into a second column (VAR00002) to identify the sample or treatment condition corresponding to each of the scores. For example, enter a 1 beside each score from sample #1 and enter a 2 beside each score from sample #2.

Data Analysis

1. Click **Analyze** on the tool bar, select **Nonparametric Tests,** and click on **2 Independent Samples.**

2. Verify that the box for **Mann-Whitney** is checked.

3. Highlight the column label for the set of scores (VAR00001) and move it into the **Test Variable** box by clicking on the arrow.

4. Highlight the column label containing the sample numbers (VAR00002) and move it into the **Group Variable** box by clicking on the arrow.

5. Click on **Define Groups.**

6. Assuming that you used the numbers 1 and 2 to identify the two sets of scores, enter the values 1 and 2 into the appropriate group boxes.

7. Click **Continue.**

8. Click **OK.**

SPSS Output

The program produces a table reporting the number of scores, the mean rank, and the sum of the ranks for each group. A second table reports the outcome of the test, including the value for the Mann-Whitney $U,$ the value for the z-score approximation, and the level of significance for both the z-score approximation (Asymp. Sig.) and for the Mann-Whitney U (Exact Sig.).

The Wilcoxon Test

Data Entry
Note: SPSS will accept either raw scores (X values before ranking) or ranks. If you enter raw scores, the program will transform the scores into ranks before conducting the Wilcoxon test.

1. Enter the data into two columns (VAR00001 and VAR00002) in the data editor with the first score for each participant in the first column and the second score in the second column.

Data Analysis

1. Click **Analyze** on the tool bar, select **Nonparametric Tests,** and click on **2 Related Samples.**

2. Verify that the box for **Wilcoxon** is checked.

3. Highlight both of the labels for the two data columns in the left box (click on one, then click on the second) and move the two labels into the **Paired Variables** box by clicking on the arrow.

4. Click **OK.**

SPSS Output

The program produces a table reporting the number (N), the mean rank, and the sum of the ranks for both the positive and the negative ranks. (*Note:* The Wilcoxon T is the smaller of the two sums.) A second table reports the outcome of the test, including the *z*-score approximation and the level of significance (*p* value or alpha level for the test).

The Kruskal-Wallis Test

Data Entry

Note: SPSS will accept either raw scores (*X* values before ranking) or ranks. If you enter raw scores, the program will transform the scores into ranks before conducting the Kruskal-Wallis test.

1. The scores (or ranks) are entered in a *stacked format* in the data editor, which means that all the scores from all the different treatments are entered in a single column (VAR00001). Enter the scores for treatment #2 directly beneath the scores from treatment #1 with no gaps or extra spaces. Continue in the same column with the scores from treatment #3, and so on.
2. In a second column (VAR00002), enter a number to identify the treatment condition for each score. For example, enter a 1 beside each score from the first treatment, enter a 2 beside each score from the second treatment, and so on.

Data Analysis

1. Click **Analyze** on the tool bar, select **Nonparametric Tests,** and click **k Independent Samples.**
2. Verify that the box for **Kruskal-Wallis** is checked.
3. Highlight the column label for the scores (VAR00001) and move it into the **Test Variable List** box by clicking on the arrow.
4. Highlight the column label for the treatment numbers (VAR0002) and move it into the **Grouping Variable** box by clicking on the arrow.
5. Click on **Define Range** and enter the minimum and maximum values for your grouping variable. For example, if you used 1, 2, and 3 to identify three treatments, enter a minimum of 1 and a maximum of 3.
6. Click **Continue.**
7. Click **OK.**

SPSS Output

The program produces a table showing the number of scores and the mean rank for each of the treatment conditions. A second table reports that outcome of the test, including the value for chi-square (the *H* statistic), degrees of freedom, and the level of significance (the *p* value or the alpha level for the test).

The Friedman Test

Data Entry

Note: SPSS will accept either raw scores (*X* values before ranking) or ranks. If you enter raw scores, the program will transform the scores into ranks before conducting the Friedman test.

1. Enter the scores (or ranks) for each treatment condition in a separate column, with the scores for each individual in the same row. All the scores for the first treatment go in the VAR00001 column, the second treatment scores in the VAR00002 column, and so on.

Data Analysis

1. Click **Analyze** on the tool bar, select **Nonparametric Tests,** and click on **k Related Samples.**
2. Verify that the box for **Friedman** is checked.
3. One by one, highlight the column labels for the treatment conditions and move them into the **Test Variables** box by clicking on the arrow.
4. Click **OK.**

SPSS Output

The program produces a table showing the mean rank for each treatment condition. A second table reports that outcome of the test, including the value for chi-square, degrees of freedom, and the level of significance (the *p* value or the alpha level for the test).

FOCUS ON PROBLEM SOLVING

1. In computing the Mann-Whitney *U,* it is sometimes obvious from the data which sample has the smaller *U* value. Even if this is the case, you should still compute both *U* values so that your computations can be checked using the formula

$$U_A + U_B = n_A n_B$$

2. Contrary to the previous statistical tests, the Mann-Whitney *U* and the Wilcoxon *T* are significant when the obtained value is *equal to or less than* the critical value in the table.

3. For the Wilcoxon *T,* remember to ignore the signs of the difference scores when assigning ranks to them.

DEMONSTRATION 20.1

THE MANN-WHITNEY *U*-TEST

A local police expert claims to be able to judge an individual's personality on the basis of his or her handwriting. To test this claim, 10 samples of handwriting are obtained: 5 from prisoners convicted of violent crimes and 5 from psychology majors at the college. The expert ranks the handwriting samples from 1st to 10th, with 1 representing the most antisocial personality. The rankings are as follows:

Ranking	Source
1	Prisoner
2	Prisoner
3	Student . . . 3 points
4	Prisoner
5	Prisoner
6	Student . . . 1 point
7	Prisoner
8	Student . . . 0 points
9	Student . . . 0 points
10	Student . . . 0 points

STEP 1 The null hypothesis states that there is no difference between the two populations. For this example, H_0 states that the police expert cannot differentiate the handwriting for prisoners from the handwriting for students.

　　　　　　The alternative hypothesis says there is a discernible difference.

STEP 2 For $\alpha = .05$ and with $n_A = n_B = 5$, the Mann-Whitney table gives a critical value of $U = 2$. If our data produce a U less than or equal to 2, we reject the null hypothesis.

STEP 3 We designate the students as sample A and the prisoners as sample B. Because the students tended to cluster at the bottom of the rankings, they should have the smaller U. Therefore, we identified the points for this sample. The students' point total is $U_A = 4$.

　　　　　　Because we found the smaller of the two U values, it is not necessary to compute U for the sample of prisoners. However, we continue with this calculation to demonstrate the formula for U. The sample of prisoners has ranks 1, 2, 4, 5, and 7. The sum is $\Sigma R_B = 19$, so the U_B value is

$$U_B = n_A n_B + \frac{n_B(n_B + 1)}{2} - \Sigma R_B$$

$$= 5(5) + \frac{5(6)}{2} - 19$$

$$= 25 + 15 - 19$$

$$= 21$$

To check our calculations,

$$U_A + U_B = n_A n_B$$

$$4 + 21 = 5(5)$$

$$25 = 25$$

The final U is the smaller of the two values, so the Mann-Whitney U statistic is $U = 4$.

STEP 4 Because $U = 4$ is not in the critical region, we fail to reject the null hypothesis. With $\alpha = .05$, these data do not provide sufficient evidence to conclude that there is a discernible difference in handwriting between the two populations.

DEMONSTRATION 20.2

THE WILCOXON SIGNED-RANKS TEST

A researcher obtains a random sample of $n = 7$ individuals and tests each person in two different treatment conditions. The data for this sample are as follows:

Participant	Treatment 1	Treatment 2	Difference
1	8	24	+16
2	12	10	−2
3	15	19	+4
4	31	52	+21
5	26	20	−6
6	32	40	+8
7	19	29	+10

STEP 1   *State the hypotheses, and select alpha.* The hypotheses for the Wilcoxon test do not refer to any specific population parameter.

H_0:   There is no systematic difference between the two treatments.

H_1:   There is a consistent difference between the treatments that causes the scores in one treatment to be generally higher than the scores in the other treatment.

We use $\alpha = .05$.

STEP 2   *Locate the critical region.* A small value for the Wilcoxon T indicates that the difference scores were consistently positive or consistently negative, which indicates a systematic treatment difference. Thus, small values will tend to refute H_0. With $n = 7$ and $\alpha = .05$, the Wilcoxon table shows that a T value of 2 or smaller is needed to reject H_0.

STEP 3   *Compute the test statistic.* The calculation of the Wilcoxon T is very simple, but requires several stages:

a. Ignoring the signs ($+$ or $-$), rank the difference scores from smallest to largest.

b. Compute the sum of the ranks for the positive differences and the sum for the negative differences.

c. The Wilcoxon T is the smaller of the two sums.

For these data, we have the following:

Difference	Rank	
$(+)$ 16	6	$\Sigma R_+ = 6 + 2 + 7 + 4 + 5 = 24$
$(-)$ 2	1	$\Sigma R_- = 1 + 3 = 4$
$(+)$ 4	2	
$(+)$ 21	7	
$(-)$ 6	3	
$(+)$ 8	4	
$(+)$ 10	5	

The Wilcoxon T is $T = 4$.

STEP 4   *Make a decision.* The obtained T value is not in the critical region. These data are not significantly different from chance. Therefore, we fail to reject H_0 and conclude that there is not sufficient evidence to suggest a systematic difference between the two treatment conditions.

DEMONSTRATION 20.3

THE KRUSKAL-WALLIS TEST

The following data represent the outcome from a research study using three separate samples to compare three treatment conditions. After treatment, the individuals from all three samples were ranked and the sum of the ranks, T, was computed for each treatment condition.

	Treatments		
I	II	III	
8	1	11	$N = 15$
2	7	5	
4	6	9	
10	3	12	
14	13	15	
$T_1 = 38$	$T_2 = 30$	$T_3 = 52$	
$n_1 = 5$	$n_2 = 5$	$n_3 = 5$	

STEP 1   *State the hypotheses and select alpha.* The null hypothesis states that there are no differences among the three treatment conditions.

H_0:   There is no tendency for the ranks in any treatment condition to be systematically higher or lower than the ranks in any other treatment condition. There are no differences among the three treatments.

H_1:   The ranks in at least one treatment condition are systematically higher (or lower) than the ranks in another treatment condition. There are differences among the treatments.

STEP 2   *Locate the critical region.* The Kruskal-Wallis test statistic is equivalent to chi-square with degrees of freedom equal to the number of treatment conditions minus one. In this case, $df = 3 - 1 = 2$. With $\alpha = .05$, the critical value is 5.99. If we obtain an H value greater than 5.99, our decision will be to reject H_0.

STEP 3   *Compute the test statistic.* We have already calculated the sum of the ranks, T, for each treatment condition. For these data, $n = 5$ for each treatment and the three samples combine for a total of $N = 15$ scores. Therefore, the Kruskal-Wallis statistic is

$$H = \frac{12}{N(N + 1)}\left(\Sigma\frac{T^2}{n}\right) - 3(N + 1)$$

$$H = \frac{12}{15(16)}\left(\frac{38^2}{5} + \frac{30^2}{5} + \frac{52^2}{5}\right) - 3(16)$$

$$= \frac{1}{20}(288.8 + 180 + 540.8) - 48$$

$$= \frac{1}{20}(1009.6) - 48$$

$$= 50.48 - 48$$

$$= 2.48$$

STEP 4   *Make a decision.* The H value for these data is not in the critical region. Therefore, we fail to reject H_0 and conclude that the data are not sufficient to show any significant differences among the three treatments.

DEMONSTRATION 20.4

THE FRIEDMAN TEST

A researcher obtains a sample of $n = 7$ college students and asks each student to rank-order the following three possibilities in terms of "the closest personal relationship in your life":

a. with a friend of the same gender
b. with a friend of the opposite gender
c. with a family member of either gender

The data are as follows:

Student	Same Gender	Opposite Gender	Family Member
A	1	3	2
B	1	2	3
C	2	1	3
D	2	3	1
E	2	1	3
F	3	2	1
G	1	2	3
Totals	12	14	16

STEP 1 *State the hypotheses and select alpha.* The null hypothesis states that there are no differences among the three kinds of relationships.

> H_0: There are no differences among the three relationships. There is no tendency for the ranks for any one relationship to be systematically higher or lower than the ranks in any other.

> H_1: There are differences among the three relationships. The ranks for at least one are systematically higher or lower than the ranks for another.

STEP 2 *Locate the critical region.* With $df = 2$, a chi-square value of at least 5.99 is needed to reject the null hypothesis.

STEP 3 *Compute the test statistic.* Adding the ranks for each of the three categories produces R values of 12, 14, and 16. With $n = 7$ participants and $k = 3$ treatments, the chi-square statistic is

$$\chi_r^2 = \frac{12}{nk(k + 1)} \Sigma R^2 - 3n(k + 1)$$

$$= \frac{12}{7(3)(4)} (12^2 + 14^2 + 16^2) - 3(7)(4)$$

$$= \frac{12}{84} (144 + 196 + 256) - 84$$

$$= \frac{1}{7} (596) - 84$$

$$= 1.14$$

STEP 4 *Make a decision.* With chi-square = 1.14, we fail to reject H_0 and conclude there are no significant differences among the three kinds of relationships.

PROBLEMS

1. Following is a set of numerical scores for a sample of $n = 10$ subjects. Find the rank for each subject.

Subject:	A	B	C	D	E	F	G	H	I	J
Score:	2	4	7	7	7	12	15	15	19	22

2. Following is a set of numerical scores for a sample of $n = 8$ subjects. Find the rank for each subject.

Subject:	A	B	C	D	E	F	G	H
Score:	12	12	19	19	19	24	27	30

3. Under what circumstances is it necessary or advisable to convert interval or ratio scale measurements into ranked data before attempting a hypothesis test?

4. Problem 22 from Chapter 10 described an experiment in which humorous sentences were presented to one group and nonhumorous sentences to another group (Schmidt, 1994). Both groups were then tested to see how many sentences they could recall. The data are as follows:

Number of Sentences Recalled	
Humorous Sentences	Nonhumorous Sentences
4 5 2 4	5 2 4 2
6 7 6 6	2 3 1 5
2 5 4 3	3 2 3 3
3 3 5 3	4 1 5 3

The independent-measures t test showed a significant difference between the two types of sentences. Transform the scores to ordinal data (ranks) and use a Mann-Whitney test with $\alpha = .05$ to determine whether the ordinal test produces the same conclusion.

5. Problem 23 from Chapter 10 described a research study comparing health (number of doctor visits) for elderly participants who own a dog with those who do not own a dog (Siegel, 1990). The independent-measures t test showed a significant difference between the two groups. Transform the scores to ordinal data (ranks) and use a Mann-Whitney test with $\alpha = .05$ to determine whether the ordinal test produces the same conclusion. The data are as follows:

Control Group	Dog Owners
10	7
8	4
7	9
9	3
13	7
7	
6	
12	

6. One of the advantages of using ranks (ordinal data) instead of interval or ratio scores is that ranks can reduce the effect of one extreme individual score. The following data represent results from an independent-measures study comparing two treatments. Notice that most subjects in treatment 1 have scores around 40, and most subjects in treatment 2 have scores around 15. However, there is one odd individual in treatment 2 with a score of $X = 104$. This extreme score distorts the data so that an independent-measures t test indicates no significant difference between treatments, $t(10) = 0.868$, $p > .05$. When the scores are converted to ranks, however, the effect of the extreme score is much less dramatic. Transform the score into ranks and use a Mann-Whitney test to evaluate the difference between treatments. Test with $\alpha = .05$.

Treatment 1	Treatment 2
22	16
39	10
37	14
30	16
47	17
41	12
57	104
$M = 39.00$	$M = 27.00$
$s = 11.30$	$s = 34.04$

7. Occasionally, researchers measure a behavior that has extremely large variability. In these situations, the variability of the scores can overwhelm any mean differences in the data so that it is essentially impossible to find significant differences. If the scores are converted to ranks, however, the problems of large variability can be eliminated. Consider the following example. A researcher is examining how success or failure on a task can influence an individual's future performance on the same task. A group of participants is given 10 minutes to work on a problem-solving task. Based on their performance, they are classified as having either succeeded or failed. The participants are then given a second chance to work on a similar task with no time limit. The researcher simply records how long each person persists on the task before stopping. The scores (in minutes) for the two groups are as follows:

    Initial success: 13, 20, 14, 7, 19, 23, 38, 10

    Initial failure: 2, 5, 7, 3, 6, 5, 4, 32

With the large variability in the scores, a *t* test would conclude that there is no significant difference between the two groups. However, you can rank the scores and use a Mann-Whitney test to evaluate the difference between the two groups. Perform the Mann-Whitney test using the .05 level of significance.

8. One assumption for parametric tests with independent-measures data is that the different treatment conditions have the same variance (homogeneity-of-variance assumption). However, a treatment effect that increases the mean often also increases the variability. In this situation, the parametric *t* test or ANOVA is not justified, and a Mann-Whitney test should be used. The following data represent an example of this situation:

Treatment 1 (Sample A)	Treatment 2 (Sample B)
1	8
5	20
0	14
2	27
4	6
2	10
3	19

a. Compute the mean and variance for each sample. Note the difference between the two sample variances.
b. Use a Mann-Whitney test, with $\alpha = .05$, to test for a significant difference between the two treatments.

9. Example 10.1 (page 318) reported research results demonstrating that high school students who regularly

watched Sesame Street as 5-year-old children had higher grades than their peers who did not watch the program. The results of an independent-measures *t* test indicated a significant differenced between the two groups. The data are repeated here.
a. Does a Mann-Whitney U test also indicate a significant difference? Test at the .05 level of significance.
b. The Kruskal-Wallis test is usually intended for evaluating the differences among three or more separate samples. However, the test can be used with only two samples. Use the Kruskal-Wallis test with $\alpha = .05$ to evaluate the difference between the two groups of students. Is the result consistent with the outcome of the Mann-Whitney test?

Average High School Grade			
Watched Sesame Street		Did Not Watch Sesame Street	
86	99	90	79
87	97	89	83
91	94	82	86
97	89	83	81
98	92	85	92
$n = 10$		$n = 10$	
$M = 93$		$M = 85$	
$SS = 200$		$SS = 160$	

10. Problem 22 from Chapter 11 described a research study comparing the accuracy of Olympic marksmen for shots fired during a heartbeat with the accuracy for shots fired between heartbeats (Pelton, 1983). The repeated-measures *t* test showed a significant difference between the two conditions. Transform the scores to ordinal data (ranks) and use a Wilcoxon test with $\alpha = .05$ to determine whether the ordinal test produces the same conclusion. The data are as follows:

Participant	During Heartbeats	Between Heartbeats
A	93	98
B	90	94
C	95	96
D	92	91
E	95	97
F	91	97

11. Problem 24 from Chapter 11 described a study evaluating the effect of a drug (MAO inhibitor) for treating nightmares for veterans with post-traumatic stress disorder (PTSD). The repeated-measures *t* test showed significant improvement after patients took the

drug for 1 month. Transform the scores to ordinal data (ranks) and use a Wilcoxon test with $\alpha = .05$ to determine whether the ordinal test produces the same conclusion. The data are as follows:

Number of Nightmares	
One Month Before Treatment	One Month During Treatment
6	1
10	2
3	0
5	5
7	2

12. In a repeated-measures research study comparing two treatments, if all of the subjects show a decrease in scores for treatment 2 compared with treatment 1, then what is the minimum number of subjects needed to demonstrate a significant difference between treatments at the .05 level of significance?

13. Hyperactive children often are treated with a stimulant such as Ritalin to improve their attention spans. In one test of this drug treatment, hyperactive children were given a boring task to work on. A psychologist recorded the amount of time (in seconds) each child spent on the task before becoming distracted. Each child's performance was measured before he or she received the drug and again after the drug was administered. Because the scores on this task are extremely variable, the psychologist decided to convert the data to ranks. Use a Wilcoxon test with $\alpha = .05$ to determine whether the following data provide evidence that the drug has a significant effect:

Child	Treatment 1 (Without the Drug)	Treatment 2 (With the Drug)
1	28	135
2	15	309
3	183	150
4	48	224
5	30	25
6	233	345
7	21	43
8	110	188
9	12	15

14. The school psychologist at an elementary school is conducting a counseling program for disruptive

students. To evaluate the program, the students' teacher is asked to rate the change in behavior for each student using a scale from $+10$ (maximum positive change) to -10 (maximum negative change). A rating of 0 indicates no change in behavior. Because the teacher's ratings are subjective evaluations, the psychologist feels more comfortable treating the scores as ordinal values instead of interval or ratio scores. Convert the following data to ranks, and use a Wilcoxon T to evaluate the effectiveness of the counseling:

Student	Change in Behavior As Rated by Teacher
A	+1
B	+6
C	0
D	+8
E	-3
F	+10
G	+4
H	+3

15. Problem 21 in Chapter 11 reported the results of a study showing that time out can be an effective reinforcement (punishment) for adults. Participants had to classify colored shapes into two categories. Correct responses were rewarded with a nickel and incorrect responses were punished with a time out period during which the participant did not have a chance to earn more nickels. A repeated-measures t test showed that a 10-second time out produced significantly better performance than a 5-second time out. The performance scores for a sample of $n = 9$ participants are shown in the following table. Does a Wilcoxon T test also indicate a significant difference between the two time-out conditions? Test at the .05 level of significance.

5-Second Time Out	10-Second Time Out
71	86
68	80
91	88
65	74
73	82
81	89
85	85
86	88
65	76

16. A new cold remedy contains a chemical that causes drowsiness and disorientation. Part of the testing for this drug involved measuring maze-learning performance for rats. Using a repeated-measures design, individual rats were tested on one maze with the drug and on a similar maze without the drug. The dependent variable is the number of errors before the rat solves the maze. In the drug condition, two rats failed to solve the maze after 200 errors and were simply marked as "failed."

Subject	No Drug	Drug
1	28	125
2	43	90
3	37	(Failed)
4	16	108
5	47	40
6	51	75
7	23	91
8	31	23
9	26	115
10	53	55
11	26	(Failed)
12	32	87

  a. Do the data indicate that the drug has a significant effect on maze-learning performance? Use a Wilcoxon test with $\alpha = .05$.

  b. The Friedman test is usually intended for evaluating the differences among three or more separate treatments in a repeated-measures study. However, the test can be used with only two treatments. Use the Friedman test with $\alpha = .05$ to evaluate the difference between the drug and no-drug conditions. Is the result consistent with the outcome of the Wilcoxon test?

17. In Chapter 11, we noted that one advantage of repeated-measures studies compared to independent-measures studies is that the repeated-measures studies reduce the variance by removing individual differences. With smaller variance, the t test is more likely to produce a significant effect. The following data demonstrate this phenomenon. If the data are treated as two separate samples, the independent-measures t test shows no significant difference. However, if the data are treated as the same sample in two treatments, the repeated-measures t test shows a significant difference between treatments. The data are as follows:

Treatment	
I	II
7	10
9	12
10	17
6	9
12	13
4	5
14	16

  a. Treat the scores as two separate samples, transform the scores to ordinal data (ranks), then use a Mann-Whitney test with $\alpha = .05$ to evaluate the difference between treatments.

  b. Treat the data as a repeated-measures study with the same sample in both treatments, then transform the scores to ordinal data (ranks) and use a Wilcoxon test with $\alpha = .05$ to evaluate the difference between treatments.

  c. Do the ordinal tests show the same pattern as the t tests? That is, is there a significant difference for the repeated-measures design but no significant difference for the independent-measures data?

18. Problem 26 from Chapter 13 described a study comparing brain sizes for birds that migrate with brain sizes for birds that do not migrate (Sol, Lefebvre, & Rodriguez-Teijeriro, 2005). The results from the ANOVA indicate that birds that do not migrate have significantly larger brains than birds that do migrate. The researchers suggest that birds with small brains are not smart enough to find food during the winter and must move to a warmer climate where food is easily available. Transform the scores to ordinal data (ranks) and use a Kruskal-Wallis test with $\alpha = .05$ to determine whether the ordinal test also concludes that there are significant difference between the different types of birds. The data are as follows:

Non-migrating	Short-Distance Migrants	Long-Distance Migrants
18	6	4
13	11	9
19	7	5
12	9	6
16	8	5
12	13	7

19. Problem 27 from Chapter 13 described a study examining language skill development for single children compared to twins and triplets (Davis, 1937). The results from the ANOVA indicate significant differences among the three groups, with single children showing the highest skills and triplets the lowest. Transform the scores to ordinal data (ranks) and use a Kruskal-Wallis test with $\alpha = .05$ to determine whether the ordinal test also concludes that there are significant differences between the three groups. The data are as follows:

Single Child	Twin	Triplet
8	6	5
7	4	5
10	6	8
6	7	3
9	4	5
8	9	4

20. A researcher who is interested in the relationship between health and personality obtains a sample of $n = 25$ men between the ages of 30 and 35. The men are all given a physical exam by a doctor who then rank-orders the men in terms of overall health. The men are also given personality tests that classify them as either alpha, beta, or gamma personality types. Alphas are cautious and steady, betas are carefree and outgoing, and gammas tend toward extremes of behavior. The health rankings for the three groups of participants are as follows:

Alphas: 2nd, 6th, 7th, 10th, 13th, 15th, 19th, 21st, 23rd

Betas: 1st, 3rd, 5th, 8th, 11th, 12th, 16th, 17th

Gammas: 4th, 9th, 14th, 18th, 20th, 22nd, 24th, 25th

Do these data indicate significant differences in health among the three personality types? Test at the .01 level of significance.

21. Problem 23 from Chapter 14 described a study examining changes in heart rate in response to embarrassment (Harris, 2001). Participants were asked to sing the "Star Spangled Banner" in front of a video camera while their heart rates were recorded. The results show a quick increase in heart rate followed by a quick return to normal. The data are shown in the following table. The repeated-measures ANOVA showed significant differences between the three measures. Transform the scores to ordinal data (ranks) and use a Friedman test with $\alpha = .05$ to determine whether the ordinal test also concludes that there are significant differences between the three measures.

Person	Baseline	1 Minute	2 Minutes
A	74	88	75
B	76	90	77
C	78	89	76
D	76	85	76

22. The cook in the college dining hall prepared three different receipes of macaroni and cheese. A sample of $n = 7$ students was invited to taste all three and then rank-order the recipes in order of preference. The data are presented in the following table. Use a Friedman test with $\alpha = .05$ to determine whether there are significant differences between the four recipes.

	Recipe		
Student	I	II	III
A	3	2	1
B	2	1	3
C	1	2	3
D	1	2	3
E	3	1	2
F	3	2	1
G	2	3	1

23. A repeated-measures ANOVA found significant differences between treatments for the following data in problem 7 from Chapter 14, $F(2, 10) = 7.78$. Does Friedman also produce a significant result? Test with $\alpha = .05$.

	Treatments				
Person	I	II	III	Person Totals	
A	0	4	2	P = 6	
B	1	5	6	P = 12	N = 18
C	3	3	3	P = 9	G = 48
D	0	1	5	P = 6	$\Sigma X^2 = 184$
E	0	2	4	P = 6	
F	2	3	4	P = 9	
	M = 1	M = 3	M = 4		
	T = 6	T = 18	T = 24		
	SS = 8	SS = 10	SS = 10		

24. In Chapter 14 we noted that one advantage of repeated-measures studies compared to independent-measures is that the repeated-measures studies reduce the variance by removing individual differences. With smaller variance, the ANOVA is more likely to produce a significant effect. The following data were used in problem 14 from Chapter 14 to demonstrate this advantage. If the data are treated as three separate samples, the independent-measures ANOVA shows no significant difference. However, if the data are treated as the same sample in three treatments, the repeated-measures ANOVA shows significant differences between the treatments.

a. Treat the scores as three separate samples, transform the scores to ordinal data (ranks), then use a Kruskal-Wallis test with $\alpha = .05$ to evaluate the differences between treatments.

b. Treat the data as a repeated-measures study with the same sample in all three treatments, then

transform the scores to ordinal data (ranks) and use a Friedman test with $\alpha = .05$ to evaluate the differences between treatments.

c. Do the ordinal tests show the same pattern as the *t* tests? That is, is there a significant difference for the repeated-measures design but no significant difference for the independent-measures data?

Treatments		
A	B	C
0	1	2
2	5	5
1	2	6
5	4	9
2	8	8

Basic Mathematics Review

Preview

A.1   Symbols and Notation

A.2   Proportions: Fractions, Decimals, and Percentages

A.3   Negative Numbers

A.4   Basic Algebra: Solving Equations

A.5   Exponents and Square Roots

Preview

This appendix reviews some of the basic math skills that are necessary for the statistical calculations presented in this book. Many students already will know some or all of this material. Others will need to do extensive work and review. To help you assess your own skills, we include *a* skills assessment exam here. You should allow approximately 30 minutes to complete the test. When you finish, grade your test using the answer key on page 723.

Notice that the test is divided into five sections. If you miss more than three questions in any section of the test, you probably need help in that area. Turn to the section of this appendix that corresponds to your problem area. In each section, you will find a general review, examples, and additional practice problems. After reviewing the appropriate section and doing the practice problems, turn to the end of the appendix. You will find another version of the skills assessment exam. If you still miss more than three questions in any section of the exam, continue studying. Get assistance from an instructor or a tutor if necessary. At the end of this appendix is a list of recommended books for individuals who need a more extensive review than can be provided here. We stress that mastering this material now will make the rest of the course much easier.

SKILLS ASSESSMENT PREVIEW EXAM

SECTION 1

(corresponding to Section A.1 of this appendix)

1. $3 + 2 \times 7 = ?$
2. $(3 + 2) \times 7 = ?$
3. $3 + 2^2 - 1 = ?$
4. $(3 + 2)^2 - 1 = ?$
5. $12/4 + 2 = ?$
6. $12/(4 + 2) = ?$
7. $12/(4 + 2)^2 = ?$
8. $2 \times (8 - 2^2) = ?$
9. $2 \times (8 - 2)^2 = ?$
10. $3 \times 2 + 8 - 1 \times 6 = ?$
11. $3 \times (2 + 8) - 1 \times 6 = ?$
12. $3 \times 2 + (8 - 1) \times 6 = ?$

SECTION 2

(corresponding to Section A.2 of this appendix)

1. The fraction $\frac{3}{4}$ corresponds to a percentage of
   _____.
2. Express 30% as a fraction.
3. Convert $\frac{12}{40}$ to a decimal.
4. $\frac{2}{13} + \frac{8}{13} = ?$
5. $1.375 + 0.25 = ?$
6. $\frac{2}{5} \times \frac{1}{4} = ?$
7. $\frac{1}{8} + \frac{2}{3} = ?$
8. $3.5 \times 0.4 = ?$
9. $\frac{1}{5} \div \frac{3}{4} = ?$
10. $3.75/0.5 = ?$
11. In a group of 80 students, 20% are psychology majors. How many psychology majors are in this group?
12. A company reports that two-fifths of its employees are women. If there are 90 employees, how many are women?

SECTION 3

(corresponding to Section A.3 of this appendix)

1. $3 + (-2) + (-1) + 4 = ?$
2. $6 - (-2) = ?$
3. $-2 - (-4) = ?$
4. $6 + (-1) - 3 - (-2) - (-5) = ?$
5. $4 \times (-3) = ?$

6. $-2 \times (-6) = ?$
7. $-3 \times 5 = ?$
8. $-2 \times (-4) \times (-3) = ?$
9. $12 \div (-3) = ?$
10. $-18 \div (-6) = ?$
11. $-16 \div 8 = ?$
12. $-100 \div (-4) = ?$

SECTION 4

(corresponding to Section A.4 of this appendix)
For each equation, find the value of X.

1. $X + 6 = 13$
2. $X - 14 = 15$
3. $5 = X - 4$
4. $3X = 12$
5. $72 = 3X$
6. $X/5 = 3$
7. $10 = X/8$
8. $3X + 5 = -4$
9. $24 = 2X + 2$
10. $(X + 3)/2 = 14$
11. $(X - 5)/3 = 2$
12. $17 = 4X - 11$

SECTION 5

(corresponding to Section A.5 of this appendix)

1. $4^3 = ?$
2. $\sqrt{25 - 9} = ?$
3. If $X = 2$ and $Y = 3$, then $XY^3 = ?$
4. If $X = 2$ and $Y = 3$, then $(X + Y)^2 = ?$
5. If $a = 3$ and $b = 2$, then $a^2 + b^2 = ?$
6. $(-3)^3 = ?$
7. $(-4)^4 = ?$
8. $\sqrt{4} \times 4 = ?$
9. $36/\sqrt{9} = ?$
10. $(9 + 2)^2 = ?$
11. $5^2 + 2^3 = ?$
12. If $a = 3$ and $b = -1$, then $a^2 b^3 = ?$

The answers to the skills assessment exam are at the end of the appendix (pages 723–724).

A.1   SYMBOLS AND NOTATION

Table A1 presents the basic mathematical symbols that you should know, along with examples of their use. Statistical symbols and notation are introduced and explained throughout this book as they are needed. Notation for exponents and square roots is covered separately at the end of this appendix.

Parentheses are a useful notation because they specify and control the order of computations. Everything inside the parentheses is calculated first. For example,

$$(5 + 3) \times 2 = 8 \times 2 = 16$$

Changing the placement of the parentheses also changes the order of calculations. For example,

$$5 + (3 \times 2) = 5 + 6 = 11$$

ORDER OF OPERATIONS   Often a formula or a mathematical expression will involve several different arithmetic operations, such as adding, multiplying, squaring, and so on. When you encounter these situations, you must perform the different operations in the correct sequence. Following is a list of mathematical operations, showing the order in which they are to be performed.

1. Any calculation contained within parentheses is done first.

2. Squaring (or raising to other exponents) is done second.

3. Multiplying and/or dividing is done third. A series of multiplication and/or division operations should be done in order from left to right.

4. Adding and/or subtracting is done fourth.

The following examples demonstrate how this sequence of operations is applied in different situations.

To evaluate the expression

$$(3 + 1)^2 - 4 \times 7/2$$

first, perform the calculation within parentheses:

$$(4)^2 - 4 \times 7/2$$

TABLE A1

Symbol	Meaning	Example
$+$	Addition	$5 + 7 = 12$
$-$	Subtraction	$8 - 3 = 5$
$\times$, ( )	Multiplication	$3 \times 9 = 27$, $3(9) = 27$
$\div$, /	Division	$15 \div 3 = 5$, $15/3 = 5$, $\frac{15}{3} = 5$
$>$	Greater than	$20 > 10$
$<$	Less than	$7 < 11$
$\neq$	Not equal to	$5 \neq 6$

Next, square the value as indicated:

$$16 - 4 \times 7/2$$

Then perform the multiplication and division:

$$16 - 14$$

Finally, do the subtraction:

$$16 - 14 = 2$$

A sequence of operations involving multiplication and division should be performed in order from left to right. For example, to compute $12/2 \times 3$, you divide 12 by 2 and then multiply the result by 3:

$$12/2 \times 3 = 6 \times 3 = 18$$

Notice that violating the left-to-right sequence can change the result. For this example, if you multiply before dividing, you will obtain

$$12/2 \times 3 = 12/6 = 2 \qquad \text{(This is wrong.)}$$

A sequence of operations involving only addition and subtraction can be performed in any order. For example, to compute $3 + 8 - 5$, you can add 3 and 8 and then subtract 5:

$$(3 + 8) - 5 = 11 - 5 = 6$$

or you can subtract 5 from 8 and then add the result to 3:

$$3 + (8 - 5) = 3 + 3 = 6$$

A mathematical expression or formula is simply a concise way to write a set of instructions. When you evaluate an expression by performing the calculation, you simply follow the instructions. For example, assume you are given these instructions:

1. First, add 3 and 8.
2. Next, square the result.
3. Next, multiply the resulting value by 6.
4. Finally, subtract 50 from the value you have obtained.

You can write these instructions as a mathematical expression.

1. The first step involves addition. Because addition is normally done last, use parentheses to give this operation priority in the sequence of calculations:

$$(3 + 8)$$

2. The instruction to square a value is noted by using the exponent 2 beside the value to be squared:

$$(3 + 8)^2$$

3. Because squaring has priority over multiplication, you can simply introduce the multiplication into the expression:

$$6 \times (3 + 8)^2$$

4. Addition and subtraction are done last, so simply write in the requested subtraction:

$$6 \times (3 + 8)^2 - 50$$

To calculate the value of the expression, you work through the sequence of operations in the proper order:

$$6 \times (3 + 8)^2 - 50 = 6 \times (11)^2 - 50$$
$$= 6 \times (121) - 50$$
$$= 726 - 50$$
$$= 676$$

As a final note, you should realize that the operation of squaring (or raising to any exponent) applies only to the value that immediately precedes the exponent. For example,

$$2 \times 3^2 = 2 \times 9 = 18 \qquad \text{(Only the 3 is squared.)}$$

If the instructions require multiplying values and then squaring the product, you must use parentheses to give the multiplication priority over squaring. For example, to multiply 2 times 3 and then square the product, you would write

$$(2 \times 3)^2 = (6)^2 = 36$$

LEARNING CHECK

1. Evaluate each of the following expressions:

   a. $4 \times 8/2^2$
   b. $4 \times (8/2)^2$
   c. $100 - 3 \times 12/(6 - 4)^2$
   d. $(4 + 6) \times (3 - 1)^2$
   e. $(8 - 2)/(9 - 8)^2$
   f. $6 + (4 - 1)^2 - 3 \times 4^2$
   g. $4 \times (8 - 3) + 8 - 3$

ANSWERS   **1. a.** 8   **b.** 64   **c.** 91   **d.** 40   **e.** 6   **f.** −33   **g.** 25

A.2   PROPORTIONS: FRACTIONS, DECIMALS, AND PERCENTAGES

A proportion is a part of a whole and can be expressed as a fraction, a decimal, or a percentage. For example, in a class of 40 students, only 3 failed the final exam.

The proportion of the class that failed can be expressed as a fraction

$$\text{fraction} = \frac{3}{40}$$

or as a decimal value

$$\text{decimal} = 0.075$$

or as a percentage

$$\text{percentage} = 7.5\%$$

In a fraction, such as $\frac{3}{4}$, the bottom value (the denominator) indicates the number of equal pieces into which the whole is split. Here the "pie" is split into 4 equal pieces:

If the denominator has a larger value—say, 8—then each piece of the whole pie is smaller:

A larger denominator indicates a smaller fraction of the whole.

The value on top of the fraction (the numerator) indicates how many pieces of the whole are being considered. Thus, the fraction $\frac{3}{4}$ indicates that the whole is split evenly into 4 pieces and that 3 of them are being used:

A fraction is simply a concise way of stating a proportion: "Three out of four" is equivalent to $\frac{3}{4}$. To convert the fraction to a decimal, you divide the numerator by the denominator:

$$\frac{3}{4} = 3 \div 4 = 0.75$$

To convert the decimal to a percentage, simply multiply by 100, and place a percent sign (%) after the answer:

$$0.75 \times 100 = 75\%$$

The U.S. money system is a convenient way of illustrating the relationship between fractions and decimals. "One quarter," for example, is one-fourth $\left(\frac{1}{4}\right)$ of a dollar, and its decimal equivalent is 0.25. Other familiar equivalencies are as follows:

	Dime	Quarter	50 Cents	75 Cents
Fraction	$\frac{1}{10}$	$\frac{1}{4}$	$\frac{1}{2}$	$\frac{3}{4}$
Decimal	0.10	0.25	0.50	0.75
Percentage	10%	25%	50%	75%

FRACTIONS

1. Finding Equivalent Fractions. The same proportional value can be expressed by many equivalent fractions. For example,

$$\frac{1}{2} = \frac{2}{4} = \frac{10}{20} = \frac{50}{100}$$

To create equivalent fractions, you can multiply the numerator and denominator by the same value. As long as both the numerator and the denominator of the fraction are multiplied by the same value, the new fraction will be equivalent to the original. For example,

$$\frac{3}{10} = \frac{9}{30}$$

because both the numerator and the denominator of the original fraction have been multiplied by 3. Dividing the numerator and denominator of a fraction by the same value will also result in an equivalent fraction. By using division, you can reduce a fraction to a simpler form. For example,

$$\frac{40}{100} = \frac{2}{5}$$

because both the numerator and the denominator of the original fraction have been divided by 20.

You can use these rules to find specific equivalent fractions. For example, find the fraction that has a denominator of 100 and is equivalent to $\frac{3}{4}$. That is,

$$\frac{3}{4} = \frac{?}{100}$$

Notice that the denominator of the original fraction must be multiplied by 25 to produce the denominator of the desired fraction. For the two fractions to be equal, both the numerator and the denominator must be multiplied by the same number. Therefore, we also multiply the top of the original fraction by 25 and obtain

$$\frac{3 \times 25}{4 \times 25} = \frac{75}{100}$$

2. Multiplying Fractions. To multiply two fractions, you first multiply the numerators and then multiply the denominators. For example,

$$\frac{3}{4} \times \frac{5}{7} = \frac{3 \times 5}{4 \times 7} = \frac{15}{28}$$

3. Dividing Fractions. To divide one fraction by another, you invert the second fraction and then multiply. For example,

$$\frac{1}{2} \div \frac{1}{4} = \frac{1}{2} \times \frac{4}{1} = \frac{1 \times 4}{2 \times 1} = \frac{4}{2} = \frac{2}{1} = 2$$

4. Adding and Subtracting Fractions. Fractions must have the same denominator before you can add or subtract them. If the two fractions already have a common denominator, you simply add (or subtract as the case may be) *only* the values in the numerators. For example,

$$\frac{2}{5} + \frac{1}{5} = \frac{3}{5}$$

Suppose you divided a pie into five equal pieces (fifths). If you first ate two-fifths of the pie and then another one-fifth, the total amount eaten would be three-fifths of the pie:

If the two fractions do not have the same denominator, you must first find equivalent fractions with a common denominator before you can add or subtract. The product of the two denominators will always work as a common denominator for equivalent fractions (although it may not be the lowest common denominator). For example,

$$\frac{2}{3} + \frac{1}{10} = ?$$

Because these two fractions have different denominators, it is necessary to convert each into an equivalent fraction and find a common denominator. We will use $3 \times 10 = 30$ as the common denominator. Thus, the equivalent fraction of each is

$$\frac{2}{3} = \frac{20}{30} \quad \text{and} \quad \frac{1}{10} = \frac{3}{30}$$

Now the two fractions can be added:

$$\frac{20}{30} + \frac{3}{30} = \frac{23}{30}$$

5. Comparing the Size of Fractions. When comparing the size of two fractions with the same denominator, the larger fraction will have the larger numerator. For example,

$$\frac{5}{8} > \frac{3}{8}$$

The denominators are the same, so the whole is partitioned into pieces of the same size. Five of these pieces are more than three of them:

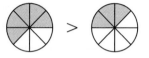

When two fractions have different denominators, you must first convert them to fractions with a common denominator to determine which is larger. Consider the following fractions:

$$\frac{3}{8} \quad \text{and} \quad \frac{7}{16}$$

If the numerator and denominator of $\frac{3}{8}$ are multiplied by 2, the resulting equivalent fraction will have a denominator of 16:

$$\frac{3}{8} = \frac{3 \times 2}{8 \times 2} = \frac{6}{16}$$

Now a comparison can be made between the two fractions:

$$\frac{6}{16} < \frac{7}{16}$$

Therefore,

$$\frac{3}{8} < \frac{7}{16}$$

DECIMALS

1. Converting Decimals to Fractions. Like a fraction, a decimal represents part of the whole. The first decimal place to the right of the decimal point indicates how many tenths are used. For example,

$$0.1 = \frac{1}{10} \qquad 0.7 = \frac{7}{10}$$

The next decimal place represents $\frac{1}{100}$, the next $\frac{1}{1000}$, the next $\frac{1}{10,000}$, and so on. To change a decimal to a fraction, just use the number without the decimal point for the numerator. Use the denominator that the last (on the right) decimal place represents. For example,

$$0.32 = \frac{32}{100} \qquad 0.5333 = \frac{5333}{10,000} \qquad 0.05 = \frac{5}{100} \qquad 0.001 = \frac{1}{1000}$$

2. Adding and Subtracting Decimals. To add and subtract decimals, the only rule is that you must keep the decimal points in a straight vertical line. For example,

$$
\begin{array}{r}
0.27 \\
+1.326 \\
\hline
1.596
\end{array}
\qquad
\begin{array}{r}
3.595 \\
-0.67 \\
\hline
2.925
\end{array}
$$

3. Multiplying Decimals. To multiply two decimal values, you first multiply the two numbers, ignoring the decimal points. Then you position the decimal point in the answer so that the number of digits to the right of the decimal point is equal to the total number of decimal places in the two numbers being multiplied. For example,

$$
\begin{array}{r}
1.73 \\
\times 0.251 \\
\hline
173 \\
865 \\
346 \\
\hline
0.43423
\end{array}
\begin{array}{l}
\text{(two decimal places)} \\
\text{(three decimal places)} \\
\\
\\
\\
\text{(five decimal places)}
\end{array}
\qquad
\begin{array}{r}
0.25 \\
\times 0.005 \\
\hline
125 \\
00 \\
00 \\
\hline
0.00125
\end{array}
\begin{array}{l}
\text{(two decimal places)} \\
\text{(three decimal places)} \\
\\
\\
\\
\text{(five decimal places)}
\end{array}
$$

4. Dividing Decimals. The simplest procedure for dividing decimals is based on the fact that dividing two numbers is identical to expressing them as a fraction:

$$0.25 \div 1.6 \text{ is identical to } \frac{0.25}{1.6}$$

You now can multiply both the numerator and the denominator of the fraction by 10, 100, 1000, or whatever number is necessary to remove the decimal places. Remember that multiplying both the numerator and the denominator of a fraction by the *same* value will create an equivalent fraction. Therefore,

$$\frac{0.25}{1.6} = \frac{0.25 \times 100}{1.6 \times 100} = \frac{25}{160} = \frac{5}{32}$$

The result is a division problem without any decimal places in the two numbers.

PERCENTAGES

1. Converting a Percentage to a Fraction or a Decimal. To convert a percentage to a fraction, remove the percent sign, place the number in the numerator, and use 100 for the denominator. For example,

$$52\% = \frac{52}{100} \qquad 5\% = \frac{5}{100}$$

To convert a percentage to a decimal, remove the percent sign and divide by 100, or simply move the decimal point two places to the left. For example,

$$83\% = 83. = 0.83$$

$$14.5\% = 14.5 = 0.145$$

$$5\% = 5. = 0.05$$

2. Performing Arithmetic Operations with Percentages. There are situations in which it is best to express percent values as decimals in order to perform certain arithmetic operations. For example, what is 45% of 60? This question may be stated as

$$45\% \times 60 = ?$$

The 45% should be converted to decimal form to find the solution to this question. Therefore,

$$0.45 \times 60 = 27$$

LEARNING CHECK

1. Convert $\frac{3}{25}$ to a decimal.

2. Convert $\frac{3}{8}$ to a percentage.

3. Next to each set of fractions, write "True" if they are equivalent and "False" if they are not:

   a. $\frac{3}{8} = \frac{9}{24}$ _____   **b.** $\frac{7}{9} = \frac{17}{19}$ _____

   c. $\frac{2}{7} = \frac{4}{14}$ _____

4. Compute the following:

    a. $\frac{1}{6} \times \frac{7}{10}$ **b.** $\frac{7}{8} - \frac{1}{2}$ **c.** $\frac{9}{10} \div \frac{2}{3}$ **d.** $\frac{7}{22} + \frac{2}{3}$

5. Identify the larger fraction of each pair:

    a. $\frac{7}{10}, \frac{21}{100}$ **b.** $\frac{3}{4}, \frac{7}{12}$ **c.** $\frac{22}{3}, \frac{19}{3}$

6. Convert the following decimals into fractions:

    a. 0.012   **b.** 0.77   **c.** 0.005

7. $2.59 \times 0.015 = ?$

8. $1.8 \div 0.02 = ?$

9. What is 28% of 45?

ANSWERS   **1.** 0.12   **2.** 37.5%   **3. a.** True   **b.** False   **c.** True

    4. a. $\frac{7}{60}$ **b.** $\frac{3}{8}$ **c.** $\frac{27}{20}$ **d.** $\frac{65}{66}$ **5. a.** $\frac{7}{10}$ **b.** $\frac{3}{4}$ **c.** $\frac{22}{3}$

    6. a. $\frac{12}{1000} = \frac{3}{250}$ **b.** $\frac{77}{100}$ **c.** $\frac{5}{1000} = \frac{1}{200}$ **7.** 0.03885   **8.** 90   **9.** 12.6

A.3   NEGATIVE NUMBERS

Negative numbers are used to represent values less than zero. Negative numbers may occur when you are measuring the difference between two scores. For example, a researcher may want to evaluate the effectiveness of a propaganda film by measuring people's attitudes with a test both before and after viewing the film:

	Before	After	Amount of Change
Person A	23	27	+4
Person B	18	15	−3
Person C	21	16	−5

Notice that the negative sign provides information about the direction of the difference: A plus sign indicates an increase in value, and a minus sign indicates a decrease.

    Because negative numbers are frequently encountered, you should be comfortable working with these values. This section reviews basic arithmetic operations using negative numbers. You should also note that any number without a sign (+ or −) is assumed to be positive.

    1. Adding Negative Numbers. When adding numbers that include negative values, simply interpret the negative sign as subtraction. For example,

$$3 + (-2) + 5 = 3 - 2 + 5 = 6$$

When adding a long string of numbers, it often is easier to add all the positive values to obtain the positive sum and then to add all of the negative values to obtain the negative sum. Finally, you subtract the negative sum from the positive sum. For example,

$$-1 + 3 + (-4) + 3 + (-6) + (-2)$$

positive sum = 6   negative sum = 13

Answer: $6 - 13 = -7$

2. Subtracting Negative Numbers. To subtract a negative number, change it to a positive number, and add. For example,

$$4 - (-3) = 4 + 3 = 7$$

This rule is easier to understand if you think of positive numbers as financial gains and negative numbers as financial losses. In this context, taking away a debt is equivalent to a financial gain. In mathematical terms, taking away a negative number is equivalent to adding a positive number. For example, suppose you are meeting a friend for lunch. You have $7, but you owe your friend $3. Thus, you really have only $4 to spend for lunch. But your friend forgives (takes away) the $3 debt. The result is that you now have $7 to spend. Expressed as an equation,

$$\$4 \text{ minus a } \$3 \text{ debt} = \$7$$

$$4 - (-3) = 4 + 3 = 7$$

3. Multiplying and Dividing Negative Numbers. When the two numbers being multiplied (or divided) have the same sign, the result is a positive number. When the two numbers have different signs, the result is negative. For example,

$$3 \times (-2) = -6$$

$$-4 \times (-2) = +8$$

The first example is easy to explain by thinking of multiplication as repeated addition. In this case,

$$3 \times (-2) = (-2) + (-2) + (-2) = -6$$

You add three negative 2s, which results in a total of negative 6. In the second example, we are multiplying by a negative number. This amounts to repeated subtraction. That is,

$$-4 \times (-2) = -(-2) - (-2) - (-2) - (-2)$$

$$= 2 + 2 + 2 + 2 = 8$$

By using the same rule for both multiplication and division, we ensure that these two operations are compatible. For example,

$$-6 \div 3 = -2$$

which is compatible with

$$3 \times (-2) = -6$$

Also,

$$8 \div (-4) = -2$$

which is compatible with

$$-4 \times (-2) = +8$$

LEARNING CHECK

1. Complete the following calculations:

   a. $3 + (-8) + 5 + 7 + (-1) + (-3)$

   b. $5 - (-9) + 2 - (-3) - (-1)$

   c. $3 - 7 - (-21) + (-5) - (-9)$

   d. $4 - (-6) - 3 + 11 - 14$

   e. $9 + 8 - 2 - 1 - (-6)$

   f. $9 \times (-3)$

   g. $-7 \times (-4)$

   h. $-6 \times (-2) \times (-3)$

   i. $-12 \div (-3)$

   j. $18 \div (-6)$

ANSWERS   1. a. 3   b. 20   c. 21   d. 4   e. 20

f. -27 g. 28   h. -36 i. 4   j. -3

A.4   BASIC ALGEBRA: SOLVING EQUATIONS

An equation is a mathematical statement that indicates two quantities are identical. For example,

$$12 = 8 + 4$$

Often an equation will contain an unknown (or variable) quantity that is identified with a letter or symbol, rather than a number. For example,

$$12 = 8 + X$$

In this event, your task is to find the value of X that makes the equation "true," or balanced. For this example, an X value of 4 will make a true equation. Finding the value of X is usually called *solving the equation*.

To solve an equation, there are two points to keep in mind:

1. Your goal is to have the unknown value (X) isolated on one side of the equation. This means that you need to remove all of the other numbers and symbols that appear on the same side of the equation as the X.

2. The equation remains balanced, provided you treat both sides exactly the same. For example, you could add 10 points to *both* sides, and the solution (the X value) for the equation would be unchanged.

FINDING THE SOLUTION FOR AN EQUATION

We will consider four basic types of equations and the operations needed to solve them.

1. **When X Has a Value Added to It.** An example of this type of equation is

$$X + 3 = 7$$

Your goal is to isolate X on one side of the equation. Thus, you must remove the $+3$ on the left-hand side. The solution is obtained by subtracting 3 from *both* sides of the equation:

$$X + 3 - 3 = 7 - 3$$
$$X = 4$$

The solution is $X = 4$. You should always check your solution by returning to the original equation and replacing X with the value you obtained for the solution. For this example,

$$X + 3 = 7$$
$$4 + 3 = 7$$
$$7 = 7$$

2. When X Has a Value Subtracted From It. An example of this type of equation is

$$X - 8 = 12$$

In this example, you must remove the -8 from the left-hand side. Thus, the solution is obtained by adding 8 to *both* sides of the equation:

$$X - 8 + 8 = 12 + 8$$
$$X = 20$$

Check the solution:

$$X - 8 = 12$$
$$20 - 8 = 12$$
$$12 = 12$$

3. When X Is Multiplied by a Value. An example of this type of equation is

$$4X = 24$$

In this instance, it is necessary to remove the 4 that is multiplied by X. This may be accomplished by dividing both sides of the equation by 4:

$$\frac{4X}{4} = \frac{24}{4}$$
$$X = 6$$

Check the solution:

$$4X = 24$$
$$4(6) = 24$$
$$24 = 24$$

4. When X Is Divided by a Value. An example of this type of equation is

$$\frac{X}{3} = 9$$

Now the X is divided by 3, so the solution is obtained by multiplying by 3. Multiplying both sides yields

$$3\left(\frac{X}{3}\right) = 9(3)$$

$$X = 27$$

For the check,

$$\frac{X}{3} = 9$$

$$\frac{27}{3} = 9$$

$$9 = 9$$

SOLUTIONS FOR MORE COMPLEX EQUATIONS

More complex equations can be solved by using a combination of the preceding simple operations. Remember that at each stage you are trying to isolate X on one side of the equation. For example,

$$3X + 7 = 22$$

$$3X + 7 - 7 = 22 - 7 \qquad \text{(Remove } + 7 \text{ by subtracting 7 from both sides.)}$$

$$3X = 15$$

$$\frac{3X}{3} = \frac{15}{3} \qquad \text{(Remove 3 by dividing both sides by 3.)}$$

$$X = 5$$

To check this solution, return to the original equation, and substitute 5 in place of X:

$$3X + 7 = 22$$

$$3(5) + 7 = 22$$

$$15 + 7 = 22$$

$$22 = 22$$

Following is another type of complex equation frequently encountered in statistics:

$$\frac{X + 3}{4} = 2$$

First, remove the 4 by multiplying both sides by 4:

$$4\left(\frac{X + 3}{4}\right) = 2(4)$$

$$X + 3 = 8$$

Now remove the $+ 3$ by subtracting 3 from both sides:

$$X + 3 - 3 = 8 - 3$$

$$X = 5$$

To check this solution, return to the original equation, and substitute 5 in place of X:

$$\frac{X + 3}{4} = 2$$

$$\frac{5 + 3}{4} = 2$$

$$\frac{8}{4} = 2$$

$$2 = 2$$

LEARNING CHECK

1. Solve for X, and check the solutions:

   a. $3X = 18$ **b.** $X + 7 = 9$ **c.** $X - 4 = 18$ **d.** $5X - 8 = 12$

   e. $\dfrac{X}{9} = 5$ **f.** $\dfrac{X + 1}{6} = 4$ **g.** $X + 2 = -5$ **h.** $\dfrac{X}{5} = -5$

   i. $\dfrac{2X}{3} = 12$ **j.** $\dfrac{X}{3} + 1 = 3$

ANSWERS

1. **a.** $X = 6$ **b.** $X = 2$ **c.** $X = 22$ **d.** $X = 4$ **e.** $X = 45$

   f. $X = 23$ **g.** $X = -7$ **h.** $X = -25$ **i.** $X = 18$ **j.** $X = 6$

A.5   EXPONENTS AND SQUARE ROOTS

EXPONENTIAL NOTATION

A simplified notation is used whenever a number is being multiplied by itself. The notation consists of placing a value, called an *exponent,* on the right-hand side of and raised above another number, called a *base*. For example,

$$7^3 \leftarrow \text{exponent}$$
$$\uparrow$$
$$\text{base}$$

The exponent indicates how many times the base is used as a factor in multiplication. Following are some examples:

$$7^3 = 7(7)(7) \qquad \text{(Read "7 cubed" or "7 raised to the third power")}$$

$$5^2 = 5(5) \qquad \text{(Read "5 squared")}$$

$$2^5 = 2(2)(2)(2)(2) \qquad \text{(Read "2 raised to the fifth power")}$$

There are a few basic rules about exponents that you will need to know for this course. They are outlined here.

1. Numbers Raised to One or Zero. Any number raised to the first power equals itself. For example,

$$6^1 = 6$$

Any number (except zero) raised to the zero power equals 1. For example,

$$9^0 = 1$$

2. Exponents for Multiple Terms. The exponent applies only to the base that is just in front of it. For example,

$$XY^2 = XYY$$

$$a^2b^3 = aabbb$$

3. Negative Bases Raised to an Exponent. If a negative number is raised to a power, then the result will be positive for exponents that are even and negative for exponents that are odd. For example,

$$(-4)^3 = -4(-4)(-4)$$

$$= 16(-4)$$

$$= -64$$

and

$$(-3)^4 = -3(-3)(-3)(-3)$$

$$= 9(-3)(-3)$$

$$= 9(9)$$

$$= 81$$

Note: The parentheses are used to ensure that the exponent applies to the entire negative number, including the sign. Without the parentheses there is some ambiguity as to how the exponent should be applied. For example, the expression -3^2 could have two interpretations:

$$-3^2 = (-3)(-3) = 9 \quad \text{or} \quad -3^2 = -(3)(3) = -9$$

4. Exponents and Parentheses. If an exponent is present outside of parentheses, then the computations within the parentheses are done first, and the exponential computation is done last:

$$(3 + 5)^2 = 8^2 = 64$$

Notice that the meaning of the expression is changed when each term in the parentheses is raised to the exponent individually:

$$3^2 + 5^2 = 9 + 25 = 34$$

Therefore,

$$X^2 + Y^2 \neq (X + Y)^2$$

5. Fractions Raised to a Power. If the numerator and denominator of a fraction are each raised to the same exponent, then the entire fraction can be raised to that exponent. That is,

$$\frac{a^2}{b^2} = \left(\frac{a}{b}\right)^2$$

For example,

$$\frac{3^2}{4^2} = \left(\frac{3}{4}\right)^2$$

$$\frac{9}{16} = \frac{3}{4}\left(\frac{3}{4}\right)$$

$$\frac{9}{16} = \frac{9}{16}$$

SQUARE ROOTS The square root of a value equals a number that when multiplied by itself yields the original value. For example, the square root of 16 equals 4 because 4 times 4 equals 16. The symbol for the square root is called a *radical,* $\sqrt{}$. The square root is taken for the number under the radical. For example,

$$\sqrt{16} = 4$$

Finding the square root is the inverse of raising a number to the second power (squaring). Thus,

$$\sqrt{a^2} = a$$

For example,

$$\sqrt{3^2} = \sqrt{9} = 3$$

Also,

$$(\sqrt{b})^2 = b$$

For example,

$$(\sqrt{64})^2 = 8^2 = 64$$

Computations under the same radical are performed *before* the square root is taken. For example,

$$\sqrt{9 + 16} = \sqrt{25} = 5$$

Note that with addition (or subtraction) separate radicals yield a different result:

$$\sqrt{9} + \sqrt{16} = 3 + 4 = 7$$

Therefore,

$$\sqrt{X} + \sqrt{Y} \neq \sqrt{X + Y}$$
$$\sqrt{X} - \sqrt{Y} \neq \sqrt{X - Y}$$

If the numerator and denominator of a fraction each have a radical, then the entire fraction can be placed under a single radical:

$$\frac{\sqrt{16}}{\sqrt{4}} = \sqrt{\frac{16}{4}}$$

$$\frac{4}{2} = \sqrt{4}$$

$$2 = 2$$

Therefore,

$$\frac{\sqrt{X}}{\sqrt{Y}} = \sqrt{\frac{X}{Y}}$$

Also, if the square root of one number is multiplied by the square root of another number, then the same result would be obtained by taking the square root of the product of both numbers. For example,

$$\sqrt{9} \times \sqrt{16} = \sqrt{9 \times 16}$$

$$3 \times 4 = \sqrt{144}$$

$$12 = 12$$

Therefore,

$$\sqrt{a} \times \sqrt{b} = \sqrt{ab}$$

LEARNING CHECK

1. Perform the following computations:
   a. $(-6)^3$
   b. $(3 + 7)^2$
   c. $a^3 b^2$ when $a = 2$ and $b = -5$
   d. $a^4 b^3$ when $a = 2$ and $b = 3$
   e. $(XY)^2$ when $X = 3$ and $Y = 5$
   f. $X^2 + Y^2$ when $X = 3$ and $Y = 5$
   g. $(X + Y)^2$ when $X = 3$ and $Y = 5$
   h. $\sqrt{5 + 4}$
   i. $(\sqrt{9})^2$
   j. $\dfrac{\sqrt{16}}{\sqrt{4}}$

ANSWERS   1. a. -216 b. 100   c. 200   d. 432   e. 225
              f. 34   g. 64   h. 3   i. 9   j. 2

PROBLEMS FOR APPENDIX A   Basic Mathematics Review

1. $50/(10 - 8) = ?$
2. $(2 + 3)^2 = ?$
3. $20/10 \times 3 = ?$
4. $12 - 4 \times 2 + 6/3 = ?$
5. $24/(12 - 4) + 2 \times (6 + 3) = ?$
6. Convert $\frac{7}{20}$ to a decimal.
7. Express $\frac{9}{25}$ as a percentage.
8. Convert 0.91 to a fraction.
9. Express 0.0031 as a fraction.
10. Next to each set of fractions, write "True" if they are equivalent and "False" if they are not:

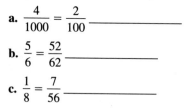

  a. $\dfrac{4}{1000} = \dfrac{2}{100}$ _____

  b. $\dfrac{5}{6} = \dfrac{52}{62}$ _____

  c. $\dfrac{1}{8} = \dfrac{7}{56}$ _____

11. Perform the following calculations:

  a. $\dfrac{4}{5} \times \dfrac{2}{3} = ?$ b. $\dfrac{7}{9} \div \dfrac{2}{3} = ?$

  c. $\dfrac{3}{8} + \dfrac{1}{5} = ?$ d. $\dfrac{5}{18} - \dfrac{1}{6} = ?$

12. $2.51 \times 0.017 = ?$
13. $3.88 \times 0.0002 = ?$
14. $3.17 + 17.0132 = ?$

15. $5.55 + 10.7 + 0.711 + 3.33 + 0.031 = ?$
16. $2.04 \div 0.2 = ?$
17. $0.36 \div 0.4 = ?$
18. $5 + 3 - 6 - 4 + 3 = ?$
19. $9 - (-1) - 17 + 3 - (-4) + 5 = ?$
20. $5 + 3 - (-8) - (-1) + (-3) - 4 + 10 = ?$
21. $8 \times (-3) = ?$
22. $-22 \div (-2) = ?$
23. $-2(-4) \times (-3) = ?$
24. $84 \div (-4) = ?$

Solve the equations in problems 25–32 for X.

25. $X - 7 = -2$
26. $9 = X + 3$
27. $\dfrac{X}{4} = 11$
28. $-3 = \dfrac{X}{3}$
29. $\dfrac{X + 3}{5} = 2$
30. $\dfrac{X + 1}{3} = -8$
31. $6X - 1 = 11$
32. $2X + 3 = -11$
33. $(-5)^2 = ?$
34. $(-5)^3 = ?$
35. If $a = 4$ and $b = 3$, then $a^2 + b^4 = ?$
36. If $a = -1$ and $b = 4$, then $(a + b)^2 = ?$
37. If $a = -1$ and $b = 5$, then $ab^2 = ?$
38. $\dfrac{18}{\sqrt{4}} = ?$
39. $\sqrt{\dfrac{20}{5}} = ?$

SKILLS ASSESSMENT FINAL EXAM

SECTION 1

1. $4 + 8/4 = ?$
2. $(4 + 8)/4 = ?$
3. $4 \times 3^2 = ?$
4. $(4 \times 3)^2 = ?$
5. $10/5 \times 2 = ?$
6. $10/(5 \times 2) = ?$
7. $40 - 10 \times 4/2 = ?$
8. $(5 - 1)^2/2 = ?$
9. $3 \times 6 - 3^2 = ?$
10. $2 \times (6 - 3)^2 = ?$
11. $4 \times 3 - 1 + 8 \times 2 = ?$
12. $4 \times (3 - 1 + 8) \times 2 = ?$

SECTION 2

1. Express $\frac{14}{80}$ as a decimal.
2. Convert $\frac{6}{25}$ to a percentage.
3. Convert 18% to a fraction.
4. $\frac{3}{5} \times \frac{2}{3} = ?$
5. $\frac{5}{24} + \frac{5}{6} = ?$
6. $\frac{7}{12} \div \frac{5}{6} = ?$
7. $\frac{5}{9} - \frac{1}{3} = ?$
8. $6.11 \times 0.22 = ?$
9. $0.18 \div 0.9 = ?$
10. $8.742 + 0.76 = ?$
11. In a statistics class of 72 students, three-eighths of the students received a B on the first test. How many Bs were earned?
12. What is 15% of 64?

SECTION 3

1. $3 - 1 - 3 + 5 - 2 + 6 = ?$

2. $-8 - (-6) = ?$

3. $2 - (-7) - 3 + (-11) - 20 = ?$

4. $-8 - 3 - (-1) - 2 - 1 = ?$

5. $8(-2) = ?$

6. $-7(-7) = ?$

7. $-3(-2)(-5) = ?$

8. $-3(5)(-3) = ?$

9. $-24 \div (-4) = ?$

10. $36 \div (-6) = ?$

11. $-56/7 = ?$

12. $-7/(-1) = ?$

SECTION 4

Solve for X.

1. $X + 5 = 12$

2. $X - 11 = 3$

3. $10 = X + 4$

4. $4X = 20$

5. $\dfrac{X}{2} = 15$

6. $18 = 9X$

7. $\dfrac{X}{5} = 35$

8. $2X + 8 = 4$

9. $\dfrac{X + 1}{3} = 6$

10. $4X + 3 = -13$

11. $\dfrac{X + 3}{3} = -7$

12. $23 = 2X - 5$

SECTION 5

1. $5^3 = ?$

2. $(-4)^3 = ?$

3. $(-2)^5 = ?$

4. $(-2)^6 = ?$

5. If $a = 4$ and $b = 2$, then $ab^2 = ?$

6. If $a = 4$ and $b = 2$, then $(a + b)^3 = ?$

7. If $a = 4$ and $b = 2$, then $a^2 + b^2 = ?$

8. $(11 + 4)^2 = ?$

9. $\sqrt{7^2} = ?$

10. If $a = 36$ and $b = 64$, then $\sqrt{a + b} = ?$

11. $\dfrac{25}{\sqrt{25}} = ?$

12. If $a = -1$ and $b = 2$, then $a^3 b^4 = ?$

ANSWER KEY   Skills Assessment Exams

PREVIEW EXAM

SECTION 1

1. 17

2. 35

3. 6

4. 24

5. 5

6. 2

7. $\dfrac{1}{3}$

8. 8

9. 72

10. 8

11. 24

12. 48

SECTION 2

1. 75%

2. $\dfrac{30}{100}$, or $\dfrac{3}{10}$

3. 0.3

4. $\dfrac{10}{13}$

5. 1.625

6. $\dfrac{2}{20}$, or $\dfrac{1}{10}$

7. $\dfrac{19}{24}$

8. 1.4

9. $\dfrac{4}{15}$

10. 7.5

11. 16

12. 36

SECTION 3

1. 4

2. 8

3. 2

4. 9

5. -12

6. 12

7. -15

8. -24

9. -4

10. 3

11. -2

12. 25

FINAL EXAM

SECTION 1

1. 6

2. 3

3. 36

4. 144

5. 4

6. 1

7. 20

8. 8

9. 9

10. 18

11. 27

12. 80

SECTION 2

1. 0.175

2. 24%

3. $\dfrac{18}{100}$, or $\dfrac{9}{50}$

4. $\dfrac{6}{15}$, or $\dfrac{2}{5}$

5. $\dfrac{25}{24}$

6. $\dfrac{42}{60}$, or $\dfrac{7}{10}$

7. $\dfrac{2}{9}$

8. 1.3442

9. 0.2

10. 9.502

11. 27

12. 9.6

SECTION 3

1. 8

2. -2

3. -25

4. -13

5. -16

6. 49

7. -30

8. 45

9. 6

10. -6

11. -8

12. 7

PREVIEW EXAM

SECTION 4

1. $X = 7$	**2.** $X = 29$	**3.** $X = 9$
4. $X = 4$	**5.** $X = 24$	**6.** $X = 15$
7. $X = 80$	**8.** $X = -3$	**9.** $X = 11$
10. $X = 25$	**11.** $X = 11$	**12.** $X = 7$

SECTION 5

1. 64	**2.** 4	**3.** 54
4. 25	**5.** 13	**6.** -27
7. 256	**8.** 8	**9.** 12
10. 121	**11.** 33	**12.** -9

FINAL EXAM

SECTION 4

1. $X = 7$	**2.** $X = 14$	**3.** $X = 6$
4. $X = 5$	**5.** $X = 30$	**6.** $X = 2$
7. $X = 175$	**8.** $X = -2$	**9.** $X = 17$
10. $X = -4$	**11.** $X = -24$	**12.** $X = 14$

SECTION 5

1. 125	**2.** -64	**3.** -32
4. 64	**5.** 16	**6.** 216
7. 20	**8.** 225	**9.** 7
10. 10	**11.** 5	**12.** -16

SOLUTIONS TO SELECTED PROBLEMS FOR APPENDIX A   Basic Mathematics Review

1. 25	**10. b.** False	**17.** 0.9	**19.** 5
3. 6	**11. a.** $\dfrac{8}{15}$ **b.** $\dfrac{21}{18}$ **c.** $\dfrac{23}{40}$	**21.** -24	**22.** 11
5. 21		**25.** $X = 5$	**28.** $X = -9$
6. 0.35	**12.** 0.04267	**30.** $X = -25$	**31.** $X = 2$
7. 36%	**14.** 20.1832	**34.** -125	**36.** 9
9. $\dfrac{31}{10,000}$		**37.** -25	**39.** 2

SUGGESTED REVIEW BOOKS

There are many basic mathematics books available if you need a more extensive review than this appendix can provide. Several are probably available in your library. The following books are but a few of the many that you may find helpful:

Gustafson, R. D., & Frisk, P. D. (2005). *Beginning Algebra* (7th ed.). Belmont, CA: Brooks/Cole.

Lial, M. L., Salzman, S.A., & Hestwood, D.L. (2006). *Basic College Mathematics* (7th ed). Reading MA: Addison-Wesley.

McKeague, C. P. (2007). *Basic College Mathematics: A Text/Workbook.* (2nd ed.). Belmont, CA: Brooks/Cole.

APPENDIX B Statistical Tables

TABLE B.1   THE UNIT NORMAL TABLE*

*Column A lists z-score values. A vertical line drawn through a normal distribution at a z-score location divides the distribution into two sections.
Column B identifies the proportion in the larger section, called the *body*.
Column C identifies the proportion in the smaller section, called the *tail*.
Column D identifies the proportion between the mean and the z-score.
Note: Because the normal distribution is symmetrical, the proportions for negative z-scores are the same as those for positive z-scores.

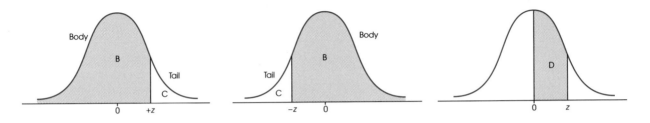

(A)	(B) Proportion in Body	(C) Proportion in Tail	(D) Proportion Between Mean and z	(A)	(B) Proportion in Body	(C) Proportion in Tail	(D) Proportion Between Mean and z
z				z			
0.00	.5000	.5000	.0000	0.25	.5987	.4013	.0987
0.01	.5040	.4960	.0040	0.26	.6026	.3974	.1026
0.02	.5080	.4920	.0080	0.27	.6064	.3936	.1064
0.03	.5120	.4880	.0120	0.28	.6103	.3897	.1103
0.04	.5160	.4840	.0160	0.29	.6141	.3859	.1141
0.05	.5199	.4801	.0199	0.30	.6179	.3821	.1179
0.06	.5239	.4761	.0239	0.31	.6217	.3783	.1217
0.07	.5279	.4721	.0279	0.32	.6255	.3745	.1255
0.08	.5319	.4681	.0319	0.33	.6293	.3707	.1293
0.09	.5359	.4641	.0359	0.34	.6331	.3669	.1331
0.10	.5398	.4602	.0398	0.35	.6368	.3632	.1368
0.11	.5438	.4562	.0438	0.36	.6406	.3594	.1406
0.12	.5478	.4522	.0478	0.37	.6443	.3557	.1443
0.13	.5517	.4483	.0517	0.38	.6480	.3520	.1480
0.14	.5557	.4443	.0557	0.39	.6517	.3483	.1517
0.15	.5596	.4404	.0596	0.40	.6554	.3446	.1554
0.16	.5636	.4364	.0636	0.41	.6591	.3409	.1591
0.17	.5675	.4325	.0675	0.42	.6628	.3372	.1628
0.18	.5714	.4286	.0714	0.43	.6664	.3336	.1664
0.19	.5753	.4247	.0753	0.44	.6700	.3300	.1700
0.20	.5793	.4207	.0793	0.45	.6736	.3264	.1736
0.21	.5832	.4168	.0832	0.46	.6772	.3228	.1772
0.22	.5871	.4129	.0871	0.47	.6808	.3192	.1808
0.23	.5910	.4090	.0910	0.48	.6844	.3156	.1844
0.24	.5948	.4052	.0948	0.49	.6879	.3121	.1879

(A)	(B) Proportion in Body	(C) Proportion in Tail	(D) Proportion Between Mean and z	(A)	(B) Proportion in Body	(C) Proportion in Tail	(D) Proportion Between Mean and z
z				z			
0.50	.6915	.3085	.1915	1.00	.8413	.1587	.3413
0.51	.6950	.3050	.1950	1.01	.8438	.1562	.3438
0.52	.6985	.3015	.1985	1.02	.8461	.1539	.3461
0.53	.7019	.2981	.2019	1.03	.8485	.1515	.3485
0.54	.7054	.2946	.2054	1.04	.8508	.1492	.3508
0.55	.7088	.2912	.2088	1.05	.8531	.1469	.3531
0.56	.7123	.2877	.2123	1.06	.8554	.1446	.3554
0.57	.7157	.2843	.2157	1.07	.8577	.1423	.3577
0.58	.7190	.2810	.2190	1.08	.8599	.1401	.3599
0.59	.7224	.2776	.2224	1.09	.8621	.1379	.3621
0.60	.7257	.2743	.2257	1.10	.8643	.1357	.3643
0.61	.7291	.2709	.2291	1.11	.8665	.1335	.3665
0.62	.7324	.2676	.2324	1.12	.8686	.1314	.3686
0.63	.7357	.2643	.2357	1.13	.8708	.1292	.3708
0.64	.7389	.2611	.2389	1.14	.8729	.1271	.3729
0.65	.7422	.2578	.2422	1.15	.8749	.1251	.3749
0.66	.7454	.2546	.2454	1.16	.8770	.1230	.3770
0.67	.7486	.2514	.2486	1.17	.8790	.1210	.3790
0.68	.7517	.2483	.2517	1.18	.8810	.1190	.3810
0.69	.7549	.2451	.2549	1.19	.8830	.1170	.3830
0.70	.7580	.2420	.2580	1.20	.8849	.1151	.3849
0.71	.7611	.2389	.2611	1.21	.8869	.1131	.3869
0.72	.7642	.2358	.2642	1.22	.8888	.1112	.3888
0.73	.7673	.2327	.2673	1.23	.8907	.1093	.3907
0.74	.7704	.2296	.2704	1.24	.8925	.1075	.3925
0.75	.7734	.2266	.2734	1.25	.8944	.1056	.3944
0.76	.7764	.2236	.2764	1.26	.8962	.1038	.3962
0.77	.7794	.2206	.2794	1.27	.8980	.1020	.3980
0.78	.7823	.2177	.2823	1.28	.8997	.1003	.3997
0.79	.7852	.2148	.2852	1.29	.9015	.0985	.4015
0.80	.7881	.2119	.2881	1.30	.9032	.0968	.4032
0.81	.7910	.2090	.2910	1.31	.9049	.0951	.4049
0.82	.7939	.2061	.2939	1.32	.9066	.0934	.4066
0.83	.7967	.2033	.2967	1.33	.9082	.0918	.4082
0.84	.7995	.2005	.2995	1.34	.9099	.0901	.4099
0.85	.8023	.1977	.3023	1.35	.9115	.0885	.4115
0.86	.8051	.1949	.3051	1.36	.9131	.0869	.4131
0.87	.8078	.1922	.3078	1.37	.9147	.0853	.4147
0.88	.8106	.1894	.3106	1.38	.9162	.0838	.4162
0.89	.8133	.1867	.3133	1.39	.9177	.0823	.4177
0.90	.8159	.1841	.3159	1.40	.9192	.0808	.4192
0.91	.8186	.1814	.3186	1.41	.9207	.0793	.4207
0.92	.8212	.1788	.3212	1.42	.9222	.0778	.4222
0.93	.8238	.1762	.3238	1.43	.9236	.0764	.4236
0.94	.8264	.1736	.3264	1.44	.9251	.0749	.4251
0.95	.8289	.1711	.3289	1.45	.9265	.0735	.4265
0.96	.8315	.1685	.3315	1.46	.9279	.0721	.4279
0.97	.8340	.1660	.3340	1.47	.9292	.0708	.4292
0.98	.8365	.1635	.3365	1.48	.9306	.0694	.4306
0.99	.8389	.1611	.3389	1.49	.9319	.0681	.4319

(A)	(B) Proportion	(C) Proportion	(D) Proportion	(A)	(B) Proportion	(C) Proportion	(D) Proportion
z	in Body	in Tail	Between Mean and z	z	in Body	in Tail	Between Mean and z
1.50	.9332	.0668	.4332	2.00	.9772	.0228	.4772
1.51	.9345	.0655	.4345	2.01	.9778	.0222	.4778
1.52	.9357	.0643	.4357	2.02	.9783	.0217	.4783
1.53	.9370	.0630	.4370	2.03	.9788	.0212	.4788
1.54	.9382	.0618	.4382	2.04	.9793	.0207	.4793
1.55	.9394	.0606	.4394	2.05	.9798	.0202	.4798
1.56	.9406	.0594	.4406	2.06	.9803	.0197	.4803
1.57	.9418	.0582	.4418	2.07	.9808	.0192	.4808
1.58	.9429	.0571	.4429	2.08	.9812	.0188	.4812
1.59	.9441	.0559	.4441	2.09	.9817	.0183	.4817
1.60	.9452	.0548	.4452	2.10	.9821	.0179	.4821
1.61	.9463	.0537	.4463	2.11	.9826	.0174	.4826
1.62	.9474	.0526	.4474	2.12	.9830	.0170	.4830
1.63	.9484	.0516	.4484	2.13	.9834	.0166	.4834
1.64	.9495	.0505	.4495	2.14	.9838	.0162	.4838
1.65	.9505	.0495	.4505	2.15	.9842	.0158	.4842
1.66	.9515	.0485	.4515	2.16	.9846	.0154	.4846
1.67	.9525	.0475	.4525	2.17	.9850	.0150	.4850
1.68	.9535	.0465	.4535	2.18	.9854	.0146	.4854
1.69	.9545	.0455	.4545	2.19	.9857	.0143	.4857
1.70	.9554	.0446	.4554	2.20	.9861	.0139	.4861
1.71	.9564	.0436	.4564	2.21	.9864	.0136	.4864
1.72	.9573	.0427	.4573	2.22	.9868	.0132	.4868
1.73	.9582	.0418	.4582	2.23	.9871	.0129	.4871
1.74	.9591	.0409	.4591	2.24	.9875	.0125	.4875
1.75	.9599	.0401	.4599	2.25	.9878	.0122	.4878
1.76	.9608	.0392	.4608	2.26	.9881	.0119	.4881
1.77	.9616	.0384	.4616	2.27	.9884	.0116	.4884
1.78	.9625	.0375	.4625	2.28	.9887	.0113	.4887
1.79	.9633	.0367	.4633	2.29	.9890	.0110	.4890
1.80	.9641	.0359	.4641	2.30	.9893	.0107	.4893
1.81	.9649	.0351	.4649	2.31	.9896	.0104	.4896
1.82	.9656	.0344	.4656	2.32	.9898	.0102	.4898
1.83	.9664	.0336	.4664	2.33	.9901	.0099	.4901
1.84	.9671	.0329	.4671	2.34	.9904	.0096	.4904
1.85	.9678	.0322	.4678	2.35	.9906	.0094	.4906
1.86	.9686	.0314	.4686	2.36	.9909	.0091	.4909
1.87	.9693	.0307	.4693	2.37	.9911	.0089	.4911
1.88	.9699	.0301	.4699	2.38	.9913	.0087	.4913
1.89	.9706	.0294	.4706	2.39	.9916	.0084	.4916
1.90	.9713	.0287	.4713	2.40	.9918	.0082	.4918
1.91	.9719	.0281	.4719	2.41	.9920	.0080	.4920
1.92	.9726	.0274	.4726	2.42	.9922	.0078	.4922
1.93	.9732	.0268	.4732	2.43	.9925	.0075	.4925
1.94	.9738	.0262	.4738	2.44	.9927	.0073	.4927
1.95	.9744	.0256	.4744	2.45	.9929	.0071	.4929
1.96	.9750	.0250	.4750	2.46	.9931	.0069	.4931
1.97	.9756	.0244	.4756	2.47	.9932	.0068	.4932
1.98	.9761	.0239	.4761	2.48	.9934	.0066	.4934
1.99	.9767	.0233	.4767	2.49	.9936	.0064	.4936

(A)	(B) Proportion in Body	(C) Proportion in Tail	(D) Proportion Between Mean and z	(A)	(B) Proportion in Body	(C) Proportion in Tail	(D) Proportion Between Mean and z
z				z			
2.50	.9938	.0062	.4938	2.95	.9984	.0016	.4984
2.51	.9940	.0060	.4940	2.96	.9985	.0015	.4985
2.52	.9941	.0059	.4941	2.97	.9985	.0015	.4985
2.53	.9943	.0057	.4943	2.98	.9986	.0014	.4986
2.54	.9945	.0055	.4945	2.99	.9986	.0014	.4986
2.55	.9946	.0054	.4946	3.00	.9987	.0013	.4987
2.56	.9948	.0052	.4948	3.01	.9987	.0013	.4987
2.57	.9949	.0051	.4949	3.02	.9987	.0013	.4987
2.58	.9951	.0049	.4951	3.03	.9988	.0012	.4988
2.59	.9952	.0048	.4952	3.04	.9988	.0012	.4988
2.60	.9953	.0047	.4953	3.05	.9989	.0011	.4989
2.61	.9955	.0045	.4955	3.06	.9989	.0011	.4989
2.62	.9956	.0044	.4956	3.07	.9989	.0011	.4989
2.63	.9957	.0043	.4957	3.08	.9990	.0010	.4990
2.64	.9959	.0041	.4959	3.09	.9990	.0010	.4990
2.65	.9960	.0040	.4960	3.10	.9990	.0010	.4990
2.66	.9961	.0039	.4961	3.11	.9991	.0009	.4991
2.67	.9962	.0038	.4962	3.12	.9991	.0009	.4991
2.68	.9963	.0037	.4963	3.13	.9991	.0009	.4991
2.69	.9964	.0036	.4964	3.14	.9992	.0008	.4992
2.70	.9965	.0035	.4965	3.15	.9992	.0008	.4992
2.71	.9966	.0034	.4966	3.16	.9992	.0008	.4992
2.72	.9967	.0033	.4967	3.17	.9992	.0008	.4992
2.73	.9968	.0032	.4968	3.18	.9993	.0007	.4993
2.74	.9969	.0031	.4969	3.19	.9993	.0007	.4993
2.75	.9970	.0030	.4970	3.20	.9993	.0007	.4993
2.76	.9971	.0029	.4971	3.21	.9993	.0007	.4993
2.77	.9972	.0028	.4972	3.22	.9994	.0006	.4994
2.78	.9973	.0027	.4973	3.23	.9994	.0006	.4994
2.79	.9974	.0026	.4974	3.24	.9994	.0006	.4994
2.80	.9974	.0026	.4974	3.30	.9995	.0005	.4995
2.81	.9975	.0025	.4975	3.40	.9997	.0003	.4997
2.82	.9976	.0024	.4976	3.50	.9998	.0002	.4998
2.83	.9977	.0023	.4977	3.60	.9998	.0002	.4998
2.84	.9977	.0023	.4977	3.70	.9999	.0001	.4999
2.85	.9978	.0022	.4978	3.80	.99993	.00007	.49993
2.86	.9979	.0021	.4979	3.90	.99995	.00005	.49995
2.87	.9979	.0021	.4979	4.00	.99997	.00003	.49997
2.88	.9980	.0020	.4980				
2.89	.9981	.0019	.4981				
2.90	.9981	.0019	.4981				
2.91	.9982	.0018	.4982				
2.92	.9982	.0018	.4982				
2.93	.9983	.0017	.4983				
2.94	.9984	.0016	.4984				

TABLE B.2   THE *t* DISTRIBUTION

Table entries are values of *t* corresponding to proportions in one tail or in two tails combined.

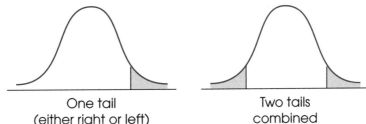

One tail
(either right or left)

Two tails
combined

| | Proportion in One Tail | | | | | |
	0.25	0.10	0.05	0.025	0.01	0.005
			Proportion in Two Tails Combined			
df	0.50	0.20	0.10	0.05	0.02	0.01
1	1.000	3.078	6.314	12.706	31.821	63.657
2	0.816	1.886	2.920	4.303	6.965	9.925
3	0.765	1.638	2.353	3.182	4.541	5.841
4	0.741	1.533	2.132	2.776	3.747	4.604
5	0.727	1.476	2.015	2.571	3.365	4.032
6	0.718	1.440	1.943	2.447	3.143	3.707
7	0.711	1.415	1.895	2.365	2.998	3.499
8	0.706	1.397	1.860	2.306	2.896	3.355
9	0.703	1.383	1.833	2.262	2.821	3.250
10	0.700	1.372	1.812	2.228	2.764	3.169
11	0.697	1.363	1.796	2.201	2.718	3.106
12	0.695	1.356	1.782	2.179	2.681	3.055
13	0.694	1.350	1.771	2.160	2.650	3.012
14	0.692	1.345	1.761	2.145	2.624	2.977
15	0.691	1.341	1.753	2.131	2.602	2.947
16	0.690	1.337	1.746	2.120	2.583	2.921
17	0.689	1.333	1.740	2.110	2.567	2.898
18	0.688	1.330	1.734	2.101	2.552	2.878
19	0.688	1.328	1.729	2.093	2.539	2.861
20	0.687	1.325	1.725	2.086	2.528	2.845
21	0.686	1.323	1.721	2.080	2.518	2.831
22	0.686	1.321	1.717	2.074	2.508	2.819
23	0.685	1.319	1.714	2.069	2.500	2.807
24	0.685	1.318	1.711	2.064	2.492	2.797
25	0.684	1.316	1.708	2.060	2.485	2.787
26	0.684	1.315	1.706	2.056	2.479	2.779
27	0.684	1.314	1.703	2.052	2.473	2.771
28	0.683	1.313	1.701	2.048	2.467	2.763
29	0.683	1.311	1.699	2.045	2.462	2.756
30	0.683	1.310	1.697	2.042	2.457	2.750
40	0.681	1.303	1.684	2.021	2.423	2.704
60	0.679	1.296	1.671	2.000	2.390	2.660
120	0.677	1.289	1.658	1.980	2.358	2.617
∞	0.674	1.282	1.645	1.960	2.326	2.576

TABLE B.3   CRITICAL VALUES FOR THE *F*-MAX STATISTIC*

*The critical values for $\alpha = .05$ are in lightface type, and for $\alpha = .01$, they are in boldface type.

$n - 1$	$k =$ Number of Samples										
	2	3	4	5	6	7	8	9	10	11	12
4	9.60	15.5	20.6	25.2	29.5	33.6	37.5	41.4	44.6	48.0	51.4
	23.2	**37.**	**49.**	**59.**	**69.**	**79.**	**89.**	**97.**	**106.**	**113.**	**120.**
5	7.15	10.8	13.7	16.3	18.7	20.8	22.9	24.7	26.5	28.2	29.9
	14.9	**22.**	**28.**	**33.**	**38.**	**42.**	**46.**	**50.**	**54.**	**57.**	**60.**
6	5.82	8.38	10.4	12.1	13.7	15.0	16.3	17.5	18.6	19.7	20.7
	11.1	**15.5**	**19.1**	**22.**	**25.**	**27.**	**30.**	**32.**	**34.**	**36.**	**37.**
7	4.99	6.94	8.44	9.70	10.8	11.8	12.7	13.5	14.3	15.1	15.8
	8.89	**12.1**	**14.5**	**16.5**	**18.4**	**20.**	**22.**	**23.**	**24.**	**26.**	**27.**
8	4.43	6.00	7.18	8.12	9.03	9.78	10.5	11.1	11.7	12.2	12.7
	7.50	**9.9**	**11.7**	**13.2**	**14.5**	**15.8**	**16.9**	**17.9**	**18.9**	**19.8**	**21.**
9	4.03	5.34	6.31	7.11	7.80	8.41	8.95	9.45	9.91	10.3	10.7
	6.54	**8.5**	**9.9**	**11.1**	**12.1**	**13.1**	**13.9**	**14.7**	**15.3**	**16.0**	**16.6**
10	3.72	4.85	5.67	6.34	6.92	7.42	7.87	8.28	8.66	9.01	9.34
	5.85	**7.4**	**8.6**	**9.6**	**10.4**	**11.1**	**11.8**	**12.4**	**12.9**	**13.4**	**13.9**
12	3.28	4.16	4.79	5.30	5.72	6.09	6.42	6.72	7.00	7.25	7.48
	4.91	**6.1**	**6.9**	**7.6**	**8.2**	**8.7**	**9.1**	**9.5**	**9.9**	**10.2**	**10.6**
15	2.86	3.54	4.01	4.37	4.68	4.95	5.19	5.40	5.59	5.77	5.93
	4.07	**4.9**	**5.5**	**6.0**	**6.4**	**6.7**	**7.1**	**7.3**	**7.5**	**7.8**	**8.0**
20	2.46	2.95	3.29	3.54	3.76	3.94	4.10	4.24	4.37	4.49	4.59
	3.32	**3.8**	**4.3**	**4.6**	**4.9**	**5.1**	**5.3**	**5.5**	**5.6**	**5.8**	**5.9**
30	2.07	2.40	2.61	2.78	2.91	3.02	3.12	3.21	3.29	3.36	3.39
	2.63	**3.0**	**3.3**	**3.5**	**3.6**	**3.7**	**3.8**	**3.9**	**4.0**	**4.1**	**4.2**
60	1.67	1.85	1.96	2.04	2.11	2.17	2.22	2.26	2.30	2.33	2.36
	1.96	**2.2**	**2.3**	**2.4**	**2.4**	**2.5**	**2.5**	**2.6**	**2.6**	**2.7**	**2.7**

TABLE B.4   THE *F* DISTRIBUTION*

*Table entries in lightface type are critical values for the .05 level of significance.
Boldface type values are for the .01 level of significance.

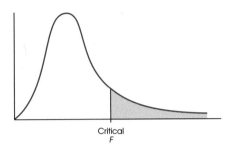

Critical
F

Degrees of Freedom: Denominator	Degrees of Freedom: Numerator														
	1	2	3	4	5	6	7	8	9	10	11	12	14	16	20
1	161	200	216	225	230	234	237	239	241	242	243	244	245	246	248
	4052	**4999**	**5403**	**5625**	**5764**	**5859**	**5928**	**5981**	**6022**	**6056**	**6082**	**6106**	**6142**	**6169**	**6208**
2	18.51	19.00	19.16	19.25	19.30	19.33	19.36	19.37	19.38	19.39	19.40	19.41	19.42	19.43	19.44
	98.49	**99.00**	**99.17**	**99.25**	**99.30**	**99.33**	**99.34**	**99.36**	**99.38**	**99.40**	**99.41**	**99.42**	**99.43**	**99.44**	**99.45**
3	10.13	9.55	9.28	9.12	9.01	8.94	8.88	8.84	8.81	8.78	8.76	8.74	8.71	8.69	8.66
	34.12	**30.92**	**29.46**	**28.71**	**28.24**	**27.91**	**27.67**	**27.49**	**27.34**	**27.23**	**27.13**	**27.05**	**26.92**	**26.83**	**26.69**
4	7.71	6.94	6.59	6.39	6.26	6.16	6.09	6.04	6.00	5.96	5.93	5.91	5.87	5.84	5.80
	21.20	**18.00**	**16.69**	**15.98**	**15.52**	**15.21**	**14.98**	**14.80**	**14.66**	**14.54**	**14.45**	**14.37**	**14.24**	**14.15**	**14.02**
5	6.61	5.79	5.41	5.19	5.05	4.95	4.88	4.82	4.78	4.74	4.70	4.68	4.64	4.60	4.56
	16.26	**13.27**	**12.06**	**11.39**	**10.97**	**10.67**	**10.45**	**10.27**	**10.15**	**10.05**	**9.96**	**9.89**	**9.77**	**9.68**	**9.55**
6	5.99	5.14	4.76	4.53	4.39	4.28	4.21	4.15	4.10	4.06	4.03	4.00	3.96	3.92	3.87
	13.74	**10.92**	**9.78**	**9.15**	**8.75**	**8.47**	**8.26**	**8.10**	**7.98**	**7.87**	**7.79**	**7.72**	**7.60**	**7.52**	**7.39**
7	5.59	4.74	4.35	4.12	3.97	3.87	3.79	3.73	3.68	3.63	3.60	3.57	3.52	3.49	3.44
	12.25	**9.55**	**8.45**	**7.85**	**7.46**	**7.19**	**7.00**	**6.84**	**6.71**	**6.62**	**6.54**	**6.47**	**6.35**	**6.27**	**6.15**
8	5.32	4.46	4.07	3.84	3.69	3.58	3.50	3.44	3.39	3.34	3.31	3.28	3.23	3.20	3.15
	11.26	**8.65**	**7.59**	**7.01**	**6.63**	**6.37**	**6.19**	**6.03**	**5.91**	**5.82**	**5.74**	**5.67**	**5.56**	**5.48**	**5.36**
9	5.12	4.26	3.86	3.63	3.48	3.37	3.29	3.23	3.18	3.13	3.10	3.07	3.02	2.98	2.93
	10.56	**8.02**	**6.99**	**6.42**	**6.06**	**5.80**	**5.62**	**5.47**	**5.35**	**5.26**	**5.18**	**5.11**	**5.00**	**4.92**	**4.80**
10	4.96	4.10	3.71	3.48	3.33	3.22	3.14	3.07	3.02	2.97	2.94	2.91	2.86	2.82	2.77
	10.04	**7.56**	**6.55**	**5.99**	**5.64**	**5.39**	**5.21**	**5.06**	**4.95**	**4.85**	**4.78**	**4.71**	**4.60**	**4.52**	**4.41**
11	4.84	3.98	3.59	3.36	3.20	3.09	3.01	2.95	2.90	2.86	2.82	2.79	2.74	2.70	2.65
	9.65	**7.20**	**6.22**	**5.67**	**5.32**	**5.07**	**4.88**	**4.74**	**4.63**	**4.54**	**4.46**	**4.40**	**4.29**	**4.21**	**4.10**
12	4.75	3.88	3.49	3.26	3.11	3.00	2.92	2.85	2.80	2.76	2.72	2.69	2.64	2.60	2.54
	9.33	**6.93**	**5.95**	**5.41**	**5.06**	**4.82**	**4.65**	**4.50**	**4.39**	**4.30**	**4.22**	**4.16**	**4.05**	**3.98**	**3.86**
13	4.67	3.80	3.41	3.18	3.02	2.92	2.84	2.77	2.72	2.67	2.63	2.60	2.55	2.51	2.46
	9.07	**6.70**	**5.74**	**5.20**	**4.86**	**4.62**	**4.44**	**4.30**	**4.19**	**4.10**	**4.02**	**3.96**	**3.85**	**3.78**	**3.67**
14	4.60	3.74	3.34	3.11	2.96	2.85	2.77	2.70	2.65	2.60	2.56	2.53	2.48	2.44	2.39
	8.86	**6.51**	**5.56**	**5.03**	**4.69**	**4.46**	**4.28**	**4.14**	**4.03**	**3.94**	**3.86**	**3.80**	**3.70**	**3.62**	**3.51**
15	4.54	3.68	3.29	3.06	2.90	2.79	2.70	2.64	2.59	2.55	2.51	2.48	2.43	2.39	2.33
	8.68	**6.36**	**5.42**	**4.89**	**4.56**	**4.32**	**4.14**	**4.00**	**3.89**	**3.80**	**3.73**	**3.67**	**3.56**	**3.48**	**3.36**
16	4.49	3.63	3.24	3.01	2.85	2.74	2.66	2.59	2.54	2.49	2.45	2.42	2.37	2.33	2.28
	8.53	**6.23**	**5.29**	**4.77**	**4.44**	**4.20**	**4.03**	**3.89**	**3.78**	**3.69**	**3.61**	**3.55**	**3.45**	**3.37**	**3.25**
17	4.45	3.59	3.20	2.96	2.81	2.70	2.62	2.55	2.50	2.45	2.41	2.38	2.33	2.29	2.23
	8.40	**6.11**	**5.18**	**4.67**	**4.34**	**4.10**	**3.93**	**3.79**	**3.68**	**3.59**	**3.52**	**3.45**	**3.35**	**3.27**	**3.16**
18	4.41	3.55	3.16	2.93	2.77	2.66	2.58	2.51	2.46	2.41	2.37	2.34	2.29	2.25	2.19
	8.28	**6.01**	**5.09**	**4.58**	**4.25**	**4.01**	**3.85**	**3.71**	**3.60**	**3.51**	**3.44**	**3.37**	**3.27**	**3.19**	**3.07**

TABLE B.4   (continued)

Degrees of Freedom: Denominator	Degrees of Freedom: Numerator														
	1	2	3	4	5	6	7	8	9	10	11	12	14	16	20
19	4.38	3.52	3.13	2.90	2.74	2.63	2.55	2.48	2.43	2.38	2.34	2.31	2.26	2.21	2.15
	8.18	**5.93**	**5.01**	**4.50**	**4.17**	**3.94**	**3.77**	**3.63**	**3.52**	**3.43**	**3.36**	**3.30**	**3.19**	**3.12**	**3.00**
20	4.35	3.49	3.10	2.87	2.71	2.60	2.52	2.45	2.40	2.35	2.31	2.28	2.23	2.18	2.12
	8.10	**5.85**	**4.94**	**4.43**	**4.10**	**3.87**	**3.71**	**3.56**	**3.45**	**3.37**	**3.30**	**3.23**	**3.13**	**3.05**	**2.94**
21	4.32	3.47	3.07	2.84	2.68	2.57	2.49	2.42	2.37	2.32	2.28	2.25	2.20	2.15	2.09
	8.02	**5.78**	**4.87**	**4.37**	**4.04**	**3.81**	**3.65**	**3.51**	**3.40**	**3.31**	**3.24**	**3.17**	**3.07**	**2.99**	**2.88**
22	4.30	3.44	3.05	2.82	2.66	2.55	2.47	2.40	2.35	2.30	2.26	2.23	2.18	2.13	2.07
	7.94	**5.72**	**4.82**	**4.31**	**3.99**	**3.76**	**3.59**	**3.45**	**3.35**	**3.26**	**3.18**	**3.12**	**3.02**	**2.94**	**2.83**
23	4.28	3.42	3.03	2.80	2.64	2.53	2.45	2.38	2.32	2.28	2.24	2.20	2.14	2.10	2.04
	7.88	**5.66**	**4.76**	**4.26**	**3.94**	**3.71**	**3.54**	**3.41**	**3.30**	**3.21**	**3.14**	**3.07**	**2.97**	**2.89**	**2.78**
24	4.26	3.40	3.01	2.78	2.62	2.51	2.43	2.36	2.30	2.26	2.22	2.18	2.13	2.09	2.02
	7.82	**5.61**	**4.72**	**4.22**	**3.90**	**3.67**	**3.50**	**3.36**	**3.25**	**3.17**	**3.09**	**3.03**	**2.93**	**2.85**	**2.74**
25	4.24	3.38	2.99	2.76	2.60	2.49	2.41	2.34	2.28	2.24	2.20	2.16	2.11	2.06	2.00
	7.77	**5.57**	**4.68**	**4.18**	**3.86**	**3.63**	**3.46**	**3.32**	**3.21**	**3.13**	**3.05**	**2.99**	**2.89**	**2.81**	**2.70**
26	4.22	3.37	2.98	2.74	2.59	2.47	2.39	2.32	2.27	2.22	2.18	2.15	2.10	2.05	1.99
	7.72	**5.53**	**4.64**	**4.14**	**3.82**	**3.59**	**3.42**	**3.29**	**3.17**	**3.09**	**3.02**	**2.96**	**2.86**	**2.77**	**2.66**
27	4.21	3.35	2.96	2.73	2.57	2.46	2.37	2.30	2.25	2.20	2.16	2.13	2.08	2.03	1.97
	7.68	**5.49**	**4.60**	**4.11**	**3.79**	**3.56**	**3.39**	**3.26**	**3.14**	**3.06**	**2.98**	**2.93**	**2.83**	**2.74**	**2.63**
28	4.20	3.34	2.95	2.71	2.56	2.44	2.36	2.29	2.24	2.19	2.15	2.12	2.06	2.02	1.96
	7.64	**5.45**	**4.57**	**4.07**	**3.76**	**3.53**	**3.36**	**3.23**	**3.11**	**3.03**	**2.95**	**2.90**	**2.80**	**2.71**	**2.60**
29	4.18	3.33	2.93	2.70	2.54	2.43	2.35	2.28	2.22	2.18	2.14	2.10	2.05	2.00	1.94
	7.60	**5.42**	**4.54**	**4.04**	**3.73**	**3.50**	**3.33**	**3.20**	**3.08**	**3.00**	**2.92**	**2.87**	**2.77**	**2.68**	**2.57**
30	4.17	3.32	2.92	2.69	2.53	2.42	2.34	2.27	2.21	2.16	2.12	2.09	2.04	1.99	1.93
	7.56	**5.39**	**4.51**	**4.02**	**3.70**	**3.47**	**3.30**	**3.17**	**3.06**	**2.98**	**2.90**	**2.84**	**2.74**	**2.66**	**2.55**
32	4.15	3.30	2.90	2.67	2.51	2.40	2.32	2.25	2.19	2.14	2.10	2.07	2.02	1.97	1.91
	7.50	**5.34**	**4.46**	**3.97**	**3.66**	**3.42**	**3.25**	**3.12**	**3.01**	**2.94**	**2.86**	**2.80**	**2.70**	**2.62**	**2.51**
34	4.13	3.28	2.88	2.65	2.49	2.38	2.30	2.23	2.17	2.12	2.08	2.05	2.00	1.95	1.89
	7.44	**5.29**	**4.42**	**3.93**	**3.61**	**3.38**	**3.21**	**3.08**	**2.97**	**2.89**	**2.82**	**2.76**	**2.66**	**2.58**	**2.47**
36	4.11	3.26	2.86	2.63	2.48	2.36	2.28	2.21	2.15	2.10	2.06	2.03	1.98	1.93	1.87
	7.39	**5.25**	**4.38**	**3.89**	**3.58**	**3.35**	**3.18**	**3.04**	**2.94**	**2.86**	**2.78**	**2.72**	**2.62**	**2.54**	**2.43**
38	4.10	3.25	2.85	2.62	2.46	2.35	2.26	2.19	2.14	2.09	2.05	2.02	1.96	1.92	1.85
	7.35	**5.21**	**4.34**	**3.86**	**3.54**	**3.32**	**3.15**	**3.02**	**2.91**	**2.82**	**2.75**	**2.69**	**2.59**	**2.51**	**2.40**
40	4.08	3.23	2.84	2.61	2.45	2.34	2.25	2.18	2.12	2.07	2.04	2.00	1.95	1.90	1.84
	7.31	**5.18**	**4.31**	**3.83**	**3.51**	**3.29**	**3.12**	**2.99**	**2.88**	**2.80**	**2.73**	**2.66**	**2.56**	**2.49**	**2.37**
42	4.07	3.22	2.83	2.59	2.44	2.32	2.24	2.17	2.11	2.06	2.02	1.99	1.94	1.89	1.82
	7.27	**5.15**	**4.29**	**3.80**	**3.49**	**3.26**	**3.10**	**2.96**	**2.86**	**2.77**	**2.70**	**2.64**	**2.54**	**2.46**	**2.35**

TABLE B.4   (continued)

Degrees of Freedom: Denominator	Degrees of Freedom: Numerator														
	1	2	3	4	5	6	7	8	9	10	11	12	14	16	20
44	4.06	3.21	2.82	2.58	2.43	2.31	2.23	2.16	2.10	2.05	2.01	1.98	1.92	1.88	1.81
	7.24	**5.12**	**4.26**	**3.78**	**3.46**	**3.24**	**3.07**	**2.94**	**2.84**	**2.75**	**2.68**	**2.62**	**2.52**	**2.44**	**2.32**
46	4.05	3.20	2.81	2.57	2.42	2.30	2.22	2.14	2.09	2.04	2.00	1.97	1.91	1.87	1.80
	7.21	**5.10**	**4.24**	**3.76**	**3.44**	**3.22**	**3.05**	**2.92**	**2.82**	**2.73**	**2.66**	**2.60**	**2.50**	**2.42**	**2.30**
48	4.04	3.19	2.80	2.56	2.41	2.30	2.21	2.14	2.08	2.03	1.99	1.96	1.90	1.86	1.79
	7.19	**5.08**	**4.22**	**3.74**	**3.42**	**3.20**	**3.04**	**2.90**	**2.80**	**2.71**	**2.64**	**2.58**	**2.48**	**2.40**	**2.28**
50	4.03	3.18	2.79	2.56	2.40	2.29	2.20	2.13	2.07	2.02	1.98	1.95	1.90	1.85	1.78
	7.17	**5.06**	**4.20**	**3.72**	**3.41**	**3.18**	**3.02**	**2.88**	**2.78**	**2.70**	**2.62**	**2.56**	**2.46**	**2.39**	**2.26**
55	4.02	3.17	2.78	2.54	2.38	2.27	2.18	2.11	2.05	2.00	1.97	1.93	1.88	1.83	1.76
	7.12	**5.01**	**4.16**	**3.68**	**3.37**	**3.15**	**2.98**	**2.85**	**2.75**	**2.66**	**2.59**	**2.53**	**2.43**	**2.35**	**2.23**
60	4.00	3.15	2.76	2.52	2.37	2.25	2.17	2.10	2.04	1.99	1.95	1.92	1.86	1.81	1.75
	7.08	**4.98**	**4.13**	**3.65**	**3.34**	**3.12**	**2.95**	**2.82**	**2.72**	**2.63**	**2.56**	**2.50**	**2.40**	**2.32**	**2.20**
65	3.99	3.14	2.75	2.51	2.36	2.24	2.15	2.08	2.02	1.98	1.94	1.90	1.85	1.80	1.73
	7.04	**4.95**	**4.10**	**3.62**	**3.31**	**3.09**	**2.93**	**2.79**	**2.70**	**2.61**	**2.54**	**2.47**	**2.37**	**2.30**	**2.18**
70	3.98	3.13	2.74	2.50	2.35	2.23	2.14	2.07	2.01	1.97	1.93	1.89	1.84	1.79	1.72
	7.01	**4.92**	**4.08**	**3.60**	**3.29**	**3.07**	**2.91**	**2.77**	**2.67**	**2.59**	**2.51**	**2.45**	**2.35**	**2.28**	**2.15**
80	3.96	3.11	2.72	2.48	2.33	2.21	2.12	2.05	1.99	1.95	1.91	1.88	1.82	1.77	1.70
	6.96	**4.88**	**4.04**	**3.56**	**3.25**	**3.04**	**2.87**	**2.74**	**2.64**	**2.55**	**2.48**	**2.41**	**2.32**	**2.24**	**2.11**
100	3.94	3.09	2.70	2.46	2.30	2.19	2.10	2.03	1.97	1.92	1.88	1.85	1.79	1.75	1.68
	6.90	**4.82**	**3.98**	**3.51**	**3.20**	**2.99**	**2.82**	**2.69**	**2.59**	**2.51**	**2.43**	**2.36**	**2.26**	**2.19**	**2.06**
125	3.92	3.07	2.68	2.44	2.29	2.17	2.08	2.01	1.95	1.90	1.86	1.83	1.77	1.72	1.65
	6.84	**4.78**	**3.94**	**3.47**	**3.17**	**2.95**	**2.79**	**2.65**	**2.56**	**2.47**	**2.40**	**2.33**	**2.23**	**2.15**	**2.03**
150	3.91	3.06	2.67	2.43	2.27	2.16	2.07	2.00	1.94	1.89	1.85	1.82	1.76	1.71	1.64
	6.81	**4.75**	**3.91**	**3.44**	**3.14**	**2.92**	**2.76**	**2.62**	**2.53**	**2.44**	**2.37**	**2.30**	**2.20**	**2.12**	**2.00**
200	3.89	3.04	2.65	2.41	2.26	2.14	2.05	1.98	1.92	1.87	1.83	1.80	1.74	1.69	1.62
	6.76	**4.71**	**3.88**	**3.41**	**3.11**	**2.90**	**2.73**	**2.60**	**2.50**	**2.41**	**2.34**	**2.28**	**2.17**	**2.09**	**1.97**
400	3.86	3.02	2.62	2.39	2.23	2.12	2.03	1.96	1.90	1.85	1.81	1.78	1.72	1.67	1.60
	6.70	**4.66**	**3.83**	**3.36**	**3.06**	**2.85**	**2.69**	**2.55**	**2.46**	**2.37**	**2.29**	**2.23**	**2.12**	**2.04**	**1.92**
1000	3.85	3.00	2.61	2.38	2.22	2.10	2.02	1.95	1.89	1.84	1.80	1.76	1.70	1.65	1.58
	6.66	**4.62**	**3.80**	**3.34**	**3.04**	**2.82**	**2.66**	**2.53**	**2.43**	**2.34**	**2.26**	**2.20**	**2.09**	**2.01**	**1.89**
∞	3.84	2.99	2.60	2.37	2.21	2.09	2.01	1.94	1.88	1.83	1.79	1.75	1.69	1.64	1.57
	6.64	**4.60**	**3.78**	**3.32**	**3.02**	**2.80**	**2.64**	**2.51**	**2.41**	**2.32**	**2.24**	**2.18**	**2.07**	**1.99**	**1.87**

Table A14 of *Statistical Methods*, 7th ed. by George W. Snedecor and William G. Cochran. Copyright © 1980 by the Iowa State University Press, 2121 South State Avenue, Ames, Iowa 50010. Reprinted with permission of the Iowa State University Press.

TABLE B.5   THE STUDENTIZED RANGE STATISTIC (q)*

*The critical values for q corresponding to α = .05 (lightface type) and α = .01 (boldface type).

df for Error Term	k = Number of Treatments										
	2	3	4	5	6	7	8	9	10	11	12
5	3.64	4.60	5.22	5.67	6.03	6.33	6.58	6.80	6.99	7.17	7.32
	5.70	**6.98**	**7.80**	**8.42**	**8.91**	**9.32**	**9.67**	**9.97**	**10.24**	**10.48**	**10.70**
6	3.46	4.34	4.90	5.30	5.63	5.90	6.12	6.32	6.49	6.65	6.79
	5.24	**6.33**	**7.03**	**7.56**	**7.97**	**8.32**	**8.61**	**8.87**	**9.10**	**9.30**	**9.48**
7	3.34	4.16	4.68	5.06	5.36	5.61	5.82	6.00	6.16	6.30	6.43
	4.95	**5.92**	**6.54**	**7.01**	**7.37**	**7.68**	**7.94**	**8.17**	**8.37**	**8.55**	**8.71**
8	3.26	4.04	4.53	4.89	5.17	5.40	5.60	5.77	5.92	6.05	6.18
	4.75	**5.64**	**6.20**	**6.62**	**6.96**	**7.24**	**7.47**	**7.68**	**7.86**	**8.03**	**8.18**
9	3.20	3.95	4.41	4.76	5.02	5.24	5.43	5.59	5.74	5.87	5.98
	4.60	**5.43**	**5.96**	**6.35**	**6.66**	**6.91**	**7.13**	**7.33**	**7.49**	**7.65**	**7.78**
10	3.15	3.88	4.33	4.65	4.91	5.12	5.30	5.46	5.60	5.72	5.83
	4.48	**5.27**	**5.77**	**6.14**	**6.43**	**6.67**	**6.87**	**7.05**	**7.21**	**7.36**	**7.49**
11	3.11	3.82	4.26	4.57	4.82	5.03	5.20	5.35	5.49	5.61	5.71
	4.39	**5.15**	**5.62**	**5.97**	**6.25**	**6.48**	**6.67**	**6.84**	**6.99**	**7.13**	**7.25**
12	3.08	3.77	4.20	4.51	4.75	4.95	5.12	5.27	5.39	5.51	5.61
	4.32	**5.05**	**5.50**	**5.84**	**6.10**	**6.32**	**6.51**	**6.67**	**6.81**	**6.94**	**7.06**
13	3.06	3.73	4.15	4.45	4.69	4.88	5.05	5.19	5.32	5.43	5.53
	4.26	**4.96**	**5.40**	**5.73**	**5.98**	**6.19**	**6.37**	**6.53**	**6.67**	**6.79**	**6.90**
14	3.03	3.70	4.11	4.41	4.64	4.83	4.99	5.13	5.25	5.36	5.46
	4.21	**4.89**	**5.32**	**5.63**	**5.88**	**6.08**	**6.26**	**6.41**	**6.54**	**6.66**	**6.77**
15	3.01	3.67	4.08	4.37	4.59	4.78	4.94	5.08	5.20	5.31	5.40
	4.17	**4.84**	**5.25**	**5.56**	**5.80**	**5.99**	**6.16**	**6.31**	**6.44**	**6.55**	**6.66**
16	3.00	3.65	4.05	4.33	4.56	4.74	4.90	5.03	5.15	5.26	5.35
	4.13	**4.79**	**5.19**	**5.49**	**5.72**	**5.92**	**6.08**	**6.22**	**6.35**	**6.46**	**6.56**
17	2.98	3.63	4.02	4.30	4.52	4.70	4.86	4.99	5.11	5.21	5.31
	4.10	**4.74**	**5.14**	**5.43**	**5.66**	**5.85**	**6.01**	**6.15**	**6.27**	**6.38**	**6.48**
18	2.97	3.61	4.00	4.28	4.49	4.67	4.82	4.96	5.07	5.17	5.27
	4.07	**4.70**	**5.09**	**5.38**	**5.60**	**5.79**	**5.94**	**6.08**	**6.20**	**6.31**	**6.41**
19	2.96	3.59	3.98	4.25	4.47	4.65	4.79	4.92	5.04	5.14	5.23
	4.05	**4.67**	**5.05**	**5.33**	**5.55**	**5.73**	**5.89**	**6.02**	**6.14**	**6.25**	**6.34**
20	2.95	3.58	3.96	4.23	4.45	4.62	4.77	4.90	5.01	5.11	5.20
	4.02	**4.64**	**5.02**	**5.29**	**5.51**	**5.69**	**5.84**	**5.97**	**6.09**	**6.19**	**6.28**
24	2.92	3.53	3.90	4.17	4.37	4.54	4.68	4.81	4.92	5.01	5.10
	3.96	**4.55**	**4.91**	**5.17**	**5.37**	**5.54**	**5.69**	**5.81**	**5.92**	**6.02**	**6.11**
30	2.89	3.49	3.85	4.10	4.30	4.46	4.60	4.72	4.82	4.92	5.00
	3.89	**4.45**	**4.80**	**5.05**	**5.24**	**5.40**	**5.54**	**5.65**	**5.76**	**5.85**	**5.93**
40	2.86	3.44	3.79	4.04	4.23	4.39	4.52	4.63	4.73	4.82	4.90
	3.82	**4.37**	**4.70**	**4.93**	**5.11**	**5.26**	**5.39**	**5.50**	**5.60**	**5.69**	**5.76**
60	2.83	3.40	3.74	3.98	4.16	4.31	4.44	4.55	4.65	4.73	4.81
	3.76	**4.28**	**4.59**	**4.82**	**4.99**	**5.13**	**5.25**	**5.36**	**5.45**	**5.53**	**5.60**
120	2.80	3.36	3.68	3.92	4.10	4.24	4.36	4.47	4.56	4.64	4.71
	3.70	**4.20**	**4.50**	**4.71**	**4.87**	**5.01**	**5.12**	**5.21**	**5.30**	**5.37**	**5.44**
∞	2.77	3.31	3.63	3.86	4.03	4.17	4.28	4.39	4.47	4.55	4.62
	3.64	**4.12**	**4.40**	**4.60**	**4.76**	**4.88**	**4.99**	**5.08**	**5.16**	**5.23**	**5.29**

TABLE B.6   CRITICAL VALUES FOR THE PEARSON CORRELATION*

*To be significant, the sample correlation, r, must be greater than or equal to the critical value in the table.

| | Level of Significance for One-Tailed Test | | | |
	.05	.025	.01	.005
	Level of Significance for Two-Tailed Test			
$df = n - 2$	.10	.05	.02	.01
1	.988	.997	.9995	.9999
2	.900	.950	.980	.990
3	.805	.878	.934	.959
4	.729	.811	.882	.917
5	.669	.754	.833	.874
6	.622	.707	.789	.834
7	.582	.666	.750	.798
8	.549	.632	.716	.765
9	.521	.602	.685	.735
10	.497	.576	.658	.708
11	.476	.553	.634	.684
12	.458	.532	.612	.661
13	.441	.514	.592	.641
14	.426	.497	.574	.623
15	.412	.482	.558	.606
16	.400	.468	.542	.590
17	.389	.456	.528	.575
18	.378	.444	.516	.561
19	.369	.433	.503	.549
20	.360	.423	.492	.537
21	.352	.413	.482	.526
22	.344	.404	.472	.515
23	.337	.396	.462	.505
24	.330	.388	.453	.496
25	.323	.381	.445	.487
26	.317	.374	.437	.479
27	.311	.367	.430	.471
28	.306	.361	.423	.463
29	.301	.355	.416	.456
30	.296	.349	.409	.449
35	.275	.325	.381	.418
40	.257	.304	.358	.393
45	.243	.288	.338	.372
50	.231	.273	.322	.354
60	.211	.250	.295	.325
70	.195	.232	.274	.302
80	.183	.217	.256	.283
90	.173	.205	.242	.267
100	.164	.195	.230	.254

Table VI of R. A. Fisher and F. Yates, *Statistical Tables for Biological, Agricultural and Medical Research*, 6th ed. London: Longman Group Ltd., 1974 (previously published by Oliver and Boyd Ltd., Edinburgh). Copyright ©1963 R. A. Fisher and F. Yates. Adapted and reprinted with permission of Pearson Education Limited.

TABLE B.7   CRITICAL VALUES FOR THE SPEARMAN CORRELATION*

*To be significant, the sample correlation, r_s, must be greater than or equal to the critical value in the table.

	Level of Significance for One-Tailed Test			
	.05	.025	.01	.005
	Level of Significance for Two-Tailed Test			
n	.10	.05	.02	.01
4	1.000			
5	0.900	1.000	1.000	
6	0.829	0.886	0.943	1.000
7	0.714	0.786	0.893	0.929
8	0.643	0.738	0.833	0.881
9	0.600	0.700	0.783	0.833
10	0.564	0.648	0.745	0.794
11	0.536	0.618	0.709	0.755
12	0.503	0.587	0.671	0.727
13	0.484	0.560	0.648	0.703
14	0.464	0.538	0.622	0.675
15	0.443	0.521	0.604	0.654
16	0.429	0.503	0.582	0.635
17	0.414	0.485	0.566	0.615
18	0.401	0.472	0.550	0.600
19	0.391	0.460	0.535	0.584
20	0.380	0.447	0.520	0.570
21	0.370	0.435	0.508	0.556
22	0.361	0.425	0.496	0.544
23	0.353	0.415	0.486	0.532
24	0.344	0.406	0.476	0.521
25	0.337	0.398	0.466	0.511
26	0.331	0.390	0.457	0.501
27	0.324	0.382	0.448	0.491
28	0.317	0.375	0.440	0.483
29	0.312	0.368	0.433	0.475
30	0.306	0.362	0.425	0.467
35	0.283	0.335	0.394	0.433
40	0.264	0.313	0.368	0.405
45	0.248	0.294	0.347	0.382
50	0.235	0.279	0.329	0.363
60	0.214	0.255	0.300	0.331
70	0.190	0.235	0.278	0.307
80	0.185	0.220	0.260	0.287
90	0.174	0.207	0.245	0.271
100	0.165	0.197	0.233	0.257

Zar, J. H. (1972). Significance testing of the Spearman rank correlation coefficient. *Journal of the American Statistical Association*, 67, 578. Reprinted with permission from the *Journal of the American Statistical Association*. Copyright © 1972 by the American Statistical Association. All rights reserved.

TABLE B.8   THE CHI-SQUARE DISTRIBUTION*

*The table entries are critical values of χ^2.

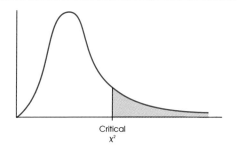

Critical
χ^2

df	Proportion in Critical Region				
	0.10	0.05	0.025	0.01	0.005
1	2.71	3.84	5.02	6.63	7.88
2	4.61	5.99	7.38	9.21	10.60
3	6.25	7.81	9.35	11.34	12.84
4	7.78	9.49	11.14	13.28	14.86
5	9.24	11.07	12.83	15.09	16.75
6	10.64	12.59	14.45	16.81	18.55
7	12.02	14.07	16.01	18.48	20.28
8	13.36	15.51	17.53	20.09	21.96
9	14.68	16.92	19.02	21.67	23.59
10	15.99	18.31	20.48	23.21	25.19
11	17.28	19.68	21.92	24.72	26.76
12	18.55	21.03	23.34	26.22	28.30
13	19.81	22.36	24.74	27.69	29.82
14	21.06	23.68	26.12	29.14	31.32
15	22.31	25.00	27.49	30.58	32.80
16	23.54	26.30	28.85	32.00	34.27
17	24.77	27.59	30.19	33.41	35.72
18	25.99	28.87	31.53	34.81	37.16
19	27.20	30.14	32.85	36.19	38.58
20	28.41	31.41	34.17	37.57	40.00
21	29.62	32.67	35.48	38.93	41.40
22	30.81	33.92	36.78	40.29	42.80
23	32.01	35.17	38.08	41.64	44.18
24	33.20	36.42	39.36	42.98	45.56
25	34.38	37.65	40.65	44.31	46.93
26	35.56	38.89	41.92	45.64	48.29
27	36.74	40.11	43.19	46.96	49.64
28	37.92	41.34	44.46	48.28	50.99
29	39.09	42.56	45.72	49.59	52.34
30	40.26	43.77	46.98	50.89	53.67
40	51.81	55.76	59.34	63.69	66.77
50	63.17	67.50	71.42	76.15	79.49
60	74.40	79.08	83.30	88.38	91.95
70	85.53	90.53	95.02	100.42	104.22
80	96.58	101.88	106.63	112.33	116.32
90	107.56	113.14	118.14	124.12	128.30
100	118.50	124.34	129.56	135.81	140.17

Table 8 of E. Pearson and H. O. Hartley, *Biometrika Tables for Statisticians*, 3rd ed. New York: Cambridge University Press, 1966. Adapted and reprinted with permission of the Biometrika trustees.

TABLE B.9A   CRITICAL VALUES OF THE MANN-WHITNEY U FOR $\alpha = .05$*

*Critical values are provided for a *one-tailed* test at $\alpha = .05$ (lightface type) and for a *two-tailed* test at $\alpha = .05$ (boldface type). To be significant for any given n_A and n_B, the obtained U must be *equal to* or *less than* the critical value in the table. Dashes (—) in the body of the table indicate that no decision is possible at the stated level of significance and values of n_A and n_B.

n_B \ n_A	1	2	3	4	5	6	7	8	9	10	11	12	13	14	15	16	17	18	19	20
1	—	—	—	—	—	—	—	—	—	—	—	—	—	—	—	—	—	—	0	0
	—	—	—	—	—	—	—	—	—	—	—	—	—	—	—	—	—	—	—	—
2	—	—	—	—	0	0	0	1	1	1	1	2	2	2	3	3	3	4	4	4
	—	—	—	—	—	—	—	**0**	**0**	**0**	**0**	**1**	**1**	**1**	**1**	**1**	**2**	**2**	**2**	**2**
3	—	—	0	0	1	2	2	3	3	4	5	5	6	7	7	8	9	9	10	11
	—	—	—	—	**0**	**1**	**1**	**2**	**2**	**3**	**3**	**4**	**4**	**5**	**5**	**6**	**6**	**7**	**7**	**8**
4	—	—	0	1	2	3	4	5	6	7	8	9	10	11	12	14	15	16	17	18
	—	—	—	**0**	**1**	**2**	**3**	**4**	**4**	**5**	**6**	**7**	**8**	**9**	**10**	**11**	**11**	**12**	**13**	**13**
5	—	0	1	2	4	5	6	8	9	11	12	13	15	16	18	19	20	22	23	25
	—	—	**0**	**1**	**2**	**3**	**5**	**6**	**7**	**8**	**9**	**11**	**12**	**13**	**14**	**15**	**17**	**18**	**19**	**20**
6	—	0	2	3	5	7	8	10	12	14	16	17	19	21	23	25	26	28	30	32
	—	—	**1**	**2**	**3**	**5**	**6**	**8**	**10**	**11**	**13**	**14**	**16**	**17**	**19**	**21**	**22**	**24**	**25**	**27**
7	—	0	2	4	6	8	11	13	15	17	19	21	24	26	28	30	33	35	37	39
	—	—	**1**	**3**	**5**	**6**	**8**	**10**	**12**	**14**	**16**	**18**	**20**	**22**	**24**	**26**	**28**	**30**	**32**	**34**
8	—	1	3	5	8	10	13	15	18	20	23	26	28	31	33	36	39	41	44	47
	—	**0**	**2**	**4**	**6**	**8**	**10**	**13**	**15**	**17**	**19**	**22**	**24**	**26**	**29**	**31**	**34**	**36**	**38**	**41**
9	—	1	3	6	9	12	15	18	21	24	27	30	33	36	39	42	45	48	51	54
	—	**0**	**2**	**4**	**7**	**10**	**12**	**15**	**17**	**20**	**23**	**26**	**28**	**31**	**34**	**37**	**39**	**42**	**45**	**48**
10	—	1	4	7	11	14	17	20	24	27	31	34	37	41	44	48	51	55	58	62
	—	**0**	**3**	**5**	**8**	**11**	**14**	**17**	**20**	**23**	**26**	**29**	**33**	**36**	**39**	**42**	**45**	**48**	**52**	**55**
11	—	1	5	8	12	16	19	23	27	31	34	38	42	46	50	54	57	61	65	69
	—	**0**	**3**	**6**	**9**	**13**	**16**	**19**	**23**	**26**	**30**	**33**	**37**	**40**	**44**	**47**	**51**	**55**	**58**	**62**
12	—	2	5	9	13	17	21	26	30	34	38	42	47	51	55	60	64	68	72	77
	—	**1**	**4**	**7**	**11**	**14**	**18**	**22**	**26**	**29**	**33**	**37**	**41**	**45**	**49**	**53**	**57**	**61**	**65**	**69**
13	—	2	6	10	15	19	24	28	33	37	42	47	51	56	61	65	79	75	80	84
	—	**1**	**4**	**8**	**12**	**16**	**20**	**24**	**28**	**33**	**37**	**41**	**45**	**50**	**54**	**59**	**63**	**67**	**72**	**76**
14	—	2	7	11	16	21	26	31	36	41	46	51	56	61	66	71	77	82	87	92
	—	**1**	**5**	**9**	**13**	**17**	**22**	**26**	**31**	**36**	**40**	**45**	**50**	**55**	**59**	**64**	**67**	**74**	**78**	**83**
15	—	3	7	12	18	23	28	33	39	44	50	55	61	66	72	77	83	88	94	100
	—	**1**	**5**	**10**	**14**	**19**	**24**	**29**	**34**	**39**	**44**	**49**	**54**	**59**	**64**	**70**	**75**	**80**	**85**	**90**
16	—	3	8	14	19	25	30	36	42	48	54	60	65	71	77	83	89	95	101	107
	—	**1**	**6**	**11**	**15**	**21**	**26**	**31**	**37**	**42**	**47**	**53**	**59**	**64**	**70**	**75**	**81**	**86**	**92**	**98**
17	—	3	9	15	20	26	33	39	45	51	57	64	70	77	83	89	96	102	109	115
	—	**2**	**6**	**11**	**17**	**22**	**28**	**34**	**39**	**45**	**51**	**57**	**63**	**67**	**75**	**81**	**87**	**93**	**99**	**105**
18	—	4	9	16	22	28	35	41	48	55	61	68	75	82	88	95	102	109	116	123
	—	**2**	**7**	**12**	**18**	**24**	**30**	**36**	**42**	**48**	**55**	**61**	**67**	**74**	**80**	**86**	**93**	**99**	**106**	**112**
19	0	4	10	17	23	30	37	44	51	58	65	72	80	87	94	101	109	116	123	130
	—	**2**	**7**	**13**	**19**	**25**	**32**	**38**	**45**	**52**	**58**	**65**	**72**	**78**	**85**	**92**	**99**	**106**	**113**	**119**
20	0	4	11	18	25	32	39	47	54	62	69	77	84	92	100	107	115	123	130	138
	—	**2**	**8**	**13**	**20**	**27**	**34**	**41**	**48**	**55**	**62**	**69**	**76**	**83**	**90**	**98**	**105**	**112**	**119**	**127**

TABLE B.9B   CRITICAL VALUES OF THE MANN-WHITNEY U FOR $\alpha = .01$*

*Critical values are provided for a *one-tailed* test at $\alpha = .01$ (lightface type) and for a *two-tailed* test at $\alpha = .01$ (boldface type). To be significant for any given n_A and n_B, the obtained U must be *equal to* or *less than* the critical value in the table. Dashes (—) in the body of the table indicate that no decision is possible at the stated level of significance and values of n_A and n_B.

n_B \ n_A	1	2	3	4	5	6	7	8	9	10	11	12	13	14	15	16	17	18	19	20
1	—	—	—	—	—	—	—	—	—	—	—	—	—	—	—	—	—	—	—	—
2	—	—	—	—	—	—	—	—	—	—	—	—	0	0	0	0	0	0	1	1
	—	—	—	—	—	—	—	—	—	—	—	—	—	—	—	—	—	—	**0**	**0**
3	—	—	—	—	—	—	0	0	1	1	1	2	2	2	3	3	4	4	4	5
	—	—	—	—	—	—	—	**0**	**0**	**0**	**1**	**1**	**1**	**2**	**2**	**2**	**2**	**2**	**3**	**3**
4	—	—	—	—	0	1	1	2	3	3	4	5	5	6	7	7	8	9	9	10
	—	—	—	—	—	**0**	**0**	**1**	**1**	**2**	**2**	**3**	**3**	**4**	**5**	**5**	**6**	**6**	**7**	**8**
5	—	—	—	0	1	2	3	4	5	6	7	8	9	10	11	12	13	14	15	16
	—	—	—	—	**0**	**1**	**1**	**2**	**3**	**4**	**5**	**6**	**7**	**7**	**8**	**9**	**10**	**11**	**12**	**13**
6	—	—	—	1	2	3	4	6	7	8	9	11	12	13	15	16	18	19	20	22
	—	—	—	**0**	**1**	**2**	**3**	**4**	**5**	**6**	**7**	**9**	**10**	**11**	**12**	**13**	**15**	**16**	**17**	**18**
7	—	—	0	1	3	4	6	7	9	11	12	14	16	17	19	21	23	24	26	28
	—	—	—	**0**	**1**	**3**	**4**	**6**	**7**	**9**	**10**	**12**	**13**	**15**	**16**	**18**	**19**	**21**	**22**	**24**
8	—	—	0	2	4	6	7	9	11	13	15	17	20	22	24	26	28	30	32	34
	—	—	—	**1**	**2**	**4**	**6**	**7**	**9**	**11**	**13**	**15**	**17**	**18**	**20**	**22**	**24**	**26**	**28**	**30**
9	—	—	1	3	5	7	9	11	14	16	18	21	23	26	28	31	33	36	38	40
	—	—	**0**	**1**	**3**	**5**	**7**	**9**	**11**	**13**	**16**	**18**	**20**	**22**	**24**	**27**	**29**	**31**	**33**	**36**
10	—	—	1	3	6	8	11	13	16	19	22	24	27	30	33	36	38	41	44	47
	—	—	**0**	**2**	**4**	**6**	**9**	**11**	**13**	**16**	**18**	**21**	**24**	**26**	**29**	**31**	**34**	**37**	**39**	**42**
11	—	—	1	4	7	9	12	15	18	22	25	28	31	34	37	41	44	47	50	53
	—	—	**0**	**2**	**5**	**7**	**10**	**13**	**16**	**18**	**21**	**24**	**27**	**30**	**33**	**36**	**39**	**42**	**45**	**48**
12	—	—	2	5	8	11	14	17	21	24	28	31	35	38	42	46	49	53	56	60
	—	—	**1**	**3**	**6**	**9**	**12**	**15**	**18**	**21**	**24**	**27**	**31**	**34**	**37**	**41**	**44**	**47**	**51**	**54**
13	—	0	2	5	9	12	16	20	23	27	31	35	39	43	47	51	55	59	63	67
	—	—	**1**	**3**	**7**	**10**	**13**	**17**	**20**	**24**	**27**	**31**	**34**	**38**	**42**	**45**	**49**	**53**	**56**	**60**
14	—	0	2	6	10	13	17	22	26	30	34	38	43	47	51	56	60	65	69	73
	—	—	**1**	**4**	**7**	**11**	**15**	**18**	**22**	**26**	**30**	**34**	**38**	**42**	**46**	**50**	**54**	**58**	**63**	**67**
15	—	0	3	7	11	15	19	24	28	33	37	42	47	51	56	61	66	70	75	80
	—	—	**2**	**5**	**8**	**12**	**16**	**20**	**24**	**29**	**33**	**37**	**42**	**46**	**51**	**55**	**60**	**64**	**69**	**73**
16	—	0	3	7	12	16	21	26	31	36	41	46	51	56	61	66	71	76	82	87
	—	—	**2**	**5**	**9**	**13**	**18**	**22**	**27**	**31**	**36**	**41**	**45**	**50**	**55**	**60**	**65**	**70**	**74**	**79**
17	—	0	4	8	13	18	23	28	33	38	44	49	55	60	66	71	77	82	88	93
	—	—	**2**	**6**	**10**	**15**	**19**	**24**	**29**	**34**	**39**	**44**	**49**	**54**	**60**	**65**	**70**	**75**	**81**	**86**
18	—	0	4	9	14	19	24	30	36	41	47	53	59	65	70	76	82	88	94	100
	—	—	**2**	**6**	**11**	**16**	**21**	**26**	**31**	**37**	**42**	**47**	**53**	**58**	**64**	**70**	**75**	**81**	**87**	**92**
19	—	1	4	9	15	20	26	32	38	44	50	56	63	69	75	82	88	94	101	107
	—	**0**	**3**	**7**	**12**	**17**	**22**	**28**	**33**	**39**	**45**	**51**	**56**	**63**	**69**	**74**	**81**	**87**	**93**	**99**
20	—	1	5	10	16	22	28	34	40	47	53	60	67	73	80	87	93	100	107	114
	—	**0**	**3**	**8**	**13**	**18**	**24**	**30**	**36**	**42**	**48**	**54**	**60**	**67**	**73**	**79**	**86**	**92**	**99**	**105**

TABLE B.10   CRITICAL VALUES OF *T* FOR THE WILCOXON SIGNED-RANKS TEST*

*To be significant, the obtained *T* must be *equal to* or *less than* the critical value. Dashes (—) in the columns indicate that no decision is possible for the stated α and *n*.

	Level of Significance for One-Tailed Test					Level of Significance for One-Tailed Test			
	.05	.025	.01	.005		.05	.025	.01	.005
	Level of Significance for Two-Tailed Test					Level of Significance for Two-Tailed Test			
n	.10	.05	.02	.01	*n*	.10	.05	.02	.01
5	0	—	—	—	28	130	116	101	91
6	2	0	—	—	29	140	126	110	100
7	3	2	0	—	30	151	137	120	109
8	5	3	1	0	31	163	147	130	118
9	8	5	3	1	32	175	159	140	128
10	10	8	5	3	33	187	170	151	138
11	13	10	7	5	34	200	182	162	148
12	17	13	9	7	35	213	195	173	159
13	21	17	12	9	36	227	208	185	171
14	25	21	15	12	37	241	221	198	182
15	30	25	19	15	38	256	235	211	194
16	35	29	23	19	39	271	249	224	207
17	41	34	27	23	40	286	264	238	220
18	47	40	32	27	41	302	279	252	233
19	53	46	37	32	42	319	294	266	247
20	60	52	43	37	43	336	310	281	261
21	67	58	49	42	44	353	327	296	276
22	75	65	55	48	45	371	343	312	291
23	83	73	62	54	46	389	361	328	307
24	91	81	69	61	47	407	378	345	322
25	100	89	76	68	48	426	396	362	339
26	110	98	84	75	49	446	415	379	355
27	119	107	92	83	50	466	434	397	373

Adapted from F. Wilcoxon, S. K. Katti, and R. A. Wilcox, *Critical Values and Probability Levels of the Wilcoxon Rank-Sum Test and the Wilcoxon Signed-Ranks Test*. Wayne, NJ: American Cyanamid Company, 1963. Adapted and reprinted with permission of the American Cyanamid Company.

Solutions for Odd-Numbered Problems in the Text

Note: Many of the problems in the text require several stages of computation. At each stage there is an opportunity for rounding answers. Depending on the exact sequence of operations used to solve a problem, different individuals will round their answers at different times and in different ways. As a result, you or your students may obtain answers that are slightly different from those presented here. To help minimize this problem, we have tried to include the numerical values obtained at different stages of complex problems rather than presenting a single final answer.

CHAPTER 1 INTRODUCTION TO STATISTICS

1. **a.** The population is the entire set of adolescent boys.
   b. The sample is the group of 30 boys who were tested in the study.

3. Descriptive statistics are used to simplify and summarize data. Inferential statistics use sample data to make general conclusions about populations.

5. A correlational study has only one group of individuals and measures two different variables for each individual. Other research evaluating relationships between variables compares two (or more) different groups of scores.

7. The independent variable is holding a pen in your teeth versus holding the pen in your lips. The dependent variable is the rating given to each cartoon.

9. This is a correlational study. The researcher is simply observing, not manipulating, two variables.

11. This is not an experiment because there is no manipulation. Instead, the study is comparing two preexisting groups (American and Canadian students).

13. A discrete variable exists as indivisible categories such as the number of children in a family. For a continuous variable the categories are infinitely divisible such as a 1-inch interval on a ruler, which can be divided in half, in quarters, in eighths, and so on.

15. **a.** The independent variable is humorous versus nonhumorous.
   b. The independent variable is measured on a nominal scale.
   c. The dependent variable is the number of sentences recalled.
   d. The dependent variable is measured on a ratio scale.

17. **a.** The independent variable is whether the rats receive the antioxidant supplements, and the dependent variable is amount of time needed to find the submerged platform.
   b. Time is measured on a ratio scale.

19. **a.** $\Sigma X = 11$
   b. $\Sigma X^2 = 45$
   c. $\Sigma(X + 1) = 15$
   d. $\Sigma(X + 1)^2 = 71$

21. **a.** $\Sigma X = 22$
   b. $\Sigma Y = 20$
   c. $\Sigma XY = 79$

23. **a.** $\Sigma X^2 = 80$
   b. $(\Sigma X)^2 = 144$
   c. $\Sigma(X - 3) = 0$
   d. $\Sigma(X - 3)^2 = 44$

CHAPTER 2   **FREQUENCY DISTRIBUTIONS**

1.

X	f
9	2
8	3
7	5
6	4
5	3
4	2
3	1

3. a. $n = 11$
  b. $\Sigma X = 32$
  c. $\Sigma X^2 = 106$

5.

X	f
450–499	2
400–449	0
350–399	2
300–349	1
250–299	5
200–249	3
150–199	4
100–149	2
50–99	2

7. A regular table reports the exact frequency for each category on the scale of measurement. After the categories have been grouped into class intervals, the table reports only the overall frequency for the interval but does not indicate how many scores are in each of the individual categories.

9. a. $N = 17$
  b. $\Sigma X = 55$

11. a. A bar graph should be used for measurements from an ordinal scale
  b.

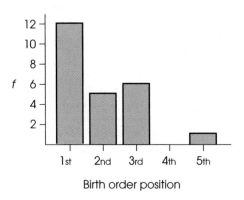

Birth order position

13. a.

X	f
9	2
8	5
7	7
6	3
5	1
4	1

  b. negatively skewed

15. a.

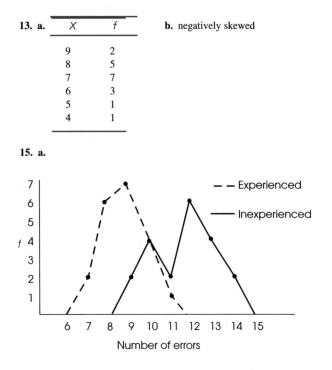

Number of errors

  b. Yes, it appears that the inexperienced rats tend to make more errors than the experienced rats do.

17. a.

Section I

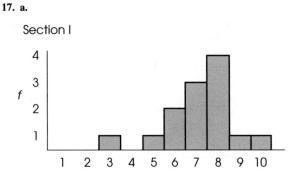

b. The scores in Section I are centered around $X = 8$ and form a negatively skewed distribution. In Section II, the scores are lower (centered around $X = 6$) and the distribution is more symmetrical with some tendency toward a positively skewed distribution.

Section II

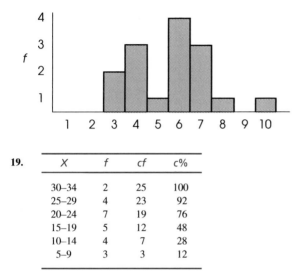

19.

X	f	cf	c%
30–34	2	25	100
25–29	4	23	92
20–24	7	19	76
15–19	5	12	48
10–14	4	7	28
5–9	3	3	12

a. The percentile rank for $X = 14.5$ is 28%.
b. The percentile rank for $X = 29.5$ is 92%
c. The 48th percentile is $X = 19.5$.
d. The 92nd percentile is $X = 29.5$.

21. a. The percentile rank for $X = 5$ is 16%.
b. The percentile rank for $X = 12$ is 83%.
c. The 30th percentile is $X = 6.5$.
d. The 72nd percentile is $X = 10.5$.

23. a. The 30th percentile is $X = 79$.
b. The 80th percentile is $X = 88.25$.
c. The percentile rank of $X = 60$ is 2.2%.
d. The percentile rank for $X = 78$ is 26%.

25. a. 4
b. 72, 71, 78, and 74
c. 2
d. 46 and 40

CHAPTER 3 CENTRAL TENDENCY

1. The purpose of central tendency is to identify a single score that serves as the best representative for an entire distribution, usually a score from the center of the distribution.

3. The mean is $\frac{27}{10} = 2.7$, the median is 2.5, and the mode is 2.

5. The mean is $\frac{40}{12} = 3.33$, the median is 3.50, and the mode is 4.

7. a. Median = 2.25
b. Median = 2

9. $\Sigma X = 48$

11. The original sample has $n = 8$ and $\Sigma X = 80$. The new sample has $n = 9$ and $\Sigma X = 81$. The new mean is $M = 9$.

13. The new mean is $\frac{198}{9} = 22$.

15. The new mean is $\frac{48}{6} = 8$. Changing $X = 14$ to $X = 2$ subtracts 12 points from ΣX.

17. The original sample has $n = 7$ and $\Sigma X = 35$. The new sample has $n = 8$ and $\Sigma X = 48$. The new score must be $X = 13$.

19. a. The new mean is $M = 6$.
b. The new mean is $(12 + 56)/10 = 6.8$.
c. The new mean is $(28 + 24)/10 = 5.2$.

21. The combined mean is $(56 + 36)/10 = 9.2$.

23. With a skewed distribution, the extreme scores in the tail can displace the mean. Specifically, the mean is displaced away from the main pile of scores and moved out toward the tail. The result is that the mean is often not a very representative value.

25. a. median (a few extreme scores would distort the mean)
b. mode (nominal scale)
c. mean

27. a. The histogram shows that 18 out of 20 participants had positive scores, indicating more pain before treatment than after treatment.
b. The mean difference in pain level is $M = 59/20 = 2.95$. On average, pain was greater by an average of 2.95 points before treatment.
c. Yes, the pain level was higher before treatment for nearly all the participants.

CHAPTER 4 **VARIABILITY**

1. a. The range is 7 points and the interquartile range is 2 points (from 2.5 to 4.5).
  b. The range is 27 points and the interquartile range is 2 points (from 2.5 to 4.5).
  c. The range is completely determined by the extreme scores. The interquartile range is not influenced by the most extreme scores.

3. The standard deviation, $\sigma = 20$ points, measures the standard distance between a score and the mean.

5. A standard deviation of zero indicates there is no variability. In this case, all of the scores in the sample have exactly the same value.

7. a. The range is 5 points, the interquartile range is 1 point (from 2.5 to 3.5), and the standard deviation is $\sqrt{1.25} = 1.12$.
  b. After adding 2 points to every score, the range is still 4, the interquartile range is still 1, and the standard deviation is still 1.12. Adding a constant to every score does not affect variability.

9. a. Median $= 3.5$. Semi interquartile range $= 2$ (Q1 $= 2.5$ and Q3 $= 6.5$)
  b. The median is still 3.5 and the semi-interquartile range is still 2 points.
  c. One extreme score does not influence the median or the semi-interquartile range.

11. a. With a standard deviation of 10 points, a score of $X = 38$ would not be considered extreme. It is within one standard deviation of the mean.
  b. With a standard deviation of only 2 points, a score of $X = 38$ is extreme. In this case, the score is located above the mean by a distance equal to four times the standard deviation.

13. a. The new mean is $\mu = 35$ and the standard deviation is still $\sigma = 5$.
  b. The new mean is $\mu = 90$ and the new standard deviation is $\sigma = 15$.

15. a. The definitional formula is easy to use when the mean is a whole number and there are relatively few scores.
  b. The computational formula is preferred when the mean is not a whole number.

17. a. Mean $= \frac{21}{7} = 3.0$

b. and c.

X	Deviation	Squared Deviation
3	0	0
2	−1	1
5	+2	4
0	−3	9
1	−2	4
2	−1	1
8	+5	25
	0	44 = SS

19. a.

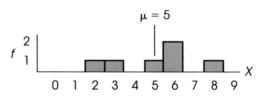

$\mu = 5$

b. The mean is $\frac{30}{6} = 5$. The score $X = 5$ is exactly equal to the mean. The scores $X = 2$ and $X = 8$ are farthest from the mean (3 points). The standard deviation should be between 0 and 3 points.
  c. $SS = 24$, $\sigma^2 = 4$, $\sigma = 2$

21. a.

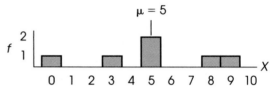

$\mu = 5$

b. The mean is $\mu = \frac{30}{6} = 5$ and the standard deviation appears to be about 3 points.
  c. $SS = 54$, $\sigma^2 = 9$, $\sigma = 3$

23. $SS = 80$, the sample variance is 16 and the standard deviation is 4.

25. $SS = 9$, the sample variance is 3, and the standard deviation is $= 1.73$.

27. $SS = 32$, the population variance is 4, and the standard deviation is 2.

29. a. For the younger woman, the variance is $s^2 = 0.786$. For the older woman, the variance is $s^2 = 1.696$.
  b. The variance for the younger woman is only half as large as for the older woman. The younger woman's scores are much more consistent.

CHAPTER 5   *z*-SCORES

1. A *z*-score describes a precise location within a distribution. The sign of the *z*-score tells whether the location is above ($+$) or below ($-$) the mean, and the magnitude tells the distance from the mean in terms of the number of standard deviations.

3. a. above the mean by 8 points
   b. above the mean by 2 points
   c. below the mean by 8 points
   d. below the mean by 2 points

5.

X	z	X	z	X	z
65	0.56	70	1.11	53	−0.78
48	−1.33	75	1.67	58	−0.22

7. a.

X	z	X	z	X	z
43	0.25	49	0.75	52	1.00
34	−0.50	28	−1.00	64	2.00

   b.

X	z	X	z	X	z
49	0.75	58	1.50	16	−2.00
37	−0.25	34	−0.50	55	1.25

9.

X	z	X	z	X	z
84	0.80	86	1.20	90	2.00
78	−0.40	77	−0.60	71	−1.80

11. a. $X = 49$
   b. $X = 70$
   c. $X = 86$
   d. $X = 92$

13. $\sigma = 8$

15. $M = 56$

17. $\sigma = 3$

19. $\mu = 60$ and $\sigma = 4$. The distance between the two scores is 10 points, which is equal to 2.5 standard deviations.

21. $\sigma = 2$

23. $X = 43$ corresponds to $z = 1.50$, and $X = 60$ corresponds to $z = 0.50$. $X = 43$ has a higher position in its distribution and should receive the higher grade.

25. a. $X = 80$ ($z = -1.00$)
   b. $X = 40$ ($z = -3.00$)
   c. $X = 140$ ($z = 2.00$)
   d. $X = 110$ ($z = 0.50$)

27. a. $\mu = 3$ and $\sigma = 2$
   b. and c.

Original X	z-score	Transformed X
6	1.00	60
1	−1.50	35
3	−0.50	45
4	0	50
7	1.50	65
3	−0.50	45

29. a.

X	z
10	+1.00
0	−1.50
4	−0.50
8	+0.50
4	−0.50
10	+1.00

   b. For the sample of $n = 6$ *z*-scores, $\Sigma z = 0$ and $M = 0$. Also for the *z*-scores, $SS = 5$, $s^2 = \frac{5}{5} = 1$, and $s = 1$.

CHAPTER 6   PROBABILITY

1. a. $p = \frac{30}{50} = 0.60$
   b. $p = \frac{20}{50} = 0.40$
   c. $p = \frac{10}{50} = 0.20$

3. The two requirements for a random sample are: (1) each individual has an equal chance of being selected, and (2) if more than one individual is selected, the probabilities must stay constant for all selections.

5. a. tail to the right, $p = 0.0668$
   b. tail to the left, $p = 0.0401$
   c. tail to the right, $p = 0.4207$
   d. tail to the left, $p = 0.3085$

7. a. $p(z > 1.00) = 0.1587$
   b. $p(z > 0.80) = 0.2119$
   c. $p(z < -0.25) = 0.4013$
   d. $p(z < -1.25) = 0.1056$

9. a. $z = 1.28$
   b. $z = 0.84$
   c. $z = -0.84$
   d. $z = -0.52$

11. a. $p = 0.3830$
   b. $p = 0.6826$

c. $p = 0.4332$
d. $p = 0.5586$

13. a. tail to the right, $p = 0.4013$
   b. tail to the left, $p = 0.3085$
   c. tail to the right, $p = 0.0668$
   d. tail to the left, $p = 0.1587$

15. a. $z = 2.00$, $p = .0228$
   b. $z = 0.50$, $p = 0.3085$
   c. $z = 1.28$, $X = 628$
   d. $z = -0.25$, $X = 475$

17. a. tail to the right, $p = 0.2266$
   b. tail to the right, $p = 0.4013$
   c. tail to the left, $p = 0.3085$
   d. ttail to the left, $p = 0.0668$

19. a. $z = 0.82$, $p = 0.2061$
   b. $z = -0.78$, $p = 0.2117$

21. a. $z = -2.00$, $p = 0.0228$
   b. Yes, only about 2% of regular rats perform as well on the maze. The rat with the supplement is smarter than 98% of regular rats.

c. Yes, the supplement seems to work. We either picked a very unusual rat or the supplement works.

23. a. $p = \frac{1}{2}$
   b. $\mu = 25$
   c. $\sigma = \sqrt{12.5}, = 3.54$ and for $X = 30.5$, $z = 1.55$, and $p = 0.0606$
   d. For $X = 29.5$, $z = 1.27$, and $p = 0.1020$

25. a. With five options, $p = \frac{1}{5}$ for each trial, $\mu = 20$ and $\sigma = 4$. For $X = 20$, $z = \pm0.13$ and $p = 0.1034$.
   b. For $X = 30.5$, $z = 2.63$, and $p = 0.0043$.
   c. $\mu = 40$ and $\sigma = 5.66$, for $X = 49.5$, $z = 1.68$ and $p = 0.0465$

27. a. With $n = 50$ and $p = q = \frac{1}{2}$, you may use the normal approximation with $\mu = 25$ and $\sigma = 3.54$. Using the upper real limit of 30.5, $p(X > 30.5) = p(z > 1.55) = 0.0606$.
   b. The normal approximation has $\mu = 50$ and $\sigma = 5$. Using the upper real limit of 60.5, $p(X > 60.5) = p(z > 2.10) = 0.0179$.
   c. Getting 60% heads with a balanced coin is an unusual event for a large sample. Although you might get 60% heads with a small sample, you should get very close to a 50-50 distribution as the sample gets larger. With a larger sample, it becomes very unlikely to get 60% heads.

CHAPTER 7 THE DISTRIBUTION OF SAMPLE MEANS

1. a. The distribution of sample means consists of the sample means for all the possible random samples of a specific size (n) from a specific population.
   b. The expected value of M is the mean of the distribution of sample means (μ).
   c. The standard error of M is the standard deviation of the distribution of sample means ($\sigma_M = \sigma/\sqrt{n}$).

3. a. $\mu = 80$, and $\sigma_M = 20/\sqrt{4} = 10$.
   b. $\mu = 80$, and $\sigma_M = 20/\sqrt{16} = 5$.

5. a. Standard error $= 20/\sqrt{4} = 10$ points
   b. Standard error $= 20/\sqrt{16} = 5$ points
   c. Standard error $= 20/\sqrt{100} = 2$ points

7. a. $n \geq 4$
   b. $n \geq 25$
   c. $n \geq 100$

9. a. $\sigma = 40$
   b. $\sigma = 20$
   c. $\sigma = 8$

11. a. $\sigma_M = 2$ points and $z = 2.00$
   b. $\sigma_M = 4$ points and $z = 1.00$
   c. $\sigma_M = 8$ points and $z = 0.50$

13. a. $z = 1.00$. This is a typical sample.
   b. $z = 1.00$. This is a typical sample.
   c. $z = 2.00$. This is an extreme sample.

15. a. $z = 0.25$ and $p = 0.5987$
   b. $\sigma_M = 10$, $z = 0.50$ $p = 0.6915$
   c. $\sigma_M = 5$, $z = 1.00$ $p = 0.8413$

17. a. $z = \pm0.50$ and $p = 0.3830$
   b. $\sigma_M = 5$, $z = \pm1.00$ and $p = 0.6826$
   c. $\sigma_M = 2.5$, $z = \pm2.00$ and $p = 0.9544$

19. a. $z = \pm0.50$ and $p = 0.3830$
   b. $z = \pm1.00$ and $p = 0.6826$
   c. $z = \pm1.50$ and $p = 0.8664$

21. a. With $n = 36$, the standard error is 2 and the sample mean corresponds to a z-score of $z = -0.95$. This is a reasonable outcome for a z-score. There is nothing unusual about the sample.
   b. With $n = 16$, the standard error is 3 and the sample mean corresponds to a z-score of $z = -2.73$. This is an extreme value. This sample is very unlikely to have occurred just by chance.

23. a.

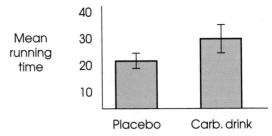

   b. Even considering the standard error for each mean, there is no overlap between the two groups. The carbohydrate-electrolyte does seem to have increased endurance, even allowing for the standard error.

CHAPTER 8   INTRODUCTION TO HYPOTHESIS TESTING

1. a. $M - \mu$ measures the difference between the sample mean and the hypothesized population mean.

   b. A sample mean is not expected to be identical to the population mean. The standard error measures how much difference, on average, is reasonable expect between M and μ.

3. The alpha level is a small probability value that defines the concept of *very unlikely*. The critical region consists of outcomes that are very unlikely to occur if the null hypothesis is true, where *very unlikely* is defined by the alpha level.

5. a. The null hypothesis states that the herb has no effect on memory scores.

   b. H_0: $\mu = 80$ (even with the herbs, the mean is still 80). H_1: $\mu \neq 80$ (the mean has changed)

   c. The critical region consists of z-scores beyond ± 1.96. For these data, the standard error is 3 and $z = \frac{4}{3} = 1.33$. Fail to reject the null hypothesis. The herbal supplements do not have a significant effect on memory scores.

7. a. With a standard error of 2.5, a sample mean of $M = 54$ produces $z = 1.60$. This is not beyond the critical value of 1.96, so we conclude that there is not a significant effect.

   b. With $M = 56$, $z = 2.40$ and we conclude that there is a significant effect.

9. a. H_0 $\mu = 40$. The sample mean $M = 42$ corresponds to $z = 1.00$. This is not sufficient to reject the null hypothesis. You cannot conclude that the program has a significant effect.

   b. H_0 $\mu = 40$. The sample mean $M = 44$ corresponds to $z = 2.00$. This is sufficient to reject the null hypothesis and conclude that the program does have a significant effect.

   c. The 2-point difference between the sample and the hypothesis in part a was not sufficient to reject H_0, however, the 4-point difference in part b is significantly more than chance.

11. a. With $n = 4$, the standard error is 6, and $z = \frac{6}{6} = 1.00$. Fail to reject H_0.

   b. With $n = 36$, the standard error is 2, and $z = \frac{6}{2} = 3.00$. Reject H_0.

   c. Cohen's $d = \frac{6}{12} = 0.50$ for both tests.

   d. A larger sample increases the likelihood of rejecting H_0 but has no influence on Cohen's d.

13. a. The null hypothesis states that there is no increase in REM activity, $\mu \leq 110$. The critical region consists of z-scores beyond $z = 2.33$. For these data, the standard error is 12.5 and $z = 33/12.5 = 2.64$. Reject H_0. There is a significant increase in REM activity.

   b. Cohen's $d = 33/50 = 0.66$.

15. a. H_0: $\mu = 50$. The critical region consists of z-scores beyond $z = \pm 1.96$. For these data, $\sigma_M = 2.74$ and $z = 2.92$. Reject H_0 and conclude that only children are significantly different.

17. H_0: $\mu \leq 100$ (performance is not increased). The critical region consists of z-scores beyond $z = +2.33$. For these data, $\sigma_M = 3$ and $z = 1.33$. Fail to reject H_0 and conclude that performance is not significantly higher with the easy questions added.

19. H_0: $\mu \leq 12$ (no increase during hot weather). H_1: $\mu > 12$ (there is an increase). The critical region consists of z-score values greater than $+1.65$. For these data, the standard error is 1.50, and $z = 2.33$, which is in the critical region so we reject the null hypothesis and conclude that there is a significant increase in the mean number of hit players during hot weather.

21. a. The z-score increases (farther from zero).

   b. Cohen's d is not influenced by sample size.

   c. Power increases.

23. a. The critical boundary, $z = 1.96$, corresponds to $M = 167.84$. With a 5-point effect, this mean is located at $z = 0.71$ and the power is 0.2389 or 23.89%.

   b. With a 10-point effect, $M = 167.84$ is located at $z = -0.54$ and the power is 0.7054 or 70.54%.

25. a. With $n = 16$, the standard error is 2 points and the critical boundary, $z = 2.33$, corresponds to a sample mean of $M = 44.66$. In the treated distribution, this sample mean corresponds to $z = -0.17$ and power $= 0.5675$ or 56.75%.

   b. With $n = 64$, the standard error is 1 point and the critical boundary, $z = 2.33$, corresponds to a sample mean of $M = 42.33$. In the treated distribution, this sample mean corresponds to $z = -2.67$ and power $= 0.9962$ or 99.62%.

CHAPTER 9   INTRODUCTION TO THE *t* STATISTIC

1. A z-score is used when the population standard deviation (or variance) is known. The t statistic is used when the population variance or standard deviation is unknown. When the population values are unknown, you use the sample data to estimate the variance and the standard error.

3. a. The sample variance is 144 and the estimated standard error is 4.

   b. The sample variance is 100 and the estimated standard error is 2.

   c. The sample variance is 36 and the estimated standard error is 1.

5. As the value of df increases, the t distribution becomes less variable (less spread out) and more like a normal-shaped distribution. For a two-tailed test with $\alpha = .05$, the critical t values decrease in magnitude and move toward ± 1.96 as df increases.

7. a. $M = 2$ and $s = 2$

   b. $s_M = 1$

9. a. With $n = 4$, $s_M = 4$ and $t = \frac{6}{4} = 1.50$. This is not greater than the critical value of 3.182, so there is no significant effect.

   b. With $n = 16$, $s_M = 2$ and $t = \frac{6}{2} = 3.00$. This value is greater than the critical value of 2.131, so we reject the null hypothesis and conclude that there is a significant treatment effect.

11. a. 6 points

 b. The sample variance is 64 and the estimated standard error is $s_M = 2$.

 c. For these data, $t = 3.00$. With $df = 15$ the critical value is $t = \pm2.131$. Reject H_0 and conclude that there is a significant effect.

13. With $df = 15$, the one-tailed critical value is 2.602. For these data, the sample variance is 400, the estimated standard error is 5, and $t = \frac{21}{5} = 4.20$. Reject the null hypothesis and conclude that the sample mean is significantly greater than would be expected from a population with $\mu = 100$.

15. The null hypothesis is H_0: $\mu \le 20$ years. With $df = 29$, the one-tailed critical value is 2.462. For these data, the estimated standard error is 1.5, and $t = \frac{3.8}{1.5} = 2.53$. Reject the null hypothesis and conclude that the average age of the library books is significantly greater than 20 years.

17. The estimated standard error is 1.50, and $t = \frac{7.7}{1.50} = 5.13$. For a one-tailed test, the critical value is 2.602. Reject the null hypothesis, children with a history of day care have significantly more behavioral problems.

19. a. The standard deviation is $\sqrt{400} = 20$, and Cohen's $d = \frac{4}{20} = 0.20$. With $n = 16$, the estimated standard error is 5 and $t = \frac{4}{5} = 0.80$. $r^2 = \frac{0.64}{15.64} = 0.041$.

 b. Cohen's $d = \frac{4}{20} = 0.20$. With $n = 25$, the estimated standard error is 4 and $t = \frac{4}{4} = 1.00$. $r^2 = 1.00/25.00 = 0.04$.

 c. The sample size does not have any influence on Cohen's d and has only a minor effect on r^2.

21. a. The estimated standard error is 0.20 and $t = 2.2/0.2 = 11.00$. The t value is well beyond the critical value of 2.492. Reject the null hypothesis.

 b. Cohen's $d = 2.2/1 = 2.20$ and $r^2 = 121/145 = 0.8345$

23. H_0: $\mu = 40$. With $df = 8$ the critical values are $t = \pm2.306$. For these data, $M = 44$, $SS = 162$, $s^2 = 20.25$, the standard error is 1.50, and $t = 2.67$. Reject H_0 and conclude that depression for the elderly is significantly different from depression for the general population.

25. H_0: $\mu \le 20$. With $df = 8$, the critical region consists of t values greater than 1.860. For these data, $M = 23$, $s^2 = 9$, the standard error is 1, and $t(8) = 3.00$. Reject H_0 and conclude that participation in the interview significantly increases life satisfaction.

CHAPTER 10   THE *t* TEST FOR TWO INDEPENDENT SAMPLES

1. An independent-measures study requires a separate sample for each of the treatments or populations being compared.

3. The size of the two samples influences the magnitude of the estimated standard error in the denominator of the t statistic. As sample size increases, the value of t also increases (moves farther from zero), and the likelihood of rejecting H_0 also increases.

5. The homogeneity of variance assumption specifies that the variances are equal for the two populations from which the samples are obtained. If this assumption is violated, the t statistic can cause misleading conclusions for a hypothesis test.

7. a. The first sample has a variance of 7, the second sample variance is 9, and the pooled variance is 8 (halfway between).

 b. The first sample has a variance of 7, the second sample variance is 3, and the pooled variance is $\frac{80}{20} = 4$ (closer to the variance for the larger sample).

9. a. The pooled variance is 120.

 b. The estimated standard error is 4.00.

 c. A mean difference of 8 would produce $t = \frac{8}{4} = 2$. With $df = 28$ the critical values are ±2.048. Fail to reject H_0.

 d. A mean difference of 12 would produce $t = \frac{12}{4} = 3$. With $df = 28$ the critical values are ±2.048. Reject H_0.

 e. With a mean difference of 8 points, $r^2 = 0.125$. With a difference of 12 points, $r^2 = 0.24$.

11. a. The estimated standard error for the sample mean difference is 4 points.

 b. The estimated standard error for the sample mean difference is 2 points.

 c. Larger samples produce a smaller standard error.

13. The pooled variance is 24, the estimated standard error is 2, and $t = 1.50$. With $df = 22$, the critical boundaries are $t = \pm2.074$. Fail to reject H_0. Larger variance reduces the likelihood of rejecting H_0.

15. a. The pooled variance is 1.2, the estimated standard error is 0.4, and $t = 0.9/0.4 = 2.25$. With $df = 28$ the critical value is 2.048. Reject the null hypothesis and conclude that there is a significant difference between group performance and individual performance.

 b. The estimated Cohen's $d = 0.9/\sqrt{1.2} = 0.822$.

17. The pooled variance is 32, the estimated standard error is 2, and $t = \frac{6}{2} = 3.00$. With $df = 30$ the critical value is 2.042. Reject the null hypothesis. There is a significant difference between the two treatments. Larger samples increase the likelihood of rejecting H_0.

19. a. The pooled variance is 90, the estimated standard error is 3, and $t = \frac{11}{3} = 3.67$. Because $df = 38$ is not listed in the table, use $df = 30$ and obtain critical boundaries of ±2.750. Reject the null hypothesis. There is a significant difference between the boys and girls.

 b. Cohen's $d = 11/\sqrt{90} = 1.16$ and $r^2 = 0.26$.

21. The pooled variance is 64, the estimated standard error is 4.00, and $t = \frac{11}{4} = 2.75$. With $df = 14$ the critical value is 2.145. Reject the null hypothesis and conclude that there is a significant difference between the two sleep conditions.

23. a. The null hypothesis states that owning a pet does not reduce the number of doctor visits. For a one tailed test, the critical boundary is $t = 1.796$. For the control group, $M = 9$ and $SS = 44$. For the dog owners, $M = 6$ and $SS = 24$. The pooled

variance is 6.18 and $t(11) = 2.11$. Reject H_0. The data show that the dog owners have significantly fewer doctor visits.
   b. For these data, $r^2 = 0.288$ (28.8%).

25. **a.** The null hypothesis states that there is no difference between the two sets of instructions, $H_0: \mu_1 - \mu_2 = 0$. With $df = 6$ and $\alpha = .05$, the critical region consists of t values beyond ± 2.447. For the first set, $M = 6$ and $SS = 16$. For the second set, $M = 10$ with $SS = 32$. For these data, the pooled variance

is 8, the estimated standard error is 2, and $t(6) = 2.00$. Fail to reject H_0. The data are not sufficient to conclude that there is a significant difference between the two sets of instructions.
   b. For these data, the estimated $d = 4/\sqrt{8} = 1.41$ (a very large effect) and $r^2 = \frac{4}{10} = 0.40$ (40%)

CHAPTER 11 **THE *t* TEST FOR TWO RELATED SAMPLES**

1. **a.** This is an independent-measures experiment with two separate samples.
   b. This is a repeated-measures study. The same individuals are measured twice.
   c. This is a repeated-measures study. The same individuals are measured twice.

3. For a repeated-measures design the same subjects are used in both treatment conditions. In a matched-subjects design, two different sets of subjects are used. However, in a matched-subjects design, each subject in one condition is matched with respect to a specific variable with a subject in the second condition so that the two separate samples are equivalent with respect to the matching variable.

5. **a.** The standard deviation is 8 points and measures the average distance between the individual scores and the sample mean.
   b. The estimated standard error is 2 points and measures the average distance between a sample mean and the population mean.

7. **a.** The estimated standard error is 2 points and $t(8) = 1.50$. With a critical boundary of ± 2.306, fail to reject the null hypothesis.
   b. With $M_D = 12$, $t(8) = 6.00$. With a critical boundary of ± 2.306, reject the null hypothesis.
   c. The larger the mean difference, the greater the likelihood of finding a significant difference.

9. The sample variance is 9, the estimated standard error is 0.75, and $t(15) = 4.33$. With critical boundaries of ± 2.131, reject H_0

11. **a.** The null hypothesis says that the judgments for smiling are not funnier than for frowning. For these data, the sample variance is 64, the estimated standard error is 2, and $t = 3.6/2 = 1.80$. For a one-tailed test with $df = 15$, the critical value is 1.753. Reject the null hypothesis.
   b. $r^2 = 3.24/18.24 = 0.178$ (17.8%)

13. The null hypothesis states that there is no difference in the perceived intelligence between attractive and unattractive photos. For these data, the estimated standard error is 0.4 and $t = 2.7/0.4 = 6.75$. With $df = 24$, the critical value is 2.064. Reject the null hypothesis.

15. **a.** The difference scores are 3, 7, 3, and 3. $M_D = 4$.
   b. $SS = 12$, sample variance is 4, and the estimated standard error is 1.
   c. With $df = 3$ and $\alpha = .05$, the critical values are $t = \pm 3.182$. For these data, $t = 4.00$. Reject H_0. There is a significant treatment effect.

17. The null hypothesis states that the images have no effect on performance. For these data, the sample variance is 12.6, the estimated standard error is 1.45, and $t(5) = 2.97$. With $df = 5$ and $\alpha = .05$, the critical values are $t = \pm 2.571$. Reject the null hypothesis, the images have a significant effect.

19. **a.** The pooled variance is 6.4 and the estimated standard error is 1.46.
   b. For the difference scores the variance is 24 and the estimated standard error is 2.

21. The null hypothesis says that the duration of the time out has no effect, $H_0: \mu_D = 0$. With $df = 8$ and $\alpha = .05$, the critical values are $t = \pm 2.306$. For these data, $M_D = 7$, $SS = 288$, the standard error is 2, and $t(8) = 3.50$. Reject H_0 and conclude that the length of the time out has a significant effect on accuracy.

23. The null hypothesis says that fatigue has no effect, $H_0:$ $\mu_D = 0$. With $df = 11$ and $\alpha = .05$, the critical values are $t = \pm 2.201$. For these data, $M_D = -5.17$, $SS = 105.67$, the standard error is 0.89, and $t(11) = -5.81$. Reject H_0 and conclude that fatigue has a significant effect on perceptual speed performance.

CHAPTER 12 **ESTIMATION**

1. The general purpose of a hypothesis test is to determine whether a treatment effect exists. A hypothesis test always addresses a yes/no question. The purpose of estimation is to determine the size of the effect. Estimation addresses a how-much question.

3. A narrower interval provides a more precise estimate than a wider interval. As confidence is increased, the interval becomes wider and, therefore, less precise.

5. a. Hypothesis test (is there any difference?), repeated measures

b. Estimation (how much effect?), independent measures

c. Hypothesis test, independent measures

d. Estimation, repeated measures

7. a. Using a single-sample t statistic with $df = 8$, the standard error is 2, the boundaries for 80% confidence are $t = \pm 1.397$, and the interval extends from 43.206 to 48.794.

b. For 90% confidence the boundaries are $t = \pm 1.860$, and the interval extends from 42.28 to 49.72

c. For 95% confidence, the boundaries are $t = \pm 2.306$, and the interval extends from 41.388 to 50.612.

d. As the percentage of confidence increases, the width of the interval also increases.

9. a. For the point estimate, use the sample mean $M = 87$. The daycare children score 6 points higher than the overall population mean of $\mu = 81$.

b. For the interval estimate the standard error is 1.6 and with $df = 24$ the boundaries for 95% confidence are $t = \pm 2.064$. The interval extends from 83.698 to 90.302.

11. a. $M = 7.00$ and $s^2 = 9$

b. Use $M = 7.00$ as the point estimate for μ.

c. With $df = 8$ and an estimated standard error of 1.00, the boundaries for 80% confidence are $t = \pm 1.397$, and the interval extends from 5.603 and 8.397.

13. Use the sample mean difference, $M_1 - M_2 = 8$, as the point estimate of the population mean difference. For the confidence interval, the pooled variance is 20 and the estimated standard error is 2 points. With $df = 18$, the boundaries for 90% confidence are $t = \pm 1.734$, and the interval extends from 4.532 to 11.468.

15. Use the sample mean difference, 11.2 points, as the point estimate of the population mean difference. With $df = 28$, the t values for

the 90% interval are ± 1.701. The pooled variance is 187.5, the standard error is 5 points, and the interval extends from 2.695 to 19.705 points.

17. a. Use the sample mean difference, $M_1 - M_2 = 14$ points for the point estimate.

b. For the interval estimate, the pooled variance is 90 and the standard error is 5. With $df = 13$ the boundaries for 90% confidence are $t = \pm 1.771$. The interval extends from 5.145 to 22.855.

19. a. For the control group, $M = 9$ and $SS = 44$. For the dog owners, $M = 6$ and $SS = 24$. Use the sample mean difference, $M_1 - M_2 = 3$ points for the point estimate.

b. The pooled variance is 6.18 and the estimated standard error is 1.42. With $df = 11$, the boundaries for 90% confidence are $t = \pm 1.796$. The interval extends from 0.45 to 5.55.

21. a. Use the sample mean difference, $M_D = 6$, for the point estimate.

b. Using a repeated-measures t statistic with $df = 8$, the estimated standard error is 1.73, the boundaries for 95% confidence are $t = \pm 2.306$, and the interval extends from 2.011 to 9.989.

23. Use the sample mean difference, $M_D = 21$, as the point estimate. For the interval estimate, the sample variance is 81 and the estimated standard error is 2.25. With $df = 15$, the boundaries for 80% confidence are $t = \pm 1.341$. The interval extends from 17.98 to 24.02.

25. For these data, $M_D = 3$, $SS = 36$, $s^2 = 5.14$, the standard error is 0.80. Use the sample mean difference, $M_D = 3$, as the point estimate. With $df = 7$, the boundaries for 80% confidence are $t = \pm 1.415$. The interval extends from 1.87 to 4.13.

CHAPTER 13 INTRODUCTION TO ANALYSIS OF VARIANCE

1. When there is no treatment effect, the numerator and the denominator of the F-ratio are both measuring the same sources of variability (random, unsystematic differences from sampling error). In this case, the F-ratio is balanced and should have a value near 1.00.

3. a. As the differences between sample means increase, $MS_{between}$ also increases, and the F-ratio increases.

b. Increases in sample variability cause MS_{within} to increase and, thereby, decrease the F-ratio.

5. a. Post tests are used to determine exactly which treatment conditions are significantly different.

b. If there are only two treatments, then there is no questions as to which two treatments are different.

c. If the decision is to fail to reject H_0, then there are no significant differences.

7. a. The three means produce $SS = 14$.

b. $n(SS_{means}) = 140$

c. $SS_{between} = 140$

9. a. Reducing the differences between treatments should reduce the value of $MS_{between}$ and should lower the value of the F-ratio.

b.

Source	SS	df	MS	
Between treatments	10	2	5	$F(2, 12) = 2.50$
Within treatments	24	12	2	
Total	34	11		

With $\alpha = .05$, the critical value is $F = 3.88$. Fail to reject the null hypothesis and conclude that there are no significant differences among the three treatments. Notice that the F-ratio is substantially smaller than it was in problem 8.

11. a.

Source	SS	df	MS	
Between treatments	72	2	36	$F(2, 9) = 9.00$
Within treatments	36	9	4	
Total	108	11		

With $\alpha = .05$, the critical value is $F = 4.26$. Reject the null hypothesis and conclude that there are significant differences among the three treatments. Increasing the mean differences between treatments produced a larger F-ratio.

b. For these data, eta squared is $\frac{72}{108} = 0.667$ or 66.7%. A larger difference between treatments produces a larger effect size.

13. a. Increasing the variance within each treatment will increase $MS_{\text{within treatments}}$ and will lower the value of the F-ratio.

b.

Source	SS	df	MS	
Between treatments	32	2	16	$F(2, 9) = 4.00$
Within treatments	36	9	4	
Total	68	11		

With $\alpha = .05$, the critical value is $F = 4.26$. Fail to reject the null hypothesis and conclude that there are not significant differences among the three treatments. As predicted, the F-ratio is smaller than in problem 12.

15. a. With $df_{\text{between}} = 3$, there must be 4 treatments.
b. Adding the two df values (3 and 36) gives $df_{\text{total}} = 39$. Therefore, the total number of subjects must be $N = 40$.

17. $SS_{\text{total}} = 40$ (same as SS_{between} from problem 16)

19.

Source	SS	df	MS	
Between treatments	72	2	36	$F(2, 21) = 7.20$
Within treatments	105	21	5	
Total	177	23		

With $\alpha = .05$, the critical value is $F = 3.47$. Reject the null hypothesis and conclude that there are significant differences among the three types of teachers.

21.

Source	SS	df	MS	
Between treatments	30	2	15	$F = 7.50$
Within treatments	54	27	2	
Total	84	29		

23.

Source	SS	df	MS	
Between treatments	23	2	11.5	$F(2, 11) = 3.83$
Within treatments	33	11	3	
Total	56	13		

With $df = 2, 11$, the critical value for $\alpha = .05$ is 3.98. Fail to reject the null hypothesis.

25. a. The pooled variance is 2, the estimated standard error is 0.89 and $t(8) = 4.49$. With $df = 8$, the critical value is 2.306. Reject the null hypothesis.

b.

Source	SS	df	MS	
Between treatments	40	1	40	$F(1, 8) = 20$
Within treatments	16	8	2	
Total	56	9		

With $\alpha = 1, 8$, the critical value is $F = 5.32$. Reject the null hypothesis. Within rounding error, note that $F = t^2$.

27. a. The means and SS values are:

Single	Twin	Triplet
$M = 8$	$M = 6$	$M = 5$
$SS = 10$	$SS = 18$	$SS = 14$

Source	SS	df	MS	
Between treatments	28	2	14	$F(2, 15) = 5.00$
Within treatments	42	15	2.8	
Total	70	17		

With $df = 2, 15$ the critical value is 3.68. Reject the null hypothesis.
b. For single versus twin, the Scheffé $F = 6/2.8 = 2.14$ (not significant). For single versus triplet, the Scheffé $F = 13.5/2.8 = 4.82$ (significant). For twin versus triplet, the Scheffé $F = 1.5/2.8 = 0.54$ (not significant).

CHAPTER 14 REPEATED-MEASURES ANALYSIS OF VARIANCE

1. For an independent measures design, the variability within treatments is the appropriate error term. For repeated measures, however, you must subtract out variability due to individual differences from the variability within treatments to obtain a measure of error.

3. a. A total of 60 participants is needed; three separate samples, each with $n = 20$. The F-ratio has $df = 2, 57$.

b. One sample of $n = 20$ is needed. The F-ratio has $df = 2, 38$.

5. a. 4 treatments
b. 13 participants

7.

Source	SS	df	MS	
Between treatments	28	2	14	$F(2, 10) = 7.78$
Within treatments	28	15		
Between subjects	10	5		
Error	18	10	1.8	
Total	56	17		

With $df = 2, 10$, the critical value is 4.10. Reject H_0. There are significant differences among the three treatments.

9. a. $SS_{total} = 100$ and $SS_{between\ treatments} = 64$.

   b. For the modified scores, $\Sigma X^2 = 436$ and both treatments have $T = 40$.

   c. For the modified scores, $SS_{total} = 36$.

   d. The difference between the original SS_{total} and the modified SS_{total} is 64 points.

11. a. $SS_{total} = 96$ and $SS_{between\ subjects} = 18$.

   b. For the modified scores, all five participants have $P = 9$ and $\Sigma X^2 = 213$.

   c. For the modified scores, $SS_{total} = 78$.

   d. The difference between the original SS_{total} and the modified SS_{total} is 18 points.

13. a. The null hypothesis states that there are no differences among the three treatments, $H_0: \mu_1 = \mu_2 = \mu_3$. With $df = 2, 6$, the critical value is 5.14.

Source	SS	df	MS	
Between treatments	8	2	4	$F(2, 6) = 0.27$
Within treatments	94	9		
Between subjects	4	3		
Error	90	6	15	
Total	102	11		

Fail to reject H_0. There are no significant differences among the three treatments.

   b. In problem 12 the individual differences were relatively large and consistent. When the individual differences were subtracted out, the error was greatly reduced.

15.

Source	SS	df	MS	
Between treatments	56	2	28	$F(2, 6) = 10.49$
Within treatments	28	9		
Between subjects	12	3		
Error	16	6	2.67	
Total	84	11		

Reject H_0 and conclude that quality of life changes significantly over the three measurements.

17.

Source	SS	df	MS	
Between treatments	60	3	20	$F(1, 14) = 5.00$
Within treatments	158	36		
Between subjects	50	9		
Error	108	27	4	
Total	218	39		

19. a. $SS_{within\ treatments} = 36$

   b. $SS_{within\ subjects} = 36$

21. a. The null hypothesis states that there is no difference between the two treatments, $H_0: \mu_D = 0$. The critical region consists of t values beyond ± 3.182. The mean difference is $M_D = 4.5$. The variance for the difference scores is 9, the estimated standard error is 1.50, and $t(3) = 3.00$. Fail to reject H_0.

   b. The null hypothesis states that there is no difference between treatments, $H_0: \mu_1 = \mu_2$. The critical value is $F = 10.13$.

Source	SS	df	MS	
Between treatments	40.5	1	40.5	$F(1, 3) = 9.00$
Within treatments	31	6		
Between subjects	17.5	3		
Error	13.5	3	4.5	
Total	71.5	7		

Fail to reject H_0. Note that $F = t^2$.

23.

Source	SS	df	MS	
Between treatments	384	2	192	$F(2, 6) = 96.00$
Within treatments	24	9		
Between subjects	12	3		
Error	12	6	2	
Total	408	11		

Reject H_0 and conclude that heart rate changes significantly over the 2-minute period.

25. The means and SS values are as follows:

Days Before Giving Birth				
9	7	5	3	1
$M = 36$	$M = 39$	$M = 42$	$M = 46$	$M = 55$
$SS = 20$	$SS = 10$	$SS = 22$	$SS = 18$	$SS = 14$

The ANOVA produces:

Source	SS	df	MS	
Between treatments	1086	4	271.5	$F(4, 16) = 120.67$
Within treatments	84	20		
Between subjects	48	4		
Error	36	16	2.25	
Total	1170	14		

Reject the null hypothesis. There are significant differences in pain threshold.

CHAPTER 15 TWO-FACTOR ANALYSIS OF VARIANCE

1. a. In ANOVA, an independent variable (or a quasi-independent variable) is called a *factor*.
  b. The values of a factor that are used to create the different groups or treatment conditions are called the *levels* of the factor.
  c. A research study with two independent (or quasi-independent) variables is called a *two-factor study*.

3. During the second stage of the two-factor ANOVA the mean differences between treatments are analyzed into differences from each of the two main effects and differences from the interaction.

5. a. The main effect for treatment is the 1-point difference between the overall column means, $M = 7$ and $M = 8$.
  b. The main effect for gender is the 7-point difference between the overall row means, $M = 4$ and $M = 11$.
  c. There is an interaction. The effect of the treatment depends on gender. The treatment decreases scores by an average of 2 points for the males and increases scores by an average of 4 points for the females.

7. a. $M = 20$
  b. $M = 50$

9. a. There is a main effect for factor A, but no main effect for factor B, and no interaction.
  b. There is a main effect for factor A, a main effect for factor B, but no interaction.
  c. There is a main effect for factor A, a main effect for factor B, and an interaction.

11. a. 2
  b. 3
  c. 2, 42

13. a.

Source	SS	df	MS	
Between treatments	400	5		
A	120	1	120	$F(1, 24) = 15.00$
B	260	2	130	$F(1, 24) = 16.25$
A × B	20	2	10	$F(1, 24) = 1.25$
Within treatments	192	24	8	
Total	592	39		

The F-ratio for factor A has $df = 1, 24$ and the critical value is $F = 4.26$. Factor B and the interaction have $df = 2, 24$ and the critical value is 3.40. The main effect is significant for both factors but not for the interaction.
  b. For level B_1, $F = 10/8 = 1.25$, for B_2, $F = 40/8 = 5.00$, and for B_3, $F = 90/8 = 11.25$. For all tests, $df = 1, 24$ and the critical value is 4.26. There are significant differences at levels B_2 and B_3 but not at B_1.

15. The null hypotheses state that there is no difference between levels of factor A (H_0: $\mu_{A1} = \mu_{A2}$), no difference between levels of factor B (H_0: $\mu_{B1} = \mu_{B2}$), and no interaction. All F-ratios have $df = 1,28$ and the critical value is $F = 4.20$.

Source	SS	df	MS	
Between treatments	148	3		
A (safety)	98	1	98	$F(1, 28) = 13.52$
B (skill)	32	1	32	$F(1, 28) = 4.41$
A × B	18	1	18	$F(1, 28) = 2.48$
Within treatments	203	28	7.25	
Total	351	39		

There is no significant interaction, but the A effect and the B effect are significant at the .05 level.

17.

Source	SS	df	MS	
Between treatments	144	8		
Achievement need	36	2	18	$F(2, 72) = 6.00$
Task difficulty	24	2	12	$F(2, 72) = 4.00$
Interaction	84	4	21	$F(4, 72) = 7.00$
Within treatments	216	72	3	
Total	360	80		

19. a.

Source	SS	df	MS	
Between treatments	128	3		
Gender	0	1	0	$F(1, 12) = 0$
Treatments	64	1	64	$F(1, 12) = 7.53$
Gender × treatment	64	1	64	$F(1, 12) = 7.53$
Within treatments	102	12	8.5	
Total	230	15		

With $df = 1, 12$, the critical value for all three F-ratios is 4.75. The main effect for the treatments and the interaction are significant, but there is no gender difference.
  b. The main effect for treatments and the interaction both have $\eta^2 = \frac{64}{166} = 0.3855$ and $\eta^2 = 0$ for gender.

21. a.

Source	SS	df	MS	
Between treatments	360	5		
Gender	72	1	72	$F(1, 12) = 9.00$
Treatments	252	2	126	$F(1, 12) = 15.75$
Gender × treatment	36	2	18	$F(1, 12) = 2.25$
Within treatments	96	12	8	
Total	456	17		

With $df = 1, 12$, the critical value for the gender main effect is 4.75. The main effect for gender is significant. With $df = 2, 12$, the critical value for the treatment main effect and the interaction is 3.88. The main effect for treatments is significant but the interaction is not.
  b. For treatment I, $F = 0$; for treatment II, $F = \frac{54}{8} = 6.75$; and for treatment III, $F = \frac{54}{8} = 6.75$. With $df = 1, 12$, the critical

value for all three tests is 4.75. The results indicate a significant difference between males and females in treatments II and III, but not in treatment I.

23. a. The null hypotheses state that counseling has no effect (H_0: $\mu_{YES} = \mu_{NO}$), the drug has no effect (H_0: $\mu_{YES} = \mu_{NO}$), and there is no interaction. All F-ratios have $df = 1,36$ and the critical value is $F = 4.11$.

Source	SS	df	MS	
Between treatments	200	3		
A (counseling)	0	1	0	$F(1, 36) = 0$
B (drug)	160	1	160	$F(1, 36) = 32.00$
A × B	40	1	40	$F(1, 36) = 8.00$
Within treatments	180	36	5	
Total	380	39		

b. The data indicate that overall the drug has a significant effect. The main effect for counseling is not significant. However, there is a significant interaction meaning that the effect of the drug is significantly affected by counseling. Without counseling the drug has a minimal effect, but with counseling the drug has a big effect.

25. The null hypotheses state that gender has no effect (H_0: $\mu_{Male} = \mu_{Female}$), the drug dosage has no effect

(H_0: $\mu_{None} = \mu_{Small} = \mu_{Large}$), and that there is no interaction. For $df = 1,24$, the critical value is $F = 4.26$ and for $df = 2,24$, the critical value is $F = 3.40$.

Source	SS	df	MS	
Between treatments	130	5		
A	30	1	30	$F(1, 24) = 6$
B	20	2	10	$F(2, 24) = 2$
A × B	80	2	40	$F(2, 24) = 8$
Within treatments	120	24	5	
Total	250	29		

The significant interaction indicates that the hormone affects the eating behavior of males differently than females. For males, small doses appear to increase eating relative to no drug or a large dose. The test of simple main effects indicates that the drug effect is significant for males, $F(2,24) = 7.00$, $p < .05$. For females, the drug (in any dose) appears to inhibit eating, but a test of simple mean effect indicates that the effect is not significant, $F(2,24) = 3.00$, $p > .05$.

CHAPTER 16 **CORRELATION**

1. A positive correlation indicates that X and Y change in the same direction: As X increases, Y also increases. A negative correlation indicates that X and Y tend to change in opposite directions: As X increases, Y decreases.

3. $SP = -11$

5. a. The scatter plot shows points widely scattered around a line sloping up to the right.
b. The correlation is small but positive; around 0.4 to 0.6.
c. For these scores, $SS_X = 8$, $SS_Y = 32$, and $SP = 8$. The correlation is $r = \frac{8}{16} = 0.50$.

7. a. The scatter plot shows points scattered around a line sloping up to the right.
b. This should be a moderate, positive correlation; around 0.70.
c. For these scores, $SS_X = 10$, $SS_Y = 40$, and $SP = 14$. The correlation is $r = \frac{14}{20} = 0.70$.

9. a. The scatter plot shows scattered points with no obvious trend.
b. The correlation should be near zero.
c. For these scores, $SS_X = 18$, $SS_Y = 18$, and $SP = 0$. The correlation is $r = 0$.

11. a. For these scores $SS_X = 18$, $SS_Y = 32$, and $SP = -16$. The correlation is $r = -\frac{16}{24} = -0.667$.
b. After adding 2 points to each score, the correlation is still $r = -\frac{16}{24} = -0.667$.
c. Adding a constant to each score does not change the value of the correlation.

13. a. For the children, $SS = 32$, and for the birth parents, $SS = 14$. $SP = 15$. The correlation is $r = 0.709$.
b. For the children, $SS = 32$, and for the adoptive parents, $SS = 16$. $SP = 3$. The correlation is $r = 0.133$.

c. The children's behavior is strongly related to their birth parents and only weakly related to their adoptive parents. The data suggest that the behavior is inherited rather than learned.

15. a. For these data, $SS_7min = 98$, $SS_{cognitive} = 236$, and $SP = 127$. $r = 0.835$.
b. With $df = 9$, the critical value is 0.735. The correlation is significant.
c. $r^2 = 0.697$ or 69.7%

17. a. not significant
b. significant
c. not significant
d. significant

19. The Pearson correlation measures the degree of linear relationship. The Spearman correlation measures the degree to which the relationship is monotonic or one-directional independent of form.

21. a. $r_s = +0.907$
b. Yes, there is a strong positive relationship between the grades assigned by the two instructors. The critical value for $\alpha = .05$ is 0.618, so the data indicate a statistically significant Spearman correlation.

23. a. Both the independent-measures t test and the point biserial correlation evaluate the relationship between two variables when one of the variables is dichotomous. The correlation measures the strength of the relationship and the t test determines whether the relationship is significant (more than can be reasonably explained by chance alone).
b. The sample size can influence the t test but has no effect on the correlation. Specifically, the same correlation can be significant with a large sample but not significant with a smaller sample.

25. Using the eating concern scores as the X variable and coding males as 1 and females as 0 for the Y variable produces $SS_X = 1875.6$, $SS_Y = 3.6$, and $SP = -50.4$. The point-biserial correlation is $r = -0.613$. (Reversing the codes for males and females will change the sign of the correlation.)

27. a. The converted data show 5 individuals with scores of $X = 1$ and $Y = 1$, 3 individuals with $X = 1$ and $Y = 0$, 1 individual with $X = 0$ and $Y = 1$, and 3 individuals with $X = 0$ and $Y = 0$.

   b. $r = 0.354$ (the sign is meaningless)

CHAPTER 17   INTRODUCTION TO REGRESSION

1.

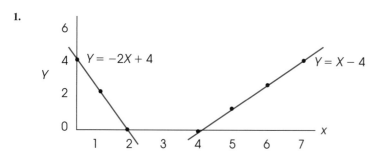

3. a. $r = 0.625$

   b. $\hat{Y} = 0.5X + 1$

5. The standard error of estimate is a measure of the average distance between the predicted Y points from the regression equation and the actual Y points in the data.

7. a. $SS_X = 20$, $SS_Y = 24$, $SP = -20$. The regression equation is $\hat{Y} = (-1)X + 10$.

9. $SS_{\text{regression}} = r^2 SS_Y = 90.02$ with $df = 1$. $MS_{\text{residual}} = 18/4 = 4.5$. $F = 90.02/4.5 = 20.00$. With $df = 1, 4$, the F-ratio is significant with $\alpha = .05$.

11. a. The predicted portion is determined by $r^2 = 0.36$ or 36%. The unpredicted (residual) portion is determined by $(1 - r^2) = 0.64$ or 64%.

   b. $SS_{\text{regression}} = 3.6$ with $df = 1$, and $SS_{\text{residual}} = 6.4$ with $df = 18$. $F = 3.6/0.356 = 10.11$. With $df = 1, 18$ the F-ratio is significant.

   c. $SS_{\text{regression}} = 36$ with $df = 1$, and $SS_{\text{residual}} = 64$ with $df = 18$. $F = 36/3.56 = 10.11$. With $df = 1, 18$ the F-ratio is significant. *Note:* The F-ratios are exactly the same. SS_Y has no effect on the significance of the regression equation.

13. a. $b = \frac{122}{172} = 0.709$ and $a = 9 - 0.709(7) = 4.037$.

   b. $r = 0.904$ and $r^2 = 0.817$ or 81.7%.

   c. $SS_{\text{regression}} = 86.602$ with $df = 1$, and $SS_{\text{residual}} = 19.398$ with $df = 6$. $F = 86.602/3.233 = 26.79$. With $df = 1, 6$ the F-ratio is significant with $\alpha = .05$ or with $\alpha = .01$.

15. a. The standard error of estimate is $\sqrt{27/13} = 1.44$.

   b. With $n = 30$, the standard error of estimate is $\sqrt{27/28} = 0.98$.

17. a. $df = 1, 16$

   b. $n = 26$ pairs of scores

19. a. $F = 1.85/0.604 = 3.06$. With $df = 2, 27$, the critical value is 3.35. The equation does not account for a significant portion of the variance.

   b. $F = 18.50/6.04 = 3.06$. Again, the equation is not significant.

21. The extra variance predicted by the multiple regression is 38.5% − 27.2% = 11.3% and has $df = 1$. The residual variance from the multiple regression is $1 - R^2 = 61.5\%$ and has $df = 30 - 3 = 27$. $F = 11.3/2.28 = 4.96$. With $df = 1, 27$, this is significant at the .05 level.

23. Using the biological parents as a single predictor accounts for $r^2 = 0.503$ or 50.3% of the variance. The multiple regression equation accounts for $R^2 = 58.7\%$ (see problem 22). The extra variance predicted by adding the adoptive parents as a second predictor is 58.7 − 50.3 = 8.4% and has $df = 1$. The residual from the multiple regression is $1 - R^2 = 41.3\%$ and has $df = 6$. The F-ratio is 8.4/(41.3/6) = 1.22. With $df = 1, 6$ the F-ratio is not significant.

25. $n = 37$

CHAPTER 18 **CHI-SQUARE TESTS**

1. Nonparametric tests make few if any assumptions about the populations from which the data are obtained. For example, the populations do not need to form normal distributions, nor is it required that different populations in the same study have equal variances (homogeneity of variance assumption). Parametric tests require data measured on an interval or ratio scale. For nonparametric tests, any scale of measurement is acceptable.

3. The null hypothesis states that there is no preference among the three photographs; $p = \frac{1}{3}$ for all categories. The expected frequencies are $f_e = 50$ for all categories, and chi-square $= 20.28$. With $df = 2$, the critical value is 5.99. Reject H_0 and conclude that there are significant preferences.

5. The null hypothesis states that wins and losses are equally likely. With 64 games, the expected frequencies are 32 wins and 32 losses. With $df = 1$ the critical value is 3.84, and the data produce a chi-square of 6.25. Reject the null hypothesis and conclude that home team wins are significantly more common than would be expected by chance.

7. The null hypothesis states that couples with the same initial do not occur more often than would be expected by chance. For a sample of 200, the expected frequencies are 13 with the same initial and 187 with different initials. With $df = 1$ the critical value is 3.84, and the data produce a chi-square of 2.96. Fail to reject the null hypothesis.

9. The null hypothesis states that the grade distribution for last semester has the same proportions as it did in 1985. For a sample of $n = 200$, the expected frequencies are 28, 52, 62, 38, and 20 for grades of A, B, C, D, and F, respectively. With $df = 4$, the critical value for chi-square is 9.49. For these data, the chi-square statistic is 6.68. Fail to reject H_0 and conclude that there is no evidence that the distribution has changed.

11. **a.** The null hypothesis states that there is no advantage (no preference) for red or blue. With $df = 1$, the critical value is 3.84. The expected frequency is 25 wins for each color, and chi-square $= 2.88$. Fail to reject H_0 and conclude that there is no significant advantage for one color over the other.
    b. The null hypothesis states that there is no advantage (no preference) for red or blue. With $df = 1$, the critical value is 3.84. The expected frequency is 50 wins for each color, and chi-square $= 5.76$. Reject H_0 and conclude that there is a significant advantage for the color red.
    c. Although the proportions are identical for the two samples, the sample in part b is twice as big as the sample in part a. The larger sample provides more convincing evidence of an advantage for red than does the smaller sample.

13. The null hypothesis states that the distribution of preferences is the same for both groups (the two variables are independent). With $df = 2$, the critical value is 5.99. The expected frequencies are:

	Design 1	Design 2	Design 3
Students	24	27	9
Older Adults	24	27	9

Chi-square $= 7.94$. Reject H_0.

15. The null hypothesis states that there is no relationship between personality and heart disease. For $df = 1$ and $\alpha = .05$, the critical value for chi-square is 3.84. The expected frequencies are:

	No Disease	Heart Disease
Type A	40	10
Type B	120	30

For these data, chi-square $= 10.67$. Reject H_0 and conclude that there is a significant relationship between personality and heart disease.

17. **a.** The null hypothesis states that the distribution of opinions is the same for those who live in the city and those who live in the suburbs. For $df = 1$ and $\alpha = .05$, the critical value for chi-square is 3.84. The expected frequencies are:

	Favor	Oppose
City	60	40
Suburb	120	80

For these data, chi-square $= 6.25$. Reject H_0 and conclude that opinions in the city are different from those in the suburbs. The larger sample produces a significant relationship.
    b. The phi coefficient is still 0.144. The sample size has no effect on the strength of the relationship.

19. The null hypothesis states that fear of falling is independent of church attendance. The distribution across fear categories is no different for those who attend church and those who do not. With $df = 3$ and $\alpha = .05$, the critical value is 7.81. The expected frequencies are as follows:

	No Fear	Somewhat Afraid	Fairly Afraid	Very Afraid	
Church	44	64	54	38	200
No Church	22	32	27	19	100

$$n = 300$$

Chi-square $= 10.71$. Reject the null hypothesis and conclude that fear of falling is related to church attendance.

21. **a.** The null hypothesis states that there is no relationship between allergies and childhood pets. With $df = 2$ and $\alpha = .05$, the critical value is 5.99. The expected frequencies are:

	Number of Pets		
	0	1	2 or More
No Allergies	30	45	15
Allergies	10	15	5

The chi-square statistic is 13.16. Reject H_0 with $\alpha = .05$ and $df = 2$.
    b. $V = 0.331$ (a medium effect)

23. a. The null hypothesis states that littering is independent of the amount of litter already on the ground. With $df = 2$, the critical value is 5.99. The expected frequencies are:

Amount of Litter

	None	Small	Large
Litter	31.33	31.33	31.33
Not Litter	58.67	58.67	58.67

Chi-square = 25.88. Reject H_0.

b. $V = 0.310$ (a medium effect)

CHAPTER 19 THE BINOMIAL TEST

1. H_0: p(home team win) = .50 (no preference). The critical boundaries are $z = \pm 1.96$. With $X = 42$, $\mu = 32$, and $\sigma = 4$, we obtain $z = 2.50$. Reject H_0 and conclude that there is a significant difference. Home teams win significantly more than would be expected by chance.

3. H_0: $p = q = \frac{1}{2}$ (right and left are equally common). The critical boundaries are $z = \pm 2.58$. With $X = 104$, $\mu = 72$, and $\sigma = 6$, we obtain $z = 5.33$. Reject H_0 and conclude that right- and left-handed rats are not equally common.

5. H_0: $p = q = \frac{1}{2}$ (no preference). The critical boundaries are $z = \pm 1.96$. With $X = 28$, $\mu = 20$, and $\sigma = 3.16$, we obtain $z = 2.53$. Reject H_0. There is evidence of a significant preference between bottled water and tap water.

7. H_0: $p = .25$ (the general population has the same proportion of belief as the psychotherapists). The critical boundaries are $z = \pm 1.96$. With $X = 65$, $\mu = 48$, and $\sigma = 6$, we obtain $z = 2.83$. Reject H_0 and conclude that the proportion of belief is significantly different for the general population than for psychotherapists.

9. H_0: $p = .20$ (correct predictions are just chance). The critical boundaries are $z = \pm 1.96$. With $X = 27$, $\mu = 20$, and $\sigma = 4$, we obtain $z = 1.75$. Fail to reject H_0 and conclude that this level of performance is not significantly different from chance.

11. H_0: $p = \frac{1}{4} = p$ (guessing correctly). The critical boundaries are $z = \pm 1.96$. With $X = 32$, $\mu = 25$, and $\sigma = 4.33$, we obtain $z = 1.62$. Fail to reject H_0 and conclude that this level of performance is not significantly different from chance.

13. a. H_0 $p = q = \frac{1}{2}$ (positive and negative correlations are equally likely). The critical boundaries are $z = \pm 1.96$, With $X = 25$, $\mu = 13.5$, and $\alpha = 2.60$, we obtain $z = 4.42$. Reject H_0, positive and negative correlations are not equally likely.

b. The critical boundaries are $z = \pm 1.96$. With $X = 20$, $\mu = 13.5$, and $\sigma = 2.60$, we obtain $z = 2.50$. Reject H_0, positive and negative correlations are not equally likely.

15. a. With $p = \frac{1}{2} = p$(improving). The critical boundaries are $z = \pm 1.96$. Discarding the 12 patients showing no change, we have n = 100. The binomial distribution has $\mu = 50$ and $\sigma = 5$. With $X = 64$ we obtain $z = 2.80$. Reject H_0 and conclude that there is significant evidence of improving symptoms.

b. H_0: $p = \frac{1}{2} = p$(improving). The critical boundaries are $z = \pm 1.96$. Dividing the 12 patients between the two groups, we have $n = 112$. and $X = 70$. The binomial distribution has $\mu = 56$ and $\sigma = 5.29$. With $X = 70$ we obtain $z = 2.65$. Again, we reject H_0 and conclude that there is significant evidence of improving symptoms.

17. H_0: $p = \frac{1}{2} = p$(no preference). The critical boundaries are $z = \pm 1.96$. The binomial distribution has $\mu = 40$ and $\sigma = 4.47$. With $X = 47$ we obtain $z = 1.57$. Fail to reject H_0 and conclude that there is no preference for either of the two cages.

19. a. H_0: $p = \frac{1}{3}$ and $q = \frac{2}{3}$ (just guessing). The critical boundaries are $z = \pm 1.96$. The binomial distribution has $\mu = 12$ and $\sigma = 2.83$. With $X = 25$ we obtain $z = 4.59$. Reject H_0 and conclude that the children with autism are performing significantly better than chance.

b. H_0: $p = \frac{1}{3}$ and $q = \frac{2}{3}$ (just guessing). The critical boundaries are $z = \pm 1.96$. The binomial distribution has $\mu = 12$ and $\sigma = 2.83$. With $X = 16$ we obtain $z = 1.41$. Reject H_0 and conclude that the children with SLI are not performing significantly better than chance.

21. H_0: $p = q = \frac{1}{2}$ (any change in grade point average is the result of chance). The critical boundaries are $z = \pm 2.58$. The binomial distribution has $\mu = 22.5$ and $\sigma = 3.35$. With $X = 31$ we obtain $z = 2.54$. Fail to reject H_0 and conclude that there is no significant change in grade point average after the workshop.

23. H_0: $p = q = \frac{1}{2}$ (any difference between the two paragraphs is the result of chance). The critical boundaries are $z = \pm 1.96$. The binomial distribution has $\mu = 18$ and $\sigma = 3$. With $X = 25$ we obtain $z = 2.33$. Reject H_0 and conclude that there is a significant change in performance.

CHAPTER 20   **TESTS FOR ORDINAL DATA**

1. The ranks for the 10 subjects are as follows:

Subject:	A	B	C	D	E	F	G	H	I	J
Rank:	1	2	4	4	4	6	7.5	7.5	9	10

3. If one or more set of scores is extremely variable, it may violate the homogeniety of variance assumption and can create a very large error term for an F-ratio or t statistic. Also, if the original scores include an undetermined value, it is impossible to compute a mean or variance. In these cases, ranking often can eliminate the problem.

5. The null hypothesis states that there is no systematic difference between doctor visits for dog owners versus non-owners. The critical value is 6. For the control group, $\Sigma R = 67.5$ and for the dog owners, $\Sigma R = 23.5$. $U = 8.5$. Fail to reject H_0 and conclude there is no significant difference between the two groups.

7. The null hypothesis states that there is no systematic difference between success and failure for future performance. The critical value is 13. For the success group, $\Sigma R = 92.5$ and for the failure group, $\Sigma R = 43.5$. $U = 7.5$. Reject H_0 and conclude there is a significant difference between the two groups.

9. **a.** The null hypothesis says that there is no systematic difference in grades for students who watched Sesame Street and those who did not watch. The critical value is $U = 23$. For the students who watched $\Sigma R = 144.5$ and for those who did not watch, $\Sigma R = 65.5$ $U = 10.5$. Reject H_0.

   b. For the Kruskal-Wallis test, $df = 1$ and the critical value is 3.84. The data produce $H = 8.92$. Reject H_0.

11. With $n = 5$, there is no value for T that is significant with $\alpha = .05$. Even $T = 0$ is not sufficient to reject the null hypothesis. Therefore, there is no need to conduct the test. However, these data produce $\Sigma R = 10$ for the decreases, and $\Sigma R = 0$ for the increases. However, there is no significant difference.

13. The null hypothesis states that the drug has no effect. The critical value is $T = 5$. For the increases $\Sigma R = 39$ and for the decreases $\Sigma R = 6$. Wilcoxon $T = 6$. Fail to reject H_0. These data do not provide sufficient evidence to conclude that the drug works.

15. The null hypothesis states that the duration of the time out has no effect on performance. Discarding the zero difference leaves $n = 8$. The critical value is $T = 3$. For the increases $\Sigma R = 34$ and for the decreases $\Sigma R = 2$. Wilcoxon $T = 2$. Reject H_0 and conclude that the length of the time out does affect performance.

17. **a.** The null hypothesis states that there is no systematic difference between the two treatments. The critical value for the Mann-Whitney test is 8. The ranks for treatment I total $\Sigma R = 42.5$ and the ranks for treatment II have $\Sigma R = 62.5$. The data produce $U = 14.5$. Fail to reject H_0 and conclude that there is no significant difference between the two treatments.

   b. The critical value for the Wilcoxon test is 2. The positive differences have $\Sigma R = 28$ and the negative differences have $\Sigma R = 0$. With $T = 0$, reject H_0 and conclude that there is a significant difference between the two treatments.

   c. Yes, the repeated-measures test finds a significant difference but the independent-measures test does not.

19. The null hypothesis states that there are no systematic differences in language skill among the three classifications of children. With $df = 2$, the critical value is 5.99. The single children have $\Sigma R = 83$, the twins have $\Sigma R = 52$, and triplets have $\Sigma R = 36$. For these data, the Kruskal-Wallis $H = 6.68$. Reject H_0 and conclude that there are significant differences among the three types of children.

21. The null hypothesis states that there are no systematic differences in heart rate among the three measures. With $df = 2$, the critical value is 5.99. The baseline measures have $\Sigma R = 5.5$, the 1-minute measures have $\Sigma R = 12$, and the 2-minute measures have $\Sigma R = 6.5$. For these data, chi-square $= 6.125$. Reject H_0 and conclude that there are significant differences among the three measurements.

23. The null hypothesis states that there are no systematic differences among the three treatments. With $df = 2$, the critical value is 5.99. The first treatment has $\Sigma R = 7$, the second treatment has $\Sigma R = 13$, and the third treatment has $\Sigma R = 16$. For these data, chi-square $= 7$. Reject H_0 and conclude that there are significant differences among the three treatments.

APPENDIX D — General Instructions for Using SPSS

The Statistical Package for the Social Sciences, commonly known as SPSS, is a computer program that performs statistical calculations, and is widely available on college campuses. Detailed instructions for using SPSS for specific statistical calculations (such as computing sample variance or performing an independent-measures *t* test) are presented at the end of the appropriate chapter in the text. Look for the SPSS logo in the Resources section at the end of each chapter. In this appendix, we provide a general overview of the SPSS program.

SPSS consists of two basic components: A data editor and a set of statistical commands. The **data editor** is a huge matrix of numbered rows and columns. To begin any analysis, you must type your data into the data editor. Typically, the scores are entered into columns of the editor. Before scores are entered, each of the columns is labeled "var." After scores are entered, the first column becomes VAR00001, the second column becomes VAR00002, and so on. To enter data into the editor, the **Data View** tab must be set at the bottom left of the screen. If you want to name a column (instead of using VAR00001), click on the **Variable View** tab at the bottom of the data editor. You will get a description of each variable in the editor, including a box for the name. You may type in a new name using up to 8 lowercase characters (no spaces, no hyphens). Click the **Data View** tab to go back to the data editor.

The **statistical commands** are listed in menus that are made available by clicking on **Analyze** in the tool bar at the top of the screen. When you select a statistical command, SPSS typically asks you to identify exactly where the scores are located and exactly what other options you want to use. This is accomplished by identifying the column(s) in the data editor that contain the needed information. Typically, you are presented with a display similar to the following figure. On the left is a box that lists all of the columns in the data editor that contain information. In this example, we have typed values into columns 1, 2, 3, and 4. On the right is an empty box that is waiting for you to identify the correct column. For example, suppose that you wanted to do a statistical calculation using the scores in column 3. You should highlight VAR00003 by clicking on it in the left-hand box, then click the arrow to move the column label into the right hand box. (If you make a mistake, you can highlight the variable in the right-hand box and the arrow reverses so that you can move the variable back to the left-hand box.)

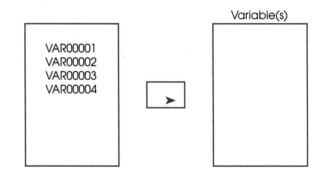

SPSS DATA FORMATS

The SPSS program uses two basic formats for entering scores into the data matrix. Each is described and demonstrated as follows:

1. The first format is used when the data consist of several scores (more than one) for each individual. This includes data from a repeated-measures study, in which each person is measured in all of the different treatment conditions, and data from a correlational study where there are two scores, X and Y, for each individual. Table D1 illustrates this kind of data and shows how the scores would appear in the SPSS data matrix. Note that the scores in the data matrix have exactly the same structure as the scores in the original data. Specifically, each row of the data matrix contains the scores for an individual participant, and each column contains the scores for one treatment condition.

TABLE D1

Data for a repeated-measures or correlational study with several scores for each individual. The left half of the table (a) shows the original data, with three scores for each person; and the right half (b) shows the scores as they would be entered into the SPSS data matrix. Note: SPSS automatically adds the two decimal points for each score. For example, you type in 10 and it appears as 10.00 in the matrix.

(a) Original data

Person	I	II	III
A	10	14	19
B	9	11	15
C	12	15	22
D	7	10	18
E	13	18	20

(Treatments)

(b) Data as entered into the SPSS data matrix

	VAR0001	VAR0002	VAR0003	var
1	10.00	14.00	19.00	
2	9.00	11.00	15.00	
3	12.00	15.00	22.00	
4	7.00	10.00	18.00	
5	13.00	18.00	20.00	

2. The second format is used for data from an independent-measures study using a separate group of participants for each treatment condition. This kind of data is entered into the data matrix in a *stacked* format. Instead of having the scores from different treatments in different columns, all of the scores from all of the treatment conditions are entered into a single column so that the scores from one treatment condition are literally stacked on top of the scores from another treatment condition. A code number is then entered into a second column beside each score to tell the computer which treatment condition corresponds to each score. For example, you could enter a value of 1 beside each score from treatment #1, enter a 2 beside each score from treatment #2, and so on. Table D2 illustrates this kind of data and shows how the scores would be entered into the SPSS data matrix.

TABLE D2

Data for an independent-measures study with a different group of participants in each treatment condition. The left half of the table shows the original data, with three separate groups, each with five participants, and the right half shows the scores as they would be entered into the SPSS data matrix. Note that the data matrix lists all 15 scores in the same column, then uses code numbers in a second column to indicate the treatment condition corresponding to each score.

(a) Original data

Treatments		
I	II	III
10	14	19
9	11	15
12	15	22
7	10	18
13	18	20

(b) Data as entered into the SPSS data matrix

	VAR0001	VAR0002	var
1	10.00	1.00	
2	9.00	1.00	
3	12.00	1.00	
4	7.00	1.00	
5	13.00	1.00	
6	14.00	2.00	
7	11.00	2.00	
8	15.00	2.00	
9	10.00	2.00	
10	18.00	2.00	
11	19.00	3.00	
12	15.00	3.00	
13	22.00	3.00	
14	18.00	3.00	
15	20.00	3.00	

Statistics Organizer

The following pages present an organized summary of the statistical procedures covered in this book. This organizer is divided into four sections, each of which groups together statistical techniques that serve a common purpose. The four groups are

   I. Descriptive Statistics

  II. Parametric Tests for Mean Differences

 III. Nonparametric Tests for Systematic Differences

 IV. Evaluating Relationship Between Variables

Each of the four sections begins with a general overview that discusses the purpose for the statistical techniques that follow and points out some common characteristics of the different techniques. Next, there is a decision map that leads you, step by step, through the task of deciding which statistical technique is appropriate for the data you wish to analyze. Finally, there is a brief description of each technique and a reference to a complete example in the text.

I   DESCRIPTIVE STATISTICS

The purpose of descriptive statistics is to simplify and organize a set of scores. Scores may be organized in a table or graph, or they may be summarized by computing one or two values that describe the entire set. The most commonly used descriptive techniques are as follows:

A. Frequency Distribution Tables and Graphs

A frequency distribution is an organized tabulation of the number of individuals in each category on the scale of measurement. A frequency distribution can be presented as either a table or a graph. The advantage of a frequency distribution is that it presents the entire set of scores rather than condensing the scores into a single descriptive value. The disadvantage of a frequency distribution is that it can be somewhat complex, especially with large sets of data.

B. Measures of Central Tendency

The purpose of measuring central tendency is to identify a single score that represents an entire data set. The goal is to obtain a single value that is the best example of the average, or most typical, score from the entire set.

Measures of central tendency are used to describe a single data set, and they are the most commonly used measures for comparing two (or more) different sets of data.

CHOOSING DESCRIPTIVE STATISTICS: A DECISION MAP

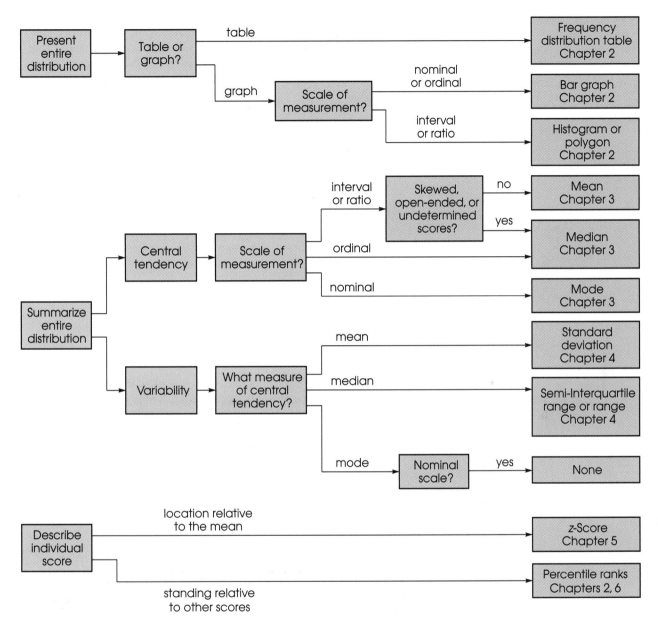

C. Measures of Variability

Variability is used to provide a description of how spread out the scores are in a distribution. It also provides a measure of how accurately a single score selected from a distribution represents the entire set.

D. z-Scores

Most descriptive statistics are intended to provide a description of an entire set of scores. However, z-scores are used to describe individual scores within a distribution. The purpose of a z-score is to identify the precise location of an individual within a distribution by using a single number.

1. **The Mean** (Chapter 3)

   The mean is the most commonly used measure of central tendency. It is computed by finding the total (ΣX) for the set of scores and then dividing the total by the number of individuals. Conceptually, the mean is the amount each individual will receive if the total is divided equally. See Demonstration 3.1 on page 100.

2. **The Median** (Chapter 3)

   Exactly 50% of the scores in a data set have values less than or equal to the median. The median is the 50th percentile. The median usually is computed for data sets when the mean cannot be found (undetermined scores, open-ended distribution) or when the mean does not provide a good, representative value (ordinal scale, skewed distribution). See Demonstration 3.1 on page 100.

3. **The Mode** (Chapter 3)

   The mode is the score with the greatest frequency. The mode is used when the scores consist of measurements on a nominal scale. See Demonstration 3.1 on page 100.

4. **The Range** (Chapter 4)

   The range is the distance from the lowest to the highest score in a data set. The range is considered to be a relatively crude measure of variability. See the example on page 107.

5. **The Semi-Interquartile Range** (Chapter 4)

   The semi-interquartile range is half of the range covered by the middle 50% of the distribution. The semi-interquartile range is often used to measure variability in situations where the median is used to report central tendency. See the example on page 108.

6. **Standard Deviation** (Chapter 4)

   The standard deviation is a measure of the standard distance from the mean. Standard deviation is obtained by first computing SS (the sum of squared deviations) and variance (the mean squared deviation). Standard deviation is the square root of variance. See Demonstration 4.1 on page 133.

7. *z*-**Scores** (Chapter 5)

   The sign of a *z*-score indicates whether an individual is above $(+)$ or below $(-)$ the mean. The numerical value of the *z*-score indicates how many standard deviations there are between the score and the mean. See Demonstration 5.1 on page 159.

PARAMETRIC TESTS: EVALUATING MEAN DIFFERENCES BETWEEN TREATMENT CONDITIONS OR BETWEEN POPULATIONS

All of the hypothesis tests covered in this section use the means obtained from sample data as the basis for testing hypotheses about population means. Although there are a variety of tests used in a variety of research situations, all use the same basic logic, and all of the test statistics have the same basic structure. In each case, the test statistic (z, t, or F) involves computing a ratio with the following structure:

$$\text{test statistic} = \frac{\text{obtained difference between means}}{\text{mean difference expected by chance}}$$

The goal of each test is to determine whether the observed mean differences are larger than expected by chance. In general terms, a *significant result* means that the results obtained in a research study (mean differences) are more than would be expected by chance. In each case, a large value for the test statistic ratio indicates a significant result; that is, when the actual difference between means (numerator) is substantially larger than chance (denominator), you will obtain a large ratio, which indicates that the sample difference is significant.

The actual calculations differ slightly from one test statistic to the next, but all involve the same basic computations.

1. A set of scores (sample) is obtained for each population or each treatment condition.

2. The mean is computed for each set of scores, and some measure of variability (*SS,* standard deviation, or variance) is obtained for each individual set of scores.

3. The differences between the sample means provide a measure of how much difference exists between treatment conditions. Because these mean differences may be caused by the treatment conditions, they are often called *systematic* or *predicted*. These differences are the numerator of the test statistic.

4. The variability within each set of scores provides a measure of unsystematic or unpredicted differences due to chance. Because the individuals within each treatment condition are treated exactly the same, there is nothing that should cause their scores to be different. Thus, any observed differences (variability) within treatments are assumed to be due to error or chance.

With these considerations in mind, each of the test statistic ratios can be described as follows:

$$\text{test statistic} = \frac{\text{differences (variability) between treatments}}{\text{differences (variability) within treatments}}$$

The hypothesis tests reviewed in this section apply to three basic research designs:

1. Single-Sample Designs. Data from a single sample are used to test a hypothesis about a single population.

2. Independent-Measures Designs. A separate sample is obtained to represent each individual population or treatment condition.

3. Related-Samples Designs. The most common example of related-samples designs are known as repeated-measures designs in which the same group of individuals participates in all of the different treatment conditions. Related-samples designs also include matched-subjects designs, which use separate groups for each treatment condition with the requirement that every individual in one group is matched with an "equivalent" individual in each of the other groups.

CHOOSING A PARAMETRIC TEST TO EVALUATE MEAN DIFFERENCES BETWEEN TREATMENT CONDITIONS OR BETWEEN POPULATIONS: A DECISION MAP

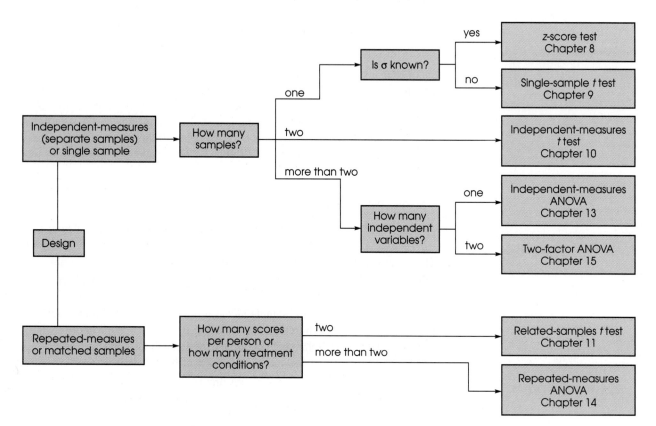

Finally, you should be aware that all parametric tests place stringent restrictions on the sample data and the population distributions being considered. First, these tests all require measurements on an interval or a ratio scale (numerical values that allow you to compute means and differences). Second, each test makes assumptions about population distributions and sampling techniques. Consult the appropriate section of this book to verify that the specific assumptions are satisfied before proceeding with any parametric test.

1. **The z-Score Test** (Chapter 8)

   The z-score test is used to evaluate the significance of a treatment effect in situations in which the population standard deviation (σ) is known. A sample is selected from a population, and the treatment is administered to the individuals in the sample. The test evaluates the difference between the sample mean and a hypothesized population mean. The null hypothesis provides a specific value for the unknown population mean by stating that the population mean has not been changed by the treatment. See Demonstration 8.1 on page 274.

   Effect Size for the z-Score Test

   In addition to the hypothesis test, a measure of effect size is recommended to determine the actual magnitude of the effect. Cohen's d provides a measure of the mean difference measured in standard deviation units. See Demonstration 8.2 on page 275.

2. **The Single-Sample *t* Test** (Chapter 9)

The single sample *t* test is used to evaluate the significance of a treatment effect in situations in which the population standard deviation (σ) is not known. A sample is selected from a population, and the treatment is administered to the individuals in the sample. The test evaluates the difference between the sample mean and a hypothesized population mean. The null hypothesis provides a specific value for the unknown population mean by stating that the population mean has not been changed by the treatment. The *t* test uses the sample variance to estimate the unknown population variance. See Demonstration 9.1 on page 302.

Effect Size for the Single-Sample *t*

The effect size for the single-sample *t* test can be described by either estimating Cohen's *d* or by r^2, which measures the percentage of variability that is accounted for by the treatment effect. See Demonstration 9.2 on page 303.

3. **The Independent-Measures *t* Test** (Chapter 10)

The independent-measures *t* test uses data from two separate samples to test a hypothesis about the difference between two population means. The variability within the two samples is combined to obtain a single (pooled) estimate of population variance. The null hypothesis states that there is no difference between the two population means. See Demonstration 10.1 on page 333.

Effect Size for the Independent-Measures *t*

The effect size for the independent-measures *t* test can be described by either estimating Cohen's *d* or by r^2, which measures the percentage of variability that is accounted for by the treatment effect. See Demonstraton 10.2 on page 335.

4. **The Related-Samples *t* Test** (Chapter 11)

This test evaluates the mean difference between two treatment conditions using the data from a repeated-measures or a matched-subjects experiment. A difference score (*D*) is obtained for each subject (or each matched pair) by subtracting the score in treatment 1 from the score in treatment 2. The variability of the sample difference scores is used to estimate the population variability. The null hypothesis states that the population mean difference (μ_D) is zero. See Demonstration 11.1 on page 358.

Effect Size for the Related-Samples *t*

The effect size for the repeated-measures (related samples) *t* test can be described by either estimating Cohen's *d* or by r^2, which measures the percentage of variability that is accounted for by the treatment effect. See Demonstration 11.2 on page 360.

5. **Single-Factor, Independent-Measures ANOVA** (Chapter 13)

This test uses data from two or more separate samples to test for mean differences among two or more populations. The null hypothesis states that there are no differences among the population means. The test statistic is an F-ratio that uses the variability between treatment conditions (sample mean differences) in the numerator and the variability within treatment conditions (error variability) as the denominator. With only two samples, this test is equivalent to the independent-measures t test. See Demonstration 13.1 on page 436.

Effect Size for ANOVA

The effect size for an ANOVA is described by a measure of the percentage of variability that is accounted for by the treatment effect. In the context of ANOVA, the percentage is called η^2 (eta squared) instead of r^2. See Demonstration 13.2 on page 438.

6. Single-Factor, Repeated-Measures ANOVA (Chapter 14)

This test is used to evaluate mean differences among two or more treatment conditions using sample data from a repeated-measures (or matched-subjects) experiment. The null hypothesis states that there are no differences among the population means. The test statistic is an F-ratio using variability between treatment conditions (mean differences) in the numerator exactly like the independent-measures ANOVA. The denominator of the F-ratio (error term) is obtained by measuring variability within treatments and then subtracting out the variability between subjects. The research design and the test statistic remove variability due to individual differences and thereby provide a more sensitive test for treatment differences than is possible with an independent-measures design. See Demonstration 14.1 on page 469.

Effect Size for Repeated-Measures ANOVA

The effect size for the repeated-measures ANOVA is described by η^2 (eta squared), which measures the percentage of variability that is accounted for by the treatment effect. The percentage is computed after other explained sources of variability have been removed. See Demonstration 14.2 on page 471.

7. Two-Factor, Independent-Measures ANOVA (Chapter 15)

This test is used to evaluate mean differences among populations or treatment conditions using sample data from research designs with two independent variables (factors). The two-factor ANOVA tests three separate hypotheses: mean differences among the levels of factor A (main effect for factor A), mean differences among the levels of factor B (main effect for factor B), and mean differences resulting from specific combinations of the two factors (interaction). Each of the three separate null hypotheses states that there are no population mean differences. Each of the three tests uses an F-ratio as the test statistic, with the variability between samples (sample mean differences) in the numerator and the variability within samples (error variability) in the denominator. See Demonstration 15.1 on page 510.

Effect Size for the Two-Factor ANOVA

The effect size for each main effect (A and B) and for the interaction are described by η^2 (eta squared), which measures the percentage of variability that is accounted for by the specific treatment effect. The percentage is computed after other explained sources of variability have been removed. See Demonstration 15.2 on page 513.

III NONPARAMETRIC TESTS: EVALUATING SYSTEMATIC DIFFERENCES BETWEEN TREATMENT CONDITIONS OR BETWEEN POPULATIONS

Although the parametric tests described in the previous section are the most commonly used inferential techniques, there are many research situations in which parametric tests cannot or should not be used. These situations fall into three general categories:

1. The data involve measurements on nominal or ordinal scales. In these situations, you cannot compute the means and variances that are an essential part of parametric tests.

2. The data do not satisfy the assumptions underlying parametric tests.

3. The data have extremely high variance, which can undermine the likelihood of significance for a parametric test. In this case, the scores can be converted to categories or ranks, and a nonparametric test can be used as an alternative.

When a parametric test cannot be used, there is usually a nonparametric alternative available. In general, parametric tests are more powerful than their nonparametric counterparts, and they are preferred over the nonparametric alternatives. However, when a parametric test is not appropriate, the nonparametric tests provide researchers with a backup statistical technique for conducting an analysis and statistical interpretation of research results.

1. **The Chi-Square Test for Goodness of Fit** (Chapter 18)

   This chi-square test is used in situations in which the measurement procedure results in classifying individuals into distinct categories. The test uses frequency data from a single sample to test a hypothesis about the population distribution. The null hypothesis specifies the proportion or percentage of the population for each category on the scale of measurement. See Example 18.1 on page 614.

2. **The Chi-Square Test for Independence** (Chapter 18)

   This test serves as an alternative to the independent-measures *t* test (or ANOVA) in situations in which the dependent variable involves classifying individuals into distinct categories. The sample data consist of frequency distributions (proportions across categories) for two or more separate samples. The null hypothesis states that the separate populations all have the same proportions (same shape). That is, the proportions across categories are independent of the different populations. See Demonstration 18.1 on page 637.

3. **The Binomial Test** (Chapter 19)

   When the individuals in a population can be classified into exactly two categories, the binomial test uses sample data to test a hypothesis about the proportion of the population in each category. The null hypothesis specifies the proportion of the population in each of the two categories. See Demonstration 19.1 on page 660.

4. **The Sign Test** (Chapter 19)

   This test evaluates the difference between two treatment conditions using data from a repeated-measures or matched-subjects research design. The null hypothesis states that there is no difference between the two treatment conditions.

CHOOSING A NONPARAMETRIC TEST USING NONNUMERICAL SCORES (RANKS OR NOMINAL CATEGORIES) TO EVALUATE
SYSTEMATIC DIFFERENCES BETWEEN TREATMENT CONDITIONS OR BETWEEN POPULATIONS: A DECISION MAP

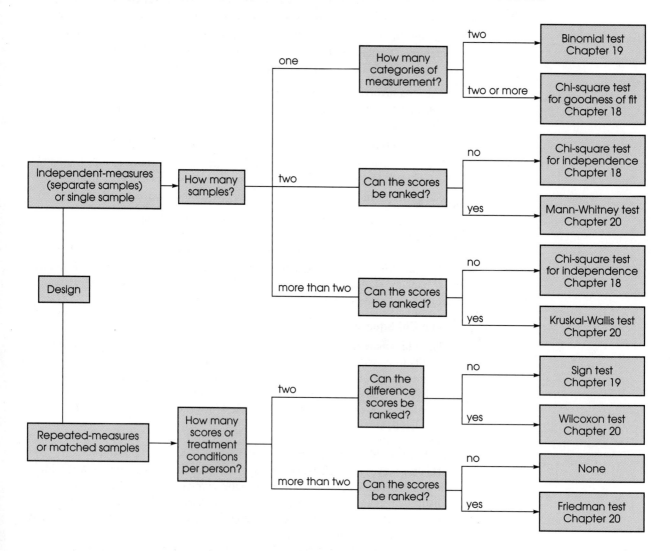

The sign test is a special application of the binomial test and requires only that the difference between treatment 1 and treatment 2 for each subject be classified as an increase or a decrease. This test is used as an alternative to the related-samples *t* test or the Wilcoxon test in situations where the data do not satisfy the more stringent requirements of these two more powerful tests. See Examples 19.2 and 19.3 on pages 655–656.

5. **The Mann-Whitney *U* Test** (Chapter 20)

This test uses ordinal data (rank orders) from two separate samples to test a hypothesis about the difference between two populations or two treatment conditions. The Mann-Whitney test is used as an alternative to the independent-measures *t* test in situations where the data can be rank-ordered but do not satisfy the more stringent requirements of the *t* test. See Demonstration 20.1 on page 692.

6. **The Wilcoxon *T* Test** (Chapter 20)

The Wilcoxon test uses the data from a repeated-measures or matched-samples design to evaluate the difference between two treatment conditions. This test is used as an alternative to the related-samples *t* test in situations in which the sample data (difference scores) can be rank-ordered but do not satisfy the more stringent requirements of the *t* test. See Demonstration 20.2 on page 693.

7. **The Kruskal-Wallis Test** (Chapter 20)

This test uses ordinal data (ranks) from three or more separate samples to test a hypothesis about the differences between three or more populations (or treatments). The Kruskal-Wallis test is used as an alternative to the independent-measures ANOVA in situations in which the data can be rank-ordered but do not satisfy the more stringent requirements of the ANOVA. See Demonstration 20.3 on page 694.

8. **The Friedman Test** (Chapter 20)

This test uses ordinal data (ranks) to test a hypothesis about the differences between three or more treatment conditions using data from a repeated-measures design (the same group participates in all treatment conditions). The Friedman test is used as an alternative to the repeated-measures ANOVA in situations in which the data can be rank-ordered but do not satisfy the more stringent requirements of the ANOVA. See Demonstration 20.4 on page 696.

IV MEASURES OF RELATIONSHIP BETWEEN VARIABLES

As we noted in Chapter 1, a major purpose for scientific research is to investigate and establish orderly relationships between variables. The statistical techniques covered in this section all serve the purpose of measuring and describing relationships. The data for these statistics involve two observations for each individual—one observation for each of the two variables being examined. The goal is to determine whether a consistent, predictable relationship exists and to describe the nature of the relationship.

Each of the different statistical methods described in this section is intended to be used with a specific type of data. To determine which method is appropriate, you must first examine your data and identify what type of variable is involved and what scale of measurement was used for recording the observations.

1. **The Pearson Correlation** (Chapter 16)

The Pearson correlation measures the degree of linear relationship between two variables. The sign (+ or −) of the correlation indicates the direction of the relationship. The magnitude of the correlation (from 0 to 1) indicates the degree to which the data points fit on a straight line. See Demonstration 16.1 on page 555.

2. **Partial Correlation** (Chapter 17)

A partial correlation measures the Pearson correlation between two variables while a third variable is controlled or held constant. See Examples 17.6, 17.7, and 17.8 on pages 589-593.

CHOOSING A METHOD FOR EVALUATING RELATIONSHIPS BETWEEN VARIABLES: A DECISION MAP

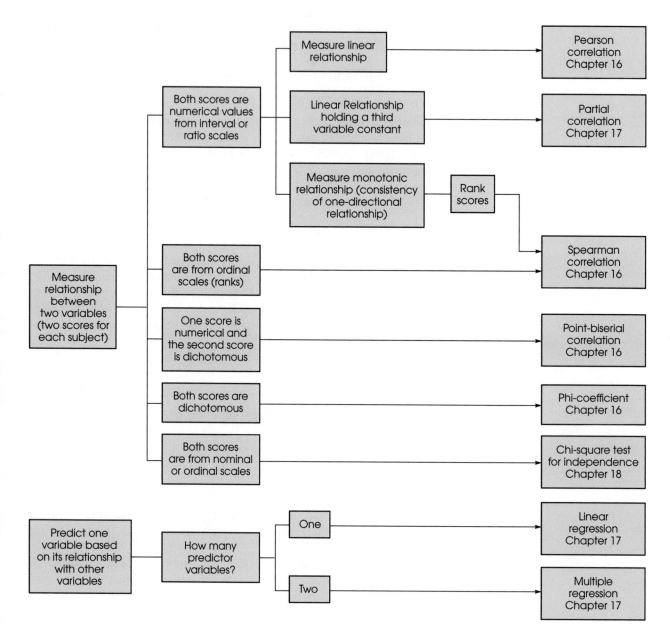

3. **The Spearman Correlation** (Chapter 16)

The Spearman correlation measures the degree to which the relationship between two variables is one-directional or monotonic. The Spearman correlation is used when both variables, X and Y, are measured on an ordinal scale or after the two variables have been transformed to ranks. See Demonstration 16.2 on page 557.

4. **The Point-Biserial Correlation** (Chapter 16)

The point-biserial correlation is a special application of the Pearson correlation that is used when one variable is dichotomous (only two values) and the second variable is measured on an interval or ratio scale. The value of the correlation measures the strength of the relationship between the two variables. The point-biserial correlation often is used as an alternative or a supplement to the independent-measures t hypothesis test. See the example on pages 548–549.

5. **The Phi-Coefficient** (Chapter 16)

The phi-coefficient is a special application of the Pearson correlation that is used when both variables, X and Y, are dichotomous (only two values). The value of the correlation measures the strength of the relationship between the two variables. The phi-coefficient is often used as an alternative or a supplement to the chi-square test for independence. See Example 16.12 on page 550.

6. **The Chi-Square Test for Independence** (Chapter 18)

This test uses frequency data to determine whether there is a significant relationship between two variables. The null hypothesis states that the two variables are independent. The chi-square test for independence is used when the scale of measurement consists of relatively few categories for both variables and can be used with nominal, ordinal, interval, or ratio scales. See Demonstration 18.1 on page 637.

7. **Linear Regression** (Chapter 17)

The purpose of linear regression is to find the equation for the best-fitting straight line for predicting Y scores from X scores. The regression process determines the linear equation with the least-squared error between the actual Y values and the predicted Y values on the line. The standard error of estimate provides a measure of the standard distance (or error) between the actual Y values and the predicted Y values. Analysis of regression determines whether the regression equation predicts a significant proportion of the variance for the Y scores by comparing the predicted portion (r^2) with the residual portion, ($1 - r^2$). See Demonstration 17.1 on page 597.

8. **Multiple Regression** (Chapter 17)

Multiple regression determines the equation that produces the most accurate predictions of one variable (Y) using two predictor variables (X_1 and X_2). Analysis of regression determines whether the regression equation predicts a significant proportion of the variance for the Y scores by comparing the predicted portion (R^2) with the residual portion, ($1 - R^2$). See Demonstration 17.2 on page 598.

References

Ackerman, P. L., & Beier, M. E. (2007). Further explorations of perceptual speed abilities in the context of assessment methods, cognitive abilities, and individual differences during skill acquisition. *Journal of Experimental Psychology: Applied, 13,* 249–272.

Albentosa, M. J., & Cooper, J. J. (2005). Testing resource value in group-housed animals: An investigation of cage height preference in laying hens. *Behavioural Processes, 70,* 113–121.

American Psychological Association (APA). (2001). *Publication manual of the American Psychological Association* (5th ed.) Washington, DC: Author.

Anderson, D. R., Huston, A. C., Wright, J. C., & Collins, P. A. (1998). Initial findings on the long term impact of Sesame Street and educational television for children: The recontact study. In R. Noll and M. Price (Eds.), *A communication cornucopia: Markle Foundation essays on information policy* (pp. 279–296). Washington, DC: Brookings Institution.

Arden, R., & Plomin, R. (2006). Sex differences in variance of intelligence across childhood. *Personality and Individual Differences, 41,* 39–48.

Athos, E. A., Levinson, B., Kistler, A., Zemansky, J., Bostrom, A., Freimer, N., & Gitschier, J. (2007). Dichotomy and perceptual distortions in absolute pitch ability. *Proceedings of the National Academy of Science of the United States of America, 104,* 14795–14800.

Bartus, R. T. (1990). Drugs to treat age-related neurodegenerative problems: The final frontier of medical science? *Journal of the American Geriatrics Society, 38,* 680–695.

Belsky, J., Weinraub, M., Owen, M., & Kelly, J. (2001). Quality of child care and problem behavior. In J. Belsky (Chair), *Early childcare and children's development prior to school entry.* Symposium conducted at the 2001 Biennial Meetings of the Society for Research in Child Development, Minneapolis, MN.

Blest, A. D. (1957). The functions of eyespot patterns in the Lepidoptera. *Behaviour, 11,* 209–255.

Blum, J. (1978). *Pseudoscience and mental ability: The origins and fallacies of the IQ controversy.* New York: Monthly Review Press.

Boogert, N. J., Reader, S. M., & Laland, K. N. (2006). The relation between social rank, neophobia and individual learning in starlings. *Behavioural Biology, 72,* 1229–1239.

Bradbury, T. N., & Miller, G. A. (1985). Season of birth in schizophrenia: A review of evidence, methodology, and etiology. *Psychological Bulletin, 98,* 569–594.

Broberg, A. G., Wessels, H., Lamb, M. E., & Hwang, C. P. (1997). Effects of day care on the development of cognitive abilities in 8-year-olds: A longitudinal study. *Development Psychology, 33,* 62–69.

Byrne, D. (1971). *The attraction paradigm.* New York: Academic Press.

Camara, W. J., & Echternacht, G. (2000). *The SAT 1 and high school grades: Utility in predicting success in college* (College Board Report No. RN-10). New York: College Entrance Examination Board.

Candappa, R. (2000). *The little book of wrong shui.* Kansas City: Andrews McMeel Publishing.

Cerveny, R. S., & Balling, Jr., R. C. (1998). Weekly cycles of air pollutants, precipitation and tropical cyclones in the coastal NW Atlantic region. *Nature, 394,* 561–563.

Chandra, A., & Minkovitz, C. S. (2006). Stigma starts early: Gender differences in teen willingness to use mental health services. *Journal of Adolescent Health, 38,* 754.el–754.e8.

Chaplin, W. F., Phillips, J. B., Brown, J. D., Clanton, N. R., & Stein, J. L. (2000). Handshaking, gender, personality, and first impressions. *Journal of Personality and Social Psychology, 79,* 110–117.

Chelonis, J. J., Bastilla, J. E., Brown, M. M., & Gardner, E. S. (2007). Effect of time out on adult performance of a visual discrimination task. *Psychological Record, 57,* 359–372.

Cialdini, R. B., Reno, R. R., & Kallgren, C. A. (1990). A focus theory of normative conduct: Recycling the concept of norms to reduce littering in public places. *Journal of Personality and Social Psychology, 58,* 1015–1026.

Cohen, J. (1988). *Statistical power analysis for the behavioral sciences.* Hillsdale, NJ: Lawrence Erlbaum Associates.

Cohn, E. J., & Rotton, J. (2000). Weather, disorderly conduct, and assaults: From social contact to social avoidance. *Environment and Behavior, 32,* 651–673.

Collins, R. L., Elliott, M. N., Berry, S. H., Kanouse, D. E., Kunkel, D., Hunter, S. B., & Miu, A. (2004). Watching sex on television predicts adolescent initiation of sexual behavior. *Pediatrics, 114,* e280–e289.

Cook, M. (1977). Gaze and mutual gaze in social encounters. *American Scientist, 65,* 328–333.

Cowles, M., & Davis, C. (1982). On the origins of the .05 level of statistical significance. *American Psychologist, 37,* 553–558.

Craik, F. I. M., & Lockhart, R. S. (1972). Levels of processing: A framework for memory research. *Journal of Verbal Learning and Verbal Behavior, 11,* 671–684.

Davis, E. A. (1937). *The development of linguistic skills in twins, singletons with siblings, and only children from age 5 to 10 years.* Institute of Child Welfare Series, No. 14. Minneapolis: University of Minnesota Press.

Downs, D. S., & Abwender, D. (2002). Neuropsychological impairment in soccer athletes. *Journal of Sports Medicine and Physical Fitness, 42,* 103–107.

Eagly, A. H., Ashmore, R. D., Makhijani, M. G., & Longo, L. C. (1991). What is beautiful is good but . . . : A meta-analytic review of research on the physical attractiveness stereotype. *Psychological Bulletin, 110,* 109–128.

Evans, S. W., Pelham, W. E., Smith, B. H., Bukstein, O., Gnagy, E. M., Greiner, A. R., Atenderfer, L. B., & Baron-Myak, C. (2001). Dose-response effects of methylphenidate on ecologically valid measures of academic performance and classroom behavior in adolescents with ADHD. *Experimental and Clinical Psychopharmacology, 9,* 163–175.

Flynn, J. R. (1984). The mean IQ of Americans: Massive gains 1932 to 1978. *Psychological Bulletin. 95,* 29–51.

Flynn, J. R. (1999). Searching for justice: The discovery of IQ gains over time. *American Psychologist, 54,* 5–20.

Friedman, M., & Rosenman, R. H. (1974). *Type A behavior and your heart.* New York: Knopf.

Frieswijk, N., Buunk, B. P., Steverink, N., & Slaets, J. P. J. (2004). The effect of social comparison information on the life satisfaction of frail older persons. *Psychology and Aging, 19,* 183–190.

Fung, C. H., Elliott, M. N., Hays, R. D., Kahn, K. L., Kanouse, D. E., McGlynn, E. A., Spranca, M. D., & Shekelle, P. G. (2005). Patient's preferences for technical versus interpersonal quality when selecting a primary care physician. *Health Services Research, 40,* 957–977.

Geiser, S., & Studley, R. (2002). UC and the SAT: Predictive validity and differential impact of the SAT I and SAT II at the University of California. *Educational Assessment, 8,* 1–26.

Gibson, E. J., & Walk, R. D. (1960). The "visual cliff." *Scientific American, 202,* 64–71.

Gintzler, A. R. (1980). Endorphin-mediated increases in pain threshold during pregnancy. *Science, 210,* 193–195.

Green, L., Fry, A. F., & Myerson, J. (1994). Discounting of delayed rewards: A lifespan comparison. *Psychological Science, 5,* 33–36.

Güven, M., Elaimis, D. D., Binokay, S., & Tan, O. (2003). Population-level right-paw preference in rats assessed by a new computerized food-reaching test. *International Journal of Neuroscience, 113,* 1675–1689.

Harlow, H. F. (1959). Love in infant monkeys. *Scientific American, 200,* 68–86.

Harris, C. R. (2001). Cardiovascular responses of embarrassment and effects of emotional suppression in a social setting. *Journal of Personality and Social Psychology, 81,* 886–897.

Harris, R. J., Schoen, L. M., & Hensley, D. L. (1992). A cross-cultural study of story memory. *Journal of Cross-Cultural Psychology, 23,* 133–147.

Hays, W. L. (1981). *Statistics,* 3rd ed. New York: Holt, Rinehart, and Winston.

Hill, R. A., & Barton, R. A. (2005). Red enhances human performance in contests. *Nature, 435,* 293.

Hunter, J. E. (1997). Needed: A ban on the significance test. *Psychological Science, 8,* 3–7.

Hylands-White, N., & Derbyshire, S. W. G. (2007). Modifying pain perception: Is it better to be hypnotizable or feel that you are hypnotized? *Contemporary Hypnosis, 24,* 143–153.

Ijuin, M., Homma, A., Mimura, M., Kitamura, S., Kawai, Y., Imai, Y., & Gondo, Y. (2008). Validation of the 7-minute screen for the detection of early-stage Alzheimer's disease. *Dementia and Geriatric Cognitive Disorders, 25,* 248–255.

Johnston, J. J. (1975). Sticking with first responses on multiple-choice exams: For better or worse? *Teaching of Psychology, 2,* 178–179.

Jones, J. T, Pelham, B. W., Carvallo, M., & Mirenberg, M. C. (2004). How do I love thee, let me count the Js: Implicit egotism and interpersonal attraction. *Journal of Personality and Social Behavior, 87,* 665–683.

Joseph. J. A., Shukitt-Hale. B., Denisova. N. A., Bielinuski, D., Martin, A., McEwen. J. J., & Bickford, P. C. (1999). Reversals of age-related declines in neuronal signal transduction, cognitive, and motor behavioral deficits with blueberry, spinach, or strawberry dietary supplementation. *Journal of Neuroscience, 19,* 8114–8121.

Katona, G. (1940). *Organizing and memorizing.* New York: Columbia University Press.

Keppel, G. (1973). *Design and analysis: A researcher's handbook.* Englewood Cliffs, NJ: Prentice-Hall.

Keppel, G., & Zedeck, S. (1989). *Data analysis for research designs.* New York: W. H. Freeman.

Kerr, D. P., Walsh, D. M., & Baxter, D. (2003). Acupuncture in the management of chronic low back pain: A blind randomized controlled trial. *Clinical Journal of Pain, 19.* 364–370.

Khan, A., Brodhead. A. E., Kolts. R. L., & Brown, W. A. (2005). Severity of depressive symptoms and response to antidepressants and placebo in antidepressant trials. *Journal of Psychiatric Research, 39,* 145–150.

Killeen. P. R. (2005). An alternative to null-hypothesis significance tests. *Psychological Science, 16,* 345–353.

Kling, K. C., Hyde, J. S., Showers, C. J., & Buswell, B. N. (1999). Gender differences in self-esteem: A meta-analysis. *Psychological Bulletin, 125,* 470–500.

Kosfeld, M., Heinrichs, M., Zak, P. J., Fischblacher, U., & Fehr, R. (2005). Oxytocin increases trust in humans. *Nature, 435,* 673–676.

Kramer, S. E., Allessie, G. H. M., Dondorp. A. W., Zekveld, A. A., & Kapteyn, T. S. (2005). A home education program for older adults with hearing impairment and their significant others: A randomized evaluating short- and long-term effects. *International Journal of Audiology, 44,* 255–264.

Kuo, M., Adlaf. E. M., Lee, H., Gliksman, L., Demers, A., & Wechsler, H. (2002). More Canadian students drink but American students drink more: Comparing college alcohol use in two countries. *Addiction, 97,* 1583–1592.

Langewitz, W., Izakovic, J., & Wyler, J. (2005). Effect of self-hypnosis on hay fever symptoms—a randomized controlled intervention. *Psychotherapy and Psychosomatics, 74,* 165–172.

Laughlin, P. R., Zander, M. L., Knievel. E. M., & Tan, T. K. (2003). Groups perform better than the best individuals on letters-to-numbers problems: Informative equations and effective strategies. *Journal of Personality and Social Psychology, 85,* 684–694.

Linde, L., & Bergstroem, M, (1992). The effect of one night without sleep on problem-solving and immediate recall. *Psychological Research, 54,* 127–136.

Loftus, E. F., & Palmer, J. C. (1994). Reconstruction of automobile destruction: An example of the interaction between language and memory. *Journal of Verbal Learning & Verbal Behavior, 13,* 585–589.

Loftus, G. R. (1996). Psychology will be a much better science when we change the way we analyze data. *Current Directions in Psychological Science, 5,* 161–171.

McGee, R., Williams, S., Howden-Chapman, P., Martin, J., & Kawachi. I. (2006). Participation in clubs and groups from childhood to adolescence and its effects on attachment and self-esteem. *Journal of Adolescence, 29,* 1–17.

Montarello, S., & Martens, B. K. (2005). Effects of interspersed brief problems on students endurance at completing math work. *Journal of Behavioral Education, 14,* 249–266.

Morse, C. K. (1993). Does variability increase with age? An archival study of cognitive measures. *Psychology and Aging, 8,* 156–164.

Mulvihill, B. A., Obuseh, F. A., & Caldwell, C. (2008). Healthcare providers' satisfaction with a State Children's Health Insurance Program (SCHIP). *Maternal & Child Health Journal, 12,* 260–265.

Murdock. T. B., Miller, M., & Kohlhardt, J. (2004). Effects of classroom context variables on high school students' judgments of the acceptability and likelihood of cheating. *Journal of Educational Psychology, 96,* 765–777.

Pelton, T. (1983). The shootists. *Science83, 4*(4), 84–86.

Persson, J., Bringlov, E., Nilsson, L., & Nyberg. L. (2004). The memory-enhancing effects of Ginseng and Ginkgo biloba in health volunteers. *Psychopharmacology, 172,* 430–434.

Plomin, R., Corley, R., DeFries, J. C., & Fulker, D. W. (1990). Individual differences in television viewing in early childhood: Nature as well as nurture. *Psychological Science, 1,* 371–377.

Raine, A., Reynolds, C., Verables, P. H., & Mednick, S. A. (2002). Stimulation seeking and intelligence: A prospective longitudinal study, *Journal of Personality and Social Psychology, 82,* 663–674.

Reed, E. T., Vernon, P. A., & Johnson, A. M. (2004). Confirmation of correlation between brain nerve conduction velocity and intelligence level in normal adults. *Intelligence, 32,* 563–572.

Reifman, A. S., Larrick, R. P., & Fein, S. (1991). Temper and temperature on the diamond: The heat–aggression relationship in major league baseball. *Personality and Social Psychology Bulletin, 17,* 580–585.

Reyes-Ortiz, C. A., Ayele, H., Mulligan, T., Espino, D. V., Berges, I. M., & Markides, K. S. (2006). Higher church attendance predicts lower fear of falling in older Mexican-Americans. *Aging and Mental Health, 10,* 13–18.

Rogers, T. B., Kuiper, N. A., & Kirker, W. S, (1977). Self-reference and the encoding of personal information. *Journal of Personality and Social Psychology, 35,* 677–688.

Rozin, P., Bauer, R., & Cantanese, D. (2003). Food and life, pleasure and worry, among American college students: Gender differences and regional similarities. *Journal of Personality and Social Psychology, 85,* 132–141.

Scaife, M. (1976). The response to eye-like shapes by birds. I. The effect of context: A predator and a strange bird. *Animal Behaviour, 24,* 195–199.

Schachter. S. (1968). Obesity and eating. *Science, 161,* 751–756.

Schmidt, S. R. (1994). Effects of humor on sentence memory. *Journal of Experimental Psychology: Learning, Memory, & Cognition, 20,* 953–967.

Segal, S. J., & Fusella, V. (1970). Influence of imaged pictures and sounds on detection of visual and auditory signals. *Journal of Experimental Psychology, 83,* 458–464.

Shrauger, J. S. (1972). Self-esteem and reactions to being observed by others. *Journal of Personality and Social Psychology, 23,* 192–200.

Shulman, C., & Guberman, A. (2007). Acquisition of verb meaning through syntactic cues: A comparison of children with autism, children with specific language impairment (SLI) and children with typical language development (TLD). *Journal of Child Language, 34,* 411–423.

Siegel, J. M. (1990). Stress life events and use of physician services among the elderly: The moderating role of pet ownership. *Journal of Personality and Social Psychology, 58,* 1081–1086.

Siegel, S. (1956). *Nonparametric statistics for the behavioral sciences.* New York: McGraw-Hill.

Slater, A., Von der Schulenburg, C., Brown, E., Badenoch, M., Butterworth, G., Parsons, S., & Samuels, C. (1998). Newborn infants prefer attractive faces. *Infant Behavior and Development, 21,* 345–354.

Smith, C., & Lapp, L. (1991). Increases in number of REMs and REM density in humans following an intensive learning period. *Sleep: Journal of Sleep Research & Sleep Medicine, 14,* 325–330.

Smyth, J. M., Stone, A. A., Hurewitz, A., & Kaell, A. (1999). Effects of writing about stressful experiences on symptom reduction in patients with asthma or rheumatoid arthritis: A randomized trial. *Journal of the American Medical Association, 281,* 1304–1309.

Sol, D., Lefebvre, L., & Rodriguez-Teijeiro, J. D. (2005). Brain size, innovative propensity and migratory behavior in temperate Palaearctic birds. *Proceedings [Proc Biol Sci], 272,* 1433–441.

Stickgold, R., Whidbee, D., Schirmer B., Patel, V., & Hobson, J. A. (2000). Visual discrimination task improvement: A multi-step process occurring during sleep. *Journal of Cognitive Neuroscience, 12,* 246–254.

Strack, F., Martin, L. L., & Stepper, S. (1998). Inhibiting and facilitating conditions of the human smile: A non-obtrusive test of the facial feedback hypothesis. *Journal of Personality & Social Psychology, 54,* 768–777.

Sümer, N., Özkan, T., & Lajunen, T. (2006). Asymmetric relationship between driving and safety skills. *Accident Analysis and Prevention, 38,* 703–711.

Trockel, M. T., Barnes, M. D., & Egget, D. L. (2000). Health-related variables and academic performance among first-year college students: Implications for sleep and other behaviors. *Journal of American College Health, 49,* 125–131.

Tryon, R. C. (1940). Genetic differences in maze-learning ability in rats. *Yearbook of the National Society for the Study of Education, 39,* 111–119.

Tukey, J. W. (1977). *Exploratory data analysis.* Reading, MA: Addison-Wesley.

Tversky, A., & Kahneman, D. (1973). Availability: A heuristic for judging frequency and probability. *Cognitive Psychology, 5,* 207–232.

Tversky, A., & Kahneman, D. (1974). Judgment under uncertainty: Heuristics and biases. *Science, 185,* 1124–1130.

Twenge, J. M. (2000). The age of anxiety? Birth cohort change in anxiety and neuroticism, 1952-1993. *Journal of Personality and Social Psychology, 79,* 1007–1021.

von Hippel, P. T. (2005). Mean, median, and skew: Correcting a textbook rule. *Journal of Statistics Education, 13.*

Wegesin, D. J., & Stern, Y. (2004). Inter- and intra-individual variability in recognition memory: Effects of aging and estrogen use. *Neuropsychology, 18,* 646–657.

Welsh, R. S., Davis, M. J., Burke, J. R., & Williams, H. G. (2002). Carbohydrates and physical/mental performance during intermittent exercise to fatigue. *Medicine & Science in Sports & Exercise, 34,* 723–731.

Wilkinson, L., and the Task Force on Statistical Inference. (1999). Statistical methods in psychology journals. *American Psychologist, 54,* 594–604.

Winget, C., & Kramer, M. (1979). *Dimensions of dreams.* Gainesville: University Press of Florida.

Yapko, M. D. (1994). Suggestibility and repressed memories of abuse: A survey of psychotherapists' beliefs. *American Journal of Clinical Hypnosis, 36,* 163–171.

Index

$A \times B$ interaction, 493
Algebra, 715–718
Alpha level, 235, 243, 244–245, 260, 270
Alternative hypothesis (H_1), 233
Analysis of regression, 577–579
Analysis of variance. *See* ANOVA
ANOVA, 392–443
    between-treatment variance, 399–400
    conceptual view, 420
    degrees of freedom, 408–410
    effect size, 417
    η^2, 417
    F distribution table, 413
    F-ratio, 397–398, 401–402, 412
    factorial design. *See* Two-factor ANOVA
    higher-order factorial designs, 505–506
    hypothesis testing, 414–417
    in the literature, 419
    independent-measures, 432
    $MS_{between}/MS_{within}$, 422
    notation, 403
    planned/unplanned comparisons, 427
    pooled variance, 422–423
    post hoc tests, 425–427
    r^2, 417–419
    repeated-measures. *See* Repeated
      measures ANOVA
    Scheffé test, 428–430
    statistical hypotheses, 396
    sum of squares (*SS*), 405–408
    summary table, 410
    t tests, compared, 431–432
    terminology, 394–396
    test statistic, 397–398, 401–402
    Tukey's HSD test, 427–428
    type I errors, 398
    variances, 410–411
    within-treatment variance, 400–401
ANOVA formulas, 404
ANOVA summary table, 410
Apparent limits, 43

Bar graph, 46–47, 94
Between-subjects design, 309. *See also*
  Independent–measures *t* test
Between-treatment variance, 399–400
Between-treatments degrees of freedom
  ($df_{between}$), 409
Between-treatments sum of squares
  ($SS_{between\ treatments}$), 406
Between-treatments variability, 451

Between-treatments variance, 449
Biased, 122
Binomial data, 645
Binomial distribution, 183–186, 647
Binomial test, 644–663
    assumptions, 652
    binomial *z*-score, 648
    chi-square, 653–654
    data, 647
    defined, 646
    hypotheses, 646–647
    in the literature, 652
    overview, 645
    real limits, 650–652
    sign test, 654–657
    steps in process, 649–650
    test statistic, 647–648
Binomial *z*-score, 648
Body, 173

Central limit theorem, 204
Central tendency, 70–103
    defined, 72
    mean, 74–81
    median, 82–86
    middle, 86–87
    mode, 88–90, 93
    selecting a measure, 90–93
    skewed distribution, 97
    symmetrical distribution, 96–97
Chi-square distribution, 611–612, 737
Chi-square formula, 611
Chi-square statistic, 604–643
    binomial test, 653–654
    expected frequencies, 629
    goodness of fit. *See* Chi-square test for
      goodness of fit
    independence of observations, 628–629
    independent-measures *t* and ANOVA, 630
    median test for independent samples,
      631–633
    Pearson correlation, 629–630
    phi-coefficient, 631
    statistical table, 737
    test for independence. *See* Chi-square test
      for independence
Chi-square test for goodness of fit, 607–617
    chi-square distribution, 611–612
    chi-square formula, 611
    chi-square statistic, 610–611
    defined, 607

    degrees of freedom, 612, 613
    example test, 614–616
    expected frequencies, 610
    in the literature, 616
    null hypothesis, 608–609
    observed frequencies, 609
    single-sample *t* test, 616–617
Chi-square test for independence,
  617–628
    Cramer's *V*, 627–628
    defined, 618
    degrees of freedom, 622
    effect size, 626–628
    example test, 622–625
    expected frequencies, 620–622
    independent variables, 619
    null hypothesis, 618
    observed frequencies, 620
    phi-coefficient, 626–627
Coefficient of determination (r^2), 534
Cohen's *d,* 262–264, 291–292
Confidence interval, 367, 375, 380, 382
Construct, 18
Continuous variable, 19
Control condition, 15
Correlation, 519–561
    causation, 531
    coefficient of determination, 534
    degrees of freedom, 538
    form of relationship, 522–523
    hypothesis test, 537–539
    in the literature, 539–540
    outliers, 533
    overview, 520–521
    Pearson. *See* Pearson correlation
    phi-coefficient, 550–551
    point-biserial correlation, 548–550
    positive/negative, 521–522
    prediction, 524
    reliability, 524–525
    restricted range, 531–532
    Spearman. *See* Spearman correlation
    strength/consistency of relationship,
      523
    strength of relationship, 534–535
    validity, 524
Correlational method, 11
Cramer's *V,* 627–628
Critical region, 236
Cumulative frequencies, 52
Cumulative percentages, 53

Data, 5
Data set, 5
Datum, 5
Decimals, 711–712
Degrees of freedom (*df*)
    ANOVA, 408–410
    chi-square test for goodness of fit, 612, 613
    chi-square test for independence, 622
    correlation, 538
    defined, 284
    independent-measures *t* test, 316
    variability, 120–121
Dependent variable, 15
Descriptive statistics, 6
Deviation, 110
Deviation score, 142
df. See Degrees of freedom (*df*)
$df_{between}$, 409
df_{total}, 408
df_{within}, 408
Dichotomous variable, 183, 548
Difference scores, 342–343
Directional hypothesis test, 255–257. *See also* One–tailed test
Discrete variable, 18
Distribution of sample means
    defined, 200
    inferential statistics, 219–222
    mean, 205
    probability, 201, 210–211
    shape of, 204
    z-score for sample means, 211–212
Distribution-free tests, 606
Distributions of scores, 49
Dunn test, 427

Effect size
    ANOVA, 417
    chi-square test for independence, 626–628
    estimation, 382
    hypothesis testing, 260–264, 267
    independent-measures *t* test, 320–323
    repeated-measures ANOVA, 456–457
    repeated-measures design, 348–349
    t statistic, 291–297
    two-factor ANOVA, 496
Envelope, 523
Environmental variable, 14
Error term, 402
Error variance, 128, 451
Estimated Cohen's *d*, 262, 291
Estimated population standard deviation, 120
Estimated population variance, 120
Estimated standard error, 282
Estimation, 365
    confidence interval, 367, 375, 380, 382
    effect size, 382
    hypothesis test, 382
    hypothesis test, compared, 368, 369
    independent-measures studies, 376–379

interval estimate, 367, 374
    logic of, 369–371
    overview, 366–372
    point estimate, 367, 380
    repeated measures studies, 379–381
    single-sample studies, 373–376
    statistical significance *vs.* practical significance, 369
    t statistic, 372
    when used, 369
η^2, 417
Expected frequencies, 610, 620–622, 629
Expected value of *M*, 205
Experimental condition, 15
Experimental method, 14
Experimentwise alpha level, 398, 426
Exponents, 718–720

f distribution, 731–733
F distribution table, 413
F-max, 730
F-max test, 328–330
F-ratio
    ANOVA, 397–398, 401–402, 412
    repeated-measures ANOVA, 448–450, 455–456
    two-factor ANOVA, 494
Factor, 395
Factor *A*, 492, 501–502
Factor *B*, 493, 502–504
Factorial design, 395, 451
Fractions, 709–711
Frequency distribution
    bar graph, 46–47
    defined, 37
    graphs, 44–50
    grouped table, 41
    histogram, 44–45
    interpolation, 54–58
    polygon, 46
    probability, 168–169
    real limits, 43
    shape, 50–51
    stem and leaf display, 59–60
    tables, 37–43
Frequency distribution graphs, 44–50
Frequency distribution tables, 37–43
Friedman test, 684–687

Graphs, use/misuse, 49
Grouped frequency distribution table, 41

H_0, 233
H_1, 233
Hartley's *F*-max test, 328–330
Higher-order factorial designs, 505–506
Histogram, 44–45, 94
Homogeneity of variance, 328
Honestly significant difference (HSD), 427
HSD, 427
Hypothesis test, 230

Hypothesis testing, 229–279
    alpha level, 235, 243, 244–245, 260, 270
    alternative hypothesis, 234
    analogy, 239–240
    ANOVA, 414–417
    Cohen's *d*, 262–264
    correlation, 537–539
    critical region, 236
    effect size, 260–264, 267
    elements, 259–260
    estimate of error, 259
    estimation, compared, 368, 369, 382
    example test, 246–248
    hypothesized population parameter, 259
    in the literature, 248–250
    independent-measures *t* test, 318–320
    independent observations, 253
    level of significance, 235
    Mann-Whitney *U*, 671–672
    normal sampling distribution, 254
    null hypothesis, 233
    number of scores in sample, 252
    one-tailed test, 255–257
    one-tailed/two-tailed tests, compared, 257–258, 270
    overlapping distributions, 284
    Pearson correlation, 537–539
    power, 265–270
    random sampling, 252–253
    reject null hypothesis *vs.* prove alternative hypothesis, 238–239
    repeated-measures ANOVA, 451
    sample in research study, 232
    sample statistic, 259
    significant/statistically significant, 249
    size of mean difference, 251
    step 1 (state the hypothesis), 232–234
    step 2 (set criteria for a decision), 234–237
    step 3 (collect data/compute sample statistics), 237–238
    t statistic, 287–291
    test statistic, 259–260
    two-tailed test, 255
    type I error, 242
    type II error, 243–244
    underlying logic, 231
    unknown population, 231–232
    value of standard error unchanged, 253–254
    variability of scores, 251–252
    Wilcoxon *T*, 677–678
    z-score statistic, 240–241

In the literature
    ANOVA, 419
    binomial test, 652
    central tendency, 93
    chi-square test, 616
    correlation, 539–540
    Friedman test, 687
    independent-measures *t* test, 323–324

Kruskal-Wallis test, 683
Mann-Whitney U test, 672
repeated-measures ANOVA, 457
repeated-measures t test, 349–350
results of statistical test, 248–250
standard deviation, 126
standard error, 217
t test, 296
two-factor ANOVA, 496–497
Wilcoxon T test, 678
Independent-measures ANOVA, 432
Independent-measures t test, 307–338
   degrees of freedom, 316
   effect size, 320–323
   estimated standard error, 311–313, 316
   final formula, 316
   Hartley's F-max test, 328–330
   hypotheses, 310
   hypothesis test, 318–320
   in the literature, 323–324
   one-tailed tests, 324–325
   overall t formula, 311
   pooled variance, 313–316
   repeated-measures design, contrasted, 353–354
   sample size, 315
   sample variance, 326
   underlying assumptions, 327–328, 329
Independent random sample, 168
Independent variable, 14
Individual differences
   repeated-measures ANOVA, 448, 460–461, 463–465
   repeated-measures design, 353–354
Inferential statistics, 6
Interactions, 482–485, 504–505
Interpolation, 54–58
Interquartile range, 107–108
Interval estimate, 367, 374
Interval scale, 22, 23

Kruskal-Wallis test, 680–683

Law of large numbers, 206
Least-squared error solution, 567
Level of significance, 235
Levels, 395
Line graph, 94
Linear equations, 564–565
Linear relationship, 564
Lower real limit, 20

Main effects, 480–482, 498
Mann-Whitney U, 669
Mann-Whitney U test, 668–674
   assumptions/cautions, 673–674
   calculation of Mann-Whitney U, 669
   computing U for large samples, 670–671
   critical values, 738–739

hypothesis tests, 671–672
in the literature, 672
normal approximation, 672–673
null hypothesis, 669
ranking without scores, 670
Matched-subjects study, 341
Mathematics, 703–721
   algebra, 715–718
   decimals, 711–712
   exponents, 718–720
   fractions, 709–711
   negative numbers, 713–714
   order of operations, 705–707
   percentages, 712
   proportions, 707–712
   skills assessment preview exam, 704
   solving equations, 715–718
   square roots, 720–721
Mean
   alternate definitions, 74–76
   changing a score, 79
   defined, 74
   frequency distribution table, 78
   graphs, 94–96
   multiplying/dividing each score by constant, 81
   new score, 79
   population, 74
   removing a score, 79
   sample, 74
   weighted, 76–77
Mean square (MS), 490, 493–494
Median
   continuous variable, 84–86
   defined, 82
   graphs, 94–96
   middle, 86–87
   when to use it, 90–91
Median test for independent samples, 631–633
Middle, 86–87
Mode, 88–90, 93
Modified histogram, 44
MS, 490, 493–494
$MS_{between}/MS_{within}$, 422
Multiple regression, 580–586

Negative correlation, 522
Negative numbers, 713–714
Negative skew, 50, 51
Nominal scale, 21
Nonequivalent groups, 15
Nonexperimental methods, 15
Nonparametric tests, 606
Normal approximation to binomial distribution, 186–188
Normal curve, 48
Normal distribution, 170–172
Null hypothesis (H_0), 233

Observed frequencies, 609, 620
One-tailed test
   hypothesis testing, 255–257
   independent-measures t test, 324–325
   repeated-measures design, 351
   t distribution, 297–299
Open-ended distributions, 92
Operational definition, 18
Order effects, 355
Order of mathematical operations, 25, 705–707
Ordinal data, 664–702
   Friedman test, 684–687
   hypotheses tests, 666
   Kruskal-Wallis test, 680–683
   Mann-Whitney. *See* Mann-Whitney U test
   ordinal measurements, 667
   tied scores, 667–668
   Wilcoxon. *See* Wilcoxon signed-ranks test
Ordinal measurements, 667
Ordinal scale, 21–22
Outliers, 533
Overview. *See* Statistics organizer

Pairwise comparisons, 426
Parameter, 5
Parametric tests, 606
Partial correlation, 588
Participant variable, 13
Pearson correlation
   calculation, 528
   chi-square statistic, 629–630
   critical values, 735
   defined, 525
   hypothesis test, 537–539
   sum of products (SP), 526, 527
   z-score, 529
Percentage, 39
Percentage of variance accounted for by the treatment (r^2), 292–295
Percentages, 712
Percentile, 52, 179
Percentile rank, 52, 179
Perfect correlation, 523
Phi-coefficient, 550–551, 626–627, 631
Planned comparisons, 427
Point-biserial correlation, 548–550
Point estimate, 367, 380
Polygon, 46
Pooled variance, 313–316, 422–423
Population, 3
Population mean, 74
Population standard variance, 115
Population variance, 111, 115, 116
Positive correlation, 521
Positive skew, 50, 51
Post hoc tests
   ANOVA, 425–427
   repeated-measures ANOVA, 457–458
Posttests, 426

Power, 265–270
Practical significance, 369
Pre-post study, 16
Probability, 163–197
    binomial distribution, 183–186
    defined, 165
    frequency distribution, 168–169
    inferential statistics, 189–190
    normal approximation to binomial
        distribution, 186–188
    normal distribution, 170–172
    proportions, 174
    random sampling, 167–168
    scores from normal distribution, 177–183
    unit normal table, 172–174
    z-scores, 174
    zero, 167
Proportion, 39
Proportions, 707–712

Quasi-independent variable, 17, 39

r^2, 417–419
    coefficient of determination, 534
    percentage of variance accented for by the
        treatment, 292–295
Random sample, 167
Random sampling, 167–168
Range, 107
Rank, 52
Ratio scale, 22, 23
Raw score, 5
Real limits, 20
Regression, 562–603
    analysis of regression, 577–579
    cuations, 570
    defined, 566
    evaluating contribution of each predictor
        variable, 586–593
    least-squared error solution, 567
    linear equations, 564–565
    multiple regression, 580–586
    partial correlation, 588
    predicted/unpredicted deviations,
        575–576
    regression equation for Y, 567–568
    significance of regression equation,
        577–579
    $SS_{regression}$, 573, 574
    $SS_{residual}$, 571, 573, 574
    standard error of estimate, 571–573,
        584–585
    standardized form of regression equation,
        570–571
    third variable, 587
    two predictor variables, 580–586
Regression equation for Y, 567–568
Regression line, 566
Regression toward the mean, 536
Relative frequencies, 39, 47

Repeated-measures ANOVA, 444–476
    advantages/disadvantages,
        461–463
    alternative view, 459–461
    assumptions, 458
    between-treatments variance, 449
    effect size, 456–457
    error variance, 449
    F-ratio, 448–450, 455–456
    hypotheses, 447
    hypothesis testing, 451
    in the literature, 457
    individual differences, 448, 460–461,
        463–465
    MS values, 455–456
    notation, 452
    post hoc tests, 457–458
    single-factor design, 446–447
    $SS_{between\ subjects}/SS_{between\ treatments}$, 454
    stage 1, 452–453
    stage 2, 453
    treatment effects, 460, 463–465
Repeated-measures design, 339–364
    analogies for H_0 and H_1, 344
    counterbalancing, 355
    defined, 341
    descriptive statistics, 350
    difference scores, 342–343
    effect size, 348–349
    hypotheses, 343–344
    hypothesis tests, 346–348
    in the literature, 349–350
    independent-measures design, contrasted,
        353–354
    individual differences, 353–354
    number of subjects, 353
    one-tailed test, 351
    order effects, 355
    study changes over time, 353
    t statistic, 344–346
    time-related factors, 354–355
    underlying assumptions, 355
    variability/treatment effect,
        350–351
Reporting. *See* In the literature
Residual variance, 451
Restricted range, 531–532

Sample, 4
Sample mean, 74
Sample size
    ANOVA, 423
    independent-measures t test, 315
    standard error, 206–208
    t statistic, 295–296
Sample standard deviation, 118, 282
Sample variance, 118, 282
Sampling distribution, 201
Sampling error, 6, 200, 367
Sampling with replacement, 168

Scales of measurement
    interval scale, 22, 23
    nominal scale, 21
    ordinal scale, 21–22
    ratio scale, 22, 23
Scheffé test, 428–430
Score, 5
Semi-interquartile range, 108–109
Sign test, 654–657
Significant, 249
Simple random sample, 168
Skewed distribution, 50
Slope, 564
Solving equations, 715–718
SP, 526, 527
Spearman correlation, 540–547
    critical values, 736
    ranking tied scores, 544–545
    special formula, 545–546
    testing the significance, 546–547
    when used, 542
SPSS, 759–761
Square roots, 720–721
SS, 113–115
$SS_{regression}$, 573, 574
$SS_{residual}$, 571, 573, 574
$SS_{between\ subjects}/SS_{between\ treatments}$, 454
$SS_{between\ treatments}$, 406
SS_{total}, 405
$SS_{within\ treatments}$, 405
Standard deviation, 111–113
    analogy, 125
    defined, 111
    descriptive statistics, 123–125
    estimated population, 120
    in the literature, 126
    population, 115
    sample, 118
    standard error, contrasted, 212
    variance for samples, 117–120
Standard error, 214–217, 282
    defined, 206
    estimation, 377
    in the literature, 217
    independent-measures t test, 311–313, 316
    measure of reliability, as, 221–222
    purposes, 205
    sample size, 206–208
    standard deviation, contrasted, 212
Standard error of estimate, 571–573, 584–585
Standardized distribution, 148
Statistic, 5
Statistical notation, 24–26
Statistical power, 265–270
Statistical significance, 369
Statistical tables
    chi-square distribution, 737
    f distribution, 731–733
    F-max, 730
    Mann-Whitney U, 738, 739

Pearson correlation, 735
Spearman correlation, 736
studentized range statistic, 734
t distribution, 729
unit normal table, 725–728
Statistically significant, 249
Statistics, 3
Statistics organizer, 762–773
descriptive statistics, 762–764
measures of relationships between
variables, 771–773
nonparametric tests, 769–771
parametric tests, 764–768
Stem and leaf display, 58–60
Studentized range statistic, 734
Sum of products (*SP*), 526, 527
Sum of squares (*SS*), 113–115, 405–408
Summation notation, 25
Symmetrical distribution, 50

t distribution, 729
Cohen's *d,* 291–292
defined, 285
effect size, 291–292
hypothesis testing, 287–291
in the literature, 296
one-tailed test, 297–299
percentage of variance explained, 292–295
proportions/probabilities, 285–286
sample size, 295–296
sample variance, 295
shape, 285
t statistic, 283
Tables. *See* Statistical tables
Tail, 50, 173
Test statistic, 240
Testwise alpha level, 398
Total degrees of freedom (*df*$_{total}$), 408
Total sum of squares (*SS*$_{total}$), 405
Total variability, 491
Treatment effects, 460, 463–465
Tukey's HSD test, 427–428
Two-factor ANOVA, 477–505
$A \times B$ interaction, 493
alternative view, 501
assumptions, 505
between-treatments variability, 451
effect size, 496
F-ratio, 494
factor *A,* 492, 501–502

factor B, 493, 502–504
graphs, 487
hypothesis tests, 488–489
in the literature, 496–497
independence of main effects/interactions,
485–486
interactions, 482–485, 504–505
interpreting the results, 497
level of arousal, 498–501
main effects, 480–482, 498
mean square (*MS*), 490, 493–494
stage 1, 491–492
stage 2, 492–493
total variability, 491
within-treatments variability, 452
Two-factor design, 395
Two-tailed test, 255
Type I error, 242, 398
Type II error, 243–244

Unbiased, 122
Unit normal table, 172–174, 725–728
Unplanned comparisons, 427
Upper real limit, 20

Variability, 104–136
bias/unbiased, 122
defined, 105
degrees of freedom, 120–121
extreme scores, 129
interquartile range, 107–108
open-ended distributions, 130
population standard variance, 115
population variance, 115, 116
range, 107
sample size, 129
sample standard deviation, 118
sample variance, 118
semi-interquartile range, 108–109
stability under sampling, 130
standard deviation. *See* Standard deviation
sum of squares (*SS*), 113–115
transformations of scale, 125–126
variance. *See* Variance
Variable, 5
Variance, 113
bias/unbiased, 122
defined, 111
error, 128
estimated population, 120

inferential statistics, 126–128
population, 111, 115, 116
sample, 118

Weighted mean, 76–77
Wilcoxon signed-ranks test
critical values, 740
hypotheses, 675–676
hypothesis tests, 677–678
in the literature, 678
normal approximations, 678
tied scores/zero scores, 676–677
Wilcoxon *T,* 676
Wilcoxon *T,* 676
Within-subject design. *See* Repeated-measures
design
Within-treatment variance, 400–401
Within-treatments degrees of freedom
(*df*$_{within}$), 408
Within-treatments sum of squares
(*SS*$_{within\ treatments}$), 405
Within-treatments variability, 452

Y-intercept, 564

z-score, 137–162
binomial *z*-score, 648
comparisons, 148–150
computing *z*-score from samples,
150–151
defined, 141
demonstration of *z*-score transformation,
148
hypothesis testing, 239–240
inferential statistics, 154–156
location in a distribution, 141–142
mean, and, 144
new distribution with predetermined
mean/standard deviation, 150–151
Pearson correlation, 529
probability, 174–176
purposes, 139–141
standard deviation, and, 144–145
standardizing a distribution, 146–148
standardizing a sample distribution,
153–154
z-score formula, 142–143
z-score formula, 142–143
Zero probability, 167